How Giraffes Work

Giraffa camelopardalis typica. Jardine's Giraffe.
From Jardine, 1838. Courtesy of Tony Ridge.

How Giraffes Work

GRAHAM MITCHELL

OXFORD
UNIVERSITY PRESS

OXFORD
UNIVERSITY PRESS

Oxford University Press is a department of the University of Oxford. It furthers the University's objective of excellence in research, scholarship, and education by publishing worldwide. Oxford is a registered trade mark of Oxford University Press in the UK and certain other countries.

Published in the United States of America by Oxford University Press
198 Madison Avenue, New York, NY 10016, United States of America.

Library of Congress Cataloging-in-Publication Data
Names: Mitchell, Graham, 1944- author.
Title: How giraffes work / Graham Mitchell.
Description: New York, NY : Oxford University Press, [2021] |
Includes bibliographical references and index.
Identifiers: LCCN 2020057172 (print) | LCCN 2020057173 (ebook) |
ISBN 9780197571194 (hardback) | ISBN 9780197571217 (epub)
Subjects: LCSH: Giraffe. | Giraffe—Conservation.
Classification: LCC QL737.U56 M58 2021 (print) | LCC QL737.U56 (ebook) |
DDC 599.638—dc23
LC record available at https://lccn.loc.gov/2020057172
LC ebook record available at https://lccn.loc.gov/2020057173

DOI: 10.1093/oso/9780197571194.001.0001

1 3 5 7 9 8 6 4 2

Printed by Integrated Books International, United States of America

For Keith, Carolyn, Andrew, and Heather
and all who follow them

Contents

Preface

No one likes an unsolved mystery. Giraffes are a mystery. And they are unique. Where did they come from? How do they work? "Magnificent in its attire, bizarre in its shape, singular in its walk, colossal in its proportions, inoffensive in character, the giraffe has long attracted the attention of man. There is no individual of the animal kingdom more worthy of a monograph." So wrote Joly and Lavocat in 1846 in their monograph. I share their sentiments. There are few creatures more beautiful, more aloof, and more fascinating than giraffes. They seem to have reached that rare combination of contentment and serenity that we all strive for. Richard Lydekker, a Victorian zoologist, was equally impressed. He wrote in 1901, "[It is} one of the most strikingly beautiful animals in the creation." However, a giraffe's life is not an easy one. "It is" in the words of Ray Lankester in 1908, another distinguished Victorian zoologist, "altogether exceptional, novel, and specialised" and they have to be to survive.

Much has been written about the public lives of giraffes, and their history in art and diplomacy, and it has been well recorded in books (e.g., Laufer, 1928; Spinage, 1968; Sherr, 1997; Allin, 1998; Williams, 2010). It is dealt with cursorily in this book. Any reader who knows it all can safely skip Chapter 2. I didn't, and had to discover it afresh. The important matter of giraffe conservation is the main focus of Julian Fennessy and his Giraffe Conservation Foundation, and is touched on here. The behavior and social biology and ecology of giraffes have been described in several books, with the first, best, and most well-known of these being that of Anne (Innes) Dagg and J. Bristol Foster. Anne Dagg is a Canadian who in the 1950s was a PhD student and had what any doctoral student with a consuming interest in wild animals, their habits, and the habitats they occupy would say is a dream job. She spent a year in the Eastern Transvaal Lowveld/ Bushveld of South Africa (now in the provinces of Limpopo and Mpumalanga), making observations on almost all visible aspects of the lives of giraffes. But so far relatively little has been written about their private lives, their physiology, and their anatomy—in short, the normal functions of a free-living animal in its natural environment. These aspects of giraffe biology have been mentioned in passing by C. A. Spinage (1968) whose book emphasized their history more than their biology, and by Ingo Krumbiegel, whose book, published in German in 1971, was more biology than physiology, as was Brian Shorrock's (2016) book. When I thought that a book like this one, on the interaction between the anatomy and physiology of giraffes and between them and their environment—their morphophysiology and ecophysiology—might fill a gap, I also thought that study of a single species could be concise and relatively simply told. In reality, it turned out that it could not be, because a species never exists in isolation from the general biological milieu in which it finds itself. Matters of tectonics, astronomical physics, climate, and purely biological factors affecting food and water resources all interact with its morphology, its internal physiological and biochemical systems, and the behavioral patterns that regulate its daily life. In short, the tale of a single species is very complex. Doing it justice in a relatively short book is and was a challenge, not least because studies on and scientific articles about giraffes are ongoing, keeping up with them a constant task, and books must have an end. Readers, both friend and foe, will judge the success of this attempt to do so. In many ways this book is a mystery tale, a tale of investigation and of analysis that tells of what we know and what we could find out if we wanted to, but I think that there comes a point in the study of animals where we should stop short of full knowledge, admire the knowledge that we have, and speculate on the knowledge we do not have. That is the point I have reached. It has been a personal obsession to reach it, and the story is personal too and written *prodesse et delectare*. I hope any reader finds that it rings true.

My earliest memory of a wild giraffe was a picture taken by my father of a giraffe that had tripped, fallen, and broken its neck while trying to leap a fence in the then Eastern Transvaal Lowveld of South Africa, a place rich in history and as romantic as any. I was to discover much later that accidents of that sort are not uncommon among giraffes, but the struggles of that giraffe and its tragedy as captured in that photograph have never left me, and why that accident happened and why its consequences were so severe have been thoughts never very far away throughout the time I have studied giraffes. Although this is a book about the physiology and anatomy of extant giraffes or, in other words, about how modern giraffes work, the most frequent questions asked about giraffes are not about how they work, but how their shape came about. All three of the scientists—Darwin, Wallace, and Lamarck—who in the nineteenth century pondered the question of how the shape of an animal develops and how a species originates used giraffes to illustrate their ideas. Their answers were theoretical. Fossils of giraffes were few and gave no idea of who the ancestors of modern giraffes were that bequeathed their characters to their offspring. When did those ancestors live, and where did they live? What were the selection pressures then that resulted in their shape now? What advantages did and does their shape give them so that they can survive successfully? Their survival is usually attributed to their great height and unique markings, but it is the functional anatomy or morphophysiology of these extraordinary creatures that has contributed most to their survival. What were/are they? Answers to that question have been the subject of many, many years of research. It is a story filled not only with a unique creature, with geography, and climate changes of great magnitude, but also by the labors of extraordinary people who put many pieces of the puzzle together. Very often the language they used to describe their observations far defeats the best efforts I could make, so in many places I have allowed them to speak for themselves, when they, like Cortez, saw a peak in Darien. Throughout this book I have tried to interpret the best scholarship, some of it difficult to understand and much of it hidden from public view in the pages of learned scientific journals. Often that scholarship is in the indigestible form much loved by scientific journals and so, despite being relevant and interesting, it is simply unknown and unappreciated. One consequence is that the information that seeps out of dusty pages is oversimplified into easily understandable myths. All this I have tried to correct, but in doing so I have almost certainly made errors of interpretation which no doubt many will detect with glee. Many references have been used to acquire the information recorded in this book. I am certain that others should have been read, marked, learned, inwardly digested, noted, and used, and others are still being published and just not read. My apologies to those authors whose work, highly relevant and useful, has been overlooked and thus not acknowledged. This book is also an opportunity to correct the many errors I and my colleagues have made in published articles. Hindsight is a source of acute embarrassment. In telling this story of giraffes, I have followed a historical approach to reveal how the knowledge progressed, and wherever possible I have used original diagrams to illustrate aspects of giraffe physiology and anatomy. This book cannot be complete or indeed accurate. Science is a demanding enterprise and will not allow a study to reach an endpoint or finality, and will not allow us to say we now know everything there is to know.

Many people have helped in gathering data or in other ways. I am especially grateful to two people. I am indebted most to the late Professor John Skinner. He was an outstanding zoologist, mammalogist, the force behind and coauthor of the definitive *Mammals of the Southern African Region*, and mentor to many students. His influence on students of zoology in Africa cannot be underestimated. Over a period of 30 years we collaborated on studies of giraffes. This book is his as much as mine. In the last years of his life he was a Professor *Extraordinaire* in the Centre for Veterinary Wildlife Studies, Faculty of Veterinary Science, University of Pretoria, South Africa, and I thank that University for the support it gave us. The second is my brother Duncan. Brothers, especially elder ones are, typically, exceptional people. He is, and also is a physiologist. He not only collaborated on some studies of giraffes but also undertook to read this book when it had passed infancy but was not yet mature. He gratefully took the opportunity to point out, with delight, all the many errors I had made and inconsistencies I had introduced. His efforts were invaluable. Several embryonic chapters were scrutinized by Peter and Margaret Ardington. Distinguished Zululand veterinarians and steadfast friends, they deserved better. I apologize for the pain I inflicted on them, which they cheerfully bore. Their advice too was invaluable. I also owe much to my wife, Jeanette, a rare companion, who for many years encouraged me to write down what I knew of giraffes. She is an expert anatomist and expert copyeditor, took on the task of copyediting the many "first drafts," contributed original data on aspects of giraffe anatomy, as well as contributing in many other ways to the study of giraffes. Friends, family, and former colleagues at the University of the Witwatersrand, Johannesburg, South Africa, made comments and took photographs, many of which appear in the book, properly acknowledged. Helpful too have been Imarie de Vaal, Sally MacRoberts, Fylyppa Meyer, Alan Garlick, and Desre Stead of the Brenthurst Library, Johannesburg, South Africa, who very kindly made available the maps of Robert Jacob Gordon's journeys, and a copy of the painting of a giraffe made by William Paterson and much else besides. I am grateful to Evelyn van Kaam and Maria Smit of the Rijksmuseum, Amsterdam, for their help in locating and obtaining copies of Gordon's drawings and providing me with much additional information about Gordon; to Lucie Strnadova, Katie Holyoak, and Agata Rutkowska for their help in obtaining copies of Davis's painting of George the IV's giraffe in the Royal Collection; to Laurie Sherwood and Sarah Broadhurst of the Zoological Society London for their help in obtaining copies of Agasse's painting of the giraffe in the park and Hill's painting of Zaida and her calf; to Caroline Lam of the Geological Society for providing the image of the C3 of *G sivalensis*; to Stephen Atkinson at the Natural History Museum, London, for the image of Richard Owen; to Antoinette Lourens in the Faculty of Veterinary Science and Susan Marsh of the University of Pretoria Library for their help in providing me with a copy of Anthony

Hall-Martin's thesis on giraffes and for arranging permission to use images and tables from that thesis and from a dissertation by Dr. Errol Ackerman; to Zhaojie Zhang, expert electron microscopist and photographer; and to my wife for turning many pictures into prints. Obtaining permission to use images was a challenging task, made easier by the many librarians, archivists, and others who with good grace have helped. There are too many to name and there are some I should have acknowledged but haven't. Here I express my thanks to them all. Permission to print pictures has been given by the Royal Collection, the Zoological Society, London, the Geological Society, London, Rijksmuseum, Amsterdam, the Linnean Society, London, the Brenthurst Library, Johannesburg, and the British Library, and the journals in which they first appeared. I thank my daughter Carolyn for her efforts to obtain illustrations from the British library for me. My son Keith, mathematician and statistician, gave me much advice and help, especially help manipulating computer programs. Last but far from least are my thanks to Jeremy Lewis, acquisitions editor at Oxford University Press, without whose (large) appetite for risk this book would not have been published and to Bronwyn Geyer, editorial assistant, Dorothy Bauhoff and Rajesh Kathamuthu who guided me through the copyediting and printing phases.

Studies like those John Skinner and I were involved in of wild, free-living giraffes require material support. John provided funds from the Don Craib Trust, and his personal research grant. Significant support also was provided by the University of Wyoming. I cannot thank that University and especially Professor Ollie Walter, Dean of the College of Arts and Sciences, enough for employing me and my wife between 2003 and 2009 and for giving us the opportunity to contribute to the University. It was a wonderful experience at an outstanding University. We acknowledge here too the support our colleagues in the Department of Zoology and Physiology there gave us. It was during that time that work on this book began. The University's library and its staff deserve special mention for their expertise and for the service they provided then and still do. No article, no matter how obscure, was beyond their reach. Material support and great expertise were provided by Blondie Leatham and staff of the Bubye Valley Conservancy, Zimbabwe, especially Kenneth Manyangadze, Mark Brewer, and Blake Wilelmi. Four postgraduate students, Louis van Schalkwyk, Sybrand van Sittert, Ian Bredin, and David Roberts, earned higher degrees arising out of the work they did. Mark Nijland and Liz Rickards and undergraduate students Glenda Hardy, Carl-Heinz Moeller, Lauren Leathem, Kyle Pearse, Declan Gallagher, and Struan Muirhead provided technical and field support. Dr. Roy Bengis, who was for over 30 years the Chief Veterinarian in the Kruger National Park, South Africa, was a source of great expertise and wisdom. One other person deserves a mention. That person is Professor Mike Brown, an inspiring teacher of physiology in the Faculty of Veterinary Science at the University of Pretoria, South Africa, from whom I learned not only the facts of physiology and biochemistry, but how the facts explained so much about how an animal works in health and disease. Many Brownisms are to be found in this book, and I think he would have approved if he had lived to read it.

GM
Devon
2020

1
What's in a Name?

When, in 1762, Mathurin Jacques Brisson, a French zoologist, classified giraffes as *Giraffa giraffa*, why did he choose that name? Why are giraffes[1] called *Giraffe*? What is the meaning of the word "giraffe"? Brisson had many other names from which to choose. In 1968, in his *Book of the Giraffe*, C.A. Spinage, for example, recorded that over the centuries giraffes have been given at least 54 common names, many of them beautifully descriptive of their anatomy, demeanor, or habits. One of the earliest and most evocative names given to giraffes was *Zemer*, a word derived from the Hebrew root *zmr* and one of its related verbs meaning "to prune," which perfectly describes a giraffe's habit of browsing as the pruning of trees. The Septuagint, the 72 religious scholars who translated the Bible from Hebrew around 250 BC, modernized *zemer* as *kamelopardalis* (Brenton, 1851), a name they attributed to an invention of the Ancient Greeks who had joined the two descriptive nouns *kamelos* (because giraffes have a long neck like a camel) with *pardalis* (because they had spots like a leopard) into *kamelopardalis*, although some 100 years before the Septuagint, the name given to giraffes, if we believe interpreters of Aristotle's writings, was *Pardium* or *Hippardium*. *Camelopardalis* was the name adopted by the Romans. It has endured, though Pliny the Elder in AD 64, in one of the very first descriptions of the shape and character of giraffes, referred to giraffes as *Ovis fera*, the wild sheep, presumably because he had been told (Pliny had never seen a giraffe) that giraffes had a sweet nature (Vosmaer, 1787). *Ovis fera* seems to have been corrupted at some stage to *Orasius* or *Oraflus*, the latter a misreading of the Roman *s* and *i* as *f* and *l*. *Orasius* returned later, but very briefly, as the name for an ancestor of modern giraffes. According to Pliny the Elder, the ancient Ethiopians called giraffes *Nabun*, subsequently corrupted by the medieval polymath, Albert the Great, to *Anabulla* and similar variants (Vosmaer, 1787). *Nabun* means "to be elevated," a word close to the Indian word *Nohna*, meaning height. Their height was the basis for the use by the Ancient Egyptians of the figure of a giraffe to mean "prophesy or foretell" (Gardiner, 1927), because the combination of their height and visual acuity alerts them early to danger. The Persians gave giraffes a name designed to show their origins from camels, the ox, and the leopard (*Ostorcamjalanca*). Modern Ethiopians call giraffes *Schiratachazin* or *Zirath-Katchin* or *Gierata-cachin*, all of which mean "long, thin tail." Beautifully descriptive too is the isiZulu name *ndlulamithi*—the "surpasser of trees."

None of these ancient names reveals the origin of the word "giraffe." Its origin seems to be the Egyptian *sr*, itself derived from the Ethiopian *Zarat*, meaning "to foretell or prophesy." Many similar sounding names are known, and presumably they have evolved from that origin. Marco Polo, for example, recorded that giraffes were called

[1] Both singular and plural forms can be either "giraffe" or "giraffes." The modern usage for both is "giraffes" unless it is clear that a single giraffe is being referred to.

How Giraffes Work. Graham Mitchell, Oxford University Press. © Oxford University Press 2021.
DOI: 10.1093/oso/9780197571194.003.0001

Graffa by the Zanzibar Islanders, *Ziraffa* in the Tigre (Turkey), and *Giraffa-kelchin* by the Amhara, Ethiopia (Latham, 1958). The Arabs of the fifteenth century called them *Seraffa*, while modern Arabs call them *Siraf*, *Seraph*, and *Zurappha*. Samuel Bochardt, a domineering religious scholar of the seventeenth century, linked the word *Zurappha* to a description of their gait and the subsequent contradictory interpretations that *Zarapha* means "one or who walks slowly or swiftly!" Bochardt also recorded that the Greeks called them *Zoraphis*, the Italians and Spanish *Girafa*, and according to Vosmaer (1787), the Germans, Polish, and Hungarians *Greif*. Modern Greek names include *Gyrafa*, *Giraffa*, and *Zirapha*. Another derivation is *Zerafa*, a variant of *Zarafa*, which is Arabic for "charming" or "lovely one" or "elegant," and it was the name given to the giraffe that arrived in France in 1826 as a gift for Charles X. Everard Home referred to the giraffe given to George IV in 1827 as *Xariffa* in his lecture on the nature of its tongue and as *Zariffa* in his scientific essay on the anatomy of its stomach.

Their modern scientific name, *Giraffa camelopardalis*, clearly is derived from these later common names, but it arrived by a very convoluted path. Carl Linnaeus, who is credited with developing the binomial classification system, attempted the first classification of giraffes. The system was actually built on the earlier ideas of John Ray, a classic English parson who had enough spare time to become a naturalist. In1693 Ray published a classification of animals and plants in a *Synopsis methodica animalium quadrupedum et serpentini generis* (*Synopsis of Animals and reptiles*). Over the many decades after Ray's death (in 1705), his method of classifying animals (and plants) incubated slowly and matured into the *Systema Naturae* first published by Linnaeus (Figure 1.1) in 1735. Its definitive tenth edition was published in 1758. In the tenth edition Linnaeus placed giraffes in the genus *Cervus* with deer, as he thought they were related to the deer family by virtue of their antler-like horns. He borrowed the species

(a) (b)

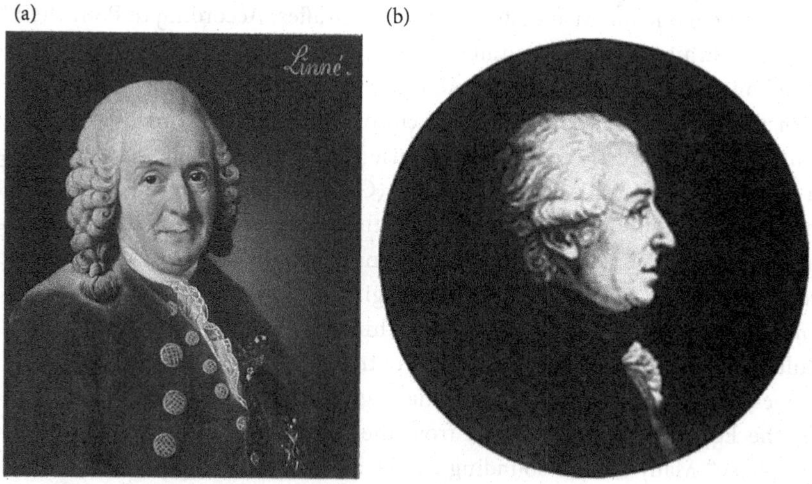

Figure 1.1. Left: Carl Linnaeus by Alexander Roslin.
With permission of the Stockholm National Museum.
Right: Mathurin Brisson. Artist unknown.
Public domain image from Wikipedia.

name of *camelopardalis* from the Roman name for giraffes. Linnaeus described the characteristics of the Cervidae as having

> *Horns: solid, soft hairy skin covers the tip, and emerge annually. Eight lower front teeth, no upper incisors (sometimes one).*

His description of the species *camelopardalis* was brief. He wrote:

> *Very simple horns, front legs very long. It lives in Ethiopia and Sennar. It lies down leaning like camels. It grazes grass by spreading its front legs, but chiefly feeds on the leaves of trees.*

Four years later Mathurin Brisson (Figure 1.1) published his own classification of animals and placed giraffes in their own genus rather than in the genus *Cervus* and selected arbitrarily, it seems, the name *Giraffa* for the genus from the several alternatives that he listed. Brisson characterized his new genus as having

> *Maxillary incisors none, mandibular eight. Cloven Hooves. Horns simple projecting upwards. Front legs much longer than the hind legs.*

He also gave giraffes the species name *giraffa* with its chief characteristics described as:

> *The length of the body, from head to tail is 10 feet. Neck 7 feet long. It is able to carry the head 16 feet from the ground: the front thighs (not true legs) are much longer than the posterior. It is not able to grasp food from the ground except by spreading its front legs wide. Two horns about 6 inches long project from the forehead, bent for a short distance towards the top. The ears are quite large. It has a mane like a horse, from the top of its head to the back. (They can be without a mane, by which perhaps the sexes are distinguished.) The tail is thin and short, and hairy at the point. The whole body is covered in spots, not indeed all round but either squarish, or round. It lives in Africa, and particularly in Ethiopia.* (Brisson, 1762)

This description is more detailed than that of Linnaeus, but does not make the case that giraffes should be placed in their own genus. What is usually given as the reason for Brisson's classification is that he thought that the horns of giraffes were permanent, unlike those of deer, which are shed each year. Perhaps in using the word "simple" to describe the horns he meant "fixed." If they were "fixed," that difference would merit placement in a genus different from that of deer. Whether the horns of giraffes are shed would not be resolved for another 20 or more years, and even then Brisson's classification did not prevail. Nevertheless, his description is remarkably accurate and impressive, as Brisson could not possibly have ever seen a giraffe. Nor had Linnaeus. The first specimens of giraffes would only be sent to Europe from southern Africa in 1780, and the first living giraffes would only arrive in Europe 46 years after that. Having never seen a giraffe, Brisson and Linnaeus had to rely on illustrations and descriptions. Both taxonomists were paraphrasing the descriptions of giraffes seen in captivity in Cairo by Pierre Belon (his description was also copied by John Ray, [Harper, 1940]), and by Petrus Gillius, a contemporary of Belon, who had seen three there. Both Belon and

Gillius were French naturalists who had traveled widely in Egypt and Arabia in the mid-1500s on behalf of Francis I of France. Four years after returning to France, in 1553, Belon published an account of his travels with illustrations: *Les observations de plusieurs singularitez et choses memorables trouvées en Grèce, Asie, Judée, Egypte, Arabie et autres pays étrangèrs* (Observations of several remarkable and memorable things found in Greece, Asia, Judaea, Egypt, Arabia, and other strange countries) "I saw," he wrote,

> *an animal at the castle of Cairo, which is commonly called Zurnapa; the Romans anciently styled it camelopardalis, a name compounded of leopard and camel, for it is sprinkled with spots like the first, and has a long neck like the latter. It is a very beautiful animal, as gentle as a lamb, and more sociable than any other wild beast. Its head is almost like that of the stag, excepting its size; on it are two small horns, about half a foot long, covered with hair; those of the male are longer than those of the female. They both have ears as large as those of a cow, and the tongue black, like that of the ox; it has no incisive teeth in the upper jaw; its neck is long, straight, and slender; its horns round; its legs thin and long, but so low behind that the animal appears to be sitting; its feet are like those of the ox. When it grazes it is obliged to spread its forelegs very wide, and even then feeds with great difficulty, therefore it rather chooses to feed on the leaves of trees than to graze in the fields, especially as its neck is exceedingly long, and can reach to a great height.* (Buffon, 1807)

Belon's drawing of one of the two giraffes he saw is shown in Figure 1.2. The Swiss physician and naturalist Conrad Gesner, who was an exact contemporary of Belon, in his *Historia Animalium* quoted Belon as saying that the horns of giraffes are shed annually, but nothing in Belon's description allows that conclusion. Perhaps Gesner had discussed the matter with Belon (they did not live all that far apart), and Belon had expressed the opinion Gesner quoted.

The Comte de Buffon (see Chapter 3) thought a better description than Belon's was one by Petrus Gillius (or Pierre Gilles). Gillius had been sent to Constantinople as a member of the French embassy with the aim of improving relations between the Ottoman Empire and France. He traveled throughout the regions abutting the Mediterranean Sea and saw

> *three camelopards at Cairo; on their heads are two horns six inches long, and in the middle of their forehead a tubercle rises to about the height of two inches, which appears like a third horn. This animal is sixteen feet high when he holds up his head. Its neck alone is seven feet, and it is twenty-two feet long from the tip of the nose to the end of the tail; its fore and hind legs are nearly of an equal height; but the thighs before are so long in comparison to those behind, that its back inclines like the roof of an house. Its whole body is sprinkled with large yellow spots which are nearly of a square form. Its feet are cloven like the ox; its upper lip hangs over the under; its tail is slender, with hair on it to the very point; it ruminates like the ox, and, like that animal, feeds upon herbage; its mane extends from the top of the head to the back. When it walks it seems as if its legs and flanks on both sides were alternately lame; and when it grazes, or drinks, it is obliged to spread its fore legs prodigiously wide.* (Buffon, 1807)

It is quite clear that Linnaeus and Brisson had used Belon's and Gillius's descriptions as their sources and that they were describing *G.c. typica* (see Plate 1.1), the variant that

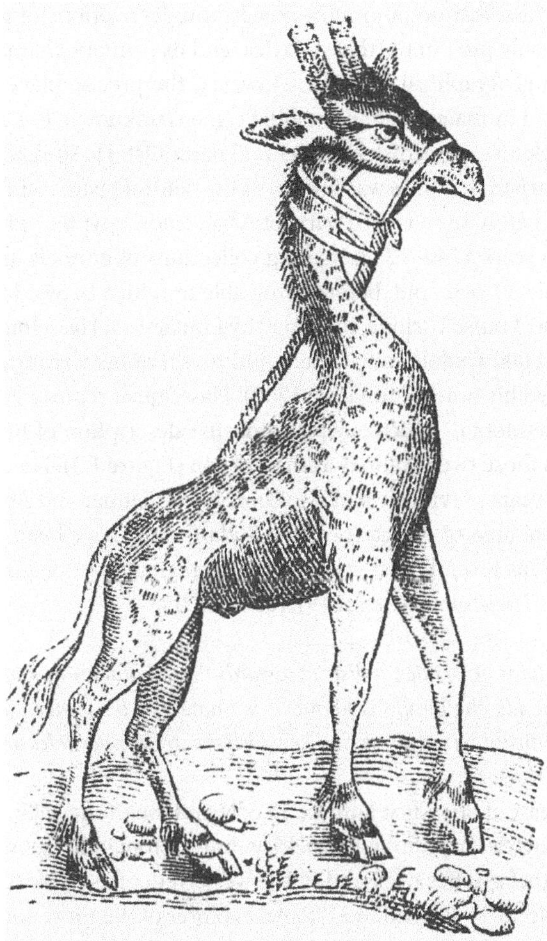

Figure 1.2. Pierre Belon's giraffe.
From Belon, 1554.

lived and still lives in northeast Africa, and whose characteristic feature is a third horn. How did Linnaeus and Brisson come to such different conclusions about the taxonomic place of giraffes? It is not easy to say. Perhaps Linnaeus believed Gesner while Brisson did not. Brisson's classification was not accepted immediately. The German naturalist Johann Friedrich Gmelin reclassified giraffes in the genus *Camelopardalis* in 1788, and Lorenz Oken, Extraordinary Professor of the Medical Sciences at the University of Jena, in 1802 reverted to the Roman form of *Orasius*. Classification was further complicated by the recognition that there were geographical variants. Nevertheless, once irrefutable proof that a giraffe's "horns" were permanent, Brisson's classification was accepted as correct, and giraffes became *Giraffa*, Brisson. However, the species name given by Linnaeus had historical precedence. Thus, Brisson's classification became *Giraffa* (Brisson) *camelopardalis* (Linnaeus), and the date of origin of this classification is usually given as 1758, the year Linnaeus described the species *camelopardalis*. And that is now the official scientific name of giraffes.

The scientific classification of giraffes is based on descriptions of giraffes that lived and still live (but only just) in northeast Africa, and its primary characters are defined, therefore, by this geographical location. However, the precise place from where the giraffes Brisson and Linnaeus were describing came is unknown, but a clue is provided by Frederic Hasselquist, a Swedish explorer and naturalist. He studied under Linnaeus at Uppsala and learned that little was known of the natural history of Palestine. He, like Belon and Gillius before him, visited parts of Asia Minor, Egypt, Cyprus, and Palestine and did so in the years 1749–1752, making collections of animals and plants he had seen. He died, only 30 years old, before being able to return to Sweden. The Queen of Sweden at the time, Louise Ulrique, informed by Linnaeus of Hasselquist's misfortunes, made available 14,000 rixdollars to obtain and preserve his writings and specimens. Linnaeus published his notes (Linnaeus, 1766). Hasselquist reported that giraffes lived in "Sennar and Aethiopia," which is why Linnaeus's description of his *Cervus camelopardalis* identifies those two localities as their origin (Figure 1.3). No doubt the giraffes seen in Cairo 200 years previously had also come from "Sennar and Aethiopia."

At that time that area of Africa (now the Sudan) must have been a highly suitable habitat for giraffes, as several other explorers of the region mention giraffes. John Lewis Burckhardt, in his *Travels in Nubia* (1819), recorded that

> *. . . in the mountains of Dender, a district towards the Atbara, and six or eight journeys south east of Shendi, the giraffe is found. It is hunted by the Arabs, Shenkorein, and Kowahel, and is highly prized for its skin, of which the strongest buckles are made.*

Atbara lies northeast of Shendi at latitude 17.5°N and longitude 34°E. Eduard Rüppell, better known for his description of Rüppell's vulture, wrote in his *Reisen in Nubien und Kordofan* (1829) that giraffes occurred in the western deserts of Northern Sudan near Dongola at latitude 18°N, longitude 31°E. An explorer of the more southern regions of Sudan, George Alexander Hoskins (*Travels in Ethiopia*, 1829), wrote that the

> *Giraffe is found in great numbers on the road from Edabbah (on the Nile) to Kordofan, between Sabrian and Jebel el Arazi and behind Kordofan on the Bahr al Abiad, in the territory of the Baggara.*

The Baggara live in Darfur and North and South Kordofan.

Giraffes clearly were plentiful in northeast Africa, but the most definite, or most restricted, locality mentioned by Hasselquist, and thus by Linnaeus, is Sennar, and Sennar therefore can be regarded as the locality of the giraffes that defined the genus and species (Harper, 1940). More precisely, the origin of giraffes that defined the species can be said to lie between latitude 10°–15°N and longitude 30°–35°E (Figure 1.3). A giraffe drawn for William Jardine to illustrate the description of giraffes that appeared in 1838 in Volume 21 (of 44) of his *Naturalists Library* (Frontispiece) was "taken in the vicinity of Senaar and Darfur," in other words, in the same general area. The giraffes that used to inhabit this part of Africa were described much later by Lydekker (1904a) as *G. camelopardalis typica* and his drawing of one of them from life (see Plate 1.1) shows features similar to those of the first giraffes taken to Europe and painted by Huet and Agasse (see Plates 3.7 and 3.8 in Chapter 3). Thus, we can be sure that the type specimen of modern

Figure 1.3. The type locality of *G.c. typica*. Sudan and the areas of Darfur, Kordofan, and Sennar where *G.c. typica* was seen in the eighteenth and nineteenth centuries.

extant giraffes, *Giraffa camelopardalis*, is defined by those that once lived in Sennar, Kordofan, and Darfur. Few live there today. Although the focus of all these explorations was northeastern Africa, there was contemporaneous exploration of southern Africa. In the 1660s two giraffes were seen at the Spoeg River mouth in northwestern South Africa, and more were to be seen in the 1770s in the northern Cape region and in southwestern Africa (Namibia; see Chapter 3), and from these sightings it can be said that the African range of giraffes, at least over the last 300 years, has been between latitude 31°S and 15°N. It is a purely African species.

Species, Subspecies, and Variants

Despite the efforts of Linnaeus and Brisson, in 1835 William Swainson, in his *Geography and Classification of Animals*, gave the name *Camelopardalis antiquorum* to the "northern giraffe" because "elaborate observations by Mr Rüppell have enabled us to characterise it as a distinct species from the giraffe of Southern Africa" to which Swainson gave the name *Camelopardalis australis*. On the basis of two giraffes shot by Francois LeVaillant (Harper, 1940; see Chapter 3), Rene Lesson, a French naturalist, who was both an author for and editor of *Nouveautableau du règneanimal* (A New Classification of Animals), published in Paris in 1842, this southern species was given the name *Camelopardalis capensis*. By 1897 eight subspecies had been identified and their names were reviewed by British zoologist William de Winton (Table 1.1). The

Table 1.1 Species Names Given to Various Subspecies/Variants of Giraffes

De Winton (1897)	Lydekker (1904b)	Daggand Foster (1976)	Current Names (Figure 1.5)	Fennessy (2016)
antiquorum	*angolensis*	*angolensis/ capensi*	*angolensis*	*Camelopardalis* (=*rothschildi* + *typica/ antiquorum* + *peralta*)
australis	*antiquorum*	*camelopardalis* (=*typica/ antiquorum*)	*antiquorum*	*giraffe* (=*angolensis* + *capensis*)
aethiopicus	*capensis*	*giraffe* (=*wardi* + *capensis*)	*camelopardalis*	*reticulata*
biturigum	*congoensis*	*peralta*	*giraffa*	*tippelskirchi* (=*thornicrofti*)
camelopardalis	*cottoni*	*reticulata*	*peralta*	
capensis	*peralta*	*Rothschildi* (=*cottoni*)	*reticulata*	
giraffa	*reticulata*	*thornicrofti*	*rothschildi*	
sennaarensis	*rothschildi*	*tippelskirchi* (=*congoensis*)	*thornicrofti*	
	tippelskirchi		*tippelskirchi*	
	typica			
	wardi			

names were based on the idea that each subtype had distinct features and the subtypes were geographically isolated (hence their specific names), although one of the given names was *antiquorum* borrowed from William Swainson. A specific name listed by de Winton was *biturigum*, a name based on a fossil supposedly found near Lyons in France, and thus near the home of the ancient Bituriges tribe, a tribe of Celts who famously battled Caesar during his conquest of Gaul. When writing on the fossil remains of the family Giraffidae, Forsythe Major (1891) said that this species "had haunted paleontological papers for 50 years from which it is high time that it should disappear." The "fossil" actually was the mandible of a modern giraffe, aged to look ancient.

A significant difference between northern and southern species emphasized by de Winton was that northern forms had a distinct third "horn." The "horn" had also been seen by Petrus Gillius nearly 350 years earlier in the giraffes he saw in Cairo. This "horn" is never seen in the southern forms (Figure 1.4), and de Winton speculated that it was the vestige of an appendage of an extinct ancestral form.

In 1904 Lydekker subdivided *G. camelopardalis* into subspecies by using variations in skin markings, ossicones, geographical distributions and the name of the first describer (Lydekker, 1904b). Physical characteristics are hereditable, and therefore subject to natural selection, so geography and characteristics become aligned (although Anne Dagg [1962c] wrote that "subspeciation should not be based on spotting, coat colour or horn characters because they are too varied"). Nevertheless, using coat and horn characters,

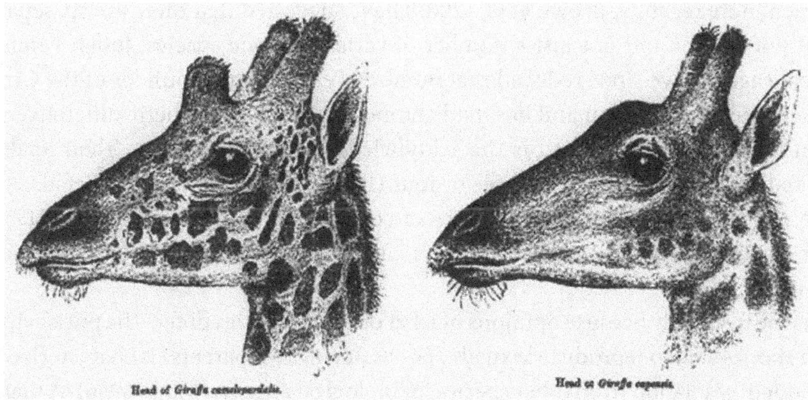

Figure 1.4. The heads of *G.c. camelopardalis* from North Africa and *G.c. capensis* from southern Africa.
From de Winton, 1897 with permission of John Wiley and Sons.
From de Winton, 1904b, with permission of John Wiley and Sons.

Lydekker distinguished 11 subspecies: *G. typica* (Ethiopia), *G. antiquorum* (Sudan), *G. peralta* (West Africa), *G. reticulata* (Somalia), *G. cottoni* (Uganda), *G. tippelskirchi* (Kenya), *G. rothschildi* (Baringo), *G. congoensis* (Congo), *G. angolensis* (Angola), *G. wardi* (Transvaal, South Africa) and *G. capensis* (Cape, South Africa) (see Table 1.1). He too noticed a progression "from south to north the gradual passage of a two-horned animal into a one with three horns," with the frontal horn of southern forms being at best "a low tuberosity or swelling."

In 1911 Lydekker added two more names. One giraffe he thought was a variant of *G. reticulata*, and because "the white lines (separating patches) are rather wider with a tinge of blackness" he gave it the name *G. reticulata nigrescens*. This name was not to last. To a second subspecies he gave the name *G. thornicrofti*. It has lasted and is the name of the small independent population of giraffes that inhabit the Luangwa Valley National Park in Zambia. Lydekker based this classification on a skin and part of a skeleton of a giraffe sent to the British Museum by H. S. Thornicroft, Native Commissioner of the Petauke district of Zambia. Lydekker's 1904 list was modified by Dagg and Foster (1976). They merged *congoensis* with *antiquorum* and *cottoni* with *rothschildi*, created *G. giraffa* to include *wardi* and *capensis*, merged *G. typica* within *G. camelopardalis* and included *G. thornicrofti*, all as additional subspecies to the remaining ones originally identified by Lydekker. Further varieties, such as *G. infumata* = *capensis* (Krumbiegel, 1951; East, 1998) or = *thornicrofti* (Fennessy, 2013), *G. renatae* (Lagos giraffe, Krumbiegel, 1971), *G. hagenbecki* = *reticulata* (Knottnerus-Meyer, 1910; Krumbiegel, 1951), *australis*=*capensis* (de Winton, 1897), and *schillingsi*= *tippelskirchi* (Krumbiegel, 1951), have also been named (Table 1.1). All-in-all, therefore, at least 21 subspecies/varieties of *G. camelopardalis* have been recognized at some point or another, given a name found to be satisfying to their creator, with each these days classified by the prefix *G. camelopardalis* (*G.c.*). The numerous varieties suggest that giraffes have a flexible, robust, and adaptable genome, with phenotypic effects that different observers have been able to identify.

Much more recently, Brown et al. (2007) have suggested that there are six separate species of giraffes and not just a number of variants of one species. Julian Fennessy and colleagues have since reduced that number. Fennessy is cofounder of the Giraffe Conservation Foundation and has made numerous studies of genetic differences between variants of giraffes and is the acknowledged expert on them. Their analyses have reduced the six species to three or four (Bock et al., 2014; Fennessy et al., 2016; Table 1.1), a conclusion that is a source of controversy (Bercovitch et al., 2017). In other words, the classification of giraffes since Linnaeus and Brisson has become very messy.

It is messy mostly because opinions of what defines a species differ. The purist view is that a species able to reproduce sexually (i.e., requiring two parents) is characterized by reproductive isolation from other species for biological reasons (Mayr, 1996). A male or female of a biological species, in other words, cannot reproduce with the opposite sex of another species because, for example, of differences in reproductive organ anatomy or because chromosomes do not match. The alternative to a biological species is a genetic species. A genetic species is one quite able to interbreed with another member of the same binomial species but is isolated from it for, say, geographical reasons (Baker and Bradley, 2006), and has a genome that differs by some threshold. So the question is whether the many types of giraffes are a single biological species with regional variants, or are they a single species made up of several isolated genetic species, some of which may be evolving into distinct biological species? If there were different biological species of giraffes, then unlike genetic species which can interbreed, each biological species could become extinct. They could not be recreated from other types of giraffes. Currently, geography and man-made barriers do separate giraffes into groups that occupy specific geographic regions (Figure 1.5), and those regional variants do interbreed easily and successfully. Hybrids between *camelopardalis* (*typica*)(the northern giraffes) and *giraffe* (the southern giraffes) (Lydekker, 1901; Beddard, 1906), between *camelopardalis* (*typica*) and *peralta* (Krumbiegel, 1951; Hassanin et al., 2007), among the several East African giraffes (Stott, 1959; Dagg, 1962; Brown et al., 2007), between *angolensis* and *giraffa* (Dagg, 1962), between *giraffa* (*wardi*) and *reticulata* (Dagg, 1968), between *rothschildi* and *tippelskirchi*, *tippelskirchi*, and *camelopardalis* (*typica*) and *typica* and *reticulata* (Krumbiegel, 1951) are well known. Thus, the criterion of reproductive isolation is not met, and modern efforts made to preserve the genomes of the current subtypes as listed in Table 1.1 as if they were distinct or "pure," seem, therefore, to be artificial. Giraffes, no matter where they live, belong to a single biological species.

However, analyses of the mitochondrial DNA that produces the enzyme cytochrome *b* have shown that giraffes can be separated into three genetically distinct groups: west African giraffes (*G.c. peralta* of Niger); north African giraffes consisting of *antiquorum*, *reticulata*, *rothschildi*, *thornicrofti*, and probably *camelopardalis* (*typica*); and southern African giraffes consisting of *angolensis* (which properly is *capensis*) and *giraffa (wardi)*, with *tippelskirchi*, the Masai giraffes, showing some affinity to the southern giraffes as well as to the northern group (Brown et al., 2007; Hassanin et al, 2007). Genetic differences range from 0.15% to 6.9%, that is, in some cases meeting the arbitrary criterion for intraspecific variation (<2%), in others for subspecies (>2<11%) but not speciation (>11%) (Bradley and Baker, 2001). The genetic differences emerged ~1 million years ago (Mya) (Brown et al., 2007) and thus after *G. camelopardalis* had

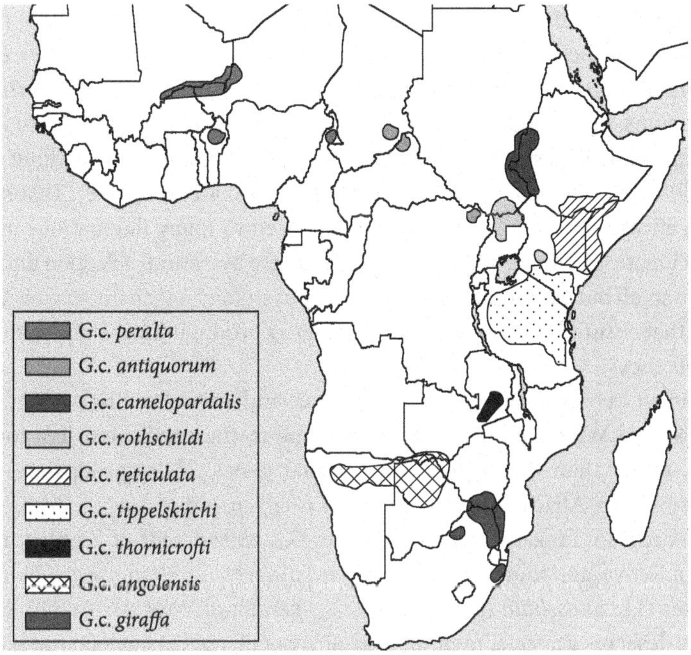

Figure 1.5. The distribution of giraffe variants in Africa.
Published with permission of its author Altaileopard.

evolved from its ancestors. The conditions that could have stimulated emergence of genetic species then were the habitat fragmentation, population isolation, and migration that most likely occurred around 800,000 years ago during the Early–Middle Pleistocene Climate Transition when the intervals between Northern Hemisphere ice ages changed from occurring every 41,000 years, coherent with changes in the obliquity of the earth's axis to every 100,000 years, coherent with its eccentricity (Imbrie et al., 1993; see Chapter 6). The effects of that transition on vegetation and climate could have led to inbreeding and fixing of genes, with distinct skin markings and ossicones the visible outcomes. The idea that there are three genetically distinct groups was modified by Bock et al. (2014), also based on mitochondrial DNA. They decided that the various types of modern giraffes were derived from a single ancestor (see Chapter 6) and could be separated into a northern group comprising *G.c. typica* from Kordofan, a group consisting of the reticulated giraffes of Kenya and *G.c. peralta* of Niger in West Africa, and a southern group comprising *G.c. capensis/angolensis and G.c. giraffa* from South Africa and *G.c. tippelskirchi* in Tanzania, and that among the southern African giraffes there is a west-east division between *G.c. capensis/angolensis* in the west and *G.c. giraffa* in the east. Mutations in mitochondrial DNA act as a clock that tells when a species might have diverged from its ancestors. Bock et al. estimated that separation into the three groups might have occurred as long ago as 2.6 million years ago (Mya), which is about the time that there was a mini-radiation of the very first African giraffes to produce the four classic Plio–Pleistocene giraffes (see Chapter 6). The division between *G.c. capensis* and *giraffa* was estimated to have begun 1.4 Mya. However, because modern variants are regarded as being variants of *G. camelopardalis*, none could have arisen before *G. camelopardalis* itself appeared in the fossil record, and that was ~1 Mya (see Chapter 6). Nevertheless, identifiable mitochondrial mutations may have originated 2.6 Mya and, if maintained in descendants, when circumstances occurred that created sub-populations, inbreeding could have produced the genetic groups, but so far has not produced any biological species. This is not to say the mutations were without effect. It is quite possible that they reflect the variance hypothesis of Potts (1998). That hypothesis posits that ancestral giraffes will have been exposed to many fluctuations in environment and climate and they will have adapted to these by natural selection and will have evolved the small but essential changes to their physiology, biochemistry, and behavior, and so on, that ensured survival without necessarily producing any significant morphological differences.

In the most recent studies of both nuclear and mitochondrial DNA, Fennessy et al. (2016) and Winter et al. (2018) have come to the conclusion that four groups of giraffes, rather than three, share very similar genes. Those groups are, first, *G.c. giraffa* of southern Africa, which includes *G.c. capensis* and *G.c. angolensis*; second, *G.c. tippelskirchi* in Tanzania, which includes *G.c. thornicrofti* of Zambia; third, *G.c. reticulata* of Kenya; and fourth, *G.c. camelopardalis/typica/antiquorum* of North Africa, and includes *G.c. rothschildi* of Kenya and *G.c. peralta* of West Africa (Table 1.1). *G.c. thornicrofti* deserves a special mention. Fennessy et al. (2013) have pointed out that its isolation in the South Luangwa Valley means that it is 400 km from the nearest northern giraffe populations in Tanzania and 450 km from the giraffe populations to the south in Zimbabwe. From studies of its DNA by Fennessy et al. (2013), this variant seems to have been an offshoot of *G.c. tippelskirchi* that has become isolated from its parental group and may in time become a true biological species. That conclusion is supported by another analysis of the genetic makeup of giraffe variants (Petzold and Hassanin, 2020). Petzold and Hassanin used a complicated array of molecular methods to identify the number of giraffe species. Depending on the method used, between two and six genetic species could be identified, but they concluded that there were, in fact, just three—*G. camelopardalis*, *G. giraffa*, and *G. tippelskirchi*, with *G. tippelskirchi* giving rise to *G.c. thornicrofti* by male dispersal.

Genetic subgroups arise by restriction or even absence of gene flow between individuals and are precipitated by barriers, so it is possible for genetic isolation to evolve without reproductive isolation being a consequence. Perhaps the best example of this process is *G.c. peralta.* It once was part of the community of giraffes that arose in northeastern Africa as the product of the radiation of giraffes that occurred there ~3 Mya (see Chapter 6) or was a variant/subspecies that arose after the appearance of *G. camelopardalis* about 1 Mya. Whatever its origin, it chose a moment—precisely when is unclear, but within the last 2.6 million years (Hassanin et al., 2007)—to migrate westward across the Sahara Desert or along the North African coast. The Sahara became a desert about 3 Mya, but intermittent periods of humidity produced intermittent habitats of grassland and trees, and lakes filled by perennial rivers, a habitat ideal for giraffes as well as other large species and one that could easily have allowed migration across it. Whatever the route *peralta* chose, it was a successful migration, for *G.c. peralta* once occupied the whole of northwestern Africa from Morocco in the far north to Niger and Nigeria in the south (Ciofolo, 1995). The most recent Sahara humid period began 10,000 years ago, coinciding with the end of the last ice age, and ended ~5000 years ago.

The Sahara became desert again quite quickly (in geological time), isolating *G.c. peralta* from *antiquorum* in northeastern Africa and *reticulata* in eastern Africa. The Sahara was not the only obstruction to gene flow between *peralta* and the giraffes of eastern Africa. Another was Mega Lake Chad (Hassanin et al., 2007), which 6,000 years ago stretched for 750–1000 km (500–600 miles) between 10°30' N to 18°N and once had a surface area of 300,000 km^2 (deMenocal et al., 2000; Leblanc et al., 2006), effectively cutting off migration between east and west. Lake Chad these days has a surface area of ~1,500 km^2 and continues to shrink, but human activity has replaced it as the barrier to migration. Another and obvious topographical barrier separating *peralta* from *capensis/angolensis* in Namibia and Angola in southwestern Africa is the tropical Congo rain forest and the Congo River that runs through it. African giraffes, apart from their primitive relative, the okapi (*Okapia johnstoni*), have never inhabited tropical forests, are very reluctant swimmers, and do not readily cross perennial rivers and especially not one as large as the Congo River. Although *peralta* is not yet reproductively isolated from giraffes to the south and east, it is possible that it could become so over time. Its mitochondrial genome currently differs by 4.6% from *capensis/angolensis* compared with 1.44% from *antiquorum* and *reticulata* (Hassanin et al., 2007). Despite the historical isolation between *antiquorum* and *peralta*, they can hybridize (Hassanin et al., 2007), and *antiquorum* and *camelopardalis* (*typica*) are contiguous, genetically indistinguishable, and certainly are not separated by impassable topographical barriers, so it is likely that *peralta* currently can hybridize with *camelopardalis* as well. An apparent *camelopardalis* (*typica*) was captured in 1907 near Timbuktu in Mali (16°45'N, 3°4'W) (Trouessart, 1908), which is at the same latitude as the *peralta* population but 500 miles further west. This giraffe was almost certainly a *peralta*, the skin markings of which, easily mistakable for those of *camelopardalis* (*typica*), led to the misdiagnosis of type.

Movement of *capensis* (*angolensis*) southward into the northwestern part of South Africa from Namibia is obstructed by the Orange/Gariep River. Apart from the sighting of two giraffes south of the Orange/Gariep River in the seventeenth century (GodeeMolsbergen, 1916) and two shot south of it in the eighteenth century (Chapter 3), none has been recorded in the Cape Province south of the river (apart, of course, from any introduced in modern times into Conservancies/National Parks). Movement between Namibia and Angola is unobstructed, but movement eastward from Namibia is, by the Kalahari Desert. Historically, movement eastward may also have been obstructed by a "cryptic rift" characterized by a perennial Okavango River and the Fish River Canyon (Bock et al., 2014), although there are no geological fault lines to suggest the existence of such an isolator (Cooke, 1978). Though giraffes do inhabit parts of the Kalahari (Deacon and Bercovitch, 2018), this desert effectively separates modern *giraffe* (*wardi*) in Zimbabwe/South Africa in the east and *capensis* (*angolensis*) in the west, and maybe it is the origin of the genetic west-east split identified by Bock et al. in 2014. The Kalahari barrier, combined with human activity and rivers that limit movement from Angola into Zimbabwe/Zambia, continues to isolate *capensis* (*angolensis*). There are no obvious natural barriers between *tippelskirchi*, *rothschildi*, and *reticulata*, or between them and northern variants, but movement through their ranges and intermingling with *thornicrofti* clearly are severely limited by distance, human activity, and the boundaries of National Parks. Nor are there insurmountable natural barriers between southern African *giraffa* and eastern African giraffes, although man-made, artificial barriers are plentiful and, in addition, giraffes would have to cross the Limpopo River, which forms the northern boundary of South Africa and the Kruger National Park, as well as the far more formidable barrier of the Zambezi River, to reach East Africa. The Zambezi encircles Zambia, separating Zambia from Angola, and runs along the northern border of Zimbabwe before emptying into the Indian Ocean, via a large delta, in Mozambique.

The "Transvaal" giraffes identified by Lydekker (1904b) as *G.c. wardi* do not seem to have occurred naturally in kwaZulu-Natal (Goodman and Tomkinson, 1987) or the southeast of the Cape Province (Gordon, 1777–1779; Barrow, 1801; Skead, 1958, 1980). Some rock portraits of them have been found in the southeast Cape, made probably by artists who had seen them further north (Barrow, 1801), but there are no written records of them or fossils. Some giraffe bones have been found at archaeological sites further north in kwaZulu-Natal dated to 1,000 to 4,000 years ago, so it is possible that giraffes did once migrate further south down South Africa's east coast but became extirpated either by human predation and/or historical habitat destruction before early explorers, like Barrow, arrived in the Cape and Natal (Bond and Lofell, 2001; Cramer and Mazel, 2007). Exactly why modern giraffes have tended to prefer the country further north is not clear. Perhaps it was a matter of ambient temperature. Giraffes, because of their shape, do not like the cold habitats that prevail further south. Apart from perennial rivers, there are no natural obstructions to prevent migration southward, and those introduced into National Parks in the Cape Province and kwaZulu-Natal in recent times flourish, but modern giraffes are not natural inhabitants of the southeast of South Africa below the great rivers of its eastern seaboard, a fact first recorded in 1778 by Gordon while on his second journey to the Eastern Cape (Chapter 3), nor are they natural inhabitants of the northwest of South Africa south of the Orange/Gariep River, as Jacobus Coetsé, Hendrik Jacobus Wikar, Robert Jacob Gordon, and William Paterson were to discover (see Chapter 3). However, although almost without exception ancestral giraffes lived to the east of the Nile and Congo Rivers and above the Zambezi River (see Figure 6.1 in Chapter 6), the first *Giraffa* to enter Africa some six million years ago, *G. jumae*, confirmed from fossils, was found at the southernmost tip of Africa, near modern-day Cape Town (see Chapter 6), and it would have faced and had to overcome the same natural obstacles that more recent giraffes would face if they tried to migrate. Migration of modern giraffes north into Zimbabwe across the Limpopo is possible when the river is dry and, theoretically, so is migration north through Mozambique up the east coast toward Kenya, if there were no human restrictions on movement and the Zambezi was dry. The overlap between the genetic profiles of *giraffe* (*wardi*)and East African giraffes shows, though, that hybridization between *giraffa* and Masai giraffes (*tippelskirchi*) and between *tippelskirchi* and the other East African giraffes occurs, but infrequently (Brown et al., 2007). The genetic differences that exist particularly between *giraffa* and the East African group of *tippelskirchi*, *rothschildi*, and *reticulata*, in the absence of any identifiable habitat or topographic barriers to gene flow between them, and the relatively little formal evidence of hybridization, strongly suggest that natural and man-made obstacles are formidable and can severely restrict the migration of modern giraffes.

Are some types of giraffes already reproductively isolated and so meet both the classic definition of biological species as well as genetic species? There is no evidence of reproductive isolation of biological origin, but apart from physical barriers, physiological and behavioral factors have been proposed as possible isolating mechanisms (Brown et al., 2007). One factor that has been suggested is that northern and southern African giraffes have evolved different breeding cycles because "the dry season occurs in July and August in the Sahelian zone north of the equator in contrast to December to March in the south" (Brown et al., 2007). In fact, December to March is the rainy season of the southern Africa populated by giraffes, not the dry season. Giraffes do have seasonal spikes of births, but they are not regarded as being seasonal breeders (Hall-Martin et al., 1975a; Skinner, 1978) and genetic isolation because of different breeding cycles is unlikely. Another isolating mechanism might be a specific mate-recognition signal, as proposed by Patterson (in Vrba, 1985a), or Koinophylia (Koeslag, 1990; Chapter 6). Another is that skin markings are sufficiently unique to stimulate pelage-specific mate selection and cause imprinting in giraffe calves for conspecific pelage patterns (Kendrick et al., 1998; Brown et al., 2007) but, at least in zoos, newly introduced adult giraffes from widely separated types and, therefore, quite different pelage patterns, cross-breed freely with preexisting captive giraffes (Beddard, 1906; Dagg, 1968; Hassanin et al., 2007), so pelage-imprinting cannot be an important isolating mechanism, despite differences in coat markings that can be startling.

From the analyses done so far, it is possible to conclude that the modern species of giraffes, *G. camelopardalis*, is a single species consisting of a combination of regional or ecological variants that are not genetically distinct, and others in which their genes have become sufficiently isolated by natural and man-made barriers to cross-breeding so that they are genetically distinct, have identifiable characteristics such as coat markings suited to their habitat, and some with morphological differences such as the median "horn" of northern species. None of the types, however, is a separate biological species, and they would interbreed if able to. A corollary is that the various types are outbred gene pools that are a valuable source of new genes. Consequently, the often-expressed view that subspecies of giraffes are threatened with extinction is too bleak. Nevertheless, and with good reason, giraffes have recently been re-classified from "least concern" to "vulnerable" by the International Union for Conservation of Nature (IUCN) that decides these things (IUCN, 2016-3). Giraffe numbers have declined by a total of about 80,000 in East Africa (Kenya) and northeastern Africa (South Sudan) over the last 40 years or so, with *G.c. tippelskirchi* declining by 52%, *G. c. reticulata* by 80%, and *G. c. typica* by 97% (Courchamp et al., 2018). However, *G.c. peralta* has increased in Niger by the determined efforts of Isabelle Ciofolo from around 100 in 1995 and on the verge of extermination (Ciofolo, 1995) to about 700 now; *G.c. thornicrofti* is stable in Zambia and, in the other southern African countries, numbers have increased by about 30,000. Declines where they occur are attributed to indifference and competition with humans for resources. Africa has the world's highest human fertility rate, with fulfillment of Malthus's prophesies the consequence. There are now some 100,000 giraffes living in the wild and many hundreds in captivity. In some National Parks and private conservancies their numbers are a calamity, as culling is the only means of population control given that their natural response to over-crowding is migration which, by definition, is constrained by boundary fences and the predatory human populations that live on the other side. This is not to say that some of the small inbred populations like *peralta*, once decimated by droughts, indiscriminate poaching, and competition with humans for resources, and *reticulata* and *thornicrofti*, isolated by human activity, are likely to survive without protection, but as they are not true biological species, their numbers could be sustained if necessary by another human activity, namely translocation of giraffes from non-threatened populations that have similar pelage patterns. The continued success of the species even in marginal zones therefore seems to be subject only to human benevolence and a supply of browse, the absence of which so often in their past has brought extinction to giraffes.

If all modern giraffes are a single biological species, is there any value in giving names to variants? I think so. Names serve the useful purposes of identifying regional subtypes and preserving their historical type localities (Figure 1.5), even if the names do not follow revealed genetic groupings. A simple nomenclature encapsulating these two ideas would be *G.c. peralta* for northwest African giraffes in Niger, *G.c. typica* for the northeast African group, *G.c. reticulata* for the East African group, *G.c. capensis* for those in southwest Africa, and *G.c. giraffa* for those in the southeast of Africa, and these names will be used wherever appropriate in the chapters in this book that follow. In these five groupings, *typica* would recognize that this group was the defining type of the giraffes imported into Europe and used as diplomatic tools among Middle Eastern countries, and would include *antiquorum* and *camelopardalis. G.c. reticulata* would include *rothschildi, tippelskirchi,* and perhaps *thornicrofti,* as these giraffes have more-or-less reticulated skin markings. A suitable name for the southeast African giraffes, first called *wardi* by Lydekker, is *giraffa*, which would recognize the first observation of them by Cornwallis Harris in 1838. *G.c. angolensis* should, properly, be *capensis* to recognize that this variant was first sighted 450 years ago in the Cape Province of South Africa by the de la Guerre and Hop expeditions of 1663 and 1761, respectively, and by the Paterson and Gordon expeditions of 1778–1779, rather than in Namibia or Angola.

2
Some Firsts

For those of us who think that there is not anything much more important than knowing about giraffes, finding out when they made their first appearances in art and in the written word are mysteries worth pursuing. Writing itself began, perhaps, around 3500 BC, and one of its first products was a history of the world as seen through the narrow perspective of the inhabitants of the Middle East—the Old Testament. In one of the great stories in the Old Testament, Moses led the Hebrews out of Egypt to the Promised Land. The exodus happened in the reign of Ramses II (1290–1224 BC), under whose rule the Hebrews had become laborers and servants. According to Exodus 12:37, "about 600,000 on foot that were men, beside children, and a mixed multitude also with them"; perhaps an improbable two and a half million people in all made the journey, at a time when the world population, according to McEvedy and Jones (1978), was 25–50 million, and they presumably did so over a period of 40 years. After 40 years of wandering in the wilderness south and east of Palestine, they came to Jordan. Then, as now, the way to escape from hopelessness was by perilous migration. Famously, some avoided capture by Egyptian pursuers by crossing the Sea of Reeds, in the marshes of which the Egyptian chariots became hopelessly bogged down.

Feeding this multitude must have been extremely difficult, but some help was at hand. The Hebrews were given clear instructions as to what animals they could eat during their travels, and these have come down to us in Deuteronomy 14:4–5:

> *These are the beasts which ye shall eat; the calf of the herd, and lamb of the sheep, and the kid of the goats, the stag, and doe, and pygarg, and wild goat, and camelopard.* (Brenton, 1851)

These words were translated from the original Hebrew into Greek by the Septuagint, the 72 Jewish scholars who began the translation in the reign of the Greek King of Egypt, Ptolemy II Philadelphius (285–247 BC). The result was to be included in the Library of Alexandria. The work was slow and went on between the third and first centuries BC.

A pygarg is the spotty-nosed addax (*Addax nasomaculatus*). The Septuagint translated the Hebrew word *zemer* or *zamar* in the original text as camelopard, the camel-leopard, and this translation places the first written reference to a giraffe in the Old Testament. So it was that for the next 2,000 years all translations of the Old Testament listed giraffes as an approved source of food. Crucially, Saint Jerome, who revised the Septuagint and translated it from the original Hebrew into Latin in the years AD 390–405, retained the word "camelopard":

> *Hoc est animal quod comedere debatis: bovem et ovem, et capram, cervum et capream, bubatum, tragelaphum, pygargum, orygem, camelopardalis.* (Tweedale, 2005)

How Giraffes Work. Graham Mitchell, Oxford University Press. © Oxford University Press 2021.
DOI: 10.1093/oso/9780197571194.003.0002

Saint Jerome's translation was called the Vulgate, and he made it after perfecting his abilities to speak and translate Hebrew with the assistance of learned Jews. Saint Jerome must have decided, therefore, that the instructions Moses had been given were accurate: giraffes were meant, they were permissible food, and it was the correct translation. Moreover, giraffes met the required general standard for edible food set out in Leviticus 2:2–3:

> *These are the living things which ye shall eat among all the beasts that are on the earth. Everything that parteth the hoof and cleaveth the cleft of the hoofs, that bringeth up the cud among beasts that ye shall eat.*

This seemingly straightforward conclusion, has, however, proved to be a highly controversial matter, as ecclesiastical arguments with little consequence often are. The translation of *zemer* into first *camelopard* and then "giraffe" has been disputed partly for zoological reasons and partly for semantic ones. The zoological doubt relates to whether giraffes existed in Egypt at the time of the exodus. There can be no doubt that they did. Figurines of giraffes are etched into a ceremonial palette dated to ~5,000 years ago, at the height of the African Humid Period when the climate of the Sahara Desert and Egypt and surrounds was similar to that of East Africa today (Yeakel et al., 2014). In 3400 BC, the time of the first Egyptian Dynasty (and of writing), images of giraffes, much better even than those of later periods, were on pottery and knife handles (Dawson, 1927) and in tombs (Figure 2.1), and were used as the hieroglyph for "to foretell or prophesy." Thereafter images appear regularly until 1225 BC (Laufer, 1928), although in the period 1580 to 1350 BC giraffes were no longer shown in hunting scenes but as a rarity, suggesting that they had been exterminated from the lower Nile by about 1600 BC, or had been displaced by a growing human population and its by-products (Yeakel et al., 2014) or by an encroaching desert. They were still to be found, however, along the shores of the Red Sea and further south in Nubia and Somalia. The last giraffe image of a captured Somali giraffe appears in 1225 BC in the tomb of Rameses the Great at the

Figure 2.1. Left: Giraffe image from Hierakonpolis, probably ca. 3000 BC. Right: From Thebes ca. 1500 BC.

From Eastman, 1915, with permission of Springer Nature.

same time that Moses was leading the Hebrews out of Egypt. Therefore, giraffes were known at that time and could have been identified as a food source because they met the requirements set out in Leviticus and, indeed, meet all modern requirements for a source of kosher meat (Zivotofsky et al., 2002).

Semantic objections were expressed most strongly by Samuel Bochardt. His objection was that giraffes, at best, were rare in Palestine, if they existed there at all. Bochardt (1599–1667) was a French religious scholar and Pastor of the French Reformed Church. He had a thorough knowledge of the main oriental languages, including Hebrew, Syriac, Chaldeic, and Arabic. His opinions, therefore, were accepted without question. In his book the *Hierozoicon*, published in 1663, which described the animals of the Bible in detail, he decided that as giraffes had not occurred in Palestine/Judaea, the word *zemer* could not mean "giraffe." Their absence in Palestine is not surprising. Suitable browse is and was absent there, and giraffes were even less likely than Egyptian chariots to have been able to cross the Sea of Reeds to reach Palestine from Egypt. Moreover, Deuteronomy was based on where the Hebrews were coming from, not where they were going, so whether or not they lived in Palestine at the time of the Exodus has no bearing on the story. Nevertheless, Bochardt's view prevailed, and the result was that the word he introduced, and which now appears in Deuteronomy in most modern Bibles, is "chamois," although oryx and goat and wild sheep are also used. Bochardt used "chamois" supposedly because the Arab word *zamara* means "to spring." It describes the agile movements and leaps of chamois. However, no such word exists in Arabic (Driver, 1902), so chamois cannot be correct, nor even its derivative, mountain sheep.

Which of the two translations is correct? In 169 BC, at the time that the Septuagint were making their translation, there appeared a book *On the Erythraean Sea* by the Greek historian Agatharchides of Cnidus. He recorded in his book:

> *In the country of the Troglodytes there is also found the animal Greeks call "camelopard," an animal that, like its name, has in a certain sense a composite nature. For it has the spotted coat of a leopard and is the size of a camel and very fast, and its neck is so long that it obtains its food from the tops of trees.* (Burstein, 1989)

Agatharchides's description leaves no doubt that he knew of giraffes, could describe them accurately, and even knew where they lived. Troglodytes is the name the Ancient Greeks gave to cave dwellers, a people of mixed Arabian and Ethiopian stock who acted as guides for Greek military expeditions. They lived (and still live) in homes excavated in sandstone hillsides, in villages spread around North Africa in Tunisia, Morocco, and Libya, but also along the shores of the Red Sea. They were mainly pastoralists but raised cattle and hunted. They were thoroughly familiar with the wild animals that lived in their regions. Thus the coincidence of the publication of the Septuagint and the publication of Agatharchides's book is evidence that giraffes were known to Moses and to the Septuagint, and that the authors of Deuteronomy were not confused. Nor was Saint Jerome. Besides, as Joly and Lavocat (1846) have pointed out,

> *the chamois has never been an inhabitant of Egypt, while the giraffe has been represented on the many ancient monuments of the country, so it would not be surprising that Moses knew this giant quadruped.*

Lastly, when Mathurin Brisson classified giraffes and gave them their scientific name *Giraffa giraffa* in 1762, he recorded all the common names he knew. Among these is the entry, "The Hebrews call it *Zamer*." And of course their use by the ancient Egyptians as a hieroglyph confirms that they were well known by them and that the Egyptians knew about their behavior patterns. Bochardt's opinions, it would seem, were wrong. Giraffes should be among the animals listed in Deuteronomy, chamois should not be, and the Old Testament is where giraffes were first mentioned in writing.

In due course, curiosity followed. In ~100 BC Artemidorus of Ephesus, a Greek geographer, who probably was quoting Agatharchides, wrote of the Troglodytic country comprising the Nile valley and the coast of the Red Sea, where, he said, among other species, there abounded elephants, leopards, rhinos, and giraffes (Leyra, 2012). An unknown illustrator drew a giraffe, which was most likely in captivity in Alexandria, on the reverse of Artemidorus's papyrus (Elsner, 2016; Figure 2.2).

Figure 2.2. Artemidorus's giraffe.
From Elsner, 2016, with permission, LED Edizioni Universitarie.

Artemidorus's books in turn were often referred to by Strabo, a Roman geographer who lived from 64 BC to AD 24 and thus would have lived through a time of great political upheaval: the decline of the Republic, the assassination of Julius Caesar, civil war, the demise of Pompey, the emergence of Roman emperors, and the dictatorship of Augustus Caesar. I suppose not surprisingly, he traveled widely and produced his famous *Geographica* in installments over many years. It was an account of his travels through the Roman world. The penultimate part, Book 16, was written in ~AD 23, the year before he died. In this Book he describes the northeastern parts of Africa around the Nile and between the Sahara Desert and the Red Sea as mostly desert with "a few palm-trees and a thorny tree" (Jones, 1930) with the "thorny tree" now presumed to be *Acacia nilotica*. Further south, toward and into Ethiopia the region was watered by summer rains. There elephants, lions, hyaenas, rhinoceroses, hippopotamuses, and crocodiles abounded, and

> *In this region, also, are found camelopards, though they are in no respect like leopards; for the dappled marking of their skin is more like that of a fawn skin, which latter is flecked with spots, and their hinder parts are so much lower than their front parts that they appear to be seated on their tail-parts, which have the height of an ox, although their forelegs are no shorter than those of camels; and their necks rise high and straight up, their heads reaching much higher up than those of camels. On account of this lack of symmetry the speed of the animal cannot, I think, be so great as stated by Artemidorus, who says that its speed is not to be surpassed. Furthermore, it is not a wild beast, but rather a domesticated animal, for it shows no signs of wildness.* (Hamilton and Falconer, 1892; Jones, 1930)

Pliny the Elder, Gaius Plinius Secundus, was a prominent Roman administrator who found the time to write copiously about law, science, military matters, and even grammar, but of all his writings his *Natural History* is his greatest achievement. He began it in AD 77, two years before his death (he was in command of the Roman fleet in the Bay of Naples when Vesuvius erupted in August AD 79. He went ashore to get a closer view and was overcome by the ash and fumes). Pliny distilled his *Natural History* from reading the books of 100 other selected authors, to which he added many of his own observations, some incorrect and others selected because of his love of the odd, a category into which giraffes clearly fell.

In Book VIII, Chapter 26, of his *Natural History*, Pliny wrote briefly on camels, and then followed in Chapter 27 with (Rackham's 1943 translation):

> *Some resemblance to these* [the camels] *is passed on to two animals.* [In this chapter he mentions only one that resembles the camel. The other mentioned later is the ostrich.] *The Ethiopians give the name of Nabun to one that has a neck like a horse, feet and legs like an ox, and a head like a camel, and is of a ruddy colour picked out with white spots, owing to which it is called camelopard. From this it has subsequently been recognized to be more remarkable for its appearance than for ferocity, and consequently it has also got the name of "wild sheep." It was first seen at Rome in the Circus games held by Caesar the Dictator.*

Pliny, obviously, had never seen a giraffe, but he had read Agatharchides, Artemidorus, and Strabo.

Given the extent of the knowledge of giraffes in the ancient world, it is surprising that they are not mentioned by Herodotus (485–425 BC) who visited Egypt and Libya, nor, apparently, by Aristotle (384–322 BC), and giraffes do not appear in Greek art of that time. However, Aristotle does mention an animal in his *Historia Animalium* that, with some imagination, could be a giraffe. In Chapter II he wrote:

> *The hipellaphus as it is called has a mane upon the point of its shoulder, and so has the pardium, though both these have a thin mane from the head to the shoulder, and the hipellaphus has a beard upon its larynx.* (Cresswell, 1878)

Only a giraffe has a mane as described by Aristotle, but it does not have a beard. In Chapter IV is:

> *The Egyptian river-horse* [which he called "pardium"] *has a mane like a horse and a cloven hoof like an ox: it has a flat face; the talus [ankle bone] is like that of other animals with cloven hooves, and it has large projecting teeth; it has a tail like a hog, and utters a sound like the neighing of a horse; it is about the size of an ass and its skin is so thick that shields are made of it; its intestines are like those of a horse or ass.*

Aspects of this description bear a vague resemblance to a giraffe (or okapi), especially the reference to the skin, but others like the reference to intestines and its "neigh" are not accurate for a giraffe (or indeed any other ruminant or animal). Could Aristotle have known about giraffes? The 500 or so animals that he described are principally those of Greece and of the countries with which the Greeks had frequent commercial dealings. They included, of course, Egypt and parts of North Africa, so he almost certainly could have heard of giraffes. A translation of Aristotle's *History of Animals* was made by Peter Simon Pallas (1741–1811). Pallas was Professor of Natural History at St. Petersburg. He substituted for *pardium* the word *hippardium*. Pallas felt so secure in his choice of words that in a book he published in 1778 he wrote, "The *Cameelparde*, whom Aristotle calls *Hippardium*, which exists among the antelopes, is a very special animal, which, as it were, holds the middle between the antelopes and the camels." It is a conclusion that has merit because to what other animal could Aristotle be referring? Aristotle's mention of a river-horse also should not be dismissed. He could not have invented it. He must have been repeating a Somali legend. Isak Dinesen in *Out of Africa* wrote thus of it:

> *Both Farah and Ismail had been to sea, for the Somali are a seafaring nation and were, I believe, in old days great pirates of the Red Sea. They explained to me how every live creature on the earth has got its replica at the bottom of the sea: horses, lions, women, and giraffes, all live down there, and from time to time have been observed by sailors. They also recounted tales of horses which live at the bottom of the rivers of Somaliland and at full-moon nights come up to the grasslands to copulate with the Somali mares grazing there and breed foals of wonderful beauty and swiftness.*

It was not just Pliny and Aristotle who had never seen a giraffe. Living giraffes were exceptionally difficult to find. The giraffe exhibited by Caesar in 46 BC and described by Pliny the Elder was a gift from Cleopatra (Balsdon, 1969). Dio Cassius Cocceianus, a Greek historian who wrote the *Annals of Rome* sometime between AD 200 and 229, described the ostentatious display of this giraffe more elaborately. According to Dio:

> *The first days of the fete he* (Caesar) *passed as was customary. On the last day, after they had finished dinner, he entered his own forum wearing fancy sandals and garlanded with all sorts of flowers; thence he proceeded homeward with the entire populace, so to speak, alongside escorting him, while many elephants carried torches. Here in honour of his daughter he had animals killed and contests between men in armour. . . . I will give an account of the so-called camelopard, because it was then for the first time introduced into Rome by Caesar and exhibited to all. This animal is in general a camel, except that it has sets of legs not of equal length. That is, its hind legs are shorter. Beginning from the rump its back grows gradually higher, appearing as if it would ascend indefinitely, until the most of its body, reaching its loftiest point, is supported on the front legs, while the neck stretches up to an unusual height. It has a skin spotted like a leopard, and for this reason bears the name common to both animals.*

This giraffe was killed by lions later that evening. Dio was paraphrasing Pliny. Dio's description is very accurate, but how he knew the details of a giraffe's shape is a mystery. What was his source? He could not have seen one, but had he lived 10 years longer until the time of the three Gordian Emperors, the father, son, and grandson, who ruled Rome over the period AD 238–244, he would have. Gordian III (the grandson) exhibited 10 giraffes at Rome at once, a remarkable feat of animal-handling. Even today, giraffes are among the most difficult of animals to capture, transport, and care for. These 10 giraffes were slain on the orders of the Emperor Philip the Arab at the millennium games in April AD 248 to obliterate the memory of the Gordians. Four giraffes were paraded at the triumph of the Emperor Aurelianus after his defeat of the Queen of Palmyra in AD 274 (Owen, 1838; Gibbon, 1897; *Historia Augusta*, 1924). While Dio and Pliny used secondary sources for their information, there was an eyewitness account of Caesar's giraffe given by a contemporary of Strabo. This eyewitness was the poet Quintus Horatius Flaccus (Horace). Horace was 19 years old at the time of Caesar's games, and he introduced giraffes into poetry. A translation of his description (in his Epistle 2.1) of a part of Caesar's triumph, from the original Latin is:

> *Nowadays a gentleman's pleasure is not of the ear anymore, but*
> *All of the eye, with gaudy spectators dazzling and empty*
> *Gauping for four or more hours at pageant displays of*
> *Squadrons of cavalry fleeing, and infantry companies routed;*
> *Then come the Kings with their hands tied behind them, deserted by Fortune,*
> *Then captured equipment, with chariots, carriages, carts, and*
> *Ship's beaks, and ivory loot, and a Corinth of bronzes in booty.*

Laughing Democritus, were he on earth, would find plenty to mock,
Watching the people gaze spellbound at a panther crossbred with a camel,
Or at a white elephant.
The entertainments themselves he would view less attentively
Than he would the spectacle far more diverting of all those spectators.

The first appearance of giraffes in fiction is in *An Aethiopian History* by Heliodorus, a third-century AD Greek author, who was born in Syria. The *Aethiopica* tells the story of Charicleia, daughter of the Queen of Ethiopia. Charicleia was born white skinned (i.e., an albino and missing the enzyme tyrosinase that synthesizes melanin) and was therefore abandoned by her mother for fear of the King's suspicions. Charicleia became a priestess at the temple of Apollo at Delphi, where she met a noble Thessalian called Theagenes. They fell in love and had numerous adventures, enduring long voyages, battles, abductions, piracy, robbery, and torture, in the best traditions of any modern penny-romance. The hero and heroine eventually live happily ever after. Although fiction, the novel included aspects of history as then known. In one interlude, Heliodorus explains why giraffes were regarded as the ultimate diplomatic gift trumping all others:

At last there came the ambassadors of the Axiomites, who brought him gifts, among which was a beast of wonderful and rare nature, as huge as a camel, whose colour was spotted, his latter part very low and lion-like, but his shoulders, forelegs and breast rose up to a height far beyond the rest. Its head was slender wherein he rolled his eyes terribly as if tinged with different colours. As soon as the beast was brought in sight all the people were afraid and suddenly called it after the fashion and principal parts of his body, Camelopardalis. At the altar of the Moon stood two bullocks and at the altar of the Sun four white horses to be sacrificed. When the monstrous and strange beast came in sight they were afraid as if they had seen a spirit and one of the bulls and two of the horses broke out of the hands that held them and ran about as fast as they could.

Giraffes in art have a longer history than in literature, with the most comprehensive and impressive surveys of that history made by Laufer (1928) and Spinage (1968). Both were unaware of the earliest illustration of a giraffe, a 6-m tall petroglyph sculpted between 9000 and 5000 BC in Niger (Figure 2.3). Obviously, *G.c. peralta* was well known to the local artist who made the petroglyph. The earliest known southern African drawings were made by the San Bushmen in the Tsodilo Hills in northern Botswana which lie near the Caprivi Strip in Namibia (see Plate 2.1), and were drawn at least in 3000 BC, in other words about the time that images of giraffes appeared in Egypt.

In the absence of living giraffes in Europe, images from observation became replaced by images of imagination. So it was that the first mention of a giraffe in a manuscript that was also accompanied by a drawing was in *A Christian Topography* by Cosmas Indicopleustes, a merchant of Alexandria, but also an explorer, theologian, and geographer. In monastic retirement, he wrote, in Greek, the 12 books that constitute the *Topography*, covering his knowledge of countries as far east as India. The books were written between AD 535 and AD 547, and their purpose was to prove the literal accuracy of the biblical universe—the Tabernacle of Moses—namely that the Earth was a rectangular plain, covered by heaven. In the center of the plain was Earth, surrounded by

Figure 2.3. Petroglyph of a giraffe (*G.c. peralta*) in Dabous, Niger.
Photo by Albert Backer for Wikimedia commons.

ocean, and beyond this was the paradise of Adam. The sun, much smaller than Earth, revolved around a conical mountain and was thereby obscured each night. The plain was inhabited by a wonderful array of beasts and plants, including giraffes, which he described, apparently, from his personal knowledge of them:

> *Cameleopards are found only in Ethiopia. They also are wild creatures and undomesticated. In the palace one or two that, by command of the King, have been caught when young, are tamed to make a show for the King's amusement. When milk or water to drink is set before these creatures in a pan, as is done in the King's presence, they cannot, by reason of the great length of their legs and the height of their breast and neck, stoop down to the earth and drink, unless by straddling with their forelegs. They must therefore, it is plain, in order to drink, stand with their forelegs wide apart. This animal also I have delineated from my personal knowledge of it.* (McCrindle, 1897)

Indicopleustes's illustration was redrawn by J.W. McCrindle, when he translated Indicopleustes's work in 1897 (Figure 2.4).

Nearly a thousand years after Indicopleustes, a Persian writer, Zakariya-al-Qazwini, or Cazurni, who lived from 1203 to 1283, in his *Marvels of Creation* tried to explain the origin of giraffes. His giraffes had cow-like horns because:

> *The giraffe is produced by the camel mare, the male hyaena and the wild cow. Its head is shaped like that of a stag, its horns like that of cattle, its legs like those of a nine year-old camel, its hoofs like those of cattle, its skin resembles that of a tiger, its tail like that of a gazelle, its neck is very long, its hands are long and its feet are short. A scholar, Timat by name, relates that in the southern equatorial region animals of various kinds congregate during the summer around water holes, being driven there by heat and thirst. If an animal of a certain species covers one of another species, strange animals like the giraffe are*

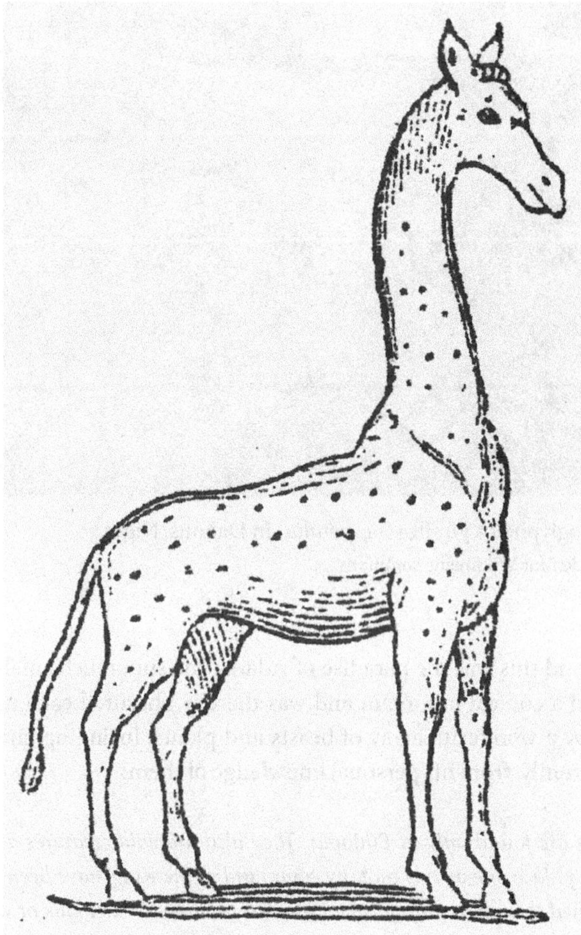

Figure 2.4. Indicopleustes's drawing of a giraffe.
From McCrindle, 1897, with permission of Cambridge University Press.

born: the male of the hyaena mates with the female Abyssinian camel; if the young one is a male and covers the wild cow it will produce a giraffe. (Joly and Lavocat, 1846)

Qazwini quite likely had seen a giraffe, as giraffes and other Western animals were introduced into Persia through trade routes as early as the thirteenth century. A Chinese painting of a giraffe (Figure 2.5) was made at about same time as, and perhaps was an interpretation of, Qazwini's description (Eastman, 1917). As a traveler along the Silk Road, the artist might have painted a giraffe he had seen in Persia. He could not have seen a giraffe in China, as it was only in the fourteenth century that the Chinese explorer Zheng He made his celebrated voyage to Africa, returning to China with, among other animals, living giraffes (Finlay, 2008).

In the Renaissance illustrations of giraffes became more accurate, and that change happened because of rising curiosity about all matters scientific. Explorers were sensitive to the need for accurate knowledge and brought back more complete

Figure 2.5. A Chinese painting of a giraffe.
From Eastman, 1917, with permission of Springer Nature.

descriptions of what they had found and seen, including giraffes. The first illustrated description of giraffes in English, a translation of Gesner's *Historia Animalium*, was in Edward Topsell's *Historie of Foure-footed Beasts*, published in 1607. The description went:

> *The pace* [walk] *of this beast differeth from all other in the world, for he doth not move his right and left foot one after another but both together, and so likewise the other, whereby his whole body is removed at every step or straine.*

Accompanying this description was a drawing by Melchior Luorigus made at Constantinople in 1559 (and so a decade or two after Petrus Gillius had been sent there) from a giraffe (Figure 2.6) owned by Suleiman the Magnificent, "the great Turke," Sultan of Constantinople, which is a fair image of a giraffe. However, the size of this animal is so extraordinary that it is doubtful if Melchior had seen it at all. From the height of the

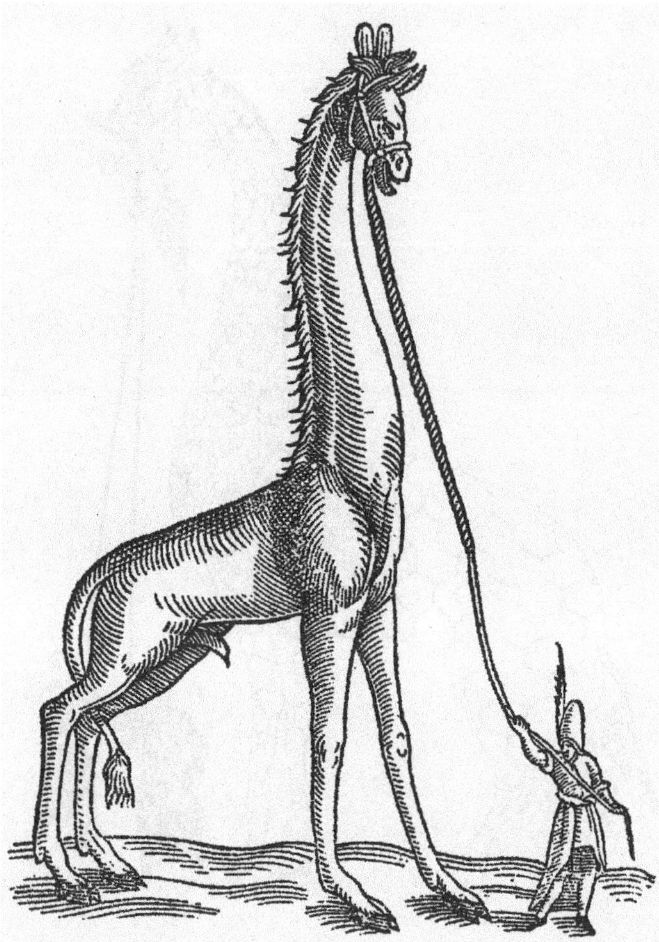

Figure 2.6. Melchior Luorigus's giraffe, drawn for Topsell, 1607.

man holding it, it would have been close to 8 m (25 ft) tall, a height no giraffe has been known to reach.

It was images such as these that suggested a fabulous, mythical creature, and as no giraffes had actually been seen in Europe since the days of Caesar and the Gordians, their existence, if they existed at all, was an enigma for any European who had a biological bent. The place a mythical creature could find a home was in the sky. Thus in the very best traditions of Greek mythology, giraffes became a constellation. In 1612, Petrus Plancius, a Flemish astronomer, imagined that he could see the shape of a giraffe in the space between Cassiopeia, the beautiful but vain Queen of Ethiopia, and the Great Bear. In 1690 yet another of the early mapmakers of the universe, Johannes Hevelius, a Pole, published his description of the heavens, which included a marvelous image of a giraffe (Figure 2.7).

There were sporadic exceptions to imaginary giraffes. In AD 1245 one animal was given to Frederick II, the Holy Roman Emperor and King of the Two Sicilies (Haskins, 1922), and possibly another to Manfred, a son, in 1261 (Laufer, 1928). Frederick had

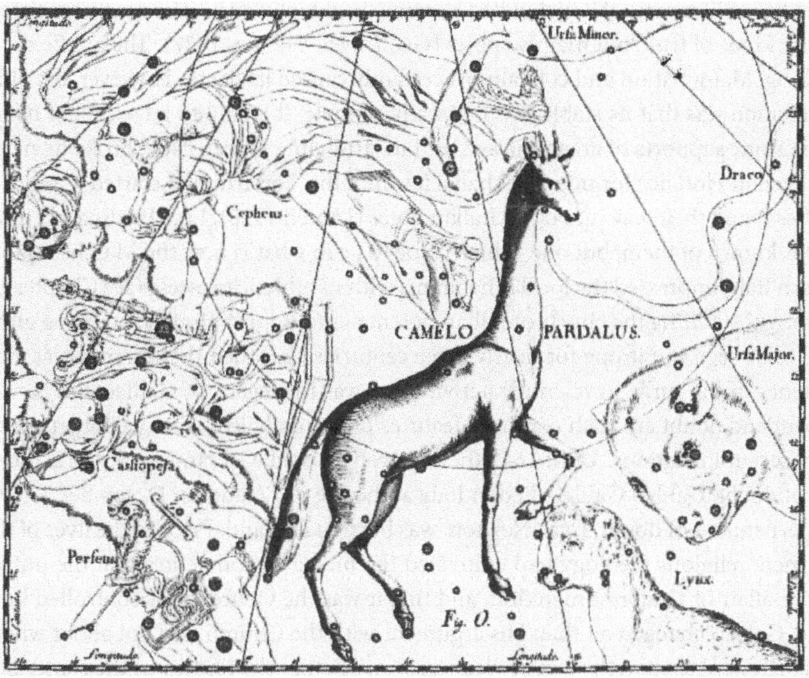

Figure 2.7. *Camelopardalis* constellation, between Cassiopeia to its left and Ursa major to its right.
From Hevelius's "Uranographia," published in 1690.

established a zoo and had obtained a giraffe from the Sultan of Egypt in exchange for a white bear. The fate of the giraffe is not recorded. In the fifteenth century two specimens (perhaps) were sent to Europe. One was sent by the Caliph of Baghdad to Ferdinand I, the King of Naples. It is possible that this giraffe was imaginary, with its purpose simply a matter of enhancing Ferdinand's reputation. If it was real, it has disappeared from history. The other one was real, and was sent by the sultan of Egypt to Lorenzo de Medici in 1486 (Dobson, 1959; Belozerskaya, 2006). Lorenzo was Machiavelli's "Prince." Lorenzo had come to power in Florence, then a city-state, in December 1469 at the age of 21. Prestige in those days depended much on the quality of gifts exchanged between potentates. He imagined himself as the equivalent of Julius Caesar, and as Caesar had showed a giraffe in Rome as evidence of his power and authority, Lorenzo had to bring one to Florence. This enterprise was not easy to achieve, as the only source of giraffes was Egypt and the Egyptian sultans needed diplomatic persuasion. In the end, though, diplomacy triumphed and a giraffe arrived in Florence in November 1487 (Joost-Gaugier, 1987). A scene of its arrival was captured in several paintings (see Plate 2.2). The paintings, possibly the first from life of a giraffe in Europe, were drawn with some artistic license: the giraffe in Plate 2.2 has no coat patches, its neck is too thin, and its head too small.

Nevertheless, Machiavelli spun the story, and we are told that the giraffe's power to impress was all that Lorenzo desired. The giraffe took daily walks through Florence, and as its head was at the height of first-floor windows, "it took meals from the hands of the ladies of Florence, of which it had become the adopted girl." The meals consisted of several kinds of fruit, but mainly apples (Geoffroy St-Hilaire, 1827). The giraffe did not live long. Malnutrition and cold almost certainly caused its death. However, the official explanation was that its stable was not really suitable. It is said to have caught its neck in the stone supports of an architrave, and in struggling to free itself, broke its neck. It had been in Florence for no more than a few months. Lorenzo had started a trend, and giraffes found their way into other Italian states (Laufer, 1928). Like Ferdinand's giraffe, little is known of them, but one, which came to be in what is now the Marche region of eastern Italy, impressed the local inhabitants with its athletic prowess (see Chapter 13).

Notwithstanding the efforts of Italian potentates to import giraffes, no living giraffes were to be seen in Europe for nearly three centuries, and then their arrival was a consequence of the surge in scientific activity that was the legacy of Galileo and Newton. Dissent and doubt are such common features of our daily lives that we forget they are fairly recent inventions. Until 1642 they scarcely existed. On January 8, 1642, the creator of doubt, Galileo Galilei, died in Italy at the age of 77, and on December 25, 1642, the perpetuator of doubt, Isaac Newton, was born in England. Prior to the lives of these two men, religious ideology had cultivated the medieval conviction that the universe was an affair of God, not mankind, and that it was the Church that controlled knowledge. Galileo changed all that. His argument with the Church was not about whether the Earth went around the sun or vice versa, or whether the Earth was the center of the universe. It was about who controlled knowledge. So strong was the vested interest of the Church to maintain its primacy over thought that the churchmen wanted Galileo's head. In 1633 he stood trial, and was found guilty of heresy. To save his life, Galileo recanted, but by then he had established that science is a powerful source of knowledge, far more powerful than belief simply because its products are more reliable. Thus the year 1642 can be identified as a watershed that changed our world from a cozy one, where all phenomena were an unchallengeable creation of God, to our raw, stress-filled, doubting one, where nothing is so until proven to be so. Newton died in 1727 at the age of 85. His legacy was profound: nothing has value unless grounded in mathematical laws, having predictable consequences that match the facts of experience.

This, then, was the background. By the mid-eighteenth century it was no longer good enough for European naturalists to imagine that giraffes existed. Incontrovertible proof of their existence was essential.

3
The Scientific Age

How did real giraffes compare with those that existed in the minds of writers like Artemidorus and Pliny, and Lorenzo's artist, and in the imaginations of Plancius and Hevelius? Did giraffes really exist? Middle and Far Eastern politicians and explorers had no doubt that giraffes were real, but the desire by European naturalists to resolve the myth of giraffes seems to have been lacking. Giraffes, after all, were still plentiful in northeastern Africa; specimens could be obtained, and so study of them, although difficult, was not impossible. We have to assume that the effort needed to obtain specimens of giraffes was greater than the interest in them. Then, in 1652 the Dutch East India Company established an agricultural settlement at the Cape of Good Hope, on the southernmost tip of Africa, under the leadership of Jan van Riebeeck. The settlement became a town and then a country and attracted adventurers. Exploration of the hinterland revealed a rich and extraordinary geography and biology. Within a few years, giraffes had been seen, and in due course around the time that European naturalists were discovering the fauna and flora of the world, southern Africa became the source of giraffe specimens, although acquiring the first few specimens took a great deal of commitment and courage. The first specimen obtained was a skin. This skin not only confirmed that giraffes were real, but also revealed that giraffes lived in southern Africa. Before this skin, giraffes had been thought to live only in North Africa. Their discovery in southern Africa was so novel that it was emphasized in all editions of Georges-Louis Leclerc's (the Comte de Buffon; see Plate 3.1) *Natural History*, which was the foremost biological encyclopedia of its time.

The skin came from a young giraffe that had died after its mother had been killed by a member of an expedition made into Namaqualand in the northwest of the Cape Province of South Africa in 1761. The previous year, a burger, Jacobus Coetsé, had been given permission by the Governor of the Cape settlement to travel inland to shoot elephants. He reached and crossed the Great (Gariep/Orange) River at Ramansdrif near Gü-daos (i.e., "sheep ford" in the local language, and now Goodhouse) and entered southern Namibia. In the narrative of his trip, he claimed that he was the first member "of any European nation" to do so. He followed the Leeuwen River northward and reached a spring from which warm water flowed. The spot was later named "Warmbad." In this vicinity he found

> *a multitude of lions and rhinoceroses as also another animal still totally unknown at the Cape, which although not as bulky as elephants is considerably taller of body, with a long neck, a humped back and long legs, and if not the real camel, is at least a kind of camel.* (Coetsé's narrative in Mossop, 1935)

In gait (Coetsé said) these animals were "slow and cumbrous." Having given chase on horseback, he was able to catch up to two of them with little difficulty and shot them.

How Giraffes Work. Graham Mitchell, Oxford University Press. © Oxford University Press 2021.
DOI: 10.1093/oso/9780197571194.003.0003

Both were females, one of which was with a calf. Coetsé kept the calf alive for 14 days on a diet of bran soaked in water. But it died and Coetsé took its skin back to the Cape. (What happened to this skin is not known for certain, but perhaps it was included in the collections of giraffe specimens that in due course were sent to Europe.) Coetsé explained, however, that "the appearance of the adult animals cannot really be pictured from this skin, because the young one is spotted and without a hump on the back; the full-grown animal on the contrary is without spots, is ruddy in colour and has massive humps. On the head of the young animal two horn-like growths are found which are really only short knobs on the adult" (Mossop, 1935).

The expedition of the following year (1761) was led by Captain Hendrik Hop with Carel Frederick Brink as notary (Mossop, 1947; Forbes, 1970). Coetsé accompanied them. On this expedition another two giraffes were shot, a male and female. This female also had a calf at foot and it too died. Coetsé's and Hop and Brink's sightings have the reputation of being the first of giraffes by Europeans in southern Africa, but the first sighting had been made nearly 100 years earlier (Bigalke, 1951; Skead, 1980), on November 28, 1663, by Sergeant Jonas de la Guerre during an expedition to the Gariep (Orange) River (Bigalke, 1951). Expeditions into the hinterland of the Cape had begun soon after the Cape settlement had been established, and their purpose was to explore the western regions of the Cape and to obtain cattle from the Namaquas, the tribe of indigenous South Africans who lived in the northwest of the Cape (Muller, 1971). Van Riebeeck's successor as Governor in 1662, Zacharias Wagenaer, commissioned the de la Guerre expedition, and Pieter van Meerhof recorded the journal. The journey lasted from October 11, 1663, to January 22, 1664. The expedition's goal to reach the Gariep (Orange) River was not accomplished, but van Meerhof recorded that, on November 28, 1663, two giraffes were seen at the Spoeg River (*in 't marcheeren is van ons twee camelen gesien welcke beesten noyt voor dezen*, translated as "on our journey two giraffes were seen creatures that have never been seen before"). No specimens were collected, and no description of the giraffes was recorded. The Spoeg River empties into the sea between the Groen River and Buffalo River mouths (see Figure 3.1) and the area around it is a beautiful, but arid, landscape filled with brightly-colored semi-succulent flora adapted to heat, but devoid of the trees so essential for the well-being of giraffes. If that was the vegetation there 350 years ago, then it is a wonder that giraffes were seen there at all. Where had they come from? Further south? Further south toward Cape Town was and is arid inhospitable country, and no modern giraffes have ever been seen there. Across equally arid country 225 km further north was the mouth of the Gariep (Orange) River, but migration southward from that river is a more likely origin for the two seen at the Spoeg River mouth. An explanation for their appearance there is that a cohort of the giraffes (*G. camelopardalis*) that had evolved in East Africa about a million years ago (see Chapter 6), through force of circumstance, migrated over many millennia not only southward (establishing *G.c. giraffa* as the giraffe of the northeastern parts of South Africa) but also westward, making their way along the Gariep (Orange) River toward what is now Namibia and moving north from Namibia into Angola to establish small but stable populations of *G.c. capensis*, a sequence of migratory events that is supported by the similar genetic makeup of *G.c. giraffa* and *G.c. capensis* (see Chapter 1).From time to time we can imagine that exploratory migrations of *G.c. capensis* southward took them across the river and further south, following browse more inland than along the

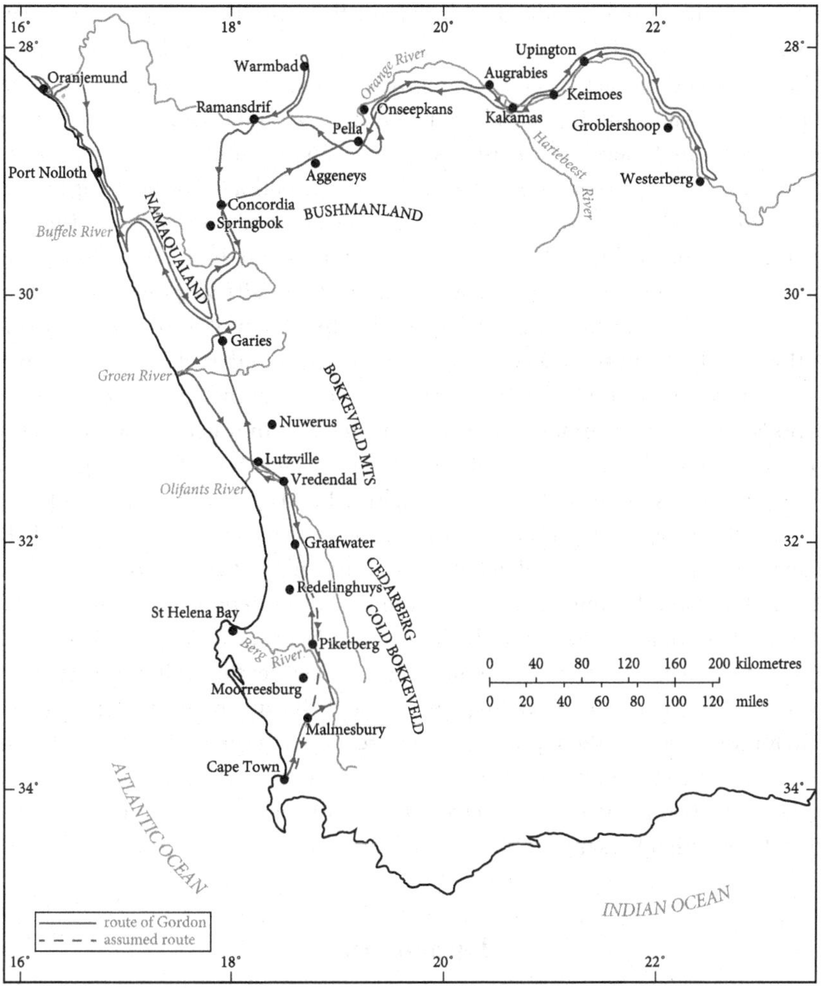

Figure 3.1. A modern map of Gordon's fourth journey to the northwest of South Africa with Paterson.

Courtesy and permission of the Brenthurst Library, Johannesburg, South Africa.

inhospitable coast to reach what are now the northwestern parts of the Cape Province. The innate urge to migrate is always present in large mammals, and allopatric speciation is a usual result, but in this case speciation seems to have been doomed. Nevertheless, for giraffes to have reached and survived, even for a short while, at the Spoeg River, supportive acacia woodland must have existed there or in the river valleys that fed the Spoeg itself.

Very few giraffes have ever been seen south of the Gariep (Orange) River and until very recent times none south of the great rivers of South Africa's eastern seaboard. Why? The most likely explanation is that they dislike soft sand and mud and cross river beds rarely and do so only, and reluctantly, when the bed is dry (Berry, 1978), and, like all terrestrial mammals, will avoid crossing rivers unless, like migrating wildebeest, they are desperate to do so. We can guess that one explanation for their reluctance is that swimming requires a dog-paddle gait which is not the natural gait of giraffes. Henderson and Naish (2010) tested the hypothesis that the avoidance of rivers by giraffes relates to their anatomy and the lack of sufficient buoyancy it confers. Lung volume is an important part of buoyancy, and giraffes have relatively small lungs (see Chapter 12), and a consequence is that their body density is higher than it would be if the lungs were the same size as they are in a mammal with a more conventional shape. Their buoyancy is, accordingly, reduced. Second, buoyancy depends on body surface area. The greater the surface area of an object, the more water it displaces and the greater is its buoyancy. Henderson and Naish estimated the surface area of a giraffe weighing ~1,600 kg and found it to be 17.7 m^2, which is much larger than the 13.3 m^2 it would be in a mammal with the same body mass but a conventional shape such as that of an ox. Nevertheless, Henderson and Naish concluded that even a surface area of the size they calculated was not enough to prevent a giraffe from becoming unstable in water that was deeper than hip height. Hip height in a small giraffe weighing 100 kg is 1.3 m from the ground, in a 1,000 kg female it is 2.6 m, and in a male weighing 1,500 kg it is 2.8 m (see Figure 18.2). In water of those depths giraffes will float, lose contact with the river bed, and roll. Plus, as Henderson and Naish pointed out, if a river is flowing fast enough to produce a current, giraffes would become unstable at even shallower depths. Their surface area is in fact less than that calculated by Henderson and Naish (see Chapter 14) so perhaps giraffes were and are aware of the implications of water depth, the limits of their buoyancy, and the consequences of wading into a strongly flowing river. Swimming is not an option for them, nor would migrating southward across the great rivers of South Africa have been, and could only have been attempted in a very dry season with a very low river flow. The two giraffes de la Guerre came across must, however, have crossed the Gariep River, with others no doubt, to enter the history books.

Hop and Brink

The expedition of Hop and Brink took place during the governorship of Ryk Tulbagh (1751–1771), and would not have taken place unless Tulbagh had had an abundance of scientific curiosity. When Tulbagh became Governor he had already been in the service of the Dutch East India Company for more than 30 years. He had come to the Cape in 1718. He had no immediate family. His wife had died in 1753 and they had no children. He put all his energy into the well-being of the Cape and its peoples. Among the visitors he entertained were Clive of India and Capt. James Cook. Johan Splinter Stavorinus, in his *Voyages to the East Indies* (1798) wrote that he was

> *the father of the country at Africa's extremity, who, after more than fifty years sojourn there really knew the true interests of the inhabitants and the nature of the country, and who was well-beloved by all for his gentle and paternal rule.*

Tulbagh was a botanist by preference and wanted to know what the flora and fauna was beyond the limits and boundaries of the by then thriving Cape Colony, and especially what it was in the region of the Gariep (Orange) River. He sent several expeditions to find out. On one of these, Tulbagh had sent his gardener, Hendrik Prisky, "to find and collect plants, vegetation, herbs and insects." The collections usually were sent to Leiden University, and they made Tulbagh well known in Europe, so well known that Carl Linnaeus, at the time Professor of Botany at Uppsala University, named a genus of lily after him, *Tulbaghia*.

The opportunity for another such expedition arose when Hop, who was a prominent, wealthy, and altruistic burgher, requested permission to explore the interior. All the support he asked for was

> *a boat, 900 pounds of powder, 2010 pounds of lead, besides various necessary and some trifling presents that were supplied to serve the purpose of the Company's Expedition.* (Mossop, 1947)

The expedition had a scientific brief. The members of the party included Jan Auge, who was a horticulturist and Superintendent of the Company's garden; Carel Rykvoet, a medical practitioner; Carel Brink as cartographer and scribe; and 10 other burghers, as well as a retinue of 85 "natives," all with Hop in charge. One of the instructions given to them was "to enquire if the natives employed giraffes as beasts of burden and if possible to bring one back" (Mossop, 1947). The expedition set out on Thursday, July 16, 1761, traveled up the west coast (see Figure 3.1), reached the Gariep River two months later on September 18, crossed it on September 29 at Ramon's drift near Gü-daos and saw and killed a giraffe, a female, on October 5 in what is now southern Namibia. It was as a historic moment in the story of giraffes as had Coetsé's been the year before. Brink described the event as follows:

> *Monday 5th October, 1761. We proceeded on, still along the river* [i.e., the Leeuwen River], *and then camped at the "warme bad." This is an ever flowing spring of water, temperately warm, and thus always suitable to bathe in. The spring lies about two hundred paces from the east bank of the (Leeuwen) river on a low and stony rise. It bubbles up like boiling water and has a somewhat briny taste, not however markedly strong so it is not unpleasant to drink. We found the soil here to be fallow and the salt beside the water lay dry for the depth of a finger. Here the first giraffe were observed by us: one of these, a female was killed and the young one with it captured. Concerning the appearance of the one killed we found the body to resemble most that of the cow although head and neck were like that of a horse. The height from hoof to head was 17 feet, the proportion of body length to height was as one to two and a half. Moreover on account of this shortness of body in comparison with its other members, and because of its extraordinary movements, it is unthinkable that this animal can be employed for any useful purpose. For the rest it is white in colour with the long neck and fore parts of the body covered with light-brown chequered and diamond shaped patches, its hindquarters likewise though with more darkly brown patches. On the forehead are two horns about a foot in length; these project forward and are encased in a hairy hide. Food is mostly sought on the tall Red-wood trees yet they can graze on level ground, although on account of the tall forepart of the body they must go on their knees. We found here the latitude to be 28°5', longitude 37°41'.* (Mossop, 1947)

This description was authentic, indisputable, and the first good one of a giraffe since the sixteenth century. The *warmebad* is the present-day Warmbad of Namibia (Figure 3.1). It is situated about 40 km (25 miles) north of the Gariep (Orange) River and its modern latitude is 28°26′S and its longitude 18°44′E, rather than the 37°41′E as recorded by Brink, a difference that can be attributed to the change in location of the prime meridian from the Canary Islands (where it was from 1643 to 1884) to Greenwich (from 1884).

The "Warmbad" spring was the only source of water in an arid, forbidding landscape. Around it a nascent tourist economy is now being built, for the brave, as its environs are hostile, hot, and dry. The surrounding vegetation is arid-adapted, and the most prominent trees are the gray camel-thorn *Acacia haemotoxylon*[1] and the true camel-thorn *A. giraffae* (its seed pods have the shape of an ear and thus its species name is now *erioloba*). Both trees have the rich red heartwood referred to by Brink, both grow in deep desert sands, and both are sparsely distributed in southern Namibia and along the Gariep River eastward toward Botswana, although their density increases the nearer they are to the Gariep River. The large pods are nutritious and are eaten by giraffes. Both trees can reach a height of ~8 m. Again, it is as surprising that giraffes were seen at Warmbad as it was that they were seen at the Spoeg River, because their preferred habitat is woodland savannah, but the vegetation around the spring must have been attractive. On October 10 the expedition left Warmbad, and traveled further north along the Leeuwen River. On this journey Brink wrote that "the young calf captured by us died despite all efforts made to preserve its life." The expedition traveled another 70 miles (110 km) further north and on October 16, 1761, another giraffe, a male, was shot by the burgher Jan Badenhorst, a member of the party. It, too, was 17 ft tall. The expedition reached 26°18′S, 37°37′E (i.e., about another 110 km [70 miles] further north; Figure 3.1) before the heat, lack of water, and scarcity of grazing compelled them to turn back, which they did on December 6, eventually arriving in Cape Town on April 27, 1762.

Allamand and Buffon

The significant products of the expedition were the skin and horns of the calf that had died and a skin and a drawing of an adult, almost certainly the male shot by Badenhorst. The drawing, the skins, and the horns of the calf were sent by Tulbagh in 1764 to Leiden University in Holland for study by the Swiss-born J.N.S. (Jean) Allamand, Professor of Natural History (see Plate 3.1). They caused great excitement, as they were the first hard evidence of southern African giraffes to arrive in Europe. Allamand was *ex officio* Director of the Natural History Cabinet at Leiden University. He had the skin of the calf stuffed and exhibited at the University, where it was seen by Arnout Vosmaer, Director of the Prince of Orange's Cabinet in The Hague, and by

[1] Disappointingly, the genus name of some *Acacia* sp. has been changed recently to *Vachellia*, and for others to *Senegalia*, no doubt for good taxonomic reasons, but the original genus of *Acacia* is deeply entrenched in all scientific studies of giraffes and, to avoid confusion, will be maintained throughout this book.

Mathew Maty, Secretary of the Royal Society of London (Carteret, 1770; Forbes, 1970; Rookmaaker, 1979). This reconstruction of a giraffe was very poor (Vosmaer, 1787). The specimens were accompanied by a short note paraphrasing the longer journal entry and contained some dimensions:

> *In an excursion from the Cape, made in 1702* [sic], *we travelled about two hundred leagues up the country, and met with the camelopardalis, a drawing of which we subjoined. Its body resembles that of an ox, and the head and neck those of the horse. All we met with were of a white colour, sprinkled with brown spots. They have two horns on the head, about half a foot long, and their feet are hoofed. We killed two of these animals, and sent their skins to Europe, one of which measured as follows: the length of the head one foot eight inches; the height, from the bottom of the fore-foot to the withers, ten feet; and from the withers to the top of the head seven feet; in all seventeen feet in height. The length from withers to reins* [= kidneys, i.e., a point roughly between the last rib and the pelvis] *is five feet six inches, and from thence to the tail one foot six; the length, therefore, of the whole body is seven feet, and the height, from the hind feet to the reins eight feet six inches. The great disproportion in the height and length of this animal seems to prevent its being of any service. It feeds on the leaves of trees, and when it wants to drink it is obliged to kneel with its forelegs* (Buffon, 1792; J. S. Barr's translation from French).

There was some confusion about the skins. Allamand had received the skin of a young giraffe, not a large one. The dimensions in the note are identical to those mentioned by Philip Carteret in a letter he wrote in 1769. Some, like the total height and the lengths of the foreleg from foot to the withers and the hind leg from foot to rump, are consistent with those of a large male with a body mass of ~1,000 kg. Others, like the length of the head, is the head length of a young giraffe weighing about 350 kg.

The measurements of the skin and the nature of the horns were important to the scientists of those days, as the thrust of biological research then was taxonomic classification of plants and animals exemplified by the binomial system developed by Linnaeus. Whether a giraffe's horns were shed each year would determine their position. Linnaeus had assumed they were. Brisson was as sure that they were not. The controversy spread. The Comte deBuffon became involved. Buffon was a French aristocrat who, despite the prevailing sentiments of the time, could not accept the idea that all creatures had been created nor that the earth was a mere 6,000 years old. He did not agree with Linnaeus's compartmentalization of animals into genuses and species, preferring to think that animal life was a continuum. He was an evolutionist before evolution had been defined. In 1739 he was appointed Director of the *Jardin des Plantes* (originally called the *Jardin du Roi*) in Paris and decided to devote his life to the description and classification of the specimens under his care and with the purpose of presenting an alternative to Linnaeus's and Brisson's classifications. The product was the *Histoire Naturelle*, an Encyclopedia of the natural world. Thirty-six of its 44 volumes were completed before his death in 1788. The other eight were added afterward, and at least 20 editions of the whole were eventually published, each one essentially the same, modified as new details became available, but each one retaining his imprint. This magnificent work put Buffon firmly at the head of the biological community.

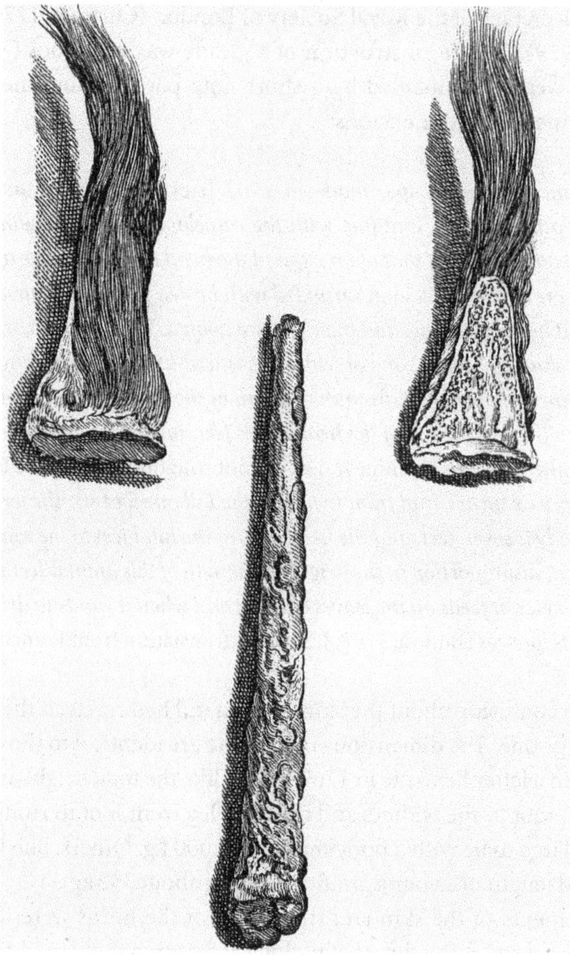

Figure 3.2. A narwhal's tusk and two giraffe ossicones.
From Buffon, 1796. Courtesy Dr. Carlos Martinez del Rio.

The descriptions of the shape, habitat, and habits of giraffes in the *Histoire Naturelle* were exhaustive, especially the descriptions in the 1835 edition. They are as good as any that can be found in a modern text and included excellent illustrations. However, in 1764 knowledge of giraffes was limited, and Allamand was eager to be associated with Buffon. The giraffe specimens gave him the means to become so, in particular by settling the dispute about the nature of giraffe horns. If they were shed, then giraffes could be linked to deer. If not, then they were a new family and genus. To add to the confusion, a narwhal's tusk had been mistaken for a giraffe's median ossicone and was illustrated as such by Buffon (1796), fueling speculation that the giraffe truly was the unicorn of fable (Figure 3.2). Saner analysis prevailed, however, with Buffon conceding that

> *I doubt very much the longest one belongs to a giraffe. It is not in proportion with the others which are thick relative to their length while it is very long for its width.* (Buffon 1796)

Allamand wrote to Louis Daubenton of the French Academy of Sciences on October 31, 1766, about the young calf. Daubenton was in charge of anatomical descriptions of animals that would appear in the *Histoire Naturelle.*

> *I have the honour to tell you that I have here a stuffed young giraffe, and you expressed the desire, as did M de Buffon, to know the nature of its horns, I determined to cut one, to send to you to give a true idea. It resembles the horns of the stag more than those of any other animal. I suspect that the animal does not shed them annually. If this conjecture be well-founded the giraffe constitutes a particular genus, perfectly distinct from the animals who shed their horns, and also from those who have hollow, but permanent horns. You observe that this giraffe is very young. The Governor of the Cape, from whom I received it, told me that it was killed lying next to its mother. Its height in effect was 6 feet.* (Buffon, 1781)

Buffon and his editors were hard to convince. In every edition of the *Encyclopaedia* between 1781 and 1835, great parts of the narrative on giraffes were devoted to the nature of their horns. In every edition the conclusion was that "even if we doubt if the horns are shed annually . . . the giraffe has to be admitted into the tribe of deer." Allamand eventually won the day, however. "The giraffe," he wrote as early as the 1775 edition of the *Histoire*, "forms a unique species, different to those that contain animals whose horns are shed, and those whose horns are hollow, but permanent." But Buffon could not bring himself to give giraffes the scientific name decided by Brisson and Linnaeus—*Giraffa camelopardalis.*

Carteret

A copy of the drawing sent by Tulbagh to Allamand was kept at the Cape in the care of Mr. Barrawke,[2] where it lay for six years until given to Captain Philip Carteret. Carteret was a British navigator whose keen interest in all matters scientific matched his excellent nautical skills. His first voyage was with the Hon. John Byron (the grandfather of the poet) on Byron's fruitless voyage of discovery to the South Seas in 1764. On the basis of his accomplishments during the 22 months of that voyage, Carteret was given the command of the second vessel in Samuel Wallis's expedition to the Southern Hemisphere in August 1766. Separated from Wallis the following April while clearing the straits of Magellan, he made his own way to the Cape. Sailing west across the Pacific, he discovered Pitcairn Island on July 2, 1767, and a number of other small Islands, one of which in the Solomon Islands bears his name. (His account of his journey, and the discovery of Pitcairn Island, had an unforeseen consequence. It was carried aboard the *Bounty* of mutiny fame. Using Carteret's description as their source of information, nine *Bounty* mutineers and their Tahitian wives and nine other Tahitians left Tahiti, for Pitcairn, arriving in January 1790).

Carteret arrived at the Cape of Good Hope in 1768, and obtained a copy of the drawing of the giraffe made on Hop's expedition. He returned to England on March

[2] A corruption of the surname of the political secretary of the Dutch East India Company at the Cape, O.M. Bergh (Mitchell, 2009).

20, 1769, and on April 20, 1769, wrote a letter to the secretary of the Royal Society of London, Dr. Mathew Maty, M.D.:

> *Sir, Inclosed* [sic] *I have sent you the drawing of a Camelopardalis, as it was taken off, from life, of one near the Cape of Good Hope. The present governor of the Cape of Good Hope has sent out parties of men on inland discoveries, some of which have been absent from eighteen months to two years. One of these parties crossed many mountains and plains, in one of which they found two of these creatures, but they only caught the young one, of which the inclosed* [sic] *is the drawing, as it was taken off by them; they endeavoured to bring him alive to the Cape Town, but unfortunately it died. They took off his skin, which they brought as confirmation of the truth, and it has been sent to Holland. These particulars, as well as the drawing, I got from Mr. Barrawke, first Secretary to the Dutch company at that place, in the presence of the governor.*

From the length of the horns and the dimensions given in the letter, the drawing (see Plate 3.2) was of the adult male shot by Badenhorst on October 16, 1761, not of the calf. It was 17 ft (5.2 m) tall and its length 14 ft (4.2 m). It is these dimensions that appear in Buffon's *Natural History*. Maty was skeptical of Carteret's story and took the trouble to go to Leiden to check that the skin and drawing had indeed arrived at the University's Cabinet of Natural History and that Carteret's story was accurate. It was, and so Carteret's letter was read to a meeting of the Royal Society in January 1770, and a drawing and letter were published in the *Transactions of the Royal Society* later that year (Carteret, 1770). The drawing that appeared in the *Transactions* was made by James Mynde. Mynde was an engraver, and responsible for numerous stock book illustrations for the Royal Society. Despite it being a copy of a copy of a copy, Mynde's drawing is good enough to identify the giraffe as what Lydekker in 1904 was to call *Giraffa camelopardalis capensis* (Plate 1.1), characterized by blotchy patches instead of, for example, the more clearly demarcated, reticulate patches of giraffes that inhabit northeastern and northwestern Africa. An outcome of the Royal Society's meeting was that one member who attended, the famous anatomist John Hunter, determined to obtain a specimen of a giraffe for his museum (Dobson, 1959). Carteret's picture appeared in the first edition of Buffon's *Histoire Naturelle* of 1770, though Buffon somewhat disparagingly remarked that "the drawing is so incorrect, that no use can be made of it" (see Plate 3.2). Nevertheless, Carteret's drawing became the iconic figure, and Buffon's descriptions definitive. The study of giraffes seemed complete. Further study of them might have stopped, had it not been for the arrival of two extraordinary men at the Cape, who were to play a pivotal role in the study of giraffe biology. Their contribution deserves a detailed description because it was through their efforts and the specimens they collected that the scientific study of giraffes really began.

Gordon and Paterson

In 1777, nine years after Carteret's visit, Captain (later Colonel) Robert Jacob Gordon (see Plate 3.1) arrived at the Cape. Gordon had been in the Cape before, in 1772, on furlough from his regiment in the Netherlands. In the late eighteenth century the Dutch had no standing army. They hired a brigade from Scotland for defense, and its commander was Gordon's father, General Jacob Gordon, who married a Dutch girl. Robert Jacob was the precocious product. His names reflected his Scottish descent, but he considered himself Dutch. He was an extrovert, had enormous energy and enthusiasm, had graduated from University at age 17, and then joined his father's brigade. After arriving in the Cape in October 1772, aged 29, he befriended a Swedish physician (and a student of Linnaeus) named Carl Peter Thunberg, and Scots horticulturist Francis Masson. Masson had arrived in September 1772, having been sent to collect plants for the Royal Botanical Garden at Kew, near London. Thunberg, who had arrived in April 1772, had been commissioned by the father and son botanists Johannes and Nicolaas Burman of Amsterdam to collect plants.

Gordon was captivated by what he saw and learned from his companions. With Masson and Thunberg he explored the mountains that surround the Cape. Alone he undertook a long journey sometime in 1773–1774 along the southeast coast of the Cape. He could not stay at the Cape; nevertheless, he was determined to return and did so in June 1777, having left the Dutch army to join the Dutch East India Company. The Company had appointed him second in command to succeed Colonel du Phren as Commander of the Garrison (which eventually he did on March 26, 1782; Rookmaaker, 1983). Gordon's arrival at the Cape was a momentous event. He became an expert on the geography and biology of southern Africa (Talbot, 1970). He had prepared himself well for this role. Between his visits to the Cape he had met with scientists, among whom were Arnout Vosmaer, Director of the Stadtholder's (Prince of Orange) Cabinet in The Hague, and Jean Allamand (Vosmaer, 1787; Rookmaaker, 1979), and it is clear that he promised to send to Vosmaer, via Allamand, any botanical and zoological specimens he collected for inclusion in the Prince's collection. Gordon was also a skilled artist, a gifted linguist, and had high military skills and leadership qualities. He was a unique individual.

In the month prior to Gordon's arrival, in May 1777, another horticulturist, William Paterson (see Plate 3.1), a Scot, arrived at the Cape to collect plants for the Countess of Strathmore. Paterson and Gordon became firm friends, sharing the common bonds of Scottish ancestry and a strong interest in natural history. Paterson and Gordon made four expeditions into the "country of the Hottentots and Caffraria" (i.e., the northwest and southeast of South Africa) between 1777 and 1779, and on three of these expeditions they traveled together for at least some of the way. Their first journey (referred to as Gordon's second journey because he had already made a journey in 1773–1774) was to the interior, traveling eastward toward Swellendam and Mossel Bay (Gordon, 1777–1779). Near Bethulie, about 200 km south of present-day Bloemfontein, Gordon came upon the origins of the Gariep (Orange) River he was to explore on his later travels to the northwest of the Cape (Axelson, 1954).

Gordon made a third journey between August 28, 1778, and January 25, 1779. His gift for languages was the reason for this trip. He was fluent in Dutch, English, and French, and quickly acquired knowledge of the Xhosa and Khoekhoen (formerly, Hottentot) languages. This journey was without Paterson, but with the new Governor of the Cape, Baron Joachim van Plettenberg. Van Plettenberg had persuaded the Dutch East India Company that he needed to journey to the "most outlying regions as far as they are occupied" (Muller, 1971); in other words, again along the southeast coast.

Gordon was needed for his knowledge of the country and the Xhosa language. En route they stayed at the farm of Hendrik Hop's widow (Mossop, 1947), where Gordon learned from her about the 1761 Hop and Brink expedition. He also learned that "the Cafres, who occupy the eastern coasts, know nothing of the giraffe" (Buffon, 1812).

Paterson, having explored the southeast Cape, resolved to make his second trip to the northwest, and took the opportunity of Gordon's obligations to the Governor to do so. It proved to be an exceptionally arduous and difficult journey. Paterson recognized his lack of knowledge and was fortunate to have had the help of the brothers Willem and Sebastiaan van Reenen, who knew the country of the northwest Cape well (Mossop, 1935) because "they possessed several farms in the interior parts of the country" (Paterson, 1789). He and his companions left the Cape on May 22, 1778, crossed the Groot Karroo, its western border, the Roggeveld, then the Bokkeveld (see Figure 3.1), and across desolate terrain toward southern Namibia. In the first week of September they reached the Gariep (Orange) River and were informed by local peoples that there were giraffes on the other side. Sebastiaan van Reenen crossed the river at Ramon's Drift, 10 km east of Goodhouse, 40 km southwest of Warmbad (Figure 3.1; Skead, 1980), where Hop and Brink also had crossed the River in 1761. Between September 15 and 19 he shot a giraffe on the north bank. Khoekhoen were sent to skin it, but when they got there they found that a lion had "disfigured it so much that nothing could be preserved except part of the skin of the neck, with the horns, and a part of the hindquarters." Paterson resolved to go and collect the skeleton the next day, but the river rose and crossing it was impossible. After a journey of six months, Paterson and van Reenen returned to the Cape empty-handed, reaching it on November 20, 1778.

Gordon returned to the Cape from his third journey on January 25, 1779, clearly obsessed with the idea of seeing a giraffe and obtaining specimens to send to Holland. On June 27, 1779, just five months after returning to the Cape, he set off up the west coast, accompanied by a cohort of Khoekhoen, Pieter Pienaar, a hunter, and Johannes Schumacher, as his artist, following essentially the same route that Hop had taken 18 years before (Figure 3.1). Paterson and van Reenen were also heading toward the west coast following the route north they had taken on Paterson's previous journey "through the most barren country I ever visited" (Paterson, 1789). A main goal was to obtain a specimen of a giraffe. Their diaries give a sense of rivalry. Paterson wanted to get his specimen of a giraffe to England before Gordon could get his to Holland.

Wikar

During July 1779, Gordon and Paterson met Hendrik Jacobus Wikar. He too contributed to the story of giraffes. Wikar was a Swede born in Gottenberg, but in the journal he wrote, in 1779, of his travels, in which he recorded in great detail all that he saw and encountered, he referred to himself as a Dutchman. Wikar had joined the Dutch East India Company in Amsterdam and arrived at the Cape in 1773. For nearly two years he worked as a clerk in the Company's hospital, but became addicted to gambling and indebted to a friend. Overcome by shame, he exiled himself to the far northwest of the Cape Province far "beyond the settled districts—not anticipating the peril and wretchedness I must encounter during the (next) 4 years and 6 months." In exile he made two trips along a 500-km stretch of the Gariep River, from Gü-daos to a place called Koegas (meaning "abundance of hippopotami" in the Nama language), which lies between Upington and Prieska on the Gariep River, and is now known as Zeekoeisteek (*Zeekoei* = hippopotamus). His journal begins with his first trip in September 1778 when he is at Gü-daos. On September 22 he saw "the fresh spoor of rhinoceros and giraffe, but we did not catch sight of them, for at this time water was till plentiful everywhere in the veld so that they were seldom seen along the river. Yet in December [1778] when I returned to these parts I saw along the river in one herd as many as 20 giraffes or more. Many black ebony trees [= *Euclea pseudebenus*] grow here and also camel-thorn trees [= *A. giraffae*] which is a very pretty red and hard. Jacobus Coetsé on his first expedition gave the tree this name because the giraffes eat mainly the tops of these trees."

On his second journey along the river, on April 11, 1779, Wikar shot a giraffe and, perhaps with Coetsé's narrative in mind, wrote a better, if brief and insensitive, description of it.

> *This giraffe was a heifer, but full grown, and you could see she had been covered. When she had been shot the long neck remained upright, so I began to make measurements from below the front feet up to the head. She was about 18 feet high, from the breast to the tail $9^1/_4$ feet and the difference in length between fore- and hind legs is about $1^1/_2$ feet.* [These measurements are very inaccurate, as having no way to measure lengths Wikar had to guess the length of a foot.]*Although I often saw giraffes singly and in big herds I never saw one eat grass; they always eat the tops of the high trees. It may be that they graze as well, but I never saw that. In my opinion and from my investigations these animals must be able to last without water for a very long time. For in the dry region where there is no water the veld is full of their spoors, but you seldom find a spoor leading to the river. If you manage to, by stalking, get within shot of this animal it is by great good fortune, and it seldom happens, for, being so tall and always eating the tree tops, it sees you from afar; but if you have a fairly good horse you can easily get it, as it cannot even run as fast as an eland. There is an easy chance of your catching the calves, for when they are a day or two old, the bushmen catch them on foot, for they know exactly when to expect the female giraffe to calve. If what I remember reading is true, that this animal was shown alive to the people in Rome in ancient times, I should imagine that an enthusiast who spared no trouble might attempt to bring one to the Cape.*

On June 16, 1779, his self-exile ended when he received the information that his misdemeanors had been pardoned and he could return to the Cape. Fortuitously for Gordon and Paterson, Wikar could not start his return journey before July. So it was that on July 25, 1779, Wikar met them and Gordon examined Wikar's "curiosities," which included the skin of the giraffe Wikar had shot, and obtained information "about some matters which might be useful to that gentleman on the journey he had begun." Paterson also questioned Wikar closely, and there can be no doubt that it was the conversations with Wikar that influenced the decisions Gordon and Paterson were going to make. Thus it was that Wikar, a much maligned person, whose place in history has been overshadowed by his apparently criminal past, came to play a significant role in the scientific history of giraffes. Apart from the biological observations his journals contain, they are a rich source of anthropological details of the many native tribes he encountered. Mossop's (1935) summary of the journal says much: "There is food for reflection in the fact that what was written by this obscure clerk is of greater importance to posterity than all that has been preserved of the writings of the august and kind-hearted Governor, who pardoned him." Just so.

On August 17, 1779, three weeks after meeting Wikar, and in some desperation from lack of food and water, Gordon and Paterson fortuitously found an ostrich nest containing 34 eggs. They cooked some to feed themselves and their companions, buried the rest, and later that day reached the mouth of the "Great River," becoming the first Europeans to do so (Mossop, 1935). Gordon was convinced that this river mouth was that of the fragment of the river he had first seen in 1777, some 1,200 km (700 miles) to the east, although how he could have known that is a mystery. Then,

> *In the evening . . . we launched Col. Gordon's boat, and hoisted the Dutch colours. Col. Gordon proposed first to drink the States' health, and then that of the Prince of Orange, and the Company; after which he gave the river the name of the Orange River, in honour of that Prince.* (Paterson, 1789)

Gordon had established from Wikar that no giraffes had been seen that far west and wanted to go further east, where Wikar had said giraffes were plentiful. Paterson decided to go to the spot where Sebastiaan van Reenen had shot a giraffe on their trip the previous year. Paterson crossed the river on October 14 and heard from local tribes that there were a number of giraffes in the vicinity, "one of which we were very desirous of shooting, as it is a beast so little known to Europeans, that even its existence has been doubted." Paterson's party traveled toward Warmbad on the route taken by Hop's expedition in 1761. Almost 18 years to the day later, Paterson saw six giraffes, and van Reenen shot one, a male, 14′9″ (4.5 m) tall. Paterson spent the day making a drawing of it (see Plate 3.3), recording a good description, preserving the skin and the skeleton, and making some measurements of its dimensions which, however, are not consistent with modern measurements.

His goal of obtaining a giraffe specimen now achieved, Paterson returned to the Cape, where he discovered that the Countess of Strathmore had stopped sending him money. He found himself deep in debt, and unable to pay a passage back to England. Nevertheless, he borrowed money from the Garrison Commander, Colonel du Phren, settled the debts, returned to England early in 1780, and gave all the specimens he had collected to the Countess. She donated them to John Hunter in 1780. Hunter, at last, had a specimen for his museum, 10 years after the Zoological Society meeting where he had first heard of Carteret's drawing (Dobson, 1959). It was the first specimen of a giraffe seen in England. The skin, the skull, cervical vertebrae, and other parts of the skeleton (the entire skeleton does not seem to have arrived in England) were stuffed, mounted, and prominently displayed in Hunter's museum. Not much seems to have been done to study or describe it, except for a note on the way the *ligamentum nuchae* "assisted the muscles of the neck to maintain equilibrium" (Dobson, 1959; see Chapter 15). When Hunter died in 1793, his collection was bought by the British government and placed in the care of the Company (later the Royal College) of Surgeons in London. In 1834 the Royal College collection was transferred to the British Museum of Natural History, but by then the skin had decayed and the specimen had disintegrated (Dobson, 1959).

Gordon in the meantime had pressed on with Pienaar and Schumacher and his Khoekhoen retinue, and on September 13 he was "given the skin of a young giraffe." Its origin is unknown, and Gordon does not mention it again. On September 30, 1779, on the south bank of the Gariep (Orange) River, they saw giraffe spoor for the first time. Thereafter spoor were plentiful. On October 12, Gordon saw a living giraffe for the first time, at a distance, near the Aughrabies Falls (28°59′S, 20°34′E) about 100 km west of present-day Upington. It was shot by Pienaar late that evening, 5 km south of the Orange River, 8 km southwest of Seekoeisteek at 28°32′S, 20°09′E (Skead, 1980). Seekoeisteek was the furthest eastern extent of Wikar's journey along the river where he too had encountered giraffes. Gordon wrote:

> *I left my horse and went on foot and came upon Pinar* [sic] *who was making signals by shooting and with fires, being an hour from the wagon. There I found this handsome and extraordinary animal, one of the most beautiful formed by nature, dead.*

October 13 was spent measuring it, and Gordon recorded everything in the great and copious detail of a man completely immersed in making a momentous discovery. He made 122 measurements of lengths and circumferences of just about every part of its body. He noted that the "skin was half an inch and one line [a line = 0.22 cm] thick" (i.e., ~13 mm). The giraffe was large and its skin thicker than any of the modern measurements made of skin thickness on younger, smaller giraffes (see Chapter 16); Gordon observed that the white stripes between patches "run like the streets on the plan of an irregularly built city." He noted the presence of a small gallbladder, a noteworthy observation, as adult giraffes do not usually have one (Cave, 1950). He also made "an accurate drawing of the animal" (see Plate 3.4). Gordon was aware of the controversy concerning whether the horns were shed annually and was emphatic that "clearly they never change or fall off." He buried the skeleton and "regretted that I could not take the skin with me." Moving further east along the river, on October 15 he saw six giraffes close by, "at last what I had for so long desired," and after crossing the river into Namibia saw many more. He traveled ~800 km (530 miles) from the mouth of the Orange River (Axelson, 1954) inland along the river, but could find no more giraffes and retraced his steps. On reaching the spot where he had buried the skeleton, he found that it had been destroyed by scavengers. This giraffe and those mentioned by Wikar were the last wild giraffes ever recorded south of the Gariep (Orange) River.

Gordon, greatly disappointed, nevertheless was determined to obtain a giraffe specimen to honor his promises to Vosmaer and Allamand. He moved west along the Orange River to Ramon's Drift, crossed the river into Namibia, and on December 9 discovered the route Hop and Brinkhad taken 18 years before, as had Paterson two months earlier. On December 10, a short distance south of Warmbad, one of the Khoekhoen in Gordon's party, a member of the Nama Afrikaanders, shot a large bull giraffe, breaking its leg, and it had to be dispatched by cutting its throat. It was 15′4″ high (5.5 m) and 13′1″ (3.9 m) from muzzle to tail, and thus weighed in the region of 1,000 kg. He prepared this specimen meticulously, taking 10 days to cure the skin and clean the bones. Schumacher made a sketch of the skeleton (see Plate 15.1 in Chapter 15). Gordon now had a giraffe specimen, and a main purpose of his journey was over.

In April 1780, after returning to the Cape, Gordon sent the skin, the skeleton, and the sketch Schumacher had made, one of his own drawings of the animal (see Plate 3.4), and his notes to Allamand for his information, but also for subsequent inclusion in the Prince of Orange's *Kabinet d'Histoire, Naturelle et des Curiosités* in De Haag (Vosmaer, 1787; Forbes, 1970; Rookmaaker, 1979; Skead, 1980). Other drawings of giraffes were also made by Gordon, as well as another of the skeleton and several others of individual bones. Gordon also made 400 watercolor drawings of the peoples he encountered, mammals, birds, reptiles, and plants, as well as plans and views of the district around Cape Town and the Gariep River (all of them can be viewed on the webpage of the Rijksmuseum, Amsterdam). They were (and are) a unique and irreplaceable record of the early biology of the Cape, and might easily have been lost, as well as much of what is known about the biology of giraffes, when Gordon committed suicide in 1795. It is said he did so in remorse over the subjugation of the Cape by the British, and/or because of serious ill-health. However, his papers, maps, and drawings were taken to England by his widow in 1796 (Raper and Boucher, 1988) or 1797 (Britten, 1914), on her return from the Cape to her native Switzerland, and were left in England on the understanding that they would be bought by the British government to provide her with a pension. That was not to be. After their release from Customs, they were placed in the custody of Edward Riou, a naval officer, who was killed at the battle of Copenhagen in 1801. The manuscripts passed to John Pinkerton, a Scottish historian. He had been tasked by Gordon's widow to find a publisher for them, but was unsuccessful, and the collection was bought by the second Marquess of Stafford (his wife was of the Scots Gordons). In 1913 the, by then, fifth Marquess sold his home and the Gordon papers were deposited for safekeeping in the Stafford County Library. The maps and drawings (but not Gordon's Diary and other papers) filled four folio volumes. They were sold at auction by Sotheby's as being "of the highest importance to students of the flora and fauna of South Africa," and were bought by Maggs Bros. of London, who added two more volumes of maps, before selling all six volumes to the Rijksmuseum, Amsterdam. Gordon's Diary and other papers remained in the Stafford Library, but were forgotten until 1964, when they were discovered by the Director of South African archives. Auctioned by Christie's in London, they were bought by Harry Oppenheimer, who was then Chairman of the Anglo-American mining corporation, for inclusion in the Brenthurst Africana Library that he had established in Johannesburg, South Africa. The diaries Gordon kept of his journey to the northwest Cape were translated from the original Dutch into English by Patrick Cullinan. They have become an invaluable resource.

Vosmaer

The amount of giraffe material collected from southern Africa and made available to the biologists in Europe was substantial. It comprised the skins, notes, and a drawing, which were obtained in October 1761 by the Hop expedition from two giraffes; a copy of this drawing and notes made by Carteret in 1770; parts of a skeleton from Gordon's first giraffe and a skin, a complete skeleton, comprehensive notes and drawings from his second giraffe, and a skin and some parts of a skeleton obtained by Paterson. No specimens from Paterson's first giraffe of October 1778 were collected. All this material, except for that of Carteret and Paterson, which stayed in England, were sent to Allamand in Holland and then passed to Arnout Vosmaer. Allamand collaborated with

Buffon, and his descriptions found their way into numerous editions of the *Histoire Naturelle*. Vosmaer studied the Gordon specimens independently and they became the subject of the first scientific article ever written on giraffes. It was written in 1786, and published in 1787 as part of Vosmaer's *Description d'un recueil exquis des animaux rares* in which, in 13 parts, he described the mammals, birds, and a snake in the Prince of Orange's collection. Each part was accompanied by an engraving. Although he was a very successful Director of the Prince of Orange's Museum of Natural History and Curiosities, Vosmaer nevertheless was on the fringes of academia and subject to some antagonism from Allamand and Buffon, and to a sharp reprimand from Gordon about the heights of giraffes. Even so, he produced an excellent analysis of the morphology of the skin and skeleton of giraffes. His article was also the first attempt to bring together all the historical descriptions of giraffes, and to correct them using the skin and horns sent to the Academy of Sciences at the University of Leiden by Tulbagh in 1764, and the skin and skeleton sent by Gordon. It was a magnificent piece of work, but Buffon ignored it, as did Richard Owen when he wrote of his dissection of three giraffes in 1838, and the French anatomists Joly and Lavocat, in 1846, shamelessly plagiarized it for their monograph based on a giraffe that had died at the Toulouse Zoo. Vosmaer's article also included an engraving of a giraffe made by Cornelis Bogerts after a drawing by Gijsbertus Haasbroek who, in turn, copied his from earlier drawings commissioned by Allamand and Buffon. Under Vosmaer's direction, those drawings were corrected by Bogerts (see Plate 3.5).

Vosmaer's illustration used Carteret's and Buffon's drawings (see Plate 3.2) as its basis, but improved them by producing the first accurate depiction of the coat patches of giraffes. The three pictures have the same background scenery and the presence of a Khoekhoen in the foreground. Vosmaer's differed not only in the shape of the skin patches (which Vosmaer described as a river of white color flowing from the head down the body and divided into an archipelago of many small islands), but also in the less equid, more giraffe-shape of the head and protruding upper lip; the shape, size, and angle of the horns; the position and size of the ears (now unstriped); the mane ending on the withers; the thoracic "hump" shaped by the thoracic vertebrae spines; the more equal (and appropriately so) heights of the fore and hindquarters; the absence of spurs at the distal end of the metapodials, and the more slender legs. Buffon's giraffe, on the other hand, captures the forward thrust of the point of the shoulder far better. Buffon's drawing clearly was of a male, Vosmaer's of a female, although in his notes he said it was a male, and its height 15′4″ (4.7 m). Vosmaer's article also included the first accurate description and drawing of the giraffe skeleton, labeled with letters, engraved by J.F. Gout in 1786, after the Dutch anatomist Petrus Camper had constructed it from the bones sent by Gordon (see Chapter 15, Plate 15.2). Vosmaer noted that it was "impossible for the horns to be shed annually as they are adhered to the frontal bone," an important if inaccurate observation that would have resolved the controversy surrounding the horns, but it was ignored by Buffon and Allamand. Earlier, in 1784, Vosmaer had had another engraving made of the skeleton by J.H. Merck, but Merck, contrary to an agreement he had with Vosmaer, had sent it, on May 6, 1785, to Joseph Banks in London before Vosmaer could publish it. Banks was President of the Royal Society and seemed to be unimpressed. He made a perfunctory acknowledgment of receiving "a plate of a *Camelopardalis*. . . . It is well-engraved and interesting." Not surprisingly, Vosmaer was angered by Merck's serious breach of scientific etiquette, and he commissioned the second engraving. As the unlabeled drawing of the skeleton appeared in several subsequent editions of Buffon's *Natural History*, the suspicion is that Merck had sold his engraving to Buffon.

LeVaillant, Burchell, and Harris

The exploits of Gordon, Paterson, and others led to South and southern Africa becoming known as a paradise, and noted zoologists, naturalists, and hunters were attracted. Among the first of these was Francois LeVaillant in 1781–1784 (LeVaillant, 1795). He exaggerated the extent of his travels and experiences and is chiefly remembered for his contribution to ornithology (six species of birds are named after him), but he did obtain a giraffe skin and made several drawings of giraffes that ended up in Paris (Geoffroy St.-Hilaire, 1827). Another visitor was William Burchell. Burchell was a Fellow of the Linnaean Society and an accomplished artist and writer whose scientific expertise included anthropology, botany, mineralogy, and zoology. Between 1811 and 1817 he collected and catalogued the skins and skulls of two giraffes, as well as 40,000 botanical specimens, and 23,000 other objects and specimens, including 120 skins of 80 different species of quadrupeds and 265 different kinds of birds. The two giraffes, a male and a female, were shot in October and December 1812 north of the Orange River (Pickering, 1997). All his specimens, including the skins and skulls of the two giraffes, were presented to the British Museum of Natural History (Axelson, 1954). It was a magnificent collection. The specimens were in high demand. He could have sold them profitably. However, he wrote:

> *At my return to England I thought proper, from feelings of patriotism, to present them to the British Museum, supposing that science would be more benefitted by placing them in our national repository, than in keeping them in any other collection.* (Burchell, 1822)

He was to be disappointed. His specimens were largely ignored.

Perhaps the most flamboyant visitor of all was Captain Cornwallis Harris, an officer in the British East India Company's Corps of Engineers. Harris had been given two years of sick leave and decided to spend it hunting and exploring in southern Africa. He arrived in 1836. His account of his adventures reveal, first, that he was not sick, and second, that he was a character of exceptional blood-thirstiness who had a wanton disregard for the pain and suffering he inflicted on hundreds of animals. He is undeserving of the admiration he still commands. His indiscriminate slaughter was mitigated partly by his ability to paint "and contribute my mite to the geography and natural history of the country" and by his very acute observation of the characteristic features of the animals he killed and painted. The account of his adventures appeared in *The Wild Sports of Southern Africa*, first published in 1839 but going to several editions.

Whereas previous explorers had ventured northwest from the Cape, Harris chose to go northeast. Having crossed the Vaal River, which is the largest tributary of the Gariep (Orange) River, Harris turned eastward into the large enclave formed by the Limpopo River (which borders modern-day Zimbabwe) to his north and the Olifants

River (which runs through what is now the Kruger National Park) to his south, where at approximately 25°30′S, 28°30′E on November 19, 1836, he shot an 18 ft (5.5 m) tall male giraffe. On subsequent days he shot several more, and says that he crossed the Limpopo River into what is now Zimbabwe before turning back. Harris made two paintings of giraffes. One appears as the frontispiece of his book and shows a giraffe stretching its neck while feeding (see Plate 3.6), and the second was entitled *Hunting the Giraffe* (see Chapter 13, Plate 13.1). These giraffes have patches with the leaf-like shape of what Lydekker (1904b) was to name the Transvaal giraffe (*G.c. wardi*). Harris's description of giraffes is a classic of such brevity and rich information that it could be an entry in any modern text on mammals:

> *In stature the tallest of mammiferous animals. Adult male twelve feet high at the shoulder, eighteen at the crown of the head. Legs slender, seven feet in length. Feet terminating in a divided hoof. No succentorial hoofs. Body short. Withers elevated, a scanty, upright, rufous mane extending along the whole neck. Back oblique. Tail thirty-four inches long, terminating in a tuft of bristly black hair about the same length, which reaches to the hocks. Head light and tapering, thirty-four inches long; provided with osseous peduncles (common to both sexes) covered with a hairy skin, and terminating in a tuft of black hair. A tuberculum on the chaffron. No muzzle, upper lip entire, hairy and pointed. Eyes large and melting. No lachrymary sinus. Ears pure white and ample. Callosities on the breast and knee. Tongue very long, pointed, and flexible. General colour deep sienna, with large angular ferruginous spots, variously disposed over the whole; each spot darker in the centre. Belly and cheeks white, with dark blotches. Female sixteen or seventeen feet in height at the crown, of a dirty white colour, with pale ferruginous spots as in the male. Mammae four. Gregarious in small troops. Inhabits the great plains of the interior.*

Living Giraffes in Europe

In the early 1800s, natural history was a developing scientific discipline. Naturalists were becoming biologists and, as exemplified by Wallace and Darwin, were studying actual and sometimes living specimens of animals and plants. Establishing zoos and museums to house specimens for study became a priority, and it was not long before giraffes themselves, rather than skins and bones, were being taken to Europe. The first living giraffe for 340 years went to France. In 1826 a French Egyptologist, Edme-François Jomard, floated the idea that a giraffe would be a worthy gift for the youngest brother of Louis XVI, who was about to be crowned King of France. The idea received the support of naturalists in France, in particular of Etienne Geoffroy St.-Hilaire, an icon of French zoology, who was one of the 12 founding professors appointed in the Museum National d'Histoire Naturelle after the French Revolution. But then as now, the capture and transport of giraffes were immensely difficult. For the idea to be realized, unique skills were needed. They were available in Bernardino Drovetti. He was a master ingratiator and an entrepreneur with wide interests: an animal trader, a tomb raider, an antiques dealer, and an Albanian mercenary in the Turkish Army, who had an ever watchful eye open for the main chance (Allin, 1998; Ringmar, 2006; Sharkey, 2015). As French Consul General in Egypt, he was also an advisor to Mehmet Ali, the Ottoman

Viceroy of Egypt. The gift of a giraffe to the future Charles X of France was arranged. The order went out to the Viceroy's military underlings in Khartoum (Sharkey, 2015) and three "Nubian" giraffes were caught in Darfur/Kordofan (precisely where is, of course, unknown), with the purpose of sending one to Austria, one to England, and one to France. To absolve himself of accusations of bias, the Viceroy hedged his bets by making the Consuls of each nation draw lots for the giraffes (*Literary Gazette*, August 25, 1827; Scherren, 1908). One was sent to Francis II of Austria, the last Holy Roman Emperor, and was kept in the Emperor's zoo in Vienna, but survived less than a year. A second was sent to George IV in England. The third was sent to France.

The story of the journey of the giraffe sent to France, a young female, has been told in several academic studies, of which the best I have come across is that by Heather Sharkey (2015), and in books, the best of which is by Michael Allin (1998) in his *Zarafa*. Zarafa's arrival in Paris was captured in a marvelous painting by Nicholas Hüet (see Plate 3.7).

Zarafa was caught as a two-month-old calf in the Ethiopian Highlands, fed cow's milk (Allin and Sharkey say 25 gallons a day, a volume the three cows that accompanied her were unlikely to be able to provide; 25 liters is more realistic. This latter volume is mentioned by Lagueux (2003) and Rigoulet (2012) in their accounts of the journey of Zarafa). Reared in Khartoum until six months old, she became very tame. Zarafa sailed from Cairo across the Mediterranean in the care of a Sudanese servant, Atir, taking three weeks to complete the journey. After landing at Marseilles on October 31, 1826, she spent two weeks in quarantine and then on November 14 was taken to winter quarters in Marseilles. St.-Hilaire described the occasion: "The prefect, Count of Villeneuve, placed her in the care of his hotel [*sic*], and gave her care which was effective, for she never ceased to enjoy, during her stay in Marseilles, the best health." An official of the Marseille branch of the French Academy of Sciences, identifiable only as M. Salze, captured the moment:

> *One can say that the Giraffe has nothing elegant or graceful in the detail of her form; her short body, her long and close together legs, the excessive length of her neck, the slope of her back, her mis-shaped rump, her long and bare tail, all these contrast in a shocking manner; she seems wrongly built, unbalanced on her feet, and yet one is seized by astonishment at the sight of her, and one finds her beautiful without being able to say why.* (Allin, 1998)

On May 20, 1827, Zarafa started out for Paris, a distance of 550 miles (~850 km). She walked all the way, being led by her Arab keepers and exciting much interest. Prominent in the organization of her trip, and reveling in the reflected glory, was Etienne Geoffroy St.-Hilaire. Zarafa took 41 days to reach Paris, a walk that she did more easily than her guardians, and lived for 18 years in the *Jardin des Plantes* to where the Royal Menagerie of Louis XVI had been moved in 1792. She died on January 12, 1845, and is now preserved in a museum in La Rochelle on the west coast of France (Allin, 1998; Sharkey, 2015).

The calf sent to George IV was sickly and not expected to live long, justifiably as it turned out. It had deformed legs. It arrived by ship at the Duchy of Lancaster Wharf near Waterloo Bridge in London on August 11, 1827, accompanied by two Egyptian cows as wet nurses, two Arab keepers, and an interpreter (*Literary Gazette*, 1827).

A sketch commemorating its arrival was published in the *Literary Gazette* "soon after it got ashore." It likely was the first living giraffe seen in England, although it has been suggested that another had been imported in 1805 (Lagueux, 2003). This earlier animal was, perhaps, a camel artificially spotted to look like a giraffe and passed off as a giraffe to a credulous public (*Literary Gazette*, 1827). George IV's giraffe had spent the winter in Malta, where some reports said that it died (Geoffroy St.-Hilaire, 1827). As the *Literary Gazette* put it, these were

> *oft reiterated statements which have appeared in the French Journals, giving the world to understand that the "Anglo-Camelopard" had died, and that Paris, in consequence, possessed the only living specimen of the creature in Europe.*

On Monday August 13, 1827, the giraffe was taken to Windsor. Like Zarafa, it was by then about 20 months old, but was somewhat smaller: it stood about 10.5 ft (3.2 m) high. Its gender is uncertain. From paintings it appears female, and both Everard Home (1828) and Geoffroy St.-Hilaire (1827) refer to it as a female, but the report of its arrival referred to it as male (*Literary Gazette*, August 25, 1827). Several paintings celebrated the arrival of this giraffe. In these pictures it is obvious that it is a female. Two paintings were made of it by Jacques-Laurent Agasse. One shows the giraffe arriving at Windsor in the company of its two Egyptian keepers and the animal dealer Edward Cross, and being offered milk. In the other, it is standing next to its keeper in a park in London (see Plate 3.8). What is clear from Agasse's paintings is that its coat markings are reticular, as they were in Zarafa, which confirms that it was caught from the same locality and likely from the same herd as Zarafa.

Another painting was by the English animal painter Richard Barrett Davis, which he completed in November 1827 (see Plate 3.9). Davis studied his subjects thoroughly before painting them. In a short article he wrote "on the habits of the giraffe" in 1827, Davis recorded that "the two outside [teeth of the lower jaw] are divided to the socket," a characteristic that Petrus Camper had seen as well (Vosmaer, 1787), which requires very close observation to be noticed and which, very much later, was identified by William Hamilton (1978) as the defining characteristic of the family *Giraffidae*. Davis also noticed that in order for a giraffe to spread its legs when bending downward it had "between the forelegs . . . folds of loose skin." He saw the deformed lower limbs and investigation revealed the reason:

> *its limbs are deformed by the treatment it experienced when in the hands of the Arabs, in its overland journey from Sennaar to Cairo. It was occasionally confined on the back of a camel, and when they huddled it together for that purpose, they were not nice in the choice of cords, or the mode of applying them.* (Davis, 1827)

As a result the joints were damaged and enlarged. However, in none of the paintings are the deformities shown.

According to Lydekker (1904a), Paterson's "Southern" giraffe, not the Windsor "Northern" one, was used by Davis as a model for his painting. Lydekker's opinion was based on the absence of a third (median) ossicone (a defining characteristic of northern variants) in Davis's giraffes, the shape of the coat markings that were blotched, not reticulated, and the lower parts of the legs were spotted. If Lydekker is correct, then Davis's impression of what the skin patches looked like is very different from those of Paterson's giraffe. Patches are variable in shape, and in Paterson's drawing (see Plate 3.3) they are distinctly, but impossibly, rectangular. It may be that Paterson, a less able artist, drew a schematic diagram. In any event, and using one's imagination, one of Davis's two giraffes has reticulated markings very similar to those of Agasse's giraffe. The other does have more "blotched" coat markings of a Cape giraffe. Davis titled his painting *Portrait of two giraffes belonging to George IV* when only one giraffe had arrived in England, so it is possible that Davis might have used Paterson's specimen in Hunter's museum to confirm details of one type of (southern) giraffe, and used George IV's (northern) giraffe for the other, achieving a painting, as had Buffon's and the two of Gordon and Patterson, that showed there are varieties of giraffes.

Owen

By the time that Zarafa and George IV's giraffes had arrived in Europe, anatomy had supplanted classification as the dominant biological science, and it required specimens that could be dissected. Zarafa survived 17 years and so was not available for an anatomical study. The first formal anatomical study of a giraffe in France was made in 1846 by Nicolas Joly, assisted by M. A. Lavocat. Joly was Professor of Zoology in the Faculty of Science at Toulouse, and Lavocat was Head of Anatomical Works at the Royal Veterinary College in Toulouse. However, George IV's giraffe died on October 11, 1829, two years and two months after its arrival at Windsor Zoo, having grown 18 inches (Scherren, 1908). After its death it was examined by Everard Home, who was the brother-in-law of John Hunter. He published a note on the tongue, a note that can be said to be the first scientific article about giraffes in England (Home, 1828). He concluded that the prehensile tongue of giraffes was propelled forward by engorgement with blood. He also made some observations on the structure of the fore-stomachs and concluded that they were "like those of all quadrupeds that chew the cud" (Home, 1830). That was the sum contribution to science of George IV's giraffe. The opportunity to make a full dissection was not taken. The giraffe's legacy nevertheless was great. Stirred by its short life, and through the intervention of George IV's successor, William IV, the Zoological Society of London acquired another four animals (three males and a female), all of them from Nubia. (The first southern giraffe arrived in London in 1895, after being caught in Mozambique, loaded onto a ship at Lorenço Marques [now Maputo] in Delagoa Bay [now Maputo Bay] and thence to Southampton [Laufer, 1928], a route others were to follow [Dinesen, 1938].) William IV's giraffes arrived on May 24, 1836. Lydekker (1901) described the occasion. Anyone, he said, in London awake early that morning would have been astonished to see these animals, accompanied by their Nubian attendants, on their way from Tilbury docks to Regents Park Zoo along New Road (now the A13 and A501). On June 19, 1839, the female, called Zaida, gave birth to the first giraffe ever born in Europe. The calf died nine days later because of what Owen said was "uncalled for aid" by zookeepers, but became the subject of a landmark article published by him in 1839. The mother and the calf were drawn by Robert Hills, whom Owen called the

Figure 3.3. *Zaida and her young one as it appeared when one day old*, by Robert Hills. The first giraffe born in Europe at Regents Park Zoo on June 19, 1839.
From Owen, 1839, with permission of John Wiley and Sons.

"celebrated delineator of the Ruminant tribe" (Figure 3.3). Zaida had a second calf in 1841, which thrived.

Richard Owen (Figure 8.1) was then the major force in English zoology. He was a great anatomist, an authority on comparative anatomy and on paleontology, and without doubt one of the greatest zoologists that have ever lived. In 1830 he was appointed assistant curator at the College of Surgeons, becoming Hunterian Professor there in 1836, and Conservator of Collections in 1849. He identified the group of animals that include ruminants and gave it the name *Artiodactyl* (Owen, 1848) and coined the term "dinosaur." He oversaw the construction of the British Museum of Natural History in South Kensington, London, and became its first Director, all the while being an active and prolific researcher, finding the time, for example, to write a three-volume compilation of the then known knowledge *On the anatomy of vertebrates*, Volume 3 of which runs to 915 pages (Owen, 1868). He developed, however, very different ideas on the origin of species from those of Darwin and, although initially supportive of Darwin, in later life became Darwin's tormentor. His cathedral-like design of the Natural History Museum better reflected his ideas on the "ordained" origin of species, although his emphasis of "ordained" is not evident in any of his anatomical descriptions. They are always precise and non-judgmental as to the origin of the animal he was dissecting. When he was at the height of his powers, Owen transformed the study of natural history and established England as the center of that universe, wresting it from Holland, which had had, in the context of giraffe research, the benefit of Allamand, Vosmaer, and Gordon's specimens, and from France, which had had Buffon, Zarafa, and St.-Hilaire. Conversely, Carteret's, Paterson's and Burchell's notes and specimens had been all but ignored. Owen changed that. In 1838 he published a very detailed and accurate description of giraffe anatomy based on a male that had died in the Zoological Society's Regents Park Zoo and a male and a female that had died at Surrey Zoo (Owen, 1838). This article, together with his description of the birth of Zaida's calf, was authoritative, and was recognized as such across Europe (Joly and Lavocat in 1846 referred to him as "the celebrated anatomist"). As this book progresses, Owen's descriptions will be seen to be a key reference for giraffe anatomy, and his articles were an important stimulus to what has become the long scientific study of the biology of this unique creature. Quite rightly, Owen's statue dominated the main staircase of the Natural History Museum until 2008, when, quite rightly, it was replaced by Darwin's statue. The occasion was the bicentenary of Darwin's birth.

4
The Antecedents

Anyone looking at a giraffe must surely wonder how its long neck could have evolved. In all contemporary zoology there are no animals that have a neck as long and just two others that have a longish neck that is held vertically. They are ostriches and llamas, both of which have a neck about one meter long. The most striking features of modern giraffes are their elongated legs and neck, distinctive coat markings and a gentle, aloof disposition, noted as long ago as AD 64 by Pliny the Elder. How did they actually come about? The shape of their antecedents was as unremarkable as it is in modern gazelles. Beautiful as modern gazelles are, there is no hint in them of a tall slender creature, yet somehow, by some miracles of evolution, the shape of the first antecedents transformed into a giraffe. It is almost impossible, at least for me, to imagine how the genes and other genetic material in all the previous species could contain the blueprint for a giraffe, but they did. Darwin had no such qualms. "By natural selection long-continued," he wrote, in a sentence that might have been written by Lamarck, "combined no doubt in a most important manner with the inherited effects of the increased use of parts, it seems to me almost certain that an ordinary hoofed quadruped might be converted into a giraffe" (Darwin, 1872). The DNA from which giraffes ultimately arose had to be in its antecedents, and it had to be that, through countless years of evolution, mutation, and indeed encryption, the DNA that ended up in the ancestors of giraffes had all the information that could be worked on to produce what we now see as a giraffe.

However, one of the enduring folklore tales about modern giraffes is that they defy the idea of a Darwinian, long, slow process of gradual change, and appeared in the African Pliocene about 6 million years ago (Mya) as if they had no ancestors, having been created by an act of God as a monument to biological structural engineering. The origin of the folklore tale is the meager fossil record of giraffes. The reason for the meager fossil record is that giraffes can constitute just 1% of all the mammals that share their habitat (Bourliere, 1963). Another is that giraffes do not live in habitats that beget fossils. Yet another is that their skeletal remains are not accumulated by scavengers. If only a tiny proportion of all deaths become a fossil, the chance of finding fossil forms of giraffes is low. Thus we are left with the impression that the origin and evolution of giraffes has been so cryptic that giraffes fulfill the requirements for creation. Evolution, however, does not work that way, and the lack of ancestral forms is implausible given that no other species seems so deprived. Every modern giraffe has an ancestor, and each ancestor yet another. So the impression of sudden appearance rests uneasily, no matter what evolutionary mechanism or paleontological reason may be invoked to explain the apparent absence of intermediate forms.

Nevertheless, the spontaneous-arrival view of the origins of the giraffe's long neck was described by Francis Hitching (1981) in his book *The Neck of the Giraffe, or, Where Darwin Went Wrong*, saying,

> *There are no intermediate forms showing a three-quarter length giraffe neck. . . . What little evidence there is, is consistent with Gould-Eldredge's punctuated*

How Giraffes Work. Graham Mitchell, Oxford University Press. © Oxford University Press 2021.
DOI: 10.1093/oso/9780197571194.003.0004

> *equilibria . . . that evolution may develop rapidly (producing) the "sudden" appearance of modern giraffe.*

Thus, his argument went,

> *[over] a period of thousands of generations a series of mutations accumulated producing stronger bronchial* [sic] *arches, greater musculature and a bigger heart. Although of no selective advantage for these at the time, they became fixed by a process of gene fixation in a small population, so that if a mutation for long neck subsequently occurred the support systems for it would be in place.*

Why Hitching focused on these three aspects of giraffe anatomy is not altogether obvious. Nor is it obvious why, in the absence of any selective pressure, these characteristics would have evolved. Branchial (not "bronchial") arches of a fish form the gills. In a mammalian embryo, they ultimately form structures related to respiration, such as the larynx and trachea, and while a giraffe's trachea is long, stronger branchial arches are not needed to form it. A *bigger* heart is also not necessary, but a *stronger* heart is. A stronger heart is needed to pump blood up the neck to the head.

In his defense, Hitching was writing when fossil evidence of giraffe prehistory was incomplete and difficult to construe. However, his idea was later supported by one of the foremost evolutionary biologists of modern times—Stephen J. Gould. Writing in 1996, Gould said of giraffes that their "ancestral species are relatively short-necked and the spotty evidence gives no insight into how the long necked modern species arose." He added that the "standard story [of access to browse] is both fatuous and unsupported . . . entirely speculative and basically rather silly story," and there are good reasons to think that that view could be correct (see Chapter 18). Nevertheless, he did reiterate the generic "standard story" of Darwin: if there is a range of neck lengths among the individuals making up a population of giraffes, and if a long neck confers even just a marginal advantage, then those with the longest necks will tend to leave more surviving offspring, each of which would inherit the potential of greater height. This slow process, continued for countless generations, will lead to a steady increase in neck length, as long as a long neck remains an advantage.

It is a commonsensible story, and if it is correct, how could natural selection build a giraffe by gradual increments? Darwin (1868) recognized the dilemma:

> *It has been proposed that all the parts must have been simultaneously modified and on the principle of natural selection this is scarcely possible. But if an animal has its neck elongated a very little without any corresponding modification in other parts of the body, animals thus modified would be enabled to browse on higher twigs and thus survive. By repetition of the same process and by occasional intercrossing of the survivors there would be some progress slow and fluctuating though it be towards the admirably coordinated structure of the giraffe.*

Darwin was writing at a time when Mendel's work on genetics had not become general knowledge and he had no idea of how characteristics might be inherited. Nevertheless, he was right. A long neck must be associated with many other modifications—an altered skeleton, muscles and ligaments, nervous, cardiovascular, and respiratory systems, and so on, all adapted to its shape, and all heritable. Did all change together and not gradually, as natural selection supposes? If a neck grows by two meters rapidly, then all other changes must be in place, as Hitching suggested. However, if the neck elongated slowly over a period of many millions of years, the change in neck length might have been about 1 μm per generation (Badlangana et al., 2009), which is a tiny, tiny amount, and an increase in neck length of that size would not impose a noticeable burden on the anatomy or physiology of an evolving giraffe.

If there was a coordinated pattern of development, then it should be both regulated by genes and be heritable. There is some evidence that this is the case. Agaba et al. (2016) compared the structure of a large number of genes in okapis with the same ones in giraffes. Okapis do not have a long neck, nor, as far as can be guessed, an unusual skeleton or cardiovascular system. Moreover, giraffes and okapi diverged from a common ancestor about 12 Mya, thus providing ample time for natural selection to work its magic and alter the necessary genes in giraffes. What Agaba et al. found was that giraffes have eight genes that are distinctly different from their counterparts in okapis. Those genes regulate aspects of cardiovascular function, some related to control of the diameter of small arteries by receptors sensitive to adrenaline and noradrenaline, and others affecting angiotensin converting enzyme (ACE). ACE synthesizes angiotensin, a small protein that is a member of a large family of similar proteins all of which regulate many physiological functions. Angiotensin's first discovered function was that it is a powerful constrictor of blood vessels and thus can regulate blood pressure. Other altered genes in giraffes regulate the shape of the cells and the contraction characteristics of cardiac muscle tissue. Agaba et al. (2016) also found, in support of the ideas of Badlangana et al. (2009) and van Sittert et al. (2010), that neck elongation is regulated by Hox genes. Overall, Agaba et al. concluded that while remarkably few genes have undergone changes, two-thirds of the genes most diverged in giraffe from those in okapis have specific roles in regulating skeletal, cardiovascular, and neural development, as well as biochemistry, all of which may have coevolved to elevate its stature, adapt its neural and cardiovascular system to the demands of its shape, and, even, adapt its metabolism to browse rich in tannins (see Chapter 7). Of course it is not just structures that had to evolve but the rhythmicity associated with them—heart rate, respiratory rate, the rate at which nerves conduct impulses—indeed, all the functions that regulate life must be in tune with the way of life of a giraffe. How rhythms might have evolved is still a mystery.

Gradual evolution of a long neck and all its supporting structures implies that intermediate forms must have existed, and Darwin (1859) again recognized the dilemma.

> *Why,* [he asked] *if species have descended from other species by insensibly fine gradations, do we not everywhere see innumerable transitional forms? Why is not all nature in confusion instead of the species being, as we see them, well defined? Innumerable transitional forms must have existed. Why do we not find them embedded in countless numbers in the crust of the earth?... Geological research has done scarcely anything in breaking down the distinction between species, by connecting them together by numerous, fine, intermediate varieties; and this not having been effected, is probably the gravest and most obvious of all the many objections which may be urged against my views.*

In other words, intermediate forms of any creature are rarely if ever found. He did have an answer, though.

> *I look at the natural geological record as a history of the world imperfectly kept, and written in a changing dialect; of this history we possess the last volume only. Of this volume, only here and there a short chapter has been preserved; and of each page, only here and there a few lines may represent the apparently abruptly changed forms of life, entombed in our consecutive, but widely separated formations. On this view, the difficulties* [arising from the poor fossil record] *are greatly diminished, or even disappear.*

Darwin need not have worried, and Hitching and Gould turned out to be wrong, because a history of intermediate forms does exist, although a neck as long as a giraffe's developed in a single ancestor from a neck about half as long over a period of millions of years. The paleontological research that has uncovered the origins of giraffes has had two main epicenters. The first was the Siwalik Hills in what are now the mountainous regions of Kashmir in Pakistan and northwest India (Figure 4.1). They were once low-lying swampland and alluvial plains, a kind of Eden that had provided, for many millions of years, a rich environment in which numerous species of plants and animals flourished. Fossils were plentiful, and study of them was driven by the search for

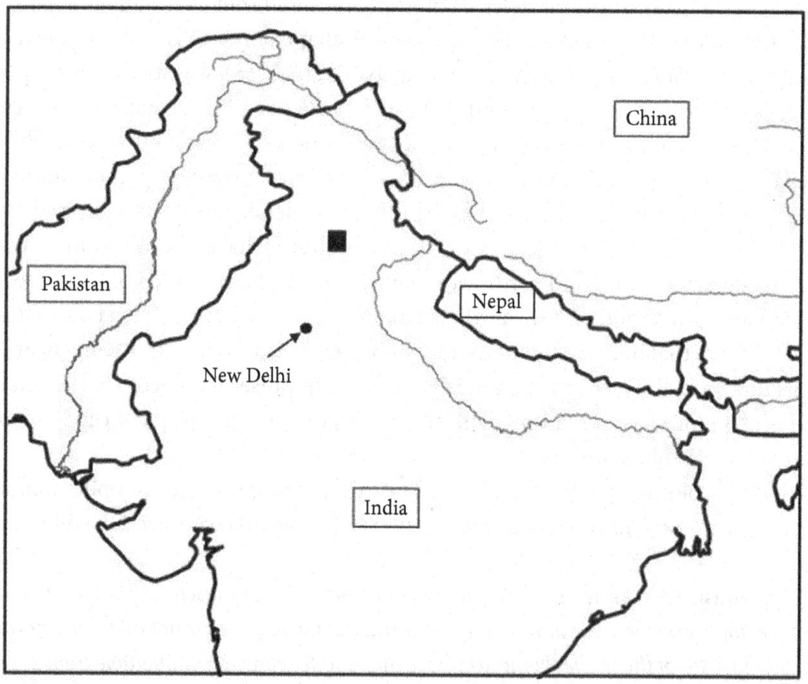

Figure 4.1. The site of the Siwalik fossil deposits. The marker (■) shows the location of Shivalik Fossil Park in the Siwalik Hills, a sub-Himalayan mountain range. This is most probably the area "*west to the River Jumna*" (currently Yamuna River) to which Falconer and Cautley (1843) referred.

After van Sittert, 2015.

intermediate forms and lineages to prove or disprove the revelations of Charles Darwin and Alfred Russel Wallace on the origin of species. This period of research began several years before Darwin and Wallace announced their discovery of the likely mechanism underlying the origin of species to the Royal Society in 1858. (Neither of them attended that historical meeting; Darwin because of acute anxiety and Wallace because he was in Indonesia. The announcement was made by Darwin's friends the geologist Charles Lyell and botanist Joseph Dalton Hooker.)

Fossil research in the Siwaliks was initiated in the 1830s by Dr. Hugh Falconer and Capt. Proby T. Cautley. Cautley was an army officer with the East India Company who later became famous for supervising the building of the 350-mile (560-km) long Ganges Canal. He had wide-ranging interests and assisted in Hugh Falconer's expeditions to, and excavations of, the Siwalik Hills fossil deposits. Falconer was extraordinarily but unassumingly accomplished. He had studied the natural sciences of zoology, geology, and botany at the University of Aberdeen before training as a medical doctor in Edinburgh. In 1830 he was appointed Assistant Surgeon in the East India Company, but on his arrival in Bengal (now mostly part of Bangladesh), he immediately began to study fossils. He recognized that evolution, while a process of continuous alteration, also seemed to move in fits and starts. The animals that were the origin of the fossils he was studying seemed to stay unchanged for long periods of time before undergoing a period of rapid change. The pattern mirrors changes in climate. When climates and habitats are stable, mutations, even favorable ones, that are occurring all the time in any species are diluted in large populations and therefore go unnoticed. When climates are so disruptive of a habitat that isolation of a subgroup of a species occurs, intense selection for any mutations, especially favorable ones, follows. The animals carrying them change and go on to become new species that have a stable, unchanging physical form, sometimes lasting for many millions of years. Falconer was the first to report this pattern, and his observations underpinned Gould and Eldredge's theory of punctuated equilibrium that emerged some 150 years later.

In 1831 Falconer was posted to the northwest of India and there met Cautley. Whenever they could spare the time, they excavated the Siwalik fossil beds. In 1838 Cautley found a perfectly preserved, elongated, third cervical (neck) vertebra (Figure 4.2) and attributed it to a giraffe species (Cautley, 1838). This and other fossils from giraffes were subsequently described by him and Falconer in an article published in 1843. They anticipated, and wishfully hoped for the demise of, the folklore of spontaneous creation of giraffes.

> *The finding of the giraffe as a fossil furnishes another link to the rapidly increasing chain which (as the discoveries of year after year evince) will sooner or later connect extinct with existing forms in a continuous series. . . . The Giraffe has hitherto been confined, like the human race, to a single species, and has occupied an isolated position in the order to which it belongs . . . it may be expected that, when the ossiferous beds of Asia and Africa are better known, other intermediate forms will be found, filling up the wide interval which now separates the giraffe from the antlered ruminants, its nearest allies in the order.*

The second epicenter of giraffe paleontological research was Africa, especially North and East Africa and mainly in Libya, Ethiopia, and Kenya, and in southern central

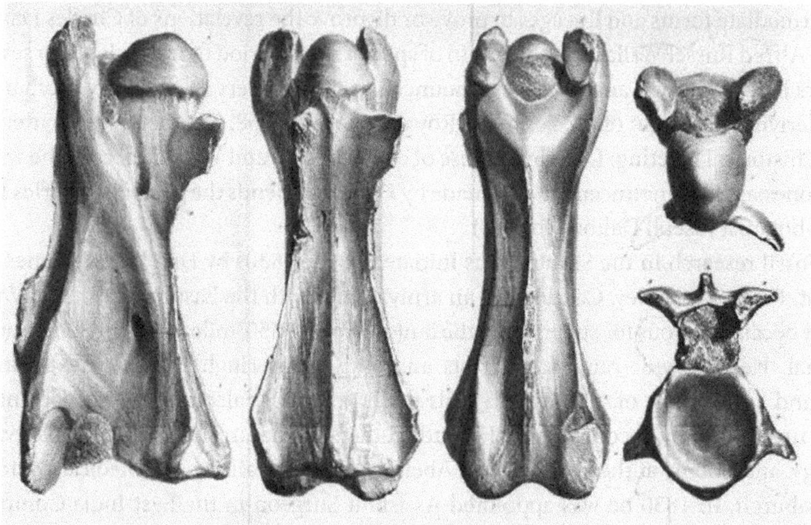

Figure 4.2. The third cervical vertebrae found in the Siwalik fossil deposits and attributed to a giraffe species.
From Falconer and Cautley, 1843. Image reproduced by permission of the Geological Society of London.

Europe, where French scientists Camille Arambourg and Denis Geraads, the Canadian palaeontologist C.S. "Rufus" Churcher, and those from England, William R. Hamilton, Alan W. Gentry, and John. M. Harris, evaluated the relatively scant giraffid material available, beginning just after the Second World War. Almost in parallel, the postwar paleontological work was complemented by a considerable volume of biological research on giraffes, to which the major contributors have been, first, Anne Innis Dagg and J. Bristol Foster, and later, John Skinner and his students, in particular Anthony Hall-Martin and Vaughan Langman. Their contributions were followed by those of Robin Pellew in the Serengeti (he later became Chief Executive of World Wildlife Fund–UK and then of the National Trust for Scotland) and Johan du Toit, now at the University of Utah, whose studies of giraffe browsing habits in the Serengeti Reserve in Tanzania and the Kruger National Park, South Africa, respectively, and much else besides, are without equal and mostly unrepeatable. This additional biological information, unavailable to the investigators of the Siwalik Hills fossil beds, and mostly to those in Africa, together with information from some other physiological studies and studies of paleoclimates, has allowed further interpretation of the origins of giraffes and their phylogeny, and provided possible explanations for their survival and for their current and historical distributions.

Two giraffids are extant. *Okapia johnstoni* (see Plate 4.1), first brought to the attention of modern science by P.L. Sclater and Ray Lankester separately in 1901, and then in 1902 by the first European to see one alive, Sir Harry Johnston (1902). Johnston was in some ways similar to Robert Jacob Gordon. He was one of those unusual people who have become contributors to the story of giraffes almost by accident. An accomplished botanist, zoologist, artist, colonial administrator who also was able to speak many African languages, he was appointed as special commissioner to Uganda in 1899 to end a war there. By 1902 he had produced a two-volume book on *The Uganda Protectorate* as "an attempt to give some description of the physical geography, botany, zoology, anthropology, languages and history." It contained 506 illustrations including drawings he made and photographs he took, 48 full-colored plates he painted, and nine maps. Most of us these days would regard that as a life's work. It took him just three years. The okapi that he described and painted (it was one of the 48 paintings) was one of only a very few large mammal species to be "discovered" in the twentieth century (see Plate 4.1). An inhabitant of the Ituri forest refuge in Zaire, its future, given the prolonged political instability that region has endured, is surprisingly good. In 2015 Ripple et al. estimated the population size of okapi as 43,000 and estimated that they still occupied ~32% of their original habitat (compared with the 11.3% of their original habitat that giraffes still occupy). The other extant species is the familiar modern African giraffe. These two species can be classified as follows:

Kingdom: Animalia
Phylum: Chordata
Class: Mammalia
Superorder: Ungulata
Order: Artiodactyla
Suborder: Ruminantia
Infraorder: Pecora
Family: Giraffidae
Genus: Giraffa; Okapia
Species: camelopardalis; johnstoni

Both giraffids are Eutherian (placental) mammals, and their origins lie in the origins of ungulates, animals characterized by their broad flat toes. An official description given to the toes by Gentry and Hooker in 1988 is "laterally broadened terminal phalanges" that can be even or odd in number. Artiodactyls, one of the orders of ungulates, have an even number of toes and include the family of animals we call ruminants and which are familiar as cattle, goats, sheep, antelope, and deer. Ruminants are herbivores, and a part of the vegetation they eat is digested in their fore-stomachs, the largest of which is the rumen. The evolutionary line from the first ruminant to modern giraffes is tortuous and far from obvious. Nevertheless, there is general agreement among the main interpreters of the paleontological and biochemical evidence (Webb and Taylor, 1980; Carroll, 1988; Gentry and Hooker, 1988; Janis and Scott, 1988; Irwin et al., 1991; Kraus and Miyamoto, 1991; Allard et al., 1992; Scott and Janis, 1993; Gentry, 1994; Hassanin and Douzery, 1999; 2003; Fernandez and Vrba, 2005) that ruminant origins can be traced back to an artiodactyl fauna that had its origins in the late Palaeocene around 55 Mya and which became recognizable in the early Eocene some 50 Mya. Their evolution into the many genera and species we recognize today is usually attributed to the great void left in the fauna of the Earth that followed the extinction of the dinosaurs ~65 Mya, caused by massive volcanic eruptions in India (Schoene et al., 2015), combined with the Chicxulub asteroid impact in Mexico (Alvarez, 1997; Bininda-Emonds et al., 2007), but it took a long time for a new, post-dinosaur world to evolve. The even-toed artiodactyls, for example, evolved 10 million years after the extinction of the dinosaurs. Throughout that long interval their existence hung by a thread, yet they survived. There was something unique that artiodactyls needed to ensure survival and success. What was it? How did some of the early artiodactyls survive to become the ancient ancestors of the modern artiodactyl ruminants with which we are so familiar today?

The oldest known artiodactyl family is the Dichobunidae. The only and almost complete skeleton of an animal from this family is that of a rabbit-sized forest-dwelling animal called *Diacodexis* (Figure 4.3).

Diacodexis lived in Pangea, a landmass that straddled the equator and comprised regions that have since then become North America, Europe, and Asia. *Diacodexis* fossils have been dated to the late Palaeocene/early Eocene of 60–55 Mya, and it lived as an identifiable species for about nine million years. What made this creature clearly an artiodactyl were the bones that form the fore- and hind limbs. They had some features that we now know are unique to artiodactyls. One of the bones, the astragulus, a part of its ankle, was formed in such a way that the foot could only move forward and backward, in other words in the vertical plane. It could not move sideways in the same way that a human ankle can. The astragulus in turn formed a joint with the main bone of the lower leg above it, the tibia, and with the small bones of the ankle below it, which are called the cuboid (shaped like a cube) and navicular (shaped like a boat) bones. The huge benefit of this arrangement was a very stable ankle joint, and the result was an improved ability to run and jump, abilities that exceeded those of any of its mammalian contemporaries (Rose, 1982). Running and jumping were also helped by long thin metatarsal bones. Those bones link the ankle to the digits. Their length increased the length of the leg and the length of the strides (Rose, 1982). Most weight-bearing was now taken by the third and fourth digits, a change that reduced the weight of the feet and reduced the amount of minerals and other constituents needed to make bones.

Diacodexis also possessed the full, archetypical mammalian complement of teeth. In each half of the upper and lower jaws it had three incisors (I), one canine (C), four

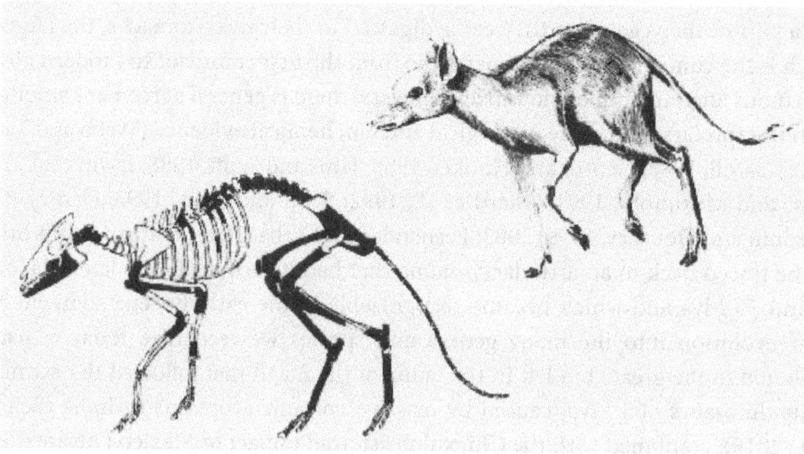

Figure 4.3. Skeleton of *Diacodexis*.

From Rose, 1982, with permission of the American Association for the Advancement of Science.

premolars (P), and three molars (M), making a total of 44 teeth when both sides of the upper and lower jaws are taken into account, an arrangement that can be described conveniently by the formula:

$$I^{3/3}\, C^{1/1}\, P^{4/4}\, M^{3/3} = 44$$

Most significantly, the molars were low-crowned (brachydont), and had knob-shaped cusps (bunodont). The general dental formula of modern ruminants is:

$$I^{0/4} C^{0/0}\, P^{3/3}\, M^{3/3} = 32$$

a formula which reveals that during their evolution ruminants lost all the incisors from the upper jaw, the canines in both jaws, and a premolar tooth in both jaws. The dental formula of giraffes is:

$$I^{0/3}\, C^{0/1}\, P^{3/3}\, M^{3/3} = 32$$

Giraffes are unique, therefore, by having the incisor furthest away from the center of the lower jaw replaced by a canine (C). In addition, their teeth are higher crowned (more hypsodont) and the molar cusps more crescent shaped (selenodont). Thus, during the long evolution of giraffes, their tooth number and character in general have changed, with the teeth of a giraffe becoming very different from those of *Diacodexis.* Tooth anatomy and number are important. Teeth are unpalatable and incredibly hard, therefore they are rarely eaten by scavengers, are resistant to the effects of erosion over time, and are often the only fossils available on which to identify a species. Justification for using such a tiny fragment of an animal to deduce a life history is that, although tooth morphology is far from constant and evolves, it is highly heritable. Richard Lydekker described their importance for understanding the evolution of giraffe, in 1891, in this way:

> [species] *are severally distinguished by the characters of their molar teeth, and that, although the teeth of one group may approximate more-or-less closely to that of another, we do not find any instances where one member of a group possesses teeth of a totally different type from those of the other representatives of the same group. These facts strongly indicate that when we meet with fossil remnants having molar teeth of the very peculiar type met with in the giraffe, we shall be justified in considering that there must be a certain amount of relationship between the owners of such teeth.*

While Lydekker emphasized giraffe molar morphology as important, much later Singer and Bonè (in 1960) and Hamilton (in 1978) determined that the tooth most diagnostic of the giraffe family was the outermost lower incisor. It had transformed into a canine and in giraffes it has two lobes. If a jaw or fragment of a jaw is found to have a canine tooth in the lower jaw that seems to be made of two halves (see Figure 6.2 in Chapter 6), then it can be assumed that the jaw came from a member of the giraffe family. This feature was first noted by Petrus Camper, the Dutch anatomist, who had reconstructed the skeletons sent to Holland from southern Africa in 1780 (Vosmaer, 1787; see

Chapter 15), and by Richard Davis in 1827 when he was preparing to draw George IV's giraffe (Chapter 3) and by Richard Owen in 1840. Tooth morphology is also important for another reason. The shape and wear of molar cusps, and the number, position, and slope of incisors, are highly correlated with the nature of the diet and with the anatomy and physiology of the gastrointestinal tract.

What made *Diacodexis* so successful that it could survive and become the ultimate ancestor of all artiodactyls? From the rounded shape and small size of its teeth, it can be deduced that *Diacodexis* was omnivorous and had a simple intestinal tract, although we cannot know for sure what the structure of *Diacodexis* intestines were because soft internal organs do not fossilize in the same way that bones do. Omnivores always can find food. Nevertheless, it does seem clear that the evolutionary success of artiodactyls in general and of *Diacodexis* in particular began not because of an ability to ferment vegetation in a rumen, which is the dietary adaptation that is most characteristic of modern ruminant artiodactyls, but because of the anatomy of their hind legs and the agility it gave them. However, *Diacodexis* must also have had other attributes as well to ensure its survival. As forest dwellers in an exceptionally hot and wet environment, they may have been able to control their body temperature better than could their competitors. Among the earliest of the structures that evolved to regulate body temperature were nasal turbinate bones. They provide a surface for the membrane in which receptors that detect odors are placed, but they also provide a large surface area that is well supplied with blood from which heat can be lost (see Chapter 14, Figure 14.4). Turbinate bones seem to have evolved 250 Mya at the start of the Triassic period, in Cynodonts, an all-embracing group of animals that includes modern mammals (Crompton et al., 2017). From the shape of its skull, it is very likely that *Diacodexis* had well-developed turbinates, a well-developed sense of smell, and a well-developed capacity to regulate its body temperature.

Hypertragulids

The Dichobunids were not on the direct evolutionary line that produced giraffes. That line began with a divergence of the first artiodactyls into non-ruminants that had four toes, simple stomachs, large canine teeth, no horns, and rounded bunodont teeth like suids (pigs) and hippopotamids, and into ruminants that had two toes, compound stomachs and complex digestive systems, horns of all types and shape and, except for camels, no incisor teeth in the upper jaw. The first family on the long journey that ends with the evolution of modern ruminants was the Hypertragulids. The first known species from this family was *Archaeomeryx* (Colbert, 1941; Figure 4.4) and it becomes the common ancestor of all ruminants. (The suffix *-meryx* that is often used to describe artiodactyl fossils is Greek for deer.)

Archaeomeryx lived in Central Asia and from it two lines of descendants arose, one of which migrated into what is now North America; the other, an offshoot, became Tragulids, which remained in Eurasia. This divergence occurred about 40 Mya. Tragulids were, like *Diacodexis* before them, small forest-dwellers feeding on a succulent, soft diet rather like their living descendants the chevrotains and mouse deer, which today live in the forests of West Africa and Asia (Figure 4.5).

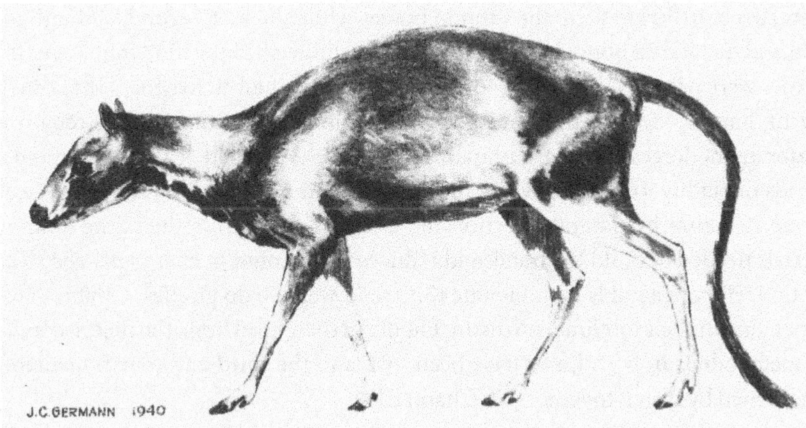

Figure 4.4. *Archaeomeryx.*
From Colbert, 1941, with permission of the American Museum of Natural History.

Figure 4.5. A chevrotain.
From Beddard, 1902b.

The teeth and jaws of the Hypertragulids differed little from those of *Diacodexis* and thus their diet differed little also, but what made them distinct can again be seen in the shape of their skeleton. The two main bones of the lower leg of the hind limb—the tibia and fibula—had begun the process of fusing into a single bone, and the cuboid and navicular bones of their ankle had fused into a cubo-navicular bone. In the front limb the two main bones of the forearm—the radius and ulna—had also begun to join together. In addition, the middle two (third and fourth) metacarpal (front limb) and metatarsal (hind limb) bones (which are completely separate, for example, in humans) had begun the process of fusion to form the cannon bones, while the first, second, and fifth metacarpal and metatarsal bones had regressed in size, more so in the hind limb (where only four toes were present) than in the forelimb, which retained all five toes. These changes confirm that selection for movement and agility and also for limb bone strength were still dominant forces driving ruminant evolution. A broken leg in an animal that depends on agility for its survival is more often than not a fatal injury. Simplification of bone structure by natural selection increases its strength, reduces the amount of materials needed to build the bone, and reduces the amount of energy needed to carry it around. Hypertragulids demonstrate this trend well. So do giraffes. Giraffes have no distinct ulna in their forelimb, no distinct fibula in their hind limb, the first, second, and fifth metapodials in both limbs have been lost, and the third and fourth metapodials strengthened by fusing together (see Chapter 15).

Leptomerycidae

The Hypertragulids were a very successful family and survived almost unchanged for many millions of years, but the evidence from fossils shows that by about 35 Mya a quite distinct new family of artiodactyls had evolved. This family was the Leptomerycidae. Their evolution seems to have started in the late Eocene, 38 Mya. They were even more successful than the Hypertragulids and lived for approximately 13 million years from the end of the Eocene era through the Oligocene era to the start of the Miocene era 20–25 Mya. Fossils of them have been found in what are now Mongolia, China, and western North America, a distribution that shows these regions were once a continuous landmass. Leptomerycids could migrate from one side to the other. They looked like a much larger version of the Hypertragulids, weighing about 8 kg compared with the 2 kg that *Archaeomeryx* weighed.

Like the artiodactyls before them, the upper (proximal) and lower (distal) parts of their astragulus formed a straight line. In addition, the large (magnum) and triangular (trapezoid) bones in the wrist had fused in the same way that the cubo-navicular bones had fused in the ankle. The bones of the lower legs—the radius and ulna of the forelimb, and tibia and fibula of the hind limb—that had begun the process of becoming one bone in the Hypertragulids, were now completely fused. Those fusions were accompanied by complete fusion of the middle two (third and fourth) metacarpals in the fore limb and the metatarsals in the hind limb to form cannon bones. The more lateral metapodial bones in both fore- and hind limbs had regressed further. What selective pressures brought about these changes? From our perspective, the changes were essential if limbs were to elongate and if body mass was increasing, because they provide the bone strength and joint stability needed to support increases in both leg length and body mass. The skeleton of Leptomerycids also shows another important development: the odontoid process of their second, cervical, neck vertebra (the axis) was elongated and spout-shaped. It forms a joint with the first cervical vertebra, the atlas. Its shape increases the stability of the upper neck and allows the head to rotate about its long axis, and foretells use of the neck as a flexible fifth limb.

Far more significantly, the Leptomerycids had molar teeth whose cusps were clearly selenodont, which means that they had become more dependent on vegetation for their nutrients. Their gastrointestinal tract also must have evolved to allow digestion of a diet containing vegetable matter. Leptomerycids were becoming ruminants. These changes in anatomy imply that the selective pressures had changed too. No longer was the evolution of artiodactyls dominated by adaptations that improved locomotion. Natural selection was now preserving mutations that improved nutrition, and those mutations not only changed the course of artiodactyl evolution, but ultimately changed the course of human development. Their origin was a dramatic change in climate. In the first 10 million years of artiodactyl evolution—from 45 to 35 Mya—the ambient temperature was far hotter than it is today and precipitation high. Most of the earth's landmass was a contiguous Pangea, covered by tropical or subtropical forest and centered around the equator (Janis, 1993). Those conditions lasted until the Pangean landmass broke apart, an event of unimaginable size that took millions of years to complete, and which brought about the precipitous fall in environmental temperature (Figure 4.6) that marked the end of the Eocene and the start of the Oligocene eras.

Tropical forest was replaced by temperate woodland, filled with broad-leafed evergreen and deciduous trees (Janis, 1993). What were to become Australia and South America separated from Antarctica, and the Antarctic icecap developed. Greenland separated from Norway. Separation allowed cold sea currents to circulate between the

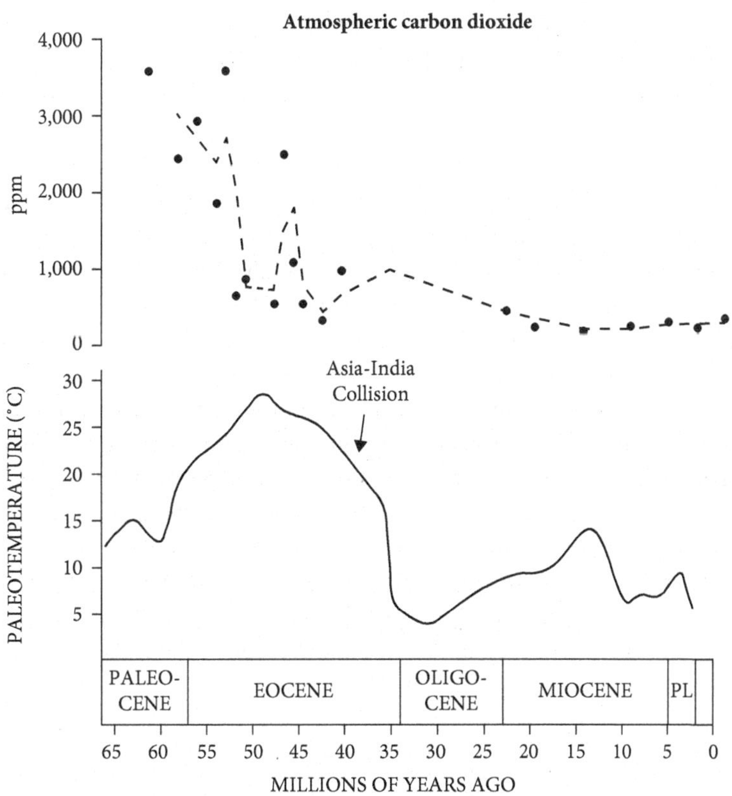

Figure 4.6. Climate change during the evolution of artiodactyls.

Redrawn using data from Kerr, 1980; Hendey, 1982; Raymo and Ruddiman, 1992; Janis, 1993; Berner, 1997; Cerling et al., 1998; Pearson and Palmer, 2000; Zachos et al., 2001.

new landmasses. The rate of cooling was accelerated by the Asia–India collision that occurred about 40 Mya, producing the still ongoing uplift of the Himalayas and the Tibetan plateau (Raymo and Ruddiman, 1992). At about the same time, the African continental plate was moving under Europe. That shift produced the Alps and a gradual emptying of the Paratethys Sea, the great inland sea of ancient Europe, the remnants of which are the Mediterranean, Black, and Caspian Seas. The uplift of the Alps and Himalayas rearranged airflow currents and effectively divided the world climates into north and south (Ramstein et al., 1997). The Paratethys Sea shrank, the Indian subcontinent was separated from China by the Himalayas, the altitude of what would become China increased, and land bridges between Africa and Asia opened. At the same time, the carbon dioxide (CO_2) concentration in air was falling (Figure 4.6) from a peak during the Palaeocene of perhaps 4,000 parts per million (ppm) (Janis, 2007), or 10–15 times as high as now, through to the mid-Eocene when they were 1,000–1,500 ppm (Pagani et al., 2005). By 51 Mya, CO_2 concentrations had fallen to 800 ppm, by 46 Mya they had dropped to 500 ppm, and are now about 400 ppm (Pearson and Palmer, 2000; Janis, 2007; Janis et al., 2004). They fell because the events that ended the dinosaur era 65 Mya, together with the later (40 Mya) Asia–India collision and the creation of the Himalayas and the Alps, exposed huge areas of bare rock. The calcium and silica in the rocks formed carbonates in a chemical reaction that consumed vast amounts of CO_2. A synchronous 17 million-year-long trend of cooling occurred, seasonal climates developed, sea temperature fell from 12°C to 4.5°C (Zachos et al., 2001), ice sheets formed, and, consequently, sea levels fell. Rather than the weather being constant, it changed into seasons. Large-scale extinction of both plant and animal species followed. It was the time when artiodactyls replaced perissodactyls like horses as the dominant ungulates. Horses almost became extinct because they digest their food inefficiently in their large intestines and they rely on having plenty of food all year round to be sure of having enough. When weather patterns became highly seasonal, with seasonal temperatures varying much like they do in the seasons we are familiar with today, there is seasonal variation in the quality and quantity of food available. Warm seasons are more favorable than cold ones for plant growth. All-year-round food was no longer available to horses. Seasonal climates affect ruminants less. Ruminants extract the nutrients from plants very efficiently, and can obtain nutrients from poor quality vegetation with the help of the trillions of micro-organisms that live in the compartments that make up their foregut. For any ungulate to have survived was a rare achievement, but Leptomerycids did, and their descendants were the ancestors of modern-day ruminants.

Toward the end of the Oligocene and into the early Miocene, it is possible to identify new characters in an artiodactyl family that had evolved from the Leptomerycidae and had survived the "deep freeze" of the Oligocene. This family, or more likely a collection or assemblage of families of similar looking animals, was the Gelocidae.

Gelocidae

In the story of giraffe evolution, the Gelocidae are a key ancestral family. They were highly successful and the "first fully progressive artiodactyls" (Carroll, 1988). Gelocid fossils appear in fossil beds dated to 25 Mya. The archetypical fossil form of the Gelocid family is *Gelocus communis* (Gentry, 1994) and, in all likelihood, it and several other members of the genus or family or assemblage showed what future ruminants might look like. Their habitat and range was the broad-leafed, temperate woodland that covered most of central southern Europe from the Chinese coast in the east to the Iberian Peninsula in the west, and lay between latitudes 30°N and 50°N. A characteristic feature of *Gelocus* was that the molar cusps were even more selenodont than those of the Leptomerycids, implying that plants were now a major source of nutrients. Their survival and success depended on herbivory. *Gelocus* also had elongated cannon bones and longer legs than their predecessors, a compact and parallel-sided astragulus, and two toes on both front and hind limbs, so speed and agility were still very important in the lives of artiodactyls. *Gelocus* was the size of and looked and functioned much like a modern gazelle, with slender legs and graceful movements, able to explode into a high-speed sprint if it was threatened by a predator. Well adapted and highly successful as they were, the Gelocids nevertheless were not so specialized that further adaptations and evolution of them was blocked. They had the supple genes needed to produce more advanced successors if the environmental conditions changed. They were at the base of the impending, expansive, adaptive radiation of artiodactyls.

From Gelocids to Giraffids

It had taken about 30 million years for Gelocids to have evolved from Dichobunids, and along the way their genes had accumulated all the information needed to produce modern pecorans (from the Latin *Pecus*, meaning "horned livestock")—the Cervidae (Old World deer), Bovidae (sheep, goats, cattle, antelopes), Antilocapriinae (pronghorn antelopes), Moschidae (musk deer), and Giraffidae (okapis and giraffes). The fossils that tell us that these five families existed have been dated to about 18–21 Mya in the early part of the Miocene era. Evolution from ancestors into families takes many millions of years, so the families had to have started evolving from their ancestral stock much earlier. From analysis of the evolution of the mitochondrial *cytochrome b* gene (Irwin et al., 1991), and from DNA sequence analysis (Kraus and Miyamoto, 1991; Allard et al., 1992) and known rates of mutation, the time of these earlier evolutionary experiments can be established with a great deal of accuracy.

Those analyses have shown that the divergence into the Pecoran families occurred over a five-million-year period between 28 and 23 Mya and thus on the border between the Oligocene and Miocene eras (Figure 4.7). Once the mutations had become established, subsequent selection of the best adapted to the ecological niche they filled, and continued selection of the best-adapted progeny produced the five distinct families that appeared 21–18 Mya. For such relatively rapid evolution of the ancestors into families and species, the environmental conditions must have been favorable for them, even if they were unfavorable for the parent stock. The period of 28–23 Mya coincides with a rising average ambient temperature (Figure 4.6) and a rising sea level (Janis, 1993) that marked the end of the Oligocene era. It was in effect a mini-duplication of the Eocene climate in which the first artiodactyls had evolved. Toward the end of the Oligocene it was becoming warmer and wetter than it had been for almost 20 million years, and the most productive times on Earth always have been

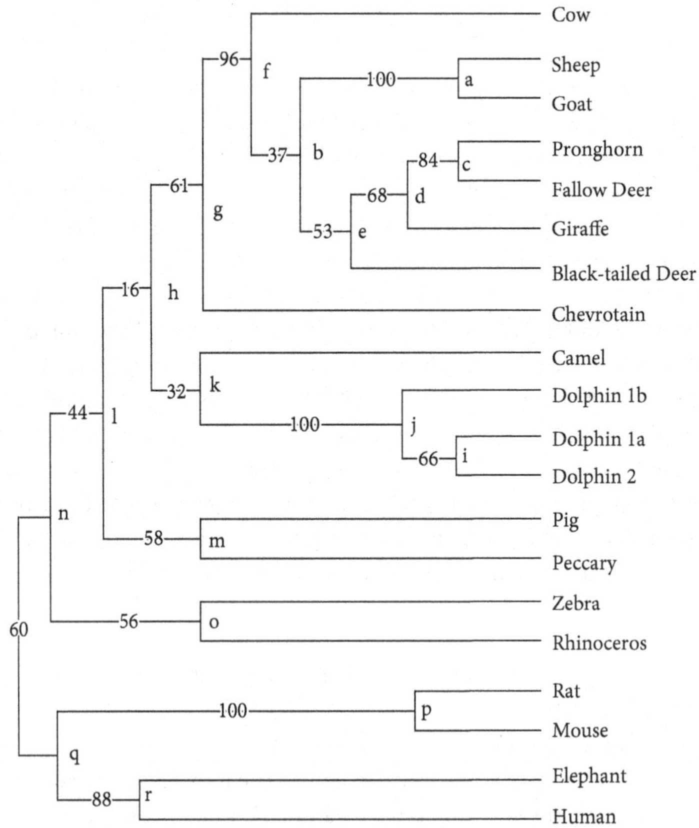

Figure 4.7. The phylogenetic relationships of 20 mammalian genera determined from analysis of the *cytochrome b* gene. The numbers refer to the number of experiments done to confirm the branch.

From Irwin et al., 1991, with permission of Springer Science.

during periods of global warming (e.g., Scafetta et al., 2017): warmth and water are ideal conditions for an evolutionary surge. Vegetation proliferated and tropical and subtropical forest zones expanded into low latitudes. The area of the Earth able to be colonized by new species enlarged.

How the various Pecoran families are related is not completely understood. From which ancestor did giraffes evolve? How did the extraordinary shape of giraffes arise from what looked like a small gazelle? The first attempts to answer those questions relied on anatomy. Similarities of giraffe and deer teeth (Colbert, 1936b) and hair structure (Ridewood, 1901) led to the conclusion that giraffids had a common origin with the deer family (cervids) rather than with bovids like antelopes and cattle. There are also similarities between the structure of cervid and giraffid horns, which famously occupied and confused the minds of Linnaeus, Brisson, Allamand, and Buffon. Cervid and giraffid types of "horn" have a dense core, are covered by skin, and receive their blood supply via skin blood vessels. Crucially, however, antlers are deciduous and are replaced annually, while those of giraffids are not. The two horn types also differ in their anatomical origin. Antlers are an outgrowth of the frontal bone of the skull. Giraffe horns develop from cartilage that lies in the skin. As the horn of a giraffe grows, it accumulates calcium, becomes hard and bone-like, and eventually fuses completely with the underlying skull on a boss that provides a large surface for attachment (see Figure 14.9). Fusion is complete at about four years of age in males and seven years of age in females (Spinage, 1968b). This type of horn was given the name "ossicone" in 1907 by its discoverer, Ray Lankester, third Director of the British Natural History Museum. Their position on the skull was first described by Owen in 1838. He wrote:

> *The two posterior or true horns are not supported exclusively by an enlarged frontal bone, but rest each upon the coronal suture which traverses precisely the middle of the expanded base.*

In other words, the horns straddle the junction between the parietal and frontal bones and so cannot be an outgrowth of one or the other. The northern varieties of giraffes also have a median ossicone (Figure 1.4). This ossicone is situated over the junction of the nasal and frontal bones and develops in the same manner as do the parietal ossicones (Spinage, 1968b, 1993).

More useful for understanding how pecorans arose has been biochemical analysis of the antigenic properties of globulins (van den Berghe and Boné, 1944), *cytochrome b* gene sequences (Irwin et al., 1991; Figure 4.7), pancreatic ribonuclease primary structure (Beintema et al., 1979), DNA sequence analysis (Kraus and Miyamoto, 1991; Allard et al., 1992), and *K-casein* gene phylogeny (Cronin et al., 1996). Irwin et al.'s analysis (Figure 4.7) shows that rapid evolution of placental mammals began about 80 Mya, and it began with divergence of the first perissodactyls (zebras and rhinoceroses) from elephants and rodents. The first artiodactyls to evolve were the pig family, followed by the divergence of the dolphin and camel families from all other artiodactyls, and finally the separation of the remaining artiodactyls into modern families. With respect to giraffid evolution, the analyses confirm the close relationship between giraffes and okapis, and confirm that giraffids have an origin closer to the deer family than to Bovid antelopes. The biochemistry brings simplicity to a complicated phylogeny. Nevertheless, existing long before the animals that we now recognize as deer or antelopes or giraffes, there were a number of intermediate species that lived all approximately at the same time as the Gelocids and probably evolved from them. Three of these species seem to have been important in the story of giraffe evolution. They were *Walangia*, *Dremotherium*, and *Teruelia*.

Given the incompleteness of the fossil record, a reasonable conclusion (but one, no doubt, that is highly disputable by expert paleontologists) is that the antelopes and domesticated ruminants arose via a *Gelocus–Walangia* ancestral line (Figure 4.8). Their further evolution (and that of Musk deer) was confined to the south of the Alps/Himalaya line with, generally, the Bovidae colonizing Africa and the Moschidae (musk deer) the Asian subcontinent. The Old World deer (the Cervidae) and a sister group, the Palaeomerycidae, evolved via a *Gelocus–Dremotherium* line with the further evolution of the cervids occurring north of the Alps/Himalaya line. The Palaeomerycidae gave rise to two subfamilies, the Dromomerycinae and the Palaeomerycinae. The Dromomerycinae had giraffoid-type horns, long, straight, probably skin-covered, non-deciduous, projecting from bone lying above the eyes and with a basal wing. An early form was *Dromomeryx* (Figure 4.9). They evolved into New World deer, and made their way to temperate North America to flourish there almost unchanged as Antilocaprid (Pronghorn) "antelope" (Figure 4.9).

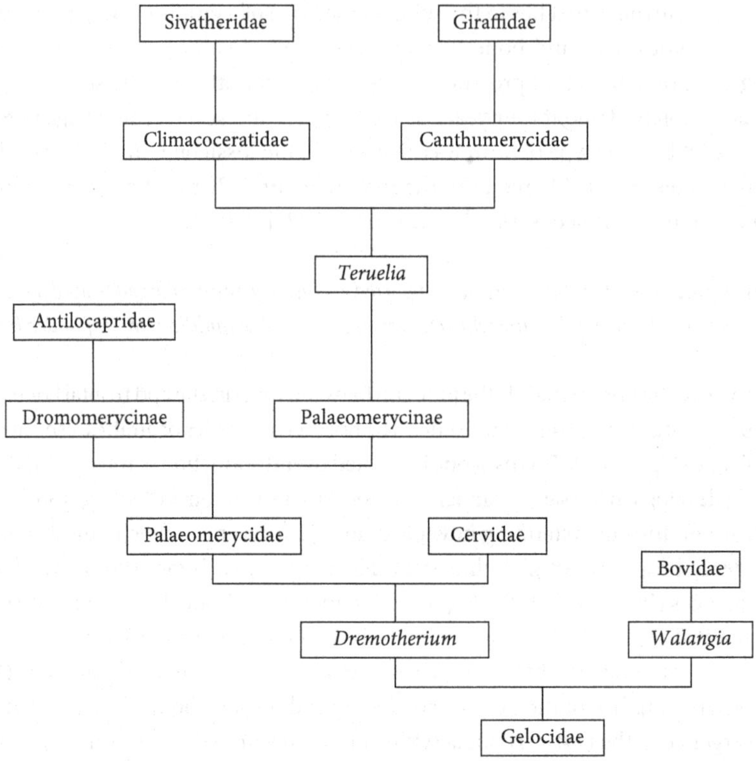

Figure 4.8. A phylogeny of giraffe based on Matthew, 1929; Colbert, 1935a; Stirton, 1944; and Gentry, 1994.

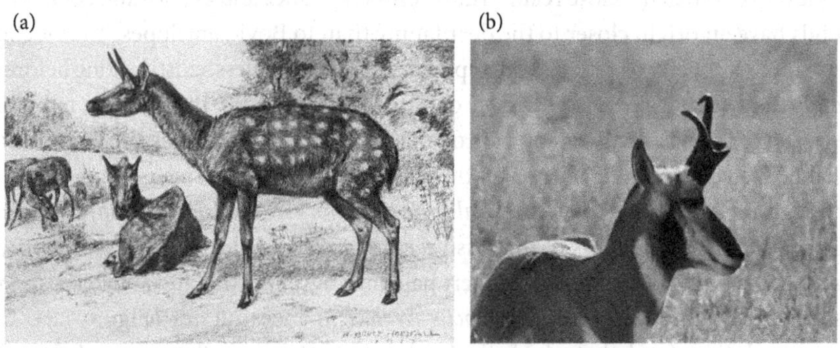

Figure 4.9. Left: *Dromomeryx borealis*. From Scott, 1913. Right: Pronghorn "antelope," Wyoming.

The Palaeomerycinae inhabited North Africa and southern Europe. Like Dromomerycinae, they had horns not antlers, but their horns could be forked and could be present as one or two pairs. These "horns," like antlers, were covered by skin and grew from a dermal cartilage rather than a bone. They were the ancestors of the family Giraffidae. *Teruelia*, a creature about the size of a small gazelle, the only fossil form of which has been found in what is now modern-day Spain, is the earliest ancestral giraffid so far discovered (Agusti and Moya Sola, 1991). From Teruelian stock, two families of giraffids arose: the Climacoceratidae and the Canthumerycidae. They are sister groups; both lived in Africa and were contemporaries. Both had the bilobed lower canine characteristic of giraffids. The giraffe lineage could have arisen from either, but modern giraffes arose from the Canthumerycids. Their story is told in Chapter 6. The Climacoceratids evolved in a completely different direction, but a direction that is relevant to the story of giraffes.

Climacoceratidae

The most primitive known member of the Climacoceratidae is *Climacoceras africanus* (Figure 4.10). MacInnes (1936) called it the "fossil deer" of Africa. From fossils found at Maboko (referred to as Kiboko by MacInnes) Island in the Kavirondo Gulf of Lake Victoria and, thus, not far from the Ethiopian highlands and the source of the Nile, it was the size of a European roe deer. The shoulder height of a roe deer is about 75 cm and its body mass is 35 kg. *C. africanus* had a single pair of permanent, forked "antlers," slender, elongated limbs, and from the length of its cervical vertebrae a neck about 56 cm long (Badlangana et al., 2009). It came onto the African scene about 15 Mya but was known to the Ancient Egyptians at least until 1778 BCE (Figure 4.10) and perhaps

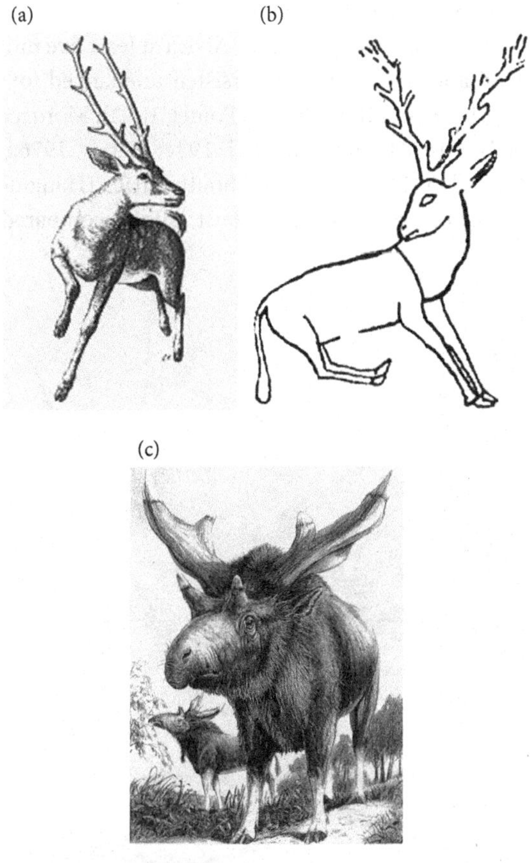

Figure 4.10. Top left: *Climacocerus africanus.*
From Hendey, 1982, with permission of Taylor and Francis.
Top right: A bas-relief on the wall in the tomb of Ukh-hotp, of the Egyptian twelfth dynasty (1991–1778 BC). The deer had been wounded and sunk onto its haunches.
From Dawson, 1943, with permission of the Linnean Society, London.
Bottom: A reconstruction of *Sivatherium giganteum.*
Savage and Long, 1986, with permission of the Natural History Museum.

until the eighteenth dynasty (1501–1480 BCE). Other Climacoceratid fossils have been found at Fort Ternan and Nyakach in Kenya (Gentry, 1970; Thomas et al., 1984), and at Arrisdrift in Namibia (Hendey, 1978). The Namibian Climacoceratid was about 20% larger than the other African species, and lived in a forested riverine habitat, isolated from its eastern and northern African relatives by the southern African rivers (Hendey, 1978). *C. africanus* could easily have been the ancestor of giraffes, but was not. The critical difference between *Climacoceras* sp. and *Giraffa* sp. is the nature of their horns. It is assumed that an unbranched simple horn such as those found in giraffes cannot evolve from a complex branched one like those found in *Climacoceras* mainly because a complex, branched horn cannot grow from an epidermal cartilage to fuse with the skull (Solounias, 1988). Branched horns must grow from the skull, and in particular from its frontal bone, upward, like antlers do. The simple, unbranched ossicones of giraffes develop from cartilage that lies in the skin—a quite different origin.

From the *Climacoceras* genetic pool there arose at least two successful species of Sivatheres. One was *Sivatherium giganteum* (Figure 4.10). Massive, heavily built, and short-legged, it was quite unlike slender, graceful, modern giraffes. It weighed between 850 and 1,800 kg (Basu et al., 2016) and perhaps up to 3,500 kg (Clauss et al., 2003a). It migrated to Asia and became extinct in the early Pleistocene (2 Mya). The other was *Sivatherium maurusium*, which was present in Africa at least five million years ago, became very widespread and common, and persisted unchanged for many millions of years. Fossils of it have been found in Algeria (Pomel, 1892), Morocco (Geraads, 2006), Egypt (Stromer, 1907), East Africa (Hopwood, 1934; Harris, 1976a, 1987a), Ethiopia (Rowan et al., 2016), Sudan (Bate, 1951), and South Africa (Haughton, 1922; Hendey, 1981, 1982). *S. maurusium* was a far less robust animal compared to *S. giganteum*,

Figure 4.11. The head of *Sivatherium maurusium* 1/12th natural size.
From Krieg, 1944, with permission of Springer Nature.

probably having a normal ruminant body shape and the size of a modern ox (Colbert, 1936a). Like *S. giganteum*, its horns were its distinguishing feature (Figure 4.11). The earliest record of *S. maurusium* in Africa is in the fossil beds at Langebaanweg, South Africa (see Figure 6.1 in Chapter 6), which are dated to 5–4 Mya. More than 500 fossils of it have been found there, clustered in a river channel, an extraordinary number given the scarcity of fossils generally. Most of those fossils came from young animals (Hendey, 1982) so *S. maurusium* must have lived in very large and fecund herds. In his wonderful analysis of life at Langebaanweg 5 Mya, Quinton Hendey speculated that the herd that provided the 500 fossils had succumbed to a catastrophic disaster, probably a flood. *S. maurusium* became extinct in southern Africa 400,000 years ago, but may have lived in Sumeria in southern Mesopotamia until 3500 BC (Colbert, 1936a). Its fossils usually are found together with giraffe fossils, whether at the bottom tip of South Africa or in East and northeastern Africa, an association similar to that between modern wildebeest and zebra. Wildebeest and zebras associate because it affords more protection—the many eyes phenomenon. Perhaps *S. maurusium* and the nascent giraffes had the same symbiotic relationship, a relationship that contributed to the survival of giraffes.

5

Notes on Allometry and Prediction of Body Mass and Age of Giraffes

A Note on Allometry

This story of giraffes so far has been of their natural history and has been descriptive. In recent decades, biology has left natural history behind, and biology (and the story of giraffes) has become much more dependent on quantification of biological structures and functions. Thus, to understand the evolution of giraffids from the Canthumerycidae, and to explain and describe how giraffes have evolved and how they work, some quantification of the arithmetical kind is essential, and to do that quantification, allometry is a tool that is often used. Allometry, as used in physiology and anatomy, is an arithmetical method that compares change in one part of the body with changes going on in another part, or to changes in a physiological function. Its literal meaning is the measurement (-metry) of other (allo-) growth. The purpose of allometry is to generate an equation that describes the relationship between an independent (X) variable, like body mass, and another, the dependent (Y) variable, like neck length, in order to see if they are related and, if they are, how closely. The closeness of the relationship is described by the coefficient of determination (which is the square of the correlation coefficient; R^2), the value of which describes how much of the variation in Y is explained by the variation in X. R^2 can range between 0 (no relationship) and 1 (a perfect relationship). In physiology, the correlation can provide very useful insights into how two systems interact. For example, are changes in the size of an animal's heart linked to changes in its body mass? If the R^2 were 0.5, then the answer is that they are linked, but changes in body mass explain only 50% of the changes in heart mass. The closer R^2 is to 1, the more likely it is, but of course does not prove, that changes in X underlie the changes in Y. Three kinds of allometric equations feature often in physiological analyses mainly because the relationships can be identified using simple computer software, an important criterion for physiologists like me. The three are linear equations, power functions, and polynomial functions. A relationship, if illustrated by a graph, often is simple, and a change in Y can be attributed to just one variable, in which case a linear equation of the form $y = ax + b$ will describe it. Wherever such a possibility can be found between variables in giraffes in all the chapters that follow this one, a linear equation will be used. However, in biology the relationship between X and Y can be more complex, often is not linear, has the shape of a curve, and is better described by a power function. Power functions have the form $y=ax^b$. A relationship that is described better (the R^2 is higher) by a polynomial function than a linear or power function usually means that the factors that affect the rate at which Y changes are more complex than a simple change in X. Polynomial equations have the form $y=ax^2+bx+c$.

How Giraffes Work. Graham Mitchell, Oxford University Press. © Oxford University Press 2021.
DOI: 10.1093/oso/9780197571194.003.0005

Recognition that allometry, and power functions in particular, is a useful way to describe patterns of growth in biology is attributed to Julian Huxley's 1932 book, *Problems of Relative Growth*. Huxley used a wonderful array of examples to show how power functions can describe the way structures or functions change as an animal grows. The essence of a power function runs thus (Huxley, 1932):

> *The law of simple allometry . . . states that the size of a part, y, is related to that of some standard, x (whether the whole body, the rest of the body without y, or some part of the body selected as standard for reasons of convenience), according to the formula* $y = ax^b$*, where a and b are constants.*

In other words, what Huxley discovered was that biological variation can come about by differences in the rate of growth of a part and/or differences in the size of a part at the start of its growth. The constant *a* is called the multiplier and it describes initial size: it is the value for *Y* when $X = 1$. The size of the multiplier reveals, for example, whether the difference in the length of a giraffe's neck compared with, say, the length of a buffalo's neck, has depended on differences in the size of neck vertebrae at birth. The constant *b* describes the rate of change in *Y* as *X* changes, and is called the scaling factor or exponent. For example, metabolic rate of an animal increases relative to increases in its body mass (Mb) raised to the power of 0.75 (i.e., metabolic rate is proportional to $Mb^{0.75}$).

Two requirements should be in place before allometry will yield meaningful results. First, the range of numerical values for *X* (and its counterpart *Y*) should be wide, and, second, there should be many of them. With these two requirements in place, it is possible to predict with confidence what the average numerical value of *Y* will be for any value of *X* within the range of values for *X*. This predictive power is especially useful when the values for the independent variable, *X*, are continuous, that is, can have any value. For example, the body mass of a giraffe may be 147 kg or 775 kg or 1,441 kg or any other mass. Specific data, for example the mass of a heart, may not have been collected from those giraffes, but that does not matter so long as the equation developed to predict *Y* (heart mass) from *X* (body mass) is strong enough. If it is, then the expected mass of the heart in each of those giraffes can be predicted with a great deal of accuracy. "Strong" in this context means that the correlation coefficient (R^2) between the two variables is high, and, consequently, the accuracy of the prediction of *Y* from *X* is high.

Two types of comparisons are often made using allometry. One is comparison of dimensions or functions between species of different sizes and especially of different body masses; this type of comparison is referred to as interspecific allometry. The other kind is comparison of dimensions or functions in the same species, like giraffes, as they grow and so have different sizes and body masses for that reason. This type of allometry is called ontogenetic allometry, and it is a more powerful predictor of size or shape or function for a species than are predictions for that species made from data collected from other species. In this book, most of the allometric equations that show how anatomy or physiology changes in giraffes as they grow are of the ontogenetic kind. However, by far the most frequently encountered type of allometric equation in the biological literature is of the interspecific type, and thus that type of equation is usually the only way that a rough idea can be obtained about the size of an anatomical or physiological variable in any mammal as, say, its body mass changes. That change often is very different from the change shown by an equation derived from the more rare ontogenetic analysis. The reason that interspecific and ontogenetic allometries can be different is that natural selection preserves the shape and functioning of a particular species, and those preserved shapes or functions are preserved because they are matched to the species' lifestyle and tend to be unique and successful and thus have their own unique allometric profile. It would be rare to find two different species that evolved under identical selective pressures. Nevertheless, interspecific analyses can be used to provide a baseline against which giraffes can be compared, and more importantly can highlight how giraffes differ from other species.

Why use allometry to describe how giraffes work? First, it allows discovery of the multipliers and scaling factors for the physiological functions or anatomical characteristics of giraffes. Does the neck of a giraffe grow at the same rate that body mass increases? If the scaling factor is 1.0, then the neck lengthens at exactly the same rate that body mass changes. However, the scaling factor may be >1 or <1, both of which are informative about the anatomy of the neck and the structures that underlie it. Second, if the allometric equations developed from actual data found from detailed studies of modern giraffes are compared with relationships between the same data obtained from other mammals, including extinct giraffes, then differences would show physiological or anatomical specialization and so begin to reveal how giraffes have evolved and survived. Third, and in the context of this book, once a reliable allometric equation for a relationship in giraffes has been developed from detailed studies, it can be used to predict the average value of Y in any giraffe. For example, we can predict what the size of the heart or the shape of a bone in a giraffe of any body mass is likely to be, even though actual data might not have been collected from that giraffe. Consequently, allometry is a productive way of building a complete picture of giraffe physiology and anatomy. In the magic of allometry you can see for the first time how a giraffe's shape and how its organs and tissues change as it grows—changes you can never see or appreciate by observation alone, no matter how many hours an observer gives to the very pleasurable task of watching wild giraffes.

Prediction of Body Mass of Giraffes

Many aspects of the life of an animal are regulated by its body mass (Mb), and therefore it is often body mass that is designated X in allometric analyses. Body mass affects metabolic rate, and metabolic rate in turn determines the amount of food the animal needs to eat and water it needs to drink, its heart rate and size, respiratory rate, the size of its skeleton, the size of its home range, and so on. Finding out an animal's body mass is easy if it is small enough to be weighed directly. It becomes increasingly more difficult as size increases, and is, for all practical purposes, impossible to measure in an adult, living, wild giraffe. It is difficult even in a dead giraffe. Body mass of a dead giraffe usually is measured by piecemeal weighing, and inherent in that method is that some fluids, blood, and tissues are lost. In addition, body mass varies with season, gender, time of day (which affects the amount of food or water present), the stage of pregnancy, and the naturally occurring differences between individuals. A way to estimate body mass that is independent of daily or seasonal variations is to assume that a giraffe (or, indeed, any other animal) has the basic shape of a cylinder, an assumption that is widely made (e.g., McMahon, 1973; Jessen, 2001). A cylinder has a volume, and its volume is calculated from its length and radius ($\pi r^2 h$). What h is can vary, but in giraffes its proxy is length measured from the tip of the nose to the tip of the last caudal vertebra (of the tail), and the proxy for radius (r) is the circumference of its chest (its girth), which is measured immediately behind the front legs. Body volume is then described by multiplying the total length of the giraffe by its girth squared (G^2), which gives a volume of m^3 or cm^3. The density of body tissues varies but, as most bodies float in water, average density is usually a little less than the density of water, which is 1 g cm^{-3}. In a theoretical analysis of giraffe buoyancy (Henderson and Naish, 2010), the average density of giraffes was estimated to be 0.96 g cm^{-3}. Theoretically, therefore, the body mass of giraffes can be estimated as volume ($L \times G^2$) multiplied by 0.96, but because giraffes are not actually a cylinder, this calculation overestimates body mass substantially. Nevertheless, there is a strong relationship between body mass of giraffes as determined by piecemeal weighing and ($L \times G^2$), and thus accurate prediction of the average body mass of a giraffe with any combination of length and girth can be made.

Anthony Hall-Martin, who had a distinguished career as a conservationist in South Africa (he died in 2014), was the first to apply the ($L \times G^2$) technique to giraffes, and he reported his findings in 1977. He measured length and girth in meters and found that the body mass (kg) of 27 male giraffes of various sizes was predicted by a linear equation: $26.1 \times (L \times G^2) + 34$, and that of 26 females, also of various sizes: $25.4 \times (L \times G^2) + 66$ (Figure 5.1). Coefficients of determination (R^2) were 0.993 for males and 0.977 for females, thus almost all of the changes in body mass (99% and 98%, respectively) were directly associated with changes in length and girth. In a similar but separate analysis of 30 males ranging in piecemeal body mass from 178 to 1,358 kg and 30 females ranging in piecemeal body mass from 141 to 1,055 kg, in which length (m) and girth (m) were also measured, the power function for males relating mass (kg) with ($L \times G^2$) was equal to $39.5 \times (L \times G^2)^{0.88}$, and for females $36.6 \times (L \times G^2)^{0.90}$ (Figure 5.1, top left) with $R^2 = 0.97$ in both cases.

There are no significant differences between the body masses predicted by these latter two equations and the body masses of the same giraffes predicted by Hall-Martin's equations for either sex. The relationship between ($L \times G^2$) and body mass is so close for both males and females that a single power function or a linear equation can be used to describe both (Figure 5.1) and covers most giraffes. The standard error of the prediction of body mass from ($L \times G^2$) is 35 kg, or about 5%. It may seem that the relationship between ($L \times G^2$) and Mb is linear, but the exponent of the power function for all giraffes (0.89) suggests otherwise. If it were linear, then the exponent would be 1. The fact that it is about 0.89, and the 95% confidence intervals of it range from 0.85 to 0.92 and thus exclude 1, indicates that as length and girth increase with growth, the body mass of giraffes increases at a slightly slower rate. The relationship between body mass, length, and girth in 21 fetuses measured by Mitchell and Skinner (2009) and Ackerman (1976) is described better by a polynomial equation (Figure 5.1, top right) than by a power equation. The way a fetus grows is not a simple matter (Chapter 17).

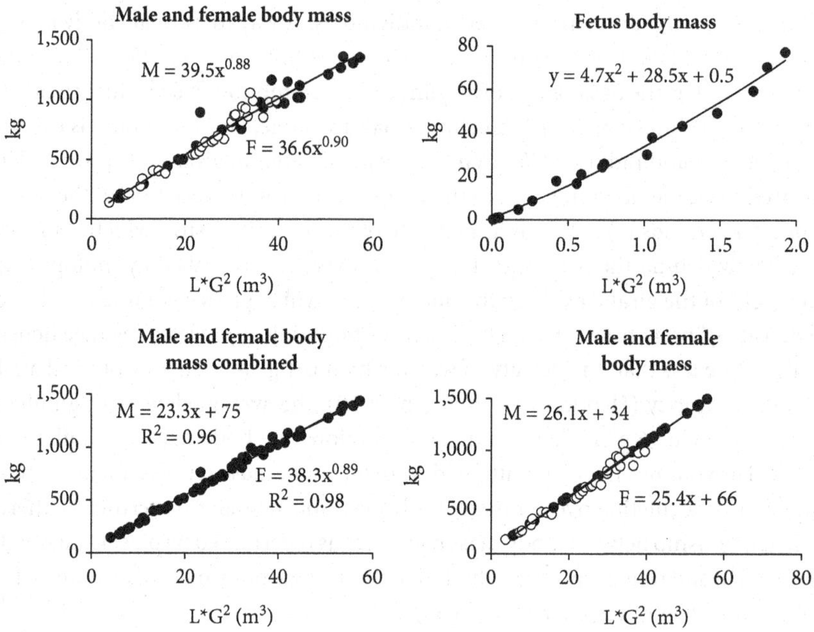

Figure 5.1. Equations for predicting body mass of male (M, ●), female (F, ○) and fetal giraffes from their length and girth. Top and bottom left data from 30 male and 30 female *G.c. giraffa* in Zimbabwe, and bottom right data from 19 male and 25 female *G.c. giraffa* in South Africa, reported by Hall-Martin, 1977.

Prediction of the Age of Giraffes

Knowing the age of an animal is a prerequisite for understanding many aspects of its life. Consider humans. Age is tied to rights, schooling, learning capacity, social and sexual maturity, legal status, and so on (e.g., Morris, 1972). For animals in general, the same principles apply. Their life span, life functions, the demography of populations, conservation, and management decisions all depend on their age. For the age of any animal to be determined accurately, the first requirement is that its birth date must be known. The birth date of a wild animal is very rarely, if ever, known. Birth date of captive animals is known, but the growth of captive animals does not necessarily reflect what happens in the wild, although it must be said that captive giraffes sometimes do seem to grow at a rate similar to wild ones (Dagg and Foster, 1976). If birth date is not known, then estimates of age can only be made by other means, of which there are several (e.g., Morris, 1972). In giraffes, correlation of age with mandible growth and eye lens mass has been investigated (see Figure 16.10 in Chapter 16), but the most precise way of estimating the age of a giraffe is from teeth eruption times, tooth wear, and the number of cementum layers in teeth (Hall-Martin, 1975a). Hall-Martin analyzed these latter characteristics in the upper first molar of giraffes, chosen by him because it is the first permanent tooth to erupt, which it does at about four months of age. Cementum lines are laid down annually, and the number of lines is highly correlated with lingual crown height and lingual occlusal surface area, in other words with the size of the tooth.

Replacement of deciduous teeth and the appearance of permanent teeth (eruption time) provide a reasonable estimate of age, but only to about age six, by which time all teeth have erupted. Richard Owen was the first to observe the pattern of tooth replacement, which he did in the first female giraffe to arrive in England in May 1836. She was then about 18 months old and had all her deciduous teeth and the first of the three permanent molar teeth. Replacement of deciduous teeth began at about three years of age with replacement of incisors, followed by replacement of premolars four months later. The emergence of the second molars and the last of the three molars began at five years of age. The large bilobed lower canines were the last to emerge (Owen, 1840).

Assessment of tooth wear is highly subjective, varies according to diet, and becomes imprecise with increasing age. The method for analyzing the number of cementum layers is exceptionally complex and time consuming and, unhelpfully, age can only be determined on dead animals, and even then it gives a good approximation rather than accuracy (Singer and Boné, 1960; Hall-Martin, 1976; Pellew, 1983b). If all three methods of eruption, wear, and cementum lines are used, then the "age of most giraffes can be determined to within approximately 1 year up to 15 years of age. Thereafter, error increases because tooth wear patterns are more variable but the error in prediction is still unlikely to exceed 2 years" (Hall-Martin, 1975a). Very helpfully, Hall-Martin (1975a) correlated age, as estimated from his careful measurements of teeth eruption, wear, and cementum lines, with the more easily measurable length, girth, shoulder height (measured as the distance from the sole of the front foot to the top of the third thoracic vertebra—the withers), and with total body mass. All of the correlations between these variables and age were described by polynomial equations (Figure 5.2), and had correlation coefficients above 0.95. All four relationships show that while males and females

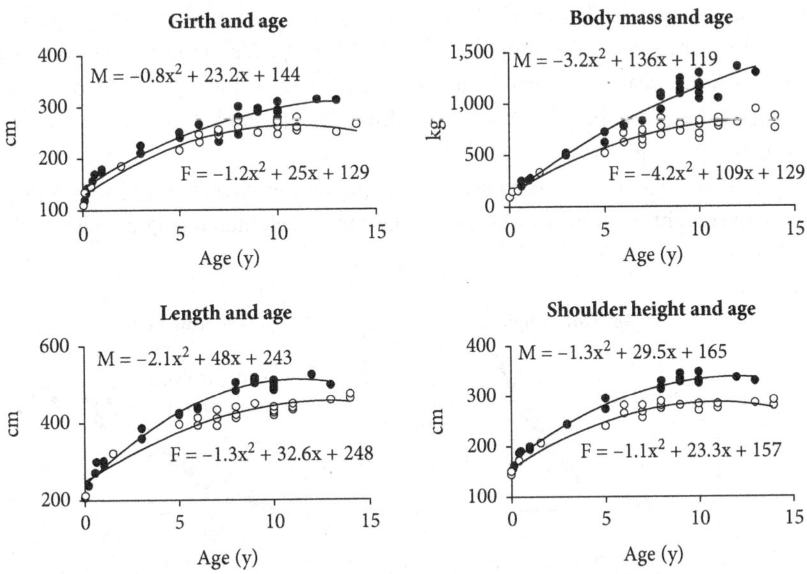

Figure 5.2. Relationships between body characteristics and age in 28 male (M, ●) and 27 female (F, ○) giraffes, *G.c. giraffa*.

Redrawn from data in Hall-Martin, 1975a.

have similar dimensions up to the age of five years, thereafter females have a slower rate of growth than males do, and their final size is smaller. Hall-Martin concluded that maximum length, girth, shoulder height, and body mass of larger males were reached at 12 years of age and at 11 in females, although from the data shown in the figures it appears that maxima may have already been reached by the age of 10 years.

An extension of the method for predicting age from body dimensions, and a useful way of determining age in living giraffes, is to correlate known age with height. This correlation has been studied in six wild giraffes of both genders and one captive male and one captive female, reported by Backhaus and recorded by Dagg and Foster (1976), and in seven males and eight females (all wild) by Pellew (1983c). In all these giraffes, the birth date was known and height measured from photographs of a giraffe against an object of known height (usually a tree). The height of a giraffe typically is measured from ground level (the sole of the foot) to the top of the head (the occiput), although Pellew measured it to the top of the ossicones. From the data obtained from these 23 giraffes, the growth patterns and equations that predict age (in years) from height (in meters) are shown in Figure 5.3. The height of a giraffe of either sex is almost identical to its length: height (m) = $1.03 \times L^{0.97}$ ($R^2 = 0.97$); and length (m) = $1.02 \times Ht^{1.0}$ ($R^2 = 0.97$). Thus height can be used as a proxy for length, which is difficult to measure in a living giraffe. The predictive equation for age from height is identical to the equation for its prediction from length. The equations constructed from Dagg and Foster's and Pellew's height data predict ages almost identical to those predicted from Hall-Martin's length data. For example, Dagg and Foster (1976) reported that a female in Nairobi National Park had reached 4 m of height in her fifth year and a male had reached 4.5 m in his seventh year. Hall-Martin's data predicts a height of 3.8 m at five years for females and a height of 4.8 m in males aged seven. Height and length correlate well with body mass (Mitchell et al., 2009b). Consequently, equations that predict age from body dimensions in male and female giraffes can be developed (Figure 5.3), and they establish important relationships given the central role of age in the life of giraffes. However, of these various correlations, the weakest is between age and body mass, especially after a body mass of 1,000 kg has been reached, because of the difficulty of establishing age and body mass accurately in older giraffes.

From the relationship between age and height, annual growth rate (meters per year) from birth (height = 1.8 m) to 12 years of age can be calculated (Figure 5.4). Figure 5.4

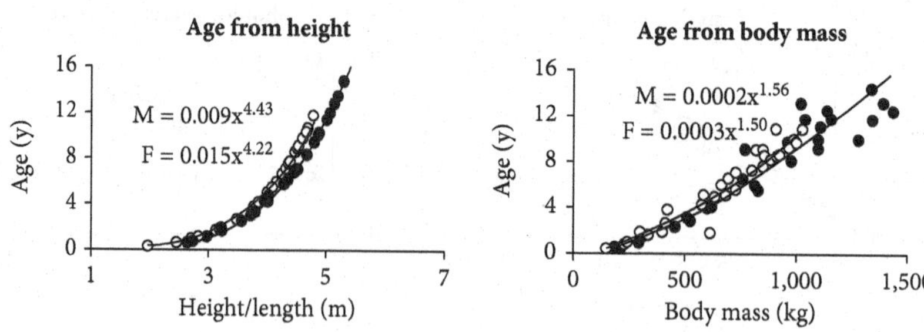

Figure 5.3. Left. Prediction of age from height or length, and right, from body mass in males (●) and females (○).

Constructed using data from Dagg and Foster, 1976, Pellew, 1983b, and from Mitchell et al., 2009b.

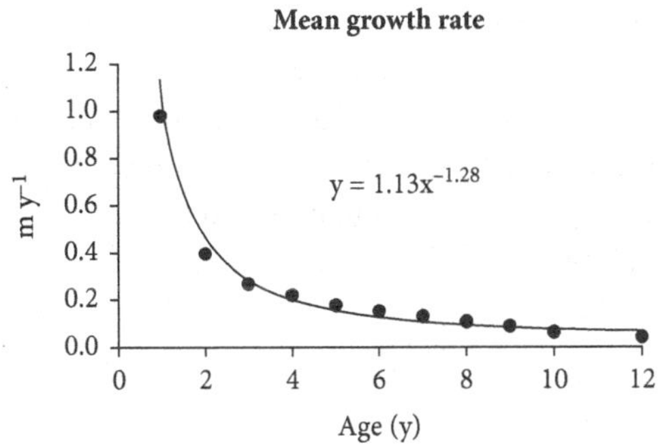

Figure 5.4. Average growth rate (meters/year) in male and female giraffes.
Birth = 0 years.
Constructed using data from Hall-Martin, 1975a, Dagg and Foster, 1976, and Pellew, 1983b.

shows that the growth rate of both sexes can be expected, on average, to be ~1.0 m in the first year, after which the annual rate declines steadily until about the age of 12 years. Females grow at a slightly slower rate than males (=1.07 × age − 1.38). Owen (1839a) described how a male calf was 1.8 m high at birth, 2.1 m at three months of age, and 2.9 m at nine months of age. It had grown 1.1 m in that time. At five years of age it was 4.4 m tall and had grown by 2.6 m. Ingo Krumbiegel (1971) reported a faster growth pattern: 180 cm tall at birth, 279 cm at one year of age, 333 cm at 1.5 years, 396 cm at two years, 450 cm at 2.5 years, and 504 cm at three years. The giraffes in both Owen's and Krumbiegel's studies were captives. By 1971 knowledge about how to look after captive giraffes was, clearly, better than it was in Owen's time.

How long do giraffes live? The expected lifespan of any animal should at least exceed the length of its reproductive period, with a safety factor that ensures that most males and females that reach adulthood produce as many young as possible and as necessary to sustain the species (see Chapter 17). For giraffes, that lifespan seems to be around 25–30 years. Foster and Dagg (1972) reported that four males in Nairobi National Park reached ages of 21 to 26 years, and two females were over 25 years old. In a later study they concluded that most giraffes die before age 30, but that reaching 25 years of age is not uncommon (Dagg and Foster, 1976). The Zambian naturalist and longstanding observer of the lives and habits of *G.c. thornicrofti* that inhabit the Luangwa Valley in Zambia, Phil Berry, and Fred Bercovitch, an expert giraffe reproductive biologist, used the blackness of coat patch color in male giraffes to estimate the age of male giraffes (Berry and Bercovitch, 2012). They found that the interval between birth (least black) and maturity (most black) ranged from 8.5 to 10.6 years and on average was 9.5 years. The period between maturity and the last time a male giraffe was seen was 7.4 ± 3.6 years (mean±SD). Thus, assuming that disappearance meant death, then death occurred on average at about the age of 16 years, with maximum longevity of 18 to 20 years. Maximum age of females in the study area was 28 years. In the Timbavati Nature Reserve in South Africa, the oldest males recorded by Hall-Martin (1975a) were 21 and 23 years of age and the oldest female 20, but he reckoned that giraffes in the wild regularly live beyond the age of 25 years. Unusually, as most captive animals live longer than their wild conspecifics because they do not have to cope with predation or the other natural hazards of nature, giraffes do not seem to live longer in captivity than they do in the wild (Hall-Martin, 1975a). Jones (1993), for example, has reported the longevity of ungulates in captivity that had died in zoos in North America, Europe, and Australia. The data set included eight giraffes, each from a different subspecies. Their age at death was on average 27.1 years, but it ranged from 17 years in a *G.c. capensis* to 32 years in a *G.c. reticulata*.

6

Antitypes, Ancestors, and the Origin of Modern Giraffes

In the far north of Africa, 200 km south of the Gulf of Sirte in Libya, lies the fossil bed of Gebel Zelten (Figure 6.1). The surrounding terrain is desert now, but 20 Mya it was a flourishing alluvial/flood plain/savannah/river basin habitat (Arambourg, 1959, 1963; Hamilton, 1973). In 1968 British geologist/paleontologist William R. Hamilton briefly visited the fossil bed and discovered the remains of ruminants. In 1973 he described the fossils he had recovered. Among them were the fossils of the earliest known ancestor of the giraffe lineage, which he named *Canthumeryx sirtensis* after the Greek mythological character Canthus, an Argonaut, who "fate had decreed . . . that he and the seer Mopsus should wander to the ends of Libya to be destroyed" (Hamilton, 1973). A few pages later in the same article, Hamilton described the fossils of a second species he found at Gebel Zelten, which he named *Zarafa zelteni*, a wonderfully evocative and appropriate name for an early giraffe ancestor that recalls the origins of the word from which *Giraffa* is derived. Hamilton thought the fossils of both species could be as old as 20–25 million years, although they are more likely to be 14–18 million years old (Gentry, 1994) and thus coincide with the time of the Miocene Climatic Optimum of 17 to 15 Mya (Holbourn et al., 2015). Then global temperature was 3°–8°C higher than the preindustrial levels of our time, a temperature caused by a period of colossal volcanic activity. In a later analysis of these fossils and others that had been found in Kenya, Hamilton (1978a) came to the conclusion that *Canthumeryx* and *Zarafa* were the same species, with the name *Canthumeryx* having precedence. Thus, in the strange world and rigid rules of binomial classification, *Zarafa zelteni*, merely because its description in 1973 followed that of *Canthumeryx* by a few pages in a journal article, became, alas, *Canthumeryx sirtensis*.

From his analysis of the dimensions of the bones, Hamilton concluded that *Canthumeryx* was about the same size as a fallow deer, *Dama dama* (Hamilton, 1973, 1978a), but, of course, was not in any way related to them. The shoulder height of fallow deer is about one meter and they weigh about 100 kg. Crucially, the lower jaw of *Canthumeryx* had the characteristic bilobed lower canines that all giraffes have (Figure 6.2), a feature that Hamilton (1978a) thought had the adaptive advantage of improving the ability of its possessor to strip leaves off trees and shrubs.

The limbs of *Canthumeryx* were about 85–100 cm long and shoulder height about 1.5 m. It had unbranched (simple) horns that projected sharply sideways and horizontally from a position above its orbits. The cervical (neck) vertebrae were short, and from the length of the cervical vertebrae in one specimen, neck length has been calculated to be 55 cm (Badlangana et al., 2009) but, more generally, was probably 60–70 cm long. Its neck, in other words, was not as long as its legs, and the ratio of leg length to neck length was identical to that of modern grazers like buffalo and wildebeest (see Chapter 18,

How Giraffes Work. Graham Mitchell, Oxford University Press. © Oxford University Press 2021.
DOI: 10.1093/oso/9780197571194.003.0006

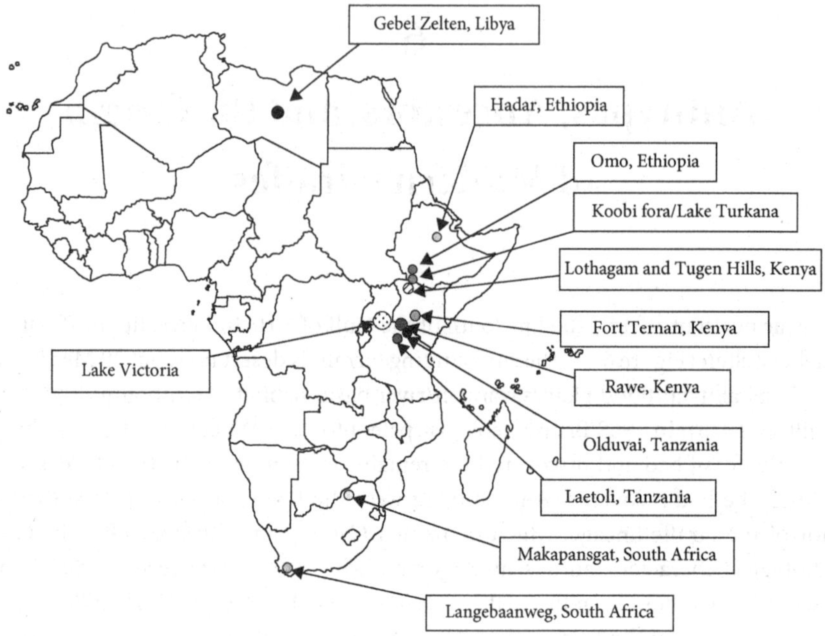

Figure 6.1. Giraffe fossil sites in Africa.
Redrawn from Cooke in Maglio and Cooke, 1978.

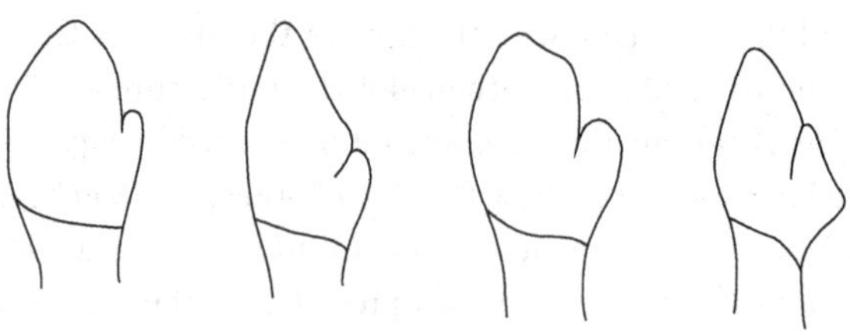

Figure 6.2. Left lower canines of giraffids illustrating the development of the accessory lobe.
From Hamilton, 1978b, with permission of the Royal Society, London.

Figure 18.5). Danowitz et al. (2015), after making detailed measurements of the shape of cervical vertebrae, concluded that the cranial part of the vertebral bodies (in front of the dorsal spine) of *Canthumeryx* had elongated more than the caudal part behind the dorsal spine, a pattern of elongation that would persist throughout the evolution of giraffids until in *Bohlinia*, the immediate ancestor of modern giraffes, the caudal part also elongated to produce the characteristic very elongated cervical vertebrae, and neck, of modern giraffes. The skull of *Canthumeryx* was wide and had large occipital condyles. The condyles form the joint between the skull and the first cervical (atlas) vertebra, and their size is correlated with the weight of the skull they carry. They are also larger in grazers than in browsers, and it is thus likely that either the skull and horns of *Canthumeryx* were heavier than they appear to be from the size of their head or they grazed more than they browsed, although rather like modern impala (*Aepyceros melampus*) they probably did both depending on the availability of fodder. Overall, and like its Gelocid ancestors, it seems to have been lightly built, medium-sized, and slender-limbed, but unlike them probably not a very agile gazelle.

Palaeotraginae

Canthumeryx became extinct about 14 or 15 Mya, an extinction that coincided with cooling of the climate that followed the warmth of the Miocene Climatic Optimum (Holbourn et al., 2005; 2015). The cooling was associated with an expansion of the Antarctic icecap. Perhaps *Canthumeryx* preferred a warm climate and its associated better food supplies. While *Canthumeryx* and its relations were the ancestral origin of giraffes, they existed millions of years before the first appearance of modern giraffes and clearly did not have the long neck and long legs of a giraffe. The evolutionary gap of 7 to 8 million years between the extinction of *Canthumeryx* and the appearance of modern giraffes has to be filled by other giraffe ancestors, and it is filled first by a family of ruminants called the Palaeotraginae, a family that could thrive in a colder world. The Palaeotraginae family appeared in mid-Miocene fossil beds dated to ~12 Mya and consists of three genera, *Giraffokeryx*, *Palaeotragus*, and *Samotherium*. However, the earliest Palaeotragine and perhaps the first of the post-*Canthumeryx* forms was *Progiraffa exigua*, first described by Guy E. Pilgrim in 1911 from a fossil that was found in the Siwalik deposits of Pakistan, a deposit that is 10–15 million years old (Matthew, 1929; Simons et al., 1971). Pilgrim was one of the most distinguished vertebrate paleontologists of his time. Born in Barbados but educated in England, he joined the Geological Survey of India in 1902, and became its superintendent in 1920. Following in the footsteps of Hugh Falconer, Proby Cautley, and Richard Lydekker, he elucidated the complicated paleontology of the Siwalik fossil deposits and the place of giraffes in them. His opinions mattered. *Progiraffa* could have been a descendant of *Teruelia* (Agusti and Moya Sola, 1991; Agusti and Anton, 2002), but its anatomy was similar to that of *Canthumeryx* (Gentry, 1994) and it lived about the time that *Canthumeryx* was becoming extinct. Its fossil remnants are scanty and little is known about its size and shape. More is known of a species that evolved from it, *Giraffokeryx punjabiensis*, which was not only an inhabitant of northern India/Pakistan (Pilgrim, 1911; Colbert, 1933; Colbert 1935a, 1935b), but also of Africa (Churcher, 1978; Gentry, 1994).

Giraffokeryx

Giraffokeryx (-keryx = herald) was a direct descendant of *Canthumeryx* (Geraads, 1986) and had all the attributes needed to be an ancestor of modern giraffes. It is what Alfred Russel Wallace would have called an antitype, "with a longer neck than usual that at once secured a fresh range of pasture over the same ground as their shorter-necked companions." It was first identified by Pilgrim (1911) from fossils again found in the Siwalik Hills and dated to the same 10–15 Mya period during which *Progiraffa* lived (Matthew, 1929; Simons et al., 1971), but the most complete description of it was

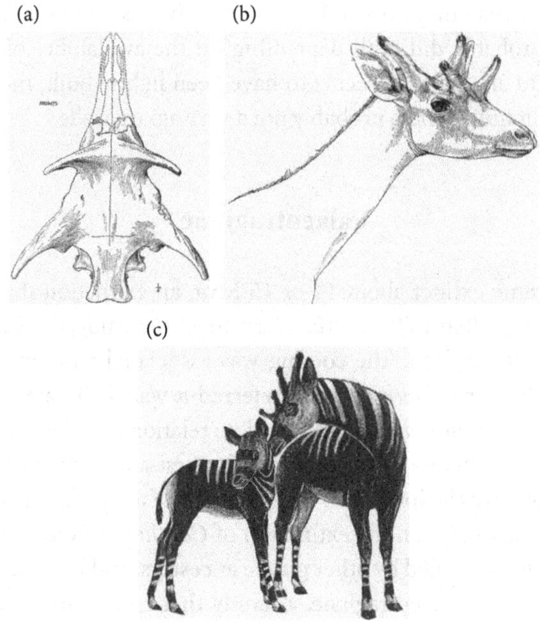

Figure 6.3. *Giraffokeryx punjabiensis.* Top left: Dorsal view of the skull showing the position of the four horns. Top right: Reconstruction of *Giraffokeryx.*
From Colbert, 1935b, with permission of the American Philosophical Society.
Bottom: Reconstruction of *Giraffokeryx.*
From Savage and Long, 1986, with permission of the Natural History Museum.

made by Edwin H. Colbert (1933, 1935b) who, for 40 years, was Curator of Vertebrate Paleontology at the American Museum of Natural History. *Giraffokeryx* was about the size of an okapi, weighing ~150 kg, and was distinguished by having two pairs of ossicones (Figure 6.3). The anterior pair arose from a precursor cartilage in the skin overlying the frontal bones of the skull, and the posterior pair from skin overlying the junction of the parietal and frontal bones and thus in the same place as the ossicones of modern giraffes. Both pairs develop from an epidermal cartilage that becomes ossified and fixed to the skull, the same process in other words by which the ossicones of modern giraffes are formed. Its skull was about 50 cm long (the head of a modern adult giraffe is about 80–90 cm long), its face was elongated, and the occipital condyles were large and thus the cervical vertebrae were heavy. The orbit, like that of modern giraffes, was conspicuously large. The site of the attachment on the occipital bone at the back of the skull for the muscle (*semispinalis capitis*) that extends and lifts the head in all mammals was shallow and small. Thus that muscle in *Giraffokeryx* could not have been very large or strong, as strong muscles require big bony ridges to which they can attach. To lift its head, *Giraffokeryx* must have relied on the *ligamentum nuchae*, which extends from the shoulders to the head (see Chapter 15). That conclusion is confirmed by the size of the insertion site on the skull for the muscle that bends the head and neck downward—the *rectus capitis ventralis* muscle. It was large so the muscle must have been strong enough to bend the neck downward whenever *Giraffokeryx* needed to drink water, against the tendency of the *ligamentum nuchae* to lift its head and neck. The bony ridges on which the internal pterygoid muscles of the jaws inserted were also large and, as these muscles move the jaw up and down and from side to side, *Giraffokeryx* had the anatomy to be a strong chewer of leaves, and probably chewed a cud just like modern ruminants do.

Reconstructions of what *Giraffokeryx* might have looked like differ. Its cervical vertebrae were elongated (Danowitz et al., 2015b), and Colbert (1935b) concluded from his analysis of its fossils that it had an elongated neck. He drew it as a small giraffe (Figure 6.3, top right). Another reconstruction by Savage and Long (1986) shows it looking more like an okapi, but with a clearly elongated neck (Figure 6.3, bottom). From data provided by Solounias (2007), its neck length can be estimated to have been approximately a meter long, and the average lengths of its legs was 115 cm, a ratio of neck length to leg length that is about the same as it is in a six-month-old modern giraffe with a body mass of 180 kg (Mitchell et al., 2009b; van Sittert et al., 2015). Krieg's (1944) drawings of the heads of *Giraffokeryx*, okapi, and giraffes (Figure 6.4) show just how similar they are. *Giraffokeryx*'s resemblance to an okapi is not fortuitous. When Harry Johnston first described okapi in 1902, he called them "living fossils." He did so because okapi evolved from the Palaeotragines and have changed little since (Joleaud, 1937; Colbert, 1938). Okapi are forest dwellers, but their ancestors might have been plains dwellers like *Giraffokeryx*, before moving into a forest biome (Thenius, 1992), where a stable unchanging environment provided no stimulus for change. Okapi today look pretty much the same as they did 12 Mya.

It always is tempting to imagine that an ancestor–descendant lineage is discontinuous: a parent produces an offspring that has the shape and form of a new species. Of course evolution does not and cannot work like that. Evolution is a very slow process, and it relies on offspring somehow becoming isolated from their parental group, finding a new niche to occupy, leaving their parents behind, and after a long period of natural selection becoming a new species. So it is with the giraffe lineage. Over many generations, descendants of *Canthumeryx* evolved into the new species *Giraffokeryx* that was bigger, had longer legs and neck than *Canthumeryx*, was better suited to its habitat, and perhaps was able to reach browse out of reach of other browsers. Success followed, but

Figure 6.4. The heads of *Giraffokeryx* (top left), okapi (bottom left), and a modern northern giraffe (right).
From Krieg, 1944, with permission of Springer Nature.

ultimately and ineluctably, change in the environment produced another subgroup that took an opportunity presented to them and evolved into a new species—*Palaeotragus.*

Palaeotragus

Numerous *Palaeotragus* spp.—the ancient goats—are known to have existed. They have resisted simple classification. Churcher (1970), one of the first to investigate the origins and phylogeny of giraffes, listed nine species, and Hamilton (1978b) concluded that the genus was derived from many different ancestors. Kostopoulos et al. (1996) pooled the many Palaeotragines into two groups—a group of small-sized animals whose type specimen was *P. rouenii*, found in Pikermi, Greece, and lived 9–7 Mya. This group also included two others—*P. microdon* and *P. primaevus*. The second group contained larger animals probably not on the giraffe lineage. Missing from these groupings is yet another—*P. germaini*—a purely African species. Hamilton thought that *P. microdon* was identical to *P. rouenii*, and that *P. primaevus* was identical to P. *tungurensis* of Asia. *P. primaevus* anatomy is indistinguishable from that of *Giraffokeryx* (Bohlin, 1926; Colbert, 1936b; Hamilton, 1978b; Geraads, 1986; Gentry 1994; Harris et al., 2010), and if so, then the neck of *Giraffokeryx* might not have been quite as long as thought by Colbert in 1935.

To a non-expert paleontologist, these consolidations and confusions bring Gelocids to mind—an awkward assemblage of similar animals out of which arose an ancestor of giraffes. What is more certain is that the *Palaeotragus* species evolved in the mid-Miocene from their ancestral Canthumerycids at the same time that bovids and antelopes were radiating into many new species, a radiation that lasted for 7million years between 23 Mya and 16–17 Mya (Allard et al., 1992), or, as analysis of changes in the cytochrome b gene has indicated, more probably somewhat later, between 15 and 12 Mya (Hassanin and Douzery, 1999), a time of rapid cooling of the Earth (Holbourn et al., 2005), but also, clearly, a climate favorable for experimentation. Apart from their skulls and a few other bones, little is known about what Palaeotragines looked like. There are no reconstructions of them (or their descendants the Samotheres) as good as that of *Giraffokeryx*, but what can be said is that they were bigger than their predecessors and had longer legs, but their neck was, at least in one case, relatively short (577 mm; Badlangana et al., 2009), and shorter than the neck of *Giraffokeryx*. The muzzle of *Palaeotragus* was elongated, the skull broad and flat, and it carried a single pair of horns that was larger in males than in females; in other words, the skull looked similar to the skull of a modern okapi. The area they inhabited ranged from East Africa (Churcher, 1970) to Mongolia (Pilgrim, 1934; Colbert, 1936b), immense distances apart, so they could survive as easily in the temperate forest habitats of Asia as in the African savannah (Heintz, 1975), partly because, while they depended on browse for nutrients, they were mixed feeders and able to exploit different sources of feed at different heights, as did their descendants the Samotheres and as do their modern counterparts, okapi (Merceron et al., 2018). The first *Palaeotragus* fossils were found in fossil deposits at Fort Ternan in Kenya, which have been dated to 14 Mya (Retallack et al., 1990) and Churcher (1970) classified the species that was their source as *Palaeotragus primaevus*. At Fort Ternan, *P. primaevus* was so common that 243 fossil specimens from it had been found there at the time Churcher was analyzing the fossils, and Churcher speculated that for so many to be found in such a confined area, it is likely that the animals that were their origin were killed by a catastrophe, perhaps poisonous gas from a volcanic eruption. The limbs of *P. primaevus* were long and slender, which gave them height to reach browse and, quite possibly, made them powerful runners and jumpers. Churcher (1970) recorded a metacarpal length of 450 mm, a tibia length of 528 mm, and a radius length of 492 mm, from which their body mass can be estimated to have been between 200 and 400 kg, compared with an okapi's body mass of 250 kg. Their dental formula (Churcher, 1970) was:

$$I^{0/3}, C^{0/1}, P^{3/3}, M^{3/3} = 32$$

which is the same as that in modern giraffes. The lower canine was bilobed (Figure 6.2), the molars narrow and short. What did *Palaeotragus* eat? From the shape of their muzzle and the wear of their teeth, both of which are similar to those found in modern giraffes (Solounias and Moelleken, 1993), they browsed on the shrubs and trees that were the dominant flora in the woodland habitat at Fort Ternan and depended almost completely on that browse for food and water (Retallack et al., 1990; Cerling, 1991, 1997a, 2005), just as modern giraffes do. The herds of *P. primaevus* coexisted with herds of antelope, and groups of rhinoceroses and elephants (Retallack et al., 1990)—a mix of species, in other words, that can be seen in any modern African National Park.

When Heintz (1975) considered the distribution of the many Palaeotragine fossils found in North and East Africa, Mongolia, and India/Pakistan, he decided that the best explanation for their distribution was that the family had originated in the Libyan alluvial plain in North Africa at least 12 Mya, migrated southward into Kenya, eastward into Mongolia and central Europe (Greece, Yugoslavia, Turkey, Crimea, Iraq), and westward to Spain and Portugal, but not into temperate northern Europe. For migration of this kind to have been accomplished, sea levels would have had to be low so that the migration routes south of the Alps/Himalaya mountain ranges were open. Expansion of polar ice caps and sequestration of water in them ensured that the routes were open. The Red Sea then was no more than an inland lake, and a land bridge between North Africa and Asia allowed migration. There must also have been a suitable habitat all the way along the southern borders of Eurasia to provide food and water. The *Palaeotragus* spp. that ended up in Asia were subject to different selection pressures from those in Africa and evolved into animals quite different in size and shape to modern giraffes (Gentry, 1994). It was from African *Palaeotragus* that modern giraffes ultimately were descended, not directly from *P. primaevus*/*Giraffokeryx*, but from *Palaeotragus germaini*—fossils of which have been found in Morocco, Algeria, and Tunisia and dated to ~10 Mya. *Palaeotragus germaini* was a descendant of *P. primaevus* (Arambourg, 1959).

For *P. germaini* to have evolved as new species, something must have precipitated a change that *P. primaevus* was unsuited to survive. What might it have been? Most likely the change started with the mid-Miocene cooling of the Earth. Cooling then had the same consequences that cooling had in the Oligocene: expansion of polar ice caps, a lowering of sea levels, less precipitation, and a more arid climate. Aridity led to reduction of forests and woodlands and expansion of the open woodland savannah that is characteristic of the modern African landscape. It was also the time that the decline in

CO_2 concentrations continued (LaRiviere et al., 2012; Holbourn et al., 2015), a decline which started the evolution of the C4 grasses we know today: C4 photosynthesis is an adaptation to low CO_2 concentrations. C4 grasses were present in Kenya 15 Mya but were an inconsequential source of food. They were a dietary source in the Siwaliks by 9 Mya and by 7 Mya were a primary resource for herbivores (Morgan et al., 1994). The selective pressures that the emergence of grasses exerted on the old order of animal species were immense. 20 Mya browsers were dominant because browse was abundant, its growth promoted by high CO_2 concentrations and efficient photosynthesis. The decline in atmospheric CO_2 concentrations reduced photosynthesis, which, together with lower rainfall and a fall in global temperatures, reduced browse productivity. Thus the proportion of browsers declined steadily throughout the middle Miocene, while expansion of C4 grasslands provided the food necessary for the evolution of grazers (Cerling et al., 1993, 1997b, 1998; Janis, 1993). About 12 Mya grazers became dominant, *P. primaevus* became extinct, and *P. germaini* evolved. *P. germaini* was not a grazer. It was a browser and it evolved probably because a new, higher, level of browse opened up in the savannah-woodland ecosystem that the grasslands brought.

Assuming that a new level of browse had appeared, to reach it *P. germaini* had to be a little taller than *P. primaevus* and taller than the co-inhabitants of the biome. How was that tallness achieved? First, *P. germaini* was bigger. Its body mass has been estimated to be 600 kg, or as much as a modern buffalo (Clauss et al., 2003a). However, there is no doubt that another component of tallness was a neck longer than the neck of *P. primaevus*. Calculation of *P. germaini*'s neck length by Badlangana et al. (2009) showed that it was probably in this genus that the genetic control of neck elongation changed, which would ultimately produce the characteristic long necks of modern giraffes. How did Badlangana and colleagues know? They compared the way that cervical vertebrae lengthen in *P. germaini* and in the *Samotherium* spp. that would follow *Palaeotragus* spp. and found that their growth patterns were similar to that found in modern giraffes. On the other hand, the pattern of vertebrae growth in *Canthumeryx* and *P. primaevus* was the same as it is in other short-necked ungulates like cattle. Both *Samotherium* (neck length ~800 mm) and *P. germaini* (neck length ~900 mm) had adult necks a third longer than the neck of *P. primaevus* (~600 mm). If *P. germaini* was a descendant of *P. primaevus*, then between the lives of *P. primaevus* and *P. germaini* was a period of 2 million years, during which there could have been some 400,000 generations, assuming a generation gap of five years as it is in modern giraffes (Badlangana et al., 2009). The neck length of *P. germaini* was about 330 mm longer than that of *P. primaevus* (570 vs. 900 mm), and thus the change in neck length was between 0.85 μm per generation, or ~0.12 μm per vertebra—infinitely small, but significant amounts. If the increase in length occurred over 50,000 to 100,000 years, the time period that is usually allowed for a punctuated evolutionary event, then the number of generations would have been 10,000 to 20,000 and the length of each cervical vertebra would have increased on average by 2.4–4.9 μm per generation, which also is a very gradual change. A longer neck would have contributed to *P. germaini*'s access to higher browse, but the very small increases in neck length over a very long period do not suggest that the selective pressures for a longer neck were strong or that a longer neck made a critical contribution to *P. germaini*'s success. Its neck length was still just half as long as it is in a modern giraffe. Much still depended on having long legs, and adult *P. germaini* had legs intermediate in length between those of *Samotherium* (~1700 mm) and modern giraffes (~2000 mm). It was leg length that ensured *P. germaini*'s access to higher browse.

Samotheres

Most but not all *Palaeotragus* species became extinct around 10–9 Mya. Samotheres lived 9–7 Mya, and the assumption is that *Palaeotragus* species were the ancestors of the last of the three genera that are classified as Palaeotragines: Samotheres. Samotheres are so-called because their first fossils were found on the Island of Samos in the Aegean Sea—the birthplace of Pythagoras. Samotheres were larger than *Palaeotragus* spp., and retained the long limbs and the mechanisms for neck elongation that *Palaeotragus* spp. had conferred on their successors. In addition, they had skulls almost identical in shape to those of modern giraffes. Their ossicones were situated above their eyes and thus further forward than those of modern giraffes, but they had an elongated muzzle, projecting lower teeth, a large diastema, a bilobed lower canine, crescent-shaped selenodont molar cusps, and a skull broadened and inflated by sinuses (Forsyth Major, 1891; Churcher, 1978). Skull sinuses are a defining characteristic of modern giraffes. It was in *Samotherium* that this adaptation seems to have evolved, so it is not unreasonable to conclude that, together with its chronological age, *Samotherium* is a predecessor of the *Giraffa* lineage. Was the neck of a Samothere longer than it was in *P. germaini*? What we know of the African Samothere—*S. africanum*—is sketchy, but we do know that a contemporary and presumably similar-looking *S. boissieri*, which was an abundant species on Samos Island (Solounias, 1988), had a neck ~800 mm long (Badlangana et al., 2009), and that in an unidentified species (but probably *S. africanum*), leg length was about 1,700 mm long (Colbert, 1938). Thus, as in *P. germaini*, the neck length of the Samotheres was still only half as long as leg length. Nevertheless, because Samotheres shared characteristics with giraffes, the comforting illusion arises that the descendants of the Samothere lineage would or should have been the origin of *Giraffa* as a continuation of an orderly progression of increasing size and development of the characteristics that modern giraffes show. However, although there is a continuous and unbroken increase in size and shape from *Canthumeryx* to *Samotherium* spp. toward the size and shape of modern giraffes, the immediate ancestor of modern giraffes was another species altogether, *Bohlinia* (Geraads, 1986; Geraads et al., 2013). What was its origin and how big was it?

Bohlinia

The first fossils of parts of a skeleton of *Bohlinia* were found at Pikermi in Greece about 20 km east-northeast of Athens. The fossils had been created from a creature that had died 7to 9million years ago (Agusti et al., 1999), and their shape and size were so similar to the bones of modern giraffes that the fossils were at first presumed to have come from a species that "was nearly of the same height as the living species but its limbs were of much more slender make" (Gaudry in Lydekker, 1883). Consequently the giraffe that was their origin was given the name *Giraffa attica* by Gaudry and Lartet (Forsyth Major,

1891). The fossils later came to be recognized as belonging not to the *Giraffa* genus but rather to an "acceptable ancestor" of modern giraffes (Forsyth Major, 1891; Pilgrim, 1911; Geraads, 1986; Geraads et al., 2013), and it was reclassified as *Bohlinia attica* to honor the discoverer of the fossils, Swedish paleontologist Birger Bohlin.

Bohlinia was not confined to Greece. Fossils of it have also been found in the shallow marine sands off the southwest coastline of Italy and in Turkey, Iran, and Bulgaria, and, although rare, are present in North Africa (Kostopoulos et al., 1996; Geraads et al., 2005; Kostopoulos 2009; Marra et al., 2011). This distribution shows that during its life *Bohlinia* was a successful, well-adapted creature, and it also shows that routes for migration to Africa from the peri-Caspian, peri-Adriatic Sea landscapes that *Bohlinia* inhabited, as well as to Asia south of the Alps and Himalayas, were open, presumably because sea levels were sufficiently low to open them (Marra et al., 2011). *Bohlinia*, like its ancestors and descendants, was a committed browser, and the necessary woodland-mingled-with-some-savannah biome necessary to support that lifestyle was wide-spread at the time, stretching through southern Russia as far east as China and as far south as the Asian subcontinent. Climatic events subsequently closed the routes to Asia. Geographical isolation of the Indo-Asian/Chinese descendants of *Bohlinia* from its African descendants followed.

What was the origin of *Bohlinia*? *Bohlinia*'s shape, if the length and diameter of their metacarpal bones are a guide, was not shared by any of the numerous giraffid co-inhabitants with which it shared the southern central Europe biome. Some dimensions of their metacarpal bones are shown in Table 6.1 and illustrate just how different the different species were. Nevertheless, *Bohlinia*'s ancestor must have been one or other of them, with a very strong likelihood that its ancestor was either a late *Paleotragus* or a *Samotherium*. The fact that *Paleotragus rouenii* had long legs and a slenderness index

Table 6.1 Metacarpal Length and Midshaft Diameter of *Canthumeryx*, Giraffids of the Late Miocene of Greece and Turkey, and *Giraffa camelopardalis*.

Species	Measured Length (mm) Mean±sd	*N*	Measured Diameter (mm) Mean±sd	*N*	Slenderness Index (see text)
Bramatherium	419	3	58±3	3	7.2±0.6
Canthumeryx	304	2	17.5	1	-
Decennatherium	391	2	40	2	10.8
Helladotherium	425±20	29	61±6	19	7.0±0.8
Samotherium sp.	417±15	41	52±5	43	8.1±0.8
Sivatherium giganteum	325±18	10	66±6	9	4.9±0.4
Okapia	315±9	15	32±1	14	9.7±0.3
Palaeotragus rouenii	414±15	9	32±5	11	13.5±3.3
Bohlinia	698±22	12	51±6	13	13.9±1.9
Giraffa	683±54	13	50±6	13	13.8±0.9

N = number of bones measured. Adapted from Rios et al., 2016.

of 13.5 (Table 6.1) (Length÷Diameter; Garcia and da Silva, 2004), like that of both *Bohlinia* and *Giraffa*, implies that *P. rouenii*, or a descendant of it, was a possible origin of *Bohlinia*. On the other hand, *Samotherium major* had a neck between 1.0 and 1.2 m long (Danowitz et al., 2015a), which is much longer than that of any other possible ancestor, and had legs as long as those of *P. rouenii*, although they were more robust. Thus *S. major* could also have been an ancestor of *Bohlinia*.

No complete skulls of *Bohlinia* have been discovered so far. From skull fragments that have been recovered, *Bohlinia* had a longer, flatter head than that of *Giraffa*, a shape in other words that was not dissimilar to that of *G. jumae* (see Figure 6.5). The ossicones arose above the orbits and were pointed and conical, rather than blunt and cylindrical as they are in modern giraffes, and *Bohlinia* did not have a median ossicone. Despite the differences between *Bohlinia*'s skull and a modern giraffe's skull, its pre-eminent status as an intermediate form between its Giraffid ancestors and *Giraffa* rests on its many common traits with *Giraffa*, particularly its very long legs, the characteristic bilobed lower canine, selenodont and rugose molars, and well-developed sinuses in the frontal and supraorbital regions (Kostopoulos et al., 1996). Most importantly, *Bohlinia* had a neck much longer than that of any of its predecessors. If *Bohlinia* had the characteristics of modern giraffes and "was nearly of the same height as the living species but its limbs were of much more slender make," as Gaudry had concluded, then its size and shape must have followed the rules of biology that govern growth patterns in modern giraffes. Consequently, estimates of its size and shape can be extrapolated from its fossils using allometric equations developed by van Sittert et al. that relate the dimensions of the six main limb bones to the body dimensions of modern giraffes, *G.c. giraffa* (Table 6.2). As for the most part, the sex of fossil giraffes is unknown, the equations shown in Table 6.2 were constructed by combining data for male and female giraffes. All the equations predict body dimensions very accurately: correlation coefficients are high and the standard errors of the predictions small.

No cervical vertebrae belonging to *Bohlinia* have so far been found, and it is their length that determines neck length most accurately. However, neck length in modern giraffes can be predicted with accuracy from the length of limb bones (Table 6.2). Using

Table 6.2 Allometric Equations That Predict Body Dimensions from Bone Lengths (L, mm) in *G. camelopardalis*

Bone	Body Mass (kg)	Height (cm)	Leg Length (cm)	Neck Length (cm)
Humerus (H)	$(1.62 \times 10^{-6}) \times HL^{3.24}$	$0.433 \times HL^{1.12}$	$1.27 \times HL^{0.80}$	$0.032 \times HL^{0.39}$
Radius (R)	$(5.94 \times 10^{-6}) \times RL^{2.85}$	$0.677 \times RL^{0.99}$	$1.74 \times RL^{0.70}$	$0.056 \times RL^{1.22}$
Metacarpal (Mc)	$(2.3 \times 10^{-7}) \times McL^{3.37}$	$0.207 \times McL^{1.18}$	$0.73 \times McL^{0.84}$	$0.015 \times McL^{1.44}$
Femur (F)	$(6.0 \times 10^{-7}) \times FL^{3.39}$	$0.322 \times FL^{1.16}$	$1.0 \times FL^{0.83}$	$0.024 \times FL^{1.43}$
Tibia (T)	$(7.8 \times 10^{-7}) \times TL^{3.25}$	$0.341 \times TL^{1.12}$	$1.06 \times TL^{0.80}$	$0.025 \times TL^{1.38}$
Metatarsal (Mt)	$(7.8 \times 10^{-7}) \times MtL^{3.19}$	$0.311 \times MtL^{1.12}$	$0.92 \times MtL^{0.81}$	$0.026 \times MtL^{1.35}$

Equations are derived from data collected from 47 giraffes ranging in body mass between 147 and 1413 kg by van Sittert et al., 2015. With permission of John Wiley and Sons.

the allometric equations listed in Table 6.2 and the length of the limb bones that have been attributed to *Bohlinia* (listed in Table 6.3), the body mass of an adult *Bohlinia* can be predicted to have been 850–900 kg, its legs ~1.8 m long, and its neck between 1.6 m and 2 m long. *Bohlinia* was thus not as heavy as modern giraffes, but its legs and neck

Table 6.3 Bone Dimensions of *Bohlinia* and Its Body Mass, Height, Neck Length, and Leg Length, Predicted from Bone Lengths

Bone	Length (mm)	Midshaft Diameter (mm)	Ref	Calculated Body Mass (kg)	Calculated Height (m)	Calculated Neck Length (m)	Calculated Leg Length (m)
R	680	63	6	702	4.31	1.60	1.67
Mc		60	7				
Mc	695	60.5	7	869	4.67	1.86	1.78
Mc	730	57	7	1026	4.95	1.99	1.86
Mc		50	7				
Mc	710	47	3	934	4.79	1.91	1.81
Mc	693	50	8	861	4.66	1.85	1.78
Mc	661	45	8	734	4.40	1.73	1.71
Mc	690		3	848	4.63	1.84	1.77
Mc	680	51.8	9	808	4.55	1.80	1.75
Mc	710	48.87	10	934	4.79	1.91	1.81
Mc	700	58.24	10	891	4.71	1.88	1.79
Mc		50.12	10				
Mc	715	48	10	957	4.83	1.93	1.82
Mc	655	47	10	712	4.36	1.70	1.69
Mc	720	40	10	979	4.87	1.95	1.83
T		60.7	9			1.70	1.79
Mt	670	55.5	9	808	4.55		
Mt		56	7			1.79	1.84
Mt	695	52.5	7	908	4.74	1.77	1.83
Mt	690		3	887	4.70		
Mt		52	7			1.73	1.81
Mt	680	46	3	847	4.63	1.66	1.76
Mt	658	49	3	762	4.46		
Mean	*691*	52		*859*	*4.65*	*1.81*	*1.78*
sd	22	6		*92*	*0.18*	*0.11*	*0.05*

Sources of the length and diameters of bones: 3 = Geraads, 1998; 6 = Theodorou et al., 2010; 7 = Geraads et al., 2005; 8 = Geraads, 1979; 9 = Kostopoulos, 1996; 10 = Rios et al., 2016.
Body dimensions calculated from allometric equations in Table 6.2.

R = radius; Mc = metacarpus; T = tibia; Mt = metatarsus.

were as long as those of modern giraffes. If so, it was slender as Gaudry had guessed, but also had, for the first time in giraffe evolution, a neck that was as long as its legs.

Can *Bohlinia*'s long neck be explained? The selection pressures operating at the time *Bohlinia* was evolving and that might have favored a long neck are unlikely ever to be known. What is certain is that a long neck was not a disadvantage. On the contrary, possessors of it were successful. Evolution of a long neck was, to use a word coined by the Russian comparative anatomist A. N. Severtsov in 1931, an "aromorph"—a "game-changing" character that results in advances unattainable in its absence, like bipedalism in mankind. How did a long neck originate? The immediate ancestors of *Bohlinia* had short necks. The neck length of one potential ancestor, *P. rouenii*, estimated from the metacarpal lengths reported by Rios et al. (2016), was ~880 mm, and that of another potential ancestor, *Samotherium* (*major*) was ~890 mm, but a length that was probably closer to 1,000 or 1,200 mm long (Danowitz et al., 2015a). Badlangana et al. (2009) calculated a length of 814 mm for *S. boisseri* and 904 mm for *P. germaini*. In other words, those predecessors had neck lengths about half the predicted length of *Bohlinia*'s neck. If *Bohlinia*'s ancestor was a *Palaeotragus*, and the mutation for a long neck originated in *Palaeotragus* as Badlangana et al. (2009) have suggested, then *Bohlinia*'s neck length increased to ~2,000 mm over a period of a few million years. It might also have done so in the relatively short period of time of perhaps 50,000 to 100,000 years of a punctuated equilibrium event. Under the intense selection pressure exerted by short geological time periods like that, neck length would have had to increase on average by 5 mm to 10 mm per generation, and each of the six main cervical vertebrae (C2–C7) by between 0.5 and 2 mm in length per generation. Is a rate like that possible? The necks of modern giraffes could easily be elongating at the same rate now, but we would never know. At that rate it would take five generations (25 years) for the neck of a modern giraffe to lengthen by 25 mm, or one inch.

It is nevertheless unlikely that natural selection alone could account for the neck elongation that occurred in *Bohlinia*, as natural selection is a drawn-out, slow process, with innumerable frustrations to its progress (e.g., Barton and Partridge, 2000). Another mechanism that could produce a long neck and do so relatively suddenly is variable expression of genes. Genes regulate metabolism, and it is generally accepted that they are switched on all the time, regulating the millions of reactions going on at any one time in the body. However, in 1961, Jacques Monod and colleagues discovered that some genes are active only when they are stimulated to be so. In the absence of a stimulus the gene is silent. The first gene of this type was discovered in a bacterium (Monod et al., 1961), but others have since been found in vertebrates, a typical example of which is the Hox gene family. The onset of Hox gene activity, at birth, is thought to be the cause of the rapid elongation of the neck in modern giraffes after birth (van Sittert et al., 2010), though what the stimulus might be that switches them on is not yet known. The general implication is that large changes in body morphology can be associated with changes in the way existing genes are regulated, rather than the more drawn-out process of evolution of new genes in the classical Darwinian manner. As a way of modifying a form, variable gene expression fits with D'Arcy Thompson's idea that new species are simply the product of different growth rates of parts of the body, an idea, it seems to me, that is as plausible an explanation for the origin of *Bohlinia*'s unique neck as any. Prior to *Bohlinia*, neck elongation (in *Palaeotragus* and *Samotherium*) had been modest and a result of elongation of the cranial part of the cervical vertebrae in front of the dorsal spines. In *Bohlinia* differential growth of the caudal part behind the dorsal spines as well as the front part of the cervical vertebrae probably was the crucial event needed for the development of the now familiar long necks of modern giraffes. Had the evolution of a long neck required a prior "series of accumulated modifications over thousands of generations" to the heart and other structures before elongating, as Hitching had concluded? It does not seem so. There would have been ample time for the evolution of the structures and functions needed to support a longer neck, especially one that was elongating slowly.

Whatever the origin of a long neck in *Bohlinia*, the evolutionary outcome with which we are familiar is the long neck of the genus *Giraffa*. Evolution of *Giraffa* from *Bohlinia* did not depend on extinction of *Bohlinia* itself. Fossils of *Bohlinia* have been dated to the 4million years between 9 Mya and 5.3 Mya (Kostopoulos et al., 1996) and those of the first unmistakable *Giraffa*, the Asian giraffe *G. punjabiensis*, are dated to 7 Mya. Consequently, if *Bohlinia* was the predecessor of giraffes, then giraffes were offshoots that developed their own characteristics in the African and Asian/Indian habitats that they chose to inhabit. The earliest migration of *Bohlinia* from the Mediterranean–Balkan–Iranian milieu took place about 7–8 Mya, followed a southeasterly migration into Indo-Asia, and eventually produced three Indian *Giraffa* spp. and the Chinese giraffe named *Honanotherium schlosseri* (Bohlin, 1926). *Honanotherium* became extinct in the early Pleistocene (~3 Mya), at much the same time as its Indian contemporary *G. punjabiensis*, probably because climate change and tectonic upheavals reduced the supply of browse on which the two species depended. Later migrations of *Bohlinia* or its descendants from Europe went southwest into Africa.

Why did *Bohlinia* migrate? About 10 Mya cold conditions prevailed worldwide, and average ambient temperature had dropped by 7°C, but between 9 Mya and 5 Mya, in the late Miocene, a pronounced warming took place (van Zinderen-Bakker and Mercer, 1986) that lasted several millions of years, during which *Bohlinia* and its co-inhabitants of southern central Europe evolved and flourished. Warmth did not last. An episode of abrupt cooling occurred around 6 Mya, sea levels dropped 70 m as water became sequestered in polar ice caps, rainfall declined, the world became arid, the Messinian crisis and desiccation of the Mediterranean Sea were triggered, and habitats fragmented. The Messinian crisis lasted a million years, from 6 to 5 Mya (Hsiu et al., 1977; van Zinderen-Bakker and Mercer, 1986), and changed the climate of the surrounding 2.6 million km^2 which included the area in which *Bohlinia* had evolved and inhabited. No doubt *Bohlinia* migrated into more hospitable habitats to escape the consequences.

Indo-Asian Giraffes

Less is known of the evolution of Indo-Asian giraffes than of African giraffes, but Indo-Asia was the region that was the home of the earliest known species of *Giraffa*: *G. priscilla*. It is thought to have evolved in the late Miocene (8 Mya). Its fossils were found in the Siwalik deposits and comprise three molar teeth and a fragment of a mandible (Pilgrim, 1911). No other parts of its skeleton have been found, so whether it was a true giraffe is uncertain. Features of *G. priscilla*'s teeth and mandible led William Diller Matthew to conclude that it was an intermediate species between *Giraffokeryx* and *Bohlinia* (Matthew, 1929) and that it migrated from Asia to Africa to become the origin of African *Giraffa*. As that origin for African giraffes was favored by Matthew, it has to be taken seriously. He was for many years the Curator of the American Museum of Natural History, was a distinguished paleontologist, and had an extraordinarily complete knowledge of *Artiodactyla*. However, the origin of African giraffes is better explained by a direct migration of *Bohlinia* or its descendants into Africa, and thus they evolved independently of any evolutionary activity going on in Asia (Gentry, 1994). *G. priscilla* was, though, the likely ancestor from which subsequent Asian giraffes evolved (Harris, 1976c; Churcher, 1978).

Giraffa punjabiensis

The second Asian *Giraffa* was *G. punjabiensis* and it is the earliest known unmistakable member of the *Giraffa* genus (Pilgrim, 1911; Harris, 1991). Its fossils are dated to 7.1–7.3 Mya. Its arrival in Asia had been preceded by uplift of the Himalaya-Tibetan plateau and a change from mixed needle-leaf and broad-leaf forests to grassland vegetation (Zhisheng et al., 2001). The identity of *G. punjabiensis* is based mostly on molar teeth (Pilgrim, 1911; Bhatti et al., 2012) which were brachydont with deep enamel folds and rugose enamel. Their teeth were also smaller than they were in the early African *Giraffa* (Pilgrim, 1911; Bhatti et al., 2012), but, as reminders of its origin, they had features similar to those of *Bohlinia* and *Honanotherium* (Harris, 1991). Small teeth imply a small body form. The co-inhabitants of the ecosystem *G. punjabiensis* occupied, like the large bovid *Pachyportax*, suggest that the habitat was an open woodland savannah not unlike an African savannah but with a humid climate, and some swamps and patches of dense forest. It seems an ideal environment for a giraffe, but *G. punjabiensis* had become extinct by the end of the Pliocene (~3 Mya), a moment in time that coincided with intensification of the Asian summer and winter monsoons, a rapid increase in the volume of ice sheets, and a shift in climate toward more glacial and colder conditions (Zhisheng et al., 2001).

Giraffa sivalensis

The last of the Indo-Asian giraffes, a descendant of *G. punjabiensis*, was *G. sivalensis*. It was this giraffe species that was the origin of the fossil of the third cervical vertebra (C3; see Figure 4.2 in Chapter 4) first described by Cautley in 1838, and it was this fossil that established an evolutionary history for the *Giraffa* lineage. *G. sivalensis* lived about 2 Mya (Churcher, 1978). When Cautley found two fossil remnants of this giraffe, he thought that they came from two separate species. One fossil, the neck vertebra, came from what he guessed was probably an immature individual, and Falconer and Cautley summarized its anatomy thus:

> *a true well-marked species of giraffe closely resembling the existing species in form, but one-third less in height, and with a neck proportionately more slender.*

The second fossil they said was "all but indistinguishable from . . . the Nubian giraffe."

Lydekker (1883) summarized his analysis of this fossil as:

> *the remains indicate an animal showing some considerable variations in size, but whose mean dimensions were not far from those of the existing species.*

Reconstruction of what this giraffe might have looked like, based on analysis of the dimensions of its third cervical vertebra and assuming that *G. sivalensis* did have a shape similar to the shape of a modern giraffe and grew in the much same way that modern giraffes grow, reveal that *G. sivalensis* could have weighed around 400 kg, have had a neck length of about 1.5 m, a foreleg length (hoof to axilla) of 1.7 m, a girth circumference of 2.7 m, and was able to reach browse at a height of 3.9 m (van Sittert and Mitchell, 2015). Its neck would have weighed about 50 kg, and thus been about 10% of its body mass, just as it is in modern giraffes. In other words, and as Lydekker surmised, it could have been a slender, tall giraffe. However, a girth of 2.7 m in a modern giraffe correlates with a body mass not of 400 kg but of ~900 kg (see Figure 18.1 in Chapter 18), a body mass that is also predicted from the dimensions of the humerus, radius, and metacarpal bones collected in the Siwalik deposits and attributed (mostly by Falconer in 1868) to *G. sivalensis*. Thus, if all the giraffe-like bones, including leg bones, so far found in the Siwalik fossil beds have been correctly assigned to this species, then it had a slender neck but a relatively stocky body and robust limbs, a shape that is not difficult to imagine but different from the shape of African *Giraffa*, but I can imagine it had a shape, but not the colours, identical to the dysplastic, "dwarf" giraffes recently photographed in Uganda and Namibia (see Plate 6.1; Brown and Wells, 2020). From analysis of the dimensions of its limb bones, there is a possibility that the body mass of males and females differed significantly. Males might have weighed about 700 kg and females about 400 kg, a difference larger than the difference between adult male and female modern giraffes. Both males and females would, however, have had a slender neck and stocky limbs. Whatever its shape, it inhabited a biome which, according to Lydekker (1876), "abounded with dense jungles of succulent plants suited for support of large herbivores": a "vast number of tree stems" where the fossils were found indicate that the habitat was the kind of tropical forest inhabited by modern-day okapi, and *G. sivalensis* lived there in a way different from that of *G. punjabiensis* and in a way that ensured its survival. Subsequent "gradual draining of the country" Lydekker (1876) wrote, "and consequent disappearance of a great part of the vegetation would have been of itself a power quite sufficient to have caused the total extinction of the fauna" adapted as it was to "life in the plains or in low jungle-clad hills." Among the fauna totally extinguished by the mid-Pleistocene (1 Mya) was *G. sivalensis*, and with it the genus *Giraffa* in Asia.

African Giraffes

African giraffes, by comparison, flourished. There were four recognized Plio-Pleistocene giraffes living in Africa at much the same time as the Indo-Asian giraffes were living: *G. jumae*, *G. gracilis*, *G. stillei*, and *G. pygmaea*, and one (*G. sp. indet.*; Harris, 1991) that, at the time of its discovery, could not be assigned to any of the other four species. The fossils of all five have been dated to cover a period from ~6 Mya (*G. jumae*) to ~1 Mya. Whether they were true species in the biological sense of being isolated reproductively is not clear, but the use of the word "species" by the experts implies that they were. The presence of several species living contemporaneously raises another, much bigger, puzzle. How could several giraffe species have evolved so quickly? What were the driving forces, the selective pressures? The classic species cannot, of course, all be ancestors and descendants, but they could be variants of a single parental stock that arose by Darwinian natural selection, each to occupy its own ecological niche. In that way several species could appear suddenly and simultaneously in the fossil record. Immigration into Africa by *G. punjabiensis* also could explain the appearance of African *Giraffas*, as immigration from Asia certainly explains the origin of five new antelope species that appeared in East Africa contemporaneously with the appearance of the four giraffes there (Vrba, 1985b), but as an explanation for the appearance of four giraffe species, this possibility is very unlikely. There must have been other factors at work.

The best explanation for the appearance of the four African giraffes and in fact any new species is Ernst Mayr's (1962) mechanism of allopatric speciation (Eldredge, 1971). It is readily applicable to the Plio-Pleistocene giraffes as well as their ancestors, like *Bohlinia*. Allopatric (= other place) speciation occurs when a small subgroup of a parental stock becomes permanently isolated and occupies a new ecological niche. Because the isolated population typically is small, pressure from natural selection is especially intense, new and beneficial mutations become fixed, and a new species evolves. Darwin (1872) had had a similar thought 100 years previously, when he wrote: "Although isolation is of great importance in the production of species, I am inclined to believe that largeness of area is still more important, especially for the production of species which shall prove capable of enduring for a long period and of spreading widely." When allopatric speciation occurs over a relatively short time of 50,000 and 100,000 years, evidence of gradual evolution in the fossil record is rare and the fossil record more usually is discontinuous. An ancestor seems to be followed by a descendant that looks different without there being any descendants with intermediate features. Eldredge and Gould (1972) coined a term, "punctuated equilibrium," to describe this pattern. Another term often used to describe the pattern is "adaptive radiation." It describes how, starting from a single ancestor, animals diversify rapidly into an array of species with different morphological and physiological traits that ensure survival in new environments. Any or all of allopatric speciation, adaptive radiation, and punctuated equilibrium can explain how it was that several giraffe species could appear suddenly (as measured in paleontological time).

Once a new body shape becomes established, that shape could be perpetuated by individual preference. In 1990 Johan Koeslag developed the idea of "Koinophylia" (= liking the same) to explain why it is that, superficially at least, all members of a species look alike even though each of their genomes is very different. The patches on any particular giraffe are unique to it, as are the stripes on any particular zebra, but at first glance all zebras and all giraffes look alike, as do a herd of wildebeest or gazelles. Koeslag proposed that members of a species are similar because when animals seek a mate they preferentially choose one with few if any features that make them look odd. Indeed, oddity would be strongly avoided by potential mates. Thus if a group of animals finds itself isolated from a parental group and in a small interbreeding group, and if they survive the genetic upheaval and develop their own characteristic appearance, selection for a similar-looking mate could explain why it is that a new species might appear in the relatively very short time of 50,000 to 100,000 years.

While allopatric speciation, adaptive radiation, and koinophylia can explain how a species can arise, they do not explain what circumstances predispose to, or create the right conditions for, speciation to take place. Most ideas about how a species comes into being stress the importance of isolation of a breeding subgroup from a larger, parental stock. If this was an important factor in the evolution of the four Plio-Pleistocene ancestral giraffes, why was East Africa so congenial for them and, incidentally, for the early humans that were evolving at the same time? What were the isolating factors that created new, exploitable, ecological niches? There is a consensus that East African niches arose out of massive, violent, tectonic events (e.g., Maslin et al., 2014) and the consensus exists because it is only geological upheavals that can fragment a landscape, throw up mountain ranges, destroy migration routes, and create barriers like lakes and rivers. North, Central, and East Africa, but less so southern Africa, were subject to just such unimaginably huge forces, and their origin was volcanic eruptions. Eruptions were infrequent. Volcanic activity started 45 Mya in Ethiopia, with further events occurring 33 Mya, 15 Mya, and 6 Mya, and in Kenya and Tanzania at intervals between 5.5 Mya and 1.2 Mya, and they occurred along the geological fault lines stretching from the Gulf of Aden in the north to Lake Malawi in the south. Before these eruptions, East Africa was a flat, homogeneous region covered with mixed tropical forest, and it was part of an equatorial rain forest that stretched across the whole of middle Africa from west to east. The geological upheavals split the forest into the Congo rain forest in the west and the grassland savannah of Kenya and Tanzania in the east, produced the East African Rift Valley System, high mountains like Kilimanjaro (5.9 km high) and Mount Kenya (5.2 km high), deserts in northern Kenya, rain forests in the Aberdare mountain range in Kenya, and numerous huge lakes, like Lake Victoria and Lake Malawi, that formed in the basins that the tectonic events had left behind. Superimposed on the tectonic upheavals was cyclical climate change. It was to play a central role in the radiation of African giraffes.

Climates, Ice Ages, and Giraffes

Of all the natural phenomena that determine how our planet functions, climate is by far the most complex, simply because there are an almost infinite number of variables and combinations of variables involved. There is, however, a strong relationship between climate, and specifically the origin and termination of ice ages, and the patterns of the Earth's cycles/orbits around the Sun. That the way Earth orbits the Sun is complex has been known since antiquity, but a basic understanding of how the patterns have influenced the climate and development of ice ages began with the work of Milutin Milanković, a Serbian mathematician who became an astrophysicist. He was born in 1879 and at the age of 21 decided to devote his life to finding practical uses for the mathematics in which he was expert. He chose climates as the vehicle and theorized that variations in the Earth's orbit around the Sun determined the Earth's climatic patterns and especially the origin of iceages. In the years between 1910 and 1940 he made the calculations needed to support his theory. The calculations were complex and had to be done by hand as there were no computers to make the task easier. One series of calculations, he wrote, "I did for a full one hundred days from morning to night" (Imbrie and Imbrie, 1979). Nevertheless, and despite opposition from others investigating the origins of iceages, he persisted. In the end he produced incontestable data to confirm his ideas, although elucidation of the details continues (e.g., Imbrie et al., 1993; Shackleton, 2000; Ruddiman, 2006; Lisiecki, 2010; Kent et al., 2018). The patterns Milanković studied were related to the eccentricity, axial tilt, and precession of Earth's orbit around the Sun. Eccentricity refers to the shape of the Earth's orbit. That shape ranges between circular and elliptical, with the Earth being equidistant from the Sun during circular orbits, while elliptical orbits take it, at intervals, further away. A change from an elliptical shape to a circular one takes ~100,000 years. It is approaching a circular shape now. The effect of eccentricity on climate has for most of the Earth's history been unnoticeable, but since ~1 Mya, the eccentricity cycle has coincided with a climate change. Axial tilt refers to the angle of the Earth's axis, with the axis defined as an imaginary line running between the North and South Poles. The Earth spins at a slight angle (obliquely) such that the line between the North and South Poles varies between 21 and 25 degrees off vertical. The change between the extremes of these two angles occurs over a period of about 40,000 years. Precession is caused by the effect that gravity, exerted by the Sun and Moon, has on the Earth's axis. Gravity's effect is to make the Earth's axis rotate just like a spinning top's axis does as it slows down. Rotation of the Earth's axis is slow. One complete circuit of rotation takes ~23,000 years. The word "precession" is derived from Latin and means "go before," because the wobble makes the equinoxes occur slightly earlier each year and so precede previous equinoxes.

How do the cycles influence the Earth's climate? The change in eccentricity from elliptical to nearly circular, in tilt from least to most, and in the position of the Earth's axis relative to the Sun all affect the amount of solar radiation reaching the Earth (e.g., Hays et al., 1976). While the amount of solar radiation is the primary source of climate change, it is its secondary effects, such as the changes in the growth of terrestrial vegetation and marine plankton, which alter global photosynthesis and thus the carbon cycle, that have a greater influence, with the greatest variability in the carbon cycle developing during those moments when the effects of precession, obliquity, and eccentricity align (Owen et al., 2018). Consequences are variation in atmospheric CO_2 concentrations and in wetland CH_4 production (Shackleton, 2000; Ruddiman, 2006) and in rainfall, and since ~3 Mya, the development of iceages. Growth and decline of polar ice caps have been a feature of fluctuations in the Earth's climate since the Oligocene and the prolonged period of cooling of the Earth that followed, interrupted only by the warm mid-Miocene Climate Optimum. Transition around 14 Mya from warmth into cold sent the Earth into the "ice-house" (Holbourn et al., 2005; Figure 4.6 in Chapter 4) we now occupy. But iceages became a phenomenon of the Pleistocene era, and when they have occurred they have been characterized by (mainly) growth of the Arctic polar icecap to cover most of the Northern Hemisphere land mass, but also growth of the Antarctic icecap. In 1976, James Hays and colleagues, in a watershed analysis, showed that over the past 300,000 years ice ages have occurred at intervals of ~23,000, ~40,000, and 100,000 years. Thus they decided that precession, obliquity, and eccentricity were the origin of modern iceages. They also concluded that between them, the three Milanković cycles explained

10%, 25%, and 50% of climate change, respectively (Hays et al., 1976). The presumption is that since the beginning of time the cycles have influenced climate patterns (Maslin, 2017), and there is no doubt that the melting of the ice that began ~10,000 to 15,000 years ago to bring about the end of the last iceage was a consequence of the warming of the Earth brought on by the effects of the Milanković cycles (Imbrie and Imbrie, 1979). Climate supercycles of 400,000 years (Kent et al., 2018), 1.2 My, and 2.4 My long have also been proposed (van Dam et al., 2006). Much shorter weather cycles of less than 100 years long, linked to sunspots, tectonic events, and water temperature cycles in the Pacific and Atlantic oceans have also been recognized (e.g., Scafetta et al., 2017), some even in biblical times.

Combinations of Milanković cycles and iceages, together with the products of geological events and weather cycles, had profound effects on the East African climate and vegetation beginning ~3 Mya. Habitat fragmentation, the cooling of the Earth, aridity, and variations in browse productivity caused extinction and/or migration and/or the conditions for speciation of terrestrial animals. The precise conditions that prevailed in East Africa at the time of the radiation of *Giraffa* are unresolved, but they seem to have begun around 3.5 Mya (deMenocal, 1995, 2004, 2011; Maslin et al., 2014). Then between 3.2 Mya and 2.5 Mya, the first ice sheet that covered most of the Northern Hemisphere landmass developed and its southern boundary pushed far south. A result was that the Indian Ocean near East Africa cooled. A sharp decline in rainfall followed. Trees died out and woodlands turned into savannah-grasslands. At intervals of 2.8 Mya, 2.5 Mya, 2.1 Mya, 1.7 Mya, and 1.2 Mya (deMenocal, 2004), peaks of variation of the East African climate occurred, all of them associated with periods of weak eccentricity, nearly circular orbits of the Earth, the effects of obliquity and precession, and Northern Hemisphere glaciation (Lisiecki and Raymo, 2005; Lisiecki, 2010). At 1.0 Mya and 0.8 Mya (Imbrie et al., 1993; Ruddiman, 2006), the East African climate oscillated between very wet and very dry, and very warm and very cold periods (deMenocal, 2011). At ~1.0 Mya, for example, there was an abundance of wooded grassland and a cool and moist environment (Owen et al., 2018).

The consequences of climate variation for the fauna and flora of East Africa were profound. In 1985, when Yale University paleontologist Elisabeth Vrba analyzed the evolution of East African antelopes, she found that bursts of speciation occurred at 2.7 Mya, 2.5 Mya, 1.8 Mya, and again around 1.0 Mya, periods that coincided almost exactly with the peaks of variation of the East African climate and its habitat, which in turn coincided with unique combinations and alignments of precession, axial tilt, and eccentricity (Vrba, 1985a, 1985b) that produced glaciation. At the times of these "icepeaks" the fauna, in her phrase, "turned over." Furthermore, after an "icepeak" and because further ice formation is inhibited by increased solar radiation produced by the precession cycle (e.g., Hays et al., 1976), the climate remains relatively stable. Consequently, any new species that had evolved during a "peak" changed little until the next coincidence of astronomical cycles upset the climate, and thus Vrba called their arrival a "pulse."

Whether Milanković cycles alone were the origin of turnover pulses has been challenged (e.g., Bibi and Kiessling, 2015), but there is an uncanny correlation between the timing of the iceages that began ~3 Mya and the appearance of the classic African *Giraffa* subspecies. The East African climate peaks at 3.2 Mya to 2.5 Mya; 1.7 Mya and 1.0 Mya coincide with the emergence of *G. stillei*, *G. pygmaea*, and *G. gracilis* 3.6–3.2 Mya, and their disappearance from the fossil record 1.8 Mya, 1.7 Mya, and 1.3 Mya, respectively. The first fossils of modern extant *G. camelopardalis* appeared at ~1.0 Mya and their genetic variants at 0.8 Mya, coinciding with the mid-Pleistocene Climate Transition (see Chapter 1). These moments in geological time also coincided with changes in vegetation with, according to Cerling et al. (2011), over the last 6 million years, oscillation of the East Africa vegetation between being mostly woodlands and then mostly grasslands. The appearance of a sparse woody cover and a habitat dominated by woody grasslands/savannah vegetation began in the late Miocene and lasted until the early Pliocene, that is, a period lasting from 5.7 to 4.4 Mya, which encouraged *G. jumae* to leave East Africa. This relatively open habitat was followed by a period of increasing woody cover such that by the middle Pliocene, about 3.6 Mya, woodland/savannah covered 40%–60% of East Africa and provided the habitat for a reappearance of *G. jumae* and for the radiation of *G. stillei*, *G. pygmaea*, and *G. gracilis*. From 3.6 to 1.4 Mya, as the Pliocene era became the Pleistocene era, woodland/savannah declined and open environments characterized by savannah/grasslands returned, reaching a peak at 1.8 Mya. Habitats containing more than 50% trees, shrubs, and bushes disappeared. The disappearance from the fossil record, but not necessarily extinction, at 1.8 Mya, 1.7 Mya, and 1.3 Mya, of *G. stillei*, *G. pygmaea*, and *G. gracilis*, respectively, followed.

The period beginning about 3 Mya is also the time that humankind emerged. The conversion to bipedalism that is a hallmark of human evolution is attributed to the disappearance of forests and the emergence of savannah. Thus it was at around 3.5 Mya, when an open savannah/woodland habitat dominated East Africa, *Australopithecus* appeared as a bipedal and tallish hominid. Some members of that genus migrated from East Africa northward ("Lucy"—*A. afarensis*) and others into South Africa ("Mrs. Ples"—*A. africanus*). Around 1.8 Mya another hominid genus, *Paranthopus*, emerged and occupied the by then even more open habitats. The first appearance of *Homo* (as *H. habilis*) in the fossil record occurred 1.8–1.9 Mya and was followed by *H. erectus*. At 1 Mya, *Homo* brain size peaked and concurrently *G. camelopardalis* emerged as the sole surviving member of the giraffe family. We can be certain that *Homo* hominids were among the first, if not the first, people to see the first modern giraffes—an event that was a product of the beginning of the iceage era.

Ancestral African Giraffe Species

The relationships between *G. jumae* and *G. gracilis*, *G. stillei* and *G. pygmaea* are unresolved and, thus, subject to considerable speculation. The different "species" are distinguished (mainly) by subtle differences in ossicone morphology and dental measurements. Identification is complicated by the rarity of fossil ossicones and complete dentitions (Rowan et al., 2017). No full skeletons of any ancestral species have been found, so the actual size of the ancestral giraffes is unknown and, it is very likely, always will be, but the dimensions of fossils of their limb bones, almost without exception, are regarded as having a very similar or even identical morphology to those of modern giraffes (Harris, 1976c; Geraads et al., 2013). If so, an idea of how big they were relative to modern giraffes and to each other can be estimated from dimensions of their limb bones and vertebrae using the relationships that exist between bone length and diameter and body mass, height, neck length, and leg length in common southern African giraffes, *G.c. giraffa* (Table 6.2).

The limb bones so far recovered that can be used in such an analysis are listed in Table 6.4 for *G. jumae* and in Table 6.5 for an indeterminate species *G.sp. indet.*, and for *G. gracilis*, *G. stillei*, and *G. pygmaea*, as are their body characteristics, as predicted from the lengths of each bone type. Their body dimensions can be compared with those of *Bohlinia* which are shown in Table 6.3.

Table 6.4 Bone Dimensions of *G. jumae* and the Predicted Size of Body Dimensions from Bone Length

Bone	Length (mm)	Midshaft Diameter (mm)	Ref	Calculated Body Mass (kg)	Calculated Height (m)	Calculated Neck Length (m)	Calculated Leg Length (m)
H	510		1	959	4.67	1.86	1.86
H	510		1	959	4.67	1.86	1.86
R	800	66	1	1,116	5.07	1.95	1.87
R	760	62	1	964	4.81	1.83	1.81
R		73	1				
R		72	1				
R		66	3				
Mc	644		2	672	4.27	1.66	1.67
Mc	636		2	645	4.21	1.63	1.65
Mc	645	43	1	676	4.28	1.67	1.67
Mc	757	52.6	1	1,160	5.17	2.10	1.91
Mc	795	53.8	1	1,368	5.48	2.25	1.99
Mc		51	1				
Mc	635	53	3	641		1.63	1.65
F		52.5	1				
F	550		2	1,169	4.86	1.99	1.88
F	533		2	1,051	4.69	1.90	1.83
T		67	1				
T	655	62	4	1,109	4.86	1.92	1.90
T	590		4	790	4.33	1.67	1.75
T	662	62	2	1,148	4.92	1.95	1.91
T	620		1	928	4.57	1.78	1.82
Mt		44	1				
Mt		54	1				
Mt	768	57.5	2	1,249	5.30	2.04	2.00
Mt	723	44.5	1	1,030	4.95	1.88	1.90

Table 6.4 Continued

Bone	Length (mm)	Midshaft Diameter (mm)	Ref	Calculated Body Mass (kg)	Calculated Height (m)	Calculated Neck Length (m)	Calculated Leg Length (m)
Mt	762	51	1	1,218	5.25	2.02	1.99
Mt		44	1				
Mt		49	1				
Mt		48.4	1				
Mt	650		3	733	4.40	1.63	1.75
Mt	630	49	3	664	4.25	1.56	1.70
Mean	*659*	*56*		*964*	*4.75*	*1.85*	*1.83*
sd	*91*	*9*		*227*	*0.38*	*0.18*	*0.11*

Sources of the length and diameters of bones: 1 = Geraads et al., 2013; 2 = Harris, 1976c; 3 = Geraads, 1998; 4 = Harris, 1976a.
Body dimensions calculated from allometric equations in Table 6.2.
sd = standard deviation; H = humerus; F = femur.

Table 6.5 Bone Dimensions of *G. sp. indet.*, *G. gracilis*, *G. stillei*, and *G. pygmaea* and the Predicted Size of Body Characteristics Derived from Bone Length

Species	Bone	Length (mm)	Midshaft Diameter (mm)	Ref	Calculated Body Mass (kg)	Calculated Height (m)	Calculated Neck Length (m)	Calculated Leg Length (m)
G. sp. indet.	H	498		5	888	4.54	1.80	1.83
	R	736		5	880	4.66	1.76	1.77
	R	779		5	1,034	4.93	1.89	1.84
	Mc	748		5	1,114	5.10	2.06	1.89
	Mc	744		5	1,094	5.06	2.05	1.89
	T	654		5	1,103	4.86	1.92	1.90
	Mt	748		5	1,148	5.15	1.97	1.96
	Mt	661		5	774	4.48	1.67	1.77
	Mt	637		5	688	4.30	1.59	1.72
	Mean	*689*			*969*	*4.79*	*1.92*	*1.87*
	sd	*88*			*167*	*0.30*	*0.12*	*0.06*
G gracilis	H	437		2	582	3.92	1.50	1.65
	H	445		2	617	4.01	1.54	1.67
	H	380		2	370	3.36	1.23	1.47

(continued)

Table 6.5 Continued

Species	Bone	Length (mm)	Midshaft Diameter (mm)	Ref	Calculated Body Mass (kg)	Calculated Height (m)	Calculated Neck Length (m)	Calculated Leg Length (m)
	R	674		2	685	4.28	1.58	1.66
	R	640		2	591	4.06	1.49	1.60
	Mc	649		2	690	4.31	1.68	1.68
	Mc	648		2	687	4.30	1.68	1.68
	Mc	632		2	631	4.18	1.62	1.64
	F	451		2	597	3.86	1.50	1.60
	F	477		2	722	4.12	1.62	1.67
	T	520		2	524	3.76	1.40	1.58
	T	520		2	524	3.76	1.40	1.58
	Mt	690		2	887	4.70	1.77	1.83
	Mt	690		2	887	4.70	1.77	1.83
	Mt	660		2	770	4.47	1.66	1.77
	Mean	*568*			*651*	*4.12*	*1.56*	*1.66*
	sd	*109*			*136*	*0.37*	*0.15*	*0.10*
G. stillei	H	465		1	711	4.21	1.63	1.73
	Mc		47	1		4.12		
	T	560	58	1	666	4.08	1.55	1.67
	Mt		41.5	1				
	Mt	652	40.8	1	741	4.41	1.64	1.75
	Mt		38.5	1				
	Mean	*559*	*45*		*706*	*4.23*	*1.61*	*1.72*
	sd	*94*	*8*		*37*	*0.17*	*0.04*	*0.03*
G. pygmaea	Mc	617		5	582	4.06	1.56	1.61
	Mc	573		5	454	3.72	1.41	1.51
	Mean	*595*			*518*	*3.89*	*1.48*	*1.56*
	sd	*31*			*91*	*0.24*	*0.11*	*0.07*

Sources of the length and diameters of bones: 1 = Geraads et al., 2013; 2 = Harris, 1976c; 5 = Harris, 1991. Body dimensions calculated from the allometric equations in Table 6.2.

The most common limb bones listed in these tables are metapodial bones. They make up 64% of all bones (56/88) and they produce 67% of all measurements (79/118), thus they constitute a robust sample on which to base estimates of body dimensions. When standard statistical methods are applied to the body dimensions calculated from limb bone dimensions, what emerges is that *G. jumae* was the same size as *G. sp. indet.* and

Bohlinia (except that the predicted body mass of *Bohlinia* is lower); *G. jumae*, *G. sp. indet.*, and *Bohlinia* were significantly larger than *G. gracilis*, *G. stillei*, and *G. pygmaea* in all respects, but there were no significant differences in the sizes of *G. gracilis*, *G. stillei*, and *G. pygmaea*. Thus the giraffe (*G. sp. indet.*) that John Harris in 1991 was unable to assign to a species had (by this method of analysis) body dimensions which, for all practical purposes, were identical to those of *G. jumae* and thus their data can be combined. *G. gracilis* and *G. stillei* already are considered to be one species (Harris, 1991; Geraads et al., 2013) and the data confirm that view. As the size of *G. pygmaea* is not significantly different from that of *G. gracilis*/*G. stillei*, *G. pygmaea* can be included with them. In other words, these analyses imply that there were perhaps just two, not four, fossil *Giraffa* species—*G. jumae* and *G. gracilis/stillei/pygmaea*—and the range of sizes among them can be attributed to the normal range of size found in any population of any mammal, or "when more than one species is present they must have formed a continuum in size with large individuals of smaller forms reaching the size of small individuals of larger forms" (Geraads et al., 2013). Body dimensions calculated from metapodial length for each of the giraffes in these combined groups are summarized in Table 6.6.

Although the neck length of modern giraffes is strongly correlated with the dimensions of any limb bone (Table 6.2), it is predicted best from the length of the body of cervical vertebrae (Badlangana et al., 2009; van Sittert et al. 2010). Allometric equations predicting neck length from the length of each of the six main cervical vertebrae (C2–C7, excluding C1 as it does not contribute significantly to neck length) have been developed. Body mass can also be predicted from cervical vertebra length, and the relevant equations are shown in Table 6.7.

Table 6.6 Summary of the Calculated Mean and Range (in parentheses) of Metapodial Lengths and Diameters for the Three Groups of Fossil Giraffes and the Predicted Mean and Range for Body Mass, Height, Neck Length, and Leg Length

Dimension	*Bohlinia*	*G. jumae*	*G. gracilis*
N	22	22	12
Body mass (kg)	866 (712–1026)	929 (641–1368)	703 (454–807)
Metapodial length (mm)	691 (655–730)	699 (635–795)	646 (573–690)
Metapodial diameter (mm)	51 (40–61)	50 (43–58)	43 (39–47
Height (m)	4.66 (4.36–4.95)	4.74 (4.21–5.48)	4.32 (3.72–4.70)
Neck length (m)	1.82 (1.66–1.99)	1.87 (1.56–2.25	1.64 (1.41–1.77)
Leg length (m)	1.79 (1.69–1.86)	1.83 (1.65–2.0)	1.70 (1.51–1.83
Neck:leg length ratio	1.02 (0.95–1.07)	1.02 (0.92–1.13)	0.94 (0.84–1.0)

Data from Tables 6.3, 6.4, 6.5.

Table 6.7 Neck Length and Body Mass Predicted from Fossil Cervical Vertebra Body Length (VBL)

Vertebra	Vertebra Body Length (mm)	Allometric Equation for Neck Length	Neck Length (cm)	Allometric Equation for Body Mass	Body Mass (kg)
Axis (C2)*	277	$0.93 \times VBL^{0.95}$	211	$0.0065 \times VBL^{2.14}$	1096
C5*	286	$0.93 \times VBL^{0.96}$	218	$0.0067 \times VBL^{2.16}$	1355
C5*	271	$0.93 \times VBL^{0.96}$	207	$0.0067 \times VBL^{2.16}$	1206
C7*	250	$1.92 \times VBL^{0.83}$	192	$0.021 \times VBL^{1.96}$	1052
Unknown⁰	205	$1.13 \times VBL^{0.93}$	160	$0.0013 \times VBL^{2.44}$	580
Unknown⁰	210	$1.13 \times VBL^{0.93}$	163	$0.0013 \times VBL^{2.44}$	603
Unknown⁰	230	$1.13 \times VBL^{0.93}$	178	$0.0013 \times VBL^{2.44}$	768

Source: Vertebra data from *Harris, 1991 and ⁰Geraads et al., 2013. Allometric equations derived from data collected from 39 postnatal giraffes by van Sittert et al., 2010.

Thus far, just seven cervical vertebrae have been recovered from fossils of African *Giraffa* spp. Four of them, and their position, came from *G. sp. indet.* (Harris, 1991). The position of the other three is not known. If the position is not known, neck length can be predicted quite accurately from an equation describing the relationship between the average lengths of C2 to C7 and neck length. The neck lengths and body masses of the fossil giraffes, calculated from the length of their vertebral bodies, coincide well with those in modern giraffes, and they further confirm that *G. sp. indet.* was as large as was *G. jumae*. Geraads et al. (2013) could not identify the position of the three cervical vertebrae they recovered but thought because "none reached the length of those of male *G. camelopardalis*, they could be female *G. jumae* or *G. stillei*." The data support their origin as *G. stillei*.

Modern African Giraffes

To get another idea of how big or small ancestral giraffes were, their sizes can be compared with those of modern giraffes. Modern giraffes fall naturally into three age groups (Table 6.8). Group 1 consists of juvenile giraffes (newborn to three years of age or puberty), Group 2 of young adults three to five years of age, and Group 3 of adults over five years of age. The body characteristics of each of these groups differ significantly (in the statistical sense): Group 1 giraffes are smaller than are Group 2 giraffes, which, in turn, are smaller than Group 3 giraffes. Table 6.8 also shows that until they are five years old the legs of modern giraffes are, on average, longer than their necks, but in mature giraffes the reverse is true. The predicted neck:leg length ratio of *G. jumae*/*G. sp. indet.* and of *Bohlinia* is greater than one, but the ratio is less than one in adult *G. gracilis*, *G. stillei*, and *G. pygmaea*—just as it is in juvenile modern giraffes (Tables 6.7, 6.8). Did male and female ancestral African giraffes differ in size? Insufficient data prevent a certain conclusion, but in young modern giraffes, body dimensions of males and females

Table 6.8 Summary of the Mean and Range (in Parentheses) of the Body Dimensions of Three Different Age Groups in *G. camelopardalis*

Dimension	Group1	Group2	Group3
N	12	11	16
Age (yr)	1.9 (0–3)	3.9 (3.1–5.0)	6.51 (>5)
Body mass (kg)	334 (147–508)	660 (525–829)	1063 (851–1441)
Metapodial length (mm)	512 (425–570)	628 (548–692)	715 (636–815)
Metapodial diameter (mm)	37 (29–44)	47 (45–51)	54 (49–59)
Height (m)	3.19 (2.44–3.80)	4.08 (3.72–4.51)	4.79 (4.51–5.32)
Neck length (m)	1.15 (0.74–1.44)	1.53 (1.37–1.65)	1.90 (1.73–2.17)
Leg length (m)	1.38 (1.12–1.66)	1.70 (1.55–1.85)	1.82 (1.59–2.08)
Neck: leg length	0.83 (0.64–0.95)	0.90 (0.79–1.06)	1.05 (0.89–1.19)

Data from Hall-Martin, 1975a; Mitchell et al., 2009b; van Sittert et al., 2015.

are the same. In adult modern giraffes, males are bigger than females. Presumably the same relationships existed among the ancestral African giraffes.

When the size of modern giraffes is compared with that of fossil giraffes (Tables 6.6 and 6.8), what is obvious is that adult *G. jumae* and *Bohlinia* were very much bigger than both Group 1 and Group 2 modern giraffes, but exactly the same size as mature modern giraffes, except, perhaps, for *Bohlinia*'s average body mass (~870 kg) which, probably, was lower than that of Group 3 giraffes (~1,060 kg) but similar to the body mass of a mature modern female giraffe (800–1,000 kg). The average diameter of the metapodial bones of *Bohlinia* and *G. jumae* (50 mm) was marginally narrower than it is in mature *G. camelopardalis* (54 mm), but all three had similar average slenderness indices (14 for *G. jumae*, 13.5 for *Bohlinia*, and 13.2 for adult modern giraffes). The length of the proximal and middle phalanges of *G. jumae* (107 mm and 60 mm, respectively)and those of *G. stillei* (~100 and ~50 mm, respectively) as recorded by Robinson (2011) are not different from those of modern giraffes (see Table 15.1 in Chapter 15). *G. gracilis/stillei/pygmaea* were bigger than Group 1 giraffes in all respects, exactly the same size as young, three- to five-year-old, extant giraffes, had a similar slenderness index (15 for *G. gracilis*, 13.6 for Groups 1 and 2 modern giraffes), but were smaller than adult extant giraffes. Thus, if the coat markings of fossil giraffes were the same as those of modern giraffes, the fossil giraffes would have been indistinguishable in a herd of modern giraffes. Geraads et al. (2013) thought that if they were true species, coat markings must have differed in order to allow interspecific recognition. There is, however, no evidence that modern giraffes use coat markings to identify a subspecies (see Chapter 1).

Table 6.9 Generic Allometric Equations for Predicting Body Dimensions in All *Giraffa* spp. and *Bohlinia* from Metapodial Length (L, mm)

Dimension	Allometric Equation	STEYX*	R^2
Body mass (kg)	$(6\times10^{-7})\times L^{3.24}$	88	0.9512
Height (m)	$0.002\times L^{1.18}$	0.136	0.9600
Neck length (m)	$0.0002\times L^{1.43}$	0.094	0.9071
Leg length (m)	$0.0077\times L^{0.83}$	0.061	0.9036

*Standard error of the prediction of Y from X.

The data shown in Tables 6.3, 6.4, and 6.5 allow a further extrapolation. They show that the dimensions of the various groups of modern giraffes and of the fossil "species" of giraffes lie on the continuum suggested by Geraads et al. (2013), with Group1 < Group2; Group2 = *G. gracilis*; Group 2 <*Bohlinia*; *Bohlinia* = *G. jumae* = Group 3. Figure 6.5 illustrates this continuum clearly. As the measurements for each body characteristic fall within the 95% confidence intervals of the measurements in modern giraffes, the relationships between metapodial dimensions and body characteristics for all *Giraffa* species, including the modern species and also *Bohlinia*, can be described by the single allometric equations shown in Table 6.9.

What Figure 6.5 also shows, though, is that none of the fossil metapodial bones so far found has the same length as that of juvenile giraffes less than three years old. Harris (1976a) did find three ossicones that must have come from immature giraffes, as they were not fused to a skull, but the absence of other small bones is odd. There must have been smaller, younger individuals than the fossils so far recovered reflect, and absence of their bones may be because small young giraffes are more susceptible to scavenging and thus less likely to end up as a fossil. As Langman (1977) has pointed out, modern juvenile giraffes, even up to two years of age, do not have the physical attributes needed, other than speed, to avoid predation. Typically fewer than 20% of giraffes survive past two years of age. It is modern giraffes older than three years that survive to adulthood, and they have a body mass of about 500 kg by that age (Chapters 5 and 18), and presumably ancestral giraffes of that or higher body mass were the ones that became fossils.

Figure 6.5 emphasizes that predicted sizes of the various fossil species, including *Bohlinia*, as well as the actual sizes of modern giraffes, are intermingled and cannot readily be separated on the basis of size. Identification of species then has to rely on the subtle differences between ossicones and teeth, which themselves can be somewhat arbitrary. A proposition that avoids this problem is that fossil *Giraffa* of Africa were all just one species. If so, the fossil "species" was constructed in much the same way as the modern species is: a single founding species and ecological or geographical variants derived from it. That variants rather than separate species coexisted has been considered by Harris (1976c, 1991) and, given the data now available, that is a plausible proposition. Variants, however, need a parent species. It was *G. jumae*.

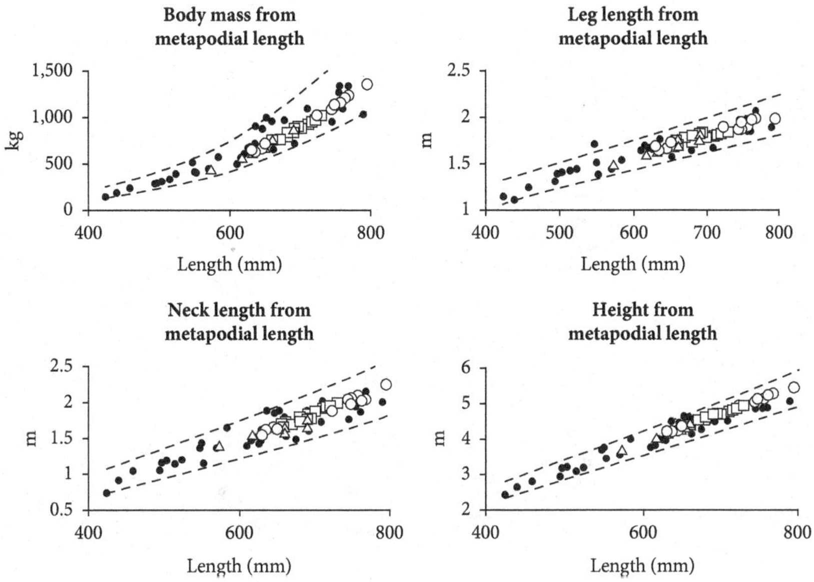

Figure 6.5. Relationships between metapodial length and body mass, height, neck and leg length in *G. jumae* (○), *G. gracilis* (△), *Bohlinia* (□), and modern giraffes (●). The dotted lines show the 95% confidence intervals for modern giraffes.
Data from Tables 6.2, 6.4, 6.5, and van Sittert et al., 2015.

Giraffa jumae

In 1965, L.S.B (Louis) Leakey, best known for his work on Kenyan paleoanthropology, found an almost perfect skull of a large giraffe (Figure 6.6) at Rawe in Kenya.

Its age was dated to the late Pliocene, about 3–4 Mya. Leakey, although cautious, decided that it was a new species, and called it *G. jumae* (because he discovered it on a Friday? *ijumaa* in Swahili means Friday). The oldest known fossils of *G. jumae*, albeit known with uncertainty, were found in the Tugen Hills near Lake Turkana (= Lake Rudolph, Figure 6.1). They are dated to ~6 Mya (Pickford, 1975), a time that coincides well with the life and times of *Bohlinia*, and with a southwestward migration of *Bohlinia* into Africa. Following the natural pattern of evolution, once in Africa the predecessors of *G. jumae* would have become isolated from a *Bohlinia* parental herd, evolved into a new, larger species to become *G. jumae*, and occupied a different ecological habitat from the Eastern Mediterranean, peri-Adriatic biome that was *Bohlinia*'s natural home. *G. jumae* became larger because when few browser species occupy a habitat competition for resources is reduced and those in it tend to increase in size (Janis et al., 2000).

G. jumae did not linger in North Africa. It was definitely present at the southernmost tip of Africa at Langebaanweg (Harris, 1976a; see Figure 6.1) in fossil deposits formed 4–5 Mya at the same time that the East African habitat was becoming grassland and the Mediterranean salinity crisis was ending (Hendey, 1970, 1981). It had arrived by migration, searching, we can imagine, for a congenial habitat, and the habitat at the Cape, fortunately, was hospitable. There was, for example, a more stable source of water (Bibi and

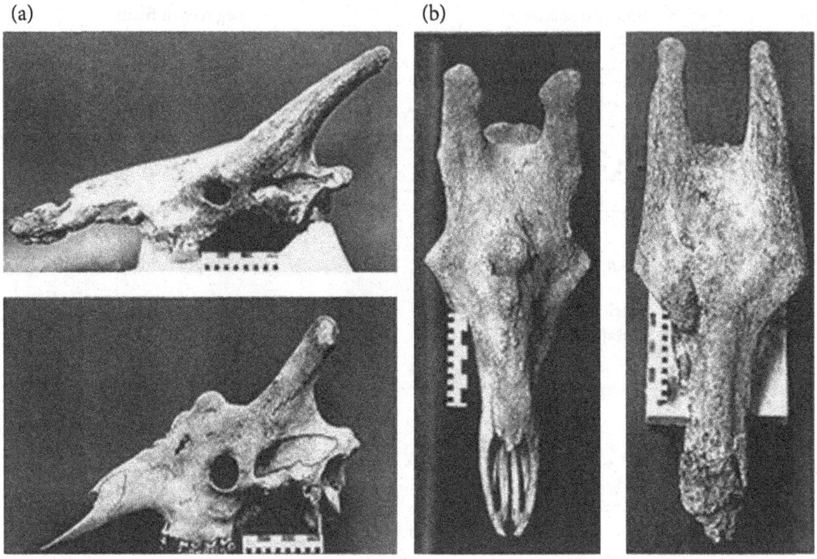

Figure 6.6. The skull of *G. jumae* (top left and far right) and *G. camelopardalis* (bottom left and near right).
From Leakey, 1970, with permission of Elsevier.

Kiessling, 2015) and there was, after all, nowhere further south to go. Was such a trek possible? Geraads (2006) has pointed out that *Sivatherium maurusium*, the common companion of *Giraffa*, migrated from the far northwest of Africa to Langebaanweg at much the same time, a distance of 8,000 km or more. Unchecked spread of large mammals can occur at a rate of 1,000 km in a century at what "would seem a moderate estimate for most larger mammals" (Kurten, 1957). Therefore, it might have taken *G. jumae* and its progeny a relatively very short time, perhaps a mere 500 years (but at least 100 generations), to have escaped the aridity caused by the Messinian crisis, to reach the Cape by traveling down the east coast of southern Africa (no doubt leaving some behind to enjoy the habitat in what is now Mozambique and the north eastern parts of South Africa) to Langebaanweg where, via the presence of its fossils, it became the earliest undoubted record of *Giraffa* in Africa (Harris, 1991).

By the late Pliocene/early Pleistocene, 3.0–2.6 Mya, as the first iceage ended, the sea levels rose once again, and the climate at Langebaanweg became more temperate. The number of mammal species there dwindled from 82 to 17 (Hendey, 1981). The woodlands became grassland and then scrubland, and browsers declined. *G. jumae* disappeared. Where did *G. jumae* go? However unlikely such a proposition seems, some may have migrated up the western coast of the Cape Province toward Namibia and Angola to become *G.c. capensis*. A more convincing origin of *G.c. capensis* is outlined in Chapter 3. Other *G. jumae* definitely migrated eastward along the southern Cape coast. There they would have found refuge in the flat valleys of the southern Cape rivers that supported an acacia-savannah type woodland that is the ideal habitat for giraffes. Along this coast, at Still Bay, 400 km east of Langebaanweg, tracks of an adult giraffe and, from the size of its footprint, a newborn calf have been preserved in the coastal dunes and dated to 125,000 years ago. Regrettably, so far no confirmatory fossils of giraffes have been found in the local fossil deposits (Helm et al., 2018). Slow migration further east and then up the east coast of South Africa could explain the rock paintings of giraffes in the southeast Cape and the archaeological evidence of giraffes in what is now Kwazulu-Natal. Some *G. jumae* might have stayed behind there to join those left behind there earlier to become the origin of *G.c. giraffa* whose current home is the mid-eastern regions of the eastern seaboard and hinterland of southern Africa stretching from the northeastern regions of South Africa through Mozambique to Zimbabwe. Much earlier eastern migration had occurred: a tooth of a large *Giraffa* has been found at Makapansgat (Singer and Boné, 1960), a fossil deposit dated to about 3 Mya (McFadden, 1979). Makapansgat is en route from the Cape to the east of southern Africa (Figure 6.1). This tooth implies that herds of *G. jumae* were making a northward migration up the eastern half of sub-Saharan Africa at that time. At the other end of the continent, in North and East Africa, the early Pliocene cold period had been followed by a period (4.6–3.6 Mya) of rapid warming, the Mediterranean Sea was re-established, the climate of the wide surroundings became humid, and a period of increasing woody cover followed (van Zinderen Bakker and Mercer, 1986; Cerling et al., 2011). It was a decidedly more attractive place for a giraffe to live. Leakey's discovery of *G. jumae* at Rawe in Kenya around 3–4 Mya confirmed that *G. jumae* had reached Rawe about one to two million years after its first appearance in northeastern Africa. The progeny of those that once had lived at the Cape had taken a long time to return to their origin.

For *G. jumae* to have been a true species, as it undoubtedly was, it must have had some characteristics quite different from those of its ancestor *Bohlinia*. What were they? Their body dimensions were almost identical, but the biggest *G. jumae* were bigger than the biggest *Bohlinia*. *Bohlinia*'s neck probably was between 1.60 and 2.00 m long and its height 4.3–4.9 m. *G. jumae*'s neck was 1.6–2.3 m long and its height 4.2–5.5 m (Tables 6.3, 6.4). They shared the primitive features of a less inflated skull—more so in *Bohlinia* than in *G. jumae*—and the absence of a median ossicone. The ossicones in *Bohlinia* were shorter, more conical, and arose above the orbits, rather than behind them as they do in *G. jumae* (Figure 6.5), and *Bohlinia*'s teeth were smaller than those of *G. jumae*. Not many differences, in other words, but combined with the chronology of *Bohlinia* living 9–5 Mya and *G. jumae* from 6 to ~1.5 Mya, they are enough to establish that these two giraffes were different species. There is no doubt, though, that *Bohlinia* was *G. jumae*'s immediate ancestor. Typically any species (including humans) has an expected lifespan of 3million years (e.g., van Dam et al., 2006). *G. jumae* had an unusually long one of 5million years, so it must have been exceptionally well adapted. It was widespread throughout Africa from its first appearance, probably 6 Mya in East Africa, to its 4–5 Mya appearance at Langebaanweg, then its reappearance at Omo and Laetoli 3.6 Mya, at Hadar in Ethiopia 3.4–3.0 Mya, and West Turkana 3.5–2.3 Mya. *G. jumae* disappears from the fossil record about 1.3 Mya, the time to which the first fossils of modern giraffes—*G. camelopardalis*—are dated.

Giraffa stillei, Giraffa gracilis, Giraffa pygmaea

I can imagine that around 3 Mya, large herds of *G. jumae* roamed the landscape of East Africa, Sudan, Somaliland, and Ethiopia—the part of Africa from where, in the early

1800s, the first modern giraffes were captured and taken to Europe. This stable community of a single *Giraffa* in balance with its environment was disrupted by tectonic events and changes in climate 4–3 Mya. Fragmented landscapes were created with many browser niches, not only for giraffes but also for artiodactyls generally (Vrba, 1985a, 1985b). Grasslands were dominant, the environment arid, and browse diminishing. The circumstances were ideal for allopatric speciation, and it is easy to imagine that with food diminishing there was contemporary emergence of herds of smaller giraffes needing less food and water and having a relatively large body surface area through which to lose body heat—*G. stillei*, *G. pygmaea*, and *G. gracilis*. Though they were almost as tall as *G. jumae*, the lower mass of these three species translates into more slender body dimensions with the thermoregulatory advantages that such a shape confers (see Chapter 14). A common response to arid conditions is a smaller size, as it reduces food and water requirements, but size differences may simply reflect that the morphology of early *Giraffa* was highly variable, sensitive to nutritional status, or that they were sexually dimorphic, or that the different fossil types reflected adaptations to different ecologies and types of vegetation (Harris, 1991). A scenario has been imagined by Harris for the four species based on the relationship between modern giraffes and their food sources in the Serengeti National Park (Pellew, 1983b), which is, of course, the area where the ancestral *Giraffa* once lived. Pellew recognized three different woodland habitats—ridge-top and upper slope *Acacia* regeneration thicket, mid-slope open *Acacia-Commiphora* woodland, and valley bottom woodland, which was further divided into riverine woodland and korongo (ephemeral river) thicket. Modern giraffes in the Serengeti at the time of Pellew's study migrated between the ridge-top areas (exploited in the wet season) and valley bottom woodland (dry season), which is a tactic giraffes practice in any arid region because they follow the availability of browse and water (e.g., Fennessy et al., 2003). The mid-slope biomass was relatively small and limited to the undersides of mature trees, thus accessible mainly to the tallest giraffes. Harris (1991) extrapolated these habits to the East African Plio-Pleistocene giraffes and thought that it was possible that the browser niche of *G. stillei* and *G. gracilis* was ridge top and riverine woodland, *G. jumae* exploited the mid-slope woodland, and *G. pygmaea* the korongo thicket.

The earliest fossils of *G. stillei*, *G. gracilis*, and *G. pygmaea* have been dated to 3.6 Mya, 3.2 Mya, and 3.2 Mya, respectively. The *Giraffa* fossils dated to 3.6 Mya were discovered at Laetoli not far from Rawe (Figure 6.1), and two species were identified from them by W.O. Dietrich in 1942. Dietrich classified them as belonging to *G. camelopardalis* and Stille's okapi (*Okapia stillei*). Some of the fossils were large enough to have come from *G. camelopardalis*, but as that species did not evolve as a species until much later, the fossils must have come from *G. jumae*, and they confirm *G. jumae*'s presence in East Africa then. Okapi have teeth that are morphologically different from those described by Dietrich, and the other fossils were, therefore, reassigned to *Giraffa* as *G. stillei* (Harris, 1976b). By this reclassification it is *G. stillei* that becomes the first of the variants or subspecies of *G. jumae* to appear. It could have had no other predecessor than *G. jumae*. *G. stillei* was the most common *Giraffa* at Laetoli, which was then savannah, a habitat that favored *G. stillei* rather than *G. jumae* (Harris, 1976c), and at Hadar in Ethiopia (see Figure 6.1) and dated to 3.4–3.1 Mya (Geraads et al., 2013). Its identity as a separate species depends on the shape and size of its teeth and ossicones (Harris, 1976c, 1991; Geraads et al., 2013).

The consensus for the origin of *G. gracilis* is that it was *G. stillei*, which, over geological time, became progressively larger as the climate and the quantity and quality of browse available improved, to become *G. gracilis*. Some fossils attributed to *G. gracilis* pre-date those of *G. stillei*, but so similar are their fossils that some mis-attribution is, I suppose, inevitable. Fossils assigned to *G. gracilis* (but are as likely to be *G. stillei*) have been found at Omo (3.75–1.84 Mya; Arambourg, 1947), at Laetoli (3.8–2.9 Mya; Harris, 1987a), at East Rudolph (3.2–1.3 Mya), in the Lower Awash Valley just four to five miles from Hadar in Ethiopia (2.8–2.6 Mya; Rowan et al., 2017), and at Olduvai (2.1–1.7 Mya; Leakey, 1965) (Figure 6.1). The increase in size of *G. stillei* may in turn have been stimulated by competition with *G. pygmaea* (Robinson, 2011) if *G. pygmaea* was a separate species. *G. gracilis* was extinct by 1.3 Mya (Churcher, 1978), but clearly it ranged far and wide in the East African Plio-Pleistocene and was a contemporary of *G. jumae* (and of *Australopithecus*). Its height, body mass, leg length, and neck length were significantly smaller than those of *G. jumae*, as were its teeth, and the size, slope, and shape of the ossicones were different. In particular, their ossicones, like the ossicones of *Bohlinia*, did not have a terminal knob. Also significantly, and unlike *G. jumae*, *G. gracilis* had evolved a third, median ossicone, but why it did and what that ossicone's selective advantage might have been is yet another mystery. Over the 3 million years that *G. stillei*/*G. gracilis* lived, the length of their neck and legs increased until they became much the same size as young, modern giraffes, and Hassanin et al. (2007) thought *G. stillei*/*G. gracilis* represented the origin of *Giraffa* as a species in Africa.

The third variant, *G. pygmaea*, is an enigma. Too few fossils of it exist to provide definitive evidence for its relationships with the other species (Geraads et al., 2013), but its chief characteristic, as its name implies, was its small size. The width of a skull attributed to it was 142 mm. A skull of that width is the width of the skull of a newborn modern giraffe with a body mass of 100 kg (Mitchell et al., 2013b), but the dimensions of the few limb bones known of presumably mature *G. pygmaea* predict an average body mass of about 500 kg (Table 6.5), which is the size of a three-year-old modern giraffe. What characters did it have that identified it as a species? Its teeth were smaller than those of *G. stillei* and *G. gracilis*, though the teeth assigned to it cannot always be distinguished from the teeth of *G. gracilis* (Harris, 1976b, 1976c, 1991; Churcher, 1978). As *G. gracilis* and *G. stillei* were the same species, and as the fossils of *G. pygmaea* were of similar size to theirs (Table 6.5), it is not unreasonable to suppose that the fossils attributed to *G. pygmaea* are just those of a young or small *G. gracilis*/*G. stillei*. *G. pygmaea* appears in the fossil record at Laetoli in the late Pliocene ~3 Mya (Robinson, 2011). At that time Laetoli must have been an ideal habitat for giraffes, as it is a rich source of fossils of *G. jumae*, *G. stillei*, and *G. pygmaea*. Thus an equally plausible origin of *G. pygmaea* is that the Laetoli habitat offered an isolated habitat for allopatric speciation of a giraffe species smaller than *G. stillei*, and a possible habitat was the riverine woodland and korongo thicket, as Harris proposed (in 1991). Classification of *G. pygmaea* as a separate species has depended in part on the size and shape of its ossicones, which were its most distinctive feature. They were slender, smaller than those of modern giraffes (Churcher, 1978), and showed secondary bone apposition (Harris, 1987a). Secondary bone apposition is a characteristic of males, not females. The ossicones of *G. stillei* and *G. gracilis* found thus far do not show secondary bone apposition; therefore, if *G. pygmaea* had that gender-specific feature, it was a different species, although the link between ossicone morphology and reproductive isolation from *G. stillei* and *G. gracilis* is not obvious, other than that it may be indicative of the inbreeding associated with the process of allopatric speciation.

Implied from its name is that *G. gracilis* (and by extension *G. stillei* and possibly *G. pygmaea*) were gracile giraffes. The nuchal crest on the skull of *G. gracilis* where the *ligamentum nuchae* is attached is larger than it is in modern giraffes; its first cervical (atlas) vertebra was small (Harris, 1987a), as were the muscles that lift the head and neck that attach to it. Small muscles imply that its neck was light, and, therefore, relatively slender. A skull width of 245 mm recorded by Harris (1976c) for *G. gracilis* is associated with a body mass of ~600 kg in a modern giraffe. Another skull measurement reported by Churcher (1978) was a width of 212 mm and it would have come from a modern giraffe with a body mass of ~400 kg, so it is quite likely that this skull was that of a juvenile or perhaps from *G. pygmaea* if *G. pygmaea* was a different species. The midshaft diameters of only five metapodial bones are known for *G. stillei*/*G. gracilis*, and they came from fossils assigned to *G. stillei*. Their average diameter (45 mm; Table 6.6) was significantly narrower than those reported for *G. jumae* (50 mm; Table 6.6). *G. stillei*/*G. gracilis* had a predicted average body mass of ~700 kg compared to ~900 kg for *G. jumae*, and predicted height about 91% of that of *G. jumae*. Thus *G. stillei*/*G. gracilis* had a body mass 75% of that of *G. jumae* and more slender limb bones, but was almost as tall as *G. jumae*. So we can say that *G. stillei*/*G. gracilis* and *G. pygmaea* were indeed tall, slender creatures. These estimates confirm the view of Arambourg (1947):

> [G. gracilis] *was comparable in length of the limb bones and neck to G. camelopardalis but with a more lightly constructed skeleton and finer proportions in all parts, with metapodials remarkable for the narrowness of their articular extremities.*

Small fossilized footprints made by a *Giraffa* in the Laetoli ash bed have been found near the spot where the famous footprints of a bipedal hominid and its child were found. When Mary Leakey asked a local tracker what might have made the spoor, the spontaneous answer was "eland" (Figure 6.7). The *Giraffa* that made them must have been about half the size of an adult *G. camelopardalis* and thus the size of a juvenile *G. stillei*/*G. gracilis* or *G. pygmaea*, about 2.5 m tall (Harris, 1987a).

Giraffa camelopardalis

It has not been easy to identify the precise moment when modern giraffes emerged as a distinct species because their morphology and that of their potential ancestors differ too little. Evolution of its ancestor into *G. camelopardalis* was a slow process. Nevertheless, fossil skeletons having features indistinguishable from the skeletons of modern giraffes have been recovered from a fossil deposit, dated to 1.0 to 1.3 Mya, at West Turkana in Kenya (Figure 6.1; Harris, 1991; Robinson, 2011). One million years ago was a moment in geological time that coincided with another period of glaciation and climate change (Lisiecki, 2010) and another burst of speciation among ruminant artiodactyls (Vrba, 1985a, 1985b), and it was in that period that *G. camelopardalis* emerged as a distinct species. It must have been a descendant of one or other of the prior forms, but which one? It

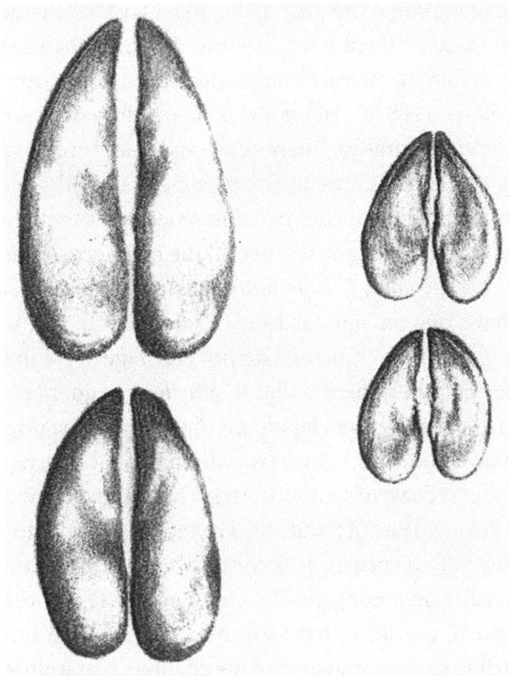

Figure 6.7. The footprints of modern giraffe (left) and those of an eland (right).
Redrawn from Skinner and Smithers, 1990.

is unlikely and probably impossible for either *G. pygmaea* or *G. stillei* to have been its ancestor. They became extinct long before *G. camelopardalis* arrived. *G. pygmaea* became extinct either because of unequal competition for browse with larger browsers or because it was a just a young *G. stillei* and *G. stillei* disappeared because it became *G. gracilis*. The most likely origin of *G. camelopardalis* is, therefore, *G. jumae* and/or *G. gracilis*. It is conceivable that one of them was extirpated by early humans, leaving the other to proliferate and change into modern giraffes. The earliest humans were co-inhabitants with giraffes in East Africa and ate animals (Faith, 2014). Were there enough hominids, and if there were, did they have the tools and skills to hunt large herbivores? One analysis (Lyons et al., 2004) concluded that large African herbivores declined in number and diversity in the early to middle Pleistocene because they were hunted by, and during, the species lifespan of *Homo erectus* (1.8 Mya to 140 kya), but would or could *Homo erectus* have extirpated either *G. jumae* or *G. gracilis* specifically? A better explanation for the apparent extinction of *G. jumae* is that if *G. jumae* were a true biological species, and were sexually isolated from a numerically dominant, more successful, and enlarging *G. gracilis*, they simply succumbed to competition, with *G. gracilis* ultimately reaching the size of *G. camelopardalis*. If *G. jumae* were not sexually isolated, then *G. camelopardalis* could have been a product of interbreeding between *G. gracilis* and *G. jumae* to produce the vigorous, successful hybrid that replaced its parents over time and is now recognized as *G. camelopardalis*. Another possibility is that the mid-Pleistocene turnover event at 1 Mya may have eliminated the larger *G. jumae* because the quantity of browse available fell below what they needed to meet their daily requirements, with

G. gracilis surviving in the same way that the younger and smaller adults and females of modern giraffes are more resilient to environment change than are larger giraffes. Yet another explanation is that the size of *G. jumae* decreased and became identical to that of *G. camelopardalis* (Harris, 1987a). However, *G. jumae* does not seem to have ever been much bigger than modern giraffes. There is a simpler and more attractive possibility. Neither *G. jumae* nor *G. gracilis* have in fact disappeared. Both, in evolutionary terms, are still with us. The single most distinctive difference between them is the presence or absence of a median ossicone. *G. gracilis*, like all the northern variants of *G. camelopardalis* (*G.c. typica, G.c. reticulata, G.c. peralta*), had a well-developed median ossicone; *G. jumae* did not have one, or one far less well-developed, just as the two southern subspecies (*G.c. giraffa* and *G.c. capensis*) do not (see Figure 1.4 in Chapter 1). A plausible origin of modern giraffes, then, is that *G. gracilis*, a resident of East and Northeast Africa, evolved into a variant *G. camelopardalis* that became the origin of northern subspecies. *G. jumae* evolved into a *G. camelopardalis* that inhabited, never left, or migrated southward, down the east coast of southern Africa into what is now Zambia, Zimbabwe, Mozambique, and the northeast of South Africa, and westward into Namibia/Angola to become *G.c. giraffa* and *G.c. capensis*. This conclusion is congruent with the most recent analyses of genetic variation among giraffes (see Chapter 1). However, for one or other or both *G. jumae* and *G. gracilis* to have survived into modern times required special attributes. These attributes are the subject of the chapters that follow.

7
The World of Browsers

A primary daily goal of any animal is to acquire sufficient energy from its environment to maintain its metabolism. Giraffes get their energy from the leaves, shoots, and flowers of trees and shrubs. They are obligate browsers. Why browse? Why not graze grass? No proven reason is known, but perhaps the best explanation is that giraffes have retained a taste for an ancient source of energy. In the Eocene era forests of ~55 Mya that was the habitat of the first artiodactyls, leaves, fruits, pods, flowers, and shoots of trees and shrubs would have been the only but very plentiful sources of food. Grasses did not grow in an Eocene forest, and in any case grasses as we know them would not appear on Earth for another 45 million years. Thus browse was the first and only source of plant energy available to the earliest artiodactyls. Learning how to graze grass was the task of the artiodactyls that have evolved relatively recently (i.e., in the last few millions of years). Modern giraffes, however, were not seduced into changing, have remained committed browsers, and still enjoy browse and browsing, as anyone who has had the privilege and patience to watch wild giraffes go through their daily routine will notice. Feeding for them is a relaxed affair that occupies a good part of their day. A nibble here, a grab there, sometimes reaching to the top of a tree, and at other times nibbling much closer to the ground. It is a leisurely activity with plenty of time to stop and stare (see Plate 7.1).

The Chemistry of Browse

What is in the browse eaten by giraffes? The first analysis of giraffe diets was done by Anne Dagg in 1960. Her hypothesis was straightforward: giraffes show definite food preferences, and those preferences must be related to the chemical composition, or to the shape, of the leaves that they eat. Dagg excluded shape because although giraffes browse mostly during the day, they also browse freely at night when discrimination of shape is less easy. Dagg concluded that giraffes must rely on taste or smell, that is, the chemistry of the browse, to make their selections. However, her analysis of leaves did not produce a clear conclusion. She analyzed the mineral, fat, and protein concentration in the leaves of 20 different browse species that she subdivided into the most-eaten and least-eaten and found no significant differences in their composition. In both types the mineral content was 12.0% ±4.7%, fat concentration 9.2% ±2.7% and crude protein 5.2% ±1.3%. The remainder was the cellulose fiber in the walls of the cells. The protein concentration she found is remarkably low and suggestive of a peculiar seasonal deficit or a measurement error, because a protein concentration of 12% or more in the diet of a herbivore is necessary for proper growth and repair of damaged tissues.

How Giraffes Work. Graham Mitchell, Oxford University Press. © Oxford University Press 2021.
DOI: 10.1093/oso/9780197571194.003.0007

In a similar study done in Kenya, Dougall and colleagues (1964) spent several years following East African herbivores, observing what they ate, and collecting and analyzing samples thereof on the grounds that knowing what herbivores eat and what giraffes in particular eat gives "a reasonably full understanding of herbivorous animals in relation to their environment." Giraffes browsed on 17 different plant species, which Dougall and colleagues divided into legumes (those able to use atmospheric nitrogen to synthesize their proteins) and non-legumes. Nitrogen is essential for the synthesis of proteins, so not surprisingly they found that the protein concentration of the legumes was 50% higher than in non-legumes (18.3% vs. 12.7%), but fiber content (25–29%), fat content (2.5%), and mineral content (10%) were similar in both. Significantly, the ratio of calcium:phosphate in the plants that giraffes chose to eat was 7:1, which is much higher than it is in grass (2:1).

Do giraffes select browse based on the nutrients it contains? How carefully do giraffes select their browse? To answer that sort of question, Leuthold and Leuthold (1972) made direct observations of feeding habits of giraffes in Tsavo National Park, Kenya, and found that, of 66 plant species eaten, the great majority were shrubs and trees. Of those species, just five supplied two-thirds of a giraffe's diet. The most common source of browse throughout the year was leaves of trees and shrubs of the Acacia family, with the "twisted trunk" umbrella thorn—*A. tortilis*—forming 12% of the browse in the green (rainy) season and 29% in the dry season. Plants of the mint family (*Premna* species) formed 35% of the rainy season browse. Overall, the preferred vegetation was flowers, fruits, leaves, especially green leaves, and new shoots of the trees and shrubs. The giraffes ate little or no grass. Pellew (1984b), in his study of giraffes in the Serengeti, noted that the species browsed by giraffes were not the most abundant. The six species most preferred by giraffes formed just 3.6% of the total available browse. Ninety-six percent of all browse eaten came from trees or shrubs, herbs provided 2.5%, vines 1.5%, and the giraffes never ate grass, except accidentally. Seeber et al. (2012), studying giraffes in the Hwange National Park in Zimbabwe, found that they spend just 1.5% of total foraging time (about 15 minutes a day) eating grass and then only in dry seasons when general nutrient availability was low, and when they did they may not have been eating grass at all but, instead, the soil among its roots to obtain micronutrients not available in browse. Other studies have confirmed the general pattern of exploratory sampling of many trees and shrubs and then selection of a few (Oates, 1972; Van Aarde and Skinner, 1975; Hall-Martin, 1974a, 1974b, 1975a; Sauer et al., 1977; Kok and Opperman, 1980; Pellew, 1984a, 1984b; Parker et al., 2003). In an exhaustive study in the Timbavati Nature Reserve that adjoins the Kruger National Park, Hall-Martin (1975a) collected 8,099 plant fragments from the rumens of culled giraffes, identified them, and concluded that over 75% of the fragments came from not more than five species, with four species of Mimosa (*Acacia*, *Dichrostachys*, *Albizia*, and *Peltophorum*, the weeping wattle), *Combretum* (bush-willow), and *Colophospermum* (Mopane) fragments making up 90% of them in the wet season from November to March and in the cool dry season from April to July. In the dry period of September–October, when those trees had lost their leaves, giraffes chewed thin branches, but mostly made up any shortfall from semi-deciduous spike thorn (*Maytenus* sp.) and evergreen Guarri trees (*Euclea* sp.). Monthly use of these species is summarized in Figure 7.1. On average throughout the year, three main species make up 88% of browse: *Acacia*, 38%;

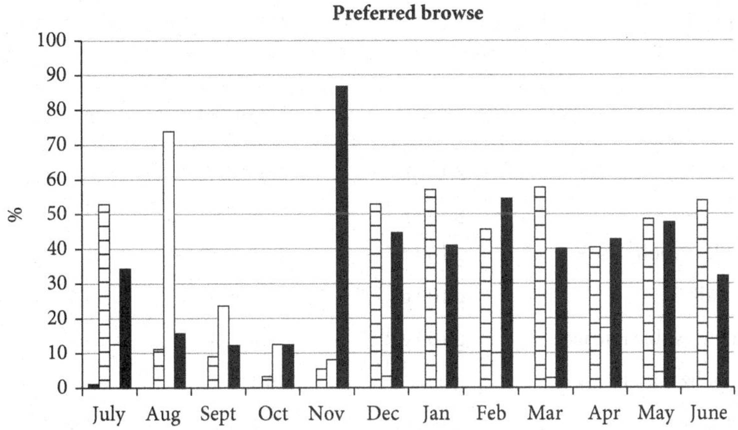

Figure 7.1. Monthly preferred browse in the Timbavati Nature Reserve. ■ = *Acacia* sp. ⊟= *Combretum* sp. □ = *Mopane* sp. In September and October the shortfall is made up of the evergreen shrub *Euclea* sp.
Constructed from data in Hall-Martin, 1975a.

Combretum, 36%; and *Mopane*, 14%. While this is an impressive analysis, it is in fact impossible to give a comprehensive list of all browse eaten by giraffes. Food preferences change from season to season, and region to region, but a common theme emerging from all the studies is that from Niger in West Africa, through Kenya and Tanzania in East Africa, to Namibia and Angola in the west of southern Africa, to the southeast of South Africa, *Acacia* spp. are the most preferred source of browse, and an obvious but possibly incorrect conclusion is that giraffes and acacias have evolved together to mutual benefit.

Giraffes would not choose particular species of plants unless the plants contained all the nutrients they need for growth and reproduction. Many analyses of the chemical composition and nutritive value of the trees and shrubs that constitute "browse" in southern Africa have been made (e.g., Lukhele et al., 2003) but there have been only five comprehensive studies made of the nutrients in the browse giraffes eat (Dagg, 1960; Dougall et al., 1964; Hall-Martin and Basson, 1975; Sauer et al., 1982; Pellew, 1984b). Dagg's original study was inconclusive, and Dougall et al.'s analyses were aimed more at the range of species browsed than their contents. Hall-Martin and Basson's study was done in the Timbavati Nature Reserve adjoining the Kruger National Park in South Africa, Sauer et al.'s mainly in nature reserves in the Western Transvaal of South Africa, and Pellew's in the Serengeti of Tanzania. Sauer et al. analyzed 54 sources of browse collected throughout the year, with eight species forming the preferred diet, and confirmed that *Acacia* spp. browse was sought after most and had more protein than other species, and that giraffes had a preference for new, succulent, protein-rich shoots. Taste was important: use of *Colophospermum mopane* as a food source correlated with protein concentration, but the tree was avoided if it could be, because when its leaves were chewed they released unpleasant aromatic compounds. As the dry season approached and trees lost their leaves, giraffes sought out evergreens, even if the protein concentration was poorer. Succulence, in other words, is attractive, and, it turns out, absolutely vital (see Chapter 11). Pellew collected samples of the 12 species most frequently browsed by

Table 7.1 The Average Concentration (g 100 g^{-1} DM) of Nutrients in the Browse Eaten by Adult Male and Female Giraffes over the Course of a Year in the Serengeti

Nutrient	Units	Mean±sd	kg Day^{-1}*
Protein	g 100 g^{-1}	14.8±2.9	2.6
Fats	g 100 g^{-1}	3.9±0.8	0.7
Soluble carbohydrate	g 100 g^{-1}	37.7±7.3	6.6
Cell wall°	g 100 g^{-1}	36.6±8.0	10.6
Total digestible nutrients**	g 100 g^{-1}	58.7±3.9	10.3
Metabolizable energy	MJ kg^{-1}	8.3±0.1	–
Total minerals	g 100 g^{-1}	7.1±1.7	1.00
Calcium	g 100 g^{-1}	1.3±0.3	0.18
Phosphorus	g 100 g^{-1}	0.18±0.05	0.03
Ca:P		7.7±2.9	

DM = dry matter; total DM intake (kg day^{-1}) = 0.103 × $Mb^{0.74}$.
* For a giraffe weighing ~1,100 kg.
** Mainly protein, fat, and soluble carbohydrates.
°Cellulose, hemicellulose, and lignin.
Constructed from data in Pellew, 1984a, 1984b.

giraffes in the Serengeti in direct proportion to their use in wet and dry seasons, and at a height above four meters, which a mature male might eat, and below three meters for cows. A summary of his analysis is shown in Table 7.1.

The total amount of dry matter eaten each day by the giraffes he studied was 19 kg for adult males and 16 kg for adult females. The energy content of their diet was 8.3 MJ kg^{-1} of dry matter consumed, and the ratio of calcium:phosphate was ~7:1, which is the same as that measured by Dougall and colleagues in their 1964 study. Hall-Martin and Basson analyzed samples every month and thus were able to develop a picture of how nutrient content of browse changed throughout a year. On average the protein concentration over the course of a year was ~18% and thus higher than it was in the Serengeti, with the highest concentration in the "wet" season from November to March, when it was 21%, falling to 17% for the months of April to July, and even further to 13% in August to October, which is the driest part of the year in the Southern Hemisphere. Significantly, the protein concentration was always above 12% and the available energy content was 8.5 MJ kg^{-1} of dry matter, the same as that of Serengeti browse. Fat content was 2%–6% (vs. 4% in Serengeti), cell wall constituents had a concentration of 25% throughout the year (vs. ~37% in Serengeti), and ash ~17% (vs. ~7% in Serengeti). The proportion of ash increased as the proportion of woody plants in the diet increased. These two studies show that whether giraffes live near the equator or below the Tropic of Capricorn, they choose browse that provides similar amounts of energy, protein, and fats, but variable amounts of residue. The perfect browse is succulent, green (high in chlorophyll and also, therefore, high in protein), rich in energy, and full of the minerals and vitamins needed. This is a somewhat trivial summary for such a complex and vital process, dependent as much on taste and palatability as it is on nutrient concentrations. Acquisition of browse by giraffes is a complex process relying on taste, touch, smell, and vision, in much the same way that we rely on these senses to decide if we like a food. We have to conclude that giraffes know far more about what they are eating, what it tastes like, and why they are eating it than can be revealed by the simple questions we can ask and the analyses that we can do. The choices giraffes make translate into excellent blood concentrations of key nutrients. For example, the concentration of essential amino acids in their blood is twice that found in the blood of domestic steers. Blood cholesterol and lipid concentrations in free-living giraffes are much lower than those in the blood of captive giraffes and one-seventh of those found in domestic cows (Schmidt et al., 2007). Not surprisingly, atherosclerosis, the hardening of the arteries common in humans who have a high-fat, high-cholesterol diet, is unknown in wild giraffes (Goetz and Budtz-Olsen, 1955).

The actual eating of browse is facilitated by a long prehensile tongue, mobile lips, short crowned molars with crescent-shaped ridges ideal for crushing leaves, and a narrowed lower jaw and face that allows giraffes to penetrate quite deeply into foliage and between branches (see Plate 7.2).

Moreover, the skin overlying the lips and chin is covered in tactile hairs that are the first hairs to develop in a growing fetus (Ackermann, 1976), and almost all of them, unlike those on other parts of the skin, are supplied by nerves (Dimond and Montagna, 1976). As they brush against leaves or branches, the hairs move and the movement is translated into a useful sensory signal. Usefully too, giraffes can distinguish colors, although the colors they recognize seem to be in the red–orange–yellow part of the spectrum (Backhaus, 1959; Chapter 16) and not in the green part, as might be expected. Perhaps our colors are not their colors. Giraffes use sight more than they do smell in selection of food. Owen, describing the interaction between giraffes and visitors to a London Zoo in, wrote how "many a fair lady has been robbed of an artificial flower which adorned her bonnet," a flower obviously lacking a natural smell. They also have an extraordinary ability to distinguish visual detail, an ability that allows giraffes to guide, with great precision, where to put their tongue and lips. This ability is related not only to the size of the retina and the number of photoreceptors present, both of which are larger in giraffes than in any other ungulate, but also to the types of photoreceptors and the way in which they are distributed in the retina (see Chapter 16). Moreover, as a giraffe grows into adulthood, stereoscopic vision increases. Stereoscopic vision, together with the retina structure, improves their ability to select preferred browse.

The rate at which browse is eaten depends on the amount of browse that can be grasped in one mouthful. Giraffes use their tongues to grasp a branch, grip the stem between their incisors and the hard maxillary pad, with the terminal portion protruding from one side of the mouth (see Plate 7.3).

A sharp tug backward or sideways follows, and the leaves are stripped off by the comb-like incisors (Woolnough and du Toit, 2001), aided, perhaps, by the bi-lobed lower canine (Hamilton, 1978). This maneuver produces a bite size of about 2.5 g (Serengeti; Pellew, 1984a), 2 g (South Africa; Woolnough and du Toit, 2001), or, in Young and Isbell's Kenyan study, 0.5 g of dry matter per bite, the amount varying

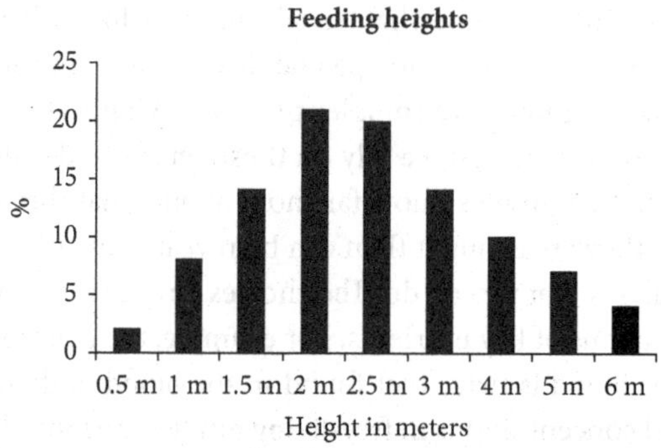

Figure 7.2. Proportion of feeding heights across all food plant species for adult male and female giraffes.
Redrawn from data in Young and Isbell, 1991.

with the species browsed and foliage density. The number of bites per minute varies with browse height. The greatest rate, at least in Young and Isbell's study, was at about 60% of adult height (between 2 m and 3 m for adults, i.e., their preferred feeding height; Figure 7.2), probably because at that height, movement of the head is easiest, stripping of branches is easiest and, therefore, giraffes can feed more quickly. For all giraffes, feeding rates are low near the ground and also low near the upper limit of feeding heights.

Giraffes, therefore, have all the necessary anatomy to be able to select good quality browse, and thus the quality of their diet should not be an issue. But quantity is an issue because giraffes live in places where seasons are well-defined, and the growth of the browse is seasonal. In southern Africa, for example, during the hot wet season (November to March) browse grows vigorously. During the cool dry season (April to July) growth declines and leaf fall sets in. During the dry season (August to October) browse is almost absent until the rains begin again. Hall-Martin and Basson were impressed by the effects of seasonal variation on browse quantity and quality. They noted that it is in the dry months of the year in South Africa, when the absolute amount of browse available has declined, that giraffes lose condition and death rates are higher, partly from starvation and malnutrition, but also because weakened giraffes are less able to flee from a predator. They lose condition because as the quantity of their preferred browse declines, they are forced to eat browse that is less easy to digest and thus is of lower nutritional value (see Chapter 8). That browse quantity is crucial was also the conclusion drawn from a rare opportunity to study the effects of drought on the demography of giraffe deaths in Zimbabwe (Mitchell et al., 2010). A severe drought caused the deaths of juveniles and large males, but left most other giraffes relatively untouched. Juveniles died out of proportion to their numbers because they had to compete for browse with other browsers such as kudu and eland. The biggest animals, usually males, died because their daily requirement for browse is highest and the quantity of browse available to them during the drought was well below the amount needed to maintain their metabolism. The survivors were young adults of both sexes that had lower requirements for browse than those of the biggest animals (see Chapter 18).

Minerals

Analyses of giraffe browse have emphasized its energy content more than its mineral content, yet the mineral content of browse is vital, especially the amount of calcium and phosphorus in it, almost all of which is used to build bones. The chemical composition of giraffe bones is ~33% minerals, 34% collagen, and 33% water (Bredin et al., 2008), and 99% of all calcium and 80% of all phosphorus in the body are in the mineral component of the bones. Giraffes have large skeletons that grow rapidly, reaching their adult height of four to five meters in four to five years (see Chapter 5), but their skeletons continue to amass the calcium and phosphorus needed for bone strength until growth has stopped at about 10–12 years of age. Thereafter minerals are needed in far smaller amounts to replace calcium or phosphate released from bone stores from time to time for use in other functions. The total mass of a giraffe's skeleton relative to its body mass is described by the linear equation in males as $0.075 \times Mb-0.65$ kg and in females by $0.064 \times Mb+3.0$ kg (see Chapter 15). These equations predict that the dry mass of the skeleton in both sexes will be 7%–8% of body mass and wet mass 10%–11%. Thus in a 1,500-kg male giraffe the dry mass of its skeleton will be about 110 kg, and in a 1,000-kg female about 70 kg (Table 7.2). The concentration of calcium and phosphate in bones of giraffes is 20.0 ± 1.4 g 100 g^{-1} bone (20%) and 9.5 ± 1.1 g 100 g^{-1} (9.5%), respectively (i.e., in the typical ratio of 2:1; Mitchell et al., 2005). The total amount of calcium in a dry skeleton thus ranges from about 3 kg in a neonate to 21 kg in an adult male weighing 1,400 kg and to 14 kg in adult females. Phosphorus content is half of that.

To provide sufficient calcium and phosphorus for bone growth, the daily average amount of calcium needed until maturity in both sexes is 6.0 ± 0.4 g and for phosphate it is 3.0 ± 0.2 g, but, as a proportion of skeleton mass, calcium and phosphorus requirements are greatest in the early years of life when bone growth is fastest (Figure 7.3). Until weaning, the main source of calcium and phosphorus is milk, which contains about 1 g L^{-1} of each (see Chapter 17), and thereafter it is browse.

Table 7.2 The Daily Calcium and Phosphorus Requirements of Giraffes

	Male	Female
Adult body mass (kg)	1,500	1,000
Mass of skeleton (kg)	110	75
Calcium in bone (%)	20	20
Phosphate in bone (%)	9.5	9.5
Calcium in an adult skeleton (kg)	23	15
Phosphorus in an adult skeleton kg)	11	7.1
Calcium required (g day^{-1})*	6.1	5.7
Phosphorus required (g day^{-1})*	2.9	2.7

Mass of the skeletons calculated from Figure 15.3.

* Calcium and phosphorus required daily from birth to maturity.

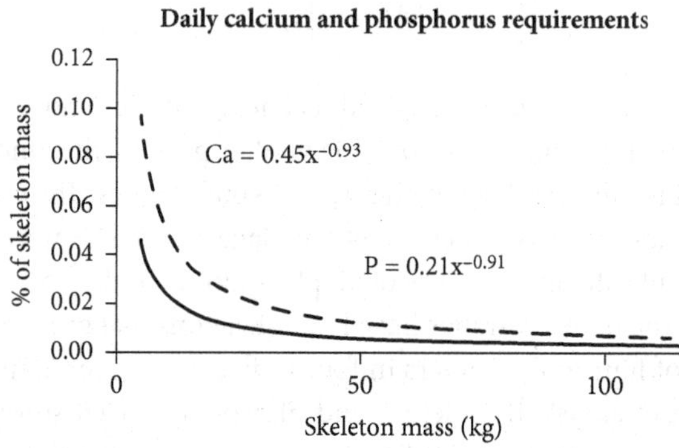

Figure 7.3. Changes in calcium (dashed line) and phosphorus (solid line) requirements as skeleton mass increases.

Data from Mitchell and van Sittert, unpublished.

Table 7.3 Mean Calcium (Ca) and Phosphorus (P) Content ($g\ kg^{-1}$ DM) of Giraffe Browse

Plant type	Ca	PO_4	Ca: P
Grass	4.2	2.3	1.8
All legumes	11.7	2.9	4.0
Giraffe legumes	14.2	2.0	7.1
Giraffe non-legumes	13.4	1.8	7.1

Constructed from data in Dougall et al., 1964, and Pellew, 1984b.

The amount of calcium and phosphorus in browse (and for comparison in grass) is shown in Table 7.3. The browse preferred by giraffes contains ~14 g of calcium and ~2 g of phosphorus per kilogram of dry matter (DM) eaten. An adult giraffe having an average body mass of 750 kg eats 14 kg of DM day^{-1} and thus the amount of calcium eaten each day by an adult male is ~200 g and ~28 g of phosphorus, which far exceed requirements, even in a lactating female that can need 10 g of calcium and 5 g of phosphorus daily (see Chapter 17)and also taking into account that ~1 g calcium and ~1 g phosphorus is excreted in urine each day (Chapter 11, Table 11.5). Additional amounts of each can, of course, be obtained by eating more browse. When Pellew (1984b) studied giraffe feeding habits, he correlated browse selectivity with its calcium or phosphate content. Giraffes did not select browse with high calcium content, but females chose browse with high phosphorus content during the wet season and males did so throughout the year.

Nevertheless, there is considerable anecdotal and formal evidence that a fine balance exists between dietary intake and the requirements for calcium and phosphorus by giraffes. Formal evidence of mineral imbalance is an apparently highly variable blood concentration of calcium and phosphorus, with phosphorus varying more than calcium. The ratio between the two in blood should be 2:1, but in giraffes it is more usually 1:1 or even reversed (Table 7.4). Another indicator is the activity of the enzyme alkaline

Table 7.4 Blood Calcium and Phosphorus Concentrations in Giraffes

N	Calcium Concentration (mmol L^{-1})	Phosphorus Concentration (mmol L^{-1})	Reference
4	3.5	1.3	Sikes, 1969
1	2.4	2.4	Rossof, 1972
4	3.0±0.6	1.6 ±0.3	Drevemo et al., 1974
Not stated	2.1±0.2	2.0±0.3	Rhodes, 1975
14	2.6±0.4	3.2±0.9	Bush et al., 1980
<1 month old; *N* not stated	3.1±0.2		Bush et al., 1980
1 month–1 year; *N* not stated	2.5±0.1		Bush et al., 1980
>1 year old; *N* not stated	2.4±0.4		Bush et al., 1980
4	0.65–2.2		Clauss et al., 1999
39	2.4±.05	3.1±0.26	Schmidt et al., 2007
10*	2.3±0.2	1.5±0.4	Sullivan et al., 2010

* Captive giraffes.

phosphatase. When it is activated, its action is to release calcium and phosphorus from bone with the purpose of replenishing usage. Its activity is 1.7 times higher in giraffes than it is in buffalo (Sikes, 1969) and, unsurprisingly, is much higher in young giraffes, whose skeletons are growing faster than they do in adults (Bush et al., 1980). Furthermore, rickets develops frequently in young captive giraffes and it can be cured by a calcium and phosphate dietary supplement (Langman, 1978); an isolated case of calcium deficiency rickets in a young giraffe has been reported (T-W-Fiennes, 1960) and bone fractures occur occasionally in low-density bones like vertebrae and the pelvis. Another sign of imbalance is the occurrence of kidney stones (urolithiasis). Kidney stones were found in 30% of 75 giraffes studied by Langman (1978). The stones ranged from 1 mm to 8 cm in diameter, and they can cause obstruction to urine outflow (e.g., Wolfe et al., 2000). Obstruction in turn causes kidney degeneration (Langman, 1978; Sullivan et al., 2010) and rupture of the bladder (Wolfe et al., 2000; Maluf, 2002), both of which usually are fatal. The stones consist mainly of magnesium (Mg^{++}) of which 60% is as carbonate ($MgCO_3$) and 40% is combined with phosphate as $MgPO_4$ (apatite) (Langman, 1978; Wolfe et al., 2000). The source of the magnesium is, of course, chlorophyll, but Langman attributed the formation of the kidney stones to a primary calcium:phosphate imbalance, and correcting that imbalance can prevent their occurrence (Wolfe et al., 2000).

Osteophagia

The anecdotal evidence of an imbalance between what is eaten and what is needed by giraffes is the frequent observations of osteophagia, their well-documented habit of chewing bones (Pattern, 1940; Nesbitt-Evans, 1970; Western, 1971; Wyatt, 1971; Hall-Martin and Basson, 1975; Langman, 1978; Kok, 1982; Hampton, 2002). It is mainly female giraffes that show osteophagia because mineral requirements are higher during pregnancy and lactation, and it is not surprising, therefore, to find that the habit is seen most often in cows that have calved in the dry months when the demand for milk by newborn calves is highest and the quantity and quality of browse lowest (see Chapter 17): 90% of osteophagia in southern Africa occurs between April and October (winter). From April to July, minerals are withdrawn from leaves and the abundance of browse starts to decline, events that coincide with the first occurrence of osteophagia (Hall-Martin and Basson, 1975; Langman, 1978). In September and October, there is very little browse left and any rain leaches minerals from the soil, and it is then that a second peak of osteophagia occurs (Langman, 1978). What causes its onset?

In 1891, a young, Swiss, newly qualified veterinarian, Arnold (later Sir Arnold) Theiler emigrated to South Africa. Theiler's life in South Africa started disastrously. Within a few weeks of arriving in South Africa his left hand was cut off by the blades of a chaff-cutter. His future as a veterinarian seemed over. The simplest task of making a blood smear or giving an injection became a nightmare of fumbling frustration. However, his career was not over. He became internationally famous as a veterinary pathologist, research scientist, vaccine developer, establisher of the world-famous Onderstepoort Veterinary School, and much else besides (Gutsche, 1979). The veterinarians he produced investigated the toxicology of poisonous plants and investigated, and developed vaccines against, insect-borne diseases. In doing so, they made livestock farming possible in South Africa. In the 1920s, Theiler made a classic study of the relationship between the phosphorus content of the diet of free-ranging cattle and the development of osteophagia (and its other correlate, botulism) and concluded that "phosphorus shortage is the only really significant factor in the development of osteophagia, and that other factors, if they exist, are quite subsidiary" (Theiler et al., 1924, 1927). Since then, it is presumed that whenever osteophagia occurs in a ruminant, the cause is a dietary phosphorus deficiency (Denton et al., 1986), including in giraffes (Langman, 1978). Its trigger is a blood concentration of phosphorus of less than 1 mmol L^{-1}, and the response is a search for bones, a response that depends on a physiological need and not, in Theiler's words, "a depravity of taste." Smell is the cue that drives the seeking out of bones (Denton et al., 1986). Cattle, for example, will not eat phosphate-rich fertilizer even though it is an excellent source of phosphorus (and if they are dosed with the same fertilizer, osteophagia is relieved rapidly), but they will eat bones.

Theiler established that a concentration of 0.2% (2 g kg^{-1}) phosphorus in fodder was the threshold concentration at which osteophagia occurred in cattle, and that is its concentration in the browse preferred by giraffes (Table 7.3). Phosphorus is one of the most abundant elements in the earth's crust, and so why browse (and grass) does not contain more phosphorus is hard to understand. What seems to be the reason is that phosphorus is highly reactive and forms insoluble complexes (phytates) with almost any other mineral, especially in acidic soils, and the phosphorus in the complexes are not available to plants. The little phosphorus that is taken up by plants is stored in them as phytate and, theoretically, that phosphorus also is not available to animals. Whatever the cause of the unavailability of phosphorus, at a concentration of 0.2% in plants, the blood concentration of phosphorus in cattle falls over time from its normal level of 2.5 mmol L^{-1} to less than the 1 mmol L^{-1} at which osteophagia is triggered. In most of those giraffes in which blood phosphorus concentration has been measured, it is above the critical 1 mmol L^{-1} level (Table 7.4). Nevertheless, if osteophagia starts at that blood concentration in cattle, we can assume it will also do so in giraffes, and even though theoretically their diets contain more than enough phosphorus to meet their physiological needs, a time comes when the minerals available are insufficient to meet their needs. Osteophagia in giraffes starts then, presumably because the proportion of ingested phosphorus that is absorbed is insufficient to maintain blood phosphorus concentrations. Moreover, and paradoxically, the high-calcium diet of giraffes is a disadvantage. Why? Calcium has ubiquitous physiological functions. Apart from being an essential constituent of bones and teeth, it controls nerve excitability, muscle contraction, the clotting of blood, and has innumerable intracellular, biochemical functions. Consequently, over evolutionary time, several mechanisms have evolved to regulate its concentrations in blood tightly, of which two are dominant: its release from bone under the influence of parathyroid hormone and increased absorption of it from the intestine, a process that is regulated by vitamin D. Synthesis of the active form of vitamin D is stimulated by parathyroid hormone. Thus, when calcium in blood is deficient, parathyroid hormone is secreted, vitamin D is activated, calcium is released from bone, increased amounts are absorbed from the diet, and calcium concentration in blood is restored to normal levels. Along with calcium release from bones is release of its associated phosphorus. No similar mechanisms to those that have evolved to regulate calcium levels have evolved to regulate blood phosphorus concentrations, other than mechanisms that reduce its excretion in urine. Phosphorus is filtered in the kidney and would be excreted in urine but, in ruminants, most filtered phosphorus is reabsorbed, the reabsorbed phosphorus is secreted in saliva, and by that route re-enters the intestinal tract for further use. Release of phosphorus from bone for physiological use, in other words, depends on bone turnover, but a diet high in calcium inhibits parathyroid hormone release, suppresses bone turnover, and prevents release of phosphorus from the bones. Under those circumstances, if dietary sources of phosphorus are low, osteophagia will follow. If an increase in browse intake by giraffes in the dry months has been stimulated by low blood phosphorus levels, then a diet high in calcium follows, and the imbalance underlying osteophagia is compounded, to be reversed only by the return of fresh spring browse rich in phosphorus.

Given that giraffe diets can be deficient in phosphorus, the most obvious reason for osteophagia is to acquire it. Living bone is rich in both calcium and phosphorus and in the correct nutritional ratio of 2:1, and contains about 200 g of calcium and 100 g of phosphorus per kilogram of bone. In weathered, dry bone, the ratio of calcium:phosphate is 1.2:1, but nevertheless the bones contain 120 g of calcium and 100 g of phosphorus per kilogram of bone (Langman, 1978). For osteophagia to be an effective source of phosphorus, the bones must be digested in order to release it. In carnivores, digestion of bone and release of calcium and phosphate is achieved by the very low pH (pH=2) in the stomach. In giraffes the pH of the true stomach—the abomasum—is 3.6, which is still sufficiently acidic to dissolve bones, but in giraffes, and ruminants generally, bones cannot reach the abomasum directly in the same way that they can in carnivores and other monogastric animals because in a ruminant the esophagus empties into the rumen, which has a pH of ~6.5. Giraffes could select easy-to-crush spongy bone, such as that found in the cervical vertebrae of animals, and reduce it to particles small enough to reach the abomasum. However, there is no evidence that they do. The many observations made of osteophagia in giraffes show that the bones selected range from fresh to weathered and from dense to highly spongy, sometimes with fatal consequences when bones get stuck between the teeth (e.g., Ferreira, 2003). If particles of bone reach the rumen, can they be digested? That question was investigated by Bredin et al. (2008) by placing a range of bone types in the rumens of sheep and analyzing the mineral content 30 days later. Disappointingly, there was no significant digestion. Presumably there also would not be in giraffes that have the same rumen chemical milieu as do sheep. The consistent source of the phosphorus that giraffes both need and obviously obtain thus remains a mystery, at least to me. It may not be a mystery at all. Giraffes get enough from browse, but, if so, their habit of osteophagia remains an enigma. Perhaps they just like chewing bones.

The Hidden Costs of Browsing: Tannins, Thorns, Ants, and Vigilance

Tannins

The nutrients in preferred browse and those in browse sampled but not eaten are not different. So why do giraffes choose particular sources of browse after sampling many more? One obvious explanation is that browse that is chosen has nothing to do with the concentration of its nutrients, but everything to do with the concentration of its plant secondary metabolites (PSM). Plants can be seen to be passive participants in the "eat or be eaten" world. They exist to supply vegetation, and animals eat them. To a large extent, that contention is true of grasses, but it is not true of browse. In the last few millions of years as atmospheric CO_2 concentrations have declined, browse has become less plentiful than grass and it has evolved methods of protecting itself from being overeaten in sometimes surprising ways. Production of PSMs is one such way. PSMs have no known function in the primary metabolic processes of plants such as photosynthesis, and have evolved to protect plants from parasites and insects, and from being eaten by herbivores: PSMs can be antibiotic, antifungal, and antiviral, as well as unpleasant to the taste (Bourgaud et al., 2001). Some also can absorb UV light, moderating the amount that penetrates to the chloroplasts, and thus preventing leaf damage. PSMs include tannins, terpenes, alkaloids, oxalates, and many metabolites that are used by humans in products such as aspirin, cocaine, nicotine, caffeine, and morphine. (Why they evolved and what role anti-inflammatory, analgesic and stimulant PSMs could have in plants is a mystery.) In rare cases, PSMs are so poisonous that they are instantly fatal if eaten: a few leaves of the South African plant *gifblaar* (*Dichapetalum cymosum*), which is a mostly subterranean, woody plant, can kill a herbivore in seconds. Its active principle is monofluoracetate, which causes irreversible inhibition of the Krebs cycle, the series of biochemical reactions in cells that are essential for the conversion of glucose and fats into energy (and which earned for Hans Krebs, its discoverer, the Nobel Prize in Physiology and Medicine in 1953). Browsers somehow know this biochemistry and strictly avoid *gifblaar* (Cooper et al., 1988).

Some PSMs are metabolites of nitrogen metabolism and others of carbon metabolism. The PSMs that trees and shrubs synthesize in infertile soils of arid regions that also have high temperatures and high light intensity, typical of the regions that giraffes inhabit, are mainly derivatives of carbon (Frutos et al., 2004), although in at least one study (Zinn et al., 2007) a non-carbon-based PSM—hydrogen cyanide—was found to be synthesized by acacias as a response to browsing by giraffes. Carbon is plentiful in arid, hot environments, and thus carbon-based metabolites such as tannins and other phenols are the defense metabolites found more frequently. Tannins are most abundant in the parts of plants that are most valuable to the plant, such as new leaves and flowers, especially so in trees, shrubs, and herbaceous leguminous plants like acacias than in grasses. Tannins, as their name implies, bind to "tan" proteins, in particular those proteins rich in the amino-acid proline. Tannins have a very bitter taste, as anyone who has had the misfortune to taste red wine or, worse, drink tea, will attest. If eaten in sufficient quantity they can inhibit digestive enzymes and depress the digestion of browse (see Chapter 8). In browsers and ruminants generally, inhibition of digestion reduces the amount of energy produced from fodder, and also reduces the emptying of the rumen, which gives the animal a feeling of fullness, which, in turn, reduces the sensation of hunger and reduces feed intake. Tannins also can be toxic. Toxic effects include damage to intestinal, kidney, liver, and nerve cells.

Acacias, the most preferred food of giraffes, are rich in tannins, and it would be logical to assume that browse containing them would be avoided, but their effect on giraffe browsing habits is not straightforward. For example, in early studies of the relationship between tannins and browsing, Sauer and colleagues (1982, 1983a, 1983b) and Cooper et al. (1988) showed that succulent browse will be eaten even if its protein concentration is low and even if tannin concentration is high. On the other hand, giraffe females seem to avoid tannin-rich foods when nursing young (Caister et al., 2003), a response presumably a result of learning that their calves do not like milk rich in tannins. Browsing itself can change the tannin concentration (du Toit et al., 1990): the leaves of heavily browsed trees have much lower concentrations of tannin than unbrowsed leaves, because in new shoots the carbon that could be used to make tannins is diverted to make carbohydrates, and it is those carbohydrates that make the leaves highly palatable. It is not only a combination of high carbohydrates and low tannin that makes browse highly acceptable, but also a combination of high protein and low tannin concentration. A combination of high protein and high tannin is, however, unpalatable (du Toit et al., 1990).

The most complete study of the relationship between tannins and giraffes was done by Furstenburg and van Hoven (1994) in the Kruger National Park in South Africa. The giraffes they observed sampled 32 species of browse of which just five made up 83% of the diet: *Acacia nigrescens*, *A. tortilis*, *A. welwitschii*, *Combretum* sp., and *Dichrostachys*. On average, in all the preferred food sources, tannins formed less than 5% of dried leaf weight. In the other plants tasted by the giraffes, tannin concentration could be as high as 36%. They were avoided unless a decline in preferred browse forced the giraffes to eat the unpalatable species. Within a tree the tannin concentration varied (Figure 7.4).

In areas of direct sunlight where metabolic rate was higher, tannin concentration was also higher. In contrast to du Toit's analyses, the tannin content of new shoots was higher than in old and mature leaves. The highest concentration of tannin that Furstenburg and van Hoven found was in leaves that were 2–3 m above the ground (Figure 7.4), which is the preferred feeding height of giraffes. High tannin concentration at that

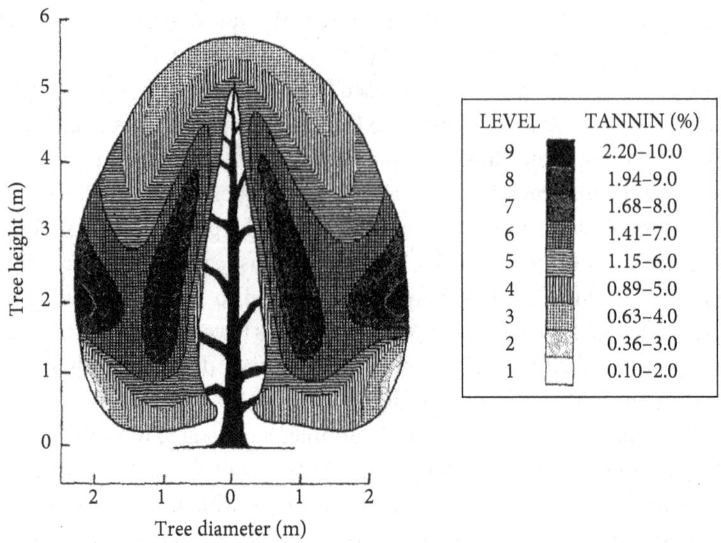

Figure 7.4. Distribution of tannin (% of dried leaf weight) in *Acacia nigrescens*.
From Furstenburg and van Hoven, 1994. With permission of Elsevier.

level is a response to browsing. Tannin concentration in leaves increases within 2 to 10 minutes from the start of browsing because chewing releases stored tannin. After 30 to 100 minutes, synthesis of new tannin by leaves increased the concentration to as much as 120% of pre-disturbed levels. What triggers this biochemical reaction is unknown, and it only occurs in leaves that are being browsed. Other leaves do not increase tannin synthesis, nor do leaves in other trees react by increasing synthesis. There is no communication between leaves or between trees that, as was once thought, would alert other leaves or trees to the presence of a browser and thus stimulate them to increase tannin synthesis, but this myth persists. In the Furstenburg and van Hoven study, tannin concentrations in browsed leaves returned to normal levels 40 to 66 hours after they had been browsed. Thus browse became unpalatable after 10 minutes and took two to three days to return to normal palatability. During that time the tree was avoided. The greater the time spent browsing on a tree, the greater the tannin increase and the longer the recovery time.

Giraffes avoid tannin-rich browse if they can, and do so presumably because of the bitter taste. However, they cannot avoid eating tannin and have had to evolve ways of neutralizing their effects. Detoxification of tannins that are absorbed from the intestines takes place in the liver. Browsers that consume tannin-rich plants are assumed to have larger livers than do grazers (Robbins et al., 1995; Owen-Smith, 1997) because grass contains very little tannin, and this proposition is certainly true for giraffes. They do have livers that are much larger, as a percentage of body mass, than are the livers of grazers (see Chapter 8), and their livers may also have evolved genes that can regulate tannin detoxification better than livers can in okapi (Agaba et al., 2016). The process of detoxification involves the synthesis of water-soluble compounds that are easy to excrete in urine or feces. Tannins could also be neutralized by rumen microbes. Rumen microbes have a legendary capacity to adapt to any diet, and it is well known that microbes able to digest tannin live in the rumen (e.g., Makkar, 2003). Thus it is entirely feasible that if a giraffe ate tannin-rich browse regularly, the organisms able to digest tannin would proliferate and neutralize the tannins. Fermentation of tannin-rich browse would proceed. This idea has been studied by Odenyo et al. (1999) and Jones et al. (2001) in giraffes. In both studies the presence of tannin did suppress digestion, but also in both, artificial removal of tannin increased digestion. The significance of the latter observation is that if the rumens of giraffes had significant numbers of tannin-digesting organisms, all the tannin would have been digested and protein digestion would have been the same before and after addition of the agent that removed tannin. There would have been no tannin left to be removed. Thus for some as yet unexplained reason, organisms that can digest tannin do not proliferate in the rumens of giraffes even though the microorganisms are constantly exposed to tannins in the browse that giraffes eat. As in the case of bone digestion, this finding is unexpected and disappointing. Proliferation of tannin-digesting organisms would be a beautiful and simple way of expanding the range of plants that a giraffe could eat, reduce the overeating of a few, and encourage the use of a wider variety to the benefit of both plants and giraffe. Evolution has chosen otherwise, for reasons not yet clear.

Tannins and other aromatic phenols make the taste of browse unpleasant, at least to humans. If giraffes cannot avoid tannin-containing plants entirely, how do they neutralize the taste? Hofmann's conclusion (in 1989) was that tannin-rich leaves stimulate saliva secretion and that saliva contains more of the amino acid proline. The proline forms complexes with tannin, and the complex neutralizes its taste. A further benefit of proline-rich saliva is that when the proline is bound to tannin, the tannin no longer can bind to dietary protein, and protein digestion by rumen microbes improves. Tannin-protein complexes are stable at pH between 5 and 7 but dissociate in acidic fluid at pH of 2.5–3.5 and in alkaline fluids at pH greater than 7. The pH of giraffe rumen fluid is kept at a remarkably constant 6.5±0.2 units (Maloiy et al., 1982) so proline–tannin complexes are not dissociated in it, can bypass rumen fermentation, and enter the abomasum. In giraffes the pH of the abomasum is 3.6±0.1 (Maloiy et al., 1982), which is not acidic enough to dissociate the complexes. However, after pancreatic secretions rich in bicarbonate have entered the small intestine, the pH of its contents rises rapidly to above 7, which is high enough to dissociate the tannin and free the protein for digestion. The tannin in turn, if it is absorbed, is detoxified in the liver. Giraffes can, therefore, eat tannin-rich browse, but not without consequences. They avoid them if they can, and compensate for any nutrient deficits arising from reduced digestion by eating more browse (see Chapter 8).

In one sense the most successful PSM would make a plant completely unpalatable, but if that were the case, then some of the benefits of being eaten would be lost. Potential benefits include pollination, seed dispersal, and stimulation of the growth of browse. Browsing is not a one-way street. Browse can benefit from browsing. For example, du Toit (1990b) studied the benefits for acacias of browsing by giraffes. Acacia flowers are an attractive food source for giraffes in the late dry season when leaves have been lost, so much so that the flowers of *A. nigrescens* are a staple food of giraffes in the Kruger National Park in South Africa. Only giraffes can reach them. In contrast to the flowers of *A. tortilis*, which are well protected, the abundant flowers of *A. nigrescens*, for a very good reason, are unprotected by thorns and are easily cropped by giraffes in large quantities (see Plate 7.4).

The good reason is that the pollen of *A. nigrescens* is too heavy to be dispersed by wind, but collects on the hair of the muzzles of giraffes while the giraffes are browsing. The giraffes carry it to other flowers. More subtly, evolution has ensured that *A. nigrescens* flower at the end of the dry season, when other sources of browse have all but vanished, thus making sure that giraffes eat them, and thereby maximizing the chance of pollen transfer. The products of fertilization of acacia flowers are seeds that are carried in aromatic and very palatable pods, not unlike those of a pea or of the honey locust tree (*Gleditsia triacanthos*), on which, famously, John the Baptist dined while enduring his 40 days in the desert. The seed coat of *A. nigrescens* is hard and impervious to water, and seeds deposited on the ground directly do not germinate. When the pods are eaten by giraffes, the warmth and water in the digestive tract soften the seed coat, the seeds are released, and later when the seeds are dispersed in droppings they germinate easily. This beautiful story has blemishes, of course. If there is a benefit for *A. nigrescens* to have its flowers eaten by giraffes, then it could be expected that the flowers would contain less PSM, but they contain far more tannin than do its leaves. So why do giraffes eat them? The answer is that the flowers also have more water (64% vs. 40%), more protein (22% vs. 13%) and less lignin (18% vs. 29%) than the leaves (Fleming et al., 2006), and these are highly desirable attributes. Potentially, giraffes could eat sufficient *A. nigrescens* flowers to have a negative, destructive effect on the trees, but obviously they do not: the trees blossom every year and giraffes have a valuable food resource.

Thorns and Ants

The idea that thorns and ants protect plants from being eaten is such a part of common knowledge these days it is hard to think that the link between them had to be discovered. Three-quarters of all browse chosen by giraffes comes from trees that are full of thorns, and this is especially true of the most common source of browse for giraffes—the acacia clade—those archetypical flat-topped trees of the African Bushveld and savannah. The thorns can be several centimeters long and are strong, sharp, formidable weapons (see Plate 7.2). Some acacias naturally produce ant domiciles at the base of thorns, an inflated bulb (see Plate 7.5). The bulbs provide a source of concentrated sugary, fatty, and protein-rich matter that the ants use for food, and the ants can live nowhere else. Plants that have them are called myrmecophytes. The ants of African acacia trees are the very aggressive *Crematogaster* sp. Biologists, including Darwin, having seen the ants and their domiciles, wondered if this phenomenon was another example of coevolution of very long standing with both browse and ants benefiting, or if the ants were just parasites exploiting an opportunity. Do myrmecophytic plants receive the benefit of protection from ants? Do plants evolve ant domiciles or thorns, or both, as defenses?

The first person to connect thorns, ants, and the possibility that they were plant defense systems was William L. Brown (Brown, 1960), a prominent twentieth-century ant biologist. He too observed the "inflated stipular spines" on acacias and their mimosa relatives and saw that they were usually inhabited by ants. Brown traveled widely, and on a visit to Australia he noted that even though the number and variety of Australian acacias are far greater than those in Africa, and Australia has a rich ant fauna, few acacias have thorns or ants. Why? What were the selective forces that induced the absence of thorns that housed ants? One possibility he considered was that Australia was too arid, but he excluded this possibility because not all myrmecophytes are found in arid biomes. He came to the conclusion that the presence or absence of large tree-browsing mammals was the "one influence that fits all of the facts available." Large tree-browsing herbivores are abundant in Africa where both thorns and ants occur. In Australia there are no extant large browsers, and Brown concluded that "the long continued absence of large browsers" was the reason that during their evolution many Australian acacias had lost their thorns, with the advantage that the resources needed to make thorns could be put to other uses. Conversely, he concluded that thorns in myrmecophytic African acacias had evolved because they encouraged the presence of ants, and the ants prevented or deterred browsing. In other words, acacias that had evolved these particular types of thorns survived better than those without them.

Brown's ideas, while logical, were unproven and, indeed, in some instances were based on incorrect information (very large herbivorous marsupials, now extinct, for example *Diprotodon*, would have browsed on Australian acacias). However, some evidence to support his hypothesis was provided by Foster and Dagg (1972), who studied the interaction of giraffes with the trees that were their favored foods. They found that all the trees of the acacia clade they studied were well armored with thorns, either straight, spear-like thorns or thorns curved to form hooks. A typical example of the latter is the *wag-'n-bietje* (wait-a-bit) tree. A human, once entangled in the hooks, finds escape to be very difficult and painful. Foster and Dagg also found that the number and size of thorns were much less at heights above 5 m—the maximum height that a giraffe can reach. Therefore, they concluded, thorns are an evolutionary reaction to browsing by giraffes or equally large other browsers or even extinct large browsers. It is a common-sense conclusion. Thorns function as defense mechanisms through pain and thus limit access to and acquisition of browse. Giraffes, though, seem to be quite impervious to thorns (see Plates 7.2 and 7.3). They tend to eat growing shoots in places where thorns have not yet hardened into spikes or claws. Giraffes also have evolved a pointed muzzle that allows them to bite very selectively, their oral mucosa and tongue are covered by a protective keratin layer, and muzzle hairs act as sense organs alerting giraffes to the presence of objects. They use their tongues to draw branches toward their mouths, grip it between their incisors and the upper jaw, and then strip it of leaves by a lateral twist. It seems likely, from our perspective, that trying to do all this when faced with a barrier of thorns, hard or soft, must be off-putting, and experimental evidence that thorns do defend acacias against browsing by giraffes comes from a study by Milewski et al. (1991) in Kenya. They removed thorns from branches of three acacias preferred by giraffes—*A. xanthophloea*, *A. seyal*, and *A. drepanolobium*—and left other branches on the trees intact. In the three weeks after thorn removal, most browsing (90%) was on thornless branches, and 10% on thorned branches. Regrowth of thornless branches and their leaves was severely retarded, but when they did regrow, the density of thorns was higher, and the thorns were longer and the leaves shorter than before they were browsed. As these changes only occurred in browsed branches, it appears as if they had been induced by browsing (Young, 1987). However, an explanation for long thorns and short leaves is that regrowth of thorns sequesters resources diverting them from leaves to thorns (du Toit et al., 1990). An explanation for greater thorn density is that thorns only grow at nodes, and the intermodal distance in heavily browsed branches is short (Milewski et al., 1991). Whatever the mechanisms, the fact is that trees browsed by giraffes respond by producing longer, more densely packed thorns and smaller leaves, and in some acacias more tannin (Furstenburg and van Hoven, 1994) and other PSMs (Zinn et al., 2007), and the trees become protected from over-browsing to the benefit of both tree and, ultimately, giraffes.

The thorn trees Foster and Dagg (1972) studied were host to *Crematogaster* ants. Foster and Dagg saw that within only one to two minutes after a browsing giraffe has put its head and all its vulnerable tissues—the eyes, nose, mouth, and ears—into a tree, its face and head became covered in ants and soon thereafter the giraffe moved on to another part of the tree or to another tree altogether. The response of the ants to giraffes has been studied in greater detail by Madden and Young (1992). They observed that the ants rushed out of their bulbs in the first 45 seconds after a giraffe put its head into the browse. The ants attacked by lifting their *gaster* (the bulbous posterior portion) upward, secreting a mixture of volatile, pungent, and irritant phenols from it, and followed that up by biting the skin and rubbing the *gaster* into the wound to deposit the phenols. Pain and irritation result, at least in humans, and because giraffes do not like it, we have to assume they feel the same. By nature, *Crematogaster* ants are aggressive, and it could be that their aggressiveness increases if they are disturbed often. However, they are not more aggressive at the heights at which a giraffe might feed, and rely for their effect purely on the number of ants that attack. More ants are found at the level that giraffes feed because, as Dagg and Foster showed, there are more thorns at that level. It would be wrong to think that the actions of the ants are what we would call altruistic, with its purpose the protection of trees. They are not. They are motivated by self-interest. The ants rely on the energy- and protein-rich secretions of the inflated bulb for food. The secretions are the product of the photosynthesis in the surrounding leaves. The evolutionary pressure on ants to protect the photosynthesis producing their food supply resulted in ants better able to do so, so they survived and thrived.

Whether something more subtle is happening has been investigated by Ward and Young (2002). It would make evolutionary sense for there to be a trade-off between the presence of ants, thorns, and tannins because producing thorns or tannins is costly and resources used to make them are diverted away from producing leaves for photosynthesis and flowers for reproduction. Therefore, if a tree is host to ants it need not produce long and plentiful thorns, and if it has long thorns it need not synthesize tannins. Regrettably, no simple relationship seems to exist, partly because the tannin concentration in leaves varies hugely for reasons unrelated to the presence of thorns or ants or browsing. Nevertheless, the study did confirm that leaves rich in tannins (and a concentration of at least 50 g kg^{-1} dry matter of leaf is needed; Frutos et al., 2004) reduced browsing. Leaves containing lower concentrations of tannin are browsed more frequently. Whether all three of ants, thorns, and tannins reduce browsing more than any one or any two of them, and hence that is why all three are produced by a tree even though the costs of producing them or supporting them are high, has, as far as I know, not yet been studied.

The effect of thorns and ants relies on what part of an animal they stab or bite. The lips, tongue, nostrils, and eyelids of a giraffe are the most sensitive part of the mouth and face, and it is these that are presented to thorns and ants. Brown (1960) speculated that browsing mammals would themselves have evolved protecting mechanisms against plant defenses. One protective mechanism giraffes have is the layer of keratin that covers the skin of the lips, mouth, and tongue. The keratin does protect giraffes from the thorns, but the dexterity with which giraffes can manipulate their lips and tongues allows them to avoid eating the hard thorns that are most likely to cause pain. Avoiding ants is not as easy. Protection of eyes generally is through the "blink" reflex. If the cornea of the eye is touched, the upper and lower eyelids close. Brown (1960) suggested that the thickened nictitating membrane, the third eyelid found in all ruminants and located at the medial corner of the eye, provided protection against ants. Unlike the upper and lower eyelids, which close from top to bottom when the eye is touched, the nictitating membrane moves horizontally across the eyeball from the inner, medial, angle near the nose to the outer rim, wiping the eye rather like a windscreen wiper. Ants will trigger both blink and membrane reflexes. What about the nostrils? When Owen made his famous dissection of giraffes in 1838 he wrote,

> *The oblique and narrow apertures of the nostrils, defended by hair which is continued to their margins, and surrounded by cutaneous muscular fibres by which the animal can close them entirely and at will, form a beautiful provision for defence of the air passages, and the irritable membrane lining the olfactory cavity, against the fine particles of sand which the storms of the desert occasionally raise in suffocating clouds, and which man, and the animals compelled through his necessities to become occasional inhabitants of the desert, find so much difficulty in excluding.*

We now know that few giraffes live in deserts and those few that do are not often exposed to sandstorms; perhaps nostril muscles and hairs have evolved not as sand traps, but to prevent ants from creeping up their noses.

Browsing and Vigilance

While they are browsing, giraffes are vulnerable. If they are watching, then they are not eating (see Plate 7.6), and if they are eating, then they are not watching. In biological terms this means that there is a vigilance cost to browsing because giraffes, while feeding, are not looking for predators, or they are looking for predators which reduces feeding time, or, more in the case of bulls than cows, they are not looking for competitors and so miss out on a mating opportunity, or they are looking out for competitors which reduces feeding time. Identification of a receptive female is another major distraction for bulls (see Chapter 17) and further reduces their browsing time. Time management in the life of a giraffe can, in other words, be a challenge.

When are giraffes most vulnerable to predation or a sneaky maneuver by a competitor? Du Toit (1990a) thought that it depended on the height at which a giraffe was feeding. He studied feeding heights and the angle of the neck preferred for feeding by bulls and cows in the Kruger National Park, and found, as did Young and Isbell (1991; Figure 7.2), that throughout the year, with no differences between seasons, the preferred feeding height was 2.7±0.2 m above ground level, which is where the highest proportion of new, protein-rich shoots is found. To place that height in perspective, the next biggest ruminant browser in the Kruger National Park, the greater kudu *Tragelaphus strepsiceros*, has a preferred feeding height of 0.9±0.1 m. When a giraffe that is four to five meters tall feeds at a height of 2.7 m, the angle of its neck relative to the ground must be about 120°–130°. Du Toit (1990a) defined a neck angle as 0° relative to the ground if a giraffe is feeding vertically down as if drinking water, as 90° if the neck is held horizontal to the ground, and 180° if vertically upright. Midway between horizontal and upright, the neck angle is 135°. Bulls spend 84% ±4% of feeding time with the neck angle greater than 90°, while cows spend 49% ±4% at 90°—a significant difference. Why? Du Toit's conclusion was that bulls are much larger than cows (final body mass of a bull is 50% more than final body mass of a cow) and thus the absolute amount of feed they need to ingest is much greater than it is for a female, and females can feed at lower levels with other benefits. Males, however, spend far less time actually feeding than females do because they spend more time searching for cows in estrus or watching other bulls and, therefore, when they do feed they need to maximize feed intake by eating at high levels where high-quality, high-quantity, competition-free browse is to be found. It is then, however, that they are most vulnerable to predation. Lions in the Kruger National Park kill one female giraffe for every 1.8 males that are killed (Pienaar, 1969). More male giraffes are killed partly because they are more solitary, partly because they prefer thickets where visibility and opportunities to escape are reduced, and partly because they are feeding at full stretch. A measurable consequence is that giraffe populations are disproportionately female. The sex ratio is 1 male:1.6 females in the Kruger National Park, and 1:1.7 in the Serengeti, Tanzania (Pellew, 1983b). A contrary opinion was expressed by Young and Isbell (1991), studying giraffes in Kenya. They concluded that feeding at high levels, which happened only in high-ranking male giraffes accompanying females, improved vigilance and ought to reduce vulnerability. This difference of opinion can, perhaps, be explained by differences in understanding of what "high level" means. If a giraffe is feeding at full stretch (a neck angle of 180°) then its eyes are directed upward and it cannot see predators or competitors. However, if "high level" means a neck angle of 135°, then giraffes would detect predators. It is long-distance vision that alerts giraffes to a threat, and giraffes seem able to detect a predator or an intruder at long distances, especially when the neck angle is in its usual position of about 135°.

Commonsense suggests that feeding level or neck angle should be minor contributors to predator avoidance. Living in large herds would be a better strategy. Large herds have "many eyes." Giraffe herds were once very large, consisting of a hundred or more individuals, with the benefits that the chance that any one of its members would detect a predator increases and the likelihood that any one of its members would become a target decreases (Bercovitch and Berry, 2010b). In modern times, most herds consist of fewer than six individuals (e.g., Dagg and Foster, 1976; van der Jeugd and Prins, 2000; Muller et al., 2018). The causes of that change in structure are not understood, but if it ever existed, the "many eyes" benefit has now been lost. Congregation by small groups to form larger groups as a response to a threat would be the way to compensate, but giraffes do not congregate (Muller et al., 2018). Increased vigilance, not aggregation, is the way giraffes respond to a threat. Perhaps "many eyes" was never the benefit of a large herd. The vigilance of individual giraffes has always been their protector.

The vigilance of individuals has been studied giraffes in the central part of the Kruger National Park by Cameron and du Toit (2005). They used the amount of time spent scanning the environment as the measure of vigilance, and found that the total time spent scanning formed a very small part of total activity time, with cows spending more time scanning than bulls, but the difference was small and probably not of any biological significance. However, the angle of the neck was important. When the neck was held at 45° (i.e., below horizontal) when a low bush is being eaten, or at 180° when it is high-level browse that is being selected, then scanning activity increased because at a low neck angle the horizontal visual field is restricted and at full stretch eyes are directed skyward. Nevertheless, the most significant scanning differences related not to predator avoidance but to the structure of the group. Giraffe groups aggregate and disaggregate freely, with visitors/intruders watched carefully, and, if they are unfamiliar males, sometimes subjected to a sparring match or a fight. Cows did not change their vigilance behavior if they were in a group of cows or even if they had calves (a surprising discovery given the high loss of calves to lions in the first year of life), but they were concerned if a bull was encroaching on their space. Scanning by bulls when other bulls were nearby was high, especially if the other bulls were large.

Browsing Ecology

Almost all the studies of the interaction between giraffes and their browse have attempted to understand whether a long neck confers browsing benefits. Long necks in many taxa have evolved to enable access to more food (Wilkinson and Ruxton, 2011), and so it is that the most widely accepted explanation for the long neck of giraffes is that their morphology coevolved with an *Acacia* sp. savannah woodland, giving them access to browse free from competition with smaller browsers. Darwin entrenched this idea in the famous passage from his 1872 edition of *The Origin of Species* mentioned earlier:

> *The giraffe, by its lofty stature, much elongated neck, forelegs, head and tongue, has its whole frame beautifully adapted for browsing on the higher branches of trees. It can thus obtain food beyond the reach of other Ungulata or hoofed animals inhabiting the same country.*

However, it is not only tallness that is important. An often overlooked part of their anatomy (but one not missed by Darwin) is the "elongated head." In a mature male giraffe, its head can be a meter long, and it thus adds a meter to the height that the giraffe can reach. Head length in a mature female head is ~80 cm and improves their reach too. Thus, although total height (from foot to head) is around 5 m in a mature male, a browse line at 5.9 m in big trees can be formed (Scheepers, 1992). However, it is not just the lengths of their necks and heads that are important. For giraffes to be able to feed comfortably at high levels, their heads and necks must also be easy to maneuver. In what other ways is their frame "beautifully adapted"? One way was noticed by Owen (1838). He saw the unique anatomy of the occipital condyles, the bones that make up the atlanto-occipital joint between the back of their skulls and the first, atlas, vertebra. These, Owen wrote,

> *are remarkable for their extent in the vertical direction; it is this structure which enables the giraffe to raise the head into line with the neck, and even to bend it a little way back upon the neck.* (see Plate 3.6)

Subsequent studies have shown several other adaptations. The skull has to be large to provide the support for the muscles of chewing, but it cannot be heavy, so it is enlarged by the presence of sinuses filled with air. Its neck needs to be flexible and not be dependent on muscles that can tire, for its support, so the weight of the neck decreases from base to apex (because the mass of the cervical vertebrae and neck muscles decrease) and the neck and head are suspended with minimal muscular effort by the *ligamentum nuchae* (see Chapter 15). It is these adaptations that make it easy for a giraffe to feed on high-level browse, and support the idea so well expressed by Darwin.

Having the anatomy to reach high browse does not mean that it is used to do so. It must also be worthwhile to do so. Is there more browse available at high levels? By far the best and most exhaustive analysis of the quantity of browse available at different heights was done by Robin Pellew in a study done in the Serengeti (1983b). He found that in the rainy season 45% of the biomass (kg hectare^{-1} of browse available) occurred below a height of 2 m. The other 55% was added by browse above this level. In the dry season 61% of the biomass of browse available occurred above 2 m. The density of browse (g m^{-3}) is greater at high levels, where light intensity is greatest: a cubic meter of the canopy above a height of 2 m produces twice as much browse in the wet season and almost three times as much in the dry season than is available at levels of less than 2 m. Taken together, having a long neck that gives access to browse 2 m above the ground doubles the potential food supply. Pellew concluded that the benefits that flow from being able to eat high-level browse included faster growth rates, shorter calving intervals, a younger age of first conception and, therefore, more calves per female and a shorter intergenerational gap, all of which favor survival of the species. Convincing as these results are for supporting Darwin's hypothesis, other studies of giraffe feeding patterns and behavior have produced more mixed results. Do giraffes prefer to eat high-level browse? In the first study of giraffe feeding habits, Leuthold and Leuthold (1972) found that giraffes in Tsavo National Park, Kenya, did use the upper layers of vegetation when it was available but on average, over a year, about 50% of browsing was below 2 m, that is, at a level that is in direct competition with smaller browsers. Two-thirds of browsing occurred at less than 2 m in the "green" season when browse was abundant, and one-third was at less than 2 m in the dry season. In South Africa, Van Aarde and Skinner (1975) found a stronger bias toward high-level browse: over the course of a year, 17% of eating was at less than 2 m and 83% at higher levels, a proportion similar to that seen by Kok and Opperman (1980; over 90% of browsing above knee level and most at head level) and by Young and Isbell (1991; Figure 7.2) in Kenya.

If giraffes can feed at low levels and often prefer to do so, when and why do they feed at high levels? Do they feed at high levels only when they have been displaced from low-level feed? Do both sexes have the same feeding patterns? Although males and females select the same browse (Bercovitch and Berry 2018), that the two sexes differ in their choice of habitat to obtain it has been recognized for some time: males prefer forests where the vegetation is taller and thicker, while females prefer more open habitats (Foster, 1966; Foster and Dagg, 1972). Why? Young and Isbell (1991) speculated that as males are taller than females, they are better suited to feeding off tall trees. Height differences are small, however: mature females are on average just 40 cm shorter than males (see Chapter 18) and, besides, males as well as females that are not rearing a young calf prefer habitats where vegetation is dense, the quantity greater, rapid acquisition of large amounts of browse is easier, time is freed for other activities, but visibility is poor. Females, especially those with young calves at foot, choose open habitats where browse is less dense, so they have to work harder to find it, and its quality is poorer, but there is better visibility, which means that avoidance of predators is better. Formation of nurseries of young, supervised by a few mothers, does, however, give other mothers the opportunity to feed on better browse. Differences in feeding habits between the sexes could also be linked to body size. If body size is a factor, then adult males should have a larger feed intake than adult females. When this prediction was tested by observing feeding behavior of giraffes in Mikumi National Park in Tanzania, Ginnett and Demment (1997, 1999) confirmed that males and females without young selected heavily wooded habitat where browse is far more plentiful and they can eat a larger quantity, more quickly. However, if both sexes chose the same supply of dense browse, body size could not have been a deciding factor. That study also confirmed that females with calves favored open habitats where food was more thinly spread but the vigilance cost was low. The females in that habitat selected browse from all heights. It is reproductive success through minimizing predation of their young, or, if there is competition for browse, where females feed is determined by whatever it takes to increase the prospects of their calves' survival (Bercovitch and Berry, 2018). For example, even when there are no predators, as in Niger (Le Pendu et al., 2000), females, especially those nursing calves, feed on more dispersed and much lower-quality browse in order to avoid tannin-rich browse (Caister et al., 2003).

Giraffes are not the only users of browse, so how are resources shared between species? Do giraffes feed above the level of other browsers because they can, or because they have to, or because they benefit from doing so? Are the feeding patterns of giraffes a result of competition, or are they merely opportunistic? Darwin had no doubts. "In South Africa," he wrote in 1872, "the competition for browsing on higher branches of the acacias and other trees must be between giraffe and giraffe, and not with other ungulate animals." The possibility that smaller browsers force larger ones to feed at higher levels was first noticed by Innes (1958), who saw tall giraffes feeding at a tree whose lower branches had been eaten. Smaller giraffes that could not reach the higher branches fed at shrubs nearby, an arrangement also noticed by Bercovitch and Berry (2018), who have studied the feeding habits of *G.c. thornicrofti* in Zambia. There, juvenile giraffes fed on bushes and small trees, while adults fed at higher levels. Woolnough and du Toit (2001) studied the possibility that it is competition that determines feeding height by recording the feeding behaviors of four browsers, ranging in size from the very smallest to the very largest—steenbok, impala, kudu, and giraffes—in the South African late dry season when competition for food is highest. They found the expected hierarchy: small browsers deplete the foliage at their level, forcing larger ones to use browse at the next higher level they can reach, and so on. In the case of giraffes, intake of browse per bite increased with height, so not only did giraffes browse at high levels because they had to, but it was more efficient for them to do so. The pressure to feed higher also applied between sexes of giraffes: smaller females removed browse above the level of smaller browsers, forcing males to feed at an even higher level.

While none of these studies proves Darwin's hypothesis that giraffes evolved tallness because of the advantage it gives for reducing competition for browse, they do provide circumstantial, supporting evidence for it. Cameron and du Toit (2007) attempted to test Darwin's hypothesis directly in the Kruger National Park. They fenced off some shrubs and trees so they could not be eaten by any browser, but did not fence other trees and shrubs of the same species and of similar size. They found that in trees and shrubs protected from being eaten there were no differences in leaf biomass or its nutrients at various heights; thus quantity and quality were constant throughout the canopy. In trees and shrubs that could be eaten, there was much less browse available at low levels because small browsers had eaten it. Therefore, they concluded, giraffes have to eat at high levels not only to avoid competition, but also to acquire the quality and quantity of food they need to sustain their large body size and its metabolism. Tallness, in other words, is a naturally selected response to competition from smaller browsers, and provides a critical advantage when lower leaves of trees have been eaten by shorter browsers. It also gives access to leaves at the center of low trees and shrubs that are out of reach to smaller browsers (see Plate 7.7).

Of course, whether access to competition-free browse was the selective pressure at the time long necks had begun evolving cannot be known for sure, simply because we can have no idea of what other browsers were eating or whether they were competing with nascent giraffes. Moreover, Darwin's hypothesis can only apply to adult giraffes. Can a unique advantage for adults be an advantage for the species generally? What of small, young giraffes? Their necks are shorter than their legs, and the height that they can reach is accordingly lower. It is only adult giraffes that have necks longer than their legs. How do young giraffes ever become big giraffes, and if small trees are heavily browsed, how do they ever grow into big trees? There are two answers to this apparent paradox. One is that smaller giraffes need much less food than larger ones do. The second is that by the time a newborn giraffe has been weaned and has to survive entirely on browse, the height it can reach is much more than 2 m; therefore, there is not really a paradox. Yet another answer is that the quantity of low-level browse that giraffes could use is actually much greater than it has been measured to be, is greater than it is at higher levels, and is abundant enough to provide browse for small browsers, small giraffes as well as large giraffes, and all the other browsers wanting it, and is also enough to allow for growth of the browse itself. Du Toit et al. (1990) have shown, for example, that after being browsed, regrowth of *A. nigrescens* shoots is exceptionally rapid and sufficient to compensate for removal and to achieve net annual growth of the tree itself. Pellew (1984a) found something similar: Acacias in the Serengeti have evolved a strategy of growth by spurts, with individual shoots having short periods of extremely rapid growth at intervals of four to six weeks. Moreover, up to 85% of young leaves can be removed without any effect on plant vigor (Pellew, 1983b). A study by Dangerfield and Modukanele in Botswana in 1996, for example, established that a once-only browsing episode produced an over-compensatory response in *A. erubescens*, and they suggested that over-compensation allows a shrub to reach a size quickly that minimizes the time a shrub is vulnerable to fire. Both tree and browser benefit. Nevertheless, the question that remains is, if a height of 2 m is adequate to avoid nutrient competition, why do giraffes grow to heights of 5 m? Dagg and Foster (1976) thought that when giraffes were evolving they were competing for browse with a number of other high-level browsers, including very large ones like Sivatheres. This hypothesis is weak, however, because for many millions of years small giraffes with shorter necks were coeval with Sivatheres, and yet survived. In modern times the herbivores most likely to compete for browse with giraffes are elephants, but there is more competition between elephants for specific browse species than between elephants and other browsers for those species (Makhabu, 2005; O'Kane et al., 2011), and little between elephants and giraffes (Makhabu, 2005), although the degree of competition between them depends on the quantity of browse available (O'Kane et al., 2011). Moreover, elephants do not feed at the same height as do giraffes (Makhabu, 2005; O'Kane et al., 2011).

In many ways, the paradoxical height to which giraffes grow is no different from the paradoxical height to which trees and shrubs grow. Analysis of the maximum height of trees and shrubs throughout the world reveals that "trees" have a mean height of 26 m and "shrubs" one of 2.8 m. Intermediate sizes apparently, and contrary to everyday observation, are relatively very rare (Scheffer et al., 2014). Why? An often given reason is that when C4 grasses became dominant, fire became an annual hazard, and evolution of fire-adapted and fire-resistant savannah trees evolved in a habitat in which the fire-sensitive and much taller trees had been killed. The fire-resistant smaller trees/shrubs did not have to compete with taller trees for access to light, and in particular the flat-topped Acacias so relished by giraffes encourage the growth of palatable grasses beneath them, which grazers consume, reducing the fire hazard while fertilizing the base of the tree with dung and urine (e.g., Midgley et al., 2001). On the other hand, within a forest habitat competition for light is fierce and trees have to grow as tall as they can if they are to expose their canopy to it. But do they have to grow so tall? If all the trees in the forest could somehow agree to grow to a much lower height, the costs of tallness would be reduced and access to light would not be different. Of course they, like giraffes, cannot agree to grow to an intermediate height, which in the case of giraffes, might be 3 m and not the 5 m they can reach commonly. This unnecessary competition is an example of the "prisoner's dilemma," a dilemma that explains why individuals do not cooperate even if it is in their best interests to do so. For trees, the final height they reach is a trade-off between the costs of supporting a long trunk and the amount of light they will gain (Falster and Westoby, 2003). For giraffes, a lower height would reduce the effects of gravity and reduce the quantity of browse required. So why grow taller? In both trees and giraffes, tallness is the consequence of an evolutionary uncertainty of how tall your conspecifics will be.

If only it were all so simple. The median height of the woody "shrubs" that are giraffe browse is 2.8 m, which is also, as numerous studies have shown, the preferred feeding height of adult giraffes (e.g., du Toit, 1990a; Young and Isbell, 1991; Figure 7.2). That feeding height is preferred because it is the height at which, on average, the weight of the head and neck are balanced by the elastic power of their *ligamentum nuchae* and browse is most abundant. No muscle work is needed to support the head and neck and thus it is the height at which the head is most maneuverable, acquisition of browse easiest, and the posture most comfortable. All this being the case, it seems that the height of a mature giraffe, measured as the sum of the lengths of the legs, trunk, and neck, and the height most giraffes are programmed to reach, of 4–5 m, is perfect to produce a browsing height of 2.8 m when the neck is at its preferred angle of carry of ~135°to the ground. Is the similarity of their tallness and that of their browse a lucky coincidence or a consequence of coevolution?

It could also be, of course, that acquisition of browse is not the main advantage that tallness confers. Pincher (1949) thought that droughts severe and frequent enough would lead to recurrent wastage of female and young giraffes, that if a long neck had utility then other ungulates would have evolved one, and therefore the value of a long neck was not to enable browsing but to offset very long forelegs and thus enable a giraffe to drink ground water. Brownlee (1963) also acknowledged the paradox of Darwin's hypothesis: if access to browse unavailable to other browsers was the only or even the main advantage of tallness, then females would tend to be eliminated in droughts, as would the very young. Instead he proposed that their "tall slender shape is an aid in achieving heat loss" and "will serve a similar purpose to young and old, male or female continuously and not merely in times of drought." Both these alternatives are perfectly reasonable and are discussed further in Chapters 14 and 18.

8

The Machinery of Digestion

Much has been discovered about the diets of wild giraffes and their browsing ecology, but how and where the browse they eat is stored, digested, and absorbed are less well known, mainly because it is so difficult to gather the necessary information. What is needed, first, is anatomical study. That could have started with the ancient Romans. They had many opportunities to study giraffe anatomy. Richard Owen (Figure 8.1) lamented their crassness and barbarity:

> *The spectacles and slaughter of the amphitheatre must have afforded ample opportunity of making observations on the form and organization of foreign animals yet it seems that these animals, once killed, were applied to no further use. Everything that such occasions could afford to debase the human mind and heart was extracted from them, but nothing to elevate or improve.* (Owen, 1838)

Anatomical study had to await the arrival of living giraffes in Europe 2,000 years later. The first giraffe to arrive was the giraffe that went to Paris in 1827. It lived to an old age, dying in 1845 without its anatomy being studied. George IV's giraffe died in 1829 and should have been a source of knowledge but was the subject of just two brief anatomical descriptions by Sir Everard Home, the brother-in-law and pupil of John Hunter. Home speculated on the anatomy of the tongue and made a brief dissection and description of the forestomachs. He concluded, albeit inaccurately, that the forestomachs were little different from those found in an ox. In 1836–1837, three of the seven giraffes then living in England died. Owen was given the opportunity to make comprehensive dissections. He did not repeat the Romans' mistakes. His dissections were remarkable, still never bettered, and described the intestinal tract (and other aspects of giraffe anatomy) in great detail.

The unique and defining anatomical feature of ruminants is their three forestomachs: the rumen, reticulum and omasum (Figure 8.2). Soft tissue, like intestine, rarely fossilizes, therefore what the forestomachs looked like in the first ruminants, and thus our understanding of the origin of the forestomachs, is based more on guesswork than data. There is little doubt, however, that the intestinal tract of the original ruminants must have evolved from the tract of the early omnivorous mammals that were their antecedents. The antecedents' digestive tracts had evolved to digest and absorb high-quality food containing concentrated nutrients. The tract would have been very like that of a modern rat—relatively short but very efficient and able to provide the nutrients to support the high metabolic rate of the small creatures that were the early artiodactyls. The tract most likely consisted of the same five basic components that modern omnivorous mammals like we humans (or rats) have: a mouth and all its associated structures, the esophagus that transports food

How Giraffes Work. Graham Mitchell, Oxford University Press. © Oxford University Press 2021.
DOI: 10.1093/oso/9780197571194.003.0008

Figure 8.1. Richard Owen, founder of the Natural History Museum.
With permission of the Natural History Museum.

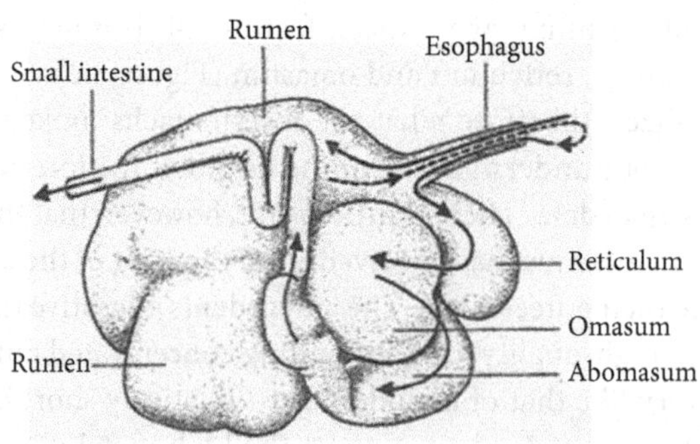

Figure 8.2. The forestomachs of a ruminant, showing the direction of food transit. The dotted line shows the path of a cud.
From Schmidt-Nielsen, 1983, with permission of Cambridge University Press.

from the mouth to a stomach, the stomach itself, the small intestine, and, lastly, the large intestine or colon. The antecedents would not have had any forestomachs and would have depended entirely on their small and large intestines, just as we do, to digest food and then absorb the products. As plant material became a greater source of nutrients, the amount of fiber in the diet increased. Fiber is rich in glucose, but it is in the inaccessible form of cellulose, and mammals do not have nearly enough of the cellulase enzyme needed to convert cellulose into glucose. However, microorganisms like bacteria do have cellulase, and some mammals (and even some birds) evolved pouches in their esophagus in which microorganisms could thrive. It is those esophageal pouches that would eventually become the forestomachs of ruminants.

Owen was very familiar with much of this anatomy, but until his dissection the detailed anatomy of the forestomachs and the rest of the intestines of giraffes was unknown. His dissection was followed by those of Joly and Lavocat in 1846, Crisp (1864a, 1864b, 1866), Murie (1872), Garrod (1877), and Richiardi (1880), all of whom were astonished by the length of giraffe intestines. A quite outstanding study was made by Edwards Crisp (1864a, 1864b, 1866). Crisp was a surgeon by profession but a comparative anatomist by preference who, among his many other duties, found time to investigate the causes of deaths of animals held in the zoos of the Zoological Society of London. Over time he measured total intestinal length in 35 ruminants, including giraffes. Crisp's giraffes included a two-month-old giraffe that weighed 153 kg and an old female giraffe that weighed 815 kg; their intestine length was 38 m and 77 m, respectively. The total intestinal length of an "old," presumably adult, bison, which might have weighed 900 kg, was 47 m and that of an ox 56 m. For more than 100 years, the observations of Owen, Crisp, and others were overlooked. During that interval it had become recognized that ruminants could be separated into browsers (concentrate selectors) and grazers, but the connection between feeding habits and intestinal tract anatomy was not made. Then, in 1989, a German veterinarian, Reinhold Hofmann, published his controversial but seminal summary of the ecophysiological adaptations of wild ruminants. The summary was based on decades of prior research. In the summary he raised the possibility that the forestomachs and the other parts of the intestinal tract were not the same in all ruminants. Hofmann proposed that feeding strategies were a continuum, with archetypical browsers, like giraffes, at one end of the spectrum and archetypical grazers, like cattle and their wild counterparts such as buffalo, at the other (Figure 8.3). The anatomy of their digestive tracts was also a continuum in which Hofmann saw "the beauty and the richness of forms in harmony with highly complicated functions." Left undisturbed, each ruminant species thrived in its environment in perfect balance with the vegetation it consumed. Each species had evolved the structures that optimized the use of its preferred vegetation. In between the two extremes were what Hofmann called intermediate feeders, the diets of which varied seasonally between browse and grass, according to quality and quantity. In the intermediate feeders he saw the plastic genetic pool that could adapt to changes in the ecosystem, whatever their origin.

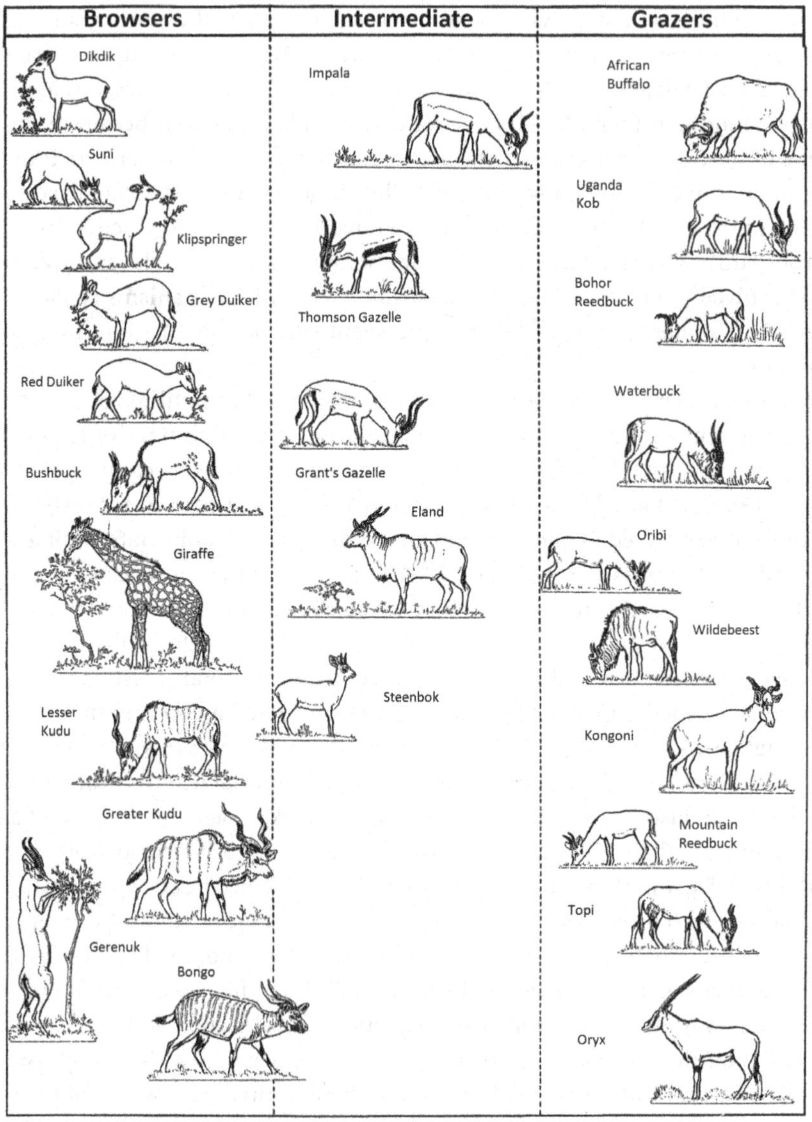

Figure 8.3. African ruminants categorized by feeding type.
From Hofmann, 1989, with permission of Springer Nature.

Giraffe Digestive Anatomy

Mouth and Teeth

Hofmann's classification, with modifications, has stood the test of time (see Clauss et al., 2010), but Hofmann never made as complete a dissection of a giraffe that Owen had, and it is Owen's that is the benchmark. Owen began his dissection at the beginning. He noted that the upper lip of giraffes is not divided as it is in camels, and compared with camels it tapers to a relatively sharp point. This shape is the product of evolution: giraffes use their narrowed, pointed lower jaw, tapered upper lip, and tongue as forceps for collecting desired food items (Solounias and Moelleken, 1993; Plates 7.2 and 7.3 in Chapter 7). Owen saw that the inner surface of the lips was covered in numerous papillae located in two fields, with 15–20 papillae in each of them. Sonntag (1922) counted a total of 50 oral papillae in a giraffe. The papillae point backward and, according to Owen, are better developed in giraffes than they are in other ruminants. The posterior part of the hard palate had 16 transverse ridges that face backward, and its anterior part was also covered in papillae. The ridges and papillae, Owen surmised, were needed to detain food "as giraffe have a small head and jaws compared to the body," and as the cud of "soft slimy comminuted vegetable substances" is expelled into the mouth with great force, the ridges and papillae prevent its escape and direct it to the molar teeth. "The papillae," he wrote, "appear to be of sufficient importance to found an argument of special adaptation or design." The papillae also contain receptors for taste. Owen did not comment on the structure of the teeth of giraffes until 1840, when he did so in a monograph: *Odontography: A Treatise on the Comparative Anatomy of the Teeth*, an awe-inspiring work of 800 pages of text and ~150 plates of illustrations, with many of the illustrations by Owen himself, including the very first depiction of the anatomy of a giraffe's molar tooth (Figure 8.4).

The first complete, modern analysis of the dentition of mature giraffes was made by Singer and Boné in 1960, and that analysis showed that the dental formula of mature giraffes is:

$$I^{0/3}\,C^{0/1}\,P^{3/3}\,M^{3/3} = 32$$

In his analysis of the relationship between age and dentition in 1976, Hall-Martin confirmed Singer and Boné's formula for adults and established that the formula for deciduous teeth in juvenile giraffes is:

$$I^{0/3}\,C^{0/1}\,P^{3/3} = 20$$

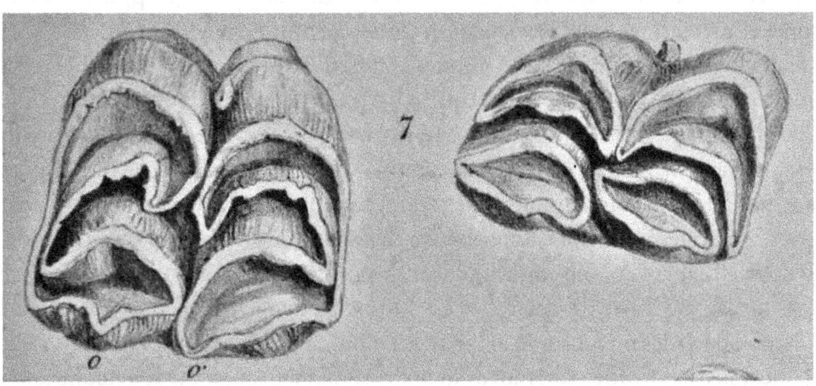

Figure 8.4. The second upper (left) and lower (right) permanent molar, of a giraffe, showing how the crescent-shaped ridges on the molars enclose a hollow space or basin.
From Plate 134 of *Odontography* (Owen, 1840).

The molar teeth of adult giraffes have a relatively low crown (more brachy- than hypsodont), and the molar cusps are crescent shaped (selenodont) (Figure 8.4). All the teeth are covered by a coat of unusually rugose, wrinkled, corrugated enamel, which is harder than smooth enamel. Deciduous and permanent teeth of giraffes erupt in a specific sequence, as they do in humans. Soon after birth, the incisors and premolars that form the deciduous dentition erupt and are replaced by permanent teeth from the age of three years onward. The molars are not part of the deciduous dentition, are the first permanent teeth to erupt, and begin to erupt in the first year of life. By four years of age, all 12 molars have erupted. By five to six years of age all deciduous teeth have been replaced by permanent teeth (Owen, 1840; Singer and Boné, 1960; Hall-Martin, 1976). All permanent teeth are firmly attached to the jaw by roots: the incisors and canines by a single root, all the lower premolars and molars by two roots, and all the upper premolars and molars by three roots, from which we can deduce that the stresses and strains of chewing are borne more by the upper teeth than the lower teeth.

The lower jaw incisors of giraffes slope forward, a slope that makes it easier to bite leaves and fruits off trees and shrubs. In most ruminants the lower incisors are arranged in a transverse line at the extremity of the lower jaw, but in giraffes that straight line has evolved into "a semi-circle" (Owen, 1840), a shape that clearly improves their ability to acquire the browse they prefer. In non-ruminant mammals, like us, a bite takes place between the upper and lower incisors, but giraffes, like all ruminants, have no incisor teeth in the upper jaw. All that is there is a hard pad. How and why have the upper incisors disappeared? Each side of the upper jaw of non-ruminant mammals is made up of a premaxilla and a maxilla, and the upper incisors are anchored on the premaxilla bone. In ruminants, over evolutionary time, the premaxilla gradually became smaller with the consequences of, first, a decrease in size of the upper incisors and then their disappearance. Was loss of upper incisors a benefit? It must have been, otherwise their absence would not have persisted. An explanation for the persistence is that damage to their tongues by upper jaw incisors is avoided (Janis, 2007). Ruminants depend on their tongues to acquire food, and if a large prehensile tongue emerged early in ruminant evolution, then those ruminants that had small or no incisors would have had an advantage and that advantage may have been the selection pressure that ultimately produced a small premaxilla and loss of incisor teeth. Alternatively, loss of the upper incisor teeth for some other reason would have made evolution of a large tongue easier.

The obvious purpose of teeth in giraffes is to reduce browse to a digestible form or, as Owen put it in 1840, "[Their] essential character is the complex structure of teeth produced by an interblending of the dental tissues in the crown, whereby they are made more efficient instruments for the mastication of vegetable substances." Unlike grazers, browsers obtain most of their nutrients from what is inside plant cells, but cells are part of leaves. Microbes in the forestomachs can digest cellulose, but they cannot digest the waxy cuticle that covers the surface of leaves, nor can they digest the woody parts of browse and, in any case, the microbes are most effective when there is a large surface area of food particles on which to act. Furthermore, the food particles must be reduced to a critical size before they can move out of the rumen, through the reticulum, and into the omasum, and eventually reach the abomasum and the intestines beyond. Thus, the cell walls of browse must be crushed, the cells burst, and their contents released. The teeth of giraffes, like those of all ruminants, have evolved in a way that achieves these goals efficiently (Archer and Sanson, 2002; see Figure 8.4).

Browse is crushed by the molar teeth; Figure 8.4 shows how the crescent-shaped ridges on the molars enclose a hollow space or basin. The ridges cut the browse being eaten and the residue fills the space. Compression of the residue when the upper and lower jaws close bursts the plant cells, releasing their contents, and also crushes the cell wall, exposing the cellulose. If the browse is woody, this process is abrasive; herbivores that eat hard, abrasive vegetation, such as modern C4 grasses, have evolved teeth with higher crowns that allow for wear and tear. Giraffes do not have teeth with a crown as high as that of grazers and, overall, their teeth, in their ancestors as much as in modern giraffes, are disproportionately small for such a large animal (Harris, 1991). The browse that modern giraffes select is soft and succulent, and their cud consists of browse that has been partially digested in the rumen and so is even softer, and we have to conclude that the quality of the browse giraffes have eaten over evolutionary time has always been good enough not to have required evolution of large, hard-wearing teeth.

The Muscles of Eating

Feeding, chewing, and rumination depend to a large extent on the tongue, the masseter muscles (from the Greek meaning "one who chews") that lie on the outer surface of the lower jaw (the mandible), and the pterygoid, wing-shaped muscles on the inside of the mandible. Like everyone else who has seen a giraffe's tongue in action, Owen was impressed by the "extraordinary command and power [they have] over their tongue." The tongue is prehensile and very flexible. Everard Home had concluded, not from dissection but from observation of George IV's giraffe, that the tongue could extend 17 inches (43 cm), but could also be fully enclosed within the mouth. Home decided that as "a muscle contracted and relaxed occupies the same volume," the giraffe tongue "does not appear to obey this rule," so there had to be another mechanism for extension and retraction. From what he was seeing, the change in size was "effected by the organ containing a reservoir which could be filled with blood at the will of the animal" with "the increase and decrease in size of the tongue arising from the blood vessels being at one time loaded with blood and at another empty" (Home, 1828). Owen summarized this explanation by saying that "the movements of the Chameleon's tongue have been explained on a similar theory of vascular or erectile action."

Owen did not agree with Home's explanation, and as they were contemporaries and rivals, it is easy to imagine that their disagreement would have occasioned heated argument. Owen noted that the structure of the tongue was no different from that in other ruminants. It was covered, like the entire oral mucosa, by a thin but tough integument. The muscles of the tongue, the lingual muscles, were, however, larger than those in other ruminants, and the styloglossus muscle—the principal retractor of the anterior part of the tongue—was, from its size, stronger than it was in other ruminants. The tongue Owen dissected had a short but high torus (the swollen mass at the back that functions to compress food against the hard palate prior to it being swallowed) and a long free end (Figure 8.5).

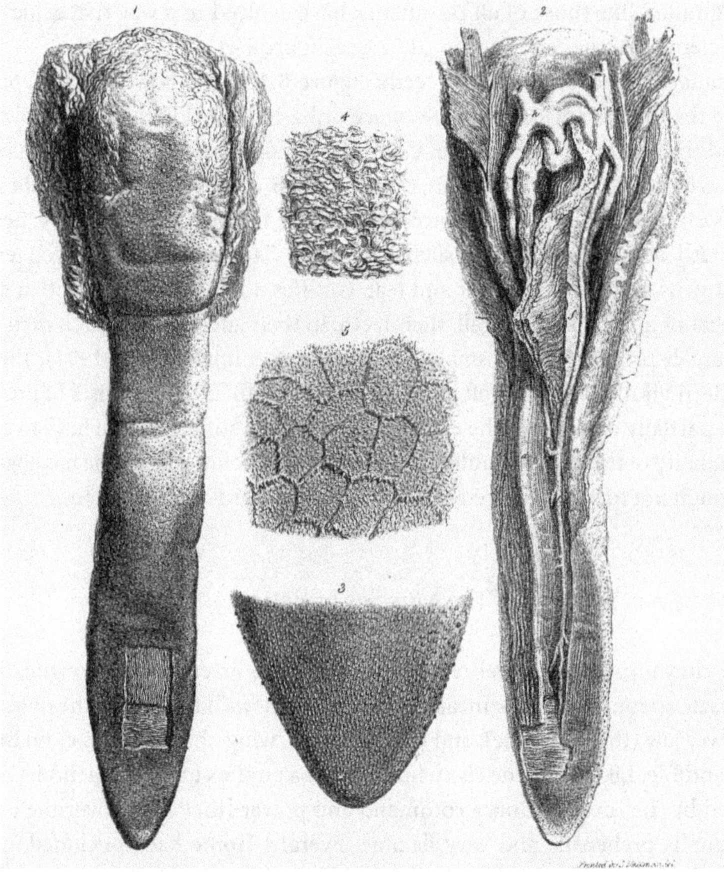

Figure 8.5. The tongue of a giraffe (1) showing the torus, and (2) the blood vessels and the concertinaed lingual nerve. The middle figures show rumen papillae (4), the reticulum cells (5), and the tip of the tongue (3).
From Owen, 1838, with permission of John Wiley and Sons.

The relaxed length of the tongue reaches its maximum of ~70 cm relatively early in a giraffe's life, but its mass, which is mainly muscle, continues to increase. In a newly weaned six-month-old giraffe weighing 180 kg, tongue length is 40 cm and tongue mass about 0.5 kg (Table 8.1). Maximum length is reached in both genders when males and females weigh about 700 kg. From Home's observation, the length of the tongue can almost double when it is being used to collect browse. Its mass in a 700-kg giraffe is only about 1 kg, which is two-thirds of what it is in a mature adult male (Table 8.1).

Owen found no vascular reservoir and decided that the protrusion of the tongue was not driven by blood, as thought by Home, although the tongue was exceptionally well supplied with blood via the two lingual arteries (Figure 8.5). Extension of the tongue, Owen explained, was achieved by contraction of transverse fibers of the lingual muscle, which narrowed the tongue and which, together with relaxation of the retractor muscle, increased its length. Retraction was brought about by contraction of the longitudinal lingual muscle fibers and the styloglossus muscle. Of course, many, many more intricate

Table 8.1 Masseter, Mandible, and Tongue Mass, and Tongue Length in Nine Giraffes

Body Mass (kg)	Gender	Masseter Mass (g)	Masseter Mass (% body mass)	Mandibles Mass (g)	Mandibles Mass (% body mass)	Masseter: Mandible ratio	Tongue Length (cm)	Tongue Mass (g)
181	M	240	0.13	338	0.19	0.71	40	438
227	F	264	0.12	396	0.17	0.67	42	502
602	M	848	0.14	1,254	0.21	0.68	63	1,056
727	F	1,056	0.15	1,274	0.18	0.83	66	1,278
770	M	1,132	0.15	1,634	0.21	0.69	68	1,194
801	F	1,304	0.16	1,406	0.18	0.93	64	1,160
813	M	1,184	0.15	1,420	0.17	0.83	63	1,194
1110	M	1,322	0.12	1,672	0.15	0.79	71	1,424
1396	M	1,642	0.12	1,706	0.12	0.96	71	1,824
736			*0.14±0.02*		*0.18±0.03*	*0.79±0.11*		

From Mitchell et al., 2015, with permission of Elsevier.

movements than those can be produced by the tongue muscles under the control of the lingual nerve. The lingual nerve supplying these muscles was (in Owen's dissection) concertinaed when the muscles were relaxed, which allowed it to elongate when the tongue was protruded (Figure 8.5). The anterior part of the tongue is covered in small papillae containing taste buds. The tip is also pigmented, sensibly Owen thought, but improbably, because of its frequent exposure to the sun, an idea, improbable as it is, that has nevertheless entered giraffe mythology. The tongue's grey color is the same color as the skin of a giraffe and has nothing to do with exposure to the sun. Why are the tongue muscles so well supplied with blood? Surprisingly, given the importance of the tongue in ruminant nutrition, no one seems to have studied the physiology and biochemistry of ruminant tongue muscles, but it is possible to guess. The muscles cannot afford to fatigue and must be able to contract and relax and twist with precision, not unlike the muscles in the fingers of a pianist. Muscles with those characteristics are built of muscle fibers that are fatigue resistant. They depend on oxygen for their metabolic processes, and the oxygen is provided by the excellent blood supply.

The masseter muscles are the most important chewing muscles, and typically they consist of superficial and deep layers of fibers that have their origin on the facial bone and on the zygomatic arch of the skull. The zygomatic arch forms the lower edge of the eye socket (Figure 8.6). The fibers of the muscle are directed obliquely downward and backward to insert on the outer surface of the mandible, and in giraffes they attach to the mandible further back than they do, for example, in a cow, a position that allows giraffes to open their mouths somewhat wider "permitting sideways stripping of twigs" (Hofmann, 1989). Their action is assisted by the pterygoid muscles, which originate from the sphenoid bone of the skull and insert on the medial, inner surface of the mandible. When the masseters and the pterygoids on both sides of the jaw contract together, they draw the mandible forward and upward. If the masseter on one side contracts and its partner pterygoid on that side relaxes and, simultaneously, the opposite pterygoid contracts and its partner masseter relaxes, the lower jaw is pulled toward the side of the contracting muscles. If this cycle of contraction and relaxation occurs rhythmically, first on one side and then the other, the result is the ruminating, grinding movement that is

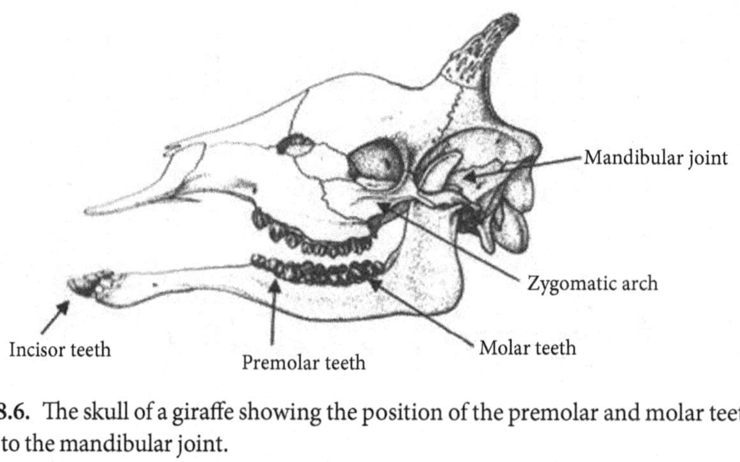

Figure 8.6. The skull of a giraffe showing the position of the premolar and molar teeth relative to the mandibular joint.
Redrawn from Skinner and Smithers, 1990.

so commonly seen in, and characteristic of, ruminants. Of the two muscle groups, the mass of the masseters in giraffes is about twice that of the pterygoids (Furuuchi et al., 2013), which implies that the masseter muscles provide most of the force needed to crush browse.

In giraffes, the typical anatomy of the masseter and pterygoid muscles is followed, except that their masseter has three, not two, layers: superficial, middle, and deep (Sasaki et al., 2001). Sasaki et al. (2001) also described it as a multipennate muscle (i.e., a muscle that is made up of discrete bundles of fibers that slope obliquely like the plumules on a feather, rather than lying parallel to the direction of contraction). Depending on the angle that the bundles of pennate muscles form with the direction of contraction, the muscles can shorten quickly with weak force or contract slowly with a stronger force. The masseters of giraffes follow this latter combination: their masseters have a strong force of contraction and a slow contraction rate, a combination that suits their leisurely approach to browsing.

Are the masseter muscles of giraffes smaller or bigger than those in the ruminants that share their habitat? Axmacher and Hofmann (1988) did not think so. In their study, the mass of masseter muscles in grazers and browsers formed 0.20%±0.04% and 0.21% ±0.05% of body mass, respectively. In giraffes (*G.c. giraffa*) in Zimbabwe, the combined (left and right sides together) masseter mass increases from about 250 g in a six-month-old giraffe to over 1.5 kg in a mature adult, and it forms a constant 0.14% ±0.02% of body mass (Table 8.1), which is lower than the 0.2% Axmacher and Hofmann found in other ruminants. Why? An answer can be derived from the ratio of masseter mass to mandible mass. The mass of the two mandibles making up the lower jaw in the nine giraffes shown in Table 8.1 increases fivefold from 338 g in a young giraffe to 1.7 kg in an adult and on average they form a fairly constant 0.18% ±0.03% of body mass (Table 8.1). The ratio of masseter mass:mandible mass in giraffes is, therefore, 0.79±0.11, compared with 1.3 in ruminant herbivores generally, 2.4 in non-ruminant herbivores, such as a horse, and 3.0 in carnivores (Clauss et al., 2008). The feed of horses is coarser, and carnivores tear and chew flesh and bones, so their masseters are much larger than are the masseters of a ruminant. If the ratio in ruminants typically is 1.3 and that of giraffes is 0.79, and the mandible mass (as a proportion of body mass) is similar in both, the lower ratio of muscle mass:mandible mass in giraffes must be because the masseter muscles of giraffes are smaller. Their size is determined by the force needed to chew browse, and the force needed to chew the succulent, soft, leaves that giraffes like is correspondingly less. In addition, evolution has produced a rearrangement of the positions of the teeth and muscles in giraffes. The molars of giraffes are situated further back in the mouth and close to the mandibular joint (Figure 8.6), so the force needed to crush browse is far less than if the molar teeth were at the other end of the jaw. Consequently, over evolutionary time the size of the masseter muscles could reduce and yet exert the same or even a larger chewing force. It does make evolutionary sense not to have heavy muscles in a head perched on the end of a long lever. Being choosy about what you eat and having teeth placed quite far back so there is no need to have large muscles to grind food may just be part of that whole, subtle evolutionary process needed if a long neck was going to be a success. Of course the converse is also plausible: a long neck allows giraffes to be choosy, and masseter muscles do not have to be large.

Salivary Glands

The main salivary glands are named for their positions in the mouth and neck: the parotid (near the ear), mandibular (behind the mandible) and sublingual (below the tongue). Of these glands the parotid gland is regarded as being the most important, and in giraffes it has two lobes (Perez et al., 2012; Figure 8.7).

The function of salivary glands in mammals is well known. They secrete saliva. Saliva contains enzymes which digest and thus remove food particles that remain in the mouth at the end of a meal. Before toothpaste, toothbrushes, mouthwashes, and chewing gum were invented—all invented more to achieve a sense of cleanness than oral health—saliva is and was the way that evolution has ensured that teeth are kept free of the detritus that encourages the growth of bacteria, and it still is the only way that non-human animals keep their gums and teeth healthy. Saliva also contains mucins that lubricate food, antibodies against bacteria and viruses, bicarbonate ions to neutralize acids, and nerve and epidermal growth factors. The presence of growth factors is the reason why oral tissues that constantly are being damaged during eating are rapidly repaired, and it is why licking of skin wounds promotes their healing.

Humans secrete about 1.5 L of saliva each day. Ruminants can secrete a volume of saliva equivalent to five times their plasma volume each day (Silanikove, 1994), and in a giraffe weighing 750 kg that saliva volume amounts to ~200 L each day (its blood

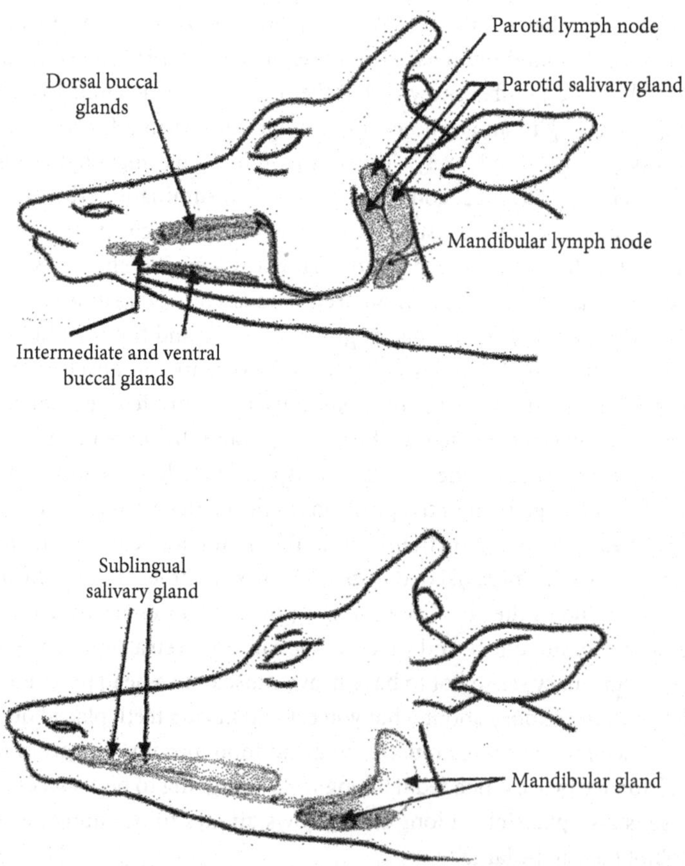

Figure 8.7. The superficial salivary glands of giraffes (above) and deep glands (below).
Redrawn from Perez et al., 2012.

volume is ~60 L and the proportion of blood that is plasma is ~60%). The saliva of giraffes contains all the constituents that human saliva has, and it has similar functions. However, it differs in that it is particularly rich in the bicarbonate and phosphate buffers that neutralize the acids produced by fermentation of food in the rumen, and it contains the proteins that neutralize tannins, and it is also the main way that urea is recycled to the rumen. Urea is rich in nitrogen, which can be used by rumen microbes to synthesize the proteins they need. It is produced by the urea cycle in the liver from the products of protein catabolism in an energy-expensive process. Its recycling for protein synthesis conserves energy and provides a huge amount of inorganic nitrogen for use by microbes. It would be a sin to excrete it, and evolution has ensured that animals, especially ruminants, that were able to recycle much of it via their saliva, thrived.

Hofmann (1989) raised the possibility that salivary glands in browsers and grazers functioned differently. The mass of all the salivary glands in the browsers and grazers he studied formed 0.36% of body mass in browsers and 0.18% in grazers (and 0.26% in intermediate feeders). The differences in mass, he surmised, was related to the volume of saliva produced: browsers had large glands because they needed to secrete large volumes of buffers in order to neutralize the large amount of acid produced by fermentation in their rumens. Do giraffes fit this proposition? There have been three measurements of salivary gland mass in giraffes: by Robbins et al. (1995), Hofmann et al. (2008), and Mitchell et al. (2015). Robbins and colleagues provided no details of salivary gland mass, but their conclusion was that their size depended on the amount of tannin-neutralizing proteins they secreted. Hofmann et al.'s study in 2008 covered 62 species of ruminant, and included data for the East African browsers listed in Table 8.3 and for three giraffes. The giraffes had an average body mass of 515 kg, and their parotid gland weighed an average of 309 g (0.06% of body mass). Mitchell et al.'s data are shown in Table 8.2; in those giraffes, average parotid gland mass

Table 8.2 The combined Mass of Both Left and Right Parotid Glands in Nine Giraffes

Body Mass (kg)	**Gender**	**Parotid Gland Mass (g)**	**Parotid Mass (% body mass)**
181	M	90	0.050
227	F	74	0.033
602	M	162	0.027
727	F	258	0.035
770	M	340	0.044
801	F	288	0.036
813	M	242	0.030
1,110	M	272	0.025
1,396	M	422	0.030
736			*0.034±0.01*

From Mitchell et al., 2015, with permission of Elsevier.

Table 8.3 Parotid, Mandibular, and Buccal Salivary Gland Mass in African Browsers

Species	Grass in Diet (%)	Body Mass (kg)	Parotid Gland Mass (% body mass	Mandibular Gland Mass (% body mass)	Buccal Gland Mass (% body mass)	Sublingual Gland Mass (% body mass)
Suni	0	3.7	0.22	0.16	0.03	0.03
Blue duiker	1	4	0.18	0.12	0.04	
Dik-dik	5	4.4	0.28	0.10	0.03	0.03
Red duiker	0	8.3	0.19	0.13	0.04	
Steenbok	10	10.8	0.16	0.06	0.02	0.01
Common duiker	5	13	0.29	0.16	0.03	0.02
Gerenuk	0	33.5	0.18	0.10*		
Bushbuck	10	45.3	0.13	0.10		
Lesser kudu	10	46.5	0.11	0.10	0.02	
Greater kudu	5	102.3	0.11	0.10	0.02	0.01
Mean	*4.6±4.3*		*0.18±0.06*	*0.11±0.03*	*0.03±0.01*	*0.02±0.01*
Okapi	0	231	0.07			
Giraffe	0	515	0.06	0.07••	0.02••	0.01••
Giraffe•••	0	736	0.034			
Giraffe••••	0	570	0.035	0.024		
Mean	*0*		*0.05±0.02*			

*Mb = 27 kg.
••Mb = 395 kg.
••• From Table 8.2.
•••• From Sauer et al., 2015.
Browsers have less than 10% grass in their diet.
Republished from Mitchell et al., 2015, with permission of Elsevier.

was 0.034% of body mass. Thus the parotid glands of giraffes do not fit Hofmann's prediction of being 0.36% of body mass, nor of being much larger than they are in grazers. They are also much smaller than those of other browsers (Table 8.3), so the data do not support the idea that the size of the salivary glands of browsers is related to saliva volume or, as Rǫbbins et al. suggested, to the amount of tannin in their diets. Hofmann et al.'s 2008 data also included an okapi; its parotid gland weighed 0.07% of body mass. The small parotid gland in okapi and giraffes, he wrote "seems to be complemented by comparatively large mandibular, buccal and sublingual glands," but that conclusion also is not supported by the data. One of the three giraffes Hofmann et al. studied had a body mass of 395 kg, and three salivary glands were weighed. The mass of its mandibular gland was 272 g (0.07% of body mass), its ventral buccal gland 77.1 g (0.02%) and its sublingual gland 38.8 g (0.01%). These relative masses are not different from those found in other African browsers (Table 8.3).

So what determines the size of salivary glands of browsers generally and giraffes specifically? The data in Table 8.3 show that the mass of the glands, as a proportion of body mass, tends to decline as the body mass of the various species increases. Parotid gland mass, for example, is described allometrically as PG mass (g) = $0.36 \times Mb^{-0.28}$ (R^2=0.82). The size of that exponent strongly supports the idea that the mass of the glands is correlated with mass-specific metabolic rate (kj kg^{-1} body mass min^{-1}), which declines proportionately with body mass ($Mb^{-0.25}$). Because their metabolic rate is higher, small ruminants need to eat relatively large amounts of food to sustain it, and have to secrete larger amounts of saliva to manage the food. The absolute mass of the parotid gland in individuals of a specific ruminant species that have different body masses has been measured only in giraffes (Mitchell et al., 2015; Table 8.2), and the glands' absolute mass does not decline as body mass increases; it increases linearly with increases in body mass (PG mass (g) = $0.27x + 40$; R^2= 0.83). Thus, it seems that the size of the glands in different species is related to their metabolic rate, but within a species, size is simply related to body mass and the quantity of browse eaten: small giraffes have small glands and eat less browse daily than do large giraffes, and large giraffes have proportionately larger salivary glands.

Esophagus

Much has been written about the apparently anomalous length of the recurrent laryngeal nerve in giraffes (see Chapter 12) and its corollary that no designer would have made it so, but another anomalous relationship with a potentially more serious outcome is that between the esophagus and trachea at the back of the throat. It would be sensible if the trachea lay above the esophagus and was aligned with the nasal passages, as that is the direct route of air on its way to the lungs, and if the esophagus was below it and aligned with the back of the tongue as that is the direct route of food on its way to the rumen. However, the arrangement that mammals have inherited from past evolutionary experiments is the opposite. The esophagus lies above the trachea in line with airflow, and the trachea below it is in line with the mouth. That anatomy predisposes mammals to choking on food. Swallowed food is directed toward the opening of the trachea, and can enter the trachea, sometimes with disastrous consequences. No designer would have made it this way. Evolution, for mysterious reasons, has.

The esophagus, as might be expected, is very long in giraffes, and much longer than the trachea. The trachea ends in the thorax, the esophagus ends in the abdomen. The trachea in situ in an adult giraffe can be as much as 2 m long , and it can be expected then that esophagus length will be some 50 cm or more longer. The few measurements of its length confirm that idea. In the young giraffe Joly and Lavocat dissected in 1846, they found it to be 1.4 m long, and Crisp (1866) found it to be 1.85 m long in a 12-year-old female. In three giraffes that Murie dissected in 1872, the esophagus was 1.5

m long in a two-month-old giraffe, 1.85 m long in a female 11 years and 7 months old, and 2.5 m long in a giraffe that was 21 years and three months old when it died. Owen did not measure esophagus length, but found its diameter to be a uniform one and three-quarters of an inch (4–5 cm) throughout its length. This measurement was made in a three- to four-year-old giraffe and *postmortem*. The diameter is variable and changes according to the volume of browse being swallowed or the volume of cud being regurgitated. The wall of the esophagus consists of two layers of muscle. The external layer is the thickest, and its fibers lie at right angles to the length and so they circle the esophagus. The fibers of the internal layer lie more obliquely but "approach the longitudinal disposition," in Owen's words. This arrangement of fibers is essential to bring about peristaltic waves of contraction. The transverse fibers squeeze and the longitudinal fibers shorten the esophagus, a combination that propels a food bolus downward toward the rumen or a cud upward for rumination. The mechanism is highly efficient: the speed at which a cud goes up and down the esophagus, an event that can easily be seen in a giraffe, is astonishing. Owen, and later Richiardi (1880), noted that the esophageal muscles resembled skeletal muscle, having a deep red color—a color that we now know indicates the presence in muscle cells of myoglobin and mitochondria—to supply the oxygen-dependent energy needed for "the increased number and force of the contractions they have daily to perform." Peristalsis is a coordinated series of contractions that are regulated by autonomic nerves (see Chapter 16), which respond to mechanical and chemical signals coming from sensory nerves. The sensory nerves lie between the two muscle layers, where they detect stretch caused by a food bolus and the chemicals in it. In humans, most sensory nerves are found in the upper two-thirds of the esophagus. In ruminants, the nerves are distributed along the entire esophagus, an arrangement necessary to regulate swallowing and its reverse of regurgitation of a cud for rumination.

The Forestomachs

All ruminants have three forestomachs—the rumen, reticulum, and omasum (Figure 8.2). All three evolved from esophageal pouches. The first and by far the largest of the forestomachs is the rumen. It is this organ that gives the name to the whole tribe of animals that possess one. The word "rumen" is surprisingly old and probably originated some 8,000 or more years ago from the Sanskrit word *romantha*, meaning "throat" or "gullet" or "paunch," and so can be traced back to the earliest times of the domestication of wild ruminants and the investigation of their internal workings by the herders of Mesopotamia. The entire surface of the rumen of a giraffe is covered by the papillae shown in Figure 8.5, a distribution that is characteristic of browsers: as the proportion of grass in a ruminant's diet increases, the number and surface area covered by papillae decreases (Clauss et al., 2010). Papillae increase the surface area for absorption of nutrients in particular volatile fatty acids (VFAs) and water. The second forestomach is the reticulum. The rumen and reticulum are often described as two distinct organs but, in reality, the reticulum is just a ventral pouch of the rumen (Figure 8.2). They share the functions of fermentation of browse into particles and sorting of the particles into those requiring further fermentation and those that can be moved on to the stomach and intestines for further processing. Both rumen and reticulum have muscular walls and can contract strongly. The leaves and shoots that form the diet of giraffes have been thoroughly chewed before they enter the rumen, mixing is easy, and, consequently, the rumen muscles of giraffes are half as thick as they are in a grazer like a buffalo (Hofmann, 1989): unlike browse, grass forms a thick mat in the rumen that is difficult to move, and moving it requires bigger muscles (Clauss et al., 2003b). The inner surface of the reticulum is covered by a network of hexagonal cells about 5 cm in diameter, and their walls are 1–3 mm high (Clauss et al., 2006b; Sauer et al., 2016). It is these cells that give the reticulum its name (*reticulum* = network; Figure 8.5). The height of the cell walls tends to increase along with increases in a giraffe's body mass (Clauss et al., 2006b). A small outflow channel or orifice connects the reticulum to the third forestomach, the omasum, sometimes called the psalterium because its internal surface is highly folded to form leaves (laminae) that resemble the pages of a book (Figure 8.8). In 1877 Garrod excellently described the omasum as a

> *subglobose dilatation of the upper wall of the canal leading from the reticulum to the abomasum, the lumen of which is much reduced by the development from all but its lower wall of longitudinal, villous-covered folds of its mucous lining. Of the laminae there are a certain number (frequently ten) of a definitely greater depth than any of their neighbours; and these may be termed primary laminae. Between each two of these there are secondary laminae of smaller size. When the organ is more complex a still shallower tertiary lamina appears on each side of each secondary one; and there may be quaternary laminae on each side of each tertiary.*

The omasum empties its contents into the abomasum, which is identical to the single stomach that is found in humans and other non-ruminant mammals.

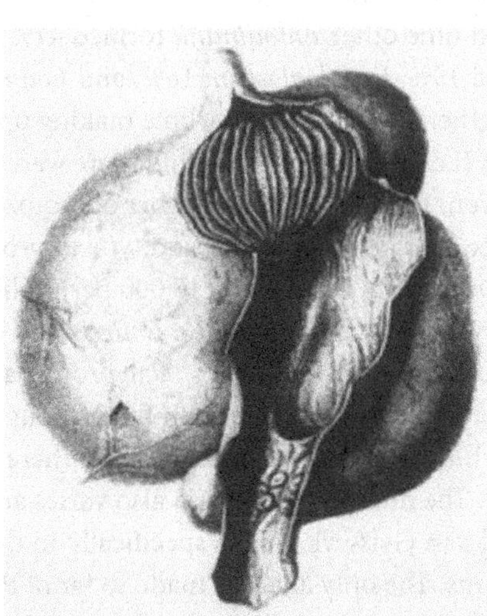

Figure 8.8. The lamellae of the omasum of King George IV's giraffe.
From Home, 1830, with permission of John Wiley and Sons.

What do the forestomachs do? The rumen is home to many trillions of bacteria, fungi, and protozoa, and they make fats and proteins as by-products of their metabolism and, living in happy symbiosis with their host, these by-products can be used by their host as nutrients. Animals that have a rumen can harvest large amounts of plant material quickly, store it, retire to digest it in security and, in doing so, reduce the risk of predation. Later, through regurgitation and rumination, they can chew it at leisure. The rumen is also a water reservoir, thus allowing foraging long distances away from water. All these are very valuable assets, and just how valuable can be gauged from the success of ruminants over time. They have become one of the most successful and diverse groups of mammals and, consequently, the evolution of a rumen can be regarded as being one of the great mutations of all time, and not just for ruminants. Without the evolution of ruminants, human civilization would not have evolved. A very large proportion of human nutrients and other products are still, as they were historically, derived from ruminants.

The rumen is, basically, a fermentation tank filled by a soup of saliva, salts, fats, proteins, carbohydrates, water, gases (mainly carbon dioxide and methane), and microorganisms, many of which in the rumen of giraffes have not yet been identified or classified (Roggenbuck et al., 2014). Rumen pH in giraffes is 6.5±0.2 and thus on the slightly acidic side of neutral (pH=7). The reactions that underpin fermentation produce heat: the temperature of the soup in giraffes is 39°–40°C, which, apart from keeping the body of its possessor warm, makes the broth a warm, nutritious medium for growing microbes and, when it is filtered to remove solid material, the soup is not unpleasant to human taste and just as nutritious for humans as it is for ruminants. The main rumen microbes are protozoans and bacteria. The rate at which the fluid soup moves through the forestomachs is much faster in browsers than in grazers, so it was once thought that protozoa, which have a longer life cycle than bacteria, could not multiply fast enough to stay in a giraffe's rumen long enough to be able to participate in fermentation. However, Kleynhans and van Hoven (1976) found 26 different species of protozoa in the rumen of wild giraffes (*G.c. giraffa*) including two new species, *Entodinium longicorpum and E. bispinosum*. These and nine other *Entodinium* formed 41% of total protozoa population. *Holotrichs* formed 19%, *Enoploplastron* 16%, and *Eodinium* 11%, with the balance made up by three other species and the whole making up a rich protozoal fauna. In each cubic centimeter (i.e., 1 mL) of rumen fluid there were about 340,000 protozoa (Kleynhans and van Hoven, 1976). A similar number of protozoa species occurs in *G.c. peralta* (Noirot-Timothée, 1963). The number and type of protozoa present vary with the diet. Dehority and Odenyo (2003) found 110,000 per milliliter of rumen fluid and 54,000 per milliliter in two East African giraffes. *Entodinium* was again the dominant protozoan present, accounting for 97% and 61% of all protozoa, respectively, in the two giraffes. It is bacteria that predominate in rumen fluid, though, and there are at least 1,000 times more in each milliliter of rumen fluid than there are protozoa (Giesecke and van Gylswyk, 1975). The number of bacteria also varies according to diet and, according to Giesecke and van Gylswyk, varies specifically to the availability of soluble carbohydrates and proteins. The only analysis made so far of the types of bacteria present in a giraffe's rumen has been made by Roggenbuck and colleagues in 2014. They found that there were large numbers of *Firmicutes* and *Bacteroidetes*—the main fiber-degrading bacteria—and of *Prevotella*, *Succinclasticium*, and *Methanobrevibacter.*

However, 21% of the bacteria could not be assigned to any known bacterial phyla, and 70% not to a known genus.

Contractions of the rumen and reticulum occur regularly at intervals of about 30 s, and their purpose is to mix the contents, cause regurgitation of the cud, expel gases, and move digested food out of the rumen and reticulum into the rest of the intestinal tract. All the while, the microorganisms are fermenting plant material, synthesizing proteins and vitamins, releasing minerals, sugars, fats, and proteins from the plant cells, producing methane and carbon dioxide, and turning the glucose of cellulose into VFAs. The VFAs are absorbed directly into the blood through the papillae in the rumen wall (Figure 8.5) and, beautifully, the size of the rumen papillae increases and decreases in accordance with the amount of fatty acids being made. The microbes are constantly multiplying and dying. The dead ones are digested and are themselves a major source of nutrients, especially protein. As fermentation proceeds, the size of the food particles in the rumen decreases, their density increases, and the smallest particles become trapped within the hexagonal cells of the reticulum. Larger, less dense particles remain suspended in fluid. When the reticulum contracts, the large, less dense particles regurgitate into the rumen for further fermentation, and the small, denser particles are emptied out of the cells and are washed into the omasum.

Resolving the Problems of Infection and Dilution

The fluid that leaves the rumen and reticulum is watery and full of microbes. The potential consequences are, first, that the nutrients that reach the intestines, and the chemicals and enzymes needed for digestion in the intestines, are diluted and, second, that microorganisms enter the intestines. The evolution of the third forestomach, the omasum, resolved the problem of dilution. The omasum's main function is to reabsorb water, and the number of laminae the omasum contains and the surface area they provide are proportional to the amount of water coming from the reticulum. Garrod (1877) found that the omasum of giraffes have 18 laminae and, that being the case, they are mainly primary and secondary in structure. Sauer et al. (2016), in contrast, reported 57 laminae in the omasum of giraffes, and it is difficult to reconcile these two very different numbers. The diameter of the omasum is ~17 cm (Sauer et al., 2016) and, therefore, its volume (assuming that it is a sphere) is ~2.5 L in a giraffe weighing 550 kg, and in an adult the volume is about 7 L. The amount of digesta it contains is about one-tenth of the amount in the rumen/reticulum (Clauss et al., 2006a). The surface area that the omasal laminae provide has been estimated to be ~5,000 cm^2 in a giraffe weighing 550 kg and, as it increases linearly with body mass, in a giraffe weighing 1,200 kg its surface area is ~9000 cm^2 (Sauer et al., 2016). The size of the omasum in giraffes is, however, small compared to its size in grazers. The evolutionary reason for the difference is that ingesta moving out of the rumen–reticulum of giraffes contain far less water than they do in grazers.

Infection of the intestinal tract by bacteria and protozoa is to a large extent prevented in all mammals by hydrochloric acid secreted by the abomasum, or true stomach. In mature carnivorous mammals like humans, the pH of abomasum secretions is between 1.5 and 2 pH units, which is not far off the pH of pure hydrochloric acid, and that pH sterilizes food. In mature ruminants, fluid entering the abomasum from the rumen has a pH of 6.5, contains great numbers of living microbes, and unless the pH is reduced the risk of developing an intestinal infection is high. Hofmann (1989) thought that the potential for infection is increased in browsers because he was convinced that large volumes of saliva, which also has a pH of 6.5, could bypass the rumen and enter the abomasum directly. The structure that has evolved to allow fluid to bypass the rumen is a groove in the rumen wall that lies at the point where the esophagus opens into the rumen and connects the esophagus to the abomasum. The groove is quite short in okapis (2–5 cm) and giraffes (8 cm) compared to its length in cattle (20 cm; Burne, 1939). The edges of the groove are formed of muscle. When the muscles contract, the two edges of the groove close to form a watertight tube that connects the esophagus to the abomasum. Closing of the groove is its usual response when a calf drinks milk. The presence of milk in the esophagus, the negative pressure generated by sucking, and extension and tilting of the head and neck of the calf upward toward its mother's udder all stimulate closure. Evolution of the groove has occurred because if milk enters the rumen, it ferments, prodigious volumes of gas are produced, and potentially fatal bloat results. Thus, bypassing the developing rumen is essential for the health of young ruminants. Hofmann (1989) proposed that in adult giraffes the browsing posture was similar to the drinking posture, and voluminous saliva filled with soluble carbohydrates, fats, and proteins released from browse had the same effect as milk. The advantages conferred if that mixture did enter the abomasum directly are prevention of the loss of food energy as heat during fermentation (~8% of food energy is lost as heat; Czerkawski, 1980) or in the production of methane (~9% of food energy; Crutzen et al., 1986). Does the groove form in adult giraffes? The general conclusion from a study done in North American captive browsers (muskoxen, reindeer, and moose) fed an artificial diet was that the groove probably does not form in adult browsers (Lechner et al., 2009), but an active groove must have existed in Zarafa, the giraffe that arrived in France in 1827. She was fed 20–25 L of milk a day throughout her life, and that milk could not have entered the rumen. The groove must also have functioned in George IV's giraffe, as it arrived in England accompanied by two cows to provide milk. Apart from these two giraffes, there is no direct evidence that the groove-reflex works in adult giraffes, but there is some circumstantial evidence that it does. If the fluid entering the abomasum comes not only from the omasum but also from saliva washed directly into the abomasum via the esophageal groove, then large volumes of fluid with a high pH would enter the abomasum. Reducing the pH of that large volume of fluid sufficiently to kill microbes would require large amounts of hydrochloric acid. The source of hydrochloric acid is acid-secreting cells in the mucosa of the abomasum, and the mucosa of giraffes is twice as thick as it is in grazers and 50% thicker than it is in intermediate feeders (Hofmann, 1989) with the thickness due to the presence of acid-secreting cells.

The Intestines

In all mammals the intestines have the same functions and the same parts—a small intestine so named because its diameter is narrow (in giraffes, according to Owen, it is "pretty uniform in size measuring in circumference four inches"; that is, it had a diameter of just 3 cm in those giraffes) consisting of a duodenum, jejunum, and ileum, and a large intestine (which, according to Joly and Lavocat, had an average diameter of 9 cm in their giraffe) consisting of a cecum, and a proximal and distal colon. The basic functions of the parts are similar in all mammals, and giraffes are no exception to the general rule. The intestines start at the abomasum. The abomasum forms a continuous tube with the first part of the small intestine, the duodenum, which is short. At the junction of the abomasum and duodenum, a ring of muscular tissue forms the pyloric valve. It controls the rate at which abomasal contents leave the abomasum: in a beautiful, self-regulating reflex, the contents of the abomasum are released into the duodenum only when what is in the intestine has been digested and the products absorbed. The length of the duodenum in giraffes has not yet been measured, but in humans it is 12 finger breadths (*duodenum digitorum*) long. The pH of its contents is the (acid) pH of abomasal fluid. Its low pH is essential. The acid reduces minerals like calcium, magnesium, and phosphorus in the diet to their divalent, ionic form (the form in which they have to be if they are to be absorbed). The duodenum ends at the point where the common bile duct enters the small intestine. At its opening, the duct has a muscular valve, famously required to be remembered by physiology and anatomy students as the sphincter of Oddi and Boyden. (The valve was first identified in 1887 by Ruggero Oddi, an Italian anatomist, and its functions were elaborated in the mid-twentieth century by Edward Boyden, an American anatomist.) The common bile duct carries the digestive secretions of both the pancreas and the liver into the small intestine. Apart from digestive enzymes, the secretions of the pancreas are rich in bicarbonate, which neutralizes the acid contents arriving from the abomasum, and raises the pH to above 7, the pH at which the enzymes that reduce carbohydrates, fats, and proteins to their simple, absorbable, components are active.

The sphincter of Oddi and Boyden regulates the rate at which the products of the pancreas and liver reach the intestine. The sphincter is relaxed by the presence of digesta and constricts when there are none. In many mammals the synthesis of pancreatic secretions is stopped between meals, and those from the liver (bile) are stored in a gallbladder, but giraffes do not seem to store bile. Murie (1872) found none in the three giraffes he dissected, nor did Garrod (1877), and in 17 adult giraffes examined by Cave in 1950 no gallbladders were found. Cave did find gallbladders in giraffe fetuses, and concluded that the gallbladder develops *in utero* but undergoes progressive atrophy, disappearing at or about the time of birth. Keen and Goetz (1957) found a very small gallbladder in a seven-month-old fetus. In contrast, Dagg and Foster (1976) reported that gallbladders were found in three adult captive, zoo giraffes dissected in Japan, and remarkably both Gordon in 1779 and Owen in 1838 (in one of three giraffes) found gallbladders in the adult giraffes they dissected. Grazers like cattle always have a gallbladder, and the implication of a usually absent gallbladder in giraffes is that bile is not stored, and the pancreas and liver empty their products continuously into the intestine in concert with a continuous flow of digesta and/or saliva into the small intestine.

Bile contains many substances destined for excretion, including the products of detoxification of plant secondary compounds like tannins. Bile's origin is the liver, and as browsers are exposed to more tannins than are grazers, Hofmann (1989) concluded that browsers must have relatively larger livers than do grazers, although he provided no numerical data to support his claim. Owen-Smith (1997), in the context of how tannins are detoxified, said "the relatively smaller liver mass of grazers compared with browsers probably restricts detoxification capacity." Data compiled by Crile and Quiring (1940) support the idea that the livers of grazers are smaller than they are in browsers. In a variety of grazers, liver mass formed 1.3%±0.3% of body mass, while in browsers, including a giraffe, liver mass was significantly higher (1.9%±0.2%). In one of the giraffes he dissected, a three-year-old, Owen (1838) found the liver to weigh 3.04 kg. Crisp (1864a) weighed the liver in three giraffes and found its mass to be 1.36 kg, 7.0 kg, and 5.5 kg, with size differences roughly following body mass and forming 0.8% ±0.1% of body mass, a proportion about half that reported by Crile and Quiring. Murie in 1872 reported that the liver weighed 1.4 kg in a juvenile giraffe weighing 130 kg (1.1% of body mass) and 5.3 kg in a giraffe weighing 827 kg (0.6% of body mass). Cave (1950) found a 9.3-kg liver in a 17-year-old giraffe, but did not disclose the animal's body mass. In another study (Mitchell, unpublished), giraffe liver mass ranged from 3.9 to 17.0 kg in six giraffes that weighed between 289 and 1,282 kg and was 1.4% ±0.2% of body mass. Liver mass in these six giraffes increased linearly with body mass, and in any giraffe with body masses within that range, liver mass can be calculated from the equation liver mass (kg)=0.012*Mb+1.26 kg (R^2=0.62). Very few studies have been made of liver mass in other ruminants, and we have to go back to 1883 and a study by one Joseph Rowe Smith Jr. to find a comprehensive study of liver mass in grazers. Smith was a medical military man, a Colonel and surgeon, but also a member of the Texan Committee on Cattle Diseases. That committee wanted to know what the "normal liver and spleen weight and temperature of Texas cattle" was. Texan cattle at that time would have been about as close to free-range grazers as is possible to find. From the liver mass reported for 814 cattle, ranging in body mass from 79 to 232 kg, average liver mass can be calculated to be 0.6% ±0.1% of body mass, which meets the criterion of being significantly smaller than those in giraffes and supports Owen-Smith's and Hofmann's conclusions that browsers have livers that are larger than they are in grazers.

Small Intestines

Digestion of carbohydrates, fats, and proteins takes place in the next and longest section of the small intestine, the jejunum. Its archaic, medieval name means "empty," as it is usually empty when seen at autopsy. In life it is anything but empty, because it is here that digestion by enzymes of ingesta coming out of the abomasum reduces carbohydrates to glucose, fats to fatty acids, and proteins to amino acids, almost all of which are absorbed, by complicated processes, through the jejunum wall. Nutrients that are absorbed enter the capillaries of the intestinal blood supply. The capillaries coalesce into intestinal veins and ultimately the veins become the hepatic portal system, which takes the nutrients to the liver for subsequent catabolism. The volume of blood (L min^{-1}) flowing through the portal system of mammals at rest (i.e., not exercising) fits a regression equation developed by Norman Edwards (1975; Flow = $0.04 \times Mb^{0.88}$) and amounts to about 3 L min^{-1}

in a juvenile giraffe to 24 L min^{-1} in a mature adult giraffe, or 30% to 40% of the total amount of blood pumped out of the heart each minute.

What remains after digestion in the jejunum has been completed passes into the ileum, an obscure word for the rest of the small intestine given to it by the medical men of ancient times, and loosely meaning "entrails," most functions of which are as obscure. One function of some importance is absorption of vitamin B_{12}. Omnivores like humans cannot synthesize vitamin B_{12} and the only significant source (an insignificant source is comfrey) of Vitamin B_{12} for humans is animal products (and sometimes the bottles that haunt the shelves of health food shops). In ruminants it is synthesized in the rumen by bacteria, in part by the replacement of Mg^{++} at the center of the chlorophyll molecule with cobalt. Vitamin B_{12} is an essential cofactor for the synthesis of DNA and for the synthesis of myelin, the protective sheath that insulates nerve fibers. Anemia and irreversible degeneration of the nerves are consequences of a deficiency—deficiencies that are well-recognized in non-ruminants. Unlike the other B vitamins, B_{12} is a relatively large molecule and in all mammals its absorption requires special transport mechanisms found only in the ileum. It is easy to see why evolution has ensured its absorption, but not why the ileum is the site.

Large Intestines

The large intestines in all mammals consist of the cecum, and proximal and distal colons. They augment the fermentation that occurs in the rumen. Fermentation in the abomasum and small intestine of giraffes is negligible, accounting for 5.5% and 3.4%, respectively, of total VFAs produced (Clemens and Maloiy, 1983). The cecum of giraffes has a pH of 7.1±0.1 units, and the concentration of VFAs in it is almost as high (146 mmol L^{-1} of fluid) as they are in the rumen (158.3±3.5 mmol L^{-1}; Maloiy et al., 1982) but its volume is small. Table 8.4 lists all the measurements so far made of cecum length of giraffes, and it varies from 44 to 96 cm. Owen (1838) measured its diameter to be 4.8 cm, and Joly and Lavocat (1846) 9.0 cm. If its average diameter is 7 cm and its average length 63 cm (Table 8.4), then cecum volume would be ~2.4 L, but Clemens and Maloiy (1983) found the weight of the contents of the cecum to be 9.2 kg, so its volume can be larger than the dimensions suggest.

The proximal and distal colons ferment any fiber still remaining, and between them in giraffes they contain 8% of all intestinal matter. Their important contribution to energy supply is fermentation of hemicellulose. Hemicellulose is a mixture of sugars (unlike cellulose, which consists only of glucose) and it can form 50% of total carbohydrate eaten. It resists fermentation in the rumen, where the pH is too high to prepare hemicellulose for fermentation. It has to be exposed to the acid in the abomasum before fermentation of it can occur. Apart from hemicellulose digestion, the large intestine is also the last chance for the absorption of water, an essential function in animals like giraffes that have evolved to live in hot, dry, arid habitats (see Chapter 11). The various sections of the giraffe's intestinal tract are illustrated in Figure 8.9.

Figure 8.9 is from the monumental study of the intestinal tracts of virtually all mammals made by Dr. P. Chalmers Mitchell, who was secretary of the Zoological

Table 8.4 The Length of Intestines (meters) in Giraffes

Reference	Age (y)	Body Mass (kg)	Sex	Small Intestine (m)	Large Intestine (m)	Ratio of Small I: Large I Length	Cecum (m)	Total Intestine (m)
Owen, 1838	~3	Not known	F	27.7	13.2	2.1	0.67	40.9
Owen, 1838	~3	Not known	M	26.8	13.1	2.0	0.61	39.9
Owen, 1838	~3	Not known	M	25.0	12.2	2.1	0.61	37.2
Joly and Lavocat, 1846	Young	Not known	F					62.3
Crisp, 1864a	18	815	F					77.4
Crisp, 1864a	4	815	M					63.7
Crisp, 1864a	0.17	153	F					32.6
Crisp, 1864b	0.58	153	M	27.4	10.1	2.7		37.5
Crisp, 1866	12	Not known	F	39.5	25.2	1.6		64.7
Murie, 1872	0.17	Not known	M	24.7	5.8	4.3	0.48	31.2
Murie, 1872	11.6	Not known	F	39.2	24.9	1.6	0.53	64.8
Murie, 1872	21.3	Not known	M	39.8	27.7	1.4	0.46	68.1
Garrod, 1877	Adult	Not known	F	59.7	22.9	2.6	0.79	83.4
Richiardi, 1880	Old	Not known	M	52.2	24.2	2.2	0.72	76.4
Perez et al., 2009	2.1	754	M	27.8	24.3	1.1	0.44	52.5
Perez et al., 2009	17	800	F	47.4	23.8	2.0	0.96	72.2
Sauer et al., 2015	3–4	532–559	–	39.1	16.6	2.4	0.50	56.2
Mean±sd				*36±12*	*19±8*	*2.2±0.7*	*0.63±0.16*	*57±17*

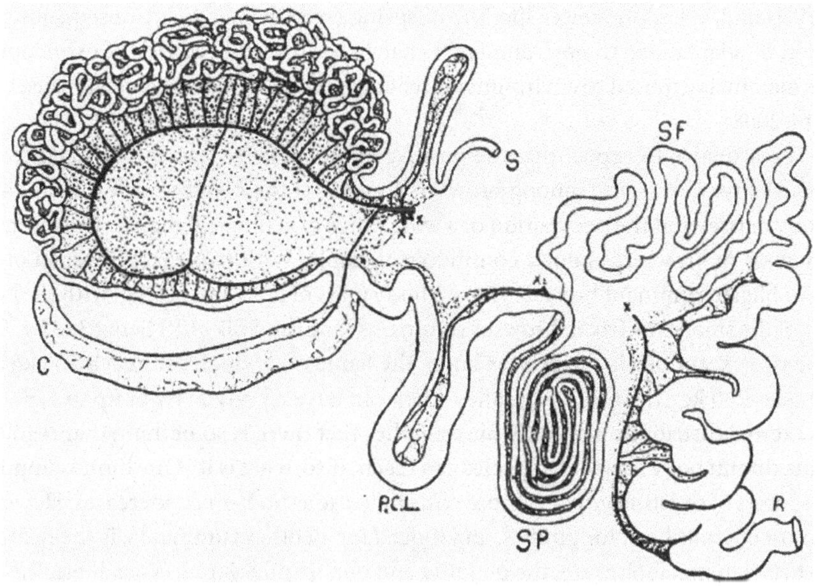

Figure 8.9. The intestinal tract of a giraffe. S: cut end of the duodenum near the abomasum; R: cut end of the colon near the anus; C: cecum; PCL: postcecal loop of the hindgut; SP: spiral loop of the proximal colon; SF: distal colon; X, X_2: cut ends of veins.
From Mitchell, 1905, with permission of John Wiley and Sons.

Society of London for 32 years. In situ in the abdomen, the intestines of a giraffe are marvelously arranged so as to accommodate their impressive length within the limited space they can occupy. The duodenum can be seen to form a well-marked loop within which the pancreas nestles, followed by the highly coiled jejunum and ileum, the last part of which, leading into the elongated cul-de-sac of the cecum, is a simple tube. The first part of the large intestine forms a loop that leads into the proximal, watch-spring-like, spiral colon that acts to delay movement of ingesta, prolonging the time for fermentation and providing the surface area and time for water and electrolyte absorption. The distal colon reabsorbs what little water remains, and it is where the characteristic hard pellets of giraffe droppings are formed.

Body Size and the Size of the Intestinal Tract

Studies of giraffe-feeding ecology consistently support the notion that a benefit of their large size and slender, tall shape is access to competition-free browse (see Chapter 7), but resistance to predation, vigilance, and better control of body temperature are other benefits. On the other hand, large size requires large absolute amounts of feed to support it, space in the abdominal cavity is required to house the organs needed to digest it, and a large home range is needed to acquire it. In a large home range, resources become shared, competition for browse increases, and small declines in browse abundance causes populations to crash, with the most vulnerable being the largest (Owen, 1851). Large size is also associated with longer gestation times and longer generation intervals and, therefore, fewer lifetime offspring (see Chapter 17). Consequently, evolution of adaptations to environmental change is slow, and the risk of extinction of large mammals exposed to environmental change is correspondingly high (Clauset and Erwin, 2008).

It is reasonable to expect that the positive effects of large size will be balanced by their negative effects and among browsers in general the result will be a wide range of body masses and the occupation of a wide range of ecological niches. However, the body mass of browsers is not a continuous variable. With two exceptions, all of the many obligate ruminant browsers have a body mass of less than 100 kg, with the body mass of the smallest African browser ruminants (suni and dik-dik) being 3–4 kg. One of the two exceptions is the greater kudu, the females of which can reach 200 kg and the males 250 kg. The other is giraffes. They can have a body mass of up to 1,500 kg. This skewed distribution of body mass implies that there is something that regulates the maximum body size that a species can reach. If so, what is it? One limit is imposed by the costs of resisting gravity. Those costs increase as body mass increases. However, the most obvious limit for giraffes, and indeed for all other ruminants, is the relationship between metabolic rate, the quantity and quality of vegetation available, the size of the rumen to hold and ferment it, and the size of the intestines needed to digest and absorb it.

In ruminants the most logical way of ensuring a continuous supply of energy is to evolve an as-large-as-possible rumen–reticulum complex, which is where energy extraction begins and where food is stored. Analysis of this idea by Demment and van Soest in 1985 found that in many ruminants, rumen capacity does increase in line with increases in body mass ($Mb^{1.0}$) and in some cases even faster ($Mb^{1.16}$), although Clauss et al. (2013) and Müller et al. (2013) have suggested that rumen capacity increases at a rate closer to $Mb^{0.75}$, i.e., at a rate proportional to metabolic rate. Large rumens slow the rate at which fodder moves through it, with the advantage that the time available for fermentation is prolonged. However, a slowly emptying rumen gives a sensation of being full, dulls the sensation of hunger, and reduces the amount of food eaten. Consequently, rumen size should be large enough to allow fermentation without making the animal feel full and large enough to provide its energy needs (Clauss et al., 2003a). Rumen (and intestine) size, though, is limited by the space available for it in the abdominal cavity, a space that also has to accommodate other organs. This limiting nature of the shape of giraffes was first recognized by James Murie in 1872. He wrote "the great depth and capricious enclosure by the ribs limit considerably the loose abdominal area. The form of the body of the giraffe is indeed remarkable in this respect." Giraffes have a short trunk. The effect of gestation on rumen volume in giraffes is particularly marked. Hall-Martin (1975a) established that for every 100 days of gestation, rumen fill as a percentage of body mass decreased by 2%, from 14% of body mass at the time of conception to 6% at full term. Taking all these factors into account, the maximum body size of an animal, and of ruminants in particular, is determined by the balance between the maximum possible size of the digestive tract, the size of the rumen complex specifically, and the quantity and quality of the food available. In most browsers this balance is reached at relatively low body mass.

The Size of the Rumen in Giraffes

Do giraffes, which are the very largest of ruminants and by far the largest browsers, have small or large rumen-reticulum complexes for their size? Measurements of its linear dimensions are rare and equivocal. Joly and Lavocat (in 1846) reported a measurement of 1.6 m for the rumen, and Crisp (1866) of 1.74 m, but in neither case was the method of measurement nor the dimension that was measured explained. In all likelihood they measured rumen circumference (they could not have measured rumen length, as the whole length of an adult giraffe's trunk from withers to tail base is about one meter). Murie in 1872 reported "forestomach length" to be 0.6 m in a juvenile giraffe, 1.7 m in a mature female, and 2.0 m in a mature male, and presumably Murie meant the distance that browse would travel from rumen to duodenum. Rumen circumference in giraffes weighing 532 kg is ~2 m, as rumen diameter is ~70 cm in giraffes of that body mass (Sauer et al., 2016), and diameter would have been ~55 cm in Crisp's and Joly and Lavocat's giraffes. Like the rumen, the reticulum is spherical and had a diameter of ~30 cm in the giraffes studied by Sauer et al. (2016). The first quantitative study of the volume (rather than of linear dimensions) of the rumen complex in wild African grazers and browsers was made by Giesecke and van Gylswyk in 1975. They found that rumen complex volume was much smaller in browsers than it was in African grazers, but their study did not include giraffes. Hofmann (1989) concluded that browsers, because they eat a concentrated form of nutrients that did not require prolonged fermentation, should have relatively small rumens, an idea that Clauss and colleagues (2003b) have supported: coarse grass requires a large rumen and softer browse a smaller one. Analysis of rumen complex mass (kg) in a group of 32 wild giraffes did find that it is lower than it can be in grazer ruminants (Mitchell et al., 2015; Figure 8.10). In those giraffes the scaling factor for the relationship between rumen mass and body mass was 0.64 ($Mb^{0.64}$; Figure 8.10). The rumen complex of giraffes is, therefore, small and forms a smaller and smaller proportion of body mass as body mass increases. Consequently, giraffes must have evolved ways other than a large rumen to solve the problem of providing enough energy and nutrients to support their body mass.

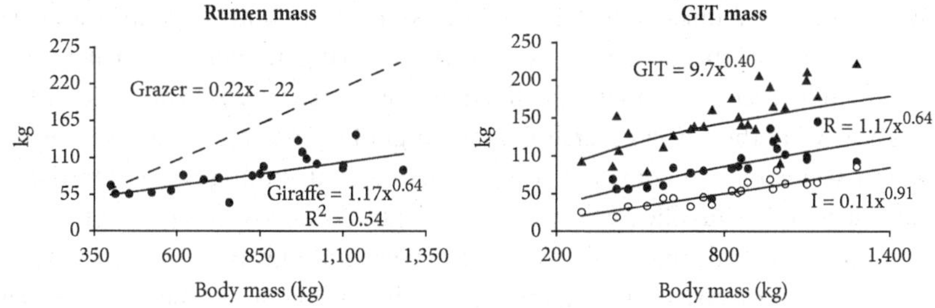

Figure 8.10. Left: The change in rumen mass in giraffes (●) and grazing ruminants (dashed line), as body mass increases.

Grazer equation from Demment and van Soest, 1985.

Right: Changes in intestine mass (I, ○), rumen complex mass (R, ●) and total gastrointestinal tract mass (GIT, ▲) in 32 wild giraffes (*G.c. giraffa*).

Rumen Energy Production

Aspects of digestion that are amenable to evolution and would improve energy provision by the rumen are the length of time food is in the rumen (which theoretically at least would optimize the time needed to extract energy from browse by fermentation), the size of the food particles deposited in the rumen and hence their surface area (which determines the efficiency of microbial action—the smaller the particles, the larger the surface area), and an ability to select browse with a high energy content, or combinations of these (Demment and van Soest, 1985; van Soest, 1996; Clauss et al., 2003a, 2007, 2009; Muller et al., 2013). Have giraffes evolved the means of optimizing energy extraction from browse in their rumens? In all ruminants the product of fermentation of fiber is VFAs. In all ruminants the main VFAs are acetic, propionic, and butyric acids, and the total energy content of VFAs varies according to the ratio of the acetic:propionic:butyric acid mixture present. In the one study thus far that has measured VFA production in wild giraffes, the ratio of VFAs was 76% acetic:14% propionic:9% butyric (Maloiy et al., 1982), proportions not dissimilar to those in other ruminants including grazers. How much energy do the VFAs produce? In eight browsers, energy produced was 1,109±49 kJ mol^{-1} of VFA (1 mol of VFA = 60 g); in eight grazers, it was 1,151±47 kJ mol^{-1}; and in four intermediate feeders, 1,110±51 kJ mol^{-1} (references in Gordon and Illius, 1994), with the minor differences reflecting the small differences in the proportions of the three main VFAs. The average value found in six giraffes by Maloiy et al. in 1982 was 1,095 kJ mol^{-1}. Thus, although the proportions of the three main VFAs varies according to diet, their energy content is remarkably constant in all ruminants. Maloiy et al. also found that in giraffes 4.06 moles of VFA were produced from each kg of dry matter (DM) in the rumen. Fermentation rate in the hindgut is much lower—1.2 mol VFA $kgDM^{-1}$ day^{-1}. What determines the rate at which VFAs are produced? From data reported by Gordon and Illius (1994), the number of moles of VFA produced from a kg of browse in the hindgut is the same in all ruminants and is independent of body mass. In contrast, the fermentation rate in the rumen is associated with body mass, and fits the regression equation: VFA (mol $kgDM^{-1}$ day^{-1}) $=13\ Mb^{-0.23}$ (R^2=0.68). The exponent of this equation shows that the amount of energy able to be extracted from a kg of DM in the rumen declines as body mass increases, and the rate of decline matches the decline in mass specific metabolic rate (kj kg^{-1} body mass min^{-1}) as body mass increases, a correlation that suggests that somehow the rate at which rumen microbes ferment browse is determined by their host's mass specific metabolic rate. Microbes are, of course, unable to know what their host's metabolic rate might be. They are what they are, so the correlation seems to be a coincidence more than a contributing factor. On the other hand, the absolute metabolic rate (kj min^{-1}) of a giraffe, or any other mammal for that matter, increases as body mass increases, and correspondingly there is an increasing requirement for energy, so a decline in the rate of VFA production is the opposite to what common sense indicates it should be. It would be far better for large browsers if they were able to extract more energy from browse than can a small browser. Extraction of more energy per kilogram of browse would mean that the amount of browse they had to eat would be reduced, an advantage that would be especially useful for giraffes when they are faced with the predicament of a seasonal decline in the quantity of browse available, but the opposite is true. The giraffe listed in Gordon and Illius's dataset weighed

703 kg and, in contrast to the fermentation rate of 4.06 mol VFA kg^{-1}DM day^{-1} reported for giraffes weighing between 500 and 750 kg by Maloiy et al., its fermentation rate was 3.21 mol VFA kg^{-1}DM day^{-1}. In a suni that had a body mass of 3.8 kg, fermentation produced 8.6 mol VFA kg^{-1} day^{-1}. As the amount of energy generated from each mol of VFA (kj mol^{-1}) in giraffes is the same as that in all other wild ruminants, it is not as if the VFAs produced in the rumens of giraffes are inferior in some way, it is that giraffes produce less of them from each kg of DM they eat. And that outcome poses a dilemma. Why has evolution produced what seems to be an anomaly?

The reasons why large browsers and specifically giraffes are unable to extract the same amount of energy from browse as can small browsers are not all that clear. Marcus Clauss and colleagues, in an excellent review of the nutritional problems faced by large ruminants, concluded that high energy intake can be achieved by having a high food intake to balance the low digestive efficiency (Clauss et al., 2010). High food intake does correlate with body mass. From the dataset reported by Gordon and Illius (1994, 1996), which included a giraffe, the mass of rumen DM fits the regression equation: rumen DM (kg) = 0.008 $Mb^{1.19}$ (R^2=0.98). In other words, as the exponent illustrates, DM intake increases at a rate higher than the rate at which body mass itself increases. Digestive efficiency in turn can be achieved by selection of high-quality browse having a low proportion of low cell wall constituents and prolonged mastication of browse before it is swallowed, which reduces particle size. This description of feeding habits fits giraffes exactly, but their VFA production rate is nevertheless very low. One explanation for that low rate is that giraffe diets contain tannin, and tannins depress digestion of browse (see Chapter 7). Depression of digestion occurs when the concentration of tannin in browse is at least 50 g kg^{-1} of DM (Frutos et al., 2004). Some Acacias (at least in Kenya) can have a concentration of ~140 g kg^{-1} DM (Abdulrazak et al., 2000). Another reason, and one that has been accepted for a long time (e.g., Jarman, 1974) is that as large browsers need a large absolute amount of browse to provide the energy they need, and have less time to be selective about quality, they inevitably have to include low quality (i.e., low energy content) browse in their diets. This explanation also is hard to reconcile with giraffe feeding behavior. Giraffes spend more time browsing than other browsers and, in contrast to their classification as "unselective browsers" by van Soest (1996), they are highly selective of the browse they prefer (see Chapter 7). Browse low in energy content is browse that is rich in cell wall constituents—cellulose, lignin, and hemicellulose—the so-called neutral detergent fiber, or NDF. It is the soluble cell contents that are the source of most energy, and a high proportion of cell wall constituents in a diet reduces the availability of the digestible cell contents (Pellew 1984b). Giraffes have a diet containing 61% NDF (Pellew 1984b), which is high but not markedly higher than the NDF in the diet of small browsers (e.g., Clauss and Dierenfeld, 2008) that have fermentation rates twice or more that of giraffes. But a high concentration of cell wall constituents does explain why it is that hindgut fermentation rate in giraffes is three to four times lower than the rate in the rumen. Dry matter in the hindgut contains the difficult-to-digest hemicellulose. When Maloiy and colleagues (1982) pondered the problem of low VFA production, they concluded that the rate at which VFAs are produced in the rumen is determined by the type and density of rumen microbes. The composition of the rumen flora is determined by the diet, and over the many millions of years of giraffe evolution we have to conclude that the diet giraffes choose has evolved in a way that satisfies all their nutrient requirements, and the microbes that are able to thrive in a giraffe's rumen happen not to be very efficient fermenters of fiber and are less able to digest the cell wall constituents than are the microbes that inhabit the rumens of smaller browsers. That proposition has yet to be tested because analysis of the correlation between microbe type and their ability to ferment fiber has yet to be done. Even so, what we can say is that as the microbe population is an adaptive population in the sense that it is only those microbes that are adapted to, or can cope with, the diet eaten that exist in the rumen. If so, exactly why that population becomes less efficient at digesting browse simply because body mass has increased is an enigma, unless it is because inefficient digestion is associated with an increase in the amount of browse that has to be eaten to supply the necessary nutrients, and that increase is correlated with an unavoidable increase in browse containing tannins, to which the microbes cannot adapt (see Chapter 7). In the case of giraffes, that too is a proposition difficult to accept, as the voluminous saliva giraffes secrete is able to neutralize the tannins they eat. None of the explanations for the low rate of VFA production in giraffes is, therefore, compelling, and the reason or reasons remain, at least for me, one of the great puzzles of giraffe digestion.

Apart from eating more browse, more energy could be extracted if the browse were to be retained in the rumen for a longer period of time and thus exposed to a long fermentation time. Measurement of retention time is difficult, but Clauss and colleagues (1998) have managed to measure retention time in three captive giraffes with estimated body masses of 1,000 kg, 850 kg, and 700 kg. In these giraffes, fluid and particle retention time averaged 33.7±5.0 and 40±7.2 hours, respectively. In other studies, retention time in a giraffe of estimated body mass of 1,000 kg was 48 hours, which was less than the retention time (52.7 hours) of an Asian buffalo that weighed 417 kg (Clauss et al., 2003a). In yet other studies, mean particle retention time has been found to be 22.5±7.0 hours in the rumen complex (Clauss et al., 2006c), and 49.8 hours in a giraffe that weighed 656 kg (Schaub in Müller et al., 2013). For ruminants in general, an equation that predicts retention time has been developed (RT (days) = $0.695 \times Mb^{0.22}$; Demment and van Soest, 1985; Illius and Gordon, 1992), and applied to giraffes, gives a retention time of 3.2 days (76 h) for a 1,000 kg giraffe, which is much longer than the time actually measured by Clauss and colleagues. Clauss et al. (2003b) also developed an equation from studies that included a giraffe (RT (days) =$13.7 \times Mb^{0.18}$). This equation gives a retention time that could be as short as 1.4 days (33 h) in a young giraffe to 2.1 days (50 h) in a mature adult male. These times are much closer to those that are observed, and they do not suggest that giraffes have evolved a long retention time as a way of improving digestion of browse. A short retention time might, however, explain why it is that VFA production rate is so low in giraffes. Perhaps there is just not enough time to extract all the energy present by fermentation in the browse giraffes eat.

Short retention times of a few days imply that the particle size of the food that reaches the rumen must be small because if small particles enter the rumen the passage of food through the rumen is quick, intake of food is not limited by a full rumen, and more browse can be consumed and digested per unit time. Particle size usually is determined by the amount of time spent ruminating, rather than browsing or grazing. Buffalo (*Syncerus caffer*), for example, ruminate for 7.1 hours a day on average and graze for 9.2 hours a day (Grimsdell and Field, 1976; Beekman and Prins, 1989; Ryan and Jordaan, 2005). Giraffes on average spend 4.5 hours a day ruminating and up to 13 hours a day feeding (Innis 1958; Leuthold and Leuthold, 1978b; Pellew, 1984b; du Toit and Yetman, 2005), although these proportions can change with season and the quality and quantity of browse. Giraffe spend more time feeding than ruminating because they need more food, because they are particular about what they eat, and the browse they eat is shoots and leaves and florets that cannot be eaten in bulk like grass. Thus more time is needed to acquire it (du Toit and Yetman, 2005), and more time is spent crushing browse at the time of foraging. The consequence is that browsing is prolonged, particle size is small, and long periods of rumination are unnecessary. Perhaps with less chewing at the time of browsing, fermentation rate and VFA production might be even lower than it has been found to be. Giraffes nevertheless are efficient ruminators. When a giraffe ruminates, it chews the cud for about 50 seconds during which time some 40 to 50 rotatory movements of the lower jaw are made, after which the cud is swallowed. Three of four seconds later, a second cud is regurgitated (Owen, 1838; Innis, 1958).

What proportion of browse is digested? In other words, how efficient are giraffes at turning browse into energy? Is their capacity to digest the DM in browse higher than it is in other browsers? If it was, then how they obtain the necessary energy and nutrients would be explained. A mole of VFA has a mass of about 60 g, and because the fermentation rate of giraffe browse in the rumen produces 3–4 mol VFA per kilogram of DM eaten each day, about 180–240 g of VFA are produced from each kilogram of rumen DM. How many kilograms of DM are available for fermentation in the rumen at any one moment? Clemens and Maloiy (1983) established that water makes up 86% and DM ~14% of the total ingesta mass in a giraffe's gastrointestinal tract (in their study, total ingesta mass in giraffes weighing 500–750 kg was, on average, 116 kg, of which 16.2 kg was DM). The total digestible nutrients (TDN) in giraffe browse, which is made up mainly by protein, carbohydrate, and fats, amount to about 600 g kg^{-1} of DM (Pellew, 1984b). Of the total DM present, the rumen complex contained 12 kg (72%) and, thus, 7,200 g of digestible nutrients. The hindgut contained 3 kg (20%) of DM and 1,800 g of digestible nutrients. The abomasum (7%; 1.2 kg) and small intestine (4%; 0.7 kg) contained the remaining DM. Fermentation of the rumen DM would produce between 36 and 48 mol VFA (2,160–2,880 g VFA) and in the hind gut could produce 1.2 mol VFAkg^{-1} (3.6 mol VFA; 216 g VFA). Thus between 30% and 40% of the 9,000 g of TDN can be expected to be fermented and would produce between 40,000 and 60,000 kj of energy. The distribution of DM outlined here changes as the body mass and the proportions of the intestines in a giraffe change as a giraffe grows, as does the amount of energy produced from fermentation. The consequence of those changes is that of all energy produced by fermentation, the proportion produced by the rumen complex declines from 90% in young giraffes to 80% in adults (=108 $Mb^{-0.04}$; R^2=0.53), and the amount produced by fermentation in the hindgut increases from 10% in young giraffes to 20% in adults (=2.9 $Mb^{0.24}$; R^2=0.58).

Fermentation is the not the main or only source of energy for giraffes (Table 8.5). Indeed, the energy required by very large herbivores cannot be met by rumen fermentation, even if rumen fermentation assisted by hindgut fermentation is efficient, because there is regular emptying of the small nutrient-rich particles from the rumen. Thus, on average, of all the energy needed to support metabolism in giraffes, fermentation provides just 43% ±5%. Where does the other 57% of the energy needed come from? It has to come from another source, and that source is the straightforward digestion of

Table 8.5 Summary of the Origins of Energy Supply for Giraffes

Variable	Allometric Equation	Value for a 750 kg Giraffe	Range in Giraffes of Body Mass 289–1,441 kg
FMR (kJ day^{-1})*	$862 \times Mb^{0.74}$	115,600	57,415–188,832
Rumen fermentation (kJ day^{-1})	$602 \times Mb^{0.64}$	41,600	22,624–64,171
Hind gut fermentation (kJ day^{-1})	$16 \times Mb^{0.92}$	7,100	2,938–12,885
Total fermentation (kJ day^{-1})	$555 \times Mb^{0.68}$	50,000	26,163–78,016
Digestion (kJ day^{-1})	$365 \text{ x } Mb^{0.79}$	68,200	32,090–114,192

*Field Metabolic Rate.

the kind that we humans subsist on—digestion of fats, carbohydrates, and proteins by enzymes secreted into the small intestine with the products absorbed from the small intestine. In juvenile but weaned giraffes, that source provides 46% of the energy needed, i.e., fermentation provides 54%; and in an adult giraffe, digestion provides 62%, i.e., fermentation provides 38%. The amount of energy digestion needs to supply each day in a juvenile giraffe is equivalent to the amount of energy found in ~1 kg day^{-1} of soluble carbohydrate, and in an adult ~6 kg day^{-1}, amounts easily met by the amount consumed by giraffes (up to 7 kg day^{-1}; Pellew, 1984b). Moreover, their prolonged chewing of browse presumably provides large volumes of a saliva and soluble nutrient mixture. It is highly likely but not yet proven that much of that mixture bypasses the rumen and enters the lower intestines directly via closure of the esophageal grove. What these analyses do reveal are the ways giraffes have evolved to cope with a low VFA production rate from fermentation. They rely less and less on it, and more and more on straightforward digestion. When the straightforward digestion component of DM digestion is added to the proportion digested by fermentation, then one estimate has found that the apparent digestion of DM is 83.5% (Schaub in Müller et al., 2013), a proportion that is significantly higher than the 56.8% for all the other ruminants in that study, and higher than it is in the okapi (65.4%) or camel (54.2%) in the same study. Apparent digestion is calculated as the percentage of DM intake that is excreted in fecal pellets. The calculation assumes that the fecal pellets consist only of the undigested portions of browse eaten, which is almost certainly not the case. Fecal pellets also contain bacteria and intestinal cells that together may make up ~50% of fecal DM. Even so, if the daily DM intake of a 750-kg giraffe is ~13 kg (Table 8.6) and the DM excreted daily in the pellets of a 750-kg giraffe is ~2 kg (Kearney, 2005), then using those data, apparent digestibility is 85% ($111 * Mb^{-0.04}$; R^2=.99), much the same as it was reported to be in Müller et al.'s study. The conclusion must be that using both fermentation and digestion, giraffes are very efficient digesters of the browse they eat. That conclusion explains why they can acquire all the nutrients they need, even though the rumen gets smaller and smaller as a giraffe matures. However, if rumen mass declines, the intestines must get bigger and bigger as

Table 8.6 Metabolic Rate, Food Consumption, and Home Range of Giraffes

Variable	Allometric Equation	Value for a 750 kg Giraffe	Range in Giraffes of Body Mass 289–1,441 kg
BMR (kJ d^{-1})	$279.4 \times Mb^{0.77}$	45,711	21,933–75,578
FMR (kJ d^{-1})	$862.0 \times Mb^{0.74}$	115,625	57,415–188,832
Mean dry browse intake (kg d^{-1} DM)*	$0.1 \times Mb^{0.74}$	13	6.8–22.5
Mean fresh browse intake (kg d^{-1})	$0.34 * Mb^{0.74}$	46	23–74
Core home range (km^2)*	$0.0054 \times Mb^{0.74}$	0.7	0.4–1.2
Predicted home range (km^2)	$0.014 \times Mb^{1.27}$	63	8.2–150.5

Data from Stahl, 1967; Nagy 1994, 2005; Hall-Martin and Basson, 1975; Pellew 1984b; and Mitchell et al., 2015.
BMR = basal metabolic rate; FMR = field metabolic rate.
*Minimum home range to supply FMR.

a giraffe grows, and they do (Figure 8.11). They are much longer than they are in any other ruminant of equivalent body mass (Crisp, 1864a, 1864b).

The Size of the Intestines of Giraffes

How big are the intestines in giraffes? In the historical literature, intestine length has been reported in 14 giraffes (8 females and 6 males), of age two months to 21 years and 3 months (Table 8.4). Perez et al. (2009) measured intestinal lengths in a two-year-old male weighing 754 kg and a 17-year-old female weighing 800 kg, and Sauer et al. (2016) measured the lengths in 21 giraffes of unspecified sex and average body mass of 548 kg (Table 8.4). One other measured length is known in a giraffe. Its total intestinal length was recorded as being 280 ft in a giraffe with an estimated body mass of 545 kg (Goetz and Budtz-Olsen, 1955). However, this length:body mass relationship is not consistent with the other data, and it is probable that the estimate of body mass or the length, or both, were incorrect. From its reproductive history, the body mass of the giraffe was probably at least ~700 kg, not ~500 kg. From data in Table 8.4, an allometric equation can be developed that describes the relationship between total intestine length and body mass: $L (m) = 3.6 \times Mb^{0.42}$ (Figure 8.11), and for a giraffe weighing 700 kg, predicted intestine length is ~56 m. Thus the length measured by Goetz and Budtz-Olsen was most likely 180 ft (~55 m) not 280 ft (85 m). In some texts, Beddard (1902) is credited with the measurement of an intestine length in a giraffe. However, he was quoting measurements made by Garrod in 1877. When the data in Table 8.4 are used to illustrate the growth patterns of the intestines (Figure 8.11), they show that the length of both the small and large intestines increases rapidly in the early years of life, with the small intestine reaching its maximum length at about five years of age. The large intestine continues to elongate, but more slowly. The average length of the small intestine

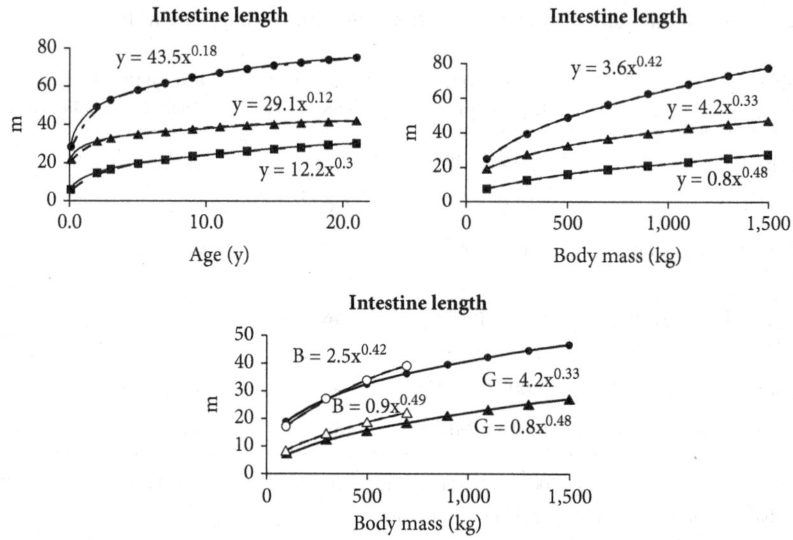

Figure 8.11. Left: Relationship between age and total (●), small (▲) and large (■) intestine length. Right: Relationship between body mass and total (●), small (▲) and large (■) intestine length.

Data from Table 8.4.

Bottom: Length of the small intestine in African browsers (○) and giraffes (●) and the large intestine in African browsers (△) and giraffes (▲).

Data from Table 8.4 and Woodall and Skinner, 1993.

in these giraffes was 36 m and of the large intestine was 19 m, and the average ratio of small:large intestine length was 2.2±0.7:1. This ratio is the same as it is in eight species of wild African browser/intermediate feeders of body mass 8.5–411 kg, as are the relative lengths (Woodall and Skinner, 1993; Figure 8.11).

The internal surface area of the small and large intestines is far more important than their length, because it is the size of the surface area that determines the efficiency of absorption of the products of digestion. The absorptive surface area of the intestines always is much larger than the surface area of a simple tube because the inner membrane of the intestinal tract is convoluted and covered in villi, a combination that enlarges the surface area to ~200 times the surface area of a smooth tube. The absorptive surface area of giraffe intestines has not been measured. However, it has been measured in other African browsers (Woodall and Skinner, 1993), and in them the absorptive surface area of the small intestine was described by the equation: SA (m^2) = 0.02 × length$^{0.17}$, and that of the large intestine: SA (m^2) = 0.25 × length$^{1.45}$. Assuming that the relationship between absorptive surface area and length of the intestines in giraffes follows the pattern of other African browsers, the absorptive surface area in a young giraffe is 5.1 m^2 and in a mature giraffe 23.0 m^2. In other words, small intestine length increases 2.4-fold, but its surface area could increase 4.5-fold during growth (Figure 8.12). Similarly, the length of the large intestine in giraffes increases from 5.8 m in a two-month-old giraffe to 27.7 m in a 21-year-old and its absorptive surface area could increase 10-fold from 3.2 to 30.9 m^2 (Figure 8.12).

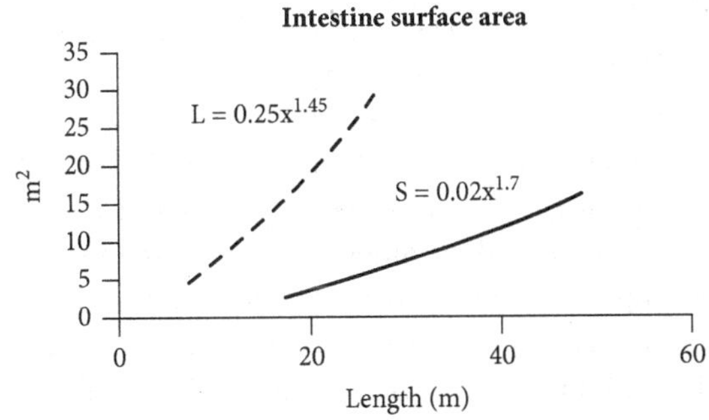

Figure 8.12. The predicted change in the surface area for absorption of nutrients as intestine length increases in the small (solid line) and large (dashed line) intestines of giraffes.

Extrapolated from data in Woodall and Skinner, 1993, for African browsers.

A Summary of Digestion in Giraffes

Is the digestive system of giraffes unique and special? In many ways, its structure is the opposite to what is expected and predicted from studies of grass-eating ruminants. A giraffe's digestive system has to supply the energy to support a body mass that is many times bigger than the body mass of the next biggest ruminant browser and bigger than any ruminant grazer, yet compared to grazer ruminants with a similar body mass, such as a Cape buffalo, their forestomachs are smaller and, as a proportion of body mass, decline with growth. The time giraffes allocate to rumination of browse and the length of time browse is retained in their rumen for fermentation are short; the muscles of mastication are small, the salivary glands are small, and giraffes can extract far less energy from browse by fermentation than can smaller ruminants. How do they manage their daily lives? What makes them special is that they have evolved large intestines with large surface areas, and have evolved all the paraphernalia needed to select and eat good-quality browse. They eat large amounts of browse each day, and any browse that has escaped fermentation in the rumen is fermented in the hindgut. It is also highly likely that a proportion of the sugars, proteins, and fats released from the cells of browse bypass the rumen and so avoid being fermented and losing some of their energy as heat or methane, and most of them are digested. Most, in other words, of the DM in the browse that is eaten is turned into nutrients either by fermentation or straightforward digestion. In short, giraffes have evolved a very efficient digestive system and waste very little of what they eat. It is an exemplarily frugal lifestyle.

Energy Balance and Food Intake

If the conversion of solar energy into food energy by photosynthesis in plants is the most fundamental biological reaction known, conversion of the sources of energy generated by digestive processes into biological energy is the most fundamental of all metabolic processes in animals. That conversion takes place in the mitochondria of cells and uses oxygen. Adenosine triphosphate (ATP) is the product. ATP is biological energy. The total oxygen required every minute or hour or day not unexpectedly increases as the body mass of an animal does because the rate of ATP use increases in proportion to increases in body mass. The rate at which ATP is used by an animal is, typically, regulated by the action of thyroid hormones. They bind to DNA and initiate gene activity, and the most important consequence of their action is an increase in the activity of sodium–potassium ion pumps in cell membranes. The pumps maintain a balance of sodium ions outside and potassium ions inside cells and probably use a half of all ATP synthesized daily to do so. If the pumps fail, ion imbalances result and death follows, so expending half of all energy on them is prudent. It would be nice to know something about thyroid hormones in giraffes, what their concentrations are in blood, and whether they regulate ATP production in the same way that they do in other mammals, but in fact nothing is known about them. A little is known about the thyroid glands that secrete them. Crisp (1864) found that the glands weighed 22.7 g in a young giraffe that weighed 153 kg. Crile and Quiring (1940) recorded thyroid gland weight as 64.7 g in a giraffe weighing 1,220 kg. Frank (1960) described their gross anatomy in a three-and-a-half-year-old captive giraffe that died in Zagreb Zoo. The main lobes were 15 cm long, 4 cm wide and 3.5 cm high, and there were two accessory thyroid glands (1.5 cm and 5.5 cm long) on the left side of the neck, and one 3 cm long on the right. The main lobes were supplied by branches of the cranial thyroid artery and the accessory thyroid glands by branches of the carotid artery. The microscopic structure of the glands in giraffes was similar to that of bovine thyroid glands. The glands in giraffes lie on the trachea just below the larynx (Erdogan and Perez, 2013), just as they do in all other mammals.

How much energy do giraffes need each day? Most ATP is made by biochemical reactions that consume oxygen; therefore, to discover how much energy a giraffe requires, all that needs to be done is to measure its oxygen consumption. Consumption of one liter of oxygen by metabolic reactions is associated with the production of 20 kilojoules of energy, so daily oxygen consumption can be easily converted into kilojoules of energy per day (kJ day^{-1}). The procedure for finding out how much energy is required by a giraffe sounds simple but, in fact, the measurement of how much oxygen is being used by any animal is very difficult and sometimes impossible. Three commonly used measures of oxygen consumption are basal metabolic rate (BMR), resting metabolic rate (RMR) and field metabolic rate (FMR). BMR is defined as the oxygen consumption in an adult animal that is not pregnant or lactating, is not exercising, is not eating or digesting food, and is in an environmental temperature such that it is not using energy to maintain body temperature (White and Seymour, 2002), a set of constraints which makes it impossible to measure in large animals in general, wild animals in particular, and ruminants specifically: ruminants are never, for example, not digesting food. Nevertheless, in the 1930s, Max Kleiber studied the metabolic rates of animals and formulated what has since become known as Kleiber's Law (Kleiber, 1947, 1961). It predicts that the BMR (kJ day^{-1}) of mammals will be equal to $294 \times Mb^{0.75}$ with body mass in kg.

RMR is easier to measure. It is the oxygen consumed by an animal resting quietly and not using energy to regulate its body temperature, criteria that are less restrictive but nevertheless difficult to achieve in a wild animal. FMR is the daily energy consumption of an animal enjoying all the activities of daily life. It is a true cost of living (Nagy, 1987). FMR is measured by analyzing the rate of incorporation of radioactive isotopes of oxygen and hydrogen into the end products of metabolism, namely carbon dioxide and water. The cost of isotopes is high and, so far, the FMR of quite small animals only has been determined. The method has been tested in reindeer having a body mass of ~250 kg, but FMR itself was not measured (Fancy et al., 1986), and the largest ruminant in which FMR has been measured by this method had a body mass of 39 kg, which is a third of the size of a newborn giraffe. As FMR has not been measured in giraffes of any size, we do not really know how much energy giraffes need to get through their day, but RMR has been measured twice in giraffes (Patterson et al., 1965; Langman et al., 1982) with very different results. In both studies the body mass of the giraffes was ~600 kg. Patterson's giraffes were wild, restrained, and anxious, and RMR in them was approximately 6 ml O_2 kg^{-1} of body mass min^{-1}. The three giraffes in Langman's study were accustomed to daily handling and RMR was measured in them 50 times. It was found to be 2.6±0.1 ml O_2 kg^{-1} min^{-1}, which is the same as the basal metabolic rate predicted by equations developed by Kleiber (1961), Stahl (1967), Hayssen and Lacy (1988), and Schmidt-Nielsen (1997). Those equations can be concatenated as $9.7 \times Mb^{-0.23}$ (mlO_2kg^{-1} min^{-1}), or as $279.4 \times Mb^{0.77}$ (kJ day^{-1}; Mitchell and Skinner, 2011). According to that latter equation, a giraffe weighing 1,000 kg will have a resting/basal metabolic rate of 57,000 kJ day^{-1}, which is equivalent to consumption of 2 ml of oxygen per kilogram of body mass every minute. The metabolic rate of an average human weighing 70 kg is about 7 000 kJ day^{-1}, or 3–5 ml O_2 kg^{-1} min^{-1}.

Kleiber's equation gives the most commonly used estimate of BMR in biology, and in his study of giraffes in the Serengeti, Pellew (1984a) used it to estimate giraffe BMR and then FMR. He concluded that in the wet season the FMR of males was 1.4 times and of females 1.75 times BMR. In the dry season in males FMR was 1.55 times and in females 1.91 times BMR. Many, many analyses have found that the main determinant of FMR in any animal is its body mass (e.g., Nagy, 2005), and an equation that summarizes the relationship between body mass and FMR in a ruminant is $FMR = 862 \times Mb^{0.74}$ kJ day^{-1} (McNab 2002; Nagy, 1994, 2005; Calder, 1996). If that equation is used to estimate FMR in giraffes, the slightly higher ratio of 2.5 times BMR (equivalent to 7 ml O_2 kg^{-1} min^{-1}) results. Translated into kJ day^{-1}, FMR ranges from around 40,000 kJ day^{-1} in a six-month-old weaned giraffe to nearly 200,000 kJ day^{-1} in a mature male (Mitchell et al., 2015). The minimum amount of browse that must be eaten by giraffes to support their FMR can be calculated by dividing FMR by the energy content of browse (Nagy, 1987), and the energy content of giraffe browse is 8.4 kJ g^{-1} of DM eaten (see Chapter 7). Thus a young giraffe needs to eat 5 kg day^{-1} of DM equivalent each day, and an adult weighing 1,500 kg must eat 24 kg day^{-1}. These amounts correspond to 2.5% of body mass in a young giraffe and to 1.6% of body mass in an adult. As the browse giraffes eat contains 60% water, the quantity of fresh browse that must be eaten is two and a half times more, or about 20 kg day^{-1} in a young giraffe and 70 kg day^{-1} in an adult. These relationships are summarized in Table 8.6 and assume that all browse eaten is converted into energy. However, as explained earlier, while there is almost complete digestion of browse by giraffes, not all of it is digested, and therefore the actual amount of browse that has to be eaten is slightly higher than the predicted minimum.

Home Range

To acquire sufficient browse throughout the year, giraffes need a large home range. What should the size of this home range be? The answer to that question is not as straightforward as it might seem, because the size of the home range of giraffes varies between a minimum (the core home range; McNab, 1963; du Toit, 1990c) and a maximum that is set by the long-distance meanderings of a giraffe outside the core area. The minimum home range is set by the relationship between food requirements and the productivity of browse, and productivity of browse is determined by climate and rainfall, offset by the size of the population of giraffes using it (Leuthold and Leuthold, 1978; Pellew, 1983a), but ultimately home range size should be determined mainly by body mass and metabolic rate (McNab, 1963).

The average population density of giraffes in Africa reported by Innis (1958; South Africa), Dagg and Foster (1976; 30 sites across Africa), Hall-Martin (1974b, 1975b; South Africa), Pellew (1983c; Serengeti), van der Jeugd and Prins (2000; Tanzania), and Cumming and Cumming (2003; Zimbabwe) is 1.3±0.7 km^{-2}, although the range is very wide (0.01–14.1 km^{-2}), and in specific areas can fluctuate according to the availability of food and water (e.g., Fennessy et al., 2003). If the average population density of giraffes is 1.3 km^{-2}, then the average core or minimum home range available for each giraffe is 0.8 km^2. Theoretically, therefore, a giraffe of average size living in Africa is able to obtain the energy it needs to meet its metabolic requirements from a habitat area of about 900 × 900 m, which seems very small. McNab (1963) developed two equations to estimate the core home range needed to support metabolism. One was derived from all mammals and is $0.027 \times Mb^{0.63}$ km^2 and the other, derived for browsers and grazers, is $0.012 \times Mb^{0.69}$ km^2, with body mass in kg in both cases. The latter equation seems more valid for giraffes, but it is based on fewer data points. If it is used to estimate core home range (in km^2) for a giraffe weighing 750 kg, the core home range is ~1.2 km^2, i.e., somewhat larger than the home range derived from population density. The core home range (km^2) of a giraffe can also be calculated by dividing FMR by the energy being produced by browse in a square kilometer. The only measurement made so far of the rate at which the browse that is eaten by giraffes is produced has been made by Pellew in the Serengeti (1983b). He defined available browse as the total plant material produced by all woody species potentially edible by giraffes, and found that browse production rate was approximately 7,000 kg DM per hectare per annum. However, it seems likely that the units of area were km^{-2}, not $hectare^{-1}$. A productivity of 7000 kg DM km^{-2} $annum^{-1}$ is equivalent to 1,866 watts km^{-2}, which is similar to the 2,188 watts km^{-2} that Jetz et al. (2004) established as a typical energy supply rate for herbivores in a savannah–woodland biome, taking into account the energy in both browse and grass. Assuming a metabolizable energy content of 8,400 kJ kg^{-1}DM, and complete assimilation of the energy, browse productivity of 7,000 kg DM km^{-2} $annum^{-1}$ theoretically can provide sufficient energy so that the FMR of an average giraffe weighing 750 kg of ~115,000 kj d^{-1} (~1330 watts) can be acquired from 0.71 km^2 of savannah–woodland, which is more or less the size expected from population density (Table 8.6).

However, direct measurement of the actual home range occupied by giraffes gives a size more in keeping with the observed daily activities of a giraffe. For example, the home range used by a single giraffe in the Kruger National Park in South Africa tracked over a period of 10 months was found to be 282 km^2 (du Toit, 1990c). Observation of 60 males and 50 females over a year in the arid Tsavo National Park in Kenya showed that their average home range was 162.8 km^2 but could be up to 650 km^2 in males and 480 km^2 in females. In 50% of males and 38% of females, home range was less than 100 km^2. In 16% of males and 2% of females, it was less than 20 km^2 (Leuthold and Leuthold, 1978). In the Tarangire Ecosystem of Northern Tanzania, giraffe home range is 90 km^2. In Nairobi National Park in Kenya, home range was 62–85 km^2 (Foster and Dagg, 1972), 68.2±7.0 in the O.I. Pejeta Conservancy in Kenya (VanderWaal et al., 2014), 79±27 km^2 in the Luangwa National Park in Zambia (Berry, 1978), 90 km^2 for males and 50 km^2 for females in Niger (Le Pendu and Ciofolo, 1999, but a home range of 1,559 km^2 has also been reported there), and 41 km^2 for a single cow and calf in Timbavati Nature Reserve in South Africa (Langman, 1977), but an average of 23 km^2 for adult males, 25 km^2 for adult females, and 13 km^2 for juveniles in that Reserve (Langman, 1973b). In the Kalahari Desert, average home range is 206 km^2 (Deacon and Smit, 2017) and in the Northern Namib Desert the average size of the home range for males is ~500 km^2 (range 27–1,950 km^2), ~400 km^2 for males (range 13–1,098 km^2) and the population density is 0.01 km^{-2} (Fennessy, 2003; 2009). All these measured home range sizes are much larger than the core home range necessary to fulfill metabolic needs. Why? The mathematics of fractal geometry (Haskell et al., 2002), the physics of particle collisions (Jetz et al., 2004), the inclusion of the concept of biological time (Lindstedt et al., 1986), the influence of social organization (VanderWaal et al., 2014), and phylogenetic factors (Tucker et al., 2014) have been used to explain the difference. The equations generated in these studies and others can be condensed into a generic equation (expanded home range (km^2) =0.014 × Mb$^{1.27}$; Mitchell et al., 2015). It predicts a home range of 63 km^2 for a giraffe weighing 750 kg, compared to the 0.7 km^2 needed to provide its daily energy (Table 8.6). Theoretically, then, an area of 65 km^2 (8,000 hectares) could, under ideal conditions, be shared by 90 giraffes without there being any shortage of browse, and their population density would be 1.4 km^{-2}, which is the average population density for giraffes in Africa. Ideal conditions, however, rarely prevail, and the home range will expand (or contract) according to circumstance.

None of the studies actually explains why home range measured directly is more than 100 times larger than is expected from calculations based on FMR and the energy in browse, but numerous reasons have been proposed. One reason is that because digestibility of browse is low, a far larger area is needed to provide the browse required to meet energy needs than the basic equations predict. Another complicating factor is that their tallness and the access tallness gives them to competition-free energy are not taken into account when the minimum home range of giraffes is predicted from equations developed from the ecology of other mammals. Perhaps because giraffes are fussy feeders, they have to wander widely to look for and find the ephemeral and patchily distributed browse on which they depend (du Toit, 1990c) or, to put it another way, being a large species, giraffes are able to use a wider range of habitat patches with the benefit of more even use of environmental resources (du Toit and Owen-Smith, 1989). Other reasons that have been suggested (so far) include the ecological pressure of having to share local resources with conspecifics (Jetz et al., 2004) or with elephants or other artiodactyl browsers (Berry, 1978; du Toit, 1990c), or the time it takes for browse to recover (du Toit, 1990c), or a search, in the case of males, for receptive females; or age, with older animals having a larger home range than younger ones (Langman, 1973b; Berry, 1978; VanderWaal et al., 2014); and edaphic and physiographic factors like aridity (Berry, 1978; Fennessy, 2009), with "homeless" sub-adult males wandering most widely. It is clearly a very complex set of interactions, the integration of which for any individual wild giraffe can only be guessed, and humbly guessed at that. Furthermore, while it is usually assumed that the meanderings are purpose driven, they may be whimsical. Giraffes wander because they can, and the energetic cost to them of doing so is low. If a human was in their position, he/she would enjoy the freedom that living well, and well inside their ability to make a living, brings. Perhaps giraffes do so, too.

9
The Blood Pressure of Giraffes

Giraffes have very high blood pressure. In humans, abnormally high blood pressure, or to give it its clinical name, hypertension, potentially is fatal. High blood pressure exposes all arteries and arterioles throughout the body to mechanical stress and the consequences are weakening of blood vessel walls, the accumulation of atherosclerotic plaques, and the formation of blood clots in blood vessels. Rupture of small blood vessels or their obstruction by clots can follow, and the consequences are especially serious if brain, heart, and kidney blood vessels are affected. In humans, most of the time the damage goes unnoticed until the person has a heart attack or develops heart failure, or has a stroke or develops dementia or renal failure. Giraffes never develop any of these complications, ever, even though they have a blood pressure that is two to three times higher than the blood pressure in a normal, healthy human. Hypertension in giraffes is necessary and not harmful. It is physiological, not pathological. What mechanisms have they evolved to prevent the pathological consequences of hypertension?

The story of how the mysteries of the giraffe cardiovascular system have been unraveled, how the first measurements of giraffe blood pressure were made, and of the discovery of the unique mechanisms that giraffes have evolved, is a story worth telling. It begins in 1935. That was when the University of Cape Town in South Africa was bequeathed funds to establish the J.S. Marais Surgical Research Laboratory. The laboratory was to become famous when Chris Barnard performed the world's first human heart transplant in Cape Town in December 1967. Some of the basic research for the transplant had been done in this laboratory. In 1937 the University recruited Robert Goetz as the laboratory's first research officer. It was an inspired choice. Goetz had graduated *summa cum laude* in medicine at the University of Frankfurt in Germany in 1935, but because he had resigned from the National Socialist Student Organization, he was declared politically undesirable and was refused a medical license (Goetz, 1988; Konstantinov, 2001). (He eventually received the license in 1997, at the age of 87, three years before he died.) He left Germany and became an assistant in the Department of Anatomy at the University of Berne in Switzerland. There he studied the placental circulation of the Tenrec (*Centetes ecaudatus*), was awarded an MD, but was not allowed to practice medicine as he was a foreigner. He went to Scotland to the Department of Physiology of the University of Edinburgh. Goetz applied to stay in the United Kingdom as a refugee. However, as he was not Jewish, the United Kingdom Home Office did not consider him to be a *bona fide* refugee, refused permission, and limited the length of his stay in the United Kingdom to one year.

As is so often the case with political rather than merit-inspired decisions, all these obstructions resulted in great cost to those countries, but they proved to be to South Africa's gain. As the end of his year at Edinburgh loomed, Goetz applied, successfully, for a research fellowship in the Department of Surgery at the University of Cape Town. In Germany, Switzerland, and Scotland he had developed a strong

How Giraffes Work. Graham Mitchell, Oxford University Press. © Oxford University Press 2021.
DOI: 10.1093/oso/9780197571194.003.0009

interest in vascular disease, and soon after his arrival in South Africa he was fully occupied in developing diagnostic tests for patients, though he could not register as a medical practitioner as the South African Medical Council did not recognize his German or Swiss degrees. Moreover, by moving to South Africa, Goetz had not escaped persecution. At the start of the Second World War, being a German, he was declared an enemy alien and had to report daily to the South African Police. This new restriction was not unwelcome. It provided an opportunity for uninterrupted study. In 1941, to gain registration with the South African Medical Council, Goetz enrolled as a part-time medical student at the University of Cape Town, although he was of course already highly qualified. As he put it, he was a *pro forma* student (Goetz, 1988), but in exchange for being a "student" he had to give up one-third of his salary. He had time for both research and for medical studies and graduated as a Bachelor of Medicine and Bachelor of Surgery (MBChB) in 1944. For the first time, nine years after graduating in Germany, and as the holder of three medical degrees, he was permitted to practice medicine.

Goetz left Cape Town in 1957 to join the Department of Surgery at the Albert Einstein College of Medicine in New York. The College had opened in 1955, and its founding charter must have been attractive to him. Einstein had agreed to attach his name to the College only if its

> *admission would not be based upon race, religion, creed, color, national origin, sex, age, disability, veteran or disabled veteran status, marital status, sexual orientation or citizenship status.*

In other words, anyone could apply; no one could be excluded for non-academic reasons. But a liberal entry charter did not mean that Goetz was able to practice medicine, as none of his now-numerous medical degrees was recognized in the United States. Hence, at the age of 47, he had to sit the American Medical Board examinations, which, in 1958, he passed at the first attempt. On May 2, 1960, he performed the first ever successful human coronary bypass operation using a cannula made of tantalum to connect an internal thoracic artery to the coronary artery. Making the connection took him 17 seconds (Konstantinov, 2001). This achievement was disregarded, and he was even denied the opportunity of presenting the case at a meeting of the American Association for Thoracic Surgery because the President of the Society declared that "the subject of coronary artery bypass surgery was unimportant and would soon be forgotten" (Haller and Olearchyk, 2002). By all accounts, Goetz retained his sense of humor despite the slight, but he surely would have had a moment of *schadenfreude* when bypass surgery was recognized as one of cardiology's 10 greatest achievements (Haller and Olearchyk, 2002). Goetz remained at Albert Einstein until 1982, when he retired to engage in the higher calling of breeding Aberdeen Angus bulls.

Goetz made many other contributions to cardiovascular physiology and pathology and became particularly interested in hypertension, which for humans is defined by the World Health Organization as a systolic pressure at or above 140 mmHg and a diastolic pressure of 90 mmHg or more (equivalent to a mean arterial pressure[MAP] of ~110 mmHg). Normal human blood pressures usually are given as ~120 mmHg systolic and ~80 mmHg diastolic, and normal MAP as 90–100 mmHg. By 1940, Goetz's thoughts had turned to the blood pressure of giraffes. In 1988, reminiscing about his research, he wrote,

> *In a fully grown giraffe standing 20 feet high the distance between heart and head is roughly 10 feet* [see Figure 9.1]. *A column of blood that high exerts a pressure of 300 mmHg and so left ventricular pressure would have to exceed that to open the aortic valve and higher pressures to perfuse the brain.*

The pressure to which Goetz was referring is the blood pressure in the carotid artery, the great artery that takes blood from the heart to the head. The artery's name comes from the Greek *karōtis*, which means "stupefy" or "sleep," as pressure on it reduces blood flow to the brain and induces short-term unconsciousness, a procedure the ancient Greeks used to induce the anesthesia needed for ritual circumcision. The size of carotid blood pressure is due to a combination of hydrostatic pressure and the pressure generated by the heart. Hydrostatic pressure is the pressure exerted by the water in blood and in a carotid artery it is determined by gravity acting on the column of blood in the carotid artery. If a giraffe's carotid artery were to be attached to a tube containing mercury, the pressure in the artery would push the mercury to a height of 250 mm up the tube. Attached to an adult human's carotid artery, the mercury would rise to 100 mm. Goetz was not the first to consider the physiological problems imposed by a long neck on the heart and blood vessels of the giraffe's cardiovascular system. When, in 1838, Owen made his famous dissection of three giraffes, he wrote,

> *it might be allowed to speculate on the influence which was due to the mechanical obstacles to the flow of blood up the singularly long and slender carotids in retarding in the Giraffe the development of antlers beyond the point at which they characterize the pricket age of the Deer.*

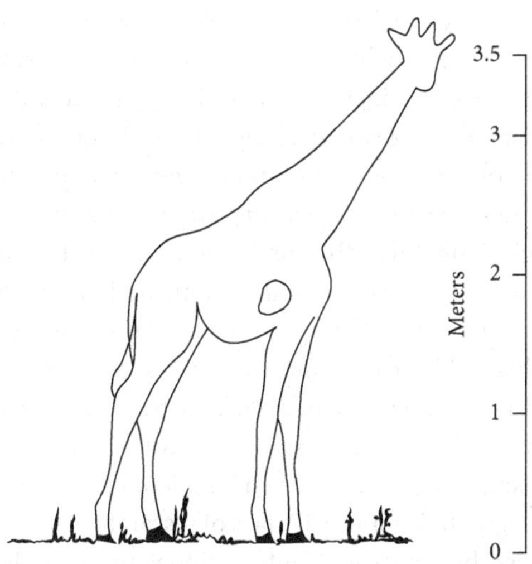

Figure 9.1. The position of the heart in a giraffe.
Redrawn after Patterson et al., 1965, with permission of John Wiley and Sons.

In 1924 the great Danish physiologist and winner of the 1920 Nobel Prize for Physiology and Medicine, August Krogh, also pondered the very high blood pressure of giraffes, and wrote,

> *it might be imagined that a circulatory system like that of a mammal might be reproduced in any desirable dimensions, but it is at least not improbable that the giraffe is not very far removed from the limit in an animal living on land.* (Krogh, 1924)

In short, Owen, Goetz, and Krogh thought that giraffes might live at the limits of cardiovascular (not to mention respiratory and skeletal) physiological possibility, and how they did so was a mystery.

Owen and Krogh did not have the opportunity to test their ideas. But Goetz could test his, and in August 1954, nearly 15 years after he had first imagined what a giraffe's blood pressure might be, he planned what he called a "scientific safari." First, he intended to study "the anatomy and physiology of the giraffe circulatory system (and especially its blood pressure) and its respiratory system." A second purpose was to investigate "the morphology and chemistry of the blood," and a third was to work out the "methods of capturing and handling this animal" (Goetz and Keen, 1957). In the end, Goetz carried out two scientific safaris (the other took place in 1956) and it would have been a logistical nightmare to mount them. In articles published in 1955 and 1957, Goetz and his colleagues described some of the problems. Permission had not been given to conduct the studies in the Kruger National Park or other National Parks in South Africa where giraffes were plentiful and assistance at hand. So they were obliged to carry them out on a cattle ranch more than 2,000 km (1,300 miles) from Cape Town in what was then the far northeastern Transvaal, which is about as far as you can get from Cape Town and still be in South Africa. This was a place "where giraffe were wild, had multiplied and become a menace to agriculture." The South African Defence Force provided jeeps, lorries, and a generator unit. The generator was a crucial piece of equipment. As Goetz noted, "the development of electric physiology equipment made it possible to take the lab to the giraffe" (Goetz and Keen, 1957), an idea that was as novel as it was original. Another essential item was formalin, to preserve specimens. A formalin tank was constructed by digging a ditch "8 ft long, 3 ft wide, and 3 ft deep," and lining it with a new material—plastic sheeting. The tank was filled with 1,000 L of 4% formalin. The hazards of field research soon became apparent. One night a small antelope, probably an impala (*Aepyceros melampus*), jumped the fence surrounding the tank and in its struggles punctured the tank's lining. All the formalin was lost, but the antelope, it seems, survived. A second tank was made. Funding was provided by the University of Cape Town, the South African Defence Force, the South African Council for Scientific and Industrial Research, and admirers from around the world, including C.J.F. Heymans. Heymans was later to become a Nobel Prize winner for his studies on the carotid sinus, the blood pressure sensing system of mammals. Such was his interest in the matter that he sent 5,000 francs, a considerable sum. Total contributions from all sources provided enough funds for just 12 days of research.

The teams assembled by Goetz were international and included O.H. Gauer from West Berlin, J.V. Warren and J.L. Patterson, both then at the Department of Surgery at the Albert Einstein College of Medicine in New York, which Goetz was about to join, and Maurice McGregor, a South African cardiologist who was then in the Department of Medicine at the University of the Witwatersrand in Johannesburg, South Africa. In 1957, the same year that Goetz had left South Africa to go to Albert Einstein College of Medicine, McGregor left South Africa to go to McGill University in Canada, where he has had a distinguished clinical and administrative career. In 2008 I asked him how he became involved. He answered:

> *I was sitting in my home one night* [in Johannesburg, South Africa] *when I got a crackly trunk call from somewhere in the Bushveld from Jim Warren. He explained that he was part of an international expedition to study the haemodynamics and respiratory function of giraffes. The team was waiting for the arrival of their equipment. They knew of me as a cardiologist who had published hemodynamic studies (on Homo sapiens) and wondered if I would like to come and join them and bring the necessary electronic equipment with me. At the time we had a four channel polygraph made by Sanborn, and in the morning I contacted the agent who was a very good guy to see what could be done. In the end, we went up in his truck with the appropriate recorders and manometers and stuff. We found the team in a state of considerable international tension, living in tents under fairly rough conditions. Gauer had, of course, brought his own, battery-operated equipment and wished to proceed with the studies. The Americans did not trust him to share his data and said they must wait until all their equipment had arrived (which it never did!). Goetz fumed in the background. The next morning we went into action, I as operator of the recording equipment. It did not go well. One of the animals died. But we managed to get some data.*

The data obtained from these studies were the first physiological measurements made on a living giraffe.

Organization was difficult, but not the main obstacle to a successful study. Rather it was capturing wild giraffes and holding them unanesthetized in a pen suitable for making measurements. As Goetz said, "no one had any experience of catching these animals and there were no references we could have followed." Physical capture was out of the question in the thick bush of the area, and there were no modern aids such as helicopters and capture pens. They decided that injecting an immobilizing agent was the best option, and the drug they selected was curare. The choice of curare reveals just how primitive capture of wild animals then was. Curare is a muscle relaxant and neither an anesthetic nor a tranquilizer. It is a substance that can be extracted from many plants and its first use is attributed to the Amazonian Indians, who still use it to kill monkeys for food. It kills by paralyzing the respiratory muscles, thereby causing suffocation. A more inappropriate drug to use to catch a giraffe scarcely can be imagined. Nevertheless, Goetz judged the dose for an adult giraffe to be 200 mg, "which turned out to be correct." However, to inject a solution containing that amount of curare was not feasible and, therefore, curare powder was used. To turn the powder into a solid injectable pellet, they experimented with several carriers before discovering one that was perfect—icing sugar. The mixture of curare and icing sugar hardened quickly, but was also very hygroscopic. Thus, once in the body, it absorbed body fluids very quickly, the body fluids turned it into a solution, and the curare became readily available for absorption and distribution around the body. But how was the curare to be administered to a running giraffe? According to Goetz, they tried using a bow and arrow but "our archer did not find it easy to shoot a giraffe racing at 30–40 miles per hour over veld and passing at about 40–50 yards distance" (Goetz and Budtz-Olsen, 1955). Moreover, the arrows did not have the force to penetrate the giraffe skin. Thus, a 0.303 (7.6 mm) rifle cartridge was taken apart, some of the lead removed to make room for the mixture, and three-fifths of the cordite sticks removed to compensate for the lower weight (relative to lead) of the sugar-curare mixture. This "hit-or-miss" approach is a far cry from modern techniques, but it worked. Giraffes could be caught 45 minutes after being shot, were blindfolded, and a tubular steel crush erected around them at the site of capture.

Goetz's studies of giraffe cardiovascular anatomy were excellent, were a classic of their kind, and remain a cornerstone for giraffe researchers. Some of his physiological results have not been easy to confirm. He did establish, however, that giraffe MAP is extraordinarily high (~250 mmHg, in an adult) and therefore two and a half times higher than that found in a person with entry-level hypertension. However, for those of us who study the physiology of free-living animals, Goetz's pioneering study had two greater legacies than the physiological data it produced (Mitchell, 2008). First, he showed that physiological research on large wild animals was feasible by "taking the lab to the animal." Since then, a vast amount of knowledge has been accumulated based on this principle and it has changed policies, generated legislation, and influenced the management of wild animals all over the world. Increasingly sophisticated and complex technology has been developed, and is being developed, to study the physiology and behavior of free-living wild animals in their natural habitats. Goetz's second great legacy arises from his use of a chemical "dart" to capture a wild animal successfully. He was the first to do so, and his achievement has burgeoned into the massive worldwide industry that it is today. Darts have brought capture, management, translocation, and scientific study of large wild animals within reach, rather than these being a daunting prospect fraught with difficulty and high mortality.

The Blood Pressure of Giraffes

A unicellular animal living in the primordial sea in which life is presumed to have begun was surrounded by an infinite volume of water, rich in nutrients that could be absorbed and a massive sink into which waste products could be excreted. Fluid-, nutrient-, and waste product–exchange were easy and life flourished. The many trillions of cells that make up the organs and tissues of a land animal also need huge amounts of fluid to function, but they are not surrounded by an infinitely large sea. Terrestrial animals solved the problem by taking the sea with them, and as far as we can tell, the concentration of the salts in the extracellular body fluids of a mammal (for example) is the same as it was in the primordial sea. The two extracellular fluids are the plasma in blood and the fluid in the interstitial space between blood vessels and the cells that make up organs. Interstitial fluid volume makes up about 15% of a mammal's body mass. The blood volume (cells and plasma) of any mammal is $0.07 \times Mb^{1.02}$ liters (L) (Stahl, 1967) and as each liter weighs ~1 kg, blood forms about 7%–8% of its body mass. The plasma part makes up 60% of blood, or 4%–5% of body mass. It is interstitial fluid and plasma that interchange with the fluid inside cells (intracellular fluid), and their combined volume in a 750-kg giraffe is ~150 L, a tiny volume compared to the volume of fluid that surrounded a unicellular animal in the primordial sea and interchanged with the intracellular fluid inside it. To compensate for the small volume in a terrestrial mammal, the fluids are circulated to cells, cleaned and refreshed by the liver and kidneys, and oxygenated by the lungs once every few minutes. So efficient is this system that, in a human, an estimated 100,000 L of fluid is able to be exchanged between the intracellular and extracellular fluids each day, and in a giraffe weighing 750 kg that volume will be ~800,000 L d^{-1}, a circulation and exchange of fluids that depends utterly on blood pressure.

What is the origin of blood pressure? Why is it so much higher in giraffes than it is in a human? Its primary, but not its only, origin is contraction of heart muscle. Contraction requires energy in the form of adenosine triphosphate (ATP). ATP is synthesized in heart muscle by (mainly) oxidation of fatty acids and glucose, and the muscle converts the chemical energy in ATP into the mechanical energy of contraction. Blood in the heart chambers, like any liquid, is incompressible, so any pressure applied to it by muscle contraction is transmitted to it, and thus the contraction generates blood pressure. The relationship between blood pressure (mean arterial blood pressure) and heart function can be described (though not, strictly speaking, accurately) by the formula

Mean arterial pressure (MAP) = cardiac output (CO) × total peripheral resistance (TPR).

Cardiac output is the product of heart muscle contraction and is the volume of blood pumped per minute (in L min^{-1}). The actual volume pumped each minute depends on the number of heart muscle contractions per unit time (usually recorded as beats per minute) and the volume of fluid (usually recorded as mL) ejected during each beat (stroke volume). The combination can be summarized as

Cardiac output = heart rate × stroke volume.

Resistance refers to the factors that affect the flow of fluids (like blood) through tubes (like blood vessels), and they were derived experimentally by Jean Poiseuille in 1838. He showed that resistance to flow is related to the viscosity of the fluid and the length and diameter of the tubes. Viscosity of blood depends on the proportion of blood that is cells, and in healthy animals that stays more or less constant (in giraffes that proportion is ~40% of blood volume). The length of the arteries, however, increases as an animal grows; thus resistance to blood flow also increases, and the consequence is that MAP increases as size increases (Figure 9.2). However, at any one moment in the life of a giraffe (or other mammal), viscosity of blood and the length of the blood vessels do not change, and thus resistance to the flow of blood through blood vessels is determined mainly by their diameter. Resistance is high if diameters are small and low if they are large.

Why was the blood pressure Goetz measured in his giraffes ~250 mmHg? The pressure that a heart has to generate must be high enough to satisfy two requirements. First, it must be high enough to overcome the characteristics of blood and the blood vessels that resist its flow through the vessels, and thus ensure that tissues are supplied with sufficient blood. Second, the pressure must be high enough to lift blood to all organs, but mainly the brain, that are above the level of the heart. The size of the pressure needed to overcome resistance is an unresolved aspect of physiology, but it is assumed to be unrelated to body mass (White and Seymour, 2013). However, large mammals do have higher blood pressure than do small ones because the distance between the heart and the tissues is large and the blood vessels are long, but giraffes have an even higher blood pressure than can be explained by size (Figure 9.2).

Their high blood pressure is a consequence of having a long neck. In a mammal whose head is at, or more or less at, heart level, like the head of a mouse or human, MAP is 75–100 mmHg, although it does rise slightly above 100 mmHg in larger mammals (Figure 9.2). As very little of the blood pressure of a mouse or human is needed to lift blood to the head, it is a pressure of 75–100 mmHg that, over evolutionary time, seems to have become the pressure the heart must generate in most if not all mammals to overcome resistance to blood flow, and provide tissues, especially the brain, with sufficient blood to sustain their metabolism. In a giraffe, whose head may be two meters above the heart (Figure 9.1), the pressure generated by the heart also has to lift blood to the head as well as being sufficient to overcome resistance. The pressure that then has to be overcome is the hydrostatic pressure exerted by blood in the carotid arteries. Its origin is gravity acting on the weight of the blood. The magnitude of the hydrostatic pressure thus depends on the vertical height of the column of blood and on the specific gravity of blood. Giraffe blood has a specific gravity of 1.055; that is, a given volume of blood will weigh 1.055 times that same volume of water, so the hydrostatic pressure generated by the column of blood in a carotid artery of a giraffe when the head is two meters above the heart is 1.055 times that of a 2-m column of water (i.e., equivalent to the weight of a column of water 211 cm high). Converted to the standard units of blood pressure (mmHg), the hydrostatic pressure generated is 155 mmHg (1 mm of Hg = 1.36 cm of water). To this pressure must be added the ~100 mmHg of pressure needed to overcome the resistance in blood vessels, and, therefore the total pressure that must be generated by the heart of a giraffe is ~255 mmHg, and that was the pressure measured by Goetz and his colleagues.

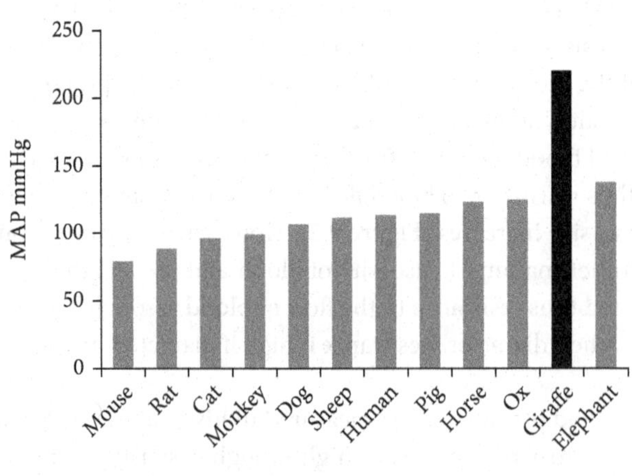

Figure 9.2. The effect of body size on mean arterial pressure (MAP).
Giraffe datum from Mitchell and Skinner, 2009. Other mammal data from Seymour and Blaylock, 2000.

Mean arterial pressure at the level of the heart has now been measured or reported many times in different giraffes, and an average MAP calculated from these measurements is 203±41 mmHg (Goetz, 1955; Goetz and Keen, 1957; Warren et al., 1957; Goetz et al., 1960; Patterson et al., 1957, 1965; van Citters et al., 1966, 1968, 1969; Warren, 1974; McCalden et al., 1977; Hargens et al., 1987a, 1987b; Mitchell and Skinner, 1993; Linton et al., 1999; Brondum et al., 2009). Using the same sources, average pressure at the head can be calculated to be 100±21 mmHg, the expected pressure needed to supply blood to the brain and the tissues of the head. The quite large variability of MAP (ranging from 160 to 240 mmHg) in giraffes can be explained by the variation in size of the animals, the difficulty of making measurements, whether the animals were active, lying, or standing, and whether they were anesthetized, sedated, or neither. In one of the few measurements ever made of blood pressure in an unanesthetized giraffe standing quietly, MAP fluctuated by ~15 mmHg (van Citters et al., 1969). This fluctuation reflects the up-and-down movement of a giraffe's head and neck as it goes about its daily chores, as blood pressure also depends on the angle that the neck is held relative to the trunk and, therefore, the height of the head above the heart and the height of the column of blood in the carotid artery. When the neck is held midway between horizontal and vertical, which is a typical angle of carry, the average value of MAP in a giraffe weighing 750 kg will be 207 mmHg. When the neck lies horizontally, MAP is reduced to 178 mmHg, and when the neck is completely vertical, then the MAP will be 241 mmHg (Mitchell and Skinner, 2009). The height of the head above the heart is of course the origin of a giraffe's high blood pressure, so in addition to neck angle, MAP is determined by neck length and increases in neck length are related to increases in body mass. Mean arterial pressure at birth in a giraffe (i.e., a body mass of 100 kg; Skinner and Hall-Martin, 1975) is the same as it is in other mammals because their necks have not yet elongated, but as giraffes grow and their body mass increases, their neck elongates, the heart muscles enlarge, and blood pressure rises. By taking into account neck angle and neck length, the way blood pressure increases as a giraffe grows can be demonstrated (Figure 9.3): for each 15 cm of increased neck length, MAP increases by ~17 mmHg and the heart muscles thicken by ~5 mm.

Despite changes in posture or changes in neck length, the pressure at the head remains the same (i.e., 100 mmHg), which means that blood pressure is regulated. In mammals, blood pressure is regulated by baroreceptors (baro- = Greek for "weight"). They usually are found (as C.J.F. Heymans discovered) in the carotid sinus, a part of the carotid artery near the head. In humans the sinus lies where the carotid artery on each side of the neck divides into its external (to the face) and internal (to the brain) branches, a point roughly at the level of the larynx. The baroreceptors contain elastic fibers, and when these are stretched, for example by a rise in blood pressure, the receptors are activated and they send signals via the ninth (glossopharyngeal) and tenth (vagus) cranial nerves to the heart and blood vessels, and the outcome is reduced heart rate, stroke volume, and resistance, and thus reduced blood pressure. Lawrence and Rewell (1948), who were the first to study the anatomy of a carotid artery in a giraffe, identified a carotid sinus, as did Goetz and E.N. "Teddy" Keen, Professor of Anatomy at the University of Cape Town, in the carotid artery where it passes the body of the first cervical vertebra (the atlas). Kimani and Mungai (1983) studied the

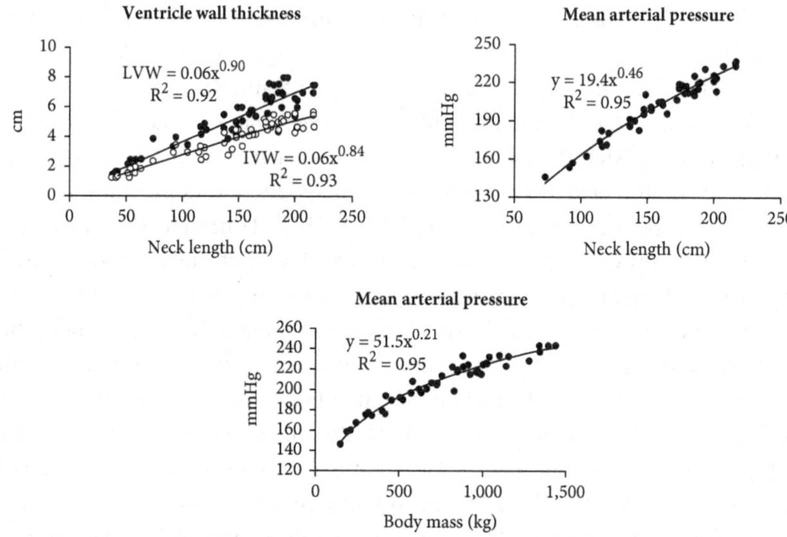

Figure 9.3. Top left: The relationship between neck length, left ventricle wall thickness (●), interventricular wall thickness (○), between neck length and mean arterial pressure (top right), and between mean arterial pressure and body mass (bottom).
Based on data obtained from 10 fetuses and 22 male and 24 female *G.c. giraffa*.

microanatomical structure of the carotid artery of giraffes and identified a segment at its upper end characterized by an elastic wall that extended 2 cm into the occipital artery; in other words, it lay where the main carotid artery divides into its occipital and external carotid branches (see Figure 10.6 in Chapter 10). This area of the artery was supplied by the glossopharyngeal nerve, and Kimani and Mungai concluded that it was the carotid sinus (Figure 9.4).

And there the matter should have ended: giraffes have an occipito-carotid sinus that detects blood pressure. Physiological experiments have confirmed that a baroreceptor reflex operates in giraffes and regulates their blood pressure (e.g., Hargens et al., 1987a). However, when Goetz raised or lowered the head of a giraffe, there was little change in heart rate (Goetz, 1955) or, as he said in a later study (1960), the changes were "small considering the large distance traversed by the head," and then he added, very surprisingly given that he had identified a carotid sinus in 1955, that "this is in keeping with anatomical findings that the giraffe does not have a carotid sinus" (Goetz et al., 1960). Perhaps, he speculated, "carotid sinus function may have been taken over by an occipital sinus." Why Goetz contradicted himself defies analysis. The contradiction had a consequence. Van Citters et al. (1969) uncritically accepted Goetz et al.'s comment, even though they had found that when giraffes lowered their heads heart rate slowed and MAP was reduced, and wrote,

> *These responses are typical of a baroreceptor reflex mechanism for blood pressure regulation. Although previous anatomic studies have failed to reveal the existence of a carotid sinus or an analogous structure in the giraffe, the physiological observations argue strongly for a similar reflex, with the receptor site located in or near the cranium.*

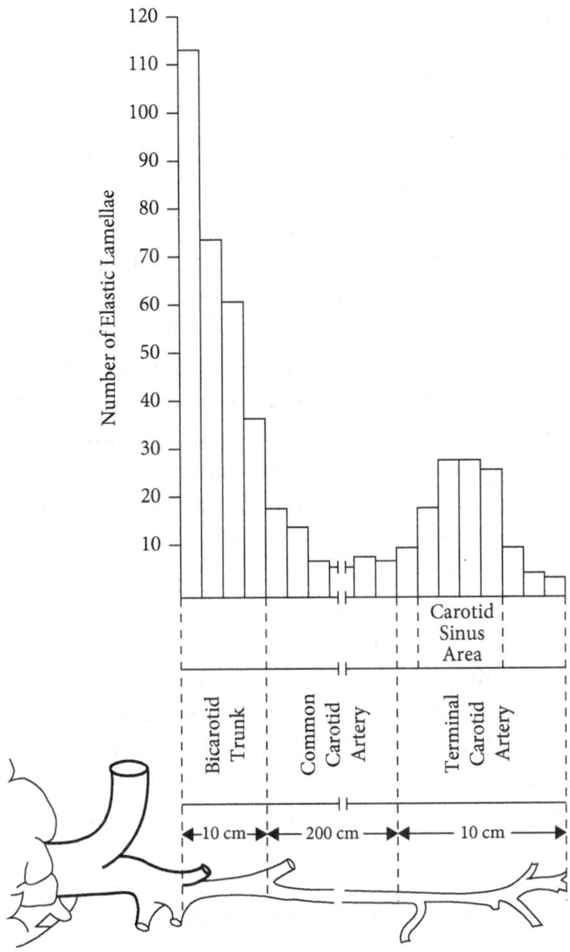

Figure 9.4. The position of the occipito-carotid sinus in giraffes.
From Kimani and Mungai, 1983, with permission of Karger Publishers.

While there are almost certainly additional blood pressure detectors in the brain, there is no doubt, despite Goetz et al.'s and van Citters et al.'s conclusions, that giraffes have baroreceptors. Giraffes could not have evolved and survived without having baroreceptors that regulate their blood pressure. What is unusual about them is that, for a very good reason (see Chapter 10), they have evolved to lie in both the occipital and carotid arteries rather than in the carotid artery alone.

The Heart of Giraffes

It has long been accepted that to generate a pressure of 200–250 mmHg, the giraffe heart must be powerful. This idea became entrenched in giraffe folklore when Goetz and colleagues reported a heart mass of 25 lbs (11.25 kg) in a giraffe that had an estimated body mass of 1,200 lbs (~500 kg) (Goetz, 1955; Goetz and Keen, 1957). In other words, the heart weighed about 2.5% of total body mass. The mammals that have the largest hearts relative to their body mass are the smallest mammals—shrews—in which heart mass is 1.7% of body weight. In large mammals heart mass typically is ~0.5% of body mass, and its actual mass can be predicted by an allometric equation: Heart mass (g) = $6^{*}Mb^{0.98}$ (Prothero, 1979). Thus the heart that Goetz weighed was five times heavier than expected. It was also 31 cm long from base to apex. Owen (1838) was the first to measure heart size; in a three-year-old giraffe he dissected, which also would have weighed about 500 kg, he found it to be 22 cm wide at its base and 22 cm long. Many years after Owen's study, Cobbold, in 1854, reported a heart of 1.82 kg taken from a giraffe of an estimated body mass of approximately 455 kg; therefore, heart mass was 0.40% of body mass. In 1864 Crisp (1864a, 1864b) dissected three giraffes and found heart mass to be 2.27 kg and 2.5 kg in two animals, each with a body mass of 800 kg, and he found a heart weighing 450 g in a young giraffe with a body mass of 153 kg. These weights correspond to a heart mass:body mass ratio of 0.3%. Altman and Dittmer (1971) reported a heart weighing 4,990 g in a giraffe with an estimated body mass of 1,220 kg for a heart mass:body mass ratio of 0.41%. In a comprehensive study done in 2009, the size of the hearts, their masses, and the body masses of 56 giraffes, ranging from an 18 kg fetus to a 1,441 kg adult male, were measured (Table 9.1).

The heart mass in those giraffes fits the regression equation Heart mass (kg) = $0.006^{*}Mb^{1.03}$, which is almost identical to the equation developed by Prothero (1979) for large mammals in general. The relative heart mass of the 56 giraffes was on average 0.51% ±0.07% of body mass, which is not significantly different from the average predicted for all mammals that have a body mass the same as that of giraffes (0.55% ±0.02%). Thus the heart mass:body mass ratios and heart masses generally in giraffes

Table 9.1 The Allometry of Heart Dimensions in Giraffes

Variable	Allometric Equation in Fetuses	Measured Fetal Range	Allometric Equation in Postnatal Giraffes	Measured Postnatal Range
Heart mass (kg)	$0.006 \times Mb^{1.03}$	0.134–0.534	$0.010 \times Mb^{0.90}$	1.07–7.75
Heart mass:body mass (%)	$0.64 \times Mb^{0.03}$	0.93–0.64	$1.0 \times Mb^{-0.10}$	0.74–0.40
Heart length (cm)	$2.6 \times Mb^{0.36}$	7–13	$3.4 \times Mb^{0.29}$	14–31
Heart diameter (cm)	$1.8 \times Mb^{0.39}$	5–10	$2.7 \times Mb^{0.29}$	12–21
Left ventricle wall thickness (cm)	$0.53 \times Mb^{0.36}$	1.5–2.5	$0.48 \times Mb^{0.38}$	3.1–8.0
Right ventricle wall thickness (cm)	$1.0 \times Mb^{0.18}$	1.2–1.8	$0.34 \times Mb^{0.28}$	1.3–2.5
Right:left ventricular wall thickness	$1.9 \times Mb^{-0.28}$	0.80–0.40	$0.71 \times Mb^{-0.1}$	0.43–0.34
Interventricular wall thickness (cm)	$0.53 \times Mb^{0.30}$	1.3–1.9	$0.42 \times Mb^{0.35}$	3.1–5.7

Based on data collected from 10 fetal giraffes and 46 postnatal giraffes by Mitchell and Skinner, 2009. Table reprinted with permission of Elsevier.

are very different from those measured by Goetz, and either the heart he weighed was unusually large or he had underestimated the giraffe's body mass, or simply made an error—an error that spawned an enduring myth. The heart from Goetz's giraffe was 31 cm long, which can be its length in giraffes that weigh 1,000 kg or more and not one that weighed 500 kg—its length in a giraffe weighing 500 kg should be ~21 cm (Figure 9.5), just as Owen had found. Goetz almost certainly underestimated the giraffe's body mass, but it is equally likely that he simply confused the units of weight. The heart probably weighed 11.25 lbs (5 kg, not 11.25 kg) and that would have been its mass if the giraffe had had a body mass of ~1,000 kg (and not ~1,000 lbs). It is from small errors like these that myths emerge.

The ratio of heart mass to body mass in giraffes is not constant throughout growth (Mitchell & Skinner, 2009; Figure 9.5). The highest ratio occurs in fetuses (0.71% of body mass) and the lowest ratio in the largest giraffes (0.41% of body mass), a decline in proportion that is probably true for all mammals (Holt et al., 1968). In fetal calves, heart mass forms 1% of body mass (Reeves et al., 1972) and in fetal kangaroos 0.7% (Snelling et al., 2015), and the reason heart mass is higher in fetuses than it is in adults is that the heart of a fetus has to circulate blood not only to fetal organs but also through the placenta (see Chapter 17). The blood supply to a placenta can add 30%–50% to the volume of the circulatory system. The decrease in relative heart mass as a giraffe grows follows the decline in the rate that heart length and heart diameter increase with growth (Figure 9.5). Even though the size of a giraffe's heart and its relative mass decline as body mass increases, its function must increase in order to supply an increasingly larger body

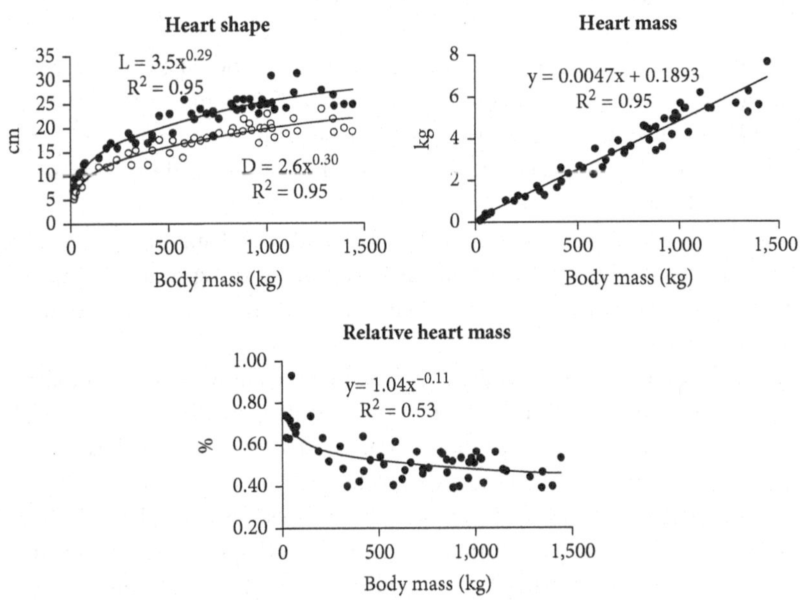

Figure 9.5. Changes in length (●) and diameter (○) of the heart (top left), changes in heart mass (top right), and changes in the heart mass:body mass ratio (bottom) as body mass increases.

Constructed using data obtained from 10 fetuses and 22 male and 24 female *G.c. giraffa*.

mass with blood. Function depends on the contraction characteristics of heart muscle, and those characteristics develop faster than does heart muscle mass.

All mammalian hearts are made up of four separate chambers—two atria and two ventricles—the walls of which are made of specialized cardiac muscle. The two atria are thin-walled and, in the sequence of contractions that make up a single beat of the heart, or what is called its cardiac cycle, they contract before the ventricles do in order to pump blood into the ventricles and, in this way, ensure that the ventricles are completely full of blood just before it is their turn to contract. The ventricles pump blood into the lungs (the right ventricle) or into the systemic circulation (the left ventricle), and are separated from each other by an interventricular wall. Ventricle wall thickness corresponds to the force of contraction required. The interventricular wall is 2.5 cm thick in a newborn giraffe and ~6 cm thick in an adult. The thickness of the right ventricle wall ranges from 1.0 to 3.5 cm but the left ventricular wall, which is the source of MAP, ranges from 3 cm thick in a newborn giraffe to 8 cm thick in a mature adult (Figure 9.3). In the heart that Goetz studied, the left ventricular wall was 7.5 cm thick. The contraction characteristics of the left ventricular wall of giraffes are impressive, and are illustrated by its ability to generate a very high pressure in a very short time (up to 6,000 mmHg s^{-1} as measured by Goetz et al. in 1960), which is two and a half times that of a cow's heart and five times that of the human heart (Patterson et al., 1965). The rate of pressure generation in the giraffe's right ventricle is 1,200 mmHg s^{-1} (Goetz et al., 1960). The origin of the left ventricle's power is the relationship between blood pressure, the tension developed by the muscle, its thickness, and the radius of the ventricle. This relationship was described by Pierre-Simon Laplace in 1806 and is encapsulated in the Laplace formula:

$$MAP = (2 \times \text{wall tension} \times \text{wall thickness}) \div \text{radius}.$$

The tension that heart muscle can produce is proportional to its cross-sectional area and it is assumed to be the same in all mammals (~20 g mm^{-2}; Sonnenblick et al., 1968); although not yet measured in them, it can be assumed to be the same in giraffes. However, because their left ventricle wall is so thick, its cross-sectional area is large and consequently the strength of the contraction it develops is high. In addition, the internal radius of the left ventricle is small (see Plate 9.1), and what Laplace's equation predicts for the left ventricle of giraffes is that its thick wall and a strong contraction, combined with the very small internal radius of the ventricle, will result in the generation of a very high pressure, and it does.

The impressive ventricle wall thicknesses of a giraffe's heart can be demonstrated if a horizontal section is made through the ventricles. Plate 9.1 shows that the left ventricular and interventricular walls of an adult giraffe are so enlarged as to obliterate the space inside the ventricle almost completely. In most mammals, wall thickness increases because the size of each muscle cell increases (hypertrophy), but in giraffes it is also caused by an increase in the number of cells (hyperplasia; Østergaard et al., 2013). The processes of hypertrophy and hyperplasia that lead to the dimensions of the heart, shown in Figure 9.5 and Plate 9.1, are gradual and start after birth. Before birth, the blood pressure generated is low and the walls of the ventricles are thin. Initially, in a fetus, the right ventricle contributes more to the circulation of blood through the body than does the left ventricle and is, therefore, thicker, but gradually the left ventricle takes on the major role (see Chapter 17). At birth, the neck of a giraffe starts to elongate rapidly (van Sittert et al., 2010) and the thickness of the left ventricle wall (and also the interventricular wall) increases in concert with increases in neck length and the associated increase in MAP (Figure 9.3). The pressure needed to circulate blood to the lungs is one-fifth of MAP, and consequently the thickness of the right ventricle wall does not increase to the same extent (see Chapter 12).

The sequence of contractions in a cardiac cycle that produce cardiac output is controlled by electrical impulses that originate in specialized areas of heart muscle called the sino-atrial and atrio-ventricular nodes. It might be expected from the thickness of the heart's muscles that the electrical events that coordinate the contraction–relaxation cycle in giraffes would be different from those found in other mammals. The only study made of the anatomy of the giraffe electrical conducting system has shown that the way that the conducting fibers are divided and distributed through the heart is not different from that found in other ruminants, but in giraffes the group of cells that is the origin of the electrical impulse that triggers ventricle contraction—the atrio-ventricular node—is bigger and the fibers that conduct impulses to the ventricles from the node are up to 50% thicker that those found in sheep and cattle (Castigli and Sacchi, 1941). A large node and large diameter fibers generate a powerful electrical impulse and conduct it fast to the muscle. The electrical activity in a giraffe's heart can be measured by an electrocardiogram (ECG), and two ECGs have been made so far in giraffes. The typical PQRST wave pattern of electrical activity is identical to that found in an ECG taken from a normal human heart with two exceptions: in a giraffe's ECG the amplitude of the waves is higher (Goetz, 1955; Jefferson, 1971; Rossof, 1972) and the T wave is inverted (Goetz and Keen, 1957; Jefferson, 1971). That the amplitude of the waves is higher than it is in humans is to be expected. Wave size is proportional to the mass of the muscles; bigger muscles discharge more electricity when they contract than do smaller ones. The T wave represents repolarization or electrical recovery of the ventricular muscles during the relaxation phase of a cardiac cycle. Jefferson (1971) thought that the inversion was a consequence of the heart having to eject blood against an already high pressure in the aorta, with the result that the direction in which the muscle fibers re-established their normal, resting, electrical activity was reversed. In humans a cause of T wave inversion is ventricular muscle hypertrophy, and this surely must be a more likely explanation in giraffes than that of Jefferson.

Normally two heart sounds can be heard during a cardiac cycle, the very familiar "lub-dub" that can be heard easily through a stethoscope. The first "lub" sound is associated with closure of the valves that separate the atria and ventricles (the atrio-ventricular valves), and their closure prevents regurgitation of blood from the ventricles into the atria when the ventricles contract. The second "dub" sound is caused by closure of the valves that lie between the left ventricle and aorta and between the right ventricle and the pulmonary artery when the heart begins its relaxation phase. They prevent regurgitation of blood from those vessels back into the ventricles. Three recordings of heart sounds in giraffes have been made (Goetz, 1955; Jefferson, 1971; Rossof, 1972), and in all three the first "lub" heart sound was so prolonged that it appeared to be split into two parts. The atrio-ventricular valves close because of the increase in pressure within the ventricles as they start to contract, and usually, as in humans, the atrio-ventricular valves on the left and right sides of the heart close simultaneously and so only one "lub" sound is heard. In giraffes it seems that the very rapid rise in pressure in the left ventricle closes its atrio-ventricular valve slightly earlier than does the valve that separates the right ventricle from its atrium, and thus a split first sound is generated. In a human a split first heart sound usually requires further investigation because it is abnormal. In giraffes a split first sound is normal.

Cardiac Output

For the blood pressure generated by the heart to be as high as it is in a giraffe, either the volume of blood pumped per minute (cardiac output) must be high, or the resistance to blood flow through the vessels (TPR) must be high, or both. Which is it? For a long time it was thought that the source of the high blood pressure of giraffes was a large heart pumping large volumes of blood. We now know that the heart is not larger than expected for a mammal of a giraffe's mass. But does it pump more blood? In adult humans that are not exercising, cardiac output is approximately 5 L min^{-1}. In other words, five liters of blood are pumped around our body every minute. Five liters is also the approximate blood volume of an adult 70 kg human at rest. Thus a volume of blood equal to our entire blood volume circulates around the body every minute, and this relationship is nearly true for all mammals. The blood volume of a giraffe has never been measured, but it is assumed to be the same proportion of body mass as it is in any other mammal and thus fits the regression equation Blood Volume (L)= $0.07{*}Mb^{1.02}$ (Stahl, 1967). For a 750-kg giraffe it is ~60 L. An extrapolation is to ask how long it takes an individual red blood cell (erythrocyte) to circulate around the body from the lungs to the body and back again. Each time a red blood cell leaves the lungs, it takes a different route through the body, perhaps to muscles or to the kidneys or any other of the vast possibilities, but on average, the time that it takes to return to the lungs in any mammal is equal to $0.35 \times Mb^{0.21}$ minutes (Stahl, 1965, 1967) and for a giraffe of average size of 750 kg, also at rest, it is 1.4 min, and if so, 60 L would also be the volume of blood that circulates every 1.4 minutes.

Cardiac output varies with metabolic rate, which varies with body mass, and thus the allometric equation developed from measurements made of cardiac output in many mammals, (CO (L min^{-1}) = $0.211 \times Mb^{0.78}$), should apply to giraffes. Accurate measurement of cardiac output is difficult and requires sophisticated equipment, but it has been measured in 14 giraffes in three different studies (Goetz et al., 1960; Linton et al., 1999; Smerup et al., 2016). In all three studies, MAP was at the level expected, but the range of cardiac output measured was wide. For example, in one of Goetz et al.'s giraffes, two measurements, each using a different technique for measuring cardiac output, gave values for cardiac output of 75 L min^{-1} and 39.1 L min^{-1}, and in another giraffe, cardiac output was measured to be 32 L min^{-1} and 22 L min^{-1}. From the same study, three other cardiac outputs have been reported: 19.4, 41, and 49 L min^{-1} (Warren et al., 1957). These giraffes were estimated to weigh approximately 500 kg, were wild, restrained, not anesthetized or sedated, and thus the variation may have reflected anxiety. Moreover, the body mass of the giraffes actually was larger and probably at least 700 kg, and predicted cardiac output in them would be 35–40 L min^{-1}. Linton et al. (1999) measured cardiac output twice in an anesthetized giraffe weighing an estimated

800 kg and found it to be 19.7 L min^{-1} and 20.3 L min^{-1}. Linton regarded these values as being low, and he speculated that the reason was that peripheral resistance was high. The predicted cardiac output of Linton's giraffe was nearer 40 L min^{-1}. In Smerup et al.'s study, the giraffes weighed on average 495 kg and their cardiac output was 16 L min^{-1}, compared to a predicted value of 27 L min^{-1}. They calculated resistance to be 16 mmHg min L^{-1}, which is about twice as high as predicted for giraffes and, as blood pressure was within the normal range for giraffes, cardiac output would have been about half the expected value. Further uncertainty about the size of cardiac output in giraffes arises because Patterson et al. (1965) developed an allometric equation that related body mass and cardiac output for a range of mammals including giraffes ($CO = 0.1017 \times Mb^{0.9988}$). This equation predicts a cardiac output for a giraffe to be twice what has been measured and twice what equations developed by Calder (1996), Stahl (1967), Holt et al. (1968), and Seymour and Blaylock (2000) predict for mammals in general (Table 9.2). When the cardiac output predicted by the general mammalian equation is applied to giraffes, the cardiac output should be 3.8–6.6 L 100 kg^{-1} of body mass, or about 40 L min^{-1} in a giraffe weighing 750 kg, a volume similar to those measured, but somewhat less than its blood volume. Thus, despite all the uncertainty and difficulties of measurement, it is reasonable to conclude, from the few data that are available, that the high blood pressure of giraffes is not a result of an unusually high cardiac output.

Table 9.2 The Allometry of Dynamic Heart Variables in Giraffes

Variable	Predictions for a 75-kg Fetus	Allometric Equation in Fetuses	Predictions for a 750 kg Giraffe	Allometric Equation in Postnatal Giraffes
Cardiac output (L min^{-1})	6	$0.211 \times Mb^{0.78}$	37	$0.211 \times Mb^{0.78}$
Predicted stroke volume (SV; mL)	114	$1.12 \times Mb^{1.07}$	860	$2.08 \times Mb^{0.91}$
SV adjusted for increases in left ventricular wall thickness (mL)		–	538	$0.55 \times Mb^{1.04}$
Predicted heart rate (min^{-1})	54	$188 \times Mb^{-0.29}$	44	$104 \times Mb^{-0.13}$
Heart rate adjusted for increases in left ventricular wall thickness (min^{-1})		–	70	$393 \times Mb^{-0.26}$
Total peripheral resistance (mmHg min L^{-1})	17	$424 \times Mb^{-0.74}$	5.5	$239 \times Mb^{-0.57}$
Mean arterial pressure (mmHg)	102	$89.5 \times Mb^{0.036}$	207	$51.5 \times Mb^{0.21}$
Systolic pressure (mmHg)	139	$128 \times Mb^{0.02}$	257	$78.0 \times Mb^{0.18}$
Diastolic pressure (mmHg)	78	$68.1 \times Mb^{0.03}$	194	$39.6 \times Mb^{0.24}$

Based on data collected from 10 fetal giraffes and 46 postnatal giraffes by Mitchell and Skinner, 2009. Table reprinted with permission of Elsevier.

Stroke Volume

Stroke volume has been calculated in nine giraffes (Smerup et al., 2016) from the difference between the volume of blood in the ventricles at the end of diastole (when the ventricles are full of blood) and at the end of systole (after the ventricles have contracted). In other giraffes it has been calculated by dividing cardiac output by heart rate. Goetz et al. (1960) calculated stroke volume by this latter method and found it to be approximately 700 mL in one giraffe, although an average of 495 mL (range 258–807 mL) can be calculated from all the data they obtained and, if those giraffes weighed 500 kg, it was 1.0 mL kg^{-1} body mass. The body mass was more likely to have been ~700 kg, and if so, the stroke volume predicted from allometric equations developed from data in other mammals (Stahl, 1967; Holt et al., 1968; Seymour and Blaylock, 2000; Table 9.2) should have been 860 mL (1.2 mL kg^{-1} body mass). Similarly for Linton's giraffe, a stroke volume of 912 mL, or 1.1 mL kg^{-1} body mass, is predicted. However, the predictions of giraffe stroke volume made from the allometric equations developed from other mammals are inaccurate. Smerup et al.'s (2016) calculations of stroke volume varied from 209 to 377 mL in giraffes weighing 495 kg, or 0.6±0.4 mL kg^{-1} body mass, which is half of predicted values. Stroke volume in giraffes is much smaller than predicted because its size decreases in proportion to increases in left ventricular wall thickness. If stroke volume is calculated assuming that its size does decrease in proportion to increases in left ventricular wall thickness, stroke volume would be equal to $0.55 \times Mb^{1.04}$ (see Table 9.2), and in a 750 kg giraffe it would be about 550 mL rather than the 860 mL that it would be in a mammal of equivalent mass with a typical blood pressure and typical ventricle wall dimensions. A stroke volume of 550 mL would be 0.73 mL kg^{-1} body mass, which is close to the stroke volume measured by Smerup et al. (2016). Stroke volume, in other words, ranges from ~60 mL per contraction in a newborn giraffe to 1.1 L per contraction in a 1,500-kg adult. For comparison, the stroke volume of an adult human is around 70 mL per contraction.

Heart Rate

As the stroke volume of giraffes is determined by heart size and thus cannot increase or decrease much, increases or decreases in a giraffe's cardiac output depend on changes in heart rate. In all mammals, heart rate tends to be directly correlated with the rate of oxygen consumption (see Chapter 13), and it is this correlation that explains why, of all the cardiovascular variables reported for giraffes, heart rate varies most widely. Heart rate of giraffes has been measured mainly in giraffes that were sedated or anesthetized, struggling or restrained, recumbent or standing, all involving different metabolic rates, and all of them circumstances that produce meaningless estimates of basal heart rate (but may give some idea of what maximal heart rates might be). For example, in one restrained giraffe, heart rate was found to be 100 min^{-1} (Goetz, 1955; Goetz and Budtz-Olsen, 1957) but in a free-ranging, unanesthetized adult giraffe, standing quietly it was about 40 min^{-1} (van Citters et al., 1969), and walking it was 33–90 min^{-1} (Hargens et al., 1987a). In four restrained but not sedated giraffes, average heart rate was 85±13 min^{-1} with a range of 65–105 min^{-1} (Goetz and Keen, 1957), and in anesthetized animals,

70–97 min^{-1} (Jefferson, 1971; Rossof, 1972) and 56±5 min^{-1} (Smerup et al., 2016). In free-ranging animals, with heart rate recorded by telemetry, heart rates of between 60 min^{-1} (prone) and 175 min^{-1} (galloping) have been recorded (van Citters et al., 1969). In nine unanesthetized giraffes accustomed to being handled, heart rate was 81±15.2 min^{-1} (Langman, 1973a). A general conclusion possible from these data seems to be that the heart rate of giraffes at rest is between 40 min^{-1} and 90 min^{-1}, rising to 175 min^{-1} during exercise (an increase mediated by the effects of noradrenaline). Heart muscle is well innervated by sympathetic nerves (Nilsson et al., 1988) which secrete noradrenaline. These values for heart rate are, however, at least twice what is predicted for a mammal of the size of a giraffe. The heart rate for giraffes at rest predicted from equations derived from measurement of heart rate in many different mammals with a wide range of metabolic rates (Stahl, 1967; Holt et al., 1968; Seymour and Blaylock, 2000; Calder, 1996; Table 9.2) should be on average 47 min^{-1}, with 29 min^{-1} in adults and 71 min^{-1} in juveniles, or approximately two-thirds of measured values. How can these differences be reconciled? The explanation for the differences is that stroke volume is far less in giraffes than it is in other mammals of comparable metabolic rate. As cardiac output is the same as it is in other mammals, and as: cardiac output = heart rate × stroke volume, if stroke volume is reduced, heart rate must be higher for the same cardiac output. When a reduced stroke volume is taken into account, the allometric equation that describes resting heart rate of giraffes is $393 \times Mb^{-0.26}$ (Table 9.2). For example, if the cardiac output of a 750-kg giraffe is predicted to be 37 L min^{-1} and its stroke volume to be 540 mL, then its heart rate will be 70 min^{-1}, which is close to what is typically measured and predicted by the equation. The high heart rate of giraffes is not, therefore, because they have a higher metabolic rate than similar-sized mammals but because the stroke volume of the heart of giraffes is small.

Blood Vessels

There are two main types of blood vessels—conducting vessels and resistance vessels. Conducting vessels are large and have the simple function of transporting large volumes of blood to organs and tissues quickly. Once in a tissue or organ, blood must be directed to those cells that at any one moment need blood most, and that is the function of resistance vessels. Not much is known about the size of blood vessels in giraffes, but Noble Maluf did provide some data that give a good idea of their size (Maluf, 2002). His data are shown in Figure 9.6.

The largest arteries are, not unexpectedly, those nearest the heart—the aorta and pulmonary artery. Their outer diameter in an adult giraffe is about 50 mm. As large arteries subdivide into branches that supply blood to organs, the diameter of the branches becomes smaller. Further subdivision of an artery within an organ ends in resistance vessels—the arterioles. They have an outer diameter of ~100 μm (see Plate 9.3); thus their diameter is about 500 times smaller than the diameter of the aorta.

Blood within the arteries is under pressure. When the left ventricle of an adult giraffe contracts it has to generate a blood pressure, if blood is to be expelled from it, higher than the pressure already in the aorta, which is between 200 and 250 mmHg. Peak left ventricle pressure has been measured by Goetz et al. (in 1957) and van Citters et al. (in

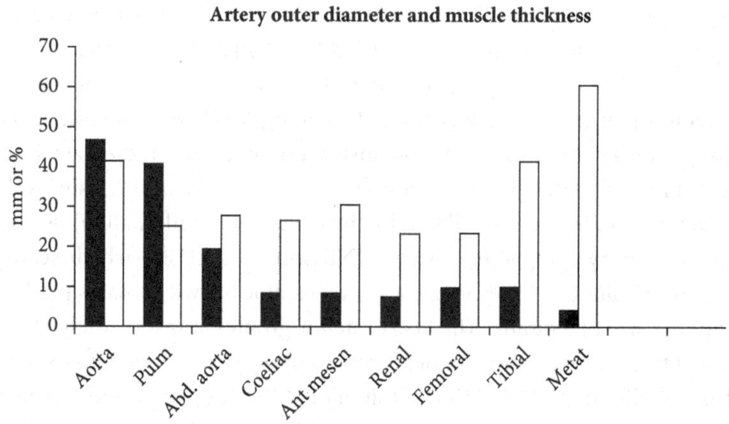

Figure 9.6. Outer diameter (■, mm) of the artery, and the thickness of the muscle layer in the artery wall as a percentage of artery outer diameter (□, %) in the arteries of a giraffe weighing ~1,100 kg.
Constructed from data in Maluf 2002.

1969) to be on average 270 mmHg. Furthermore, at the start of a contraction there is a very high spike of pressure of 284–344 mmHg that is not seen in other mammals. Goetz thought that the spike probably was caused by the large volume of blood that has just been ejected from the left ventricle colliding with the blood deposited in the aorta during the previous contraction. Whatever its origin, it is an additional stress on the aorta wall. Similar relationships exist between the right ventricle and the pulmonary artery, but the pressures involved are, of course, much lower (see Chapter 12).

To avoid rupturing under pressure, arteries (and arterioles) depend on the structure of their walls and on their thickness. The wall of an artery is made up mainly of elastic tissue and muscle tissue. What proportions of these exist in a giraffe's arteries? Study of the structure of giraffe arteries was begun by K.J. Franklin and F. Haynes in 1927. They had an opportunity to examine the carotid artery of a young giraffe, and in its upper part near the head, they found the ratio of muscle to elastic tissue to be 45% muscle:55% elastic tissue, while at the lower end near the heart, the ratio was 35% muscle:65% elastic. Since then, Goetz (1955), Goetz et al. (1955, 1957), and Kimani et al. (1983) have studied the structure of giraffe arteries and showed that the walls of all the large vessels that originate close to the heart—the carotid artery (e.g., Figure 9.4), aorta, pulmonary artery, and brachiocephalic trunk (from which the arteries to the forelegs arise)—consist mainly of elastic tissue. The further any artery is from the heart, the proportion of elastic tissue in its wall declines and the proportion of muscle tissue increases. How that proportion changes in a giraffe can be seen in the arteries studied by Maluf (Figure 9.6): as distance increases, muscle tissue increases, but most of the wall of the large conducting arteries consists of elastic tissue. Of the arteries Maluf studied, muscle tissue formed more than 50% of the wall only in the small metatarsal artery at the bottom of the hindleg, but as shown by Østergaard et al. in 2011 and Petersen et al. in 2013, muscle predominates in all the small arteries of the lower legs. The thickness of the walls, not only their structure, also changes with distance from the heart. The walls of the aorta in the giraffe dissected by Goetz and Keen (1957) were ~11 mm thick. In the thorax the wall of the aorta was 9 mm thick, 7 mm in the abdomen, and 4 mm at the point where the aorta divides into the two femoral arteries. In lower leg arteries, where blood pressure is higher than it is the femoral artery, the walls were thicker: the wall of the posterior tibial artery was 4.5 mm thick. Furthermore, wall thickness changes as a giraffe grows. Keen and Goetz (1957), for example, dissected a seven-month-old fetus and found that the wall:lumen ratio of the tibial artery was much lower than its ratio in adults. Those different ratios are illustrated in Plate 9.2. In a fetus, the diameter of the lumen of the tibial artery is ~4 times larger than wall thickness, whereas in an adult the ratio is ~1:1. In 1988 Hargens et al. made a study of wall thickness in the blood vessels of fetuses, neonates, and adults. Those in fetuses were thinner than those in neonates, and those in adults thicker than those in neonates. Fetuses are not exposed to the effects of gravity. They live in a weightless environment, their necks are short, and their heads and hearts are at the same level, and, therefore, their blood pressure is not high. After birth those conditions are changed, and the blood vessels respond. Giraffes, in other words, are not born hypertensive. They become so as they grow, and the progressive increase in wall thickness of leg arteries that Hargens et al. found correlates with increasing blood pressure. Østergaard et al. (2011) also came to the same conclusion after measuring diameters of leg vessels in newborn and adult giraffes. The changes in thickness with age apply mainly to arteries below the level of the heart. While their structure changes, the walls of arteries at or above the heart do not thicken over time. The walls of fetal and adult carotid and coronary arteries, for example, are very similar in size, but the wall of the posterior tibial artery thickens as MAP increases (see Plate 9.2).

The advantage of having more muscle tissue in the walls of distant arteries is that blood flow through them can be regulated by a muscle contraction that alters the radius of the vessel. Blood flow is inversely proportional to the fourth power of the radius, which means that even quite small changes in lumen radius can produce a very large reduction in flow. For example, if the radius of an artery is halved, then flow decreases 16-fold (2^4) (which is a phenomenon any motorist who has been held up by the closure of one lane of a two-lane motorway will recognize). The net effect is that blood flow to an organ or tissue easily can be adjusted to meet its needs, simply by muscle contractions that change the radius of the blood vessel supplying it. A second consequence of a decrease in radius and the increase in resistance that follows is that upstream arterial pressure increases, and vice versa. Changing resistance is one way that arterial pressure can be regulated.

For an artery or arteriole to be able to regulate the size of its radius, it must be not only muscular, but also innervated by vasoconstrictor sympathetic nerves, nerves which secrete noradrenaline (norepinephrine). Noradrenaline stimulates muscle cells in the walls of blood vessels to contract, and the arteriole radius narrows. The muscles that are in the wall are specialized. They do not have a striated structure like the muscles of the legs—the skeletal muscle—that is ideal for making muscles shorten in length, nor like the striated muscle of the heart—the cardiac muscle—that contracts inward like a closing net. The muscles in the wall of a blood vessel are made up of many millions of single cells. Under a microscope this type of muscle is seen to have a uniform structure and, consequently, is called smooth muscle. Contraction of each of the cells is achieved by a wave of electrical impulses that transmits from cell to cell, rather than by each cell being activated by a branch of a sympathetic nerve. Depending on the number of cells that contract, the degree of constriction of an artery can be varied or graded very precisely, and that depends on the size of the first stimulus arriving by a nerve. It can be assumed, therefore, that as arterioles are the main resistance vessels, they should be supplied by nerves. Nilsson et al. (1988) were the first to look for nerves in giraffe arteries, and found what they called "an exquisitely distributed sympathetic innervation pattern" but they did not find a clear pattern of innervation. The density of nerves did not correlate with muscle thickness, and so exactly how contraction is graded precisely in giraffes is not yet clear.

The advantage of having elastic tissue in the walls of arteries near the heart and in the conducting vessels generally is that their walls can stretch. For example, when the heart muscle contracts, it ejects a large volume of blood from each ventricle into either the pulmonary artery or the aorta. The high proportion of elastic tissue in the aorta (and pulmonary artery) allows the arteries to expand and absorb the volume of ejected blood. The expansion is sudden, and it generates a wave that is transmitted along all conducting arteries where it can be felt as a pulse. A pulse is intermittent. Its frequency is of course a measure of heart rate and specifically it indicates when contraction and emptying (systole) of the heart has occurred. Van Citters et al. (1969) measured the period of contraction and emptying to be 300 ms long in a giraffe heart beating once per second, and saw that it took another 380 ms for the heart to relax and fill (diastole). For the remaining 320 ms of the cardiac cycle the heart was at rest. Thus for about two-thirds of a cardiac cycle the heart is not contracting, yet blood flows continuously through the blood vessels. Continuous flow depends on what is often referred to as the "windkessel" effect. A windkessel is the air chamber on an oscillating water pump. The air is compressed as the chamber fills with water, and the pressure increases. The pressure drives water out of the chamber into pipes, taking it away. The windkessel effect in giraffes (and other mammals) starts when the heart contracts and the large, elastic arteries are stretched to accommodate the ejected blood. When the heart relaxes, valves between the aorta and pulmonary arteries and the left and right ventricles close, and closure prevents regurgitation of blood into the heart. At the same time, the elastic tissue contracts. As the blood cannot return to the heart, contraction of the elastic tissue pushes blood down the blood vessels in the same way that a windkessel drives water down pipes. A continuous flow of blood is the outcome. If the aorta or pulmonary artery had stiff walls, they could not expand to accumulate ejected blood, nor could they contract to squeeze blood down blood vessels while the heart is relaxing.

Carotid Arteries

In giraffes the carotid arteries are exposed to much higher blood pressures than they are in other mammals, especially when a giraffe lowers its head to drink. Kimani, in a series of studies (1981a, 1983a, 1983b), analyzed the structure of the carotid arteries of giraffes in great detail to find out how they had been modified during evolution to accommodate the stresses imposed by high blood pressure. What he found was that the innermost layer of the carotid artery contained a prominent layer of elastic tissue situated between the endothelium that lines the inner surface of the artery wall and the muscle layer. When there are elastic fibers in that region of an artery, their function is to reduce the longitudinal (i.e., in the direction of blood flow) shearing stresses that blood under high pressure exerts on the vessel wall. Thus they minimize damage, preserve the relationship between the artery wall and its muscles, and allow the muscles to contract and relax in response to contraction and relaxation of heart muscle during a cardiac cycle. Contraction of the muscles in the wall prevents over-distension of the wall (Kimani, 1987), but it is the prominent elastic layer on the inside of the vessel wall that prevents distortion of the vessel wall and thus ensures smooth blood flow when internal pressures are high. As the elastic layer is thicker in giraffes than it is in animals with a much lower blood pressure, it does seem to be a specific modification of a giraffe's carotid artery wall. At the point where the single common carotid trunk divides into its left and right branches, a division that occurs near the heart (Owen, 1838; Chapman, 1875), elastic tissue is especially prominent and forms "cushions," as it is here that the vessel wall is exposed to the highest shearing stresses, and the possibilities of distortion and the generation of abnormal stresses are greatest. The wall of the artery at this point is one centimeter thick (van Citters et al., 1969). The wall near the head is just 2 mm thick, is mainly muscle, and is not strong enough to resist pressures, for example when a giraffe lowers its head to drink. However, in the carotid artery near the head, the muscle cells are held together by stiff collagen fibers (Kimani, 1981a). They form a network, a mesh, interwoven with elastic fibers, that spreads the force exerted by pressure across many muscle cells. As collagen is not elastic and cannot stretch, its presence adds some stiffness and helps both muscle and elastic fibers resist excessive expansion when internal blood pressure threatens to do so. Another modification is that the endothelial layer of cells that lines the internal surface of the carotid artery, and is subjected to strong shear forces, is attached to the muscle layer by fibrils (Kimani 1981b). All these modifications are also found in fetuses, except that in them there is much more elastic tissue and less muscle in their carotid artery. The implication is that muscle and its support structures develop with age and are a response to increasing blood pressure.

Leg Arteries

Although the blood vessels near the heart are exposed to very high pressures, they are not the blood vessels exposed to the highest pressures found in a giraffe's cardiovascular system. Goetz and Keen (1957) speculated that the pressure in arteries of the feet of a giraffe might reach 400–500 mmHg, because added to the pressure generated by the heart is the hydrostatic pressure generated by gravity. The highest actual pressure measured by Alan Hargens and colleagues (1987b) in the feet of a free-ranging walking giraffe using radiotelemetry was 380 mmHg, but on average it was 260 mmHg (Figure 10.3 in Chapter 10). That giraffe had a body mass of about 520 kg; therefore the height from heart to feet would have been ~2.0 m, and the hydrostatic pressure gradient would have been 155 mmHg, much the same as the hydrostatic pressure generated by gravity in a carotid artery. A similar pressure gradient down the hindleg was measured by Østergaard et al. in unanesthetized giraffes in 2011. In that study, hydrostatic pressure down the leg increased by ~64 mmHg for every meter of leg length. The maximum leg pressure they recorded was 350 mmHg. Goetz and Keen (1957) predicted that the high pressure would be offset by changes in the structure of the blood vessels supplying the feet and that their structure would be unique. They saw that in the

> *vessels further away from the heart towards the head the arterial wall becomes progressively thinner while the lumen remains wide. In the legs, however, the lumen of the arteries diminishes while their walls become enormously thick.*

The classic example of thickened walls in leg arteries is the wall:lumen ratio of the tibial artery which was first discovered by Goetz and Keen in 1957. Since then, Kimani et al. (1991; hindleg), Østergaard et al. (2011; hindleg), and Petersen et al. (2013; foreleg) have all shown that the further down the leg an artery or arteriole is, the thicker the muscle layer in the wall is and the smaller the lumen is. The walls become thicker by an increase in the number of cells (hyperplasia) rather than by an increase in cell size (hypertrophy) in the same way that heart muscle thickness increases—a mechanism that seems to be unique to giraffes.

An increase in muscle wall thickness alone could be enough to protect the lower leg blood vessels from the very high blood pressure within them as Laplace's Law predicts, but in another study of artery architecture Kimani et al. (1991a) found that giraffes have evolved yet other mechanisms to resist the effects of pressure and to regulate flow to the lower legs. They found that between the knee and ankle joints of the hindleg, the arteries supplying the lower leg narrowed quite abruptly, the muscle layer became thicker (Figure 9.7), and the muscles were supplied with many more nerves than those lying above or below this region.

Østergaard et al. (2011) found, in the same 80 cm length between the knee and the ankle, similar changes to those described by Kimani et al. (1991), with the point of narrowing richly supplied by nerves. Petersen et al. (2013) found a similar arrangement in the main (median) artery of the forelimb. In their detailed analysis of artery function in the foreleg of giraffes, Petersen et al. (2013) confirmed that nerves are not uniformly distributed along an artery, but are most dense at and below a discrete "sphincter" located just below the elbow, a position similar to that of the "sphincter" identified by Kimani et al. in the hindlimb. Constriction of this section of the artery following injection of noradrenalin reduced blood flow below it. Petersen et al. also measured how powerfully the muscles in the walls of small arterioles in the lower legs can contract: they can contract against an internal pressure of 500 mmHg, a pressure much higher than is typically found. That degree of contraction had the potential to narrow the diameter of a small artery so much that blood flow in it was reduced to zero.

The presence of a "sphincter" is unique to giraffes and can explain why Hargens et al. (1987b) found a large variation of 70–380 mmHg in blood pressure in the arteries of the feet in a walking giraffe. Kimani et al. (1991a) provided an explanation for the variation. They decided that activation of the sympathetic nerves such as that which occurs during exercise, would close the "sphincter" and thereby shunt blood to the tissues that needed the blood most, namely the muscles of locomotion in the upper leg. Presumably when the "sphincter" is closed the blood pressure in the artery below it falls, and when it is open the pressure rises. However, even after maximum contraction of the "valve," the reduction in blood pressure below it is only about 70 mmHg (Petersen et al., 2013), so that alone cannot explain the variation in pressure that Hargens et al. had found. Its

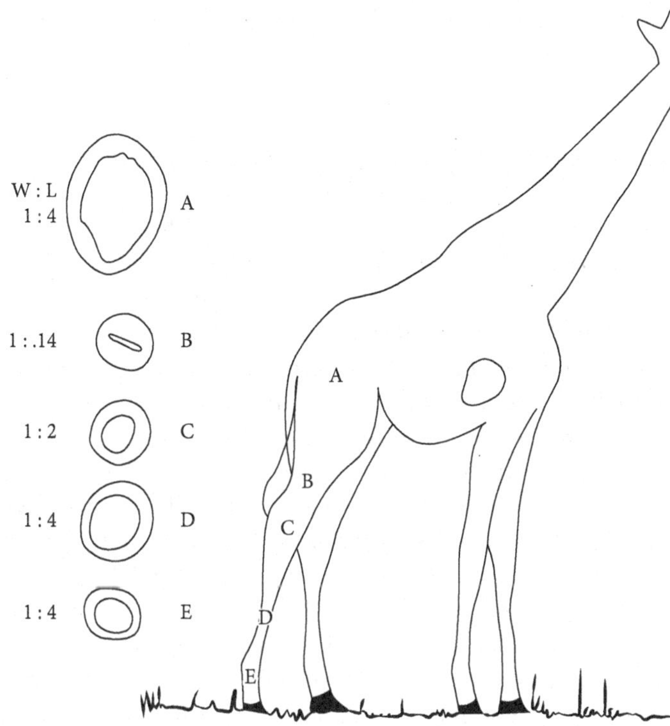

Figure 9.7. Artery diameter and the ratio of wall:lumen (W:L) thickness in the hindleg of a giraffe. A = femoral artery (1:4), B = posterior tibial artery (1:0.14), C = metatarsal artery (1:2), D, E = digital arteries (1:4).
Redrawn from Kimani et al., 1991a, with permission of John Wiley and Sons.

origin is still a mystery. Østergaard et al. (2011) discovered that the "sphincter" was well developed in a newborn giraffe. Newborn giraffes do not yet have a high blood pressure, so its structure could not be a response to high blood pressure. Østergaard and colleagues thought, therefore, that it was a "pre-adaptation" to the high blood pressure that would develop later, although this explanation for its presence in a newborn giraffe and presumably also in a fetus seems to me to be unlikely. It takes many months before blood pressure rises significantly after birth. A "pre-adaptation" might be necessary only if a giraffe's blood pressure rose rapidly after birth. A different explanation is that from the moment of birth, the newborn giraffe must be in an anatomical state to minimize predation. As Owen (1838) described it, there is "a necessity for its speedy ability to escape with the parent from its enemies." Hence, while it is a fetus, all the adaptations it needs to be able to run after birth must be in place on "day one" and are put in place by its genes. A newborn giraffe is born with legs that are over a meter long. Providing a means of short-circuiting the blood from the moment of birth so that it goes to the locomotory muscles in the upper muscles of the hind legs, as described by Kimani et al. (1991), and by a similar mechanism to the muscles of the front leg (Petersen et al., 2013), would improve a newborn giraffe's chances of escaping from a predator. An ability to bring about large changes in blood flow to the lower leg has an additional benefit and one of equal importance. It is a mechanism that can limit or promote heat loss from the legs and by doing so makes a contribution to whole body temperature regulation (see Chapter 14).

Thoracic and Abdominal Arteries and Arterioles

The arteries and arterioles that supply blood to the organs of the abdomen lie at much the same level as the heart, and thus are not exposed to the effects of gravity in the same way that leg arteries and arterioles are, but they are exposed to the pressure generated by the heart of ~200 to ~250 mmHg. How are abdominal blood vessels strengthened, and how is blood flow to abdominal organs controlled? An answer is provided by the way the walls of the small arterioles that lie deep within the tissues of the heart, lungs, intestines, liver, and kidney increase in thickness as a giraffe grows (see Plate 9.3).

For each of the tissues shown in Plate 9.3, arterioles were studied that had an outside diameter of ~100 μm. They were obtained from a fetus of 408 (of 450) days gestation and from an adult of eight years of age. The micrographs show very clearly that the muscle layer in the walls of these small arterioles in an adult giraffe is significantly thicker than it is in a fetus, and the internal diameter of the adult arteriole is significantly smaller. The thickness resists the higher blood pressure, and the smaller inner diameter, because of Poiseuille's fourth power law, reduces blood flow in the arterioles to the same relative amount that it is in a fetus. Overall what these data reveal is that the way thoracic and abdominal tissues are protected from high blood pressure and high blood flow is by hyperplasia and/or hypertrophy of the smooth muscle cells.

Peripheral Resistance

Are the hyperplasia/hypertrophy of muscle cells in blood vessel walls and the small internal radius that results a response to, or a cause of, high blood pressure? Which comes first, high blood pressure or narrow radius? What mechanism stimulates the ventricle walls to enlarge? How does each giraffe develop its own unique MAP? So far no simple answer to these questions has emerged, but there is a general relationship between pressure, flow, and resistance that must underpin any possible solution. This relationship was first worked out by the German physicist and mathematician Georg Ohm in 1827 in relation to electrical currents, but it applies equally well to blood flow, and provides the theoretical basis for the physiological relationship of

Blood Pressure = Blood Flow × Resistance.

As flow (cardiac output) is about the same in giraffes as in other mammals but MAP much higher than it is in other mammals (e.g., Figure 9.2), it follows from Ohm's Law that resistance to blood flow in the arterial tree must be higher in giraffes. Resistance is related to the number of blood vessels, their diameter, and the way that they are connected together, and giraffes are no exception to this general rule. Total peripheral resistance (the total resistance of all arteries and arterioles) is impossible to measure directly, but it can be derived from measurements of cardiac output and blood pressure using the principle of Ohm's Law (total resistance = MAP÷ cardiac output). Total

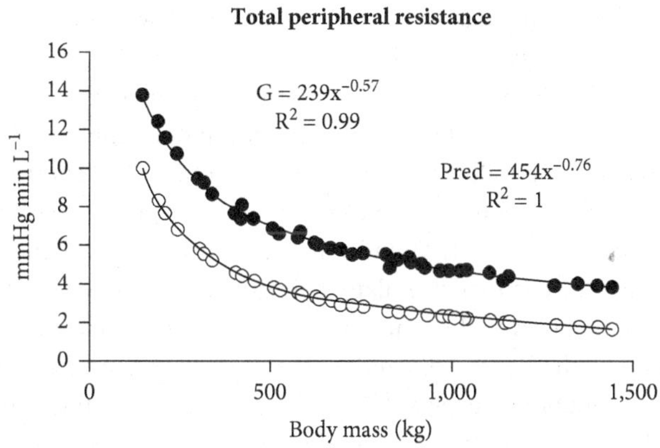

Figure 9.8. Total peripheral resistance (mmHg min L^{-1}) in giraffes (●) from birth onward, calculated from estimates of blood pressure and cardiac output in 22 male and 24 female *G.c. giraffa*, and total peripheral resistance (○) in other mammals.
Predicted from data in Holt et al., 1968; Calder, 1996; Seymour and Blaylock, 2000.

peripheral resistance in giraffes can be calculated from predicted cardiac output and blood pressure (Table 9.2), and in 56 giraffes, assuming that their necks were at the typical angle of carry of 55° to the vertical, it was 4–14 mmHg min L^{-1} (mean = 5.5 mmHg min L^{-1}) compared with an expected value of 2–10 mmHg min L^{-1} (mean = 3.5 mmHg min L^{-1}) for mammals of similar body mass (Figure 9.8). Thus, systemic resistance in giraffes is much higher than expected for mammals of its size.

How does increased resistance develop? One quite plausible theory is that as giraffes grow and their legs get longer, there is a gradual increase in the gravitational hydrostatic pressure of the blood in the limb arteries, and this pressure causes their blood vessel walls to thicken in response. This type of thickening is a passive process and it changes the structure of the vessel wall such that it becomes stiffer and less compliant. As a result, wall tension increases and internal diameter decreases, in accordance with Laplace's law (Folkow, 1991), resistance increases, and MAP will increase (Badeer, 1997). Despite their hypertrophy/hyperplasia and stiffness, the vessels retain their sensitivity to nerves so that blood pressure and blood flow can be changed as required by changes in posture and other demands such as exercise. However, if this mechanism for producing high MAP were correct, then all mammals with long legs, such as humans, elephants, and camels, would have a high MAP, but they do not (Seymour and Blaylock, 2000; Figure 9.2). Therefore, the gravitational hydrostatic pressure theory of higher resistance in giraffes does not seem to be the most likely origin of a giraffe's high MAP, although it would be unwise to rule it out completely as a contributory mechanism.

The difference between long-legged camels and elephants and giraffes is, of course, that giraffes have a long neck that is held nearly upright, unlike the neck of a camel, for example, which is long but is held horizontal and at much the same level as its heart. Accordingly, the conclusion must be that the origin of the high blood pressure of giraffes is their long upright necks rather than long legs. If so, what mechanism could exist between a long neck and high MAP? The mechanism that has been proposed, and it may well be a contributor to some types of hypertension in humans, is that it is a consequence of a reflex that starts in the brain (Paton et al., 2009; McBryde et al., 2017). The reflex has been called the Cushing's reflex after the pioneering American neurosurgeon Harvey

Cushing who, in 1901, enunciated a "simple and definite law: any cause of a decrease in blood flow to the brain causes a rise of blood pressure." The reflex is very sensitive. Very small decreases in blood flow to the brain produce large increases in MAP. The reflex activates sympathetic nerves, which secrete noradrenaline; the noradrenaline causes smooth muscle cells in the walls of blood vessels to contract, and contraction constricts the vessels, narrowing the radius and increasing total peripheral resistance, and an increase in blood pressure follows. Noradrenaline also activates biochemical pathways inside smooth muscle cells that stimulate their growth. According to this theory, any cause of reduced blood flow to the brain, such as narrowing of arteries (arteriosclerosis) that supply blood to the brain, will initiate the mechanism. Arteriosclerosis of brain blood vessels is undoubtedly a cause of hypertension in some humans but, although it has been looked for in giraffes, arteriosclerosis has not been found in any arteries (Goetz and Budtz-Olsen, 1955; Maluf, 2002), and some other way of reducing blood flow to the brain must exist for the Cushing's reflex theory to be correct.

A possible sequence of events that would link a long neck and a rise in blood pressure is that as the neck (rather than the legs) gradually elongates, so also does the height of the column of blood in the carotid artery. As height gradually increases, so does the effect of gravity, with the consequence that there is a tendency throughout growth for blood flow to the brain to diminish. As that happens, the Cushing's mechanism would be activated gradually to maintain cerebral blood flow at its required rate, in a coordinated way, through hyperplasia/hypertrophy of arterioles and of cardiac muscle, an increase in resistance, and a rise in MAP (Paton et al., 2009). The increased MAP in turn makes the muscles of the ventricles work harder, much as they have to do in someone like a top-class marathon runner, and under that stimulus the ventricle muscles enlarge. Similarly, the increase in pressure will add another stimulus for blood vessel walls in the limbs to thicken. Ultimately, when the neck of a giraffe has stopped elongating, a balance is reached such that there is a perfect match between blood pressure and neck length for each individual. This would be a very gradual process, taking many years to develop: it takes about 10 years for a giraffe to reach maturity and reach its final blood pressure. A little-known human example of this mechanism probably occurs in the giraffe-necked women of the Padaung tribe of East Myanmar. For personal adornment the women deliberately but gradually, from childhood to adulthood, lengthen their necks with brass rings. They become hypertensive at an early age and their average systolic/diastolic blood pressures are 157/101 mmHg but can be 175/110 (Roaf, 1961), pressures that easily meet the WHO definition of hypertension. These blood pressures are much higher than might be expected in healthy members of a rural community, and their elongated necks are the likely origin of the hypertension. Other examples are the MAP of llamas and elephants. Their heads and brains are about one meter above the level of the heart and in them MAP is ~152 mmHg (Llanos et al., 1998; White and Seymour, 2013), which is what they can be predicted to be. If a Cushing's mechanism is what increases blood pressure in giraffes, then it could also be expected that all arteries that are innervated by the sympathetic nervous system—not just those in the systemic circulation or those in the legs affected by gravity, but also those in the pulmonary circulation—would be affected, and they are. The resistance to blood flow in lung blood vessels in postnatal giraffes is higher than expected. It is, on average, 0.9 mmHg min L^{-1}, which is twice what it is in other mammals of similar body mass (see Chapter 12).

Activation of the Cushing's mechanism will, over time, cause hypertrophy/hyperplasia in all blood vessels innervated by the sympathetic nervous system indiscriminately and, in the case of pulmonary arterioles, without any obvious adaptive benefits.

For a Cushing's mechanism to operate in giraffes, there has to be a sensor of blood pressure at the head or of blood flow to the brain, or of the constituents of the blood going to the brain like oxygen or glucose, changes in any or all of which would result in an appropriate response of the sympathetic nervous system. It is tempting to think that the pressure sensors are the baroreceptors in the occipito-carotid sinus of giraffes, but those receptors do not seem to be involved. In humans who have hypertension, baroreceptor sensitivity to changes in blood pressure is retained even though the MAP is much higher than usual, and this must be the case in order to adjust blood pressure to changes in posture, for example when a giraffe (or a human) lifts or lowers his/her head. Baroreceptors, in other words, regulate short-term changes in blood pressure. The most likely site for a long-term sensor of blood flow to the brain is the highly vascular region of the brainstem that regulates the cardiovascular system (Dickinson, 1991). It may be a coincidence that in giraffes the C2 nucleus of this area is far more densely populated with neurons than it is in other ruminants (Badlangana et al., 2007b), but as the C2 nucleus regulates autonomic reflexes like blood pressure, it may also be that during giraffe evolution this area of the brain has become more sensitive to changes in brain blood flow. Yet another possibility is that detection of changes in cerebral blood flow resides in astrocytes. Astrocytes are specialized cells that fill all the spaces between the neurons that make up the brain, support their metabolic functions, outnumber them by fivefold, and, importantly, are also intermingled with the neurons that regulate cardiovascular mechanisms. They are able (among many other things) to detect a decline in blood flow to the brain and, having done so, activate sympathetic nerves which in turn increase MAP, maintain brain blood flow, nutrient delivery, and waste product removal (Marina et al., 2019). Thus, it is a possibility that a Cushing's reflex-type mechanism dependent on a hypersensitive C2 nucleus or astrocytes, or both, has evolved to maintain adequate blood flow to the brainstem in, possibly, all terrestrial mammals, including humans, but in giraffes it results in a substantially higher blood pressure.

10

Edema, Fainting, and Strokes

Explaining how giraffe blood pressure is as high as it is and how it is maintained at that high level has been the main aim of the studies of the giraffe cardiovascular system started by Goetz in the 1950s. It was, Goetz (in 1955) said, "a most perplexing problem." However, Goetz was also a clinician and, equally, he thought about why giraffes do not suffer from the effects of high blood pressure. Much earlier, in 1924, August Krogh had wondered about the relationship between the structure and function of capillaries and whether abnormal amounts of fluid collected in the tissues of giraffes. And "why," Goetz wondered, "does a giraffe not faint when suddenly lifting its head?" (Goetz, 1988). Perhaps "investigation of the cerebral circulation in the giraffe might reveal information useful in unravelling the problem of blackouts and fainting in crews of high-speed aircraft" (Goetz and Budtz-Olsen, 1955), a problem which, at that time, was of interest to the Royal Aircraft Establishment at Farnborough in England, as well as the US Air Force. Krogh and Goetz were right to have these thoughts. It is astonishing that despite the elaborate architecture of their heart and blood vessels, giraffes do not develop swollen legs, or suffer "head rushes" and dizziness when lifting their heads rapidly, or burst blood vessels in the head and especially in the brain when they lower their heads, all of which can happen to humans, especially humans who have high blood pressure. None of these misfortunes occurs in giraffes. Why not?

Edema

Having a head at the top of a long neck, and a neck held for the most part vertically or nearly so, requires that giraffes must have a high mean arterial pressure. It is not, however, only the arteries and arterioles of giraffes that are exposed to high pressure. Veins are as well. The blood pressure in veins is always lower than it is in arteries, because a pressure gradient between arteries and veins must exist before blood will flow from one to the other. A gradient between them does exist in giraffes. Hargens et al. (1987b), for example, measured the pressure in the veins of the legs of a free-ranging giraffe and found it to be 250 mmHg even though arterial pressure at the same level was 380 mmHg. In an adult human, arterial pressure in the legs is 270 mmHg, but venous pressure in the legs is about 180 mmHg. The difference between arterial and venous blood pressures exists because veins are separated from arteries by the network of capillaries that form the microcirculation (Figure 10.1).

The capillaries that lie between arteries and veins were first seen by Marcello Malpighi in the seventeenth century (and, at about the same time, by the Dutch microscopist, Antonie van Leeuwenhoek). Malpighi was an Italian anatomist who had the exceptional and rare skill of being able to turn a 2D image seen under a microscope into a 3D structure. He had seen that the connection between the smallest branches of arteries

How Giraffes Work. Graham Mitchell, Oxford University Press. © Oxford University Press 2021.
DOI: 10.1093/oso/9780197571194.003.0010

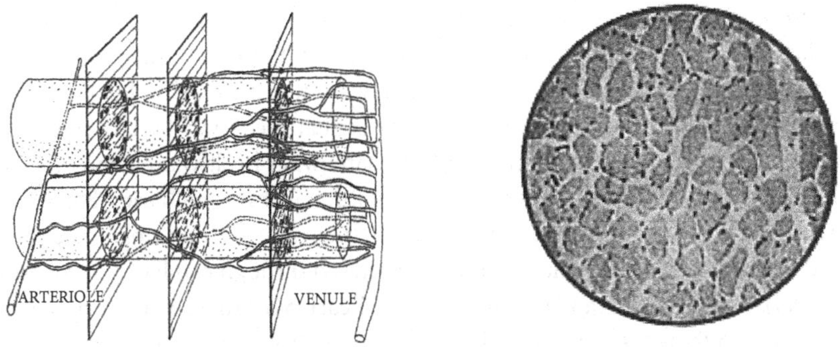

Figure 10.1. Left: Distribution of capillaries along a muscle fiber from an arteriole to a venule showing how the density of capillary branches increases toward the venule.
From Hoppeler et al., 1981, with permission of Elsevier.
Right: August Krogh's visualization of capillaries (black dots) in a muscle of a horse after India ink had been injected into the artery supplying the muscle. Magnified 120 times.
From Krogh, 1920, with permission of the Nobel Foundation.

and veins was not by random cavities between them, but by extremely fine vessels visible only under the microscope. He called the vessels capillaries, after the Latin *capillus*, meaning hair. The cross-sectional area of all the capillaries taken together typically in a mammal is ~1,000 times larger than the cross-sectional area of the aorta; this large area dissipates arterial blood pressure and, consequently, the blood pressure in the veins leaving an organ is always less than it is in the artery supplying that organ. The main origin of blood pressure in veins is gravity acting on the blood in the veins. The length of the leg veins of a giraffe is greater than it is in most other mammals and the result is that the pressure in their veins is much higher as well. The physiological problem created by the high venous blood pressure was spotted by August Krogh in 1924. He had this to say:

> *In the giraffe with a total height of 5 m the heart is at a height of about 2.5 m, and it would be extremely interesting to know just how the giraffe avoids the development of filtration oedema in its long leg.*

Edema (from the Greek, meaning "to swell") is the abnormal collection of fluid in the interstitial space, which is a conceptual more than an actual "space" between the capillaries running through tissues and the cells that make up tissues. In healthy tissues, fluid enters and almost all of it leaves the interstitial space via capillaries, and Krogh, in the address he gave at the ceremony when he received his Nobel Prize, said that capillaries

> *were, in one sense, the most important element in the whole circulatory system. All exchange between the blood and the organs, of nutrient materials and waste products, of oxygen and carbon dioxide, takes place through their walls. In them the blood fulfils its real function. One could say that the whole circulatory machine exists for their sake.* (Krogh, 1920)

There are an estimated 10 billion capillaries in an adult human, each on average ~1 mm long (i.e., there are 100,000 km of capillaries in an adult human); therefore, "if all the capillaries in a grown man formed a continuous tube, it could reach at least twice round the earth" (Krogh, 1920). It is through capillaries that the ~100,000 L of fluid needed every day to keep all the cells of their adult human owner healthy are filtered. An adult giraffe weighing 1,000 kg would have many more kilometers of capillaries than a human, and the volume of fluid exchanged daily between them and surrounding tissues is probably 10 times more. Usually, most of the fluid filtered through capillaries is reabsorbed by capillaries, enters veins, and is returned to the heart, but a tiny and important volume leaving the capillaries is not reabsorbed, enters lymph vessels, and is returned to the heart by them. The volume of lymph flow has not been measured in giraffes but has been in 15 other mammalian species, and their average volume of lymph flow (L d^{-1}) was equal to $0.015 \times Mb^{0.97}$ (Edwards, 1975). That is, it is roughly 0.02% of cardiac output in any mammal. In a small giraffe weighing 147 kg, it can be predicted to be 2 L d^{-1}. In a large one weighing 1,440 kg, it is 17 L d^{-1}.

The presence of edema implies that a more than usual amount of the fluid entering the interstitial space is not being returned to the heart either by veins or by lymph. If fluid is not being returned to the heart properly, then cells are not getting fresh nutrient-rich fluid, nor are waste products of cell metabolism being removed as they should be. If this state of affairs continues for several hours, cells eventually die. Typically, edema fluid collects in the legs and feet of standing animals as it is there that the hydrostatic pressures in veins and in the capillaries of the microcirculation are greatest. However, even though blood pressure in their leg veins is much higher than it is in humans, edema does not develop in the feet of giraffes. Why not? What prevents its formation in giraffes? Krogh's musings on the cardiovascular system of giraffes can be related to the forces affecting fluid movement in the microcirculations of mammals generally. They were analyzed first in dogs by Ernest Henry Starling, a British physiologist. In 1896 he summarized the forces involved in the equation:

$$\text{Filtration rate} = K_1\,((CHP - IHP) - K_2\,(CCOP - ICOP))$$

This looks to be a complicated equation, but it is actually quite a simple one. K_1 and K_2 are measures of the permeability of the capillary membrane; CHP is the capillary hydrostatic pressure, i.e., the blood pressure inside a capillary; IHP the interstitial hydrostatic pressure exerted by fluid in the interstitial space; CCOP is the capillary colloid osmotic pressure; and ICOP is the colloid osmotic pressure in the interstitial fluid. What the equation tells us is that the rate of filtration of fluid out of a capillary depends on how porous or permeable the capillary wall is to water and any small molecules it contains (= K_1) and to proteins (= K_2), on the difference between the hydrostatic pressures on each side of the capillary wall, and on the difference between the colloid osmotic pressure inside and outside the capillary. Capillary hydrostatic pressure is, for all practical purposes, much the same as the pressure in the veins draining a tissue, because there are no valves or other obstructions between capillaries and veins (Figure 10.1). Capillary hydrostatic pressure causes filtration of fluid out of a capillary, and capillary colloid osmotic pressure acts to reabsorb fluid. Interstitial hydrostatic pressure, if it is high enough, resists filtration, and interstitial colloid osmotic pressure resists reabsorption.

Thus, if CHP is much bigger than the IHP, fluid is filtered out of the capillary, and if the CCOP is greater than the ICOP, the filtered fluid is reabsorbed. Normally, fluid starts being filtered out of capillaries at the arterial end of the microcirculation. As fluid leaves the capillaries, CHP falls and CCOP increases, and the combination of low CHP and high CCOP is greatest where the capillaries join the vein, where not only are CHP low and CCOP high and reabsorption greatest, but capillaries are more numerous (Figure 10.1). Krogh correctly guessed (in 1924) that in giraffes, if the effect of gravity did make hydrostatic pressure in veins and capillaries of the legs and feet higher than usual, the result would be a very high filtration rate and edema would develop, unless "the unavoidable increase in hydrostatic pressure can be compensated by increasing colloid osmotic pressure of the blood." Is colloid osmotic pressure in the blood of giraffes as high as Krogh surmised it would need to be?

The source of colloid osmotic pressure in blood is the proteins in blood plasma that are unable to move through the capillary walls. One of the most important of those proteins is albumin, which is quite a small protein having a molecular weight of 70,000 units compared to the molecular weight of, say, an antibody, which may be 15 times larger, that is, one million units, but albumin is, nevertheless, too large to move across a capillary wall and out of a capillary. The size of the colloid pressure generated by proteins depends on the number of individual particles of protein that are present: the more protein molecules (per mL of fluid) there are, the fewer the water molecules, and water moves down its concentration gradient from, say, the interstitial space into capillaries. Because it is relatively small and has the highest concentration of any protein in plasma (typically ~40 g L^{-1}), albumin contributes many particles and makes up the lion's share, about 80%, of colloid osmotic pressure. Goetz and colleagues (1955) were the first to measure the concentration of all the plasma proteins, not only albumin, in giraffes and found it to be 43 g L^{-1}, which is much lower than it is in most mammals (~60–70 g L^{-1}). This low concentration of protein disproved Krogh's idea that high blood pressure in the veins must be offset by high colloid osmotic pressure if edema is to be avoided, and the paradox entered the folklore of giraffe physiology: giraffes did not have a high colloid osmotic pressure and yet did not develop edema. Goetz reasoned that substances other than protein may contribute to colloid pressure, and he measured plasma viscosity because plasma viscosity is almost exclusively dependent on the presence of colloids like proteins. However, when he measured viscosity of giraffe blood, he found it to be 4.9 times that of water, a value similar to the viscosity of human plasma. Therefore, no other colloids could be present. Then, in 1969, S.K. Sikes published a value of 130 g L^{-1} for total plasma protein concentration in a single giraffe, which is three times Goetz's value and closer to what is needed to fulfill Krogh's prophesy. The controversy needed to be resolved, and in a study of five giraffes Mitchell and Hattingh (1993) again measured plasma protein concentration and also total osmotic pressure (crystalloids, like salts, and colloids, like proteins, combined) of plasma and found them to be 61±2 g L^{-1} and 300±9 mosmol kg^{-1} of water, respectively, concentrations no different from those found in all other mammals. A plasma protein concentration of 61 g L^{-1} will generate a colloid osmotic pressure equivalent to a hydrostatic pressure of about 25 mmHg: Hargens et al. (1987) measured colloid osmotic pressure to be 27 mmHg in giraffes. That pressure is nowhere near high enough to counter capillary hydrostatic pressure in the legs of a giraffe, which may be as high as 150 mmHg (see Figure 10.3 later in the chapter).

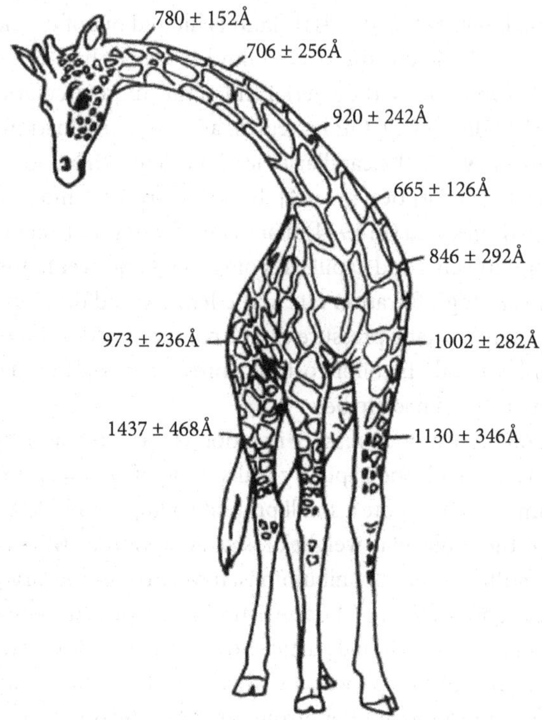

Figure 10.2. Regional variations in the width of the basement membrane of capillaries in a giraffe.
1 Angstrom (Å) = 0.1 nm.
From Williamson et al., 1971, with permission of Elsevier.

If the higher capillary hydrostatic pressure in the lower legs of giraffes is not counteracted by a higher colloid osmotic pressure, then which of Starling's other forces prevent the formation of filtration edema? Edema would be prevented if the permeability of capillaries to water and protein were low. A low permeability to water would be counterproductive, however, as it is water that carries nutrients to, and wastes from, cells, but low permeability to protein would be advantageous. Proteins would not leave capillaries when fluid did, protein concentration in them would increase as would capillary colloid osmotic pressure, and reabsorption of filtered fluid would be enhanced. Proteins leave capillaries through gaps or pores in the capillary wall. Low permeability to protein depends on the size of the pores between the endothelial cells that line the walls of capillaries and on the thickness of the basement membrane to which the endothelial cells are attached. Thus far only one study has examined capillary structure in giraffes (Williamson et al., 1971) and the thickness of the basement membrane in the hind legs (1,437±468 Å) was almost double that in the neck (780±152 Å; Figure 10.2).

Taken in isolation, this analysis looks convincing and implies that giraffes have a unique thickening of the basement membrane of capillaries in the legs but, in fact, as Williamson et al. pointed out, the same pattern is found in humans and probably in all mammals: basement membranes of capillaries in the legs are thicker than those in capillaries situated above the level of the heart, they increase in thickness with age, their

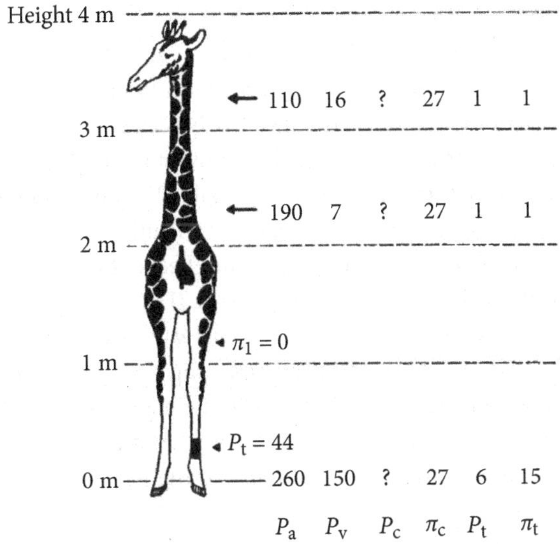

Figure 10.3. Hargens et al.'s (1987b) summary of arterial pressures (Pa) , venous pressures (Pv), tissue pressure (Pt), and colloid osmotic pressures in blood (πc), lymph (πl), and interstitial space (πt) in a standing, unanesthetized giraffe. Capillary pressure (Pc) was not measured directly and was assumed to be the same as venous pressure (Pv).
Published with permission of Springer Nature.

increase in thickness is a response to increasing hydrostatic pressure, and it is one way in which the development of edema is prevented in all healthy mammals. Nevertheless, the average thickness of the membrane lining the capillaries in the legs of giraffes is impressive. Does its thickness alter capillary permeability to protein (K_2)? If permeability to proteins was high, plasma proteins could move freely out of capillaries into interstitial fluid, the colloid osmotic pressure inside and outside capillaries would be similar, and reabsorption of fluid would be reduced. No mention was made of capillary pores by Williamson et al. in their study but, if they were there, then the concentration of proteins in the interstitial fluid would be high. Hargens et al. (1987b) measured protein concentration in the lymph draining the feet of giraffes and there was only a trace amount (Figure 10.3). Interstitial fluid protein concentration (πt) in the feet was high because, it turned out, it was contaminated with blood rather than being indicative of leaky capillaries. The absence of protein in the lymph means that K_2 is very low: the capillaries in the legs of giraffes are impermeable to protein. The consequence of a low permeability to protein, and a low interstitial protein concentration, is that ICOP is low and the colloid osmotic gradient, which favors reabsorption of fluid back into capillaries, is as high as it could be, making reabsorption of fluid as efficient as possible, but that alone cannot explain the absence of edema.

The only other force in Starling's equation that regulates filtration rate is the hydrostatic pressure in the interstitial space (IHP) and in giraffes it is high. IHP has been measured by Hargens et al. (1987b) in the lower legs of a walking giraffe and they found that it varied between −120 mmHg and +80 mmHg, but they also found that it was a constant +44 mmHg immediately above the fetlock (Figure 10.3), a pressure confirmed by Petersen et al. in 2013. This pressure is much higher than it is in upper parts of the leg or in the neck (Hargens et al., 1987b; Petersen et al., 2013) where Brondum et al. (2009) measured it to be 19 mmHg near the head and 12 mmHg at the base of the neck. For comparison, in a human, IHP in the legs usually is close to zero, thus in giraffes an IHP of 44 mmHg is extraordinarily high. The origin of the high IHP in the lower legs of giraffes is the skin, which is very stiff and inflexible because of the collagen fibers it contains (see Chapter 15). The skin functions like an anti-gravity G-suit and prevents blood vessels beneath it from expanding under the effect of high internal pressure, and, because the skin is inflexible, its effect is to increase IHP. The high IHP in turn acts to reduce the filtration rate of plasma through capillary walls. The inelastic skin, combined with the very narrow radius of the arteries supplying the legs of giraffes (Figure 9.8) and the "sphincter" of Kimani et al. (Figure 9.9 in Chapter 9), reduces blood flow to the lower leg to the same rate as it is in the upper part of the neck (~4 mL 100 g^{-1} of tissue min^{-1}) despite blood pressure being fourfold higher in the leg than in the neck (Pettersson et al., 1986).

Venous Return from the Legs and Abdomen

By itself, low blood flow is of little significance for the prevention of edema. What makes it significant in giraffes is that the volume of blood entering the microcirculation of the lower legs each minute is reduced and the volume filtered each minute falls within the capacity of the capillaries to reabsorb it. Once in the veins, the reabsorbed fluid has to be returned to the heart, and for blood to return to the heart from the legs, the gravitational hydrostatic pressure in the veins has to be overcome. Overcoming it requires a pump of some sort and also a way of preventing backflow of blood toward the feet. Backflow is prevented by non-return valves. The first mention of venous valves in giraffes was made by Edwards Crisp in 1853. His note is frustratingly brief. The valves themselves were not described, but Crisp believed they "throw some light upon the functioning of the spleen." In 1864 Crisp again mentioned valves, saying that he had found them not only in giraffes but in many other species as well and that they occurred in intestinal veins, the splenic vein, and the renal vein. The first systematic study of venous valves in the legs of giraffes was done by Emmanuel Amoroso and colleagues in 1947. They noted that the axillary and brachial veins (which carry blood from the front legs to the heart) contained a large number of three-leafed (tricuspid) and two-leafed (bicuspid) valves, and their tributary veins usually were guarded by tricuspid (large tributaries) or bicuspid (small tributaries) valves with one cusp above each tributary and one or two cusps below. Experiments they did on an isolated segment of the axillary vein confirmed that these valves were able to withstand, that is, did not rupture, when exposed to blood pressures of up to 200 mmHg, i.e., the kind of pressure that gravity would produce in the leg vein of a giraffe, and the valves closed quickly when exposed to back flow of blood.

Amoroso et al. (1947) also found that the saphenous veins draining the lower hind legs contained tri- and bicuspid valves, and that the veins were thick-walled and, therefore, not very compliant or flexible. The walls of veins containing valves must be stiff. If they were not, when a vein filled with blood it would expand and the valves would not be able to stop blood from leaking past or through them. Valves in the femoral and tibial veins draining the upper hindleg of giraffes are numerous and on average 30 mm apart (Østergaard et al., 2011). The inferior vena cava, which is the final conduit for blood returning to the heart from organs in the thorax, abdomen, and legs, and the veins draining the major internal organs, are less well supplied with valves. No valves have been found at the junction of the hepatic vein (draining the liver of giraffes) with the inferior vena cava, but the junctions with the vena cava of the left and right renal veins draining the kidneys are guarded by a single valve (Crisp, 1964; Goetz and Keen, 1957; Damkjaer et al., 2015). The relatively few valves in abdominal veins compared to the large number in leg veins illustrates well that getting blood up the legs to the central veins of the abdomen (and thorax) is the main problem giraffes have to overcome in order to ensure proper return of blood from the legs. Once in the abdominal veins, blood flow to the heart is by gravity because the heart lies toward the bottom of the thorax and thus flow toward the heart is downhill and there is no gravitational force pushing blood back up the great veins of the abdomen or those that drain its organs.

Flow of blood toward the heart in the abdominal veins is assisted by the movements of respiration by what is called the abdomino-respiratory pump. When a giraffe or any other mammal breathes in, the air pressure in the chest cavity falls to below atmospheric pressure, the veins in the thorax expand, their internal pressure falls, and blood returning to the heart from the front limbs and neck and head at a higher pressure flows into them. At the same time, the pressure in the abdomen increases because contraction of the diaphragm reduces abdominal volume (while increasing thorax volume). The pressure gradient between the abdomen (high) and the thorax (low) pushes blood that is in the abdominal veins toward the thorax. When a giraffe breathes out, the reverse occurs: pressure inside the thorax rises, pressure in the abdomen falls, and blood flows into the abdominal vena cava from abdominal organs and hind legs. This alternating sequence of high and low pressures occurs at every breath, and thus about 10 to 15 times a minute in an adult giraffe.

Valves in the leg veins, together with the respiratory pump, are not enough to ensure venous return from the legs or to ensure flow from the abdominal veins to the heart. For blood to flow toward the heart, the blood pressure in the veins of the legs has to be higher than the hydrostatic pressure generated by gravity (of about 200 mmHg), which acts in the opposite direction. When Hargens et al. measured the pressure in the leg veins of a walking giraffe the pressure ranged from −250 mmHg to +240 mmHg, and they concluded that there had to be "an effective pumping mechanism for moving blood against gravity" (Hargens et al, 1987b). What is the pump, and what generates positive blood pressure in leg veins of giraffes? In mammals, including humans, contraction of leg muscles is one source of pressure. The crucial "pump" muscle in the hind leg of humans is the soleus muscle, which, when it contracts, compresses the veins of the lower leg against the tibia, blood pressure in the vein increases, becomes greater than gravitational pressure, and as the valves in the veins prevent the blood from going back down the legs, the squeezed blood has to move up the legs toward the heart. Giraffes, however, do not have a soleus muscle, but its role in the hind leg is taken over by the plantaris muscle (Roy et al., 1988; see Chapter 16). Roy et al. studied muscles taken from a stillborn giraffe, a giraffe 17 days old, and an adult, and found that with age the fibers in the plantaris muscle enlarge markedly, especially those fibers in the part of the muscle closest to the bone. They concluded that the plantaris muscle, as it aged, acquired an unusually high percentage of fibers of the type usually found in the muscles of the body that maintain posture and also typical of the percentage found in the soleus muscle of other mammals. The plantaris muscle functions to maintain posture, but in doing so also acts as a muscle pump. It is unlikely, however, that the pressure generated in the veins of the hind leg depends only or even mainly on the plantaris muscle. The most obvious reason for saying so is that there is no similar muscle to the plantaris in the front legs, and yet venous return from them is just as efficient. Hargens et al. (1987b) thought that the variation in blood pressure in leg veins was most likely the result of "cyclical compression and decompression associated with muscular activity" acting against a tight skin. Their data is also consistent with the presence of another mechanism—a pump in the feet. Although not yet described in giraffes, it is possible that in both fore- and hind limbs a pump similar to that found in the hooves of horses may be present. The pump in horses depends on venous sinuses, which lie between the lateral cartilage of the hoof and the third bone of the foot. When the foot is lifted, a negative pressure is generated in the sinuses and blood is sucked into them. When the foot is placed on the ground, the sinuses are compressed between bone and cartilage, a positive pressure is generated, and blood is expelled from them. The sequence has the effect of "cyclical compression and decompression" that could explain the negative and positive blood pressures measured by Hargens et al. Another hindleg pump has been suggested by Kimani et al. (1991). They thought that it was the upper hindleg muscles that acted as a pump. When a giraffe is walking or running, the "sphincter" in the tibial artery is closed and blood is shunted into the thigh muscles. When the thigh muscles contract, they "squeeze the veins so that blood flows to the heart, reflux being prevented by venous valves," and "this effect may ensure not only an efficient venous return, but counteracts gravitational venous pooling." Presumably a similar mechanism exists in the front leg as well. The evolution of this array of mechanisms that return blood to the heart, and perhaps some others yet to be discovered, is why filtration edema, about which Krogh and Goetz were so extremely interested, never develops in the legs of healthy giraffes.

Venous Return from the Head

Theoretically, edema could develop in the head and brain when a giraffe lowers its head to eat from a low bush or to drink surface water because the pressure in the veins and capillaries of the head and brain will increase under the effect of gravity. In fact, edema does not develop in the head for the same reasons that it does not in feet, but also because the length of time the head is below heart level is too short for fluid to collect. Most of the time a giraffe's head is above heart level. Nevertheless, when a giraffe lowers its head to drink, there is a tendency for blood returning to the heart from the rest of the body via the inferior vena cava to flow into the superior vena cava and from there back up and into the jugular veins, and blood already in the veins of the head and neck would tend to regurgitate toward the cranial circulation and into the tributaries that empty their blood into the jugular vein. Engorgement of all these blood vessels would follow. The volume of blood that could enter the jugular veins and all its tributary veins when a giraffe lowers its head is theoretically as large as the volume of blood returning to the right atrium via the inferior vena cava, which is approximately 95% of cardiac output (Mitchell and Skinner, 1993) and, in total, could be 75 L min^{-1} in a large, mature giraffe. If blood could and did regurgitate into the jugular veins, then apart from engorging the tissues of the head and brain with blood, less blood would be available to the heart to generate cardiac output. Engorgement, however, does not happen because the jugular veins are equipped with valves. They function in the opposite direction to the valves in leg veins. When a giraffe's head is higher than heart level, the valves are open to allow blood to flow toward the heart; when the head is below heart level, they are closed to prevent flow toward the head. The studies on the jugular vein valves in giraffes made by Amoroso et al. (1947) found that

> [the jugular vein] *is a large straight vessel with 5 tricuspid valves of the usual type set at irregular intervals along it. All its tributaries are relatively very small and no valves were present at the orifices of any.*

Goetz and Keen (1957) saw in the single jugular vein they studied that it contained a "large number" of bicuspid and tricuspid valves and that each of the tributaries flowing into the jugular vein was provided with a competent set of valves. These two studies suggested that there were two kinds of valves in two different positions. One type lay at the opening of tributaries, and the other type occurred independently of tributaries and was located at irregular distances along the length of the vein. In addition, Hargens et al. (1987b) found that

> *the existence of closely spaced valves in the head and distal neck near the head as compared to sparse valves in the proximal neck near the heart indicates their importance for preventing retrograde venous flow, for example during short periods when the giraffe's head is lowered below heart level during drinking.*

This description implies that the distribution of the valves is neither random, nor irregular, and has a purpose.

In a larger study that involved the measurement of the number, location, and inter-valve distances of 396 jugular vein valves (192 left, 204 right) in 60 veins from 25 adult (11 males and 14 females) and five fetal giraffes (Mitchell et al., 2009a), the average number of valves in the left jugular veins was 6.4±2.7 (mean ± SD, range = 2–13) and in the right was 6.8±2.1 (range = 3–12). Male giraffes had 7.3±2.7, females 5.9±2.1, and fetuses 7.0±2.1 valves per vein. Brondum et al. (2009) recorded 7±4 in the jugular veins of five giraffes. None of these average values differ statistically. Most valves (88%) studied by Mitchell et al. were bicuspid in structure, very few were tricuspid, and the position of the tricuspid valves was random, so why some valves are tricuspid and others bicuspid is not clear. Most valves (36.1%) were located in the first 500-mm segment of vein nearest the heart and half in the first 800 mm (200 mm in a fetus) of jugular vein (Figure 10.4).

None of the valves was located at or near the entry of tributary veins. Each tributary had its own valves, and it is these valves that prevent regurgitation into the tributaries. As most valves in the jugular vein are in its bottom segment, they are able to prevent blood that is returning to the heart from the body from flowing into the jugular vein. This

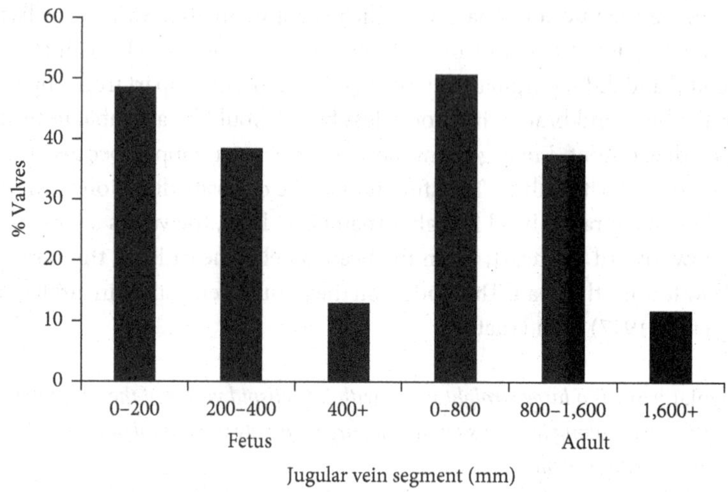

Figure 10.4. The distribution of valves in the jugular vein of fetal and adult giraffes. The segments 0–200 and 0–800 refer to the segments nearest to the heart. The data were obtained from 11 male, 14 female, and 5 fetuses of *G.c. giraffa*.

pattern is also found in kudu (*Tragelaphus strepsiceros*) and probably in all ruminants. The length of the jugular veins in a kudu is ~700 mm. In a single kudu studied (Mitchell et al., 2009a), the left jugular vein contained nine valves and the right vein eight valves. Most valves (41%) were in the bottom one-third of the veins, 35% were in the middle one-third, and 24% in the top one-third, mirroring the position in giraffes, and like giraffes, inter-valve distances were much closer in the bottom segment (35.6±11.3 mm, range = 23–53 mm) than in the middle segment (65.8±17.6 mm, range = 56–97 mm) or the top segment (206.8±109.9 mm, range = 111–360 mm).

As fewest valves in giraffes are found in the top 500-mm segment of the vein and many veins had no valves in that segment, the idea that the main function of the valves is to prevent reflux of blood into the veins of the head or brain, as suggested by Hargens et al. (1987b), is not supported by these data. In any case, despite the absence of valves at the head end of the veins, regurgitation toward the cranial circulation is unlikely because the internal and external maxillary veins that drain the tissues of the head contain valves to prevent it, and the volume of blood in the top part of the jugular vein is small. The average distance between the most rostral valve and the junction of the internal and external maxillary veins was 36 cm in the 60 veins studied by Mitchell et al., and the average diameter of this part of the jugular vein was 2.5 cm, thus the volume of blood that would fill this section of the vein would be ~200 mL and ~400 mL if both veins were taken into account, a small fraction of the blood in the cranial circulation at any moment. The capacity of giraffe jugular valves to withstand back pressure has not been measured, but the maximum back pressure to which jugular veins are likely to be exposed is, on average, 140 mmHg, given that the downward angle of the neck relative to horizontal when a giraffe is drinking is ~55°. This pressure is well within the capacity of the valves if they are as resilient as the axillary vein valves tested by Amoroso et al. in 1947. Effectiveness of the valves does not seem to depend on their number. In six veins there were just two or three valves, and these presumably were just as effective as the

10 or more valves that were present in eight veins. Why so many valves are required in some giraffes and so few in others to prevent regurgitation is, therefore, a mystery.

Jugular valves prevent blood returning to the heart in the inferior vena cava from entering the jugular veins when a giraffe has lowered its head. They do not prevent blood that is being sent to the head in the carotid artery from collecting in the jugular veins and its tributaries when a giraffe lowers its head, and that blood has to be returned to the heart. Some blood could be pumped up the veins by rhythmic contractions of the masseter muscles of the jaws that occur when a giraffe is drinking (Pedley, 1987; Brook and Pedley, 2002), and Owen (1838) noted a special mechanism for returning blood from the tongue. The lingual veins form a network or plexus at the base of the tongue and are covered by a thin layer of muscle fibers that circle the tongue. When these fibers contract, as they do when the tongue is being extended, they compress the plexus. However, the volume of blood that both of these mechanisms will move is small, and most blood is simply returned to the heart by gravity when the head is lifted.

Nevertheless, the volume of blood that could collect in the two jugular veins even in a young giraffe weighing ~400 kg can be estimated to be 400 mL (Brondum et al., 2009), and in an adult giraffe, in which each jugular vein has a diameter of 2.5 cm and is 2 m long, is 2 L (1 L in each). Under pressure the veins could distend and the volume would then become larger, but the jugular vein, unlike any other vein so far studied in giraffes, is innervated by sympathetic nerves (Nilsson et al., 1988) and if they are activated they would prevent distension. Even so, Goetz and Keen (1957) observed that in the two seconds that it takes a giraffe to lift its head after drinking, blood that has collected in the jugular veins could descend suddenly into the right atrium of the heart with considerable force, and exceed the right atrium's capacity to receive it, unless some special regulatory mechanism existed to prevent it from doing so. How is this blood flow controlled? Brook and Pedley (2002) have shown that collapse of the jugular vein when a giraffe lifts its head will "prevent blood from rushing out of the head," but Goetz and Keen found a muscular cuff in the wall of the superior vena cava at the point where it enters the right atrium. The muscle cuff in vivo is approximately 20 mm long and 5 mm thick, and its muscle almost certainly is innervated by sympathetic nerves because other parts of the jugular vein are innervated by sympathetic nerves (Nilsson et al., 1988). The outer diameter of the vena cava at this point is approximately 25 mm, thus the wall of the cuff forms ~40% of the diameter of the vena cava. The cuff's function is unknown, but the function Goetz and Keen thought it fulfilled was to contract when a giraffe lifts its head after drinking. Contraction of the cuff muscles would constrict the jugular vein, narrow its lumen, and limit the rate at which blood flowed into the heart in the same way that a collapsed jugular vein would. Constriction of the cuff (and/or collapse of the vein) would also explain why it is that when a giraffe is standing upright the pressure in the jugular vein can be positive (Figure 10.5), when it is negative or zero in other mammals.

Another possible function of the cuff is that the muscles contract and narrow the diameter of the jugular vein when a giraffe lowers its head. A constricted superior vena cava would help prevent regurgitation of blood from the inferior vena cava into the jugular vein. Indeed, this may be the main function of the cuff, as in some jugular veins there are no valves positioned close to the heart to prevent regurgitation of blood coming from the inferior vena cava into the jugular veins. A third possibility is that the cuff constricts as part of the head-raising reflex that diverts blood to the

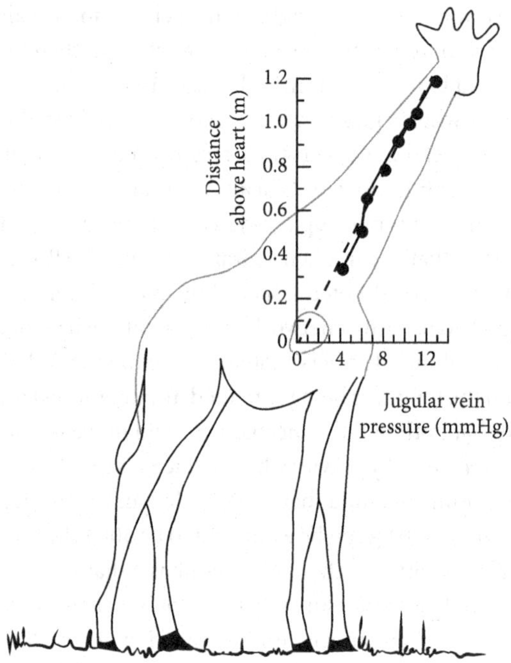

Figure 10.5. Giraffe jugular venous pressures (JVP) as measured by Hargens et al., 1987b (points and solid line) and a linear regression line calculated from the data (broken line). Figure from Mitchell et al., 2006b, reprinted with permission of the Company of Biologists.

brain. These various possibilities were tested in a mechanical model (Mitchell et al., 2006b). Mechanical "constriction" of the cuff did not produce the pressure profile typically found in a giraffe's jugular vein, as shown in Figure 10.5, but constriction did restrict flow and so could function to reduce the rate of emptying of the jugular veins, and could prevent regurgitation of blood from the inferior vena cava into the jugular veins and direct it into the right atrium. More intriguing is that in the model, constriction of the cuff increased the flow of blood to the brain, perhaps by in some way increasing blood flow in the vertebral venous sinus, although how it could do so in a living giraffe is another mystery. If the cuff did constrict as the head lifted and did improve cerebral blood flow at a time when carotid blood flow is declining, then constriction of the cuff might be part of a very elegant reflex that contributes to the prevention of fainting.

Fainting and the Cranial Circulation of Giraffes

The head of a giraffe can move through an arc four to five meters in height when it lifts and lowers its head, and when he made the first measurements of blood pressure in giraffes, Goetz concluded that, "judging by human standards, giraffe should faint when they raise their heads after drinking water"(Goetz, 1955). One of the giraffes in Goetz's 1960 study did "faint" and, because of the failure of compensatory mechanisms it suddenly collapsed and died (Goetz et al., 1960). Fainting results from insufficient blood flow to the brain, and the result is that the brain does not receive the oxygen and other nutrients it needs to maintain neuronal function. The rate at which a giraffe lifts its head complicates matters. It is, and has to be, rapid. Giraffes are vulnerable to attack by predators when they are drinking. Filmed recordings of the head-raising maneuver show that the neck swings from a drinking position to a horizontal position in half a second (Dagg and Foster, 1976). The momentum generated by the swing helps to lift the front legs to a more vertical position, after which head-raising is continued until the giraffe is upright a second later. During this maneuver the effect of gravity (the G-forces) on the column of blood in the carotid artery is increased. Goetz's conclusions about fainting in giraffes had been influenced by a study by Robert Rushmer and colleagues in 1948, who were trying to understand the phenomenon of "blacking out" in pilots as G-forces increased. Rushmer established that blood pressure at, and blood flow to, the head of pilots became progressively reduced as G-forces increased because blood accumulated in leg veins, reducing blood flow to the brain until eventually unconsciousness occurred. The problem of "blacking out" in pilots was solved by a "G-suit." What mechanisms have giraffes evolved to prevent reduced blood flow to the brain and to prevent "blacking out" when they lift their heads rapidly?

Many myths have entered giraffe folklore to explain how fainting is prevented during a head-lifting maneuver. One of the most persistent is that the left and right carotid arteries contain valves which prevent reflux of blood down the carotid artery between heart beats. The two carotids arise from the common carotid trunk, which is one of the first major arteries to branch off the aorta (Chapman, 1875). If valves are present in the carotid arteries, then all the heart has to do, this theory suggests, is to pump a volume of blood into the common carotid trunk and the new volume of blood will displace the blood already there from the preceding heart contraction. The displaced blood must move up the carotids because of the valves, and blood will reach the brain. Of course this theory is a fallacy. The only valve able to prevent backflow of blood from the carotid artery and aorta into the left ventricle during the relaxation, filling, phase of a cardiac cycle is the aortic valve. Blood moves up the carotid arteries not because of carotid valves, but because of the pressure generated by the heart, and will do so regardless of whether the blood is separated into compartments by valves. Many researchers have looked for these carotid valves, but none has ever been found; nevertheless, this myth, like the myth that the giraffe heart weighs 11 kg, persists. Another myth, but one taken much more seriously by researchers (e.g., Badeer, 1986, 1988, 1997), is that a siphon operates to suck blood up the carotid artery. A siphon would ensure a continuous flow of blood. The two arms of the siphon would be the carotid arteries on the one hand and the jugular veins on the other. Another attraction of a siphon is that it would reduce the work that the heart has to do to pump blood up the carotid arteries. It seems inevitable and even ineluctable that evolutionary mechanisms would have resulted in the disproportionate survival of giraffes that had hearts that used the least energy. In giraffe literature, reports of a siphon mechanism seem to have originated from a note by Patterson and Warren in 1952. Even the exceptionally knowledgeable Goetz (1955) imagined that a siphon might exist. He asked

> *whether the left ventricle does provide the moving force unaided or whether it is assisted in its task by other mechanisms such as a peristaltic wave along the carotid artery or a*

siphon effect of the venous blood coming down the jugular vein helping to elevate the blood in the carotid artery.

A peristaltic wave does not exist. For a siphon to operate, the arterial and venous arms of the cranial circulation must be continuous and similar to an inverted U-tube in which the fluid is returned to at least the same level at which it started. It is only when this arrangement is in place that the effect of gravity is neutralized. It is impossible to study how a siphon might work in a living giraffe, and so instead it has been studied using mechanical models (Holt, 1959; Seymour and Johansen, 1987; Hicks and Badeer, 1987; Seymour, 2000; Mitchell et al., 2006b).

Those models tested whether a siphon effect can exist in the cranial circulation of any animal that stands upright like humans and giraffes, and if it does, if the work of the heart is thereby reduced. The models have shown that if the carotid artery and jugular vein are connected directly and if both blood vessels are rigid (i.e., not able to collapse), then a strong siphon can be established and the work of the heart can be reduced by 30% (Mitchell et al., 2006b). However, if any of the blood vessels in the cranial circulation can collapse, then a siphon cannot be established, and the work of the heart is not reduced. In the cranial circulation of a giraffe, the carotid arteries can be assumed to be rigid tubes because of the structure of their walls and because blood inside them is at high pressure, but the jugular veins are not rigid and are collapsed when the giraffe is standing upright. Furthermore, the carotid arteries and jugular veins are separated by the highly collapsible capillaries that make up the extensive microcirculations of the cranial circulations. In addition, if the jugular vein was acting as the downward arm of a siphon, pressures in it should be at or below atmospheric pressure. In anesthetized giraffes lying flat or held upright, the pressure inside the jugular vein is atmospheric or slightly below it (McCalden et al., 1977, 1980; Brondum et al., 2009), but in unanesthetized giraffes standing upright, it is 16 mmHg above atmospheric pressure near the head (Hargens et al., 1987a) and 10 mmHg about one meter above the heart (Mitchell et al., 1993; Figure 10.5). The positive pressures are consistent with collapsed jugular veins offering resistance to blood flow (Pedley et al., 1996). The relationship between height (Ht) in meters above the heart and jugular venous pressure (JVP) shown in Figure 10.5 is JVP (mmHg) = 9.7 x Ht + 0.9 ($R^2 = 0.98$). This equation implies that JVP is 0.9 mmHg when height is zero (i.e., in the right atrium). The right atrial pressure has been measured by Goetz et al. (1960) to be 0 mmHg, but it fluctuates between +20 and −5 mmHg, with the positive pressure occurring during expiration and the negative pressure during inspiration (van Citters et al., 1969). The overall conclusion from these studies is that a siphon does not and cannot operate to lift blood to the giraffe's brain (Holt, 1959; Seymour and Johansen, 1987; Seymour et al., 1993; Munis and Lozada, 2000; Dawson et al., 2004; Mitchell et al., 2006b). The left ventricle, alone and unaided, provides the moving force. If so, how does enough blood reach the brain of a giraffe to prevent fainting when it lifts its head?

At first glance the circulation of blood to the head is a simple system. Blood flows to the head in the left and right carotid arteries and returns to the heart via the left and right jugular veins. However, the cranial circulation is not as simple as it seems. In mammals it can be separated into extracranial flow, which is blood flow to the muscle, skin, tongue, and other structures of the head, and intracranial flow, which is the flow to the brain (cerebral blood flow). Cerebral blood flow can further be divided into flow to the meninges surrounding the brain and spinal cord and blood flow supplying the brain tissue itself. All these blood supplies are separated from the veins that drain them by the network of capillaries that makes up the cranial and cerebral microcirculations. Blood leaving extracranial tissues returns to the heart via jugular veins. The blood leaving the brain returns to the heart via the vertebral venous plexus, which is a complex of veins found within the vertebral canal alongside the spinal cord.

Giraffe Cranial Circulation

The path that blood takes to get to the extracranial tissues is very similar in all mammals. The way blood is distributed to the brain is also similar: the brain receives blood from a ring of blood vessels that lies at the base of the brain. This ring was described first by Thomas Willis in 1664 in one of the great works of anatomy and physiology: *Cerebri anatome: Cui acccessit nervorum descriptio et uses (The Anatomy of the Brain and the Description and Function of the Nerves)*. In historical importance it ranks alongside William Harvey's *De mortu cordis*, which was published in 1628 and described how blood circulates. Since Willis's description of it, the ring of vessels at the base of the brain has been called the Circle of Willis. In most ruminant artiodactyls, interposed between the main carotid artery and the Circle of Willis is the *carotid rete mirabile*—a wonderful net—a structure that arises from multiple branching of the internal maxillary artery (see Figure 10.8 later in the chapter). In artiodactyls the final pathway for blood flow from the *rete* to the Circle of Willis is the internal carotid artery. It arises as a confluence of all the many tiny arteries that form the carotid *rete* (Figure 10.6). Willis was the first to describe the *rete* and it was drawn for him by Christopher Wren. Goetz and Keen in 1957 and Kimani et al. (1991b; Figure 10.8) have revealed its inner structure in giraffes. Willis attributed three functions to the *rete*.

> *First, that the torrent of blood being divided into small rivulets, its more rapid course may be so far dulled or broken, that it may be but leisurely instilled into the brain. For otherwise in labouring beasts, who go with their heads hanging down, and have but a weak brain, the more free influx of blood might easily overthrow the fabrick of the brain and spoil the animal spirits. Secondly, . . . another use of no less moment, to wit, that the blood may carry away some part of it to the pituitary glandula. Thirdly and lastly, for as much as the vessels on either side are mutually inoculated, there is care taken that more certain ways may be made for its passage: because, if perchance an obstruction should happen in one side of the wonderful net, the blood by being carried to the other, may find a passage that the blood might be better prevented from any impediment in its passage.*

All three functions foretold by Willis have since been shown to operate. The *rete* "reduces the impetuosity of the torrent of blood flowing to the brain" and it does provide a way of delivering hormones to the hypothalamus and pituitary gland, and a way of transferring pheromones from the nose to the pituitary gland that the *rete* surrounds (Grzegorzewski et al., 1997), and it does provide an excellent collateral circulation in the event of an obstruction in one or other carotid artery. Van Citters et al. (1966) wrote,

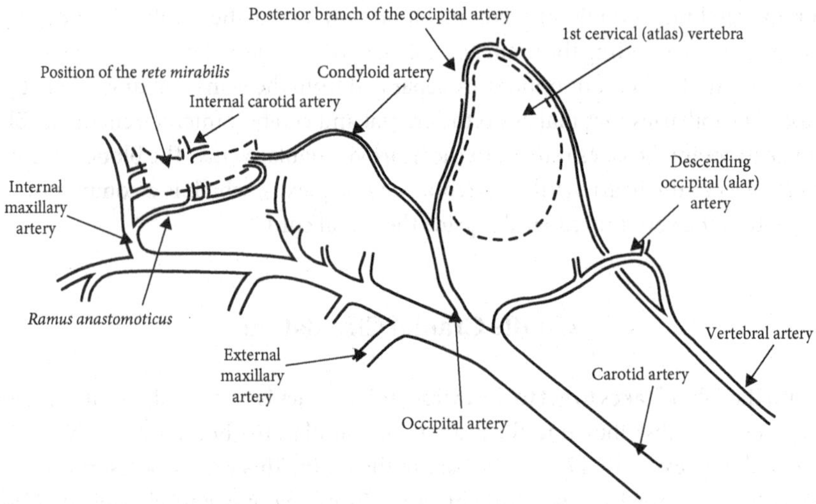

Figure 10.6. Diagram of the major arteries of the cranial circulation of giraffes.
Redrawn from Goetz and Keen, 1957, with names of arteries modified following O'Brien et al., 2016.

for example, that "despite a 10-minute continuous occlusion of the right carotid artery, there was no evidence of neurological impairment" in the giraffes they studied. There are two main ways that blood can reach the *rete* in ruminant artiodactyls (Baldwin and Bell, 1963). In sheep and goats and many antelopes, the main artery supplying blood to the *rete* is the maxillary artery and its *ramus anastomoticus* (Figure 10.6).

In cattle and other bovids like buffalo and in deer, the blood can reach the *rete* by both the maxillary artery and the vertebral artery. Which of the two ways occurs in giraffes? Lawrence and Rewell (1948) made the first dissection of giraffe cranial blood vessels and noticed two unique features. First, a main occipital artery branches off from each of the left and right carotid arteries, and it is extraordinarily large in giraffes (Figure 10.6). There is also a smaller descending occipital artery arising from each carotid artery. Second, on each side of the neck the descending occipital arteries connect to the vertebral arteries, which run up the neck alongside the vertebrae independently of the carotid arteries. Lawrence and Rewell could not find any connection between the vertebral arteries and the brain circulation in giraffes, and they concluded that the purpose of the connection between the carotid and vertebral arteries was to divert blood away from the brain when a giraffe lowered its head to drink. Arteries do not contain valves; thus blood could flow from the descending occipital arteries to the vertebral arteries and, if it did, engorgement of cerebral arteries would be avoided. After lifting their heads, the direction of blood flow would reverse, and blood would flow from the vertebral arteries into the occipital artery and then to the carotid arteries.

A more detailed study of the cranial vessels was made by Goetz and Keen (1957; Figure 10.6). They injected a radio-opaque dye into the vessels, which makes the vessels visible on an X-ray film, and reveals the routes that blood can take in a living animal. Goetz and Keen, like Lawrence and Rewell, also found junctions (anastomoses) between the carotid, vertebral, and occipital arteries, and saw that the occipital and descending occipital arteries were large and that they anastomosed with the vertebral arteries, an arrangement they concluded that could provide a second supply of blood to the *rete* in addition to the supply it received from branches of the internal maxillary artery (Figure 10.6). Importantly, at the point where the main occipital artery branches off the carotid artery, both arteries are equipped with baroreceptors and thus both can detect changes in the pressure of the blood supplying the brain. Kimani (1987) studied the anatomy of the vertebral artery and showed that its structure changed (at the level of the 5th to 7th cervical vertebrae) from being mainly composed of elastic tissue to being composed mainly of muscle. This difference led Kimani to conclude that blood in the vertebral artery is confined to the neck by contraction of its muscle, and blood in it does not reach the brain. Conversely, of course, contraction of the vertebral artery muscles would prevent blood that has been diverted from the carotid artery into the occipital artery from flowing toward the heart down the vertebral artery (Figure 10.6), making sure that diverted blood could flow toward the brain. In another study of the brain blood vessels in 12 giraffes, Frackowiakand Jakubowski (2008) confirmed that the condyloid artery, a branch of the occipital artery (Figure 10.6), anastomosed with the *rete*. Lawrence and Rewell had estimated that the cross-sectional area of the main occipital artery was at least as great as that of all the much smaller vessels supplying the *rete* via the internal maxillary branch of the carotid artery and its *ramus anastomoticus*. If so, then in giraffes the blood supply to the *rete* followed the pattern found in deer, and the vertebral arteries could carry large amounts of blood to the brain. O'Brien et al. (2016) injected a latex-based solution into the cranial blood vessels of a giraffe and, in general, their study showed a similar pattern, although in their specimen the main occipital artery was small and they could not show that it joined the *rete*.

How much blood goes to the brain? To function properly, the brain of all mammals requires an average blood flow of 50 mL 100 g^{-1} of brain tissue min^{-1} (range 25–80 mL 100 g^{-1} min^{-1}) to supply all the necessary oxygen and glucose, and to remove waste products and heat. In other words, in normal circumstances blood in the brain at any one moment makes up half of the mass of the brain. Brain mass has been found to be, on average, 0.1% of body mass in giraffes. It is proportionately higher in fetuses and newborn giraffes (0.3% of body mass) and lowest in adults (0.07%). Its absolute mass is ~500 g in neonates and ~700 g in adult giraffes (see Chapter 16). For giraffes a flow rate of 50 mL 100g^{-1} min^{-1} means that a flow of 250 mL min^{-1} is required in small giraffes and 350 mL min^{-1} in large ones for the brain to function normally. That blood has to come from the carotid arteries. The first measurements of how much blood flows in the carotid arteries of giraffes were made by van Citters et al. (1968). In a giraffe that would have weighed about 700 kg, carotid blood flow varied between 2.2 L min^{-1} and 3.2 L min^{-1}, fluctuating with the contraction and relaxing phases of the cardiac cycle, but had an average flow of 2.4 L min^{-1}. Other researchers have found similar flow rates. That volume of blood corresponds to about 5% of total cardiac output in a giraffe that has a body mass of 700 kg and will be proportionally the same (i.e., 5% of cardiac output) in all giraffes. Carotid blood flow is a measure of total blood flow to the head, i.e., not only blood flow to the brain. In a 700-kg giraffe, brain mass is ~660 g (Figure 16.10) and blood flow to the brain will be about 330 mL min^{-1}. Thus the brain of a giraffe receives only ~10%–12 % of total carotid blood flow or 0.5% of cardiac output: most blood going to the head goes to non-brain tissues.

Preventing Fainting

Making sure that the relatively small flow of blood to the brain is continuous whatever the position of the head is a complex matter because of the large variations possible in the height that a giraffe's head can be relative to the level of the heart and the changes in blood pressure that follow. The changes in blood pressure in giraffes when a giraffe lifts its head after drinking have been measured by van Citters and colleagues (van Citters et al., 1969). They measured blood pressure in the carotid artery at a point about 30–40 cm below the level of the brain and, simultaneously, in the heart. The pressures were measured in unanesthetized, free-living giraffes after a pressure sensor had been implanted in the carotid artery. The data were transmitted by telemetry and because they are the only measurements made so far in free-living giraffes, the pressures obtained represent the only true, natural, blood pressures so far recorded in giraffes. During the period of measurement that lasted several days, the position of the head was observed, and fortunately the giraffes drank water at ground level at intervals so the effect of postural change on blood pressures could be seen. The relationship between heart and brain pressures in these free-living giraffes during head lowering and raising maneuvers and immediately afterward, as extracted from a pressure trace recorded in one of them (in van Citters et al., 1969), is shown in Figure 10.7.

In the head-down position, mean arterial pressure at the head increased to nearly 325 mmHg, while heart pressure remained constant at about 175 mmHg (i.e., well

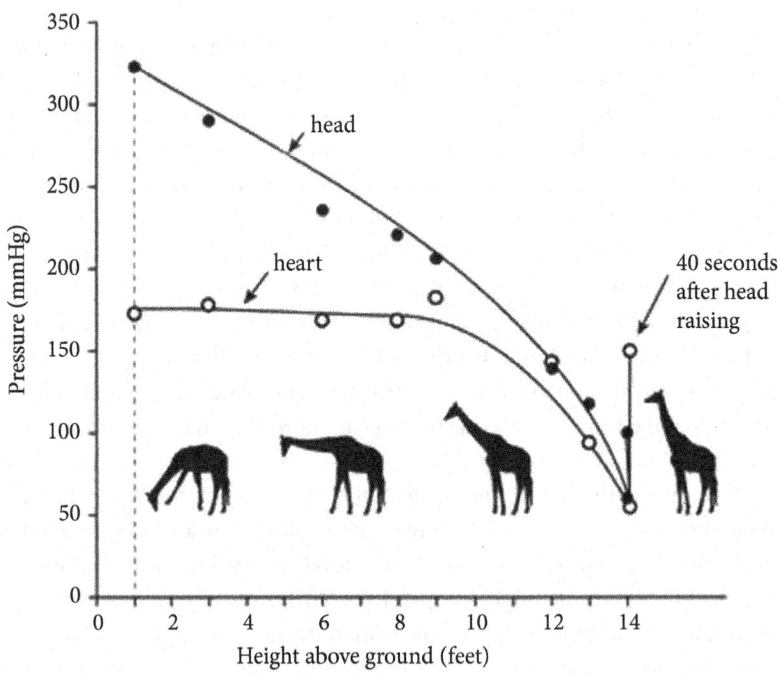

Figure 10.7. Heart and head pressures during a head-raising maneuver, drawn using data from a pressure trace in van Citters et al., 1969.
From Mitchell et al., 2008, with permission of Elsevier.

below what its pressure is when a giraffe is standing upright) and pressures of this kind also occur in sedated giraffes (McCalden and Mendelow, 1980). As the head was lifted and neared the end of its arc, pressure in both the head and heart decreased, falling below 60 mmHg despite an increase in heart rate that would normally raise blood pressure. In a similar study in sedated but mobile giraffes, Hargens et al. (1987b) found systolic/diastolic pressure to be 330/240 mmHg (mean arterial pressure = 270 mmHg) in the head-down position, and in the head-up position they were 145/55 (mean pressure = 85 mmHg). The changes in blood pressure in the blood vessels at the head are entirely due to the effects of gravity: in van Citters et al.'s study (1969), heart pressure was constant while the head moved from ground level to about 2.7 m (9 feet) above the ground, but thereafter decreased sharply, with head pressures doing the same (Figure 10.7). A decline in blood pressure will cause carotid blood flow to fall and thus a fall in blood flow to the brain. In the giraffe that they studied, McCalden et al. (1977) found, for example, that when the head was lifted, mean carotid pressure fell from 140 to 70 mmHg, and carotid blood flow simultaneously fell by half from 2.1 to 1.2 L min^{-1}. In a young giraffe, when the head was down blood flow in the carotid artery was 540 mL min^{-1}, and in the head-up position it was 360 mL min^{-1} (Mitchell and Skinner, 1993). The time taken to re-establish normal giraffe pressures (~200 mmHg at the heart and ~100 mmHg at the head) and also normal blood flow to the head after the head was raised was ~40 seconds in van Citters et al.'s study (Figure 10.7) and 60 seconds in a study done by Brondum et al. (2009), both of which are extraordinarily long times during which blood flow to the brain must be maintained in order to prevent fainting. No other animals are subjected to equivalent changes in pressures and flows, so giraffes must have evolved special mechanisms to preserve brain blood flow during this prolonged recovery phase.

One completely logical and simple mechanism for maintaining flow when the head is raised is for the blood vessels of the brain to dilate (McCalden et al., 1977). Dilation would lower resistance to flow, and flow to the brain would increase provided that blood pressure stayed constant or increased. However, blood pressure falls when the head is lifted to below the required 100 mmHg needed to sustain sufficient flow, so this possibility cannot be correct. A second possibility is that extracerebral vessels constrict while cerebral vessels do not, thus diverting whatever blood is flowing up the carotid artery to the brain. Under these circumstances, resistance to flow would increase. The only study so far that has set out specifically to measure cranial resistance found that resistance in cranial blood vessels did increase as the head lifted (Mitchell and Skinner, 1993). More circumstantial evidence for this counterintuitive response comes from changes in diastolic blood pressure in the carotid artery as posture changes. Diastolic blood pressure (i.e., the pressure in an artery when the heart is filling with blood) is a measure of resistance. The higher the diastolic pressure, the more constricted are arteries and arterioles and the higher the resistance. In a study done by Goetz et al. (1960), for example, diastolic pressure in the carotid artery increased when the head was lifted, thus the arteries and arterioles had constricted, and on lowering the head a "conspicuous fall in diastolic pressure" occurred, indicating that in the head-down position the blood vessels of the head had dilated, and the resistance reduced, a conclusion also reached by Brondum et al. (2009). Van Citters et al. in their several studies (1966, 1968, and 1969) found exactly the same changes.

For cranial resistance to increase when the head is lifted, and for blood vessels to have constricted, while at the same time allowing blood to reach the brain, there must be a way for blood to reach the brain that is not shut off by blood vessel constriction. There also must be an arrangement of nerves that brings about constriction of non-brain blood vessels but leaves the alternative routes to the brain open. An alternative pathway to the brain for blood, identified by Goetz and Keen (1957) and Frackowiak and Jakubowski (2008), is the connections between the carotid, occipital, and vertebral arteries that supply the *rete mirabile* (Figure 10.6). If this pathway is not constricted when the head is lifted, blood will reach the *rete* and the brain by its alternative route. An open pathway implies that nerves to the carotid-occipital-vertebral connection are absent or less active than the nerve supply to extracranial blood vessels. The innervation of the arteries of the cranial circulation of giraffes has been studied exhaustively by Kimani and colleagues (Kimani, 1987; Kimani and Opole, 1991; Kimani et al., 1991b), and they have found that, in sharp contrast to the rich innervation and deep penetration of nerves to the wall of the carotid artery and its extracranial branches, the *rete* itself is poorly supplied with nerves, as is, typically, the vertebral artery (Edvinsson, 1975). So impressive was this innervation pattern that Kimani et al. concluded that the extracranial arteries must function as a sphincter, shutting down the blood supply to the *rete* when a giraffe lowers its head, although if that happened the brain would be deprived of blood and the giraffe would lose consciousness. What must happen when the head is raised (if a giraffe is to remain conscious) is that non-essential, extracranial, blood vessels constrict while the alternative supply to the *rete* remains unaffected. Thus an explanation for why giraffes do not faint is that, first, the brain requires a relatively small volume of blood, and second, via the *rete*, the brain can receive blood from two sources—via the internal maxillary branch of the carotid artery and from the carotid-occipital-vertebral connection. Third, when a giraffe lifts its head the extracranial vessels, which are well supplied with nerves, constrict. The *rete* and the occipital-vertebral arteries do not, and even if blood flow up the carotid artery is reduced significantly as the head is lifted, whatever blood is still flowing up the carotid artery is diverted to the brain and is sufficient to prevent fainting.

Yet another possible way that fainting is prevented when a giraffe lifts its head is for cerebral blood flow to be maintained by an increase in cerebral perfusion pressure (CPP). Cerebral perfusion pressure is the difference between mean arterial pressure (MAP) at the head and the intracranial pressure (ICP), and the larger CPP is, the better is blood flow to the brain.

$$CPP = MAP - ICP$$

ICP is the sum of cerebral venous pressure (CVP) within the drainage veins of the skull and cerebrospinal fluid pressure (CSFP).

$$ICP = CVP + CSFP$$

With the head in the upright position, CPP is 100 mmHg and the same as MAP because CSFP and CVP are zero or even negative. Cerebral perfusion pressure could be increased at the time the head is lifted if "a siphon effect is produced through the deep non-collapsible venous channels" (Goetz et al., 1960) to produce sub-atmospheric (negative) pressure in the intracranial veins and so reduce intracranial pressure. The non-collapsible venous channel to which Goetz et al. were referring is the vertebral venous plexus, which runs to the heart inside the vertebral canal and is protected from collapse by being attached to the walls of the canal. In many animals this plexus is the major route by which cerebral venous blood leaves the skull and is a major route in humans (Eckenhoff, 1971). No one has looked for a vertebral venous drainage channel in giraffes, but as the plexus is found in other ruminants (Sisson and Grossman, 1967), it is likely that giraffes have one and, given the difficulties that must be overcome, it may be better developed in giraffes than it is in other ruminants.

It is not possible to study, in a living giraffe, how a vertebral venous plexus might function in them, but a model has been used to simulate what could happen (Mitchell et al., 2008). The model showed that when a giraffe bends to drink, ICP increases under the influence of gravity, and the size of the increase depends on the vertical height of the blood or spinal fluid column. On lifting the head, ICP becomes negative. These changes have been found in giraffes by van Citters et al. (1969); thus the model did mimic what happens in real life. The increase in ICP when the head is lowered should, however, be counteracted by a rise in MAP because of the effect of gravity on the blood in the carotid artery, and that increase in pressure should maintain CPP, but the model showed that CPP in the head-down position is half of what it is in the upright position. Arterial pressure at the head, although much lower than the 450 mmHg that it could be if it followed the effect of gravity, is nevertheless still very high when the head is down (~300 mmHg; Figure 10.7); thus the low CPP must result from a rise in intracranial pressure. It seems quite possible that the uncomfortable sensation produced by a high intracranial pressure and reduced cranial perfusion pressure partly accounts for (with a need for vigilance) the limited time that giraffes drink—about 20 s at a time (Innis, 1958). On lifting the head, MAP at the head falls from around 300 mmHg to less than 100 mmHg (Figure 10.7), but the model showed that as the head is raised, intracranial pressure and vertebral venous pressure become negative. The result is that when the head is lifted, CPP increases quickly and by a greater amount than the fall in MAP, and so cerebral blood flow is sustained. In humans this mechanism for maintaining cerebral blood flow is so effective that MAP can fall by as much as 75% without loss of consciousness (Henry et al., 1951). In giraffes, at the moment of lifting its head, MAP also falls by this proportion (Figure 10.7) and loss of consciousness also does not occur. Thus it seems very likely that an increase in CPP, combined with the diversion of blood via the occipital-vertebral anastomosis to the carotid *rete* and thence to the brain, at the expense of extracranial flow, are the reasons why a giraffe does not faint when it lifts its head.

Strokes

The signs of a stroke in a human result from a sudden reduction in blood flow to a (usually) quite small region of the brain. As far as can be judged, strokes rarely occur in animals. In humans strokes are readily recognizable, and most are caused by an obstruction of blood flow by a blood clot in a brain blood vessel. Some strokes, and the most frequently fatal form in humans, are caused by rupture of an intracerebral artery. The underlying cause of a stroke, whether caused by a clot or by a rupture, is an abnormally high blood pressure, and giraffes should be particularly susceptible to strokes. Theoretically, when a giraffe is drinking, MAP at the head could be as much as 450 mmHg (van Citters et al., 1969) to 500 mmHg (Goetz, 1957; Goetz and Budtz-Olsen, 1957) because of the combined effects of gravity and the pressure generated by the heart. There are very few blood vessels that can withstand such a pressure, and the blood vessels in the brain are not among them. However, when arterial pressure at the head of a giraffe is measured, the theoretical pressures are not reached. The measured pressure can be ~300 mmHg (van Citters et al., 1969; Figure 10.7) but is, on average, about 255 mmHg when all measurements of blood pressure at the head thus far made in giraffes when they have lowered their heads are taken into account. Although very high by human standards, this pressure is much less than theory predicts, which means that giraffes have evolved mechanisms to reduce blood pressure at the head when they bend down to drink, and, therefore, reduce the risk of blood vessel rupture (Mitchell et al, 1980).

One mechanism that reduces the blood pressure at the head is their characteristic drinking posture. When a giraffe bends to drink, it spreads its front legs in a very ungainly maneuver (Figure 11.1) that takes minutes to accomplish. Their contortions arise not because they are attempting to reduce cranial blood pressure, but because of the way in which the head and neck are supported by the strong elastic *ligamentum nuchae.* This ligament is attached to the long dorsal spines of the thoracic vertebrae, to all of the cervical vertebrae except the first one, and to the occipital crest at the back of the skull (see Chapter 15). The ligament makes lifting of the head and neck easy and mostly independent of any muscle work, but lowering of the head to ground level involves using muscle power to stretch the ligament as much as it can be, but thereafter giraffes have to spread their legs to reach the ground. How far they have to stretch is minimized by the length of their heads, which may be 80 to 100 cm long in mature giraffes, but they cannot avoid spreading their legs. Spreading of their legs is facilitated by a very loose fold of skin between their front legs and a unique arrangement of the muscles of the trunk (see Chapter 16). As a result of the bending and spreading and stretching, not just the head, but also the trunk and the heart within it, are lowered. Thus, instead of the heart being 2 m above the head, as it would be if a giraffe could bend down without adopting its drinking posture, it is 1–1.5 m above it, thus reducing the effect of gravity and blood pressure by 25%–50%.

Nevertheless, arterial pressure at the head is still very high and it triggers the baroreceptor reflex, which mediates a reduction in stroke volume, heart rate, and total peripheral resistance and, consequently, the pressure generated by the heart does not increase (Figure 10.7). Another possible cause of reduced arterial pressure suggested by Brondum et al. (2009) is that when a giraffe bends its head to drink, the jugular veins fill with the blood that would normally enter the right atrium, reducing the volume available to fill the heart, causing cardiac output and, therefore, blood pressure to fall. They found in juvenile giraffes that the MAP generated by the heart fell by 30% from 193 mmHg in the head-up position to 131 mmHg in the head-down position. It is unlikely, though, that pooling of blood in the jugular veins was the mechanism involved: the action of jugular vein valves and possibly constriction of the superior vena cava cuff prevents blood in the inferior vena cava from entering the superior vena cava and the jugular veins. In addition, the blood volume of the ~400 kg giraffes that

Brondum et al. studied can be calculated to be 22 L (see Chapter 9), and the amount of blood that they calculated to have accumulated in the jugular veins was 1.2 L, which they estimated to be 4% of blood volume. Decreases of at least 15% of blood volume are necessary before cardiac output is reduced. A more likely explanation for their data is that the baroreceptor reflex was activated to reduce cardiac output and blood pressure.

The struggles giraffes go through to lower their hearts cannot be the only way that the potential consequences of high blood pressure for cerebral vessels is controlled. Other ways must have evolved. It would be unthinkable that giraffes could have evolved so successfully if they relied solely on a contortion to prevent rupture of cerebral and cranial blood vessels. Mechanisms to control blood pressure and flow in the extracranial circulation appear to be minimal. That control relies mostly on the architecture of the carotid artery and its branches (see Chapter 9), aided by the external pressure applied by the inelastic skin of the head. The skin of the head is relatively thin by giraffe standards, at only 1.5–3.6 mm thick, but it has a very high density of inflexible collagen fibers (see Chapter 15). Consequently, tissue pressure is high and is about 19 mmHg (Brondum et al., 2009). Head skin is another G-suit. The large distributive arteries that transport blood to brain tissue and which arise from the Circle of Willis are well supplied with nerves (Kimani et al., 1991) and Kimani et al. concluded that "instantaneous vasoconstriction occurs (in them) when blood pressure increases, for example, as the animal bends its head to drink." However, the small intraparenchymal cerebral blood vessels of giraffes that supply brain tissues cannot constrict. They are composed of a thin endothelium, a single layer of elastic membrane, and a thin muscular layer (Goetz and Keen, 1957; Kimani et al., 1991b; van Citters et al., 1969), and are physically unable to constrict against a pressure of 250 mmHg or more. One way that blood pressure could be dissipated is if blood could be diverted away from the brain when a giraffe lowers its head. Lawrence and Rewell had proposed just such a mechanism: diversion of blood from the carotid artery to the occipital artery and then to the vertebral artery, but an alternative function for those connections is probable. An alternative route for removal of excess blood from the cerebral circulation is, however, provided by the basilar artery, which is a small artery that branches off the posterior part of the Circle of Willis. Its function is to carry blood from the Circle of Willis to the ventral spinal artery and thus away from the brain (Baldwin and Bell, 1963), a direction of flow that would allow decompression of blood pressure in the Circle of Willis. The basilar artery, in other words, rather than the occipital artery, would act as the pressure-release valve. Furthermore, as in all mammals, the arteries that supply intracerebral blood vessels have to pass through bony canals in the skull, and the diameter of the bony canals determines the maximum diameter of the arteries to which they can expand even if their internal blood pressure is high. The bony canals act as mechanical constrictors of the blood vessels. In addition, the carotid *rete* is surrounded by venous blood inside a stiff-walled venous sinus (Figure 10.8).

When a giraffe lowers its head, the pressure of blood in the venous sinus increases in parallel with the increasing effect of gravity, and as this pressure is outside the arteries of the *rete* it counteracts any rise in their internal pressure caused by gravity. As Willis said in 1664, one function of the *rete* is to "reduce the impetuosity of the torrent of blood flowing to the brain." Cerebrospinal fluid pressure also increases under the effect of gravity when the head is lowered. Its pressure is applied to all structures within the skull, including blood vessels. Van Citters et al. (1969) measured cerebrospinal fluid pressure

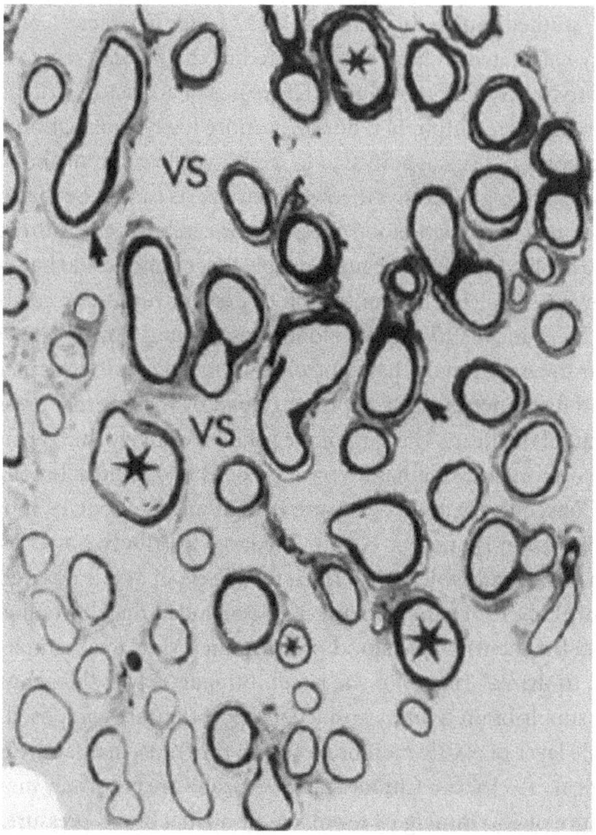

Figure 10.8. A cross section through the carotid *rete* of a giraffe showing the numerous small arteries within the cavernous venous sinus.

VS = venous sinus. Asterisks identify arterioles. Small arrows show that each arteriole is surrounded by connective tissue that separates it from the venous sinus.

From Kimani et al., 1991b, with permission of John Wiley and Sons.

to be 27 mmHg in a reclining giraffe, a pressure sufficient to increase intracranial pressure and reduce cranial perfusion pressure. These mechanisms and structures reduce the pressure of blood flowing to and within the brain (Nagel et al., 1968; Edelman et al., 1972; Lluch et al., 1985; O'Brien and Bourke, 2015). Edelman et al. (1972), for example, found that in goats arterial pressure beyond the *rete* was 20% less than it was in the arteries supplying the *rete*, and a theoretical analysis by O'Brien and Bourke calculated the pressure difference in giraffes to be 15 mmHg. None of these reductions seems significant if the blood pressure inside intracranial blood vessels is ~250 mmHg, but they will add to the effects of other mechanisms that lower the pressure. Moreover, it is obvious that giraffes do not develop strokes and thus, like prevention of fainting, giraffes must have evolved mechanisms like these, and almost certainly others that have yet to be described, which protect them from the potential consequences of the high blood pressure that develops in the blood vessels of the brain when they lower their heads to drink.

11
Water Balance in Giraffes

Even though there have been many observations of giraffes drinking water (see Plate 11.1), there are as many other convincing observations that imply that giraffes do not need to do so. One of the first such observations of this almost legendary ability of giraffes was made by Frederick Courteney Selous. Selous was born in London in 1851. While still a schoolboy at Rugby School in England, he decided to become a big-game hunter in Africa, and did so, arriving in South Africa in 1870 at the age of 19. His exploits are legendary. He recorded them in two books that made him famous throughout the world. He was a friend of Theodore Roosevelt, with whom he shared similar hunting tastes, of Cecil John Rhodes, and of Lobengula Khumalo, the King of the Matabele in what was then Rhodesia. Selous should have been killed by an elephant or a lion, but on January 4, 1917, he was killed during a minor skirmish in the war in German East Africa (now Tanzania). As a young man Selous enjoyed all the romance that surrounded the occupation of "big-game hunter," but by today's standards he was a callous slaughterer of animals, in the mold of Cornwallis Harris. Harris's slaughter was partially mitigated by his accurate descriptions and drawings of the fauna and flora he encountered. Selous's slaughter was mitigated by the knowledge he absorbed, and as an older and wiser man he became a highly regarded naturalist. He was immortalized by Sir Rider Haggard as Allan Quatermain, the protagonist of *King Solomon's Mines*, published in 1885. In 1886 Haggard published the novel *She*. *She* met her death in the ruined city of Kor, but in a later novel, published in 1921, which recounted an earlier time, Haggard interwove the stories of *She* and Quatermain. *She*'s character, it is said, was inspired by Modjaji, the Rain Queen, who then ruled and whose successors still rule the Lobedu people of Limpopo Province in South Africa. Modjaji lived (and her successors still live) on a farm once owned by my maternal grandfather. Haggard visited South Africa regularly and on one visit met my grandfather, who wrote:

> [The farm] *had a cave through the Lebaka Mountain and natives will not sleep under this hill: they say people move up and down it all night, carrying quartz on their heads. Gold is on the farm, and I told Sir Rider about it. He was attentive and made copious notes. He told me that he had woven the story of* She *out of hunter's yarns.* (Abercrombie, 1952)

The stories of *She* and Quatermain captivated me as a boy growing up in South Africa in the shadow of a grandfather, himself of mythical proportions.

In his *African Nature Notes and Reminiscences* (1908; for which Theodore Roosevelt wrote an overly long and self-congratulatory foreword), Selous wrote:

> *The belief is very general, both amongst white and native hunters in South Africa, that giraffes are capable of going for months at a time without drinking and the fact that they*

How Giraffes Work. Graham Mitchell, Oxford University Press. © Oxford University Press 2021.
DOI: 10.1093/oso/9780197571194.003.0011

are to be found during the driest season of the year in the most arid districts, far away from any place where surface water exists, lends colour to this belief.

Selous was not the first to make this observation, though. Hendrik Jacobus Wikar, the self-exiled Swede who roamed the remote, arid, hinterland of the Northern Cape (see Chapter 3), wrote in 1779,

In my opinion and from my investigations these animals must be able to last without water for a very long time. For in the dry region where there is no water the veld is full of their spoors, but you seldom find a spoor leading to the river.

The legend has found support in some modern observations. According to Foster and Dagg, (1972) even in habitats where surface water is plentiful, giraffes avoid drinking. In one of the most arid of all deserts—the Namib Desert—giraffes that live there have to be and are totally independent of surface water (Fennessy, 2009). There are saline springs in the Namib, and some intermittent, ephemeral rivers, and indeed Gordon's giraffe was shot near a saline spring, but giraffes tend to avoid drinking from them, obtaining their water from water-rich browse in riverine woodlands, or from fog that condenses on vegetation (Scheepers, 1992), or browse that has been hydrated by fog-water (Fennessy, 2004, 2009). Legends and myths are a treasured part of the life story of giraffes because they add a touch of romance to the everydayness of their lives and a bit of color to their story, but they are not, regrettably, completely accurate. Giraffes can and do drink water frequently whenever they can (see Plate 11.1) and whenever they need to, and especially when they have to eat increasingly dry browse or use body water to produce milk or to regulate body temperature and so on, but the observations suggest that under some circumstances they do not have to rely on drinking surface water to balance water losses.

How can they avoid drinking when a lack of water is a bigger threat to the everyday life of a terrestrial mammal than is a lack of food? Large mammals like giraffes can survive a lack of food for many weeks, but they cannot survive a lack of water for very long. A lack of water causes death by dehydration in a few hours for small mammals, in a few days for humans, and in a week or two for giraffes. The reasons that dehydration is fatal lie in the physiological and physical properties of water. Water's two most important physiological properties are, first, that it is a liquid at body temperature, and second, water molecules exist not as H_2O but as a positively charged hydrogen ion (H^+) and a negatively charged hydroxyl ion (OH^-). This latter property means that the many substances in body fluids that carry a charge like salts and proteins can dissolve freely in it. Water is, thus, the solvent for almost all the biochemical reactions that support life. Without water, those reactions stop. Water's physical properties, at least as they impinge on life, are as essential. Water is incompressible, so any pressure applied to it is transmitted by it, a property that is the origin of blood pressure. Water has a high latent heat of evaporation, therefore turning liquid water into water vapor requires a great deal of energy: evaporation of water from body surfaces cools the body effectively. Water also has a high heat capacity or, to put it another way, high thermal inertia, and so it takes a lot of heat to raise body temperature. Water, in other words, prevents rapid changes in body temperature. It is a prerequisite for life, yet giraffes often choose to live in arid, hot areas of Africa.

It may well be that a giraffe's need for water is less than it is for animals that live in more congenial climates. They could be, for example, as well adapted as or better adapted to an arid environment than are camels. A place to start working out if that is the case is to find out how much water giraffes need daily. How much is needed by mammals in general is determined by the rate at which body water is lost and replaced (i.e., turned over) each day. Regrettably, but understandably, water turnover has not been measured in giraffes. However, Calder (1996), based on data in Calder (1981) and Calder and Braun (1983a), calculated that for mammals generally, water turnover rate ($L.d^{-1}$) could be estimated from the equation $0.12 \times Mb^{0.80}$ with body mass in kg. Applying this equation to a medium-sized giraffe weighing 750 kg means that it would turn over ~25 L each day. What is that volume as a proportion of total body water? The total amount of water in a mammal is approximately 60% of total body mass, and in a 750-kg giraffe it is 470 L (Table 11.1), so the volume that is turned over daily in a giraffe can be expected to be about 5% of its total body water.

However, of total body water, two-thirds is situated inside cells (i.e., about 315 L in a 750-kg giraffe) and that water only leaves the cells when a giraffe is experiencing severe dehydration. Water that can easily be turned over is water outside cells in the extracellular space. Most extracellular fluid (~120 L in a 750-kg giraffe) is found in the virtual interstitial space between blood capillaries and cells, about 30 L is plasma in blood, and the rest (~10 L) is in joints, eyes, and the brain (cerebrospinal fluid). This distribution means that in a 750-kg giraffe, ~160 L of water is in the extracellular space, of which about 20% is turned over each day. Regulation of total body water does not necessarily have to be precise. Excess water can be excreted, and all mammals can tolerate a degree of dehydration before the proportion of body water that is lost becomes fatal. That fatal proportion in a ruminant is ~30%–40% of total body water (= ~25% of total body mass; Silanikove, 1994), a proportion twice as high as it is in a monogastric mammal: adult ruminants carry a supply of water in their rumens which can replace fluid lost from the body itself. Nevertheless, if the turnover rate of a giraffe weighing 750 kg is 25 $L.d^{-1}$ and the giraffe was totally deprived of any source of water, it would lose 30% of its 470 L of total body water (~150 L) in ~6 days. A newborn giraffe weighing 100 kg and having 60 L of total body water would lose 30% of that (18 L) in ~4 days: young mammals, especially those not yet weaned, are more vulnerable to dehydration than are adults. Of course, in nature, an instantaneous absence of water is a rare occurrence. Even in a drought, the amount of water available declines slowly, and giraffes invoke all the mechanisms they can to reduce water loss; thus it may take more than four days for a newborn giraffe and a few weeks for an adult giraffe to lose 30% of their total body water.

Sources of Water

From where do giraffes get the water they need? Apart from surface water, giraffes have two other sources of water. One is the water in the browse they eat, and the other is water formed as a by-product of the metabolism of the protein, fats, and carbohydrates in their diet. Of these two latter sources, it is the water in browse that is the more important. The water content of browse is detected by giraffes as succulence and is the primary cue for browse of choice for giraffes, as so many studies have shown (see Chapter 7). The proportion of water in browse varies widely. Browse eaten by giraffes in Namibia has a water content of 40%–50% (Fennessy, 2004), in southern Africa 43%–65% (Sauer et al., 1982), in East Africa 53%–62% (Taylor and Lyman, 1967), although in one Kenyan species of Acacia that giraffes prefer, *A. drepanolobium* (whistling thorn), water content is 74% (Foster and Dagg, 1972). On average, giraffes consume 35±12 kg of fresh browse daily (range 10 kg d^{-1} for juveniles to 56 kg d^{-1} for adults; see Chapter 8), which, using an average water content of 60%, provides 6–34 L of water daily and, if surface water is not available, that could represent almost 75% of their daily water intake (Table 11.1). If a giraffe is not drinking surface water, the balance of their daily water comes from metabolic water and the volume generated varies according to metabolic rate and the composition of the diet. Using Pellew's (1984b) analysis, a typical giraffe diet contains 15% protein, 4% fat, and 75% carbohydrate (the rest is minerals). The volume of water produced by metabolism of proteins is 0.41 $ml.g^{-1}$ of protein catabolized, from fats it is 1.1 $ml.g^{-1}$, and from carbohydrates it is 0.55 $ml.g^{-1}$. From the total amount of browse a giraffe has to eat to sustain its field metabolic rate and its composition as defined by Pellew, the total volume of water formed by metabolism can be calculated to be between 2 L a day for a juvenile giraffe and 12 L per day for an adult (Table 11.1). These two sources of water—browse and metabolism—can provide a giraffe weighing 750 kg with ~30 L of water every day, and any giraffe with $0.21 \times Mb^{.74}$ $L.d^{-1}$ (Table 11.1). A volume of 30 $L.d^{-1}$ exceeds the estimated turnover rate of 25 $L.d^{-1}$ of a 750-kg giraffe and explains why it is that giraffes do not need to drink surface water if succulent, good quality, browse is plentiful.

Loss of Water in Feces

If a 750-kg giraffe has a turnover rate of 25 $L.d^{-1}$, then 25 L is lost every day and has to be replaced. How is it lost? The main routes of water loss are through urine, evaporation, and in feces. The volume that could be lost in feces is large. Water enters the gastrointestinal tract when a giraffe drinks surface water or as water in browse, and, in an adult giraffe, another 200 L or more of saliva enters the gastrointestinal tract each day. The combination of saliva volume, surface water volume, and the volume of water in browse amounts to some 230 L or more of water, and if it is not reabsorbed an adult giraffe would lose half of its total body water in a day. Some of that water, perhaps 20%, is reabsorbed from the rumen–reticulum, and another 40% from the omasum. Reabsorption of most of the remaining 40% occurs throughout the small intestine tract, so that by the time intestinal contents reach the distal colon, just 2–3 L of water remain and would be lost unless they were reabsorbed. The mechanism of water reabsorption in the colon (and elsewhere) appears to be mainly by osmotic diffusion, unlike horses and similar hindgut-dependent herbivores, where water reabsorption from the hindgut is controlled by several hormones (Sneddon and Argenzio, 1998). The rate of reabsorption by osmotic diffusion depends on the surface area for reabsorption and the osmotic gradient between the fluid in the intestine (usually dilute) and the blood (more concentrated), perhaps assisted by the pumping of sodium ions from colon fluid into blood. Browsers tend to have drier fecal pellets than do grazers (Clemens and Maloiy, 1983), but even so, Clemens and Maloiy found the average water content of pellets in the distal colon/rectum of wild giraffes to be 67%, and fecal pellets of captive giraffes (the mass of the pellets was not recorded) have been found to have a water content of 56% (Clauss et al., 2004) and 63.5%–71.7% (Isobe, 2007), compared with 83% for the contents of the upper colon (Clemens and Maloiy, 1983). From data in Woodall and Skinner (1993) and Clauss et al. (2004), the water content (%) of fecal pellets in African browsers generally can be described allometrically by $51.4 \times Mb^{0.03}$; thus it is independent of body mass. So far no data on daily fecal mass and its water content in wild giraffes have been collected, and it would be very difficult to do so accurately, but Kearney (2005) collected pellets from six captive giraffes that produced 30 fecal samples, the daily mass of which was about 3 kg and represented 0.0043% ±0.0009% of body mass. Using this fecal mass and assuming fecal water content of 63%, then in a 750-kg giraffe about 2 L of the 230 L or more of water that enter its intestines each day, or about 1% of it, is lost in pellets (Table 11.1).

Evaporation

A bigger source of water loss is evaporation from the skin and the respiratory tract. From studies on a range of other and somewhat smaller mammals, the average evaporative water loss from both the skin and the respiratory tract combined during the course of a day has been calculated by Calder and Braun (1983). If the equations derived from those data are applied to giraffes (with reservation, as ambient heat load and metabolic rate of the animals concerned need to be known to arrive at an accurate estimate), total evaporative loss amounts to ~12 $L.d^{-1}$ in a giraffe that weighs 750 kg and could range from 3 $L.d^{-1}$ in a juvenile giraffe to 21 $L.d^{-1}$ in a mature adult. The volume of water lost through the respiratory tract has been measured with considerable accuracy in three giraffes (Langman et al., 1979). The three giraffes weighed about 600 kg, and Langman et al. made 35 measurements of the water content of inspired air and of exhaled air. On average, the water content of inspired air (at an ambient temperature of 21.1°C) during the study was 0.0133 $ml.L^{-1}$ of air to which, on its passage to the lungs, another 0.0313 $ml.L^{-1}$ of air was added from the water in the membranes that line the respiratory tract. This added water is a potential source of water loss from the body if all of it were to be in exhaled air. The water content of exhaled air in the three giraffes was 0.0272 $ml.L^{-1}$ of air, of which 0.0133 $ml.L^{-1}$ had already been in the inspired air, so the additional volume lost was 0.0139 $ml.L^{-1}$ of air exhaled, or 44% of the 0.0313 $ml.L^{-1}$ water added to inspired air. Thus in these giraffes 56% of the added water was recovered, and perhaps it is that proportion of added water that is recovered from all air exhaled by giraffes. The total volume of water lost per minute or per day depends on the number of liters of air breathed every minute (the minute volume; V_E; $L.min^{-1}$), and if the amount of body water in each liter of exhaled air is multiplied by the minute volume of a giraffe going quietly about its day ($V_E = 0.34 \times Mb^{0.82}$; Table 12.3 in Chapter 12), then the amount of water lost through the respiratory tract in a 750-kg giraffe is ~2 L per day, or $0.007 \times Mb^{0.82}$ $L.d^{-1}$ for any giraffe, which is much the same volume of water that is lost in feces (Table 11.1).

Table 11.1 Water Balance in Giraffes

	750-kg Giraffe	Range for Giraffes 147–1,441 kg	Allometric Equation	Reference
Total body water (L)	469	89–872	$0.605 \times Mb^{1.0}$	Calder, 1996
Water intake from browse ($L.d^{-1}$)	21	6–34	$0.154 \times Mb^{0.74}$	Taylor and Lyman, 1967; Sauer et al., 1982;
Metabolic water ($L.d^{-1}$)	7	2–12	$0.053 \times Mb^{0.74}$	Pellew, 1984a
Total browse + metabolism ($L.d^{-1}$)	28	8–46	$0.21 \times Mb^{0.74}$	See text
Respiratory evaporative loss ($L.d^{-1}$)	2	0.5–3	$0.008 \times Mb^{0.80}$	Langman et al., 1979
Sweat loss ($L.d^{-1}$)	10	2–18	$0.025 \times Mb^{0.91}$	See text
Minimum urine loss ($L.d^{-1}$)	8	2–12	$0.01 \times Mb + 0.04$	See text
Fecal water loss ($L.d^{-1}$)	2	0.5–3	$0.008 \times Mb^{0.83}$	Woodall and Skinner, 1993; Clauss et al., 2004; Kearney, 2005
Total loss ($L.d^{-1}$)	22	5–37	$0.05 \times Mb^{0.91}$	

Whether giraffes sweat, and what the total amount of water lost daily by sweating is, has not been measured in them, and it would be very difficult to do so in such a large animal, especially a free-living wild one. However, giraffes have active sweat glands (Dimond and Montagna, 1976; Mitchell and Skinner, 2004), and it would be surprising if they could not sweat. Several other artiodactyls co-inhabitant with giraffes do sweat (Robertshaw and Taylor, 1969). If giraffes do lose water by evaporation from the skin and respiratory tract at the rate predicted by Calder's equation, then despite the limitations of that equation, the volume of water lost by sweating daily by a giraffe can be estimated by subtracting respiratory water lost from total evaporative water loss. As Table 11.1 shows, evaporation by sweating could range between 2 L in a small giraffe and 18 L in an adult. Thus sweating could account for at least half of all water lost by a giraffe daily, the volume varying according to whatever thermoregulatory behaviors the giraffe calls upon to minimize it (see Chapter 14).

Kidney Function and Urine Volume

Urine is the end product of kidney function and is the biggest potential source of water loss from a mammal. Urine is a mixture of water and water-soluble waste products. Some of the waste products (particularly undesirable ones) are secreted by kidney processes into urine, but most of them enter urine by filtration of blood plasma. Urine volume depends on how much of the fluid and its contents that are filtered by the kidneys are reabsorbed. The volume filtered by the kidneys each day in a mammal varies with size and is higher, per kg of body mass, in small mammals than in large ones, but in mammals the size of a giraffe, it is about twice the volume of its total body water. Clearly, unless most filtered fluid is reabsorbed, dehydration would follow rapidly. If most filtered fluid were to be reabsorbed, achieving water balance would be easier and water balance would depend mainly on replacing the fluid lost in feces and by evaporation. However, it is not desirable for all filtered fluid to be reabsorbed because urine is an important way through which many harmful waste products are excreted. Excretion of these solutes is obligatory, and because they depend on water for their excretion, the daily amount of waste products that must be excreted determines the minimum obligatory daily volume of urine. Urine volume can be greater than the obligatory minimum because the volume of water in urine varies according to the state of hydration. A well-hydrated giraffe will excrete enough water in urine to keep total body water volume constant. A dehydrated giraffe will excrete the absolute minimum. How big is the minimum volume of urine? There have been very few studies of giraffe kidney anatomy and function, so the answer is far from certain. Nevertheless, it is a useful exercise to try to determine how much solute and how much water are filtered by the kidneys of giraffes each day and how much is reabsorbed.

Kidney Anatomy

Owen (1838), in the classic manner of his times, wrote simply that giraffe kidneys occupied the usual position in the loins, did not have the lobular appearance of bovine kidneys, the right kidney lay a little nearer to the diaphragm than did the left, and both were about 11 cm long, 10 cm wide, and 5 cm thick. He noted too that the collecting ducts that are the final part of a nephron (Figure 11.1) converge toward a single ridge. The first detailed description of the internal anatomy of a giraffe's kidney was made by an Italian comparative anatomist, Sebastiano Richiardi, in 1880. The description was detailed but brief and concerned the inverted, cone-shaped pyramids of Malpighi that can be seen if a kidney is cut in half along its longitudinal axis. The pyramids are named after the same Marcello Malpighi, who described blood capillaries (see Chapter 10). He identified the pyramids and also the many millions of "Malpighian bodies" that are found in the kidneys of mammals (Figure 11.1). The Malpighian bodies have since been renamed "glomeruli" because, under a microscope, they look like a ball of thread (*glomerulus* = thread in Latin). The "threads" are in fact the network of capillaries through which filtration of the plasma portion of blood occurs. The pyramids are made up of the many thousands of glomeruli and their tubules—the nephrons—which are the functional units of kidneys (Figure 11.1). Nephrons are collected together in bundles, with the broad outer base of the pyramid cone being filled by all the individual nephrons and the bottom, apex, of the pyramid being narrower because the nephrons join to form shared outflow channels, the collecting ducts (Figure 11.1). Richiardi counted 24 to 27 pyramids in the giraffe kidneys he studied. No further information on giraffe kidneys was to follow until 1965, when Karl-Heinz Wrobel compared, even more briefly, the gross anatomy of a giraffe's kidney with that of horses, sheep, and dogs. In all three of these latter animals, the collecting ducts emptied onto a longitudinal ridge—the renal crest—that projects into the renal pelvis where urine collects before discharge. The renal

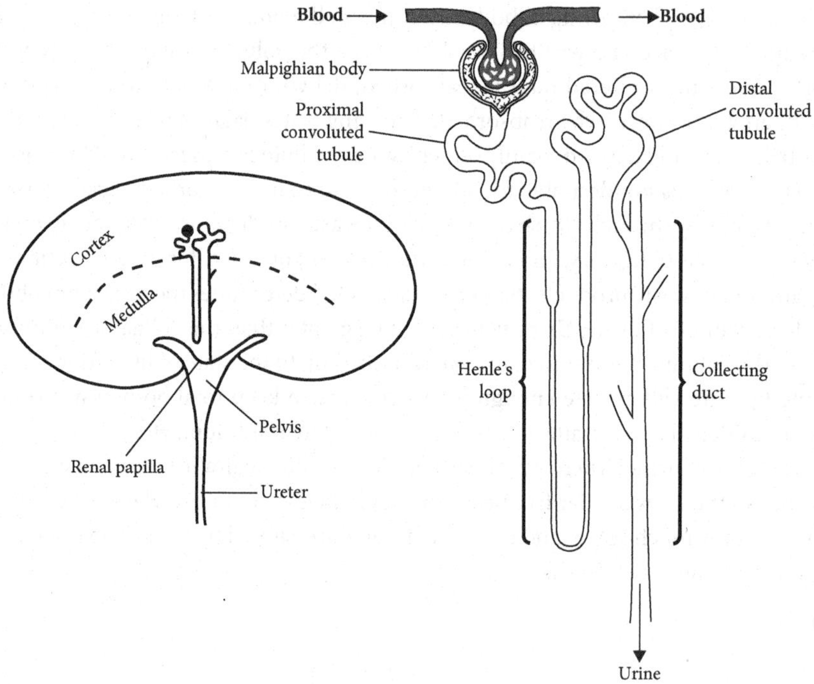

Figure 11.1. Basic anatomy of a mammalian kidney and its functional unit, a nephron.
From Schmidt-Nielsen, 1983, with permission of Cambridge University Press.

pelvis drains its contents into a ureter, and the two ureters, one for each kidney, empty their urine into the bladder and from there urine is excreted through a single urethra. Wrobel concluded that giraffe did not have a renal crest. Owen (1838) and Maluf (2002) disagreed. The collecting ducts do form a ridge in giraffes, although if some animals do not have one the value of having one is obscure. There are about 6 million nephrons in each kidney of a giraffe, and thus 250,000 to 300,000 nephrons in each pyramid, and 4,000 collecting ducts in total, or 150 draining all the nephrons in each pyramid (Maluf, 2002). For comparison in humans, there are 10 to 18 pyramids, one million nephrons and 25,000 collecting ducts in each kidney, so each pyramid contains 100,000 to 200,000 nephrons and ~2000 collecting ducts.

So far the only comprehensive study of the anatomy of giraffe kidneys has been done by Noble Maluf, an American surgeon but whose considerable contribution to giraffe kidney physiology is as a comparative nephrologist. In 2002, Maluf reported the results of a study of the gross anatomy of the kidneys in two male giraffes, one 18 years old weighing 1,172 kg and the other 22 years old weighing 1,016 kg, and also in three other giraffes that were 6 months, 11 months, and 2 years old. The kidneys in the adults, Maluf found, were covered by a tough, translucent, collagenous, and inflexible capsule, about 0.75 mm thick. Like all mammalian kidneys, the substance of giraffe kidneys consists of an outer cortex and an inner medulla (Figure 11.1), and Maluf established that in giraffes the cortex makes up ~62% of kidney mass, and the medulla the other 38% (Maluf, 2002). Maluf weighed the kidneys, and from those masses and those measured by Crisp (1864) for a single kidney obtained from a 153-kg giraffe

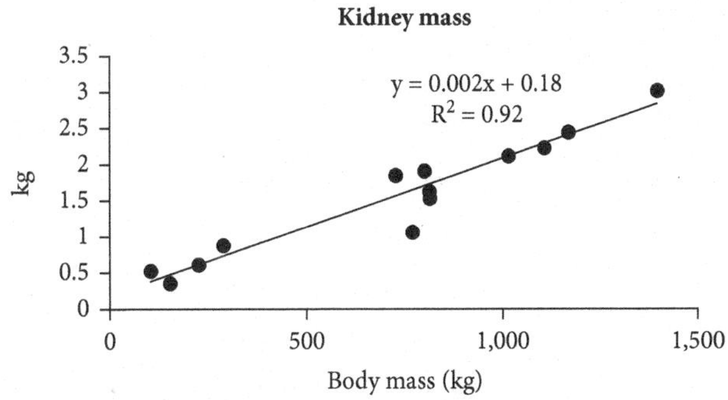

Figure 11.2. The mass of the left and right kidneys of giraffes combined.
Data from 13 measurements by Crisp, 1964a; Maluf, 2002; Mitchell, unpublished. Body mass ranged from 104 kg to 1,396 kg.

that weighed 174 g or 0.11% of body mass, 814 g in a 813-kg giraffe (0.10% of body mass), and 794 g from a giraffe weighing 815 kg (0.10% of body mass), and from the kidney masses in eight other giraffes (5 males and 3 females, Mitchell unpublished; Table 11.2), an allometric equation can be developed that predicts kidney mass in any giraffe (Figure 11.2). Together the two kidneys would weigh a substantial 1.6 kg in a 750-kg giraffe and range in mass from ~500 g in a neonate to 2.6 kg in a mature adult, masses which are 0.22% ±0.03% of body mass.

Measurement of another kidney mass in a giraffe was made by Crile and Quiring (1940); it had a mass 2,268 g (0.18% of body mass) and came from a giraffe weighing 1,220 kg, but it is not clear if that value referred to one or both kidneys. There is no difference between the measured masses of right and left kidneys (Table 11.2). Stahl (1965) derived an equation that describes the mass of kidneys in mammals generally ($0.007 \times Mb^{0.85}$ kg), and if it is applied to giraffes, the mass of giraffe kidneys should range between 0.35% of body mass in a neonate and 0.23% in a 1,500-kg adult and be 0.26% of body mass in a giraffe weighing 750 kg. By those comparisons, giraffe kidneys may be smaller than they are in other mammals.

The number of glomeruli that can be expected to be in mammalian kidneys can be estimated from the regression equation ($(9.53 \times 10^4) \times Mb^{0.62}$; Calder, 1996) and this equation predicts a number of ~7,600,000 in each kidney of the giraffe weighing 1,172 kg in Maluf's study, and ~7,000,000 in each of the kidneys of the other giraffe that had a body mass of 1,016 kg. The actual number of nephrons reported by Maluf was ~6,500,000 in each of the four kidneys he examined; thus giraffe kidneys have the number of nephrons fairly close to the number expected for mammals of comparable body mass. Calder's equation, which was developed by analysis of the kidneys of different species with different body masses, implies, correctly, that the kidneys of large mammals have more nephrons than do the kidneys of smaller mammals and that should be so because large mammals have to filter more fluid each day than does a small mammal. A progression of that kind suggests that the kidneys of large giraffes should have more nephrons than do the kidneys of a small giraffe. However, in a single species like giraffes, the number of nephrons in each kidney is laid down during fetal development. Nephrons increase in size after birth but do not increase in number as body mass

Table 11.2 Kidney Mass of Giraffes

Body Mass (kg)	Right Kidney Mass (g)	Left Kidney Mass (g)	Combined Kidney Mass (kg)	Peri-Renal Fat (kg)
104*	278	242	520	–
153°	174			–
227	300	298	598	0.066
289			870	–
727	926	924	1,850	0.442
770			1,038	0.184
801	942	958	1,900	0.290
813	814	800	1,614	0.350
815°	794	709	1,503	–
1,016*	948	1,153	2,101	–
1,110	1,066	1,138	2,204	0.392
1,172*	1,275	1,150	2,425	–
1,396	1,476	1,508	2,984	0.770

Total kidney mass (kg) excluding peri-renal fat= $0.002 \times Mb + 0.18$, $R^2 = 0.92$; left kidney mass = $0.0009 \times Mb + 0.1$, $R^2 = 0.89$; right kidney mass = $0.0009 \times Mb + 0.1$, $R^2 = 0.91$. Data from Crisp°, 1864a; Mitchell (unpublished) and Maluf*, 2002.

increases, hence the number of nephrons stays more or less the same no matter the age of the giraffe, although it is quite likely that as a giraffe becomes older, nephrons become damaged and die, just as they do in humans, and are not replaced, leading, ultimately, to a decline in kidney function with age.

Kidney Function

The main function of the kidneys is to regulate the composition of body fluids. They do so by filtering plasma, reabsorbing all useful substances, and excreting wastes. The wastes can include excess water. Damkjaer et al. (2015) have carried out the only study so far of how the kidneys of giraffes might regulate body fluid composition. There have, however, been many studies of the kidneys of other mammals, and when the data from those studies are considered together with Damkjaer et al.'s data, it is possible to piece together how giraffe kidneys might work.

Kidney function depends on how the nephrons function, and there are two main types of nephrons in a mammalian kidney. Those nephrons in the outer cortex of the kidney have small glomeruli, well-developed proximal and distal tubules, and vestigial or very short loops of Henlè. This type of nephron regulates the concentration of solutes in body fluids. Tubules situated on the border between the cortex and the inner medulla—the juxtamedullary nephrons—have large glomeruli and well-developed loops of Henlè which extend deep into the kidney medulla (Figure 11.1). The function of the loops is to establish an osmotic gradient so that water can be reabsorbed from collecting ducts when the ducts are made permeable by the action of anti-diuretic hormone. The concentration of anti-diuretic hormone in giraffe plasma is 57–97 pg.mL^{-1}, which is the same as it is in other mammals (Damkjaer et al., 2015). The ratio of cortical to juxtamedullary nephrons varies. Mammals that live in arid or semi-arid habitats and have a greater need to conserve water have more juxtamedullary nephrons and longer loops of Henlè compared to those from mesic habitats. It is not known what the relative proportions of the two types of nephron are in giraffes.

Glomerular Filtration Rate in Giraffes

The starting point of nephron function is filtration of blood plasma, and the rate at which plasma is filtered (glomerular filtration rate) has the units mL min^{-1}. Filtration rate is governed by the same forces that regulate the movement of fluids across any capillary membrane, and they are summarized in Starling's equation (see Chapter 10). Filtration across a kidney capillary wall is more complicated, though, because fluid and solutes have to cross not only the capillary wall but also a second membrane, Bowman's capsule. Bowman's capsule encloses the capillaries in the glomeruli (Figure 11.1). The proportion of the plasma that flows across these two membranes to enter a nephron is the filtration fraction.

Of the forces that determine the size of the filtration fraction, blood pressure is the most important, and several mechanisms have evolved to keep blood pressure at the glomerulus constant and at a level suitable for filtration. Perhaps the most important mechanism acts to increase mean arterial blood pressure whenever the pressure of the blood supplying the kidneys falls. If the pressure of the blood going to the kidneys was to fall below the critical level needed for proper filtration (what that blood pressure is in giraffes is not known), glomerular filtration rate would also fall and the kidneys would be unable to carry out their functions. Under those circumstances, as part of the baroreceptor reflex (see Chapter 9), a small protein—renin—is released by kidney cells; its release is stimulated, in turn, by noradrenaline released from sympathetic nerves. Renin's main action is to stimulate the synthesis of angiotensin, which is a powerful constrictor of arterioles in the systemic circulation. Constriction of arterioles increases mean arterial blood pressure, and glomerular filtration rate is returned to normal levels. Angiotensin also causes contraction of mesangial cells, which lie between glomerular capillaries. Their contraction increases the surface area through which filtration occurs and thus increases filtration rate. Renin and angiotensin are synthesized in giraffes. The concentration of renin in giraffe plasma is 2.5±0.5 mIU.L^{-1} and of angiotensin is 9.1±1.5 pg.mL^{-1}, both of which are within the range found in other mammals (Damkjaer et al., 2015). Giraffes may have evolved the ability to synthesize angiotensin more efficiently than other artiodactyls can and better than okapi can (Agaba et al., 2016; see Chapter 4) and giraffes are sensitive to the effects of angiotensin. When McCalden et al. (in 1977) infused angiotensin at a rate of 400 μg min^{-1} into the artery of an anesthetized giraffe, its aortic blood pressure increased 1.3-fold from ~210 to ~270 mmHg and carotid artery pressure increased by 1.5-fold from 95 to ~140 mmHg in a few minutes. It is likely, therefore, that the renin-angiotensin system is able to regulate blood pressure in giraffes in the same way that it does in other mammals, and thus ensures that their blood pressure always is sufficient to maintain glomerular filtration rate.

The kidneys of giraffes lie in the abdomen at much the same level as the heart does in the thorax, and so they are exposed to the blood pressure generated by the heart, which is double what it is in any other mammal. Thus, theoretically, filtration rate should be twice as high as it is in other mammals. Is it, and if it is, is the way kidneys of giraffes function unusual? To answer that question, Damkjaer and colleagues measured, in 14 anesthetized giraffes with an average body mass of 478±22 kg, three fundamental aspects of kidney function: the volume of blood plasma flowing through the kidneys (renal plasma flow; $mL.min^{-1}$), the glomerular filtration rate ($mL.min^{-1}$), and the renal interstitial fluid pressure. They also measured two important cardiovascular variables that support kidney function, namely cardiac output and mean arterial pressure. Their data are shown in Table 11.3. Damkjaer et al. concluded that the rate of plasma flow (1,270 $mL.min^{-1}$) and glomerular filtration rate (345 $mL.min^{-1}$; 0.7±0.1 $mL.kg^{-1}$ body mass min^{-1}) were, compared with other mammals, unusually low, but the filtration fraction (%) of plasma was higher than usual (30%), as was the renal interstitial pressure (47 mmHg in a giraffe lying on its side and 146 mmHg in giraffe in an upright position). The origin of the high interstitial pressure, they concluded, was the thick, collagenous, inflexible renal capsule. Its effect would be to counteract high arterial pressure, and the consequence would be that blood flow to the kidneys would be reduced in much the same way that blood flow to the feet is reduced by the collagenous skin cuff that surrounds the lower part of a giraffe's legs. In the young giraffes that Damkjaer et al. studied, the renal capsule was 0.2 mm thick (much thicker than its counterpart in cattle), but thin when compared with the capsule thickness (0.75 mm) reported by Maluf (2002) in adult giraffes, so its thickness increases as body mass and blood pressure increase. In addition, Damkjaer et al. recorded a marked pressure drop between the renal vein (high pressure) draining the kidneys and the inferior vena cava (lower pressure) taking blood to the heart, a junction guarded by a valve. Its function, they concluded, was to prevent blood from flowing into the inferior vena cava, an effect that also would increase interstitial pressure and thus counteract high blood pressure. This valve was first seen in giraffes by Crisp in 1864 and then again by Goetz and Keen (1957), and from its position and shape, Crisp, and Goetz and Keen, concluded that its function was to prevent backflow of blood from the inferior vena cava into the renal vein and not the reverse. How it could do the reverse is unclear. Nevertheless, Damkjaer et al. concluded that the combination of these mechanisms was the origin of the high interstitial pressure and that high pressure reduced glomerular filtration rate. Compensation for a reduced filtration rate was a higher filtration fraction.

All this may well be the way that sufficient filtration is maintained, but the explanation rests on the idea that glomerular filtration rate in giraffes is lower than expected. However, as Table 11.3 shows, the glomerular filtration rate for giraffes that weigh 478 kg, predicted from studies of renal physiology in many other mammals, is not different from what was measured: from data in Edwards (1975), Holt and Rhode (1976), and Singer and Morton (2000), glomerular filtration rate of mammals ranging in body mass from 100 to 1,000 kg, can be predicted to be $4 \times Mb^{-0.24}$ $mL\ kg^{-1}$ body mass min^{-1}. This equation predicts that the glomerular filtration rate in adult giraffes will be lower than it is in neonatal giraffes (0.7 $mL.kg^{-1}\ min^{-1}$ and 1.2 $mL.kg^{-1}\ min^{-1}$, respectively),

Table 11.3 Cardiovascular and Kidney Function Variables in a 478-kg Giraffe and Predicted for a 478-kg Giraffe

Variable	Damkjaer et al., 2015	Predicted	Allometric Equation	Reference
Mean arterial pressure (MAP, mmHg)	168–242	188	$51.5 \times Mb^{0.21}$	Mitchell and Skinner, 2009
Cardiac output ($L.min^{-1}$)	16*	27	$0.211 \times Mb^{0.78}$	Mitchell and Skinner, 2009
Total peripheral resistance ($mmHg.L^{-1}.min^{-1}$)	~13	7	$244 \times Mb^{-0.57}$	Mitchell and Skinner, 2009
Heart rate (min^{-1})	34–43	74	$104 \times Mb^{-.13}$	Mitchell and Skinner, 2009
Stroke volume (mL)	~ 400	365	Cardiac output÷heart rate	
Renal blood flow ($mL.min^{-1}$)°	2,300	4,568	$39.5 \times Mb^{0.77}$	Edwards, 1975
Renal blood flow as % of cardiac output	14.4	17.6		
Renal plasma flow ($mL.min^{-1}$)	1,270	2,740	$23.7 \times Mb^{0.77}$	Edwards, 1975
Glomerular filtration rate ($mL.min^{-1}$)	144–540	435	$4.0 \times Mb^{0.76}$	Edwards, 1975; Holt and Rhode, 1976; Singer and Morton, 2000
Glomerular filtration rate ($mL.kg^{-1}.min^{-1}$)	0.7	0.9	$4.0 \times Mb^{-0.24}$	Edwards, 1975; Holt and Rhode, 1976; Singer and Morton, 2000
Filtration fraction (%)	27	16	Glomerular filtration rate÷renal plasma flow	

*From Smerup et al. (2016).

°Calculated from renal plasma flow and a hematocrit of 40% (Bush et al., 1980).

filtration rates that are substantially the same as those measured by Damkjaer et al. (0.5–0.9 $mL.kg^{-1}$ min^{-1}). Filtration rate decreases as body mass increases because, as Damkjaer et al. suspected, mass-specific metabolic rate decreases as body mass increases (Figure 11.3).

Giraffe kidneys, in other words, appear to filter plasma at the same rate as do other mammals of equivalent body mass. What actually was different in Damkjaer et al.'s study was cardiac output. They measured renal function in the same giraffes in which cardiovascular functions were studied by Smerup et al. (2016), and in those giraffes cardiac output was 33 $mL.kg^{-1}$ min^{-1}, and thus 16 $L.min^{-1}$ for giraffes weighing the 478 kg of Damkjaer et al.'s giraffes, and not the 27 $L.min^{-1}$ that can be predicted for giraffes of that

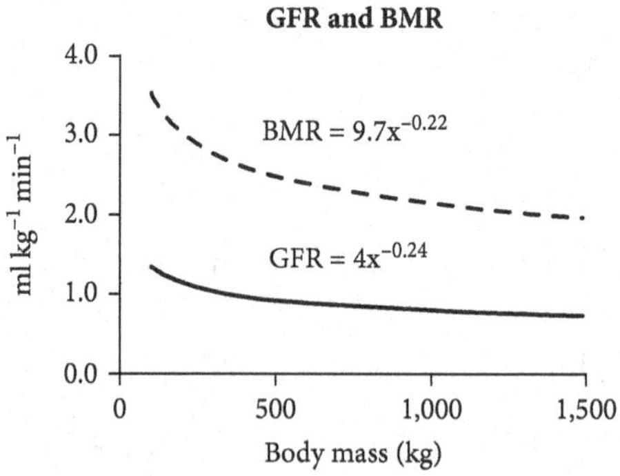

Figure 11.3. Mass specific glomerular filtration rate (GFR, ml kg^{-1} min^{-1}) predicted for giraffes from data in Edwards, 1975; Holt and Rhode, 1976; and Singer and Morton, 2000 (solid line) and mass specific basal metabolic rate (BMR, mlO_2 kg^{-1} min^{-1}) in giraffes (dashed line).

mass from other studies (see Chapter 9). The combination of a normal blood pressure (for giraffes) in their study and a low cardiac output can be achieved only if peripheral resistance is high, and in both Smerup et al.'s and Damkjaer et al.'s studies it was twice the predicted resistance. As the stroke volume was the same as predicted stroke volume, the low cardiac output resulted from a heart rate that was about half the predicted heart rate (Table 11.3), a heart rate that is expected if blood pressure is what is normal for a giraffe. The origin of high peripheral resistance is constriction of arteries and arterioles, including those in the kidney. If renal arteries and arterioles are constricted such that resistance doubles, then renal blood flow could be expected to be half of predicted values—and it was—but would be the same proportion of cardiac output—which it was (Table 11.3). Despite reduced cardiac output and renal blood flow, glomerular filtration rate in Damkjaer et al.'s giraffes was at the levels predicted for mammals of their body mass. Consequently, filtration fraction (GFR÷RBF) would be higher, which is indeed what they found (Table 11.3). Thus, it seems likely that the circumstances of Damkjaer et al.'s experiment had altered normal giraffe cardiovascular functions, but not kidney function. Giraffe kidneys function the same way that they do in any other mammal.

That conclusion, if correct, is surprising because giraffe kidneys potentially are exposed to blood pressures twice what they are in other mammals, and hence intrarenal blood flows should be twice what they are in other mammals, as would be glomerular filtration rate, and unless the nephrons are equally unusual, the consequences would be body fluid imbalances. Giraffe nephrons do not have an unusual structure (Maluf, 2002) so other mechanisms in the arterioles that supply the glomeruli must exist to counter the effects of high blood pressure and the high blood flow associated with it. One way that pressure in those small arterioles could be reduced is if the mechanical properties and radius of the main renal arteries that branch off the abdominal aorta to supply the kidneys were adapted to a high pressure. Damkjaer et al. found that their mechanical properties were the same as they are in cattle, but they also found that their internal radius was 2.3 mm compared with 3.3 mm in cattle. If blood pressure was the same in both giraffes and cattle, the smaller radius in giraffes would, because of Poiseuille's fourth power law, translate into a blood flow ~25% of what it is in cattle.

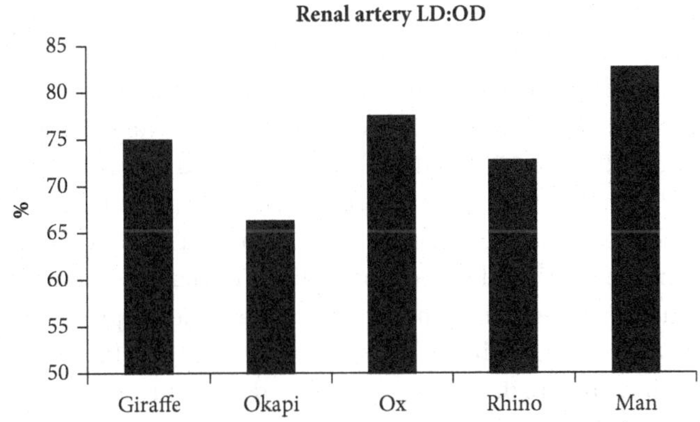

Figure 11.4. The ratio of lumen diameter (LD) to outer diameter (OD) of the main renal artery in five species.
Constructed from data in Maluf, 2002.

However, giraffe blood pressure is twice what it is in cattle, so the flow in a giraffe's renal artery would be about 50% of what it is in cattle. A narrow renal artery, in other words, would counter high blood pressure. Maluf also analyzed the dimensions of the main renal artery in giraffes and in other species (Figure 11.4).

The lumen diameter of those species including a giraffe was on average 75% ±6% of the outer diameter, which is similar to the ratio found by Damkjaer et al. There are no significant differences (in the statistical sense) between them. Consequently, the main renal artery in giraffes does not seem to be uniquely narrow. Maluf did find, however, that the wall of the interlobular renal arteries, which branch off the main artery, consisted of two layers of circular muscle, and the arteries had an outer diameter of 48 μm and an inner diameter of 25 μm, or 52% of the outer diameter. The small inner diameter and the thick muscle would increase resistance and reduce blood pressure and blood flow to the glomeruli. The structure of the interlobular arteries is similar to that of the even smaller afferent renal arterioles shown in Plate 9.2 (Chapter 9) that supply a glomerulus. In them the internal (lumen) diameter was ~45% of the outer diameter, a narrowness that also counters the effect of high blood pressure. From these data it seems probable that the filtration mechanism of the kidneys of giraffes is protected from high blood pressure by thick-walled and narrow small arteries and arterioles that supply it, rather than by a modified main renal artery. Nevertheless, the main renal artery is exposed to high blood pressure and that pressure may well be resisted by the tough renal capsule, as suggested by Damkjaer et al.

Reabsorption of Water

The volume of urine produced daily and, therefore, the volume of body fluid that could be lost as urine depend on how much water is reabsorbed from the fluid filtered into the nephrons. The relationship between the ability of a kidney to reabsorb water and kidney anatomy in mammals has been analyzed best by Carol Beuchat (1990), a vertebrate biologist and physiologist at the University of California. By reviewing the available data, she found that the maximum concentration of urine ([U]; i.e., the minimum volume of water that can be excreted or, to put it another way, how much water can be extracted from urine before the solutes in urine would precipitate) attainable by 146 different species of mammals (mainly small mammals and not including a giraffe) declined as body mass increased: $[U] = 2564 \times Mb^{-0.097}$ $mmol.L^{-1}$, and so theoretically it would be 2,800 mmol L^{-1} in a 400-g rat, 1,350 mmol L^{-1} in a 750-kg giraffe, and 600 mmol L^{-1} in a 6,000- kg elephant. Most, but not all, of the kidneys' ability to reabsorb water and concentrate urine depends on the length of the loops of Henlè. The longer their length, the higher is the osmotic gradient that they can create. Consequently, mammals that have long loops of Henlè are able to reabsorb more water and thereby produce more concentrated urine. The length of a loop of Henlè is difficult to measure, but as the loops lie in the kidney medulla, a substitute, the thickness of the kidney medulla, is used to get an idea of their length, and in the few cases where both loop length and medulla thickness have been measured, the two are correlated. Medulla thickness was measured by Maluf in three giraffes; it was 31 mm in the giraffe weighing 1,172 kg, 32 mm in the giraffe weighing 1,016 kg, and 20 mm in the giraffe weighing 104 kg. Medulla thickness (mm) in 138 mammal species is described by $8.15 \times Mb^{0.13}$ (Beuchat, 1990), and if that equation is applied to Maluf's giraffes, it predicts a medulla thickness for Maluf's adult giraffes of 20 mm and, for the small giraffe, a thickness of 15 mm. Thus, medulla thickness in Maluf's giraffes was somewhat higher than predicted for their mass and, by implication, their loops of Henlè were longer than expected and their ability to reabsorb water would have been better.

Even though absolute medullary thickness correlates quite well with loop length, it turns out that absolute medulla thickness is a poor predictor of the ability to concentrate urine in a mammal because as body mass increases so does kidney size, and increases in size are associated with increases in the diameter of the loops, and a decline in their power to establish the osmotic gradient needed for water reabsorbtion. Consequently, to take into account kidney size, relative medullary thickness (RMT) is calculated. RMT is the ratio of absolute medulla thickness (AMT) to kidney size. Kidney size (mm^3) = $(\text{length} \times \text{width} \times \text{breadth})^{0.33}$ and RMT if calculated from the formula $10 \times (\text{AMT (mm)} \div \text{kidney size})$. For 165 species of mammals, including some of the body mass of giraffes, $RMT = 5.41 \times Mb^{-0.11}$ (Beuchat, 1990). Applying that equation to Maluf's giraffes, their RMT should be 2.7 ± 0.3, and in fact RMT was 2.7 in the smallest giraffe weighing 104 kg, and the highest of 3.3 was in the giraffe weighing 1,016 kg, and thus RMT was of the size predicted for mammals of their mass (N=5; Table 11.4). The larger relative medullary thickness in larger giraffes implies that as the kidneys grow, their medulla enlarges slightly faster than does the cortex. Another way that the concentrating power of a kidney can be assessed, while at the same time compensating for changes in the overall size of the kidneys, is by using the dimensionless cortico-medullary index (cortical thickness÷medullary thickness). For Maluf's giraffes, this index was 0.5 (the medulla was twice as large as the cortex). In camels the ratio is 0.25—the medulla is four times bigger than the cortex (Abdalla and Abdalla, 1979), with the implication that giraffe kidneys are not as well-equipped to concentrate urine as are the kidneys of camels.

It would be comforting if the concentrating ability of a kidney was defined precisely by RMT but, regrettably, the correlation between concentrating ability and RMT is

Table 11.4 Relative Medullary Thickness in Giraffes

Body Mass (kg)	Kidney Mass (kg)	Length (mm)	Breadth (mm)	Width (mm)	Kidney Size (mm³)°	Absolute Medullary Thickness (AMT) (mm)	Relative Medullary Thickness (RMT)*
1,172	1.275	174	85	86	103	33	3.20
1,172	1.150	162	88	78	99	28	2.83
1,016	0.948	167	94	97	110	30	2.75
1,016	1.153	172	81	80	99	33	3.33
104	0.278	127	92	42	76	20	2.67
	0.96±0.40	*160±19*	*88±5*	*77±21*	*98±12*	*29±5*	*3.0±0.3*

°Kidney size (mm^3) = (length × width × breadth)$^{0.33}$.

*RMT = 10 (AMT ÷ ((L × W × B)$^{0.33}$).

Kidney mass and dimensions from Maluf, 2002, with permission of John Wiley and Sons.

quite weak (R2 =0.60; Beuchat, 1990) because concentrating ability depends not only on the lengths of the loops of Henlè, but also on the rate that filtrate moves along a tubule, the permeability of the collecting ducts, and tubule diameter (McNab, 2002). Nevertheless, the maximum urine concentrating ability ([U]; mosmol.L^{-1} of urine) can be estimated from RMT by [U] = (680 × RMT)–696 (Beuchat, 1990), and the maximum urine concentration that giraffes should be able to produce can be predicted to be 1,237 mosmol.L^{-1} on average (range ~1,500 mosmol.L^{-1} in juvenile giraffes to 1,000 mosmol.L^{-1} in adults). This concentration is not significantly different from the 1,350 mmol L^{-1} for a 750-kg giraffe predicted from body mass by Beuchat's equation and is four to five times the plasma concentration of solutes (= 300 mosmol.L^{-1}). In eight other East African artiodactyls in Beuchat's dataset, the average maximum urine concentration was 1,056 mosmol.L^{-1}, or 3.5 times the plasma concentration. By this comparison, giraffes may be able concentrate urine more than do the sympatric artiodactyls that share their habitat, but not as well as arid-adapted artiodactyls like camels (*Camelus dromedarius*; 3,170 mosmol.L^{-1}) or even the tiny dik-dik antelope (*Madoqua kirkii*; 4,760 mosmol.L^{-1}) (Beuchat, 1990).

Obligatory Daily Urine Volume of Giraffes

If the glomerular filtration rate of giraffes is predicted by the allometric equation 4 × $Mb^{0.76}$ mL.min^{-1}, then the kidneys of a medium-sized giraffe having a body mass of 750 kg will filter ~900 L of fluid every day, and that volume will range from ~250 L.d^{-1} in a neonate to ~1,500 L.d^{-1} in an adult. Roughly speaking, of all the water that enters a nephron daily, 75% is reabsorbed in the proximal convoluted tubule along with (mainly) the sodium ions that are reabsorbed there. Another 10% is reabsorbed in the distal convoluted tubule, along with any ions that escaped earlier reabsorption, leaving 15% of filtered water to be reabsorbed via interaction between the loops of Henlè and the collecting ducts. In a neonatal giraffe, that 15% is equal to ~40 $L.d^{-1}$, and in an adult it is equal to ~200 $L.d^{-1}$. These volumes are potentially the volumes of fluid that could be lost as urine if most of the filtered fluid were not reabsorbed. By comparison, fecal and evaporative losses at ~2 $L.d^{-1}$ each are trivial. However, the collecting duct-loop of Henlè interaction is exceptionally efficient, and almost all the water that enters a loop of Henlè is reabsorbed, except for the volume that is required for the necessary excretion of solutes. That volume is the obligatory urine volume.

The minimum, obligatory, urine volume produced daily by a giraffe depends on how much of the solutes that have been filtered end up in urine. The solutes, all of them soluble in water, cannot be excreted in any other way. How many do? The fluid that enters a nephron is, for all practical purposes, an ultrafiltrate of plasma, so the amount of solute filtered by giraffe kidneys depends on the plasma concentration of solutes which, in wild and captive giraffes, is 300 $mosmol.L^{-1}$ of water (Bush et al., 1980; Mitchell and Hattingh, 1993; Sullivan et al., 2010) and, thus, it is the same as it is in other mammals. The amount of individual solutes that is filtered each day (their filtered load) is the product of glomerular filtration rate and the concentration of the solute in blood plasma, and the filtered load for some important ones in giraffes is shown in Table 11.5.

How many of those filtered solutes are found in urine? The amount of solutes in the urine of wild giraffes has not been measured, and the difficulty of collecting urine samples from free-living, wild giraffes means that it probably never will be known. However, Damkjaer et al. measured the proportion of filtered sodium and potassium that was excreted into the urine each day in anesthetized wild giraffes and found that between 4.6% and 5.8% of filtered sodium and between 1.2% and 1.5% of filtered potassium were excreted. Sullivan et al. (2010) have reported the concentrations of solutes in the urine of captive giraffes in seven different zoos in North America. These latter data are highly variable and unreliable, and depend on the diet the giraffes had been fed, but they are the only data available for estimating what proportion (%) of the daily filtered amount of urine solutes, other than sodium and potassium, appears in urine. That proportion can be calculated from the concentration of each solute in urine and in plasma and the concentration of creatinine in urine and plasma. Almost no creatinine (a by-product of muscle catabolism) is secreted into or reabsorbed from kidney tubules, so the rate at which it appears in urine is a good measure of filtration rate. The quantities of solutes that are excreted in urine (their excreted load, as estimated from creatinine clearance), are shown in Table 11.5. The total mass of these solutes is 60 to 350 $g.d^{-1}$, and is equivalent to ~2500 $mmol.d^{-1}$ in a juvenile giraffe and 14,500 $mmol.d^{-1}$ in an adult. As the maximum concentration of giraffe urine can be estimated to be ~1,250 $mmol.L^{-1}$, in order to excrete those quantities of solute each day, the minimum amount of water needed (the obligatory urine volume) can be calculated by dividing the quantity of solute by maximum urine concentration. It ranges from 2 to 12 $L.day^{-1}$ for giraffes that have body masses between 100 and 1,500 kg ($=0.01 \times Mb + 0.04\ L.d^{-1}$). For a 750-kg giraffe, it is about 8 $L.d^{-1}$. This range of minimum urine volumes is predicted well by an equation that describes daily urine volume for mammals in general ($= 0.06 \times Mb^{0.75}\ L.d^{-1}$; Calder, 1996). Thus, the obligatory urine volume in giraffes estimated from the minimal data available for giraffe kidney function seems to be the same as it is in similar-sized mammals and does not suggest that giraffes have evolved any special mechanisms of water reabsorption or of concentrating power. That conclusion is

Table 11.5 Blood and Urine Concentrations of Sodium, Chloride, and Potassium Ions in Captive Giraffes

Solute	Blood[a] (mmol.L^{-1})	Urine[b] (mmol.L^{-1})	Filtered Load (mmol.d^{-1})	Excreted Load (mmol.d^{-1})	ExcretedLoad (g.d^{-1})
Calcium[c]	2.6±0.4	2.2	600–3,700	3–18	0.1–0.7
Phosphorus[c]	3.2±0.9	7.1	800–4,500	10–50	0.3–1.6
Sodium[d]	143±9	127	36,000–205,000	1,900–10,600	40–240
Potassium[d]	5.2±0.9	415	1,300–7,500	20–100	1–4
Chloride[c]	97±5	245	24,000–139,000	400–2,000	14–70
Creatinine	0.2±0.1	32	50–290	50–290	6–33
Total			60,000–360,000	2,500–14,500	60–350

[a]Mean±sd; [b]weighted mean; [c]excreted load calculated from creatinine clearance; [d]excreted load calculated from measured fractional excretion, after Damkjaer et al., 2015.

Blood data from 14 giraffes (Bush et al., 1980) and urine data from 11 giraffes (Sullivan et al., 2010). Ranges given for filtered and excreted loads are for giraffes of 100–1,500 kg body mass.

disappointing given the preference giraffes have for arid habitats, because the minimum urine volume is equivalent to a third of all water lost by a giraffe each day, a volume that must be replaced if water balance is to be achieved.

A Summary

Table 11.1 shows water balance in a 750-kg giraffe at rest eating high-quality succulent browse on a balmy, but hot, African summer day when the ambient temperature is 25°C–30°C. The data imply that in those circumstances, water intake from browse and metabolism exceeds water lost, and the giraffe would be in positive water balance. If ambient temperature is higher or exercise is required, or a female is lactating, and especially during the hot dry season from August to November, loss could easily exceed intake. However, water balance means just that, so whatever volume of water is taken in or lost over the course of a day or several days, giraffes find a way of balancing the volume lost or taken in. Giraffes cannot, in normal circumstances, be in anything other than water balance. If water intake is greater than water loss, dilute urine is excreted; if loss is greater than intake, urine concentrates to its maximum and/or there is reduction in evaporative loss by behavior or physiological mechanisms (see Chapter 14) to bring about balance. The volume of water ingested is related directly to the amount of browse eaten and its succulence, so water intake can be increased by eating more browse, but browse is not their only source of water. Contrary to the legend, objective observations reveal that drinking of surface water, if it is available, is frequent and occurs whenever it is needed. Giraffes' apparent independence of surface water will occur only if intake of water through succulent browse and metabolism exceeds obligatory urine loss and fecal and evaporative losses. Under most circumstances, independence of surface water will be intermittent. During the wet season, independence will predominate; in the dry season, when leaves have fallen, dependence on surface water will predominate. Lactation stimulates a similar dependency. Milk is ~90% water, and giraffe cows provide up to 10 L d^{-1}. Lactating cows drink regularly even if water is kilometers away from where they have hidden their calf (see Chapter 17). It is easy, though, for giraffes to slip into a water deficit. For example, if all metabolic heat were to be removed by sweating, dehydration would follow quite quickly unless water balance could be corrected (see Chapter 14). The only circumstance in which balance is not readily correctable is in a drought. Devastating mortality of giraffes can be the consequence.

(a) (b)

Plate 1.1. Lydekker's drawings of (left) *G.c. typica* and (right) *G.c. capensis.*

Plate 2.1. Rock painting of giraffes in Botswana ca. 3000 BC.
Courtesy Duncan Mitchell.

Plate 2.2. *The Medici Giraffe* by Francesco Bacchiacca.
With permission of the National Art Gallery, Washington, DC.

Plate 3.1. Top left: *Portrait of Colonel Robert Jacob Gordon*.
With permission of the William Fehr Art Collection, Cape Town, South Africa.
Top right: *Colonel William Paterson*.
With permission of the Dixson Galleries, State Library of New South Wales, Australia.
Bottom left: *George-Louis Leclerc, comte du Buffon* by François-Hubert Drouais.
With permission of Musée Buffon, Montbard.
Bottom right: *Jean-Nicolas-Sébastien Allamand* by Jan Palthe.
With permission of Leiden University Libraries Digital Collections.

(a) (b)

Plate 3.2. Carteret's giraffe (left) with permission of the Royal Society and Buffon's giraffe (right), drawn from Carteret's image.
With permission of the Wellcome Collection.

Plate 3.3. Paterson's giraffe.
From Paterson, 1789. Courtesy of the Brenthurst Library, Johannesburg, South Africa.

Plate 3.4. Gordon's giraffe.
Courtesy and permission of the Rijksmuseum, Amsterdam.

Plate 3.5. Vosmaer's giraffe.
From Vosmaer, 1787, with permission of the British Library Board.

Plate 3.6. Harris's giraffe, *G.c. giraffa.*
From Harris, 1852.

Plate 3.7. Zarafa. *Study of the giraffe given to Charles X of France by the Viceroy of Egypt*, by Nicholas Hüet the younger, 1827.
Public domain image from Wikimedia commons.

Plate 3.8. *George IV's giraffe in a park with its attendant,* by Jacques-Laurent Agasse, 1827.
Courtesy and permission of the Zoological Society, London.

Plate 3.9. *Portrait of two giraffes belonging to George IV*, or "*The Nubian Giraffe*" by Richard Barrett Davis, 1827.

Plate 4.1. Harry Johnston's painting of an okapi (*Okapia johnstoni*), the "living fossil."
From Johnston, 1902.

Plate 6.1. "A long neck and stocky body." The shape of *G. sivalensis* based on analysis of its bone dimensions by van Sittert and Mitchell (2015). The photograph is of a "dwarf," dysplastic *G.c. typica*.
From Brown and Wells (2020), with permission of Springer Nature.

Plate 7.1. Time to stop and stare.
Image by chuck@UPDmedia.com for Wikimedia Commons.

Plate 7.2. A mouth and head shaped for deep penetration into browse. *G.c. giraffa* in Pilanesberg National Park, South Africa, 2018.
Photograph by Andrea Fuller, with permission.

Plate 7.3. Giraffes use their tongues to grasp a branch and grip the stem between their incisors and the maxillary pad. *G.c. capensis* collecting browse in the Augrabies National Park, South Africa, 2018.
Photograph by Andrea Fuller, with permission.

Plate 7.4. *A. nigrescens* (the knoppiesdoring tree) and its flowers.
Photographed for Pretoria National Botanical Gardens by JMK.

Plate 7.5. Ant domicile in *Acacia drepanolobium*.
Image by Alastair Rae for wikmedia.commons.

Plate 7.6. Vigilant giraffes in their natural habitat of Limpopo Province (Eastern Transvaal), South Africa.
Photo courtesy of Ian Powell, December 2016, with permission.

Plate 7.7. Browsing where others cannot. *G.c. giraffa* browsing on *Dichrostachys* in the Kruger National Park, South Africa, 2010.
Photograph by Dr. S. van Sittert, with permission.

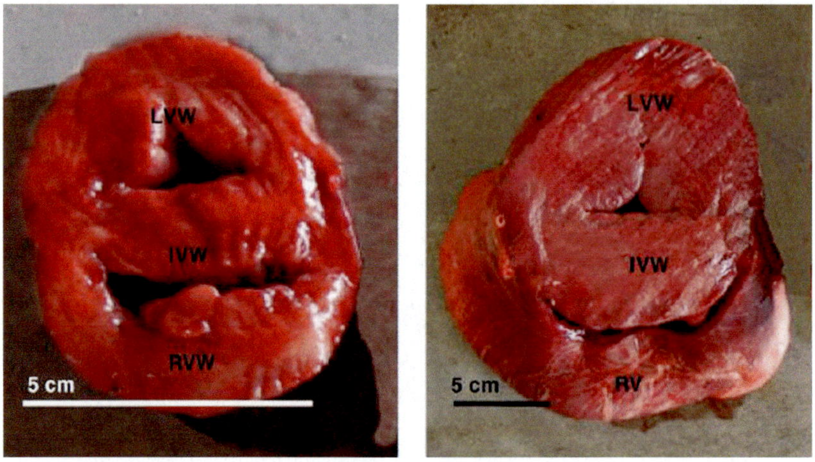

Plate 9.1. A transection through the right (RVW), left (LVW) and interventricular (IVW) walls of the ventricles of an adult (right) and fetal (left) giraffe.

Bar = 5 cm.

From Mitchell and Skinner, 2009, with permission of Elsevier.

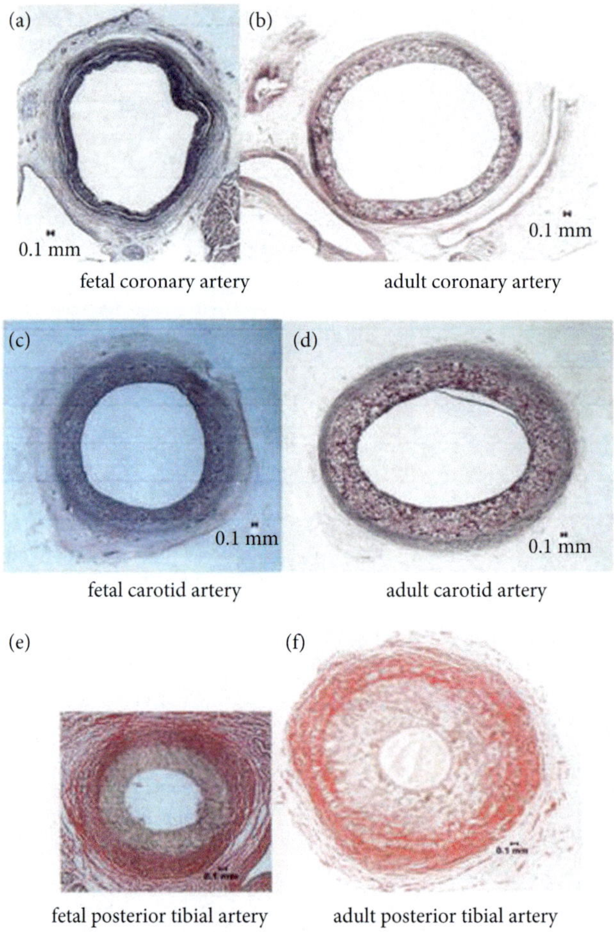

Plate 9.2. Wall thicknesses of coronary, carotid, and tibial arteries.

Bar = 100 μm.

From Mitchell JA and Mitchell G, unpublished.

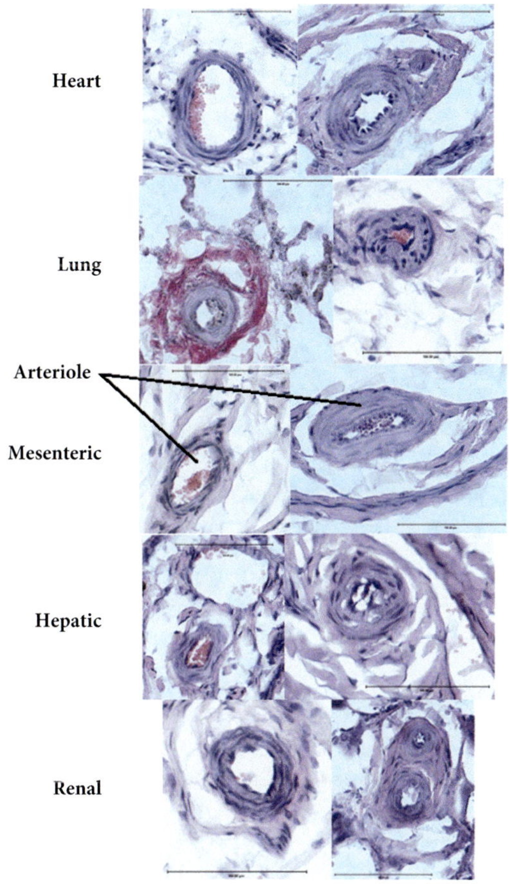

Plate 9.3. Fetal (left) and adult (right) arterioles in heart, lung, intestinal (mesenteric), liver (hepatic), and kidney (renal) tissue.

Bar = 100 µm.

From Mitchell JA and Mitchell G, unpublished.

Plate 11.1. A giraffe (*G.c. capensis*) drinking at an artificial drinking point in the Kgalagadi Transfrontier National Park, South Africa.

Photograph by Andrea Fuller, with permission.

Plate 13.1. The hunting of giraffe.
From Harris, 1852.

(a) (b)

Plate 14.1. Left: A giraffe (*G.c. giraffa*) in shade near the Kruger National Park, South Africa, January, 2018.
Photograph by Bridget Mitchell with permission.
Right: Giraffes (*G.c. capensis*) lying down in shade in the heat of the day in Kgalagadi National Park, near Mata-Mata, in the Auob River bed, on the Namibia/Botswana border.
Photograph by Andrea Fuller, 2018, with permission.

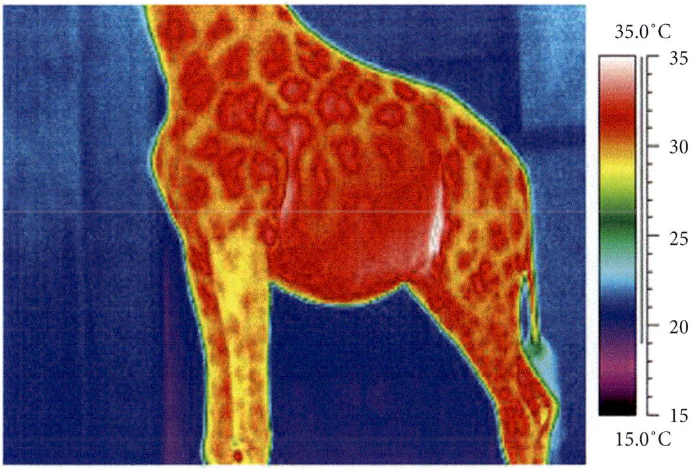

Plate 14.2. Photograph of skin temperatures in a giraffe using a thermal camera.
After Hilsberg, 2000. Reprinted with permission of Taylor and Francis.

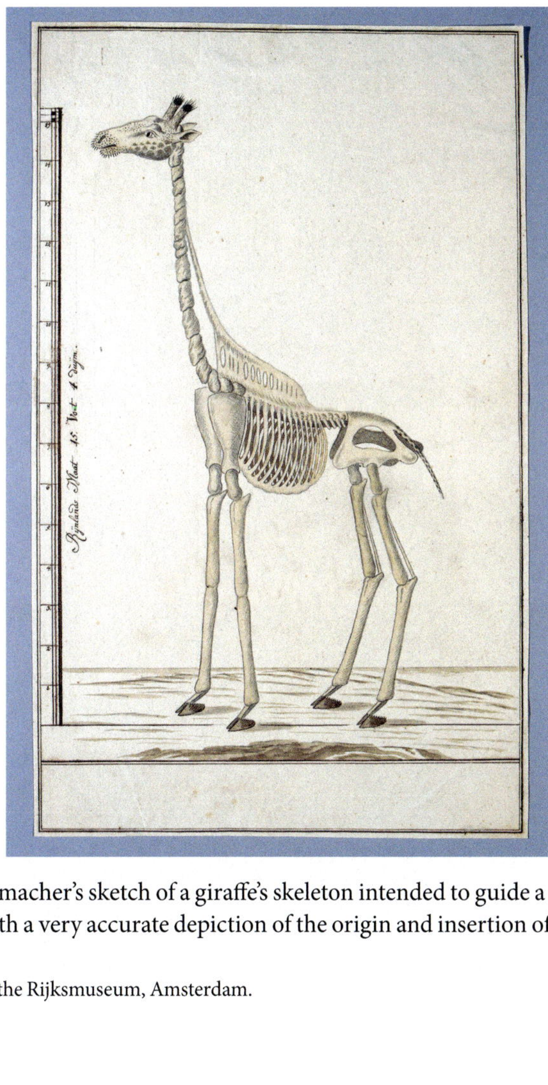

Plate 15.1. Schumacher's sketch of a giraffe's skeleton intended to guide a reassembly of the bones, but with a very accurate depiction of the origin and insertion of the *ligamentum nuchae*.
With permission of the Rijksmuseum, Amsterdam.

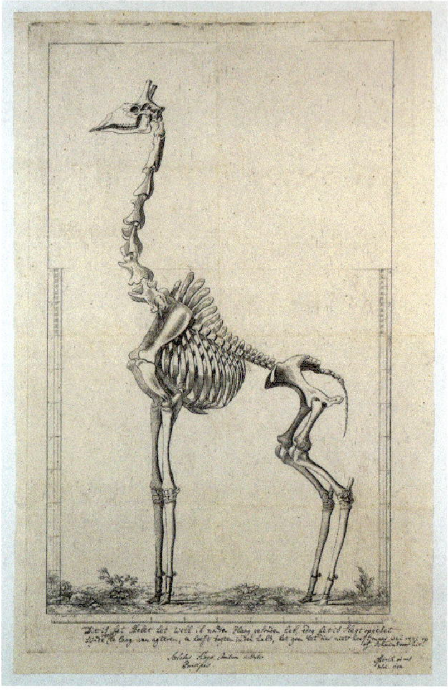

Plate 15.2. A copy of the original copper engraving of Vosmaer's *Skeleton of the Giraffe*, made by J. F. Gout in 1786.

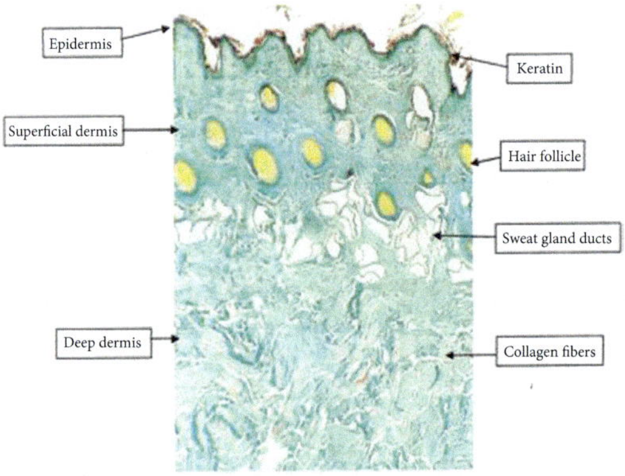

Plate 16.1. A section through the epidermis and dermis of giraffe skin stained with Masson's trichrome stain for keratin (red), collagen (green), hair follicles (yellow). Elastic tissue, if present, would stain black. Ducts of sweat glands are not stained.
From Mitchell and Skinner, 2004, with permission of Taylor and Francis.

Plate 16.2. A necessary nuisance. A giraffe (*G.c. giraffa*) and oxpeckers.
Photograph by David Roberts with permission.

Plate 16.3. A young giraffe in thick Acacia bush in Mpumalanga Province South Africa.
Photo courtesy of Andrew Mitchell, 2019.

Plate 16.4. In his prime. A mature male giraffe at least eight years old.
Photograph by David Roberts with permission.

Plate 18.1. Browsing where other browsers cannot reach. Giraffes (*G.c. giraffa*) feeding from *Acacia* sp. in Zimbabwe.
Photograph by David Roberts with permission.

Plate 18.2. The end of the line? A male *G.c. giraffa* in shade in the Bubye Valley Conservancy in Zimbabwe on November 25, 2008, during the drought referred to in the text.
Photograph by David Roberts with permission.

12
Respiration

For any mammal, including giraffes, to be able to extract oxygen from air, the air has to be delivered to a surface where the oxygen in the air can be exposed to blood. Delivery of air and its oxygen to that surface in terrestrial vertebrates would be relatively easy if the exchange surface were close to the air and its oxygen, in the same way that oxygen in water is in direct contact with a fish's gills. Gills are efficient, but they are external structures, rely on contact with water to function, are susceptible to damage, and because water contains very little oxygen the gills have to have a large surface area in order to extract it. By comparison, air is rich in oxygen, so the exchange surface can be smaller and the surface that has evolved has become protected, in lungs, in a well-defined thorax that is delimited by ribs and their muscles and in mammals is separated from the abdominal space by a diaphragm. Air reaches the lungs via an airway. Movement of air up and down the airway and in and out of a mammal's lungs follows the physical laws that govern the relationship between the pressure exerted by a gas and its volume. That relationship was discovered by Robert Boyle in 1662 and ever since has been known to countless generations of physics and physiology students as Boyle's Law. It can be stated as "the pressure exerted by a given mass of an ideal gas is inversely proportional to its volume at a constant temperature" or, in other words, provided that temperature does not change, pressure multiplied by volume is constant ($P \times V = k$). If one goes up, the other goes down. How Boyle's Law can help explain breathing in mammals is a far more recent discovery. Movement of air into the lungs of mammals is brought about by contraction of the diaphragm and, when a large breath is required, is aided in most mammals by contraction of the intercostal muscles that act to lift the ribs. Together, contraction of these two muscles increases the volume of the thorax. Boyle's Law decrees that if thorax volume increases, the pressure in it will fall. How far the pressure falls in lungs depends on the effort made to increase thorax volume, but for any air to flow into and inflate the lungs, intrathoracic pressure has to fall, at the very least, to below atmospheric pressure, which, of course and typically, it does. Healthy lung tissue is full of elastic fibers, and when the lungs expand the elastic fibers are stretched. Breathing in, therefore, requires some muscular effort. Breathing out requires none (unless forced expiratory movements are required). It is accomplished by recoil of the lung elastic fibers and relaxation of the diaphragm and the muscles of the chest wall.

Respiratory movements not only deliver oxygen to the lungs. They also remove carbon dioxide from the body. The amount of oxygen that must be delivered and the amount of carbon dioxide that has to be removed are determined by the rate at which sources of dietary energy are oxidized to produce the energy needed to sustain metabolism. At rest, we humans use 3–5 mL of oxygen per kilogram of body weight every minute, which, for a human weighing 70 kg, means that over the course of a day 500 L of oxygen are consumed. That amount is increased by exercise. In an Olympic-standard marathon runner or cross-country skier weighing, say, 60 kg and exercising at his/

How Giraffes Work. Graham Mitchell, Oxford University Press. © Oxford University Press 2021.
DOI: 10.1093/oso/9780197571194.003.00112

her peak, oxygen requirements can easily be 80 mL kg^{-1} min^{-1} or more, so if a marathon lasts for two hours, then 500–600 L of oxygen are needed to run it. The amount of carbon dioxide produced depends on the dietary source of energy. If the source of energy is a mixture of fats, carbohydrates, and proteins, then for every 1 mL of oxygen consumed, ~0.8 mL of carbon dioxide is produced, a ratio that has been confirmed in a giraffe, albeit an anesthetized one (Nyman et al., 2016). Over the course of a day in someone at rest or over the course of a two-hour marathon, some 400 L of carbon dioxide are produced. Over the course of a year, about 150,000 L of carbon dioxide are produced by every person on earth, which weighs about half a metric ton, so metabolism in the current human population on earth produces about 3–4 billion metric tons of carbon dioxide every year. An adult giraffe needs about 6,000 L of oxygen each day (or two million liters a year) and produces about 5,000 L of carbon dioxide per day (about 1.75 million liters a year or nearly six metric tons).

In most mammals, lungs are also the source of the airflow needed for making sounds. The airflow makes vocal cords vibrate. Their vibration generates sound waves that resonate in the mouth and sinuses. Sound waves are made up of a sequence of high and low air pressures, which, when they reach the tympanum (eardrum), cause it to vibrate. That vibration is magnified by the ear ossicles and ultimately generates nerve impulses in the auditory nerve, which on reaching the brain are interpreted as a sound. The loudness of a sound is correlated with the amplitude of the pressure wave (i.e., the difference in size between the low and high pressures in it) and it is measured in decibels. The pitch is the frequency of the pressure waves (Hz, sec^{-1}). Frequency is determined by the tension in the vocal cords: the greater the tension, the higher the pitch. The lower the pitch frequency, the louder a sound must be before it can be heard.

From the earliest of times, giraffes have been regarded as being mute (e.g., Selous, 1908), which means either that they cannot produce sounds, or that the sounds they produce cannot be heard by humans. It would be surprising if giraffes were mute, as they have excellent hearing. The slightest unfamiliar noise produces a characteristic ear-twitch and a penetrating look at the source (see Plate 7.6). Humans can hear sounds that have a frequency as low as 100 Hz, but only if loudness is above 40 decibels. Normal conversation has a frequency of 120 Hz to 250 Hz and a loudness of about 60 decibels. Recordings have shown that most giraffe "sounds" have a frequency of 14–250 Hz and sound pressure levels between 60 decibels at low frequency and 30 decibels at high frequency, and none of these sounds can be heard by humans (von Muggenthaler et al., 1999). Calves have been heard to bellow when seeking milk, but in older animals only a grunt or cough, or low amplitude moans or snorts, or nocturnal hums or deep long sighs, have been heard by humans or recorded, and these mostly in captive giraffes (Owen, 1838; Innis, 1958; Dagg and Foster, 1976; Kristal and Noonan, 1979; Harrison, 1981; Baotic et al., 2015; Volodina et al., 2018). The nocturnal hums have a frequency of 92±26 Hz, are thus below human hearing range, and last a few seconds. Their purpose is unknown. A long, audible hissing, caused by air being forcibly expelled through the nose, is made by suspicious cows that have a calf nearby (Langman, 1977).

Vocal cords are situated in the larynx, and the fact that the larynx of giraffes is small—smaller, for example, than it is in horses (Hahn and Mayhew, 1999)—may explain the characteristics of the "sounds" giraffes produce, but the reason most often given for the absence of more audible sounds is that the vocal cords of giraffes are rudimentary (Joly and Lavocat, 1846) or absent (Hahn and Mayhew, 1999). Donald Harrison (1980), however, who has studied the larynx of giraffes in great detail, concluded that vocal cords are present, and there is no unusual development of the larynx that could explain the apparent silence of giraffes. The more likely reason for their "silence" is that they cannot easily produce airflow of the velocity needed to vibrate their vocal folds sufficiently to produce a sound audible to humans. In humans, airflow velocity in the trachea is between 90 cm s^{-1} and 175 cm s^{-1}. In giraffes at rest, it is 80 cm s^{-1} (see Chapter 13) which, while it is insufficient to produce a sound audible to humans, will be almost certainly enough to produce sounds very audible to other giraffes, and it may well be that giraffes use sounds to stay in touch with each other, or to signal a threat, especially at night, when visibility is poorer.

The ultimate purpose of breathing is to deliver oxygen to mitochondria, where a single enzyme—cytochrome oxidase—uses almost all of it to synthesize water, a reaction that is the terminal reaction in the sequence of reactions that synthesize ATP. However, it would be wrong to think that the respiratory systems of animals and, because this is the story of giraffes, land mammals specifically is only to do with oxygen delivery and carbon dioxide removal from the body and the mitochondria in particular. Carbon dioxide produced by metabolism interacts with the water in body fluids and, in the process, releases hydrogen ions, which could and would, if they were not removed quickly, make body fluids acidic and lower their pH to levels not compatible with life. Thus, removal of carbon dioxide by the lungs is an essential part of the processes that keep body fluid pH constant. The respiratory system also has a role to play in controlling body temperature (see Chapter 14), and lung tissue activates hormones, synthesizes glucose, and traps and expels by sneezes and coughs the airborne pathogens that cause disease and the antigens that cause hay fever and asthma. Small blood clots that circulate to the lung get trapped by capillary blood vessels in the lungs and are removed from the circulation by the macrophages that populate and lurk in lung tissues. The anatomy and position of the airways and lungs and their blood supply are ideally placed to fulfill these functions. The lungs and airways are open to the environment and exposed to whatever is in the air. They monitor the air that is breathed and, for the most part, eliminate its worst constituents. They are a first line of defense. The lungs also receive more blood every minute than any other organ or tissue. Air coming into the lungs from the environment is exposed to this large blood flow and to all the defense mechanisms in it.

The Pulmonary Circulation of Giraffes

Lungs receive blood from two sources. One source is the bronchial arteries. They supply lung tissue and the lower branches of the respiratory tree with oxygen and other nutrients, and supply heat and moisture which ensure that air arriving in the lungs is at body temperature and is fully saturated with water. Bronchial veins remove carbon dioxide and other waste metabolites generated by lung tissue itself. Bronchial artery blood flow volume is 1%–2% of the cardiac output of the left ventricle, which, in an average-sized giraffe weighing 750 kg, means that it would be 400–800 mL min^{-1}. The other blood supply comes via the pulmonary artery which is the outflow channel of the right ventricle. The pulmonary artery transports deoxygenated blood from the right ventricle to the lungs, where oxygen is added to it and carbon dioxide removed from it. In other words, the pulmonary artery each minute delivers to the lungs all the blood that has returned to the heart from the body, and in the same average-sized giraffe weighing 750 kg that volume is ~40 L min^{-1}. There have been relatively few measurements of the pulmonary artery pressure that drives blood through the lungs of giraffes, mainly because it is far more difficult to measure than it is to measure systemic blood pressure, but it has been measured in giraffes with a body mass of 500–700 kg and found to be 75/25 mmHg (systolic/diastolic; Patterson et al., 1957; Warren et al., 1957), 38/13 mmHg and 48/22 mmHg (Goetz et al., 1960), and 30–50/10–30 mmHg (van Citters et al., 1969). Right ventricle systolic pressure has been measured to be 60 mmHg and its diastolic pressure to be 11 mmHg in a standing, non-anesthetized giraffe, +40 mmHg systolic to −5 mmHg diastolic when the head is up, and +60 mmHg to −5 mmHg when the head is down (Goetz et al., 1960), and 30–50 mmHg systolic and 5–10 mmHg diastolic in unanesthetized free-living giraffes (Van Citters et al., 1969). Smerup et al. (2016) recorded average systolic pulmonary artery pressure to be 45±3 mmHg in eight young anesthetized giraffes weighing on average 495 kg. The mean pulmonary pressure that can be derived from these pressures is 30–40 mmHg, that is, about one-fifth to one-sixth of the systemic arterial pressures generated by the left ventricle. Except for the difference between systolic and diastolic pressures reported by Patterson et al. and Warren et al. in 1957, the difference between pulmonary systolic and diastolic pressures (the pulmonary pulse pressure) in adult giraffes is ~20 mmHg (van Citters et al., 1969) to ~25 mmHg (Goetz et al., 1960), and if so, then systolic pressures will range between 38 and 74 mmHg and diastolic pressures between 13 and 49 mmHg, ranges that cover most of the pulmonary pressures measured thus far.

Why is pulmonary artery pressure so much lower than systemic arterial pressure? One, teleological, reason is that the right ventricle is unable to generate pressures as high as those produced by the left ventricle. The pressures that the right ventricle can generate follow Laplace's Law (Chapter 9) and so are directly proportional to the thickness of the right ventricle wall. In giraffes the right ventricle wall is much thinner than the left ventricular wall (Plate 9.1 in Chapter 9). In 11 fetal giraffes with body masses ranging from 18 to 77 kg, the right ventricular wall was found to be 68% ±12% as thick as the left ventricle wall (on average, 1.3 cm and 2.0 cm, respectively) and in postnatal giraffes weighing 147 kg to 1,441 kg, it was 38% ±9% as thick as the left ventricle wall (on average, 2.2 cm and 5.8 cm, respectively; Mitchell and Skinner, 2009). If mean pulmonary artery pressure is calculated from mean systemic arterial pressure but adjusted according to the ratio of right ventricle wall thickness:left ventricle wall thickness, mean pulmonary pressure can be shown to decline through fetal life to reach a relatively stable average value in adults of 40±9 mmHg (Figure 12.1), a pressure consistent with the pressures measured so far in the right ventricle of giraffes.

A second origin of the low pulmonary artery pressure is that lung blood vessels and especially lung capillaries distend easily, and any that are unfilled with blood when blood pressure is low can be filled if pressure tends to increase, and that adjustment dampens any potential increase in pressure. Yet another reason is that there are a very large number of blood vessels in the lungs and their collective diameter is large, so the resistance to blood flow is low. The pulmonary resistance (pulmonary artery pressure ÷ cardiac output) in an adult giraffe is ~1 mmHg min L^{-1} or roughly one-sixth of the resistance in an adult giraffe's systemic circulation. If resistance is low, then of course the right ventricle does

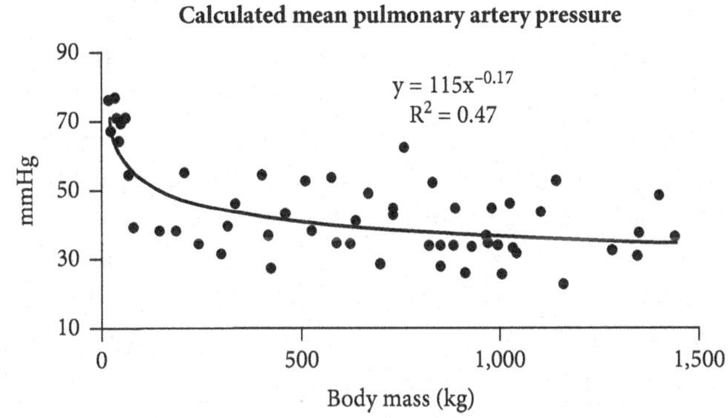

Figure 12.1. Mean pulmonary pressure calculated from measurements of left and right ventricle wall thickness in 10 fetal and 22 male and 24 female *G.c. giraffa*. The body mass of the fetuses ranged from 18 to 77 kg, that of the males from 187 to 1,441 kg, and that of the females from 147 to 1,029 kg.

not have to work hard to generate a pressure suitable to push blood through the pulmonary blood vessels and so does not have to have a thick wall. The pulmonary artery pressure of fetuses (66±11 mmHg; Figure 12.1), on the other hand, is much higher than it is in adults, and that is because the lungs of fetuses are collapsed and are filled with fluid, not air. The fluid prevents the lung and its blood vessels from expanding, so resistance to blood flow is very high. For example in a fetus weighing 75 kg and thus about three-quarters of the way through gestation, pulmonary resistance is ~9 mmHg min L^{-1}, that is, nine times higher than it is in a postnatal giraffe. The right ventricle wall is, consequently, proportionately thicker in a fetus than it is in an adult giraffe.

Even though the pulmonary artery pressure of giraffes is lower than systemic pressure, it is much higher than it is in any other mammal and it should, following Starling's equation (see Chapter 10), be sufficient to cause filtration of fluid out of pulmonary capillaries into the alveolar sacs where gas exchange takes place, and if so, it would cause pulmonary edema and prevent gas exchange. However, the hydrostatic pressure in the capillaries of the pulmonary micro-circulation in postnatal giraffes is determined mostly by the pressure in the left atrium, to which the blood is going, not by the pressure in the pulmonary artery. Left atrium pressure is 13 mmHg (Smerup et al., 2016) and thus well below the colloid osmotic pressure of giraffes (25–27 mmHg). The difference between these two pressures is sufficient to favor reabsorption of any fluid that might be filtered out of the capillaries of the pulmonary circulation.

The pulmonary circulation in all mammals is a high-volume, low-pressure, low-resistance circulation but, in giraffes, it functions at pressures much higher than they are in, for example, humans, which means that the pressures in both the pulmonary and systemic circulations in giraffes are uniquely high.

The Airways of Giraffes

Giraffes have a very long neck and to connect the atmosphere to the lungs very long airways are needed. Breathing through a long tube, as most competitive schoolboys will find out at some stage, is a task full of difficulties. If the long tube is narrow, then resistance to airflow is high and breathing in and out is hard work. If the tube is wide, then the volume of air reaching the lungs at each breath is small, the air in the lungs is not refreshed, and dizziness and fainting are the outcome of breathing through the tube. How, then, do giraffes manage to breathe through their long tube? What is its structure? In what ways is it modified to make breathing through it possible?

The airways of giraffes have the same basic structure as have all mammals. Like all mammals, in giraffes the upper part of the airway starts in the nasopharynx and ends at the larynx. Beyond the larynx, the airway becomes the trachea, which divides into large bronchi and then into many small bronchioles that connect to the respiratory airways where exchange of oxygen and carbon dioxide between blood and lungs occurs. The respiratory airways consist of respiratory bronchioles, alveolar ducts, and, finally, the alveolar sacs that have the very large surface area required for efficient gas exchange. Situated below the nasopharynx is the oropharynx. It starts in the mouth and ends at the opening of the esophagus. The esophagus itself lies between the trachea (below) and the vertebral column (above). A more useless arrangement is hard to imagine because it requires complicated interaction between the larynx and esophagus to prevent food from going down the trachea and air going down the esophagus. When a mammal breathes in, the muscles of the larynx relax and the larynx diameter increases, making inward airflow easy. When a mammal swallows food, a backward movement of the tongue forces the epiglottis over the laryngeal opening and the laryngeal muscles contract, a combination that prevents food from going down the airway. Air going down the esophagus rather than the trachea is, on the other hand, very common in all mammals, and occurs when food is swallowed. This air and any gas generated in the stomach of a monogastric mammal returns as a belch, which, for the most part, is inconsequential but should, when it happens in humans, be accompanied by "pardon." In ruminants, some air is swallowed, but a lot more gas is produced by fermentation of grass or browse in the rumen and comes up the esophagus as an eructation. Unlike a belch, this gas has the pleasant aroma of the mixture of volatile fatty acids that is its origin, so no "pardon" is necessary, but cattle I know do look guilty and surprised when a bubble emerges.

For the trachea of mammals to function as a conducting airway, it has to be a fairly rigid, non-collapsible tube. In all mammals the tube consists of a muscular wall lined with ciliated, mucus-secreting epithelium. Mucus traps pollutants, which, usually, are propelled upward by the cilia to the throat and ejected by a cough. Embedded in the wall of the trachea are incomplete rings of cartilage. The incompleteness of the rings allows trachea diameter to vary, but they prevent its complete collapse. The ends of the incomplete rings in giraffes are joined by an involuntary muscle which, when it contracts, reduces trachea diameter (Harrison, 1986). Relaxation of the muscle allows the diameter to increase. The early anatomists who counted the number of tracheal rings in the giraffes they dissected found 100 in a young giraffe weighing 153 kg (Crisp, 1864) and "more than 200" in older animals (Joly and Lavocat, 1846; Crisp, 1864), implying that the number of rings increases as the trachea elongates, but Cano and Perez (2009) counted 100 in a 754-kg giraffe. Tracheas, although flexible, are cylinders and have, therefore, a length and a diameter. Trachea length (cm) is usually regarded as the distance from the first tracheal ring below the larynx to the bifurcation into left and right bronchi (Tenney and Bartlett, 1967; Nakakuki, 1983). To determine trachea diameter (cm) in a mammal, it should be measured vertically and horizontally (dorso-ventrally and laterally) at the first tracheal ring, at the midpoint of the trachea, and just proximal to its bifurcation into the two main bronchi (to establish if it is circular or elliptical in cross-section and if its diameter changes along the length). From these six measurements of diameter, an average trachea diameter can be calculated and used to calculate trachea cross-sectional area and trachea volume. The length of the trachea in giraffes is, obviously, much longer than it is in other mammals of similar body mass, and in 46 giraffes, length ranged from 77 cm in a neonate to 2.5 m in a mature adult (Mitchell and Skinner, 2011; Figure 12.2; Table 12.1).

In those same giraffes, the trachea was circular in cross-section and had an average diameter that ranged from 2.3 cm in the smallest giraffe to 4.8 cm in the largest. Diameters at the top, middle, and bottom were not statistically different, and overall the diameters were significantly narrower than they are in other mammals (Figure 12.2; Table 12.1). Nor did the diameter increase proportionately with increase in length, so as a giraffe grows the trachea becomes relatively narrower. Cross-sectional area of the trachea calculated from average diameter increased almost threefold during growth, from 4.2 to 17.9 cm^2, but far less than it does in other mammals of comparable body mass (Figure 12.2).

A consequence of having a long and narrow trachea is that the resistance to airflow is high. As for blood flow in a blood vessel, resistance to airflow is determined mainly by the diameter of the airway. Another consequence is that the velocity at which air has to flow through it to reach the lungs will be higher than the velocity in a mammal that has a wider trachea, down which the same volume of air is moving. If the velocity of air flow is high enough, the pattern of its flow changes from being smooth (laminar) to becoming turbulent and, at least theoretically, turbulence makes it more difficult to move the air. Resistance to airflow and turbulence means that more muscle power is needed to ventilate the lungs, and the work of breathing (i.e., the amount of energy needed to

Table 12.1 Dimensions of Giraffe Conducting Airways

Variable	750-kg Giraffe	Range for 147–1,441-kg Giraffes	Allometric Equations for Giraffes	750-kg Mammal	Allometric Equations for Predicted Values*
Trachea length (m)	1.7	0.8–2.3	$0.10 \times Mb^{0.43}$	1.1	$0.15 \times Mb^{0.30}$
Trachea diameter (cm)	4.0	2.4–5.1	$0.46 \times Mb^{0.33}$	4.6	$0.35 \times Mb^{0.39}$
Trachea cross-sectional area (cm^2)	12	4–18	$0.16 \times Mb^{0.65}$	18	$0.10 \times Mb^{0.78}$
Trachea surface area (m^2)	0.22	0.06–0.35	$0.0015 \times Mb^{0.75}$	0.16	—
Dead space (L)	2.3	0.4–4.4	$0.003 \times Mb + 0.04$	2.1	0.003×-0.12

Giraffe dimensions based on data collected from 46 giraffes.

*Predicted values for other mammals from Stahl, 1967; Calder, 1981, 1996; Gehr et al., 1981.

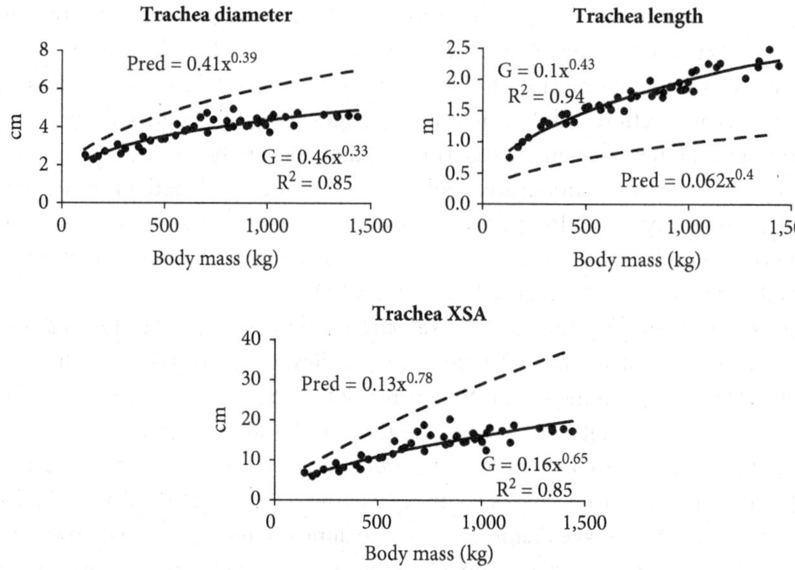

Figure 12.2. Trachea length, diameter, and cross-sectional area (XSA) in giraffes (●), and the predicted values for mammals of equivalent body mass (dashed lines). Giraffe data were obtained from 22 male and 24 female *G.c. giraffa*.
Predicted data from Tenney and Bartlett, 1967 in Calder, 1996.

breathe) increases. Any potential increase in the work of breathing could be offset if the rate at which trachea cross-sectional area increased was matched by the rate at which the volume of air (L min^{-1}) that has to be moved in and out of the lungs increased. It is in most mammals (Tenney and Bartlett, 1967), but in giraffes this ideal relationship seems to have been compromised. In giraffes, the ratio of the rate of change of cross-sectional area to ventilation is 0.65:0.82 (i.e., 0.8:1; see Tables 12.1 and 12.3) instead of the ideal 1:1, and we have to conclude that giraffes, because their trachea is narrow, have to work harder to breathe than we or other terrestrial mammals do.

Dead Space Volume

A reason why evolution has not found a way of matching cross-sectional area and ventilation and thus reducing the work of breathing in giraffes is, perhaps, because natural selection for an as-small-as-possible airway volume was/is a greater benefit. There are two kinds of airway volume. One is the volume of air that reaches the alveoli but does not take part in gas exchange because the alveoli are not being supplied with blood. The other is the volume of air that never reaches the alveoli. This latter volume is called anatomical dead space because it arises from the anatomical dimensions of the airways. The two volumes together are called physiological dead space. In all healthy mammals, alveolar dead space is negligible compared with anatomical dead space. Giraffes should have a large anatomical dead space because their tracheas are so long, and in the earliest measurements of airway volume of giraffes, Patterson et al. (1957, 1965) found that it was 2.6 mL kg^{-1} of body mass in a giraffe but just 1.0 mL kg^{-1} body mass in a cow of the

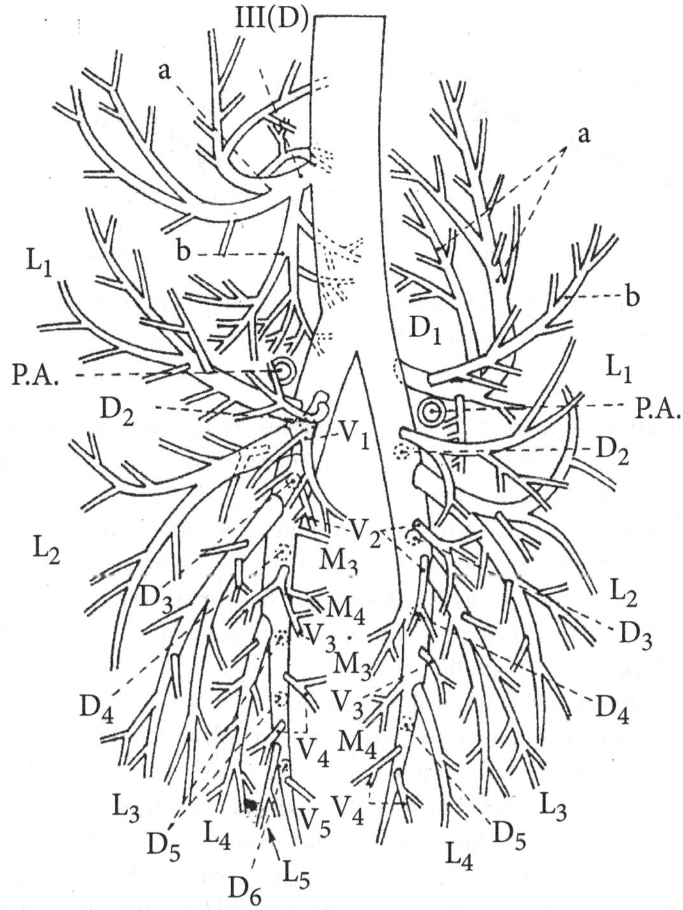

Figure 12.3. The bronchial tree of a giraffe.
From Nakakuki, 1983, with permission of Elsevier.

same body mass as the giraffe. Measuring just the size of trachea volume should, however, underestimate anatomical dead space in giraffes (and other mammals) because their bronchial tree is extensive (Nakakuki, 1983; Figure 12.3).

The total anatomical volume (trachea and bronchial tree air combined) is difficult to measure in any mammal, but has been done once in a small, unanesthetized giraffe (weighing 400 kg) by Hugh-Jones and colleagues and in a camel, a llama, and a red deer. They found that total dead space was 3 mL kg^{-1} of body mass in the giraffe, 3.2 mL kg^{-1} in the deer, 3.1 mL kg^{-1} in the llama, and 1.8 mL kg^{-1} in the camel, and thus giraffe airway volume was not unusually large, but much larger than it is in camels and larger than the trachea volume measured by Patterson et al. Langman et al. (1982), who measured but did not report dead space volume, also concluded that airway volume of giraffes was not abnormally large because the trachea, although long, was narrow. Nevertheless, Langman and colleagues decided, as had Hugh-Jones et al., that resistance to airflow would be high, especially during exercise, when larger volumes of air are moved. To meet that demand when a trachea is narrow, it is easier to move small

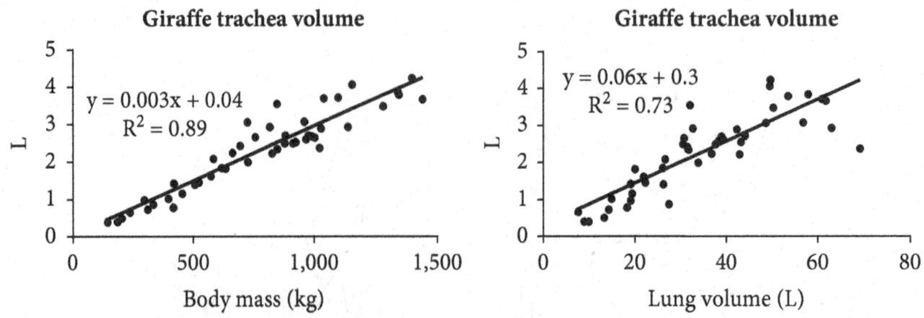

Figure 12.4. Changes in trachea volume as body mass (left) and lung volume (right) increase in 22 male and 24 female *G.c. giraffa*.

volumes often than large volumes less often: during exercise, it is easier for a giraffe to increase respiratory rate than to increase the size of each breath, which is the opposite to what we humans do if we need more oxygen.

Trachea volume calculated from trachea dimensions in 46 giraffes weighing between 147 kg and 1,441 kg increases linearly with both body mass and lung volume (Figure 12.4) and was on average 2.9±0.5 mL kg^{-1} body mass (range 1.9–4.2 mL kg^{-1}), which is much the same relative volume that Hugh-Jones and colleagues measured for total dead space in their giraffe.

Thus, at least in giraffes, tracheal volume and dead space volume are essentially the same: the size of the bronchial tree seems to add little to the dead space volume in them. If the dead space volume of giraffes is predicted from equations established from measurements made of tracheal volume in other animals of similar body mass, it would be 2.4±0.1 mL kg^{-1}, which is significantly smaller than is the measured trachea volume of giraffes. Consequently, dead space volume is bigger in giraffes than it is in mammals of similar size. However, its volume is much less than it would be if the relationship between trachea length and diameter that exists in other mammals was followed. In mammals generally, tracheal diameter is ~6% of trachea length (Tenney and Bartlett, 1967; in Calder, 1996). If that ratio were applied to giraffes, then trachea diameter of giraffes would be ~2.5 times bigger than it is, and trachea volume would be four- to sevenfold bigger than it is. A trachea volume of that potential size would mean that the size of a breath a giraffe would have to take would be bigger than is possible. In fact, trachea volume is on average just 400 mL larger (55 mL in a young giraffe and 700 mL in a large one) than it is in mammals of the same body mass (but shorter neck), a difference equivalent to about 0.5 mL kg^{-1} of body mass. So the relatively narrow tracheal diameter that giraffes have evolved has reduced the potential size of the dead space significantly, and that reduction was an essential part of the adjustments needed to make ventilation of their lungs possible when a long neck was evolving, but the cost was and is an increased work of breathing.

The Recurrent Laryngeal Nerves of Giraffes

In all mammals the first three cartilage rings in the trachea have become modified during evolution to form the larynx. In giraffes the larynx is composed of three unpaired cartilages—cricoid, thyroid, and epiglottis, and one paired cartilage, the arytenoid

(Erdogan and Perez, 2013). The cartilages can be moved by muscles—the cricothyroid, posterior and lateral cricoarytenoids, transverse arytenoids, and thyroarytenoid. In almost all mammals, contraction and relaxation of the laryngeal muscles changes the tension of vocal cords that are attached to the cartilages, and sounds of different pitch result. However, evolution also has made protection of the airways from entry of foreign bodies a high priority. In all mammals, if a particle of food enters the larynx or if there is another mechanical or artifactual stimulus, there is violent coughing, the laryngeal muscles contract, and a "laryngeal spasm" closes the larynx, completely preventing airflow. In a conscious animal, the result is distressing and even frightening. Avoiding it is best. How does a giraffe's larynx function to prevent food or a cud from entering it? Movements of the laryngeal cartilages depend on activation of the laryngeal muscles by the nerves that supply them, and few structures in the giraffe have been so talked about as the larynx and its nerve supply, especially the recurrent laryngeal nerves. Their pathway to the larynx has been used as evidence of both creation and evolution. In reality, the nerves are not that unusual and their pathway and functions are easily explained, are essential, and are not an anomaly as is often alluded. So why the fuss? The fuss is because giraffes have a very long neck, and an assumption has been made that the nerves must be twice as long as the neck (as illustrated by Berry and Hallam in 1986; see Dawkins, 2009).

In giraffes, as in all mammals, on both the left and right side of the larynx there are two laryngeal nerves (superior and inferior). Both nerves have their origin (their cell bodies) in the brain. The nerve fibers arising in the brain are carried in the vagus nerve, which is the 10th cranial nerve (see Chapter 16), and there are two of them, one on each side of the body. The nerve fibers in the vagus nerves innervate all the viscera and coordinate digestive functions, which includes preventing food entering the larynx. On each side of the larynx, the superior laryngeal nerve, after leaving the vagus on that side, innervates one of the laryngeal muscles (the cricothyroid) and typically detects sensations in the upper half of the laryngeal mucosa, such as those that arise when a particle of food, or worse, enters the larynx. The length of the superior laryngeal nerves is short: the nerves leave the vagus nerve as it passes the larynx. It is the inferior nerves that are called the recurrent nerves. The length of the fibers in the inferior nerves that innervate the larynx is long, and that would be an anomaly if the sole or main purpose of the recurrent nerves was innervation of the larynx. But that is not their main function. Their main function is to innervate the lower part of the esophagus where it passes through the thorax on its way to the rumen. In giraffes, the left inferior nerve (and presumably the right, but its anatomy has not been studied in as much detail) does not consist of a single branch but is formed by the union of four small nerves that emerge from the vagus. Two of these form the esophageal plexuses, which are nerve complexes that lie in the wall of the esophagus. The plexuses are the origin of sensory and motor nerves that supply the cervical part of the esophagus, a pattern of innervation that is found in all mammals. The rest of the esophagus is supplied by other branches of the vagus nerve and not by the recurrent nerves. Owen was the first to describe the anatomy of the recurrent nerves in giraffes. He wrote, and his description cannot be bettered:

> *The* (left) *vagus nerve as it passes down in front of the arch of the aorta sends off four small branches, which bend round the arch of the aorta on the left side of the ductus*

> *arteriosus; the two small branches on the left side pass to the oesophagus and are lost in the oesphageal plexus; the remaining two branches continue their recurrent course and ascend upon the side of the trachea, giving off filaments which communicate with branches from the neighbouring oesophageal nerves; these recurrent filaments also receive twigs from the oesophageal nerves and thus increase in size, and ultimately coalesce into a single nerve of flattened form, which enters the larynx above the cricoid cartilage and behind the margin of the thyroid cartilage.* (Owen, 1838)

On the right side of the neck, the recurrent nerve arises from the vagus nerve below the subclavian artery, which is the first major artery to branch off the aorta in giraffes, and it supplies blood to the forelimbs (Chapman, 1875). After leaving the vagus, the right recurrent nerve, like the left recurrent nerve, innervates the esophagus and its laryngeal branches and ascends to the larynx in the groove formed between the esophagus and trachea. Because both nerves pass between large arteries leaving the heart, it appears as if the nerves are pulled downward by the heart when it moves into the thorax during embryonic life, to form a U-shaped nerve with descending and ascending arms (see Dawkins, 2009). Of course the heart does not "move into the thorax"; it stays where it developed in the embryo, but embryos become fetuses and the neck elongates, especially in giraffes, and when it does, it carries with it the larynx and the nerves that innervate it. Elongation of the nerves occurs because, as expressed by Jeffrey Goldberg in 2003, in order to "wire up the nervous system properly, nerve cells must elongate their axons during development to reach their targets. They must elongate at the correct rate and stop growing at the proper time," so as to match the size of the animal. The process of elongation of a nerve is controlled by the genes in the nerve cell's nucleus. The precise mechanism that regulates the rate of elongation of a giraffe's recurrent nerves is not known, but we can be pretty sure that the general mammalian pattern will be followed: the genes controlling elongation of both neck and nerve synchronize nerve fiber elongation with the rate of neck elongation, with both neck and nerve growth slowing down as a giraffe ages, until it stops at maturity.

In two giraffes studied by Harrison (1981), the nerves that reach the larynx on the left side were 2 m long and on the right 1.7 m long, that is, they were as long as the neck, and that difference seems to be universal. In an adult human the recurrent nerve on the left is about 40 cm long and on the right is about 30 cm long. The difference in lengths arises simply because the point at which they leave the vagus nerve on each side of the neck differs slightly. The difference in lengths is a potential problem because, for the larynx to function properly and prevent foreign objects from entering it, all its muscles on both of its sides must contract simultaneously and symmetrically, and to make them do so requires that nerve impulses in both nerves reach the larynx muscles simultaneously. Thus, because the left recurrent nerve is longer than the right recurrent nerve, the impulses carried by it must travel faster than they do in the right nerve. In dogs the impulses do, at 65 m s^{-1} in the left nerve and 58 m s^{-1} in the right nerve (Harrison, 1981). The speed at which an impulse travels along a nerve fiber is proportional to its diameter: larger fibers conduct impulses faster than small ones. Is the diameter of the fibers in a giraffe's recurrent nerves large? Harrison (1981) measured the number and diameters of the fibers in the left and right recurrent nerves in humans and giraffes. He found that in humans the total number of fibers in each nerve was on average 1,242 and

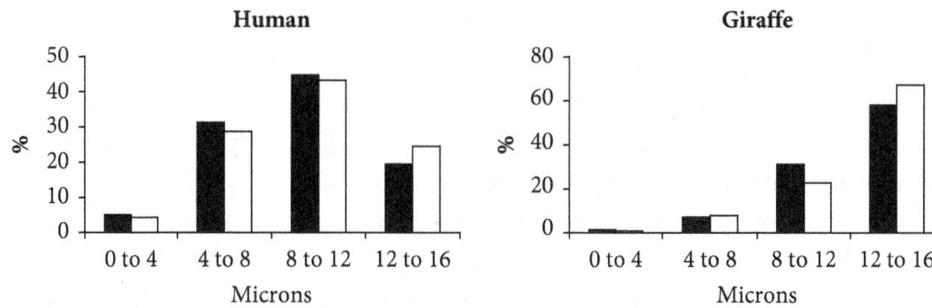

Figure 12.5. Composition of the right (■) and left (□) recurrent laryngeal nerves in humans and giraffes.
Constructed from data in Harrison, 1981.

in giraffes 911, this difference perhaps arising because the sounds humans can make are more complex. However, the nerves of giraffes had far more large fibers (>12 μm in diameter = 62% versus 22% in humans; Figure 12.5). The preponderance of large fibers in the giraffe nerves compared to the human nerves implies that during evolution, increases in nerve length were accompanied by an increase in nerve fiber diameter and faster conduction speeds. In addition, the longer left nerves had slightly more large fibers (65%) than did the right nerve (58%) in giraffes, as did the left and longer nerve in humans (24%) than the right, shorter one (19%). In both humans and giraffes, the higher number of large, faster-conducting fibers in the left nerve correlates with their slightly greater length.

Why have the inferior nerves of giraffes evolved to innervate both esophagus and larynx? The most likely reason is that while in most mammals food usually moves only down the esophagus, in a ruminant the motor nerves to the esophageal muscles not only move food down (swallowing) but also move food up (a cud) by peristalsis. When a cud is regurgitated, sensory receptors connected to the esophageal plexuses detect the stretch caused by the presence of the cud inside the esophagus. The stretch signals that a cud is coming up the esophagus and the impulses in the recurrent nerves activate the laryngeal muscles and epiglottis to close the larynx in the few seconds before a cud arrives and thus prevent the cud from entering the trachea. Ruminants propel a cud toward the mouth frequently, and they must have this early warning system, otherwise they will, inevitably, choke. In other words, during the evolution of ruminants (and perhaps in other mammals) the recurrent nerves have assumed the function of carrying sensory information from the esophagus to the larynx when a cud is regurgitated, in addition to doing their original job of innervating the larynx muscles and mucosa. This arrangement of nerves and muscles seems to have reached its apogee in giraffes. They are a creation of evolution.

Giraffe Lung Volumes

It might be expected that a giraffe's lungs are as big as they are in any mammal of equivalent body mass. In other mammals, lung mass is ~1% of body mass (Tenney and Remmers, 1963; Stahl, 1965, 1967). Are a giraffe's lungs 1% of its body mass? The length of the trunk of a giraffe is short for a mammal of their mass (Mitchell et al., 2017) because of the adjustments to the skeleton that have evolved to provide support for the neck. The effect of this shape on lung size was first observed by Joly and Lavocat (1846) and Cobbold (1854). To them, the lungs appeared small in proportion to the "caliber" of the thoracic cavity. Cobbold later confirmed this observation by measuring lung mass as an indicator of lung volume and, in three giraffes, found lung mass to be 0.56%, 0.82%, and 0.84% of body mass (Cobbold, 1860), thus much smaller than the expected 1%. Patterson et al. (1957, 1965) did find that the lung mass of a giraffe was the expected 1% of the body mass of the giraffe they studied; nevertheless, they commented that lung volume was "surprisingly small." Warren et al. (1957) made the same comment and added that total lung volume was 10 L, which is about 30% of their predicted volume, and the lungs would have weighed about 5 kg (Figure 12.6). It is quite likely, though, that the body mass of the giraffe was incorrect and was actually 700 kg or more. Thus lung mass was really 0.7% of body mass. A giraffe's shape also reduces abdominal volume (Murie, 1872). The reduced abdominal volume reduces the space available for the intestines and other abdominal organs, which are large and they push against the diaphragm, which in turn reduces the volume of the thoracic space. During his reconstruction of the skeleton of a giraffe, Vosmaer saw that the "ribs formed a right angle with the spine and are rigid, incapable of being spread to facilitate respiration" (Vosmaer, 1787). The rigidity reduces thorax flexibility and thus reduces the volume to which it and the lungs can expand. The size of the thorax of giraffes and the size of the lungs in it should, therefore, be smaller than they are in a mammal with a more conventional shape. However, Robin et al. (1960) found that the volume of the conducting airways in a giraffe of undisclosed body mass was nine times that of a "normal" adult human (i.e., 1.6 L), and they concluded that to overcome this large airway volume the lungs had to be disproportionately larger than expected for similar-sized mammals. The total lung volume they recorded was eight times (at 47 L) larger than that of human lungs (and also some five times bigger than the lung volume of giraffes recorded by Warren in 1957). Robin et al. also noted that the elastic fibers in the lung they studied were more numerous and coarser than they were in humans and "were admirably suited for extra work that might be required to ventilate a large dead space." So just how big are a giraffe's lungs?

In 1967 Walter R. Stahl, a mathematician at Oregon State University, reported an analysis of respiratory data collected from a large number of sources for mammals ranging in size from mice to horses. From those data he developed allometric equations to describe the relationship between body mass and respiratory variables. He showed that lung volume can be estimated from lung mass (Stahl, 1967). The mass of the lungs of giraffes weighing between 147 kg and 1,441 kg has been measured (Mitchell and Skinner, 2011) and the mass of the two lungs together ranged from 1.2 kg in the 147-kg giraffe (and thus 0.8% of body mass) to 9 kg in the adult (or 0.6% of body mass), which is significantly less than the 1% (1.6–15.2 kg) for the mass of the lungs in mammals of similar size (Figure 12.6). A gram of lung tissue inflates to a volume of ~7.5 mL (Stahl, 1967), and thus the lung volume of giraffes can be calculated to range from 8 to 70 L, compared to a predicted range of 10 L to 112 L (Figure 12.6). These data show that the lungs of giraffes are indeed small, and their small size is confirmed by measurements made by Gehr et al. in 1981. They determined lung volume by the method of Archimedes—by

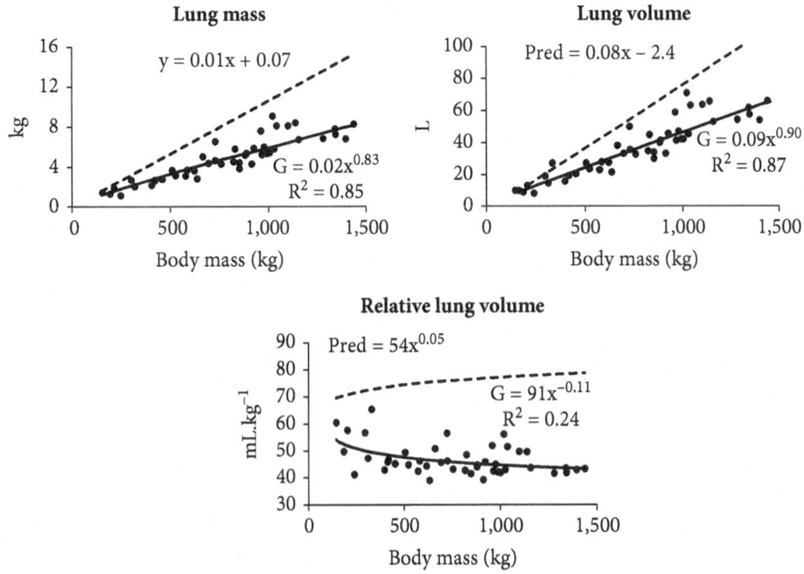

Figure 12.6. Lung mass (top left), calculated lung volume (top right) and the relative lung volume (bottom) in giraffes (●) and predicted lung mass and volumes for mammals of the same mass (dashed lines). Giraffe data were obtained from 22 male and 24 female *G.c. giraffa*. Predicted data from Stahl, 1967.

inflating lungs in a water bath. The volume of water displaced is lung volume. Gehr et al. measured the volume of the lungs of eight wild East African ruminant artiodactyls, including a small giraffe, by the volume displacement method and found that lung volume of those animals could be described by a linear equation (lung volume (mL) = 51.3 × Mb+619, R^2=0.90) or a power function (lung volume (mL) = 72.3 × $Mb^{0.97}$, R^2= 0.99). The small giraffe had a body mass of 383 kg and a lung volume of 21 L (the range predicted from the equations is 20 L to 23 L). When these two equations are applied to a giraffe weighing 750 kg, its lung volume is predicted to be 39 L–44 L, compared to the 35 L derived from extrapolation of its lung mass (5 kg) to lung volume (Figure 12.6). The predicted lung volume of a 750-kg mammal using Stahl's equation is 60 L, almost twice what it is in a giraffe of the same body mass.

Static Lung Volumes of Giraffes

Total lung volume in any mammal is made up of component volumes, which are termed static lung volumes. In humans, static volumes can be measured using a spirometer, and the results are displayed as a spirogram. The spirogram shown in Figure 12.7 shows a typical sequence of breaths in a healthy adult human starting with quiet (normal) breaths (V_T= tidal volume), followed by a deep breath in (IRV = inspiratory reserve volume) and a deep breath out (ERV = expiratory reserve volume).

These three volumes together make up vital capacity, which is the proportion of total lung volume that can actually inflate and deflate when a deep breath is taken and then

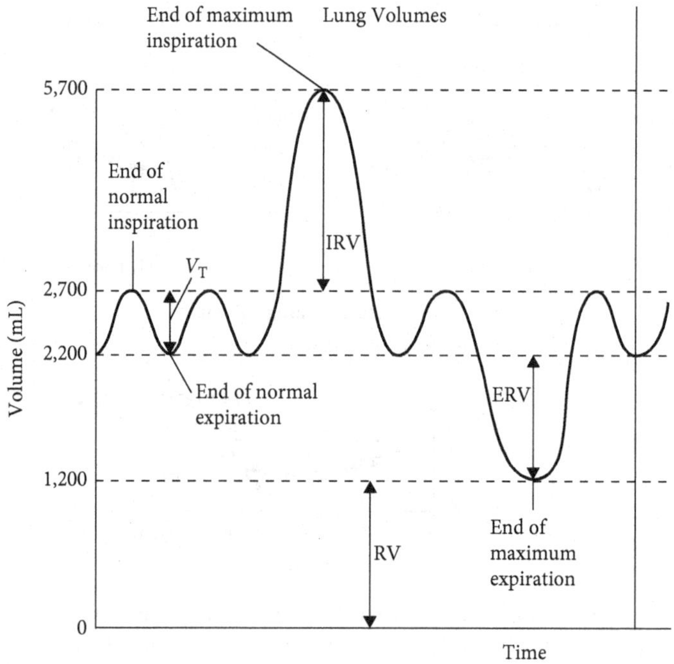

Figure 12.7. A typical human spirogram showing the main static lung volumes.
V_T = tidal volume; IRV= inspiratory reserve volume; ERV= expiratory reserve volume; RV= residual volume.

exhaled. It is impossible to breathe out all the air in the lungs because lungs are "glued" to the wall of the thorax by a viscous interpleural fluid, and because it is impossible to reduce thoracic volume to zero, it is also impossible to reduce lung volume to zero. Therefore, at the end of a maximum expiratory effort there is always some air still left in the lungs, and this volume is the residual volume (RV). Typically, in a human, it is ~20% of total lung volume. In the spirogram shown in Figure 12.7, total lung volume is the sum of V_T + IRV + ERV + RV and is 5,700 mL, of which RV is 1,200 mL.

It is not possible to measure all the static lung volumes in giraffes or any mammal other than humans, because non-human mammals cannot be persuaded to take the deep inspiratory and expiratory movements on command that are required to determine inspiratory and expiratory reserve volumes. Estimating these volumes is more feasible and depends largely on the relationships uncovered by Stahl. The vital capacity (VC) of giraffe lungs (the sum of V_T, IRV, and ERV) can be estimated from the allometric equation developed by Stahl (VC (mL) = $56.7 \times Mb^{1.03}$) and the exponent shows that it increases almost linearly with body mass, and thus forms a nearly constant proportion of lung volume (in fact, 87% ±2% in giraffes). Residual volume is the difference between lung volume and vital capacity and in giraffes forms 13% ±3% of lung volume (i.e., somewhat less than it does in humans). A third static volume that has been determined in mammals is functional reserve capacity (FRC, the sum of RV and ERV) and its volume fits the regression equation FRC (mL) = $24.1 \times Mb^{1.13}$ (Stahl, 1967). The scaling factor of this equation shows that FRC increases at a rate faster than the rate at which body mass increases, but it is, on average, 71% ±3% of lung volume in giraffes.

When these equations and proportions are applied to giraffe lungs, the volumes of vital capacity, residual volume, and functional reserve capacity and the allometric equations that define them are shown in Table 12.2. The proportions of lung volumes that vital capacity and residual volume form of lung volume are more or less similar to those predicted by Stahl's equations for mammals in general, but functional reserve capacity of giraffes is much larger. This latter difference may arise partly because Stahl's equations were not derived from giraffes or mammals with the body mass of large giraffes, but, a better explanation is that the large FRC of giraffes has evolved to optimize gas exchange.

Of the three components of vital capacity, only V_T has been measured in giraffes. V_T is relatively easy to measure, and Stahl developed an equation to describe its relationship with body mass in mammals based on 688 measurements. That equation is V_T (mL) = 7.69 × $Mb^{1.04}$, and if it is applied to giraffes ranging in body mass from 147 to 1,441 kg, then V_T would be 10.0±0.2 mL kg^{-1} of body mass. Giraffes, however, have a larger dead space than mammals in general, and their additional dead space ranges from 0.3 mL kg^{-1} body mass in a 147-kg giraffe to 1 mL kg^{-1} body mass in a 1,441-kg giraffe. If that additional volume is added to the V_T predicted for short-necked mammals by Stahl's equation, then the V_T of giraffes can be expected to be 10.7±0.4 mL kg^{-1} of body mass. In giraffes that probably weighed 700 kg to 800 kg, Patterson et al., in 1957 and 1965, measured V_T and found it to be much lower than the predicted volume at 2.7–4.1 L $breath^{-1}$ (3.5–5.5 mL kg^{-1}). Hugh-Jones et al. (1978) measured V_T to be 3.3 L in a giraffe weighing 400 kg (8.3 mL kg^{-1}). Langman et al. (1982) measured it to be 7.0±0.1 L for ~600-kg giraffes (11.7 mL kg^{-1} body mass), a volume similar to predicted volumes. Tidal volume has to exceed dead space volume comfortably to make sure that a sufficient part of inspired air reaches the lungs. The actual proportion of V_T that fills the dead space volume is illustrated by the ratio between the two (the V_D:V_T ratio). For mammals of the same body mass as giraffes, dead space volume is 24% ±0.5% of V_T. In giraffes it is 27% ±0.4% (range 22%–39%). The V_D:V_T ratio measured by Hugh-Jones et al. (1978) in a giraffe weighing 400 kg was 36%, and by Nyman et al. (2016) in a giraffe also weighing ~400 kg was 26%–33%. Thus a comfortable margin seems to be provided by a V_T that is three to four times bigger than is dead space volume, and giraffes meet that standard. Tidal volume includes not only the volume of air that fills the airways, but also the volume of air that fills the respiratory airways in the lungs. It is the parts that fill the bronchial tree and the respiratory airways that form a static lung volume, and that volume can be estimated from total V_T by subtracting tracheal volume. Tracheal volume is 2.9±0.5 mL kg^{-1} body mass in giraffes and thus the volume of air that enters the conducting airways and reaches the gas exchange surfaces during a single breath in a giraffe at rest (i.e., is not exercising) is 7.8±0.5 mL kg^{-1} body mass, or about 6 L in a 750-kg giraffe.

Expiratory and Inspiratory Reserve Volumes

Expiratory (ERV) and inspiratory (IRV) reserve volumes cannot be easily measured in non-human mammals, but they can be calculated. ERV forms, with residual volume (RV), functional reserve capacity (FRC) and, as RV and FRC of giraffes can be estimated with some degree of accuracy, ERV is the difference between them. When

Table 12.2 Approximate Volumes of Static Lung Volumes in Giraffes

Variable	Allometric Equations for Giraffes	Volume in a 147-kg Giraffe (L)	Proportion of Lung Volume in a 147-kg Giraffe (%)	Volume in a 1,441-kg Giraffe (L)	Proportion of Lung Volume in a 1,441-kg Giraffe (%)	Proportion of Lung Volume in a Human (%)
Lung volume (L)	$0.09 \times Mb^{0.90}$ $R^2 = 0.87$	8	100	63	100	100
Vital capacity (L)	$0.09 \times Mb^{0.87}$ $R^2 = 0.87$	7	86	50	80	79
Residual volume (L)	$0.003 \times Mb^{1.12}$ $R^2 = 0.92$	1	13	13	20	21
Functional reserve capacity (L)	$0.04 \times Mb^{0.97}$ $R^2 = 0.89$	5	58	45	71	39
Tidal volume (L)	$0.012 \times Mb -0.45.$ $R^2 = 0.99$	1	16	17	27	9
*Lung tidal volume (L)	$0.008 \times Mb -0.30$ $R^2 = 0.99$	1	11	11	18	6
Expiratory reserve volume (L)	$0.03 \times Mb^{0.96}$ $R^2 = 0.88$	4	45	32	51	18
Inspiratory reserve volume (L)	$3.6 \times Mb^{-0.08}$ $R^2 = 0.99$	2	30	2	3	53

Based on measurements made in 46 giraffes and on regression equations developed by Stahl, 1967 (see text).

*The volume of air reaching the lungs at each breath: tidal volume – dead space volume.

that subtraction is done, the unexpected result is that in giraffes ERV forms ~50% of lung volume (Table 12.2), and thus is a much larger component of lung volume in, presumably, all large quadrupedal mammals, than it is in a bipedal adult human (~18%; Table 12.2). IRV together with V_T and ERV form vital capacity. If V_T and ERV are known, then IRV can be estimated by subtraction, and that calculation, when it is applied to giraffe data, produces a similarly unusual result. IRV has a fairly constant volume of ~2 L in giraffes of any size but, as a proportion of lung volume, it declines from ~30% of lung volume in a juvenile giraffe to ~3% in an adult, so a small giraffe can take bigger breaths than can an adult giraffe. In an adult human, IRV forms ~50% of lung volume (Figure 12.7; Table 12.2). Thus in giraffes, and perhaps in other quadrupeds, ERV is large and IRV is small, instead of the other way around as they are in humans. If the volume of IRV is small, then use of it to increase V_T by taking a large breath, which is the easiest and the most efficient way of increasing air intake, as humans do, is not what adult giraffes can do, and this complication of their lung anatomy, together with a long and narrow trachea, means that if and when giraffes have to increase the amount of air breathed in and out over a period of time, they are obliged to increase both respiratory rate and the work of breathing. Giraffes could increase the volume of air breathed in and out by exhaling more air from the functional reserve capacity and in particular from ERV, as it is large. To use ERV would require giraffes to make forced expiratory efforts. Those efforts also would increase the work of breathing, an option mammals tend to avoid if they can. Even if giraffes could exhale more air from the ERV (and their rigid rib cage suggests that they cannot), it would be counterproductive to do so. Breathing is a cyclical process, alternating between inspiration and expiration, so there are times when no new air is being introduced into the lungs, but gas exchange must continue. Most gas exchange actually occurs during the expiration phase and between expiration and the next inspiration. Lung blood vessels are partially compressed during inspiration by the expansion of lung tissue. During exhalation the compression is reduced and it is then that the lung capillaries fill with blood and it is then that most gas exchange happens with the composition of gases in the FRC determining the efficiency of exchange. How long is the time available for gas exchange? If the respiratory rate of a 500-kg giraffe is ~10 min^{-1}, each breath should last ~6 seconds. Of those 6 seconds, 1.7 seconds is devoted to inspiration and 1.6 seconds to expiration (see Figure 12.8), with a pause of ~2.5 seconds before the next inspiratory movement begins. The longer this "respiratory pause," the more important FRC becomes for gas exchange. FRC increases as body size increases, and increases in body size are characterized, in a non-exercising giraffe, by decreases in respiratory rate and longer respiratory pauses, a combination that is ideal for optimizing gas exchange in a mammal as large as a giraffe.

The Respiratory Cycle of Giraffes

Giraffes obviously have little difficulty in supplying air to their lungs, but the dimensions of the trachea and the lungs of giraffes nevertheless do affect the delivery of air to the lungs, and the effects can be seen when the pattern of a giraffe's respiratory cycle is compared with the pattern in a mammal of similar size. That comparison has been

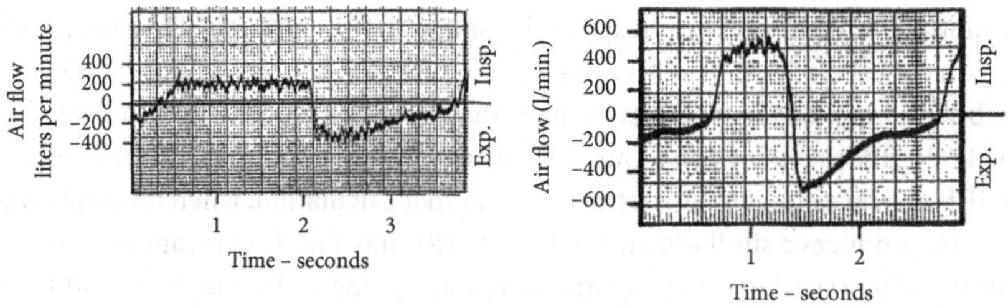

Figure 12.8. Inspiratory and expiratory patterns in a giraffe (left) and in a cow (right). Both the giraffe and the cow were assumed to have the same body mass of ~500 kg.
Reprinted from Patterson et al., 1965, with permission of John Wiley and Sons.

made with a cow of the same body mass as that of a giraffe (Patterson et al., 1965). The differences are obvious (Figure 12.8).

One difference is that even though the velocity of air flow (cm s^{-1}) is high in giraffes, the volume flow rate of ~200 L min^{-1} during inspiration was lower in the giraffe from which the tracing was obtained than it was in a cow (~500 L min^{-1}). Another difference is that the volume flow rate reached a plateau in the giraffe soon after the start of inspiration and a long, flat inspiratory effort followed. In this giraffe it lasted for 1.7 s (Figure 12.8). The shape and maximum volume flow rate during expiration was much the same in both the giraffe and the cow and lasted 1.4 sec in the cow and 1.6 sec in the giraffe. The ratio of inspiration to expiration time in giraffes was, therefore, close to 1:1 and the whole cycle took 3.3 s to complete. There was no respiratory pause because respiratory rate (calculated from the trace) was ~18 min^{-1}, and thus faster than the predicted 10 min^{-1} for a 500-kg giraffe. It was high because, as Patterson et al. acknowledged, the giraffe was under some stress and anxious at the time the measurements were made. Shortening inspiratory or expiratory time as a way of increasing respiratory rate is near to impossible for a giraffe, but reducing or eliminating the respiratory pause allows them to do so. In the cow, inspiratory time was half as long as expiratory time, the ratio of inspiration time:expiration time was 1:2, the plateau phase of flow rate was short, and the whole cycle took 2 s to complete. The cow's respiratory rate thus was ~30 min^{-1} when 10 min^{-1} was its predicted rate and the respiratory pause had disappeared, so it too must have been anxious.

Differences between the giraffe and cow inspiratory patterns arise partly because the narrow trachea of giraffes increases the resistance to airflow, but also because during inspiration there is a tendency for the trachea to narrow. It narrows because the high velocity of air flow through the trachea lowers its internal pressure. This phenomenon was described first by Daniel Bernoulli, a Swiss mathematician, and by an Italian physicist, Giovanni Venturi, in the eighteenth century in a completely different context. They were investigating why it is that a tube tends to collapse when the velocity of fluid flow through it increases, a phenomenon now known as the Venturi effect. Breathing out, despite the same velocity of airflow, has less of an effect on trachea radius, as the air pressure inside the trachea has to be greater than atmospheric pressure if air is to move out of the lungs back into the atmosphere. Consequently, trachea diameter increases, breathing out is easier than breathing in, and the length of time it takes equal-sized giraffe and cow to breathe out is very similar (1.6 s vs. 1.4 s, respectively). In all mammals at rest and breathing quietly, expelling air from the lungs is caused by recoil of the elastic fibers of the lungs. Giraffes do have relatively large and numerous elastic fibers in the lungs (Robin et al., 1960) but the fibers do not, it seems, shorten expiratory time more than do the elastic fibers in a cow's lungs.

It was once thought likely that the patterns of the respiratory cycle of giraffes were related to the anatomy of the larynx. Resistance to airflow is always highest in the upper airways, as anyone who has had a "headcold" will know. However, the larynx of giraffes is not unusually narrow (Harrison and Denny, 1985). Another (more likely) possibility is that the long inspiratory time is a consequence of low lung and/or chest wall compliance, i.e., the lungs or chest wall, or both, are more stiff. If the lungs of a giraffe were more stiff because of the more numerous and coarser elastic fibers present and the chest wall more rigid because of the anatomy of their rib cage, then their compliance would be low. Lung compliance is measured as the increase in volume produced by an increase in internal pressure. The pressure required to inflate a lung in any mammal is surprisingly low and usually is measured in units of cm of water (cm H_2O) rather than the more usual mmHg. However, if the compliance of a giraffe's lung is compared to that of a cow, no significant differences are apparent (Figure 12.9).

Over the 14 L of inflation during which the measurements shown in Figure 12.9 were made by Patterson et al. in 1965, the giraffe's lung volume increased by, on average, 475 mL for every cm of water pressure, an increase that is not statistically different from the increase in a cow (538 mL cmH_2O^{-1}). The compliance or flexibility of the chest wall of giraffes is not known, but because of its rib cage anatomy, it is likely to be much more rigid than it is in a cow, and the more rigid chest may well contribute to the prolonged time of inspiration. Intercostal muscles, which in other mammals contribute to air movement in and out of the lungs by expanding or reducing the size of the thorax, in giraffes probably play a minor, if any, role. Movement of air in and out of the lungs of a giraffe depends mainly on the muscles of the diaphragm and it is they that have to overcome the resistance to airflow that giraffe respiratory tract anatomy imposes. The characteristics of the muscles of a giraffe's diaphragm are not known, but we can assume that the muscle will be larger than it is in a mammal with a more conventional shape but

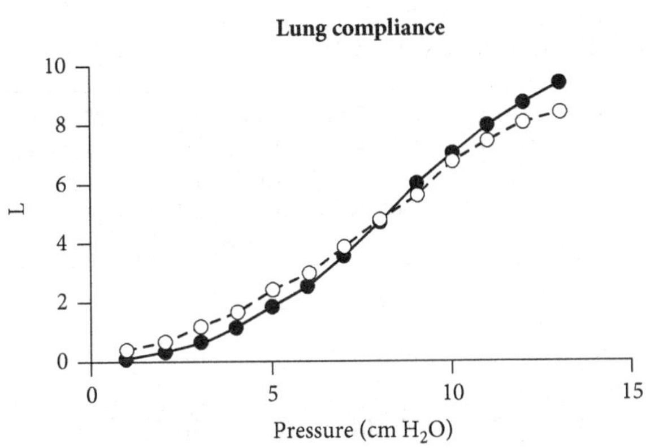

Figure 12.9. Lung compliance in a giraffe (●) and cow (○) of similar body mass.
Constructed from data in Patterson et al., 1965.

the same body mass. A longer and stronger contraction of their diaphragm and a longer time to breath in air seem to be the outcomes.

Dynamic Lung Volumes

Despite having a large dead space volume, and small lungs that are less able to expand than those of other similar-sized mammals, giraffes clearly function perfectly well, so the potential problems created by the structure of their respiratory tract have been overcome by evolution. How do giraffes ensure that delivery of oxygen to the alveoli is sufficient when the structure and size of a giraffe's respiratory system so clearly affects their breathing? The success of breathing depends on the volume of air that can be moved in and out of the lungs per unit time (L min^{-1}), a volume referred to as the dynamic (i.e., time-dependent) volume. There are two dynamic lung volumes. One is the minute ventilation (V_E) and it is the total volume of air moved in and out of the airways and lungs each minute. It is equal to V_T multiplied by respiratory rate. The size of V_E in a giraffe, measured by Hugh-Jones et al. in 1978, was 50 mL min^{-1} kg^{-1} of body mass. V_E at rest has also been measured by Langman et al. (1982) in three giraffes weighing 570 kg, 580 kg, and 640 kg, and its relative volume was 108.8±4.2 mL min^{-1} kg^{-1} of body mass. Why the volume measured by Hugh-Jones et al. was so small is not clear. The relative minute volume found by Langman et al. is the same as that calculated from the equations shown in Table 12.3 for giraffes of those body masses and indeed for giraffes ranging in body mass from 147 to 1,441 kg (1,07.7±11.3 mL min^{-1} kg^{-1} of body mass). Minute volumes of that relative size translate into absolute volumes of 16 L min^{-1} to 156 L min^{-1} for giraffes of those two body masses.

Minute volume is not the volume that reaches the gas exchange regions of the lungs and, specifically, does not reach the alveoli each minute, because about a third of inspired air stays in the dead space of the trachea and bronchi. Thus, the far more important volume for gas exchange is the volume of air reaching the alveoli each minute, or alveolar ventilation rate (V_A). Its size can be calculated from V_T minus the dead space volume (DS) multiplied by the respiratory rate (RR), and is summarized in the formula

Table 12.3 Dynamic Lung Volumes in Giraffes

Variable	750-kg Giraffe	Range for 147–1,441-kg Giraffes	Allometric Equations for Giraffes in That Mass Range	750-kg Mammal	Allometric Equations for Predicted Values
Respiratory rate (min^{-1})	9	15–8	$49 \times Mb^{-0.25}$	10	$51 \times Mb^{-0.25}$
Minute ventilation rate (L min^{-1})	78	20–132	$0.34 \times Mb^{0.82}$	73	$0.39 \times Mb^{0.79}$
Alveolar ventilation rate (L min^{-1})	59	15–101	$0.26 \times Mb^{0.82}$	58	$0.27 \times Mb^{0.81}$

Based on allometric respiratory rate and tidal volume and measured trachea volume in 46 giraffes (see text). Predicted values for other mammals from Stahl, 1967; Calder, 1981, 1996.

$V_A = (TV - DS) \times RR$. For example, in a non-exercising adult human, the volume of air in a single breath is of the order of 500 mL and respiratory rate is about 15 breaths per minute; therefore, V_E is 7.5 L min^{-1}, and as dead space volume is about 150 mL, V_A is 5.3 L min^{-1} and not 7.5 L min^{-1} (i.e., ~30% smaller). The respiratory rate of giraffes follows typical mammalian patterns, with small animals having a higher rate than large ones (Robin et al., 1960; Stahl, 1967; Hugh-Jones et al., 1978; Langman et al., 1982; Worthington et al., 1991; Calder, 1996), and is summarized by the equation $RR = 51.2 \times Mb^{-0.25}$ (Stahl, 1967) or the similar equation developed from a dataset that included data from two giraffes: $RR = 47.6 \times Mb^{-0.24}$ (Worthington et al., 1991). The respiratory rates predicted by these two equations are not significantly different, so a combined equation can be formulated, which is $RR = 49.4 \times Mb^{-0.25}$, and it predicts that resting respiratory rate in giraffes will range from 15 min^{-1} in juveniles to 8 min^{-1} in adults. Direct observations of giraffe respiration rate reported by Robin et al. (1960), Anon (1974), Hugh-Jones et al. (1978), Langman et al. (1982), Worthington et al. (1991), and Mitchell and Skinner (2011) confirm the predictions.

The purpose of alveolar ventilation is to make sure that oxygen used by metabolism is replaced. The volume of oxygen needed by a giraffe quietly going about the business of its day is proportional to field metabolic rate, and to meet that demand, a 750-kg giraffe needs about 6,000 L O_2 day^{-1} or 4 L O_2 min^{-1} (see Chapter 8). If alveolar air had the same composition as ambient air, it would contain 210 mL $O_2 L^{-1}$ at sea level if the ambient air was dry, so to provide 4 L $O_2 min^{-1}$ for a 750-kg giraffe, ~20 L of air needs to reach the alveoli each minute, or about 30 mL min^{-1} kg^{-1} of body mass. From the data shown in Tables 12.1, 12.2, and 12.3, the V_A of a 750-kg giraffe can be calculated to be ~80 mL min^{-1} kg^{-1} of body mass (~60 L min^{-1}) and the volume of oxygen delivered would be ~14 L min^{-1}, which is about three and a half times more than the 4 L min^{-1} required. The composition of air in alveoli is, however, quite different from the composition of ambient air. One reason that the compositions differ is that as air passes through the airways it becomes completely saturated with water, and the water vapor exerts a partial pressure of 47 mmHg. Following Dalton's Law of partial pressures, the presence of a non-reacting gas such as water vapor in a mixture of gases reduces the proportions of all the other gases present in the mixture and reduces the volume of oxygen in particular by 8% from 210 to 194 mL L^{-1} (in inspired air at sea level). Alveolar air also contains the carbon dioxide that has been generated by metabolism and delivered to the alveoli by blood for excretion into the atmosphere. Its partial pressure in the alveoli at sea level is, typically, 40 mmHg. The addition of water vapor and carbon dioxide to the air in the lungs reduces the partial pressure of oxygen to 110 mmHg, its proportion in alveolar air to 14%, and its volume to 144 mL L^{-1} of alveolar air. Consequently, the alveolar ventilation rate of oxygen (V_AO_2) in a 750-kg giraffe is reduced from 14 L min^{-1} to approximately 9 L min^{-1}, or about twice what is required to sustain field metabolic rate in a non-exercising giraffe.

That seems to be a sufficiently high rate. However, another complication is that alveolar air is contaminated by rumen gases. Fermentation of browse in the rumen produces carbon dioxide and methane. On average the ratio is 65% carbon dioxide, 30% methane. The other rumen gases are nitrogen and oxygen. Rumen gas is produced at a rate of around 0.75 L min^{-1} (i.e., ~1,000 L day^{-1}) and it is eliminated by eructation. Patterson et al. (1965) have estimated that of this eliminated amount, three to seven times as much enters the lungs as leaves the mouth directly, and when it is recycled through the lungs it reduces the concentration of oxygen in alveolar air. By how much oxygen concentration is reduced has so far not been established, but the presence of rumen gases in the lungs may be one reason why the oxygen content of blood is lower than it is expected to be. Discounting that effect but taking all other factors into account, oxygen delivery exceeds that required by field metabolic rate by 1.9±0.1-fold, a multiple that provides an adequate buffer for any eventuality.

The relationship between oxygen demand and oxygen delivery can of course be (and is frequently) adjusted. Increased demand, for example, stimulates an increase in respiratory rate. This response to increased demand in mammals usually is mediated by a carotid body, which contains sensors that are activated by too little oxygen or too much carbon dioxide in blood and/or a pH of blood that is too acidic. In mammals the carotid body, typically, is located in the carotid artery near the carotid sinus, and its position there presumably has evolved so that its sensors can monitor the pH and the oxygen and carbon dioxide content of blood going to the brain. Although a carotid body has been looked for in giraffes, so far one has not been found (Lawrence and Rewell, 1948; Goetz, 1955), nor has one been found in okapis (Lawrence and Rewell, 1948). (Its apparent absence in giraffes was confused by both Goetz et al. (1960) and van Citters et al. (1969) to mean the absence of a carotid sinus in giraffes, the site of the sensors of blood pressure—see Chapter 9.) Nevertheless, a sensor of oxygen in blood is present somewhere in giraffes because when Patterson et al. (1957) reduced the oxygen content of the air a giraffe was breathing from the normal ~20% to 10%, alveolar ventilation rate more than doubled from 77 L to 169 L min^{-1} within 45 seconds.

All in all, when the unique characteristics of the respiratory system of giraffes are considered, what can be shown is that giraffes react to oxygen demand in the same way that other mammals do, and despite a small lung volume and long trachea, more than enough air and its oxygen can be delivered to the lungs to support metabolism. The only apparent costs of its quite unique respiratory anatomy are that there is little scope to increase the size of a breath, and the work of breathing—the muscular effort required to breathe—is more than it is in other mammals of equivalent size.

Gas Exchange

Gas exchange between air in the alveoli and the oxygen and carbon dioxide in blood takes place across the membrane that separates the air from blood. That membrane consists of the very thin wall of the alveolus and the equally thin wall of the capillaries that surround an alveolus. Most oxygen and about 25% of carbon dioxide are transported in red blood cells (erythrocytes). An erythrocyte spends no more than 0.75 s in a capillary that surrounds an alveolus, during which time gas exchange has to be completed. How is exchange accomplished in that short time? A sensible mechanism might be one that actively pumps oxygen and carbon dioxide across the alveolar wall and erythrocyte membrane, but it turns out that in terrestrial mammals, evolution has solved the problem differently. The exchange of oxygen and carbon dioxide in the lungs is by diffusion, a passive process requiring no energy. However, diffusion requires a large surface area to be effective, and the larger the animal, the greater the diffusing surface area needed

(Hoppeler and Weibel, 1998). Thus in almost all mammals, body mass and lung volume increase linearly (Tenney and Remmers, 1963; Stahl, 1967; Gehr et al., 1981) and, consequently, so does the surface area for diffusion. However, the volume of giraffe lungs is much smaller than expected, and their size increases at a slower rate than the rate at which body mass increases (Figure 12.6), which means that increases in surface area do not match increases in body mass, yet plainly adult giraffes add oxygen and remove carbon dioxide efficiently. How?

Pressure Gradients for Diffusion

If oxygen is to enter blood from the lungs and carbon dioxide is to leave the blood by diffusion, then in addition to a large surface area, appropriate pressure gradients for diffusion must exist. The pressure of oxygen in the alveoli (P_AO_2) must be greater than it is in the deoxygenated blood (P_VO_2) arriving at the alveolus via the pulmonary artery. Similarly, if carbon dioxide is to leave the blood and diffuse into an alveolus, then its pressure in the pulmonary blood (P_VCO_2) reaching an alveolus must be greater than it is in an alveolus (P_ACO_2). Invariably in mammals, $P_AO_2 > P_VO_2$ and $P_VCO_2 > P_ACO_2$. The pressure gradient needed to ensure complete diffusion of carbon dioxide out of blood (typically 6 mmHg) is much less than that needed to get oxygen into blood (typically 60 or 70 mmHg) because carbon dioxide is many times more soluble in body fluids than is oxygen. Ideally, if the transfer of gases were perfect, then after exchange of gases has taken place the gas pressures in the alveoli should be the same as those in the arterial blood leaving the lungs. They are the same for carbon dioxide but not for oxygen; the post-exchange pressure of oxygen in arterial blood (P_aO_2) is lower than it is in the alveoli, partly because the solubility of oxygen is low, but also because the pressure of oxygen in alveolar air is lower than expected. Ewald Weibel, a Swiss physiologist famous for (among many other accomplishments) his contributions to our understanding of the mechanisms underlying oxygen delivery to muscles, showed that the size of the difference between P_AO_2 and P_aO_2 in any mammal is related to body mass and the difference (ΔPO_2) fits the regression equation ΔPO_2 (mmHg) = $25.1 \times Mb^{-0.22}$ (Weibel et al., 1981). Applying that equation to giraffes, the difference in pressure between alveolus and blood would range from ~8 mmHg in young giraffes to ~5 mmHg in adult giraffes, with the smaller difference in adults attributable to a larger surface area for diffusion. If the difference between the pressure of oxygen in the alveoli (~110 mmHg) and blood is 5–8 mmHg, the P_aO_2 of blood leaving the lungs en route to the body should be about 102–105 mmHg in a giraffe. Is it? It is hard to say, because although many measurements of blood gas pressures have been made in giraffes, most of them have been made in anesthetized animals and no matter what anesthetic agent is used to sedate a giraffe, respiration is depressed, carbon dioxide content of arterial blood and alveoli increases, and P_aO_2 falls unless supplementary oxygen is given (e.g., Bush et al., 2001; Williams et al., 2007; Brondum et al., 2009; Nyman et al., 2016). Only three measurements of arterial blood gases have ever been made in non-anesthetized giraffes, and these are shown in Table 12.4, compared with those made in a cow of similar body mass. Given the exacting field conditions under which the giraffe data were obtained, it is fair to say that in giraffes exchange of oxygen and carbon dioxide with blood follows the mammalian

Table 12.4 Blood Gas Pressures in Three Giraffes

Giraffe #	Arterial O_2 (mmHg)	Arterial CO_2 (mmHg)	Venous O_2 (mmHg)	Venous CO_2 (mmHg)
1	86	40	48	47
2	82	28	27	55
3	93	46	27	48
Mean	*87*	*38*	*34*	*50*
Cow	81	43	34	55
Ideal	*100*	*40*	*40*	*46*

Data from Patterson et al., 1965, with permission of John Wiley and Sons.

pattern, but arterial oxygen pressure (P_aO_2) in giraffes (and the cow) is much lower than expected, perhaps partly because of the presence of rumen gases in the lungs.

Surface Area for Diffusion

The surface area across which oxygen and carbon dioxide diffuse has two parts: the surface area of the alveoli, which are spherical in shape, and the surface area of the dense network of blood capillaries that surround and enclose each alveolus. Those two surface areas become functional only when an alveolus is ventilated with air and the capillaries that surround it are filled with blood. For the lungs as a whole, the relationship between ventilation and blood flow is defined as the ratio of alveolar ventilation rate (V_A; L min^{-1}) to cardiac output (Q; L min^{-1}). This ratio in free-living giraffes can be calculated from estimates of V_A and cardiac output, is independent of body mass, and V_A is 1.6±0.1 times larger than cardiac output. The V_A to Qratio has been measured in giraffes only once, and that was in an anesthetized female giraffe weighing 400 kg (Nyman et al., 2016). The ratio was ~1, but it is likely that the anesthesia had depressed respiration and the true ratio was higher. In humans at rest the ratio is 0.8 and, as the relative cardiac output is the same in humans and giraffes, the higher V_A:Q ratio of giraffes means that the small lungs of giraffes are actually better ventilated than are the lungs of humans. The actual surface area for diffusion depends on the number of alveoli. There are approximately 170 alveoli cm^{-3} of lung volume in humans (Ochs et al., 2004) and a similar number in all other mammals, including giraffes (Gehr et al., 1981). The size of each individual alveolus does not change as lung volume increases, but the number of alveoli does (Ochs et al., 2004). How the volume of the lungs of giraffes increases as they grow is known, thus the total number of alveoli (billions) in their lungs can be estimated ($= 0.015 \times Mb^{0.90}$). In a 750-kg giraffe there are ~6 billion, and the number ranges from 1.4 billion in a newborn giraffe to 11 billion in a mature adult. Measurements of the surface area (m^2) that the alveoli provide and which is available for diffusion of gases were made in a series of experiments by the legendary C. Richard Taylor and colleagues at Harvard's Museum of Comparative Zoology. They managed to persuade a variety of wild animals (in fact, 21 animals from 12 different species) to run on a treadmill and while running to have their oxygen consumption and many other respiratory variables measured. How this was achieved is truly the stuff of physiological legend. One of the species they studied was a young giraffe weighing 383 kg. It did not run on the treadmill, but it did have many other measurements made of its respiratory functions. Its lung volume was 21 L, alveolar surface area was 636 m^2, capillary surface area was 552 m^2, and the alveolar membrane thickness was 0.596 μm. The rate at which oxygen could diffuse across the membrane was ~1,400 mL min^{-1} $mmHg^{-1}$ of pressure gradient. Assuming that the maximum pressure gradient of oxygen for diffusion is ~70 mmHg, then the lungs of that 383-kg young giraffe had the capacity to diffuse ~100 L O_2.min^{-1} when just 4 L min^{-1} would be needed to sustain its field metabolic rate.

The size of all of the structures involved in gas transfer in mammals increases as body mass increases, except for the thickness of the alveolar membrane that separates alveolar air from blood. It must always be thin. Its thickness is independent of body mass and is ~0.6 μm in any mammal. The surface area realized by all the alveoli fits the regression equation SA (m^2) =3.6 × $Mb^{0.}$92 and that of the surrounding capillaries (SA (m^2) =3.2 × $Mb^{0.92}$) and are 1.5±0.1 m^2 kg^{-1} of body mass and 1.4±0.1 m^2 kg^{-1}, respectively, or ~1,000 m^2 each in a 750-kg giraffe. When Gehr and colleagues made those measurements in 1981, they wondered if the ratio of the surface area for diffusion (m^2) to the rate of oxygen consumption (L O_2 min^{-1}) was the same in neonatal mammals and adults. It should be because, like so many other anatomical structures, the lungs of a newborn giraffe must be sufficiently well developed to equip it with all that it needs to escape predation. Gehr et al. did not have data for giraffes of different ages to answer the question, but these data do exist now for growing giraffes (Mitchell and Skinner, 2011). It turns out that the ratio is not the same in neonatal and adult giraffes. It increases with age from ~27 m^2 L^{-1} O_2 min^{-1} in a 147-kg giraffe to ~33 m^2 L^{-1} O_2 min^{-1} in a 1,441-kg adult, but this smallish difference clearly makes no difference to a neonate's ability to acquire oxygen.

Gas Transport

Oxygen Transport

About 200 mL O_2 L^{-1} of blood must be carried in the arterial blood of mammals to ensure that their tissues receive enough oxygen. That concentration of oxygen happens to be the same as that found in air. In other words, mammals have evolved a way of transferring the same amount of oxygen that is in a liter of air into a liter of blood. How is that oxygen carried in blood? Oxygen is very insoluble in body fluids like the plasma of blood. Its solubility coefficient is 0.03 mL $mmHg^{-1}$ of P_aO_2 L^{-1} of plasma, and as P_aO_2 should be ~100 mmHg (Table 12.4), just 3 mL of oxygen is carried dissolved in every liter of plasma. The other 197 mL of oxygen must be carried in another way, and that way is by attachment to hemoglobin. The heme part of hemoglobin is a porphyrin and almost identical in structure to chlorophyll, but with iron as the metal ion at its center, rather than the magnesium that is at the center of chlorophyll. Each hemoglobin molecule contains four molecules of heme, and they make up about 4% of the total mass of the hemoglobin molecule. Each of the four heme molecules can bind reversibly with a molecule of oxygen, giving up its oxygen in the tissues where oxygen has been depleted and picking up oxygen from the oxygen-rich air in the lungs. It is this magical and absolutely essential property of heme that makes the collection of oxygen in lungs and its delivery to tissues possible. The other 96% of the hemoglobin molecule is made up of globin, which is a protein made up of four amino-acid chains. (Unraveling the structure of hemoglobin earned a Nobel Prize in Chemistry in 1962 for Max Perutz and John Kendrew.) The globin keeps the whole molecule soluble, binds hydrogen ions, and so helps to keep body fluid pH constant, and regulates the rate at which oxygen binds to heme. Hemoglobin absorbs all light wave lengths except red. It is this property that gives red cells, and hence blood, their color. (Chlorophyll, typically, absorbs all wavelengths except green.) Heme is not only a constituent of hemoglobin. It is also the active part of the cytochromes that synthesize ATP, and is the center molecule of myoglobin. Myoglobin is found in muscle cells. There it functions to transport oxygen arriving at the membrane of a muscle cell to the cell's mitochondria. It is also able to store oxygen inside muscle cells, which it releases when needed, for example during exercise.

The total hemoglobin molecule has a molecular mass of about 68,000 Daltons, thus it is roughly the same size as a molecule of albumin (70,000 Daltons), and it is very soluble in body fluids. Unlike a molecule of albumin, a molecule of hemoglobin is small enough to be filtered in the kidney, and if it were to be filtered, it would precipitate in the acidic filtrate and block the kidney nephrons. Furthermore, if it were to be floating free in blood plasma, it would exert a large colloid osmotic pressure in the same way that albumin in plasma does, and that pressure would alter the dynamics of fluid exchange in organs and in tissues (see Chapter 10). Evolution has solved both of these potentially fatal problems in mammals, birds, reptiles, amphibians, and fish by enclosing the hemoglobin (or equivalent oxygen transport molecule) inside erythrocytes (or equivalent cell). There it can neither be filtered nor exert an osmotic pressure. The amount of hemoglobin in blood, and hence the oxygen-carrying capacity of blood, is determined by the number of individual red cells present (erythrocyte count), their size (mean cell volume) and the amount of hemoglobin in each of them (mean cell hemoglobin). The proportion of blood that is cells is the hematocrit. The remainder is plasma, and typically the ratio of cells to plasma in mammals is 40:60. Giraffes are no exception (see Table 12.6).

It might be expected that as all mammals have a similar requirement for oxygen, a result of evolution would have been the same number and same size of erythrocytes in all mammals. However, as Table 12.5 shows, among East African ruminants no two are alike. They do not have the same red cell morphology, and their morphology differs from that in humans. The reasons why natural selection has resulted in variation are not yet resolved, but an important one is that red cell morphology is correlated with the diameter of a blood capillary. The two have evolved together. A red cell must be small and flexible enough to move along a capillary, and thus capillary diameter and red cell shape have evolved in each species in a way that optimizes delivery of oxygen to tissues. Despite the considerable variation in erythrocyte size and shape, evolution has, however, ensured that all healthy mammals have ~150 g of hemoglobin in each liter of blood (~15 g 100 mL^{-1}; Table 12.5), and as each gram of hemoglobin can carry ~1.3 mL of oxygen (or 10^{18} molecules of oxygen), ~200 mL or more of oxygen can be carried in each liter of blood.

Table 12.5 The Morphology of Red Blood Cells in Giraffes Compared with Other Species

Species	Erythrocyte Count (10^{12} L^{-1})	Hematocrit (%)	Hemoglobin (g 100 mL^{-1})	Mean Cell Volume (10^{-15} L)	Mean Cell Hemoglobin (10^{-12} g)	Mean Cell Hemoglobin Concentration (%)
Humans	5	45	15	90	30	32
Eland	8	37	14	45	17	37
Gazelle	10	43	16	43	16	38
Hartebeest	10	42	16	44	17	39
Topi	12	38	14	32	12	36
Giraffe	12	39	14	35	12	35
Wildebeest	13	40	15	30	11	37
Impala	21	40	14	19	7	36

Data from Drevemo et al., 1974, and Hawkey, 1975.

Table 12.6 The Morphology of Giraffe Red Blood Cells

Variable	Goetz, 1955 (*N*=1)	Drevemo et al., 1974 (*N*=4)	Hawkey, 1975 (*N*=3)	Bush et al., 1980 (*N*=38)	Weighted Mean (*N*=46)
Erythrocyte count (10^{12} L^{-1})	12.0	10.2±1.1	9.4±0.9	12.4±3.4	12.0±3.0
Hematocrit (%)	47.0	43.0±3.4	37.6±0.5	38.8±5.8	39.3±5.1
Hemoglobin (g dL^{-1})	17.8	14.4±3.0	13.3±0.2	13.5±2.4	13.7±2.4
Mean cell volume (10^{-15} L)	39.2	42.2	39.4±4.8	33.3±8.9	34.6±7.7
Mean cell hemoglobin (10^{-12} g)	14.8	14.1	14.1±0.6	11.4±3.1	11.9±2.6
Mean cell hemoglobin concentration (%)	35.4	33.5	34.8±2.1	34.6±2.1	34.5±1.8

There have been four systematic studies of the characteristics of giraffe red blood cells, of which the most reliable is that of Bush et al. (1980), who analyzed 38 samples from 14 giraffes of both sexes and included neonates, juveniles, and adults (Table 12.6).

Those four analyses show that giraffe blood is characterized by having a large number of small red blood cells. The shape of a mature red cell is a flattened disc with concave surfaces, a shape that increases the surface area to volume ratio and optimizes the diffusion distance between the cell's surface and the hemoglobin inside each cell. Just as important as the shape of a red cell and the amount of hemoglobin it contains is the surface area of each red cell because it is across that surface area that oxygen has to diffuse to reach hemoglobin in the lungs or to diffuse out of a red cell into tissues. What is the size of that surface area? As giraffe erythrocytes are many and small, their total surface area will be much larger than if the cells were fewer and large. The size of the surface area of a single red cell depends on its volume, diameter, and thickness. Udroiu

Table 12.7 Allometric Relationships between Red Blood Cell Size (Mean Cell Volume, MCV) and Their Diameter, Thickness, and Surface Area

Variable	Allometric Equation	R^2	Calculated Average Values from Giraffe MCV	Average for Humans Based on Human MCV
Diameter (μm)	$1.34 \times MCV^{0.37}$	0.95	5.0	7.0
Thickness (μm)	$0.68 \times MCV^{0.27}$	0.59	1.8	2.3
Surface area (μm²)	$5.92 \times MCV^{0.70}$	0.98	71	135

Equations derived from data in Udroiu, 2014. Giraffe MCV from Table 12.6.

(2014) has calculated the relationships between red cell volume, diameter, thickness, and their surface area for several mammals, and regression equations can be fitted to those data. When those equations are applied to giraffes, the surface area of each of their erythrocytes is ~70 μm^2 (Table 12.7).

Once the surface area of a single red cell is known, then the total surface area of all red cells in the blood can be calculated, and in a 750-kg giraffe it is an astonishing 50,000 m^2 and has a range of about 10,000 m^2 in a newborn giraffe to about 100,000 m^2 in a mature adult. This entire surface area passes through the lungs every one and a half minutes and is exposed to the oxygen in the alveoli, and also every one and a half minutes it passes through the capillaries that supply blood to the cells of the giraffe's body. Giraffe capillary diameter is as small as an erythrocyte and the capillaries numerous, so capillary surface area is also large (see Chapter 13). Natural selection, in other words, by producing small erythrocytes and numerous capillaries in giraffes, has provided two further adaptations which ensure that oxygen and nutrient delivery to the tissues of giraffes always is excellent.

Carbon Dioxide Transport

The main gaseous product of metabolism is carbon dioxide, and it too has to be transported in blood to the lungs for excretion to the atmosphere. At any one moment in a mammal there are about 450 mL of carbon dioxide in every liter of arterial blood and 500 mL L^{-1} of venous blood. If all this carbon dioxide were to be transported free in plasma, it would interact with water to form carbonic acid, and the plasma would become highly acidic and not compatible with life. However, just 5% is transported dissolved in plasma. Its solubility coefficient is 0.7 mL $mmHg^{-1}$ L^{-1} (vs. 0.03 mL $mmHg^{-1}$ L^{-1} for oxygen), but the pressure exerted by carbon dioxide in arterial blood is only 40 mmHg and never much more than 46 mmHg in venous blood. Thus, the total amount of dissolved carbon dioxide is around 30 mL L^{-1}. The remaining 420–470 mL of carbon dioxide has to be transported by other means, and two other ways dominate. One way is by attachment to hemoglobin. When hemoglobin gives up its oxygen in tissues, the shape of the globin chains changes and exposes the amino groups (NH_2) in the amino acids that make up the globin protein. These NH_2 groups interact with carbon dioxide to form $NHCOO^-$ (carbamino) compounds, and in this way another 75 mL L^{-1} of carbon dioxide are transported. The remaining carbon dioxide is transported as bicarbonate, which is synthesized inside red blood cells by the reaction:

$$CO_2 + H_2O \leftrightarrow H^+ + HCO_3^-$$

Theoretically, the hydrogen ion produced (H^+) will lower body fluid pH. However, hemoglobin and plasma proteins are excellent buffers and absorb the H^+ ions as fast as they are made, so the pH does not change. Of the ~400 mL of carbon dioxide carried as bicarbonate, 100 mL stays inside the red cells and the other 300 mL of HCO_3^- diffuses into the plasma, where it also functions as a buffer to mop up any stray H^+ ions that have been produced by metabolic reactions. The effectiveness of this mechanism for transporting carbon dioxide and acting as a buffer of acid can be gauged from the concentration of HCO_3^- in the plasma of giraffes. In mammals generally, it is 25–30 mmol L^{-1} and has been found to be so in giraffes (e.g., Geiser et al., 1992; Williams et al., 2007; Brondum et al., 2009). Once the carbon dioxide reaches the lungs, all three transport processes are reversed; dissolved carbon dioxide diffuses down its pressure gradient into the alveoli, which lowers the pressure exerted by it in blood. The reaction that forms HCO_3^- is reversed, carbon dioxide is generated, and it also diffuses into alveoli. In addition, when hemoglobin is oxygenated in the lungs, carbon dioxide is released from the carbamino compounds and diffuses into the alveoli and alveolar ventilation expels it into the atmosphere. These series of reactions are beautifully simple, supply the alveoli continuously with carbon dioxide, and all of them occur in the fraction of a second that a red blood cell loiters in the capillaries that surround an alveolus.

13
Exercise in Giraffes

Exercise in giraffes is an enigma. There are no experimental studies of their exercise capability. Their preferred activity is a gentle wander between trees or an amble to surface water if needed. A run is rare. There is, however, a superabundance of anecdotes about their exercise capability, the chief characteristic of which is disagreement—and disagreement with some or all of the contents of this chapter by readers thus is certain. The earliest anecdotes are most probably those of Artemidorus of Ephesus (~100 BC), who thought their speed could not be surpassed, but Strabo, the great Greek geographer, wrote in ~AD 23 "on account of their lack of symmetry the speed of the animal is not so great as stated by Artemidorus" (see Chapter 2), thus establishing the ambivalence that has persisted ever since. More recent examples are plentiful. Andre Thevet (1502–1590), a French historian, wrote in his Universal Cosmography of 1544, "They are easy to catch because they are slow runners," while at much the same time, Antonio Costanzi, in a letter to Galeas Manfredi, Prince of Faënza, told how the "inhabitants of Fano near the Metauro River have seen the giraffe run without effort, with great speed such that horsemen have not been able to follow, even when spurring them" (Vosmaer, 1787). Fano is in the Marche region of eastern Italy. Those intrepid explorers of the hinterland of South Africa Jacobus Coetsé (1760) and Hendrik Wikar (1779), who were among the first Europeans to see wild giraffes, disagreed. Coetsé told how the "gait of these animals is slow and cumbrous so that having given chase on horseback I caught up to two of them with little difficulty," and Wikar wrote, "if you have a fairly good horse you can easily get it, as it cannot even run as fast as an eland." Jean Allamand, in his supplements to Buffon's Natural History, tried, in two sentences, to satisfy both opinions. He wrote, "The movements of the giraffe are not strongly brisk, and its gallop is not equal in speed to that of the horse. Nevertheless it takes a good horse to be able to follow: the length of its legs makes it move very well." At the time of the arrival of the French giraffe in Paris in 1826, Geoffroy St.-Hilaire (1827) wrote, "The giraffe if excited to flee prevails and is soon out of sight, but it does not long maintain this effort. Fatigue results because the lungs are too small the reason being that the chest which contains them is small." One of the more trustworthy assessments of their abilities was made by Cornwallis Harris. In his 1839 memoir, he wrote that he startled a troop of some 30 giraffes and they ran off, "soon leaving me far behind," even though he was on his best horse, but "—in 5 minutes I perceived their race was run. I urged my good steed to his utmost and instantly found myself by the side of the herd" (see Plate 13.1), an observation later supported by Dagg (1962a): "giraffe dislike galloping and slow down as soon as possible."

Dagg and Foster (1976) reported that a hunter, T. S. Arbuthnot, wrote that "one might expect the giraffe with its long legs, to outrun the other plains species, but it cannot; the record for an individual is only 56 km/hr." Dagg and Foster also reported the experience of Arab hunters (with the source possibly Grzimek, 1956):

How Giraffes Work. Graham Mitchell, Oxford University Press. © Oxford University Press 2021.
DOI: 10.1093/oso/9780197571194.003.0013

> *The Arabs have hunted them for centuries on camels and horses, often using fresh relays during the chase because the giraffe are fast enough to outrun many of the small horses. The highest recommendation that an Arab can give a horse is to say that it can overtake a giraffe. Even then the horse has to be pressed forward at top speed as soon as the giraffe is sighted; if the prey is allowed any leisure, it can set up a steady pace which it can continue for hours.*

Langman et al. (1982) used almost identical words when they reported that "anecdotal evidence and personal experience indicate that giraffe can run at 60 km/h for 5 minutes," and added, "with no obvious signs of distress." Innis (1958) recorded a top speed of 54 km h^{-1} over short distances and one of 46 km h^{-1} for several miles. The mix of high speed and quick exhaustion but legendary endurance was also described by Frederick Selous in his *Narrative of Nine Years Spent Amongst the Game of the Far Interior of South Africa*, published in 1881. He had this to say:

> *They get over the ground at a great rate, and it requires a good horse to run one down. The great thing is to press them to their utmost speed at first, when, if fat, they soon get blown and can be ridden into, and, if the wind is favourable, driven for miles. At a hard gallop, however, they can spin along for miles. The giraffe is both fleet and enduring and it is a fairly good horse that can gallop clean past one. If not hard pressed they do not go at a very great pace,* [but] *how easily and with little exertion they seemed to get over the ground, with that long, sweeping stride of theirs! Yet they were going at a great rate, for I felt that my old nag was doing his best, and I could not lessen the distance between us by an inch.*

Twenty-seven years later, in his more careful and considered *African Nature Notes and Reminiscences*, he repeated these observations. Can the mix of "fleet and enduring," of high speed and staying power, be explained? What is the exercise capacity of giraffes?

Walking and Running

Generally, quadrupeds can move at three speeds—walk, trot, and gallop—but those that have sloping backs, for some not obvious biomechanical reason, cannot trot (Pennycuick, 1975; McNab, 2002). Giraffes are among them. They do not trot. They either walk or run, but they can walk and run at different speeds. There have been just three recorded estimates of the speed at which free-living giraffes amble between trees while browsing. Innes (1958) made 48 observations of the distances traveled by unhurried, unthreatened giraffes going about their daily business and the speed at which they moved was 0.06 to 0.1 km h^{-1}, that is, about 60 to 100 meters every hour. It was 0.5–0.7 km h^{-1} in Tsavo National Park (Leuthold and Leuthold, 1978b). In the Northern Namib Desert, males walk ~6 km each day (0.3 km h^{-1}) and females ~2 km daily (0.1 km h^{-1}; Fennessy, 2009). A giraffe's purposeful walk (termed its "pace") is not the amble associated with browsing, is proportional to size, and follows Heglund and Taylor's (1988) equation that describes walking speed for artiodactyls generally (i.e., $0.94 \times Mb^{0.145}$ m s^{-1}). This equation predicts a walking speed of 2 m s^{-1} for juvenile giraffes and 3 m s^{-1} for adults, speeds that match observed speeds of between 7 km h^{-1} for very

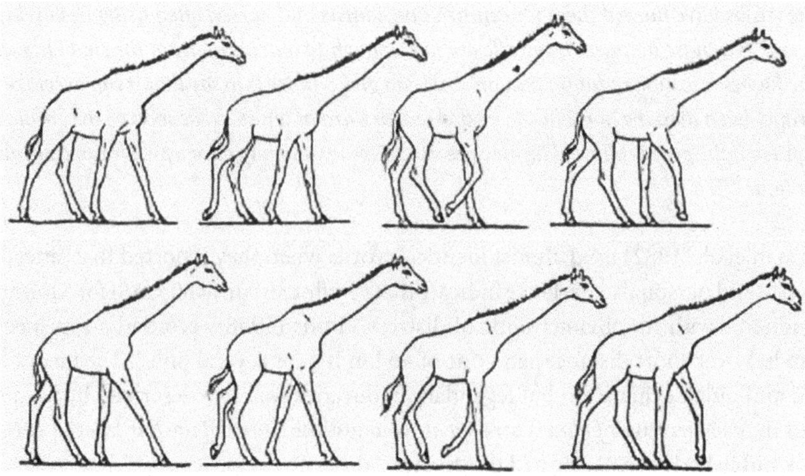

Figure 13.1. A giraffe's walk.
From Gambaryan, 1974, with permission of John Wiley and Sons.

young giraffes and 10 km h^{-1} for adults (Dagg, 1962a, 1962b). The way that giraffes (and camels) walk is unique. When they walk, the two legs on each side move together, so their body mass is alternately supported first by two left legs and then two right legs (Figure 13.1). Giraffes have relatively short trunks and long legs, and this way of walking avoids collisions between the fore and hind hoofs. Gambaryan's illustration of a giraffe walking (Figure 13.1) should also have shown that while a giraffe is walking, its neck and head move back and forth. The forward movement of the neck shifts the center of gravity, and moves the body forward. When the fore-hooves touch the ground, the head and neck recoil, forward momentum is reduced, and the step is stabilized before the next one is taken.

Dagg (1962a, b) filmed giraffes galloping at full speed; quoting Elizabeth Marshall Thomas (1959), she described the rhythm of their gallop as "with their strong legs slowly opening, then bunching as they run and their heads on their long necks stretching out and back, out and back." When a giraffe walks the legs on each side move together, but when it gallops its fore and hind legs move together, the hind legs twisting slightly outward to avoid collisions between hind and front hooves. Characteristically, too, they run with their tails twisted up and over their backs (see Plate 13.1, Figure 13.2). In almost all mammals, a run is started by a powerful thrust of the hind limbs, but giraffes seem to begin a gallop with a forward thrust of the neck that shifts the center of gravity over or in front of the forelegs, and then they push off with their hindlegs (Figure 13.2B; see also Chapter 16).

To get an idea of the size of the force generated by their neck, Dagg (1962a) drew, on the images in the films she made of giraffes running, a line down the neck and another along the back. Before a run, the neck formed nearly a right angle (115°) with the back. At the start of a run and in less than half a second, the angle formed by the neck and the back increased so that neck and back formed nearly a straight line (165°). The head and neck of a large giraffe can weigh 150 kg, and the momentum generated by that rapid forward thrust of the head and neck plunges the whole giraffe forward, overcoming its

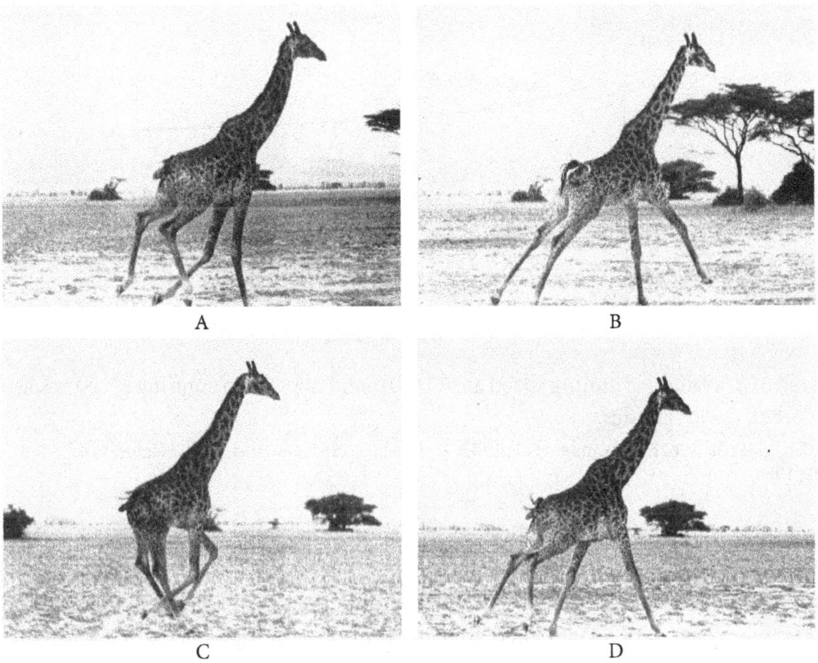

Figure 13.2. The four phases of a giraffe's gallop (see text).
From Dagg and Foster, 1976, with permission of Anne Dagg and J. Bristol Foster.

inertia and bringing about almost instantaneous movement. Thereafter, power comes mainly, but not only, from the hind quarters, and the forelegs become the fulcrum over which the body is moved. Once a gallop is in full flow, giraffes move through the air seemingly effortlessly, with fore- and hindlegs stretched far apart (Figure 13.2B), then swing their hindquarters under the body with the hindlegs spread outside the forelegs and all four hooves off the ground (Figure 13.2C), landing on the hind legs momentarily before the front legs do so, and then pushing off with the hind legs to begin the cycle again (Figure 13.2D). The oscillation of the neck during a run is even greater than at their pace.

How fast and how far can giraffes actually run? The anecdotes imply that their top speed is faster than that of a horse. The top speed of a modern-day racehorse is 70 km h^{-1}, and that speed is beyond any recorded for giraffes, but a reasonable speed for a good work horse is 45 km h^{-1}, especially a horse trying to navigate its way through savannah woodland. Giraffes can, obviously, run at many different speeds, but in the context of this analysis of their exercise capabilities, two are relevant—maximum speed and the speed at which the maximum rate of oxygen consumption (VO_2max) is reached. This latter speed has been predicted but not measured in giraffes, but maximum running speed has been measured by R. McNeil Alexander and colleagues. In 1977 they chased a number of artiodactyls, including a giraffe that weighed 1,100 kg (Alexander, 1977), for two minutes to make them run at maximum speed, and filmed them doing so. From the film they could determine the giraffe's maximum running speed, and it was 11 m s^{-1} (40 km h^{-1}), the same speed that features regularly in the anecdotes and which is roughly the top speed of a good work horse. By extrapolation

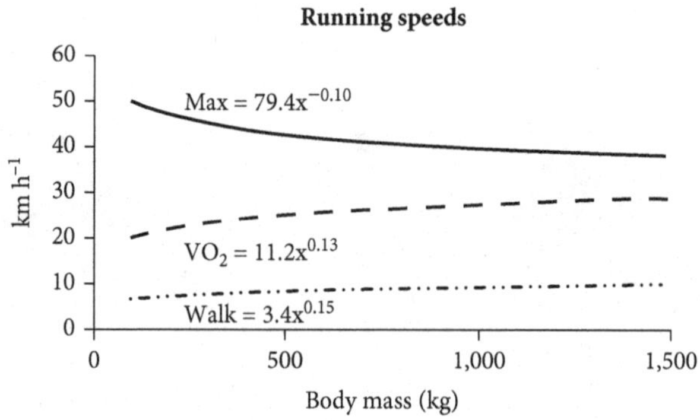

Figure 13.3. Walking, running speed at VO_2max, and maximum running speed, measured in or predicted for giraffes.

After Dagg, 1962a, 1962b; Alexander et al., 1977; Taylor et al., 1981; Garland, 1983; Heglund and Taylor, 1988.

of Alexander et al.'s 1977 data, the maximum speed (m s^{-1}) can be predicted in any giraffe by the equation $17 \times Mb^{-0.062}$ (Alexander et al., 1977). Garland (1983a), using mostly anecdotal data, established that maximum running speeds of mammals, including giraffes, can be predicted by the equation $27.2 \times Mb^{-0.13}$, which gives the same maximum speed (m s^{-1}) measured by Alexander et al. for a giraffe weighing 1,100 kg. A generic equation for maximum speed derived from these two equations is $22.1 \times Mb^{-0.10}$ m s^{-1} (or $79.4 \times Mb^{-0.10}$ km h^{-1}). The exponents of all three equations imply that maximum running speed is almost independent of body mass and, if so, a small giraffe would have the same maximum running speed as an adult. However, if the equations are applied to giraffes ranging in body mass from 100 to 1,500 kg, the calculated maximum speed of the small giraffe is 50 km h^{-1} and that of the adult is 38 km h^{-1}. Observation confirms this difference; newborn giraffes, like newborn horses, can run faster than their mothers can (Figure 13.3).

Anaerobic Exercise

The most basic need for exercise is a supply of energy for muscle contraction. The energy comes from ATP and ATP must be replenished at the rate at which it is consumed. ATP comes not only from aerobic, oxygen-dependent sources, but also from oxygen-independent, anaerobic metabolism. The ATP for short sharp bursts of exercise characteristic of an escape from a predator or, in humans, running 100 m or 200 m, comes almost completely from anaerobic glycolysis in which glucose and its stored form of glycogen are catabolized. No oxygen is needed to turn them into ATP, but oxygen is required afterward to replenish the energy used. Thus an animal's capacity for anaerobic exercise depends, first, on the size of its glycogen stores and, second, on having the kind of muscle fibers that can turn glucose into ATP anaerobically.

The muscle fibers of a good anaerobic exerciser lack all the biochemical machinery such as mitochondria and myoglobin that are needed to make use of oxygen, and the fibers consequently have a pale look to them. They are called white fibers, in contrast to the red fibers rich in mitochondria and myoglobin that are found in muscles that depend utterly on oxygen for their energy. Red (Type 1) fibers are resistant to fatigue and are the fibers that abound in the muscles of excellent endurance athletes of any species. White (Type 2) fibers fatigue more easily, but have the fast contractions and power required of sprinters. There has been no comprehensive survey of the muscle fiber types of giraffes, but Roy et al. (1988) analyzed fiber types in three muscles—the plantaris and the medial and lateral gastrocnemius (calf) muscles. These muscles act to flex the ankle upward (plantaris) and extend the ankle downward (gastrocnemius). Roy et al. found that the plantaris, which is mostly concerned with posture rather than movement, is approximately 70% red, oxidative fibers in an adult, while in the gastrocnemius muscles that play a bigger role in exercise, red fibers formed about 30%. Therefore, the remaining 70% of fibers in the gastrocnemius muscles were predominantly white Type 2 fibers.

There are two types of white Type 2 fibers in a muscle: Type 2A fibers, which have a mix of aerobic and anaerobic metabolic pathways, and Type 2X fibers, which are the classic fast-contracting, anaerobic, low in mitochondria, quickly-fatiguing fibers. What types are found in wild artiodactyls? An answer has come from a study of the fiber types of the *vastus lateralis* muscle in the quadriceps group of the hind limbs in several southern African antelopes ranging in body mass from 30-kg springbok to 150-kg wildebeest and 250-kg kudu (Curry et al., 2012; Kohn, 2014). The quadriceps group of muscles located in the upper hind leg is an important source of muscle power for exercise (see Chapter 16). Surprisingly, the muscle fiber types did not follow the classic divisions. In almost all the species, Type 2X predominated, but they had a mix of aerobic and anaerobic metabolic pathways that were similar to those in Type 2A fibers, and so were fatigue resistant and hence conferred endurance capacity. If this mix of red and white fibers is reflected in all the muscles of exercise in giraffes, then it would partly explain the ease with which they run seemingly effortlessly and also explain how they can produce short sharp bursts of speed.

Glycogen is stored in muscle cells and in liver cells. Liver glycogen makes up about 10% of liver mass, and muscle glycogen 1%–2% of muscle mass in ruminants (e.g., Perry, 2012). The glycogen content of the liver and muscles of giraffes has not been measured, but there is no reason to think that it will be significantly different from what is found in other ruminants. In six giraffes that had body masses of 289–1,282 kg, liver mass was 1.4% of body mass (see Chapter 8). Their muscle mass could be expected to be approximately 36% of body mass (Weibel et al., 2004), and the total amount of glycogen in giraffes of this range of body mass (assuming that liver glycogen is 1.5% of liver mass) can be estimated by a linear equation: total glycogen (kg) = $0.007 \times Mb + 0.124$, with liver glycogen providing about 20% and muscle glycogen the rest (Table 13.1). The energy content of the glucose in glycogen, like that of all carbohydrates, is ~20 kJ g^{-1}. However, glycogen stored in the liver and muscles is nearly 80% water, so its energy content is 4.2 kJ g^{-1} (Flatt, 1995) and, therefore, total glycogen energy in a 750-kg giraffe is ~20,000 kJ (Table 13.1). How much energy does it take for a giraffe of that body mass to run at maximum speed of 12.5 m sec^{-1} (45 km h^{-1})? The answer to that question is, of course, unknown, but Taylor et al., in 1982, measured energy consumption at various

Table 13.1 Estimates of the Glycogen Stores Available for Anaerobic Exercise in Giraffes

Variable	750 kg Giraffe	Range in Giraffes of Mb 289 to 1,282 kg	Allometric Equation
Muscle glycogen (kg)	4.4	1.6 to 7.1	0.006*Mb–0.116 R^2=1
Liver glycogen (kg)	1.0	0.4 to 1.7	0.0012*Mb+0.126 R^2=.62
Total glycogen (kg)	5.4	2.0 to 8.8	0.007*Mb+0.124 R^2=.98
Total kJ	21748	8350 to 37,000	28.3*Mb+523 R^2=.98

speeds in artiodactyls that weighed up to 245 kg and running at various speeds. If those data are extrapolated to giraffes, then a giraffe weighing 750 kg requires ~600 kJ of energy every minute to run at 45 km per hour. Thus, if all the energy in glycogen could be used by a giraffe for anaerobic exercise, it would be enough to provide about 30 minutes of running time at 45 km h^{-1}.

How long a species like an antelope or giraffe can actually run anaerobically has been estimated by measurements made on six wild African antelopes and on six other species, all of them trained to run on a treadmill (Taylor et al., 1981). All of them stopped running, not when their glycogen was depleted, but when the concentration of lactic acid (a product of anaerobic glycolysis) in blood reached a remarkably consistent 21.1±2.8 mmol L^{-1}. The time it took for that concentration to be reached was similarly consistent and was 4.0±1.5 min, which is not too dissimilar to the ~5 min that startled giraffes run for at high speed before tiring. Thus, even though giraffes and other species have the energy stores to run anaerobically for a long time, they do not do so because provision of ATP by anaerobic metabolism stops when the further catabolism of glycogen is inhibited by accumulation of its metabolic by-products. The rate at which glycogen can be used is increased by the effects of adrenaline, and it is easy to imagine, then, that a startled giraffe could run at quite a high speed for several minutes as the reports suggest, using the energy from anaerobic glycogen metabolism, even if it was not getting any additional energy from other sources.

Aerobic Exercise

The alternative energy source, and quantitatively a far more important source of ATP, is aerobic metabolism or, more accurately, oxidative phosphorylation and, as its name implies, oxygen is essential for it. It is the source of energy for endurance exercise because compared to the number of ATP molecules produced anaerobically from a molecule of glucose, aerobic metabolism produces ~20 times more. The oxygen is used by a single mitochondrial enzyme—cytochrome oxidase—in the final step of the electron transfer cascade that produces ATP. How much oxygen is used by muscles? Oxygen is distributed to tissues in proportion to demand, and Weibel et al. (2004) estimated that during strenuous aerobic exercise, over 90% of cardiac output and hence 90% of all oxygen goes to muscles. Not all of that oxygen can be used, and the maximum rate that oxygen can be delivered to the muscles of a giraffe, and can be used by them, sets an upper limit for aerobic, endurance exercise (Bassett and Howley, 2000; Weibel et al., 2004).

In their acclaimed series of experiments on wild animals running on a treadmill, Richard Taylor and Ewald Weibel and colleagues (1981, 1987, 2004) managed to measure the maximum volume of oxygen able to be consumed (VO_2max; L min^{-1}) by each species, and developed equations that allow its estimation in mammals that weigh up to ~300 kg. Taylor and colleagues showed that the mammals they studied fell clearly into athletic and non-athletic types, separated by their VO_2max. VO_2max (L min^{-1}) in all those species increased as body mass increased. A better measure of a mammal's capacity to use oxygen is their mass-specific oxygen consumption, that is, the amount of oxygen used per kg of body mass (mL kg^{-1} min^{-1}). In the non-athletic species studied by Taylor and colleagues, mass-specific aerobic capacity (mL kg^{-1} min^{-1}) was less than 60 mL kg^{-1} min^{-1}. A wildebeest (43 mL kg^{-1} min^{-1}) and an eland (36 mL kg^{-1} min^{-1}) both fell clearly into the non-athletic group. VO_2max in athletic species like sled dogs (240 mL kg^{-1} min^{-1}), horses (200 mL kg^{-1} min^{-1}), and even the best of human marathon runners (80 mL kg^{-1} min^{-1}) far exceeds the aerobic capacity of wild antelopes. Where do giraffes fall? Their actual VO_2max (L min^{-1}) and relative VO_2max (mL kg^{-1} min^{-1}) are of course unknown but, if Taylor et al.'s data are extrapolated to giraffes, a medium-sized giraffe weighing 750 kg might have a VO_2max of 25 L $min^{-1}$and a relative VO_2max of 32mL kg^{-1} min^{-1} that would range from 41 mL kg^{-1} min^{-1} in small giraffes to 28 mL kg^{-1} min^{-1} in large ones (Table 13.2).

It may be that extrapolating the oxygen consumption found in antelopes to giraffes gives an incorrect value and giraffes may have a much higher VO_2max. However, measurement of VO_2max (L min^{-1}) in wild animals running on a treadmill produces very

Table 13.2 Oxygen Consumption (VO_2) in Giraffes, Based on Measurements Made in Giraffes and Extrapolated from Measurements Made in Other Mammals

Variable	750-kg Giraffe	Range in Giraffes of 147–1,441 kg	Allometric Equation	Reference
Resting VO_2 (mL kg^{-1} min^{-1})	2.1	3.1–1.8	$9.7 \times Mb^{-0.23}$	Chapter 8
Field VO_2 (mL kg^{-1} min^{-1})	6.5	7.9--4.3	$30.0 \times Mb^{-0.23}$	Chapter 8
VO_2max (mL kg^{-1} min^{-1})	32.3	41–28	$93.1 \times Mb^{-0.16}$	Taylor et al., 1981
VO_2max L min^{-1}	24.2	7.5–42.7	$0.0931 \times Mb^{0.84}$	Taylor et al., 1981
Running speed at VO_2max (km h^{-1})	26.5	21–29	$11.2 \times Mb^{0.13}$	Taylor et al., 1981

consistent results, and there is no reason to think that giraffes would be different, and if those treadmill-derived data are extrapolated to giraffes, they indicate that giraffes are non-athletic. Taylor and colleagues also determined the running speed at which VO_2max is reached in the species they studied and, assuming that giraffes will conform to the speeds reached by other mammals of similar body masses, the speed at which VO_2max ought to be reached is 6–8 m sec^{-1} or 21 km h^{-1} in newborn giraffes and 29 km h^{-1} in adults (Table 13.2; Figure 13.3). These speeds are about half the maximum running speed measured by Alexander et al. (1977) in a giraffe, a ratio that seems to be universally true: the best human marathoners running at close to their VO_2max run at ~20 km h^{-1}, and the best human sprinters run at ~40 km h^{-1}.

Knowing, or in the case of giraffes, making an informed guess, of VO_2max is a small but very important aspect of aerobic exercise. Why is it what it is? If mass-specific VO_2max in giraffes is low, what limits their VO_2max? What makes them a non-athletic species? And if they are non-athletic, how is it possible for them to run for the long periods of time that the anecdotes suggest they can? The best way to answer those questions is to measure cardiovascular and respiratory functions in exercising giraffes, but it is very unlikely that anyone will ever be able to carry out the complex and objective measurements of those functions in a group of wild giraffes of different ages, sizes, and sexes exercising at or near their maximum capacity. Consequently, the best that can be done is to rely on Taylor et al.'s data (1981) obtained from a few wild antelopes, and applying them to giraffes while recognizing their limitations. The species studied by Taylor and colleagues were smaller than giraffes and were running on a treadmill. What might happen in giraffes running freely in the wild? No one can be sure, not only because the data have not yet been collected, but also because there are an infinite number of combinations of heart rate, stroke volume, respiratory rate, tidal volume, and oxygen content of blood and so on that could exist in a giraffe as it runs. Each giraffe will have a different combination that will depend to a large extent on how fast each one is running and how much it weighs. Nevertheless, whatever the combinations may be, they must satisfy at least two criteria. First, they must be consistent with and indeed cannot deviate from the formulae which define cardiovascular and respiratory function (see Chapters 9 and 12) and, second, they must be consistent with delivery to lungs and muscles of the volumes of oxygen needed at VO_2max. What could those combinations be in an exercising giraffe?

The combinations ultimately depend on how much oxygen is being consumed. For oxygen to be consumed, it has to be delivered to muscles and their mitochondria. The pathway that oxygen follows from the atmosphere to the mitochondria was outlined by Taylor and colleagues in 1981 and it contains a series of steps, any one or all of which can limit oxygen use. The first step involves the mechanisms responsible for oxygenation of blood in the lungs. The second depends on the carriage of oxygen to muscles by blood. The third is use of the oxygen by the muscle mitochondria themselves. As simple as all this sounds, what actually limits exercise is a surprisingly controversial topic and also one that generates an immense volume of research because of the lucrative nature of horse and dog racing and human athleticism in all its forms. The success of human or equine and canine athletes depends partly on finding ways to overcome the natural limits of oxidative energy production.

Oxygenation of Blood in the Lungs of Giraffes

Their relatively small lung volume and component volumes, a narrow trachea, and a large dead space (see Chapter 12) are the explanations usually given for the non-athletic status (i.e., low VO_2max) of giraffes. A consequence of a small lung volume is a reduced ability to increase the size of a breath, and giraffes have to rely on increased respiratory rate to produce the alveolar ventilation needed to deliver enough oxygen to the alveoli for diffusion into blood. The consequence of a narrow trachea is that giraffes have a low peak inspiratory flow rate, which is presumed to make it much harder to increase ventilation, especially under the demands of exercise (Patterson et al., 1965). When a giraffe is browsing quietly, none of this anatomy and physiology matters. Alveolar ventilation (V_A) is more than enough to deliver oxygen to the alveoli to meet the metabolic rate of a giraffe at rest (see Chapter 12). However, considering all the disadvantages that arise from their lung anatomy, it is reasonable to think that delivery of oxygen to muscles will be compromised when, during exercise, oxygen demand is greatest. Can their respiratory system provide enough oxygen at VO_2max?

Giraffes do have a very small inspiratory reserve volume (IRV) of no more than a few liters but, if most of it is recruited and respiratory rate doubles from its level at rest to ~20 breaths min^{-1}, then alveolar ventilation is sufficient to supply the oxygen needed at VO_2max. Respiratory rates of four times resting values have been recorded in anxious giraffes, so a doubling during exercise is well within what is possible (Patterson et al., 1965). Moreover, the respiratory system is sensitive to changes in oxygen demand and adjusts mainly by increasing respiratory rate (see Chapter 12). The respiratory rates actually needed to deliver sufficient oxygen to the lungs at VO_2max are 14 min^{-1} in young giraffes and 20 min^{-1} in adults (Table 13.3).

What is more, Gehr et al. (1981) have shown that in a giraffe weighing 383 kg and antelopes of lower body mass, the diffusion capacity of the lungs for oxygen (DLO_2 = the liters of oxygen diffusing into blood every minute for every mmHg of pressure

Table 13.3 Some Combinations of Respiratory Rate, Dead Space, and Tidal Volume That Exist in a 750-kg Giraffe at Rest and That Could Exist in a 750-kg Giraffe to Deliver Sufficient Oxygen to the Lungs at VO_2max

Variable	750-kg Giraffe at Rest	750-kg Giraffe at VO_2max	Range in Giraffes* at VO_2max	Allometric Equation at VO_2max
Respiratory rate (min^{-1})	9	19	14–20	$8 \times Mb^{0.13}$
Tidal volume (L)	9	12	3.6–20.1	$0.071 \times Mb^{0.77}$
Dead space (L)	2.3	2.3	0.4–4.4	$0.003 \times Mb+0.04$
Minute ventilation (V_E) (L min^{-1})	78	217	50–370	$0.56 \times Mb^{0.90}$
Alveolar ventilation (V_A) (L min^{-1})	59	174	40–300	$0.67 \times Mb^{0.84}$
Diffusion capacity (L min^{-1} $mmHg^{-1}$)	0.5	4.5	0.8–9.5	$0.004 \times Mb^{1.06}$

* Body mass between 150–1,500 kg.

gradient between alveoli and blood) increases as VO_2max increases to the power of 1.27 ($VO_2max^{1.27}$; Gehr et al., 1981), so that on average at VO_2max, diffusion rate is about 10 times higher than it is at rest (Table 13.3). Assuming that giraffes larger than the one studied by Gehr et al. will follow a similar pattern, then in an average-sized giraffe weighing 750 kg, the diffusion capacity would increase from 0.5 L min^{-1} $mmHg^{-1}$ at rest to 4.5 L min^{-1} $mmHg^{-1}$ at VO_2max (Table 13.3). As the pressure gradient for oxygen across the alveolar membrane is ~70 mmHg, the diffusion capacity (in a 750 kg giraffe would be ~350 L min^{-1} and would far exceed the volume of oxygen (25 L min^{-1}) that it is estimated a giraffe of that body mass will use at its predicted VO_2max. Why diffusion capacity has evolved to far exceed requirements is not yet known.

Despite the conclusion that giraffes can easily ventilate their lungs and can provide sufficient oxygen for exercise, Robert Schroter suggested (in 1978) that because the trachea of giraffes is so long, exercising giraffes would not be able to overcome the associated increase in the work of breathing for prolonged periods of time. Thus, they would be able to run for short bursts only. Moreover, the combination of an increase in ventilation and a narrow trachea would increase the velocity of air flow and change the pattern of airflow from laminar (i.e., smooth) to turbulent. Turbulent flow is a potential disadvantage as it increases resistance to the flow of any fluid in a tube, although by how much it increases the resistance to airflow in the trachea of a mammal is unknown. If turbulence does increase resistance to airflow through the trachea significantly, then during exercise the work of breathing would be increased even more and the length of time a giraffe could exercise would be further reduced. The velocity of air flow at which turbulence starts is predicted by the Reynolds number (*Re*). The greater the velocity, the higher the *Re* and the more likely turbulence will develop. The transition from laminar airflow to turbulent airflow occurs when *Re* is in the range 2,300–2,500, and above 2,500, turbulence is present continuously (Schroter, 1978). Hugh-Jones et al. (1978) estimated from the dimensions of the trachea they measured in a giraffe that, in giraffes quietly going about their business, air velocity would be on average 180 cm s^{-1} and *Re* would be about 1,800, but they also concluded that airflow would become turbulent as soon as giraffes begin to run and thus their endurance capacity would be diminished.

Hugh-Jones et al. (1978) did not have sufficient data on respiratory tract anatomy or dynamic lung volumes in giraffes of different body masses to calculate *Re* in them, but those data are now known (see Chapter 12), and from them *Re* that exists in the trachea can be calculated for giraffes at rest and estimated in them during exercise from the formula:

$$Re = (\text{velocity of airflow } [cm\ s^{-1}] \times \text{trachea diameter } [cm] \times (\text{air density} \div \text{air viscosity})).$$

Air density at a giraffe's body temperature of 38°C is 1.127 kg m^{-3} and its dynamic viscosity is 1.983×10^{-5} kg m sec^{-1}. The velocity at which air flows down a giraffe's trachea is proportional to minute ventilation rate and inversely proportional to the diameter of the trachea, and can be calculated from the formula:

$$\text{Air flow velocity } [cm\ s^{-1}] = (4 \times V_E) \div (\pi \times \text{diameter}^2).$$

What these two formulae reveal is that in 46 giraffes at rest, with body masses between 147 kg to 1,441 kg, and whose tracheal dimensions are known (see Table 12.1), the velocity of airflow ranges from 55 to 118 cm sec^{-1} and *Re* ranges from 800 to 2,700. At the minute ventilation that could be expected to occur at VO_2max, the velocity of airflow in those same giraffes could range from 160 to 380 $cm.s^{-1}$, and the *Re* associated with those velocities would range from 2,300 to 9,700. The turbulence threshold of 2,500 is reached at a minute ventilation of 95 L min^{-1}, which corresponds to the minute ventilation at rest of a giraffe weighing 1,331 kg. During exercise that threshold is reached in all giraffes weighing more than 330 kg and/or when running speed is ~31 km h^{-1} (Figure 13.4). Thus, during exercise, as Hugh-Jones et al. predicted, airflow is always turbulent, but at rest the *Re* is low enough for flow to be laminar and smooth, except in large, mature giraffes.

Overcoming any increased work of breathing brought on by turbulence during exercise is, from observations of their ability to exercise, achieved by giraffes. How? The most likely explanation is that the effect of turbulence is minor, inconsequential, and, if it exists at all, can easily be overcome. One quite possible but so far unproven (Alexander, 1989b) way that giraffes could overcome increased resistance to airflow caused by a narrow trachea or turbulence, or both, is by synchronizing breathing during exercise with their running speed, as proposed by Bramble and Carrier (1983) for horses. In horses, the inertia of the mass of the intestines shifts the intestines backward during a forward step, drawing the diaphragm backward and, when the forelegs hit the ground to complete the step, the intestines rebound into the diaphragm thrusting it into the thorax. The mechanism functions rather like a piston with the mass of the intestines acting as the piston itself: the movement backward helps fill the lungs, the movement

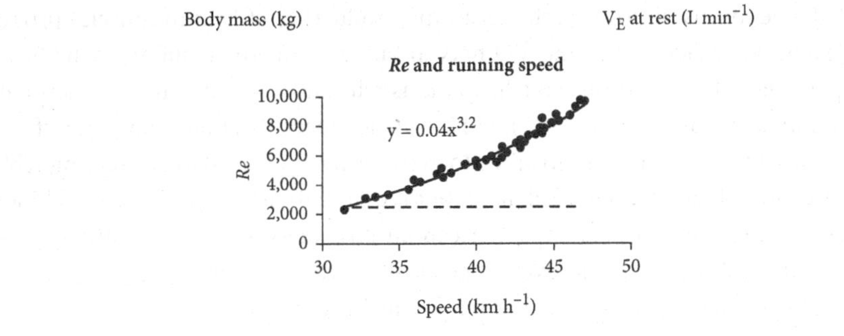

Figure 13.4. The Reynolds number (*Re*) calculated in 46 giraffes at rest (●) from the dimensions of their respiratory tract, and predicted at VO_2max (□), as body mass, minute ventilation, and running speed increases. Giraffe body mass ranged from 147 kg to 1,441 kg. The dashed line is the turbulence threshold (*Re* = 2500).

forward helps empty them, and the combination would overcome any resistance to airflow without any need for an increase in the work of breathing. There would be one breath per stride, which is what is found in horses (Bramble and Carrier, 1983). This 1:1 ratio could also arise from the contraction characteristics of the diaphragm muscle. In small mammals, its contraction frequency (s^{-1}) to achieve maximum power is equal to $4.4 \times Mb^{-0.16}$ and stride frequency (s^{-1}) equals $4.5 \times Mb^{-0.14}$ and thus they increase together. The two are linked by neural signals arising from the contracting muscles (Altringham and Young, 1991). An average-sized, galloping giraffe takes ~2 strides s^{-1} or 120 min^{-1}, so if their speed were to be linked to breathing, then their respiratory rate would be 120 min^{-1}, an impossibly high rate. However, in a giraffe being chased, van Citters et al. (1966; Warren, 1974) recorded that a drop in blood pressure occurred every 1–2 s, with each drop in pressure coinciding with the moment its front hooves hit the ground, which means the giraffe was taking a reasonable 30–60 strides min^{-1} and was running at around 30 km h^{-1} (i.e., at its expected VO_2max). However unlikely it seems that a giraffe could breathe 30 times per minute, if blood pressure is synchronized with pace, it is possible that the respiratory cycle is as well, and even if breathing is not perfectly synchronized with steps, it is hard to accept that intestinal shifts play no role in the respiratory mechanics of running giraffes.

The characteristics of the giraffe's respiratory system notwithstanding, it is possible to conclude that despite narrow conducting airways and a small lung, oxygen provision always is enough and can be adjusted upward during exercise by a combination of increased respiratory rate and increased lung diffusion capacity. A compromised supply of oxygen to the gas exchange surfaces of the lungs, or to the blood that flows past them, is thus not the origin of a giraffe's low VO_2max.

Delivery of Oxygen to Muscles

There is a lot of evidence, reviewed by Bassett and Howley (2000), that the main limiter of aerobic exercise capability and especially endurance in mammals is the capacity of the cardiovascular system to deliver to muscles the oxygen that has entered the lungs. As long ago as 1924, A.V. Hill (who in 1922 was awarded the Nobel Prize in Physiology or Medicine for his discovery relating to the production of heat in muscle) proposed that the primary factor affecting VO_2max in humans was the inability of the heart to pump enough blood, and this conclusion is as relevant to exercise in other animals as it is to humans (Taylor et al., 1987). Hill's view is especially applicable to giraffes. The structure of the giraffe heart is unsuited to exercise. It has evolved to produce high blood pressure rather than the high cardiac output required to deliver the masses of blood to muscles that racehorses, sled dogs, African hunting dogs, wolves, marathon runners, cross-country skiers, and *Tour* cyclists need, all of which, compared to giraffes, have a low blood pressure and a high cardiac output during exercise.

The relationship between the rate of oxygen use by mitochondria and the rate of delivery of oxygen to mitochondria can be summarized by the equation developed by a German physiologist, Adolf Fick, in 1870:

$$VO_2\,(mL\ min^{-1}) = cardiac\ output\ (L\ min^{-1}) \times (C_aO_2 - C_vO_2)$$

where C_aO_2 and C_vO_2 are the oxygen content of arterial and venous blood in mL L^{-1}, respectively, and of course:

cardiac output (mL min⁻¹) = heart rate (min⁻¹) × stroke volume (mL)

What these two formulae tell us is that the volume of oxygen that is being delivered and the volume being consumed at any time are determined by the combination of how fast blood is circulating (cardiac output), how much oxygen is in each liter of arterial blood (C_aO_2), and how much oxygen is extracted from each liter of blood ($C_aO_2 - C_vO_2$). For example, if, in a human, 50 mL of oxygen is being extracted from each liter of blood as it flows around the body each minute, and if cardiac output is 5 L min^{-1}, then total oxygen consumption (VO_2) must be 250 mL min^{-1} (i.e., 5 L min^{-1}× 50 mL L^{-1}). How do the equations help explain what happens in giraffes when they are running at speed? One outcome they reveal is that cardiac output is directly proportional to VO_2. As VO_2 increases, so does cardiac output. Consequently, as exercise time prolongs, giraffes have to depend increasingly on aerobic energy to supply the ATP needed for muscle contraction, and thus both VO_2 will increase and cardiac output will increase. If a giraffe weighing 750 kg is running at its VO_2max, it could be using 25 L of oxygen min^{-1} and to deliver that volume of oxygen to muscles, what is needed is an increase in heart rate and/or stroke volume to increase cardiac output and/or an increase in C_aO_2 and/or a decrease in C_vO_2. The relationship between these variables and VO_2 has been studied by Taylor et al. (1987) in both athletic and non-athletic species smaller than giraffes but from which some informed guesses can be made about how the cardiovascular system of giraffes could adapt to aerobic exercise.

Cardiac Output

Relative to increases in body mass, heart rate always decreases but can be expected to be much higher at VO_2max than at rest in giraffes in the same way that during exercise the heart rate in humans increases. The highest heart rate reported so far in giraffes came from a giraffe (which from its neck length probably weighed between 500 kg and 700 kg) that had been "vigorously chased for several minutes of hard running in which it was thoroughly exercised." Its heart rate had trebled from a base of 60 beats min^{-1} to ~170 beats min^{-1} (van Citters et al., 1966), a rate that can be assumed to be the upper limit for heart rate in giraffes of that size. Smaller giraffes would have a higher maximum heart rate during exercise, probably reaching above 200 min^{-1}, and larger ones a maximum rate somewhat less. If heart rate is two to three times higher at VO_2max than it is at rest, then that range would be encompassed by a heart rate described by $559 \times Mb^{-0.19}$ (Table 13.4).

Unlike heart rate, stroke volume always increases as body mass increases because the heart becomes larger, and during exercise it could increase if heart muscle contraction increased, but it does not increase much during exercise in non-athletic ruminants, even at VO_2max (Taylor et al., 1987). One reason for this apparent anomaly is that as heart rate increases, there is less and less time for the ventricles to fill with blood. There is an additional reason in giraffes. The walls of the ventricles thicken as

Table 13.4 Some Combinations of Heart Rate, Stroke Volume, Cardiac Output, and Peripheral Resistance and Blood Pressure That Exist in a 750-kg Giraffe at Rest and That Could Exist in a 750-kg Giraffe to Deliver Sufficient Oxygen to Muscles at VO_2max

Variable	750-kg Giraffe at Rest	750-kg Giraffe at VO_2max	Range in Giraffes* at VO_2max	Allometric Equation at VO_2max
Heart rate (min^{-1})	70	159	215–140	$559 \times Mb^{-0.19}$
Stroke volume (mL)	538	757	140–1500	$1.03 \times Mb - 16$
Cardiac output ($L\ min^{-1}$)	38	120	30–215	$0.46 \times Mb^{0.84}$
Blood pressure (mmHg)	207	240	190–270	$89 \times Mb^{0.15}$
Total peripheral resistance ($mmHg\ min\ L^{-1}$)	5.5	2	6.3–1.3	$194 \times Mb^{-0.69}$

* Body mass between 150–1,500 kg.

a giraffe grows, and the thicker walls reduce ventricle volume, which makes it difficult for stroke volume to increase significantly (see Chapter 9). However, in order to provide the necessary cardiac output at VO_2max, the combination of heart rate and stroke volume needed must include not only a higher heart rate but also a higher stroke volume. In a giraffe weighing 750 kg, an increase in heart rate from 70 to ~160 min^{-1} and an increase in stroke volume from ~ 540 mL to ~760 mL would produce a cardiac output of ~120 $L.min^{-1}$ and satisfy the requirements for oxygen delivery (Table 13.4). Cardiac output increases about threefold in other non-athletic ruminants (Taylor et al., 1987) and the higher heart rate, combined with the relatively small increase in stroke volume predicted for giraffes, produces an increase in cardiac output that is 3.2 times higher at VO_2max than it is at rest (Table 13.4). In a 750-kg giraffe, cardiac output at rest is ~40 L min^{-1}. At VO_2max it would have to be ~120 L min^{-1} if sufficient oxygen is to be delivered to muscles.

Oxygen Content of Blood during Exercise

Is an increase in cardiac output to 120 L min^{-1} enough to deliver the 25 L min^{-1} of oxygen that is potentially needed by a 750-kg giraffe at VO_2max? At rest the oxygen content of arterial blood (C_aO_2) in any giraffe is 172–193 mL L^{-1} because hemoglobin concentration is 132–142 g L^{-1} (see Chapter 12, Table 12.6) and each gram of hemoglobin can carry 1.30–1.36 mL of oxygen (depending on the pressure exerted by the oxygen available—the P_aO_2). If the C_aO_2 is 180 mL L^{-1}, then to deliver 25 LO_2min^{-1} a cardiac output of ~140 L min^{-1} (25,000 ÷ 180) is needed, even if all that oxygen could be taken up by muscles. The amount that is taken up by muscle cells is revealed by the difference between C_aO_2 and C_vO_2. At rest the difference, typically and, for example, in humans, is 45 mL L^{-1} and that volume represents about 25% of the total oxygen content of a liter of arterial blood. At VO_2max, typically 85% is given up (Weibel et al., 2004), and thus if the hemoglobin concentration did not change, then of the 180 mL L^{-1} of blood in giraffes at rest, just ~150 mL L^{-1} would move off hemoglobin into muscle cells. If so, cardiac output would need to be ~170 L min^{-1} to deliver sufficient oxygen, when the maximum cardiac output that appears to be possible in this average-sized giraffe is ~120 L min^{-1}. However, at a cardiac output of 120 L.min^{-1}, oxygen delivery would be sufficient if the C_aO_2 of blood was increased to ~250 mL L^{-1} of which 85% (210 mL L^{-1}) was released into muscles.

C_aO_2 could be increased if the hemoglobin concentration increased. The amount of hemoglobin present in blood is determined by the number and characteristics of the red blood cells in circulation. In primates like humans, the number of red cells per liter of blood changes very little from day to day but may increase as a result of exercise training at high altitude (contrary to popular opinion, that effect is moderate; Mortola & Wilfong, 2017), or dehydration or by fraudulent manipulation (a much bigger increase). In antelopes and horses, the number of red blood cells can be increased almost instantaneously at the start of exercise by contraction of the spleen, which stores a large number of them, an event that occurs at the start of exercise because of secretion of adrenalin. (The presence of red cells in blood increases blood viscosity, so storing them in the spleen when they are not needed makes it easier for the heart to pump blood around the body.) The effect of a "spleen transfusion" has been measured in both athletic and non-athletic non-primate species by Taylor et al. (1987) and they are similar: at VO_2max the C_aO_2 increases by 33% on average compared to its pre-exercise levels. If the number of red blood cells (and thus hemoglobin concentration and C_aO_2) increases by about 33% in giraffes during exercise (and from the range of hemoglobin concentrations measured in them, an increase of that proportion is easily possible; see Table 12.6 in Chapter 12), C_aO_2 will increase to around 240 mL L^{-1} of which 204 mL L^{-1} could be released, a volume that is just enough when cardiac output is 120 L min^{-1} to provide the needed 25 L min^{-1}. A fine balance thus exists between the amount of oxygen a giraffe weighing 750 kg needs when it is exercising at its VO_2max and what can be carried by, and released from, the hemoglobin in its arterial blood.

The amount of oxygen that enters muscle cells depends partly on blood filling the small capillaries supplying the muscle fibers (see Figure 10.1 in Chapter 10). As the muscles start contracting, the arterioles supplying the capillaries dilate, the capillaries fill, and their diameters increase. The consequences are that much more blood enters muscle tissue during exercise than it does at rest, blood flow through the capillaries becomes slower, and the time available for diffusion of oxygen into cells is prolonged. Just how big the surface area for diffusion can become can be estimated from the length of the capillaries (Table 13.5) and their diameter. The diameter of the capillaries in a mammal is about three-quarters of the diameter of a red blood cell (Snyder and Sheafor, 1999). As the diameter of a giraffe's red blood cell is about 5 μm, capillary diameter is ~4 μm. Thus the surface area of capillaries in exercising muscles could range from 700 m^2 in small giraffes to 6250 m^2 in adults (Table 13.5). (The surface area of a tennis court is about 200 m^2.)

A negative consequence of dilated blood vessels is that the resistance to blood flow in them falls (Table 13.4). While this fall has the advantage of making it much easier for blood to flow to and through the capillaries, it also has the potential to lower blood pressure (peripheral resistance is one of the factors that regulates blood pressure (see

Table 13.5 The Predicted Muscle Infrastructure Needed to Supply Oxygen at VO_2max in Giraffes

Variable	750-kg Giraffe	Range in Giraffes of Mb 150–1500 kg	Allometric Equation
Mitochondria volume (ml kg^{-1} muscle)	50	54–48	$72.3 \times Mb^{-0.057}$ $R^2 = 0.22$
Capillary length (km 10^5)	3.0	0.6–6.0	$0.004 \times Mb+0.02$
Capillary volume (L)	4.6	24–9.1	$0.006 \times Mb+24$
Capillary Surface Area (m^2)	3280	700–6500	$4.3 \times Mb+55$

Data from Weibel et al., 2004, with permission of Elsevier. Mb = body mass.

Chapter 9), and a lower blood pressure would reduce blood flow. During exercise, however, the mechanisms that maintain blood pressure in mammals produce a cardiac output that increases exactly enough to ensure that there is sufficient pressure to sustain blood flow to muscles whatever the resistance or level of exercise, and can be predicted to do so in giraffes (Table 13.4). The net effect of all these adjustments is that sufficient quantities of oxygen can be delivered to muscles and diffuse into the muscle cells. However, there is very little else that can improve oxygen delivery, so it seems clear that at VO_2max the cardiovascular system of giraffes is likely to be operating close to its maximum capacity and has little room for increasing oxygen delivery. Therefore, as in humans and all other mammals as far as we know, it is A.V. Hill's conjecture that limits VO_2max. It is the capacity of the cardiovascular system to deliver oxygen to muscles that is the likely limiter of VO_2max in giraffes.

Muscle Metabolism

Almost all, that is 95%, of the oxygen consumed by a mammal is used by mitochondria to generate ATP. Thus the maximum rate that a giraffe can consume oxygen during exercise ultimately is determined by the rate at which muscle mitochondria use it, and that rate depends on how many mitochondria are present and whether oxygen is being delivered to them in adequate amounts. The number of mitochondria in muscles of mammals has been worked out from their volume density (mL mitochondria mL^{-1} of muscle volume) by Mathieu et al. (1981) and Hoppeler et al. (1981), and the data have been consolidated by further research by Weibel et al. (2004) on athletic and non-athletic wild animals, including large African artiodactyls like wildebeest, waterbuck, eland, and also an ox that weighed 475 kg, but not a giraffe, so once again we have to assume that what happens in other large wild and domestic artiodactyls will also happen in small- to medium-sized giraffes and, probably, in larger giraffes. The number of mitochondria depends on muscle mass, and Weibel et al. (2004) found that muscle mass (kg) of mammals increases linearly with body mass, and was on average 36% of body mass. Thus muscle mass ranges from ~50 kg in a juvenile giraffe weighing 150 kg to ~550 kg in an adult, although these may be underestimates given the quite large amounts of muscle in the long neck. The relative volume of mitochondria in muscle (mL kg^{-1} of muscle) can be calculated from the data of Weibel et al. (2004) and is equal to 72.3 × $Mb^{-0.057}$. The correlation between body mass and mitochondrial volume is, however, poor (R^2= 0.22) but if it is applied to giraffes, the volume of mitochondria decreases moderately from 54 mL kg^{-1} of muscle in juveniles to 48 mL kg^{-1} in adults (Table 13.5). Given the poor correlation with body mass, it would be safer to assume that mitochondrial volume does not vary significantly and is ~50 mL kg^{-1} of muscle. Weibel et al. (2004) correlated VO_2max with the volume of mitochondria and found that at VO_2max, 1 mL of mitochondria consumes 5 mL O_2 min^{-1}, a consumption rate that is universal as it applies to "all mammalian species, whether small or large, athletic or non-athletic" (Weibel et al., 2004).

The VO_2 max of a 750-kg giraffe can be predicted to be ~25 L min^{-1}. If at VO_2max, 1 mL of mitochondria consumes 5 mL of oxygen each minute, the volume of mitochondria needed to consume 25 L of oxygen is 5,000 mL. In the muscles of a 750-kg giraffe there are ~270 kg of muscles and 50 mL kg^{-1} of mitochondria, thus its muscles actually contain ~13,500 mL of mitochondria in total, or about three times more than is needed, and this relationship appears to be true for all mammals studied so far (Weibel et al., 2004). Why do mammals have so many mitochondria, and why do the extra mitochondria not consume more oxygen? It is a mystery, but one reason could be that the cardiovascular system cannot supply sufficient oxygen. It would need a cardiac output of over 400 L min^{-1} in a 750-kg giraffe to provide enough oxygen for all its mitochondria when that giraffe's maximum possible cardiac output is approximately 120 L min^{-1}. If all mitochondria cannot be supplied with oxygen, then what is the selective, evolutionary advantage of having so many more mitochondria than is needed and that can be supplied with oxygen? This apparent anomaly has been studied in human athletes in whom the volume of mitochondria can increase 2.2-fold as one result of endurance training. Despite this increase, there is only a modest increase in VO_2max (Bassett and Howley, 2000). The additional mitochondria nevertheless produce a significant improvement in athletic performance because their main effect is to improve the efficiency of oxidation of fats and glucose by the muscles. If the same benefits are enjoyed by giraffes, then at their VO_2max they will be using just one-third of the maximum capacity of their muscles to consume oxygen, but will be making much better use of sources of energy.

The Economy of Running

The analyses of the combinations of interactions between the cardiovascular and respiratory systems that supply oxygen to exercising muscles apply best to giraffes using aerobic metabolism to run, and running at their VO_2max, but it is well established that no animal can run at their VO_2max for any length of time (e.g., Jones et al., 2010), and this will be as true for giraffes as it is for an elite human marathon runner. Elite marathon runners tend to have low body mass, which reduces the effects of gravity and thus reduces the effort needed to run, and they try to run at a speed that uses oxygen at a rate of about 80% of their VO_2max. They know that above that rate they begin to use anaerobic sources of energy, and fatigue sets in because the by-products of anaerobic metabolism inhibit the biochemical reactions that underpin energy supply. The fatigue serves as a warning to slow down. It is also a commonplace phenomenon that even though all the elite athletes that start a marathon have much the same or even an identical VO_2max, only one of them wins the race, and rarely does a group of them reach the finish line at the same time. Clearly other factors are important for separating the winner from the string of losers that follow him/her home, and the winner is usually credited with having better "running economy," which means that speed is maintained at a lower aerobic cost. If giraffes could achieve that state, then being able to gallop for long periods of time will be the result. Can they? How could a giraffe run for a long time at or near their VO_2max? Is there any other way that speed can be maintained and even increased at the same or lower metabolic cost?

The Cost of Transport

The essence of aerobic exercise is that oxygen is consumed to move an animal a distance, and the volume of oxygen consumed to move that distance is the cost of transport. In 1930, W.O. Fenn showed that the hard work of running in humans could be attributed to lifting body mass against gravity and to the speed at which body mass was moved forward. Both seem logical. Our own experience of exercise confirms it. Everyone knows that the faster you run, the more air you need and the more breathless you become. Large, obese people obviously use more energy to walk than do slimmer ones, and they do so because the work the muscles have to do to support their weight increases: working against gravity accounts for 40%–50% of muscle work (Pontzer 2007). For comparison, a mammal such as a sea lion uses a fraction of the energy for swimming that a terrestrial mammal uses for walking because in water sea lions are effectively weightless and, therefore, and just like fetuses, do not have to work against gravity. Taylor and colleagues in 1982 measured the contributions that body mass and speed make to oxygen consumption in 82 avian and mammalian species and concluded that the relationship they uncovered was generally applicable to all mammals (and birds) whatever their body mass was. They separated the effects of body mass and speed into two components. The first was the oxygen consumption at zero speed. It was related only to body mass and is equivalent to the Y intercept of a regression equation. For the subset of artiodactyls they studied, that component could be estimated from $0.21 \times Mb^{-0.27}$ $mLO_2\ s^{-1}\ kg^{-1}$ (Taylor et al., 1982). Applying that equation to a 750-kg giraffe, the predicted rate of oxygen consumption at zero speed (for example, a giraffe standing and browsing) would be 0.035 $mLO_2\ s^{-1}\ kg^{-1}$, or 1.6 L of oxygen every minute. The second component is the oxygen consumption while a bird or mammal is walking or running. This additional cost of transport (the milliliters of oxygen required to move a kilogram of body mass a meter; $mLO_2\ kg^{-1}\ m^{-1}$) depended on both body mass and the speed of movement. For artiodactyls ranging in body mass from a 3.5-kg suni to a 254-kg ox, that component was described by the equation $((0.79 \times Mb^{-0.41}) \times \text{speed})$ with the units of speed being $m\ s^{-1}$.

If the two parts of Taylor et al.'s equation are applied to giraffes, then the cost of walking or running for any giraffe can be estimated from its body mass and the speed at which it is moving. Unhurried, unthreatened giraffes, going about their daily business in the Eastern Transvaal lowveld of South Africa, moved at 0.06 to 0.1 km h^{-1}, or 0.02 to 0.03 m s^{-1} (Innes, 1958). They live a leisurely life. Leuthold and Leuthold (1978b) in Tsavo National Park, Kenya, where browse is more sparse, the walking speed of browsing giraffes was faster—between 0.5 km h^{-1} and 0.7 km h^{-1}, or 0.14–0.19 m s^{-1}—but still leisurely. If these latter speeds were to be maintained throughout a day by a juvenile giraffe weighing 150 kg, it would use ~130 kJ of energy for every kg of body mass to amble alongside its mother. Its mother, weighing 1,000 kg, would use ~70 kJ kg^{-1} day^{-1} to walk. The absolute amounts of oxygen (L) the mother uses would be nearly four times what its calf uses, but per unit of body mass, a small giraffe would use twice as much energy to walk as does its mother. If an occasional daily run is required, a small giraffe will use, per kg of its body mass, even more than its mother does. Female giraffes know all this physiology, of course. Giraffes have very large home ranges. On average it is ~70 km^2 (see Chapter 8), but it can be as small as 25 km^2 where browse is plentiful to ~2,000 km^2 in the Northern Namib Desert where browse is scarce (Fennessy, 2009). Apart from browse supply, many other factors play a role in determining the size of the home range, and age is one of them. Juvenile giraffes tend to have small home ranges, and large ones large home ranges. Typically, giraffes traverse their home range at a stately walk and so the size of the home range is matched to the cost of walking. When a female has a calf at foot, she knows that her calf will use more energy to cover the same distance she does, and so she confines herself to a much smaller home range than do adult males who, unencumbered by feelings of paternal duty, wander widely, and it costs them very little energy, either metabolic or emotional, to do so.

The speed at which a giraffe moves (m s^{-1}) is the product of stride length (m; the distance between successive footprints of the same foot) and stride frequency (s^{-1}). At a walk stride frequency of a (~600 kg) giraffe is 1.2 s^{-1} (Pennycuick, 1975) and its stride length would be 1.7 m to produce the speed of 2 m s^{-1} recorded by Dagg (1962a). What are stride length and frequency in a giraffe running at speed? An answer is provided from the films Alexander and colleagues made in 1977 of each of 10 different species of African ungulates, including four with body masses greater than 161 kg as well as an 1,100-kg giraffe. From the films, maximum running speed and stride length could be measured and stride frequency calculated by dividing speed by stride frequency. Stride frequency at maximum running speed fitted two allometric equations. One could be constructed from stride frequency's relationship with body mass (Frequency = $6.9 \times Mb^{-0.18}$) and the other from the distance between the foot and the hip ("hindheight" in meters; Frequency = $3.0\times HH^{-0.63}$). If those equations are used to estimate stride frequency in giraffes that weigh more than 160 kg and are running at their maximum speed, then, in an average-sized giraffe with a body mass of 750 kg and running at its maximum speed of 11.4 m s^{-1}, stride frequency will be 2.1 s^{-1}, and will range from 2.8 s^{-1} in a giraffe weighing 161 kg to 2.0 s^{-1} in a 1,100-kg giraffe, running at their maximum speeds. From stride frequency and maximum running speed, stride length in a 161-kg giraffe can be calculated to be 4.8 m, 5.4 m in a 750-kg giraffe, and 5.5 m in the 1,100-kg giraffe. These stride frequencies and lengths make sense. Small giraffes have legs about half the length of an adult's legs and can be expected to have to take more and shorter strides than would an adult to cover a distance. As legs get longer, stride frequency should fall and stride length increase. Alexander et al.'s equations were derived

Table 13.6 The Relationships between Body Mass, Maximum Running Speed, Leg Length, Stride Length, Stride Frequency and Energy Consumption Rate in Five Giraffes

Body Mass (kg)	Maximum Running Speed ($m\ s^{-1}$)[2]	Stride Frequency (s^{-1})	Leg Length (cm)[d]	Stride Length (m)[e]	Energy (kJ kg^{-1} min^{-1})[f]
160[0]	13.3	2.30[c]	131	5.8	1.48
220[0]	12.9	2.25[c]	144	5.7	1.37
720[0]	11.4	2.08[c]	203	5.5	0.82
1000[0]	11.1	2.03[c]	223	5.5	0.71
1100[a]	11.0	1.95[b]	229	5.6	0.69

[0] From height (m); body mass (kg) = $15 \times Ht^{2.7}$; $R^2 = 0.95$ (Chapter 5).
[a] From Alexander, 1977.
[b] From Alexander et al., 1977; Garland 1983a.
[c] From Dagg, 1962a.
[d] From Figure 13.5.
[e] Maximum running speed ÷ stride frequency.
[f] After Taylor et al., 1982.

from different species having different body masses or hind heights and do not necessarily describe what happens to individuals of the same species whose body masses and proportions differ through growth and, in the case of giraffes, also have a very different shape. However, when Anne Dagg (1962a) filmed three adult giraffes and two juveniles to measure changes in neck angle during a gallop, she also recorded other data from which their body mass and stride frequency can be extracted. These data are summarized in Table 13.6.

In those giraffes, stride frequency was 2.30 s^{-1} in the smallest giraffe and 1.95 s^{-1} in the largest, frequencies that are similar but not identical to the frequencies predicted by Alexander et al.'s equation but, nevertheless, showing that as body mass and leg length increase, stride frequency decreases. Stride length, surprisingly, was remarkably constant at 5.6±1 m, surprising because leg lengths increase as body mass and tallness increase (Table 13.6; Figure 13.5), and leg lengths would have been 131 cm in the smallest giraffe and 229 cm in the largest. Despite this apparent anomaly, and assuming that Dagg's giraffes were running aerobically at their maximum speed, then, as expected, the energy used by the smallest giraffe would have been 1.48 kJ kg^{-1} min^{-1}, 1.37 kJ kg^{-1} min^{-1} in the 220-kg giraffe, 0.82 kJ kg^{-1} min^{-1} in the 720-kg giraffe, and 0.71 kJ kg^{-1} min^{-1} in the 1,000-kg giraffe, amounts which emphasize that while it is speed and body mass that determine energy use, having long legs, a lower maximum running speed, and taking fewer strides per unit time are all useful for reducing energy consumption. The higher stride frequency and higher running speeds in small giraffes mean that they use more oxygen to run, and if so, then their endurance capacity, which depends partly on aerobic capacity, will be correspondingly lower. This likely lower endurance perhaps partly explains why the giraffes caught in the early nineteenth century to be taken to Europe were juveniles.

Mechanical Energy

Running economy would also be improved if some of the energy generated by movement could be captured rather than being lost as heat (e.g., Cavagna et al., 1977). When giraffes (or human marathon runners) run or walk, they generate a large amount of kinetic energy. Kinetic energy, as its name implies, is the energy generated by movement and it can be calculated by the very well-known formula

Kinetic energy = $\frac{1}{2}m \times v^2$

where the units of body mass (m) are kg, of velocity (v) are m s^{-1}, and of kinetic energy are Joules. When an adult giraffe weighing ~1,000 kg is running at 11 m s^{-1} or 40 km h^{-1}, it generates ~60,000 J of kinetic energy. When its fore-hooves hit the ground, forward movement is, for an instant, stopped, and all the kinetic energy becomes potential energy, which in turn can be reconverted into kinetic energy as soon as forward movement begins again. In a mammal, some potential energy can be stored as elastic energy. Most elastic energy is stored in muscles if they are long and their tendons are short and most in tendons if they are long and the muscles are short (Alexander, 1984) as they are in a giraffe. Tendons can store eight times more elastic energy than muscles can. In addition, the further a tendon can be stretched, the greater the amount of energy stored and the greater the force generated by recoil. The only tendons that can store and release significant amounts of energy this way during locomotion are those that extend the lower parts of the limb below the knee and ankle and the elbow and wrist, and they are stretched when those joints flex (Alexander and Bennet-Clark, 1977). About equal amounts of elastic energy are stored in the front leg and hind leg tendons (Alexander, 1984), and when downward pressure is released on stretched tendons, the tendons recoil and lift the body without muscles having to contract to do so.

Those tendons, measured in 18 adult male and 21 adult female giraffes (van Sittert et al., 2015), can be nearly two meters long (Figure 13.5; and see Figure 16.1 in Chapter 16), and like all suspensory ligaments they are exceptionally strong. The force at which they rupture is ~100 MPa (=1,000 kg.cm^2).

The amount of kinetic and potential energy that is produced at various running speeds (m s^{-1}) has been measured by Taylor and his colleagues (Fedak et al., 1982; Heglund et al., 1982) in animals with a body mass ranging from 30 g to 100 kg, so again none in the range of a giraffe's body mass, but in those animals, the kinetic energy produced (W kg^{-1} of body mass) could be calculated from $0.478 \times \text{speed}^{1.53}$. The potential and kinetic energy combined (W kg^{-1}) fitted the equation: $0.685 \times \text{speed} + 0.072$, was independent of body mass, and included elastic energy, as they assumed that all elastic energy was recovered within a stride. These equations, like those that describe the cost of transport, have very general application to all animals (Taylor et al., 1982) and when they are applied to giraffes running at their maximum speed, the combination of kinetic and potential energy produced would be ~300 kJ min^{-1} in a small giraffe and 1,800 kJ min^{-1} in an adult. Almost no additional energy would be needed to maintain a constant forward speed if all the potential and kinetic energy could be stored during a stride and used to re-accelerate the body (Cavagna et al., 1977). Of course if

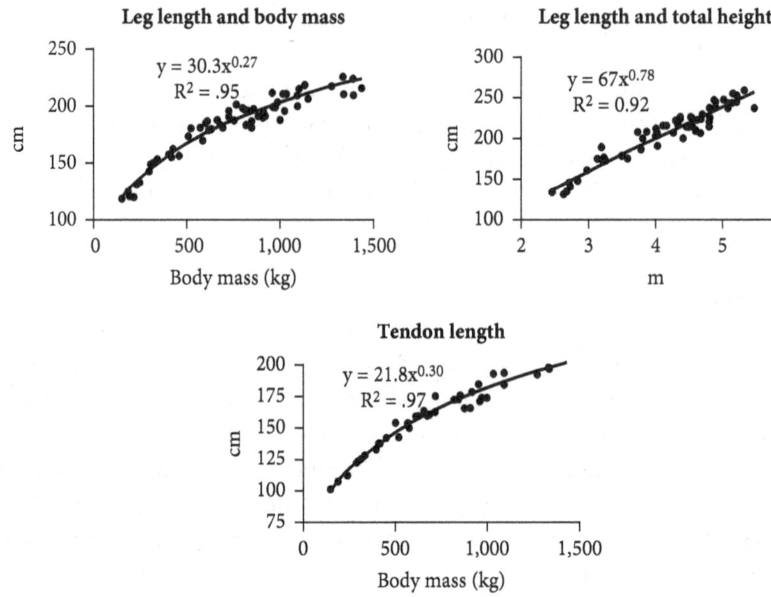

Figure 13.5. Changes in leg and tendon length as body mass increases, and leg length as total height increases in giraffes. Leg length is the mean of the fore- and hindleg lengths measured from the hoof to the axilla and groin, respectively. Tendon length is the mean of the lengths of the radius, metacarpus, and phalanges in the foreleg and the tibia, metatarsus, and phalanges in the hind leg. Giraffe data were obtained from 18 male *G.c. giraffa* and 21 female *G.c. giraffa*. The males weighed between 177 and 1,355 kg and the females between 141 and 988 kg.

Constructed using data from van Sittert et al., 2015.

this was possible, then giraffes, or indeed any other quadruped (or biped), would have discovered the secrets of a perpetual motion machine: once mechanical energy supply had been activated, they could run forever, hindered only by obstructions in their path. No such possibility exists, of course. Recovery of mechanical energy is never perfect.

Mechanical energy is not self-generating. Its origin is the expenditure of metabolic energy, and to generate 300 kJ of energy each minute, for example, requires the consumption of 15 L of oxygen min^{-1}. Nevertheless, energy that has been conserved by storing it in elastic tendons will reduce the metabolic energy needed to maintain a run. By how much? In humans, Cavagna et al. (1964) estimated that the high efficiency of running (40%–50% of energy coming from non-metabolic sources) "involves a contribution of a substantial amount of energy—from elastic recoil energy." No attempt has been made so far to measure mechanical energy in a running giraffe, and it is unlikely ever to be measured, but its contribution can, perhaps, be estimated from measurements made in other artiodactyls. Alexander (1977), for example, noted that the

> *total fluctuation of kinetic and potential energy which occurs in each stride, per unit of body mass, may be nearly the same for antelopes of a wide range of body masses galloping at top speed. The part of this which is stored as elastic energy in tendons may also be nearly constant, leaving fairly constant amounts of work to be done by muscles.*

In a 70-kg antelope galloping at maximum speed, Alexander found that recovery of energy that had been stored as elastic energy reduced the amount of oxygen used by 39%. In a camel weighing 400 kg and running at 6 m sec^{-1} (22 km h^{-1}), leg tendons stored and returned 45% of the kinetic and potential energy being generated by each stride (Alexander, 1984), thus reducing the metabolic energy required by that amount. The amount of elastic energy generated and the amount of metabolic energy saved in giraffes that have very long tendons must be at least that proportion, and if they are running fast, storage and recovery of elastic energy become even more important (Heglund et al., 1982). This conclusion has also been reached by allometric analysis of tendon and muscle shortening by Alexander and colleagues in 1981. It is very likely, then, that giraffes can reduce the metabolic energy they need for running through elastic recoil of tendons. It is equally likely that their cost of walking or running will be substantially less than expected, and their capacity to generate aerobic (and anaerobic) energy will be in excess of the metabolic energy actually used to run. If so, they would be able to sustain a gallop for a very long time, which is exactly what is recorded by many of the anecdotes.

And so it is that when Selous recorded that

> *they get over the ground, however, at a great rate, and it requires a good horse to run one down. The great thing is to press them to their utmost speed at first, when, if fat, they soon get blown and can be ridden into. At a hard gallop, however, they can spin along for miles*

this mix of exercise abilities can be explained. Their usual response to being startled is a high-speed sprint that depends mainly on anaerobic exercise and lasts only a few minutes, not because they run out of energy stores but because metabolic changes in their muscles inhibit energy production from the stores. However, the combination of a long neck and good vision usually alert them to danger well before danger is imminent. Then they can "spin along for miles—with little exertion and a long, sweeping stride" at a leisurely gallop of 40–45 km h^{-1}, sustained by a mix of anaerobic and aerobic provision of energy and the conservation of some of that energy as elastic energy in their tendons.

14
Keeping Cool

The idea that the main or even only selective advantage of tallness in giraffes is that it gave and now gives access to browse free of competition from smaller browsers has been almost undisputed from the time that Lamarck in 1809, Wallace in 1858, and Darwin in 1872 acknowledged the association between the height of giraffes and their feeding habits and used the association to explain the origin of species. However, the "access to browse" story was disputed by an agricultural scientist working in southern England, in a letter to the journal *Nature* entitled *"Evolution of the giraffes"* (Brownlee, 1963). Brownlee disputed the story on the grounds that as females and young giraffes are less tall than males, in a drought they would have to compete for food with smaller browsers and would succumb, a process that would, over time, lead to the extinction of the family. He was wrong, but he could not know that. Competition between giraffes and other browsers for browse is weak, and it is females and young adults that survive droughts best (Mitchell et al., 2010). Nevertheless, Brownlee proposed that if their shape had a more general, thermoregulatory value, that would have been a greater selective advantage because giraffes of all sizes and ages would benefit, not just the tallest, and in making that association he was absolutely correct.

Brownlee was extrapolating from an idea of Eugene Schreider, a French physical anthropologist, who had studied the geographical distribution of the shape of men (Schreider, 1950) and also the anatomy of body temperature regulation (Schreider, 1951). Schreider had found that compared to men living in temperate regions, those who lived in desert or tropical surroundings had a body surface area that was larger and limbs that were longer. "In a hot climate," he wrote, "a relative deficiency of body mass—or a relative excess of body surface—is an advantage." This idea, in turn, is an extension of Allen's Rule formulated by an American zoologist, Joel Asaph Allen, in 1877. The Rule melds thermoregulatory physiology with geographic distribution, and states that warm-blooded animals living in hot climates will have long limbs and a slender build, whereas those living in cold climates will have shorter limbs and a more compact body, and the rule is supported by observation. The advantage of a compact body for an animal living in a cold climate, for example a polar bear, is that it has a smaller surface area through which heat can be lost, and body heat is conserved. In warm climates, animals, for example giraffes, have the opposite shape, and body heat is lost more easily.

Neither Allen's nor Schreider's ideas were all that original. Schreider was rediscovering an idea first thought of more than a century earlier, in 1839, by Pierre Francois Sarrus and Jean-Francois Rameaux. Sarrus was a professor of mathematics and Rameaux a doctor of medicine and of science. Both were at the University of Strasbourg studying the temperature of endothermic (warm-blooded) animals, which they found was similar and constant no matter what the animal's size. They concluded that a constant body temperature is achieved by balancing the heat generated by metabolism with heat lost through the skin. Thus small animals that have a high metabolic rate would also have to

How Giraffes Work. Graham Mitchell, Oxford University Press. © Oxford University Press 2021.
DOI: 10.1093/oso/9780197571194.003.0014

have a relatively larger skin (body) surface area than would a large animal with a lower metabolic rate. This relationship between body surface area and heat loss has endured for the last 175 years as the Surface Law. The arithmetic of the Surface Law was established by D'Arcy Thompson (1917). He noted that "heat lost varies as the surface area, or the square of the linear dimensions, whereas heat gained by metabolism varies as the bulk of the animal, or the cube of the linear dimensions," and since then it has been accepted that the expected body surface area of a mammal is proportional to body mass raised to the power of two-thirds ($Mb^{0.67}$).

What Schreider did do, however, was to suggest that the anatomy of humans and, in particular, the relationship between their body mass and surface area had evolved as an adaptation to prevailing environmental conditions with the benefit of easier regulation of body temperature. In his 1963 article, Brownlee extended Schreider's idea to giraffes and proposed that in giraffes as:

> *In certain human races living in hot climates, dolichomorphism is an aid in achieving heat loss. So also to the giraffes, living as they do in hot climates, their dolichomorphic structure will serve a similar purpose to young and old, male or female continuously and not merely in times of drought, while at the same time enabling them to achieve that size and tallness which confers greater ability to evade, or defend against, predators and to reach a source of food otherwise unavailable to them.*

A dolichomorphic shape is tall and slender, and the human exemplars are the slender Masai tribesmen of Kenya who, like some giraffes, live in a hot and arid habitat. Slender humans do have a surface area larger than do short, stout humans of the same body mass (du Bois and du Bois, 1916), and it is certainly true that over the long course of their evolution, the shape of giraffes has become tall and slender. If giraffes do have a large surface area, then heat loss would be enhanced in them in the same way that it is in tall, slender humans, and it has been said that because of their shape, the size of the body surface area of giraffes is large, and larger than that of a rhinoceros, hippopotamus, or elephant of the same body mass (Dagg, 2014). For the shape of giraffes to have evolved to support the regulation of body temperature by enabling heat loss implies, first, that thermoregulation in a mammal is a crucial physiological function and, second, that giraffes need to lose heat more than they need to conserve heat. The second is dubious. The first is not.

In 1854, Claude Bernard, the great French physiologist and the founder of modern physiology, wrote: "*La fixité du milieu intérieur est la condition de la vie libre et indépendante*... [The stability of the internal environment is the condition for a free and independent life . . .]" (Bernard, 1974). It was a crucial insight and it has driven physiological, and a good deal of clinical, research ever since, as the numerous subsequent references to his writings confirm. He was referring to mammals and the many processes that regulate the internal environment in which the trillions of cells that make up a mammalian body exist. Temperature regulation is one of those. As a physiological variable it is unusual in that in most living things, body temperature is not actively regulated, but in the relatively few animals that do regulate it, the homeotherms (*homeo-* = constant), it is regulated quite tightly. Only birds and mammals are homeotherms. The average body temperature of birds (~41°C) is higher than it is in ruminants like giraffes (~39°C), which is higher than it is in monogastric animals like us and dogs and horses (~37°C), which is higher than that of marsupials (~35°C). Whatever the species, body temperature usually varies little over the course of a day, and its absolute value is, usually, much higher than environmental temperature.

Giraffe Surface Area

Confirmation of Brownlee's idea that a tall, slender shape is "an aid in achieving heat loss" depends partly on knowing what thermoregulatory advantages flow from that shape and partly on knowing what the surface area to body mass ratio is in giraffes. Giraffe body mass can be measured directly. Direct measurement of surface area of a giraffe or any large animal is, however, very difficult. One method used is "coating" in which the body is covered by an inelastic material that can be removed, spread out, and its surface area measured. Hogan and Skouby (1923) used gummed paper as the "coating" to measure cattle surface area. In another study, Elting (1926) used a purpose-built planimeter to measure surface area of cattle. The body mass of the 133 cattle measured by Hogan and Skouby, and Elting, ranged from 18 to 883 kg, and their surface area (m^2) fitted the allometric equation $SA = 0.14 \times Mb^{0.57}$, an equation which is the same as that derived by Brody in 1945 from 50 cattle ranging in body mass from 41 to 617 kg (Berman, 2003). Surface area can also be estimated from girth and length, assuming that an animal has the general shape of a cylinder. Walter Stahl (1967) developed an allometric equation, based on this idea, from over 100 measurements made on animals of a range of sizes from mice to horses. His equation is $SA (m^2) = 0.11 \times Mb^{0.65}$.

Giraffes do not have the shape of a cylinder, and the length of their legs and neck might lead an observer to think that their surface area is much larger than that of a standard short-necked, short-legged artiodactyl of the same body mass, such as an ox. Is it? So far there have been only two attempts to quantify the body surface area of giraffes. In the first of these, Henderson and Naish (2010) used a computerized digital slicing method based on illustrations and calculated that, in a theoretical giraffe weighing 1,611 kg, the surface area of the skin would be 17.7 m^2. If the Stahl and cattle equations are applied to a giraffe with a body mass of 1,611 kg, the resultant surface area is 13.3 m^2 (Stahl) and 9.7 m^2 (Hogan and Skouby; Elting); thus the surface area of Henderson and Naish's theoretical giraffe is much larger than the surface area of a standard mammal or an ox, and, therefore, supports Brownlee's idea that a giraffe's shape will aid heat loss.

However, in the second attempt, the surface area of 60 giraffes ranging in body mass from 141 to 1,358 kg was estimated as the sum of measurements of the surface area of subdivisions of their body shown in Figure 14.1 (Mitchell et al., 2017).

The surface area estimated this way was, very surprisingly, much smaller than the theoretical size calculated by Henderson and Naish, and was in fact similar to that of a standard mammal. The surface area calculated for the 60 giraffes fitted the regression equation: $SA (m^2) = 0.071 \times Mb^{0.70}$ (Table 14.1) and that equation should give a good idea of the surface area of any giraffe.

The surface area of each body part is summarized in Table 14.2. The trunk, which makes up 86% of body mass and provides virtually all (>95%) of the surface area in artiodactyls that have a more conventional shape, provided just 51% of total surface

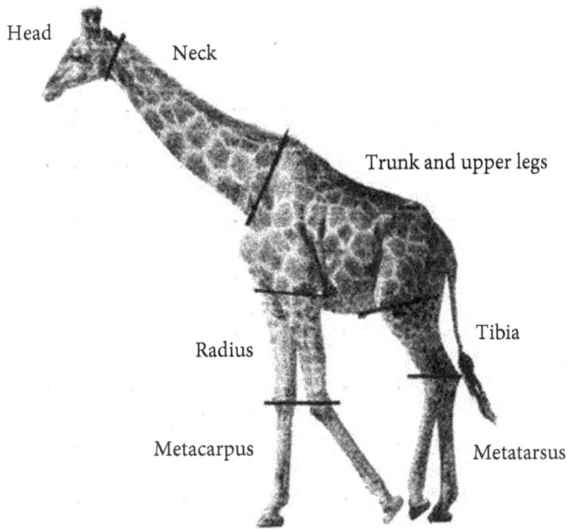

Figure 14.1. The components of the surface area of a giraffe.
Reprinted from Mitchell et al., 2017, with permission of Elsevier.

Table 14.1 Surface Area of the Four Body Parts of Giraffes Shown in Figure 14.1, and the Allometric Regression Equations with Body Mass (Mb) as the Independent Variable

Body Part	Surface Area in a 750-kg Giraffe (m^2)	Range (m^2)	Allometric Equation
Head	0.7	0.2–1.0	$0.005 \times Mb^{0.74}$
Neck	1.9	0.5–3.0	$0.011 \times Mb^{0.78}$
Trunk and upper legs	3.9	1.0–6.2	$0.027 \times Mb^{0.75}$
Lower legs	1.1	0.4–1.6	$0.027 \times Mb^{0.56}$
Total	7.3	2.1–11.8	$0.071 \times Mb^{0.70}$

Data from 60 *G.c. giraffa* with piecemeal body mass of 141–1,358 kg. Reprinted from Mitchell et al., 2017, with permission of Elsevier.

area in the 60 giraffes, the neck provided 24%, and the legs, which make up just 3% of body mass, provided 16%. The surface area of an average-sized giraffe weighing 750 kg was 7.5 m^2, and it ranged from 2.1 m^2 in the 141 kg female giraffe to 11.8 m^2 in an adult male weighing 1,358 kg.

These surface areas are much the same as those predicted by Stahl's equation (8.3 m^2 in a 750-kg giraffe; 2.7 m^2 in a 141-kg giraffe; 12 m^2 in a 1,358-kg giraffe), and similar to those of cattle (6.5 m^2, 2.4 m^2, 8.9 m^2, respectively). The change in total surface area in giraffes as their body mass increases compared to the surface area predicted from Stahl's general mammal equation is shown in Figure 14.2A. The metabolic rate (kJ kg^{-1} body mass) of young giraffes is higher than it is in adult giraffes,

Table 14.2 The Relative (%) Mass and Surface Area of the Four Body Parts of Giraffes Shown in Figure 14.1

Body Part	Mass (kg)	Mass (% of total body mass)	Surface Area (m^2)	Surface Area (% of total surface area)	Surface Area: Mass ($m^2\ g^{-1}$)
Head	6–39	2.9±0.5	0.2–1.0	8.9±0.7	0.033±0.005
Neck	14–138	9.0±0.8	0.5–3.0	24.4±1.3	0.15±0.02
Trunk and upper legs	117–1,147	85.5±1.0	1.0–6.2	51.0±2.3	0.006±0.001
Lower legs	4–34	2.6±0.1	0.4–1.6	15.6±1.5	0.06±0.02
All	*141–1,358*	*100*	*2.1–1.8*	*100*	*0.009±0.002*

Data from 60 *G.c. giraffa* with piecemeal body mass of 141–1,358 kg. Reprinted from Mitchell et al., 2017, with permission of Elsevier.

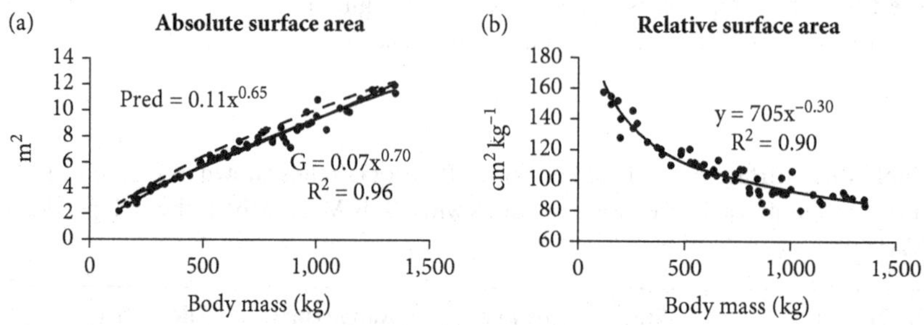

Figure 14.2. A: The change in total surface area of giraffe (●) as body mass increases and the change in surface area predicted from Stahl's general mammal equation (1967; dashed line). B: The decrease in relative surface area as body mass increases. Data are for 60 *G.c. giraffa* with piecemeal body mass of 141 kg to 1,358 kg.

Reprinted from Mitchell et al., 2017 with permission of Elsevier.

and following D'Arcy Thompson's arithmetic, the relative surface area ($cm^2\ kg^{-1}$ body mass) of young giraffes should be higher too. It is (Figure 14.2B): juvenile giraffes have a relative metabolic rate ~1.8 times that of an adult and a relative surface area ~ 1.6 times that of an adult.

In general, the ratio of surface area to mass for a giraffe's whole body ($Mb^{0.70}$) meets the expectation that it should scale with mass to the power of 0.67 (Table 14.3). The mean exponent of 0.60 for the ratio of lower leg surface area:leg mass shows that leg surface area increases at a slower rate during growth than the 0.67 expected, but that rate is offset by a faster rate of increase in head, neck, and trunk surface area (mean exponent = 0.75).

The fact that surface area of giraffes as calculated from their constituent parts is much the same as it is in standard mammals is unexpected because giraffes have a shape very different from that of a standard mammal. An explanation for this apparent anomaly is that the length of a giraffe's body, measured from the base of the neck to the base of the tail (see Figure 14.1), is much shorter than it is in a more "cylindrical" mammal of

Table 14.3 The Allometric Relationship between Surface Area (m^2) and the Mass (kg) of Each Body Part

Body Part	Allometric Equation	95% CI (exponent)
Head	$0.06 \times Mhead^{0.78}$	0.73–0.84
Neck	$0.08 \times Mneck^{0.73}$	0.71–0.75
Trunk & upper legs	$0.03 \times Mtrunk^{0.75}$	0.71–0.79
Lower legs	$0.19 \times Mleg^{0.60}$	0.58–0.62
All	$0.07 \times Mbody^{0.70}$	*0.68–0.73*

CI = confidence interval.

Data from 60 *G.c. giraffa* with piecemeal body mass of 141 to 1,358 kg reprinted from Mitchell et al., 2017, with permission of Elsevier.

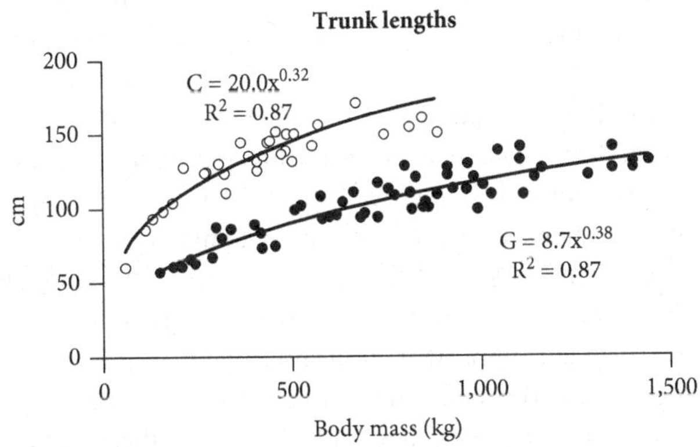

Figure 14.3. Trunk length in 60 giraffes (●) and 37 cattle (○). Cattle body mass ranged from 55 to 883 kg (Hogan and Skouby, 1923) and giraffe body mass from 147 to 1,441 kg.

similar mass, a disproportion first noticed in 1761 by Hendrik Hop, then in 1826 by M. Salze when he first observed *Zarafa* when she was in quarantine in Marseille, and in 1839 by Cornwallis Harris in his very scientific description of a giraffe's body shape (see Chapter 3). How much shorter is it? To illustrate the difference between them, the trunk's length and surface area of a giraffe can be compared with those of an ox. For example, the trunk length of a giraffe weighing 750 kg is 109 cm, and in a 750-kg ox it is 168 cm (Figure 14.3), so a giraffe's trunk is two-thirds as long as that of an ox. Their girths are similar: ~270 cm for the giraffe and ~250 cm for the ox. From these data, and treating the trunk as a cylinder, the surface area (m^2) of the trunk of a 750-kg ox is 6.6 m^2 and that of a giraffe 4.4 m^2. It is the combination of a smaller trunk surface area (51% of the total) and a larger neck and leg surface area (40%) that produces a surface area in a giraffe that is not significantly different from that of an ox, or any other mammal that conforms to Stahl's equation. These measurements of surface area imply that it is not a large surface area or a higher surface area to mass ratio that gives giraffes a thermoregulatory advantage, if indeed their shape confers such an advantage. If it is not surface area, does their shape have any other benefits for body temperature regulation that might support Brownlee's contention that "dolichomorphism is an aid in achieving heat loss"?

Giraffe Body Temperature

Giraffe are, colloquially, "warm-blooded" and homeothermic and, in accordance with that definition, they maintain a body temperature within a narrow range even though the environmental temperature of their habitat can range from being very cold to very hot. Giraffe body temperature is not easy to measure, but Table 14.4 summarizes some of the few measurements that have been made.

At first glance, one conclusion that can be drawn from these data is that giraffe body temperature is about 38.5°C and remarkably stable. But is that the case? Bligh and Harthoorn (1965) placed a thermocouple probe deep (8 cm) into the "dorsal caudal neck" of a young female giraffe weighing ~400 kg, and found that the temperature there varied between 37.8°C and 39.1°C, with an average of 38.5°C±0.5°C, over a range of environmental (ambient) temperatures of 10.5°C to 21.5°C. Langman (1973a) measured rectal temperature in 13 captive giraffes accustomed to being handled and found it to be 38.5°C±0.5°C. From data in Hetem et al. (2016) for ruminants generally, body temperature (in °C) can be calculated allometrically to be $40.3 \times Mb^{-0.008}$. Applied to a giraffe of either sex with a body mass of 150 kg, its body temperature would be 38.7°C, in a 750-kg giraffe it would be 38.2°C, and 38.0°C in a mature adult. Presumably, the decline in predicted body temperature as body mass increases is related in some way to changes in metabolic rate. These data imply that giraffe body temperature can vary, even though Bligh and Harthoorn (1965) concluded from their study that giraffes were among the most thermostable of the 20 species they studied, and that there was no hint that their body temperature varied widely under normal circumstances.

Table 14.4 Giraffe Body Temperature

Reference	Number of Animals	Number of Measurements/Method	Body Temperature (°C)
Bligh and Harthoorn, 1965	1	Continuous by telemetry for 46 hours	38.5±0.5; range 37.8 to 39.1
Langman, 1973a	13	13/rectal temperature	38.5±0.5
Langman et al., 1982	3	51/rectal temperature	37.3±0.9; range 35.7 to 38.8
Langman and Maloiy, 1989	6	Continuous by telemetry for 96 hours	No maximum and minimum reported. Diurnal range was 6.2°C
Mitchell and Skinner, 1993	3	8/rectal temperature	38.6±0.1
Hetem et al., 2016*	15 ruminants	Allometry: $40.3 \times Mb^{-0.008}$	38.3±0.2

*Predicted for *G.c. giraffa* with piecemeal body masses of 141–1,358 kg from allometry based on biologged temperatures in free-living large ruminants.

Reprinted from Mitchell and Skinner, 2004, with permission of Taylor and Francis.

The possibility that giraffe body temperature can undergo large variation was surmised from observations of the behavior of giraffes kept at the Sydney Zoo, Australia (Dagg, 1970). Over an ambient temperature range of 12°C to 37°C in the shade, giraffes appeared to be indifferent to ambient temperature and it was only at extremes of ambient temperature (for example, 54°C) that they sought shade. High body temperatures can also be a consequence of exercise. Giraffes can run at high speed over short distances and are able to run at 44–50 km h^{-1} for several kilometers, and this sort of activity will generate body temperatures of up to 42.3°C (Langman, 1973a). In 1982 Langman and colleagues made 51 measurements of rectal temperature in three giraffes and found that it varied by 3.3°C between 35.7°C and 39°C, a variation that correlated well with ambient temperature (see Figure 14.6 later in the chapter). The way to confirm if large variations of body temperature occur naturally is to measure it directly and continuously in free-living giraffes, and in 1989 Langman and Maloiy used radiotelemetry to do so. In six captive giraffes of 250–1,100 kg body mass, they found that body temperature varied by up to 6.2°C over 24 hours. In a similar study, continuous measurement of body temperature in a young giraffe showed that it varied by ~5°C over the course of a day (Mitchell and Skinner, 1993). Langman and Maloiy concluded that giraffes, like camels, were capable of storing heat and, in doing so, let their body temperatures rise in the hot part of a day and then at night when ambient temperature is colder, let their body temperature fall passively. This pattern is termed passive obligatory heterothermy. It is potentially a very valuable strategy in the hot and arid habitats that giraffes can occupy because it means that they do not have to use as much energy to keep warm or use precious body water to cool. Variation in body temperature does not mean, though, that giraffes are unable to control their body temperature. They can. How?

Body Temperature Regulation

Gaining Heat

For body temperature to be relatively stable, heat gained must equal heat lost. Body heat typically is gained in ruminants in four ways: by metabolism, by air convection if air temperature is greater than body temperature, by radiant energy from the sun, and by the heat generated by fermentation in the rumen (mainly) and the large intestine (see Chapter 8). Heat of fermentation is ~8% of total energy of fermentation—a small but significant amount that reduces the amount of heat that must be produced by a ruminant's own metabolism to maintain body temperature. In a 750-kg giraffe, total energy generated by fermentation is 66,000 kJ d^{-1}, of which 5,250 kJ d^{-1} is heat. Radiation is the transfer of heat by electromagnetic waves, and it is by far the most important way that heat could be gained or lost by an animal. While it is usually associated with heat gain, in fact the direction of heat transfer depends on the temperature difference between a source of heat and a sink for it, and as a result heat loss by radiation is just as frequent, if not more frequent, than heat gain. Animals absorb radiant energy when they are colder than the source of heat, and lose heat to surfaces that are colder than they are. During a hot cloudless day, animals will absorb radiant heat unless they seek shade and/or reduce the surface area exposed to the sun or to the radiation reflected off surroundings. At night, especially a cloudless one, they will lose body heat by radiation unless they seek shelter or invoke other heat-conservation mechanisms. The total amount of energy produced by metabolism in a medium-sized giraffe is ~115,000 kJ $^{d-1}$ and, as most of that energy is ultimately turned into heat, the heat produced is about 11 kJ min^{-1} m^{-2} of its body surface area. At the equator at noon, the amount of heat coming from the sun is around 60 kJ min^{-1} m^{-2} of body surface area. Thus, if all solar radiation were to be absorbed by a giraffe living on the equator, the heat absorbed would be equivalent to five times the amount of heat produced by metabolism. Avoiding radiant heat gain is essential if body temperature is to be kept stable.

In general, most metabolic heat is produced by muscles and the main internal, metabolically active organs (heart, kidney, liver). It might be expected that the neck of a giraffe would add metabolic mass. However, in giraffes the metabolic mass of the neck (muscle) constitutes only about 7% of body mass (Mitchell et al., 2013a). Similarly, the lower legs have virtually no metabolic tissue, but both neck and legs provide large surface areas, and offset metabolic heat without themselves being a source of metabolic heat. They function like radiators. Heat production by metabolism is not constant. Its rate is regulated by hormones that can increase metabolism, such as thyroid hormone or adrenaline, but it varies mostly because of the heat produced by movement. The mechanical efficiency of movement is low: just 20%–25% of energy used to make muscles contract is translated into movement. The other 75%–80% is released as heat. Why is this so, when evolution could have selected for high mechanical efficiency and excellent exercise ability, both of which would have been advantageous in animals that rely on speed to escape predation? A reason is that in mammals homeothermy has a stronger survival advantage. The heat produced by exercise, even if that is no more than walking, is of more benefit than an ability to exercise efficiently. Keeping warm needs a continuous supply of heat, and the inefficiency of exercise is that supply. As a low ambient temperature has prevailed throughout much of the evolutionary history of mammals, inefficient exercise was, perhaps, an outcome of natural selection.

Losing Heat

The main routes of heat loss are well known: evaporation, conduction, convection, and radiation, and the rate of loss is related to surface area. Thus, the long neck and long slender limbs that between them provide 40% of a giraffe's surface area should provide an avenue for achieving heat loss, and support Brownlee's contention.

Radiant Heat Loss

The heat unavoidably gained by a giraffe standing in the sun must be lost or its body temperature would increase. Most of the potential radiant heat is in fact not absorbed, partly because the surface area of a giraffe's body that is exposed to the sun at any moment is far smaller than the total body surface area of a giraffe; partly because the hair, especially light colored hair, acts as a barrier by reflecting the electromagnetic waves that are the source of radiation energy; and partly because a giraffe can escape the effects of radiation by seeking shade. If a giraffe is in shade, as they often choose to be (see Plates 14.1 and 18.2) and exposed to surfaces that are colder than skin or hair, heat will flow from the hair or skin to that surface. The net effect is that although radiant heat is potentially a very large source of heat for a giraffe in the habitat it prefers, that heat can be avoided or lost.

Heat Loss by Conduction

Heat loss by conduction occurs when heat flows from an animal's body to a colder surface in contact with it, like a cold concrete floor or soil. In the world of non-burrowing mammals that giraffes inhabit, heat loss by conduction is unlikely. Why? First, soil is an exceptionally poor conductor of heat. Further, assuming that a giraffe is lying down on the ground, as shown in Plate 14.1, conduction of heat from a giraffe's body to the soil will only occur if skin or fur temperature is greater than soil temperature. Soil temperatures, even in shaded areas of woodland savannahs preferred by giraffes, range between 28°C and 48°C, and even in the dense shade of an East African Forest it ranges from 25°C to 35°C (Cerling et al., 2011). A giraffe's skin temperature when it is losing heat is 32°C to 33°C (see Plate 14.2). Thus soil temperatures are usually hotter than skin or fur. Soil conducts heat at a rate of 2 W m^{-2} of surface area of contact between soil and skin/fur per degree centigrade of temperature gradient between the soil and skin/fur (Jessen, 2001), so if soil temperature is, say, 28°C, and a generous 25% of a giraffe's total skin surface area is in contact with that soil, then a 750-kg giraffe with a total skin surface area of 7 m^2 (Table 14.1) could lose 50–60 kJ of heat per hour, which is ~1% of the heat generated by metabolism (~5,000 kJ h^{-1}) and ~0.2% of potential radiant heat gain (~25,000 kJ h^{-1}). Giraffes occasionally do rest in shade and lie down to do so, and when they do their belly is in contact with the soil (see Plate 14.1; Innes, 1958), but they are far more likely to stand in shade (see Plates 14.1 and 18.2) than lie down and will avoid being in contact with a surface even slightly hotter than their skin.

Heat Loss by Convection

Heat loss by convection occurs when a fluid, which includes air, mixes with a fluid at a different temperature with heat transferred from the hotter fluid to the colder one. Heat loss by convection is greater than heat loss by conduction. In large East African antelopes (eland and hartebeest), convection from the surface of their fur accounts for 4%–6% of total body heat lost (Finch, 1972), but that proportion seems far too low to be of much use to a giraffe. How much convection contributes to heat loss in giraffes has never been quantified, but like all mammals, giraffes will transfer heat from the water or air on their skin to environmental air. The amount of heat lost by convection depends on the temperature gradient between the two fluids, the surface area exposed, the length of time of exposure, and the velocity at which the air or water is moving, circumstances that we humans try to minimize or even avoid to keep warm on a cold day. Of these factors, it is hard to be categorical about which are the most important, but two do stand out. One is surface area, especially the surface area of small animals, which is relatively much larger than it is in large animals, and is so in small giraffes (Figure 14.2). The other is air or wind velocity created by either air blowing over the body or movement of the animal through the air. The rate at which convection transfers heat is inversely proportional to the square root of the diameter of the body from which heat is being lost. The diameter of the neck of a 750-kg giraffe is ~680 mm, that of its legs is ~53 mm, and that of its thorax is ~800 mm (Mitchell et al., 2017), and it follows that convective heat transfer rate (m^{-2}) from the elongated neck and legs will be greater than it is from the trunk. The contribution of the legs and neck is amplified when a giraffe moves because movement creates a higher wind speed over them than it does over the trunk. But that is not the only benefit of long legs. Long legs elevate giraffes above the layer of low wind speed that exists close to the ground into a cooler layer, which increases convective cooling. This advantage is shared with humans, and in the first humans could have been a factor that made bipedal walking a mode of transport worth adopting (Wheeler, 1991).

The amount of heat lost through the legs and neck of a giraffe by convection depends on the amount of blood flowing to the skin because blood is at body temperature. The principle at work is that the thicker the layer of tissue between the outside of the body (the so-called shell) and the internal organs (the "core"), the better the insulation and the less likely is heat loss because body heat does not reach the skin. The ratio of shell:core can be altered by changes in blood supply. Eugene du Bois, famous for his analyses of body shape and surface area in humans, elegantly phrased it in 1916: constriction of skin blood vessels changes the skin and subcutaneous tissues into a "suit of clothes." Dilatation of skin blood vessels would achieve the opposite, and if du Bois had chosen to say so, "nakedness" would be the result. A question that arises is: how do skin cells survive if the blood supply to them is constricted during times of heat conservation? The answer is that skin cells are able to survive for long periods of time using anaerobic respiration to provide their energy. It is not only changes in blood vessel diameter that giraffes can implement, they can also direct blood away from their legs. From the moment of birth, giraffes can short-circuit leg blood flow so that the blood goes to the muscles in the upper hind legs (Kimani et al., 1991a; see Chapter 9), and by a similar mechanism to the muscles of the upper front legs (Petersen et al., 2013). This redirection of blood to muscles improves oxygen delivery to them and, thus, improves a giraffe's chances of escaping from a predator, but diversion of blood to the upper legs also and simultaneously reduces blood flow to the lower legs. The lower legs in particular have a very large surface area for their mass and a narrow diameter, both of which facilitate heat loss. Diversion of blood away from the lower legs when the ambient temperature is low would reduce heat loss not only because blood is not exposed to the large surface area, but also because the lower legs would not be warmed and their temperature would move toward the surrounding ambient temperature, thus reducing the temperature gradient between them. Diversion of blood toward the lower legs would achieve the opposite if heat loss is required.

Heat Loss by Evaporation of Water

One of the most effective ways large terrestrial mammals have of losing body heat is by evaporation of water from a body surface (e.g., Mitchell et al., 2018), which together with convection are the main ways that giraffes lose body heat. Evaporation is effective because to turn a milliliter of water into vapor requires 2.4 kJ of heat energy and some of that heat comes from the blood supplying the body surface from which evaporation is occurring. Two sites of evaporation are important—the airways and the skin. Whether evaporation occurs from the respiratory system membranes or from the skin depends on whether an animal can pant. If panting is used, as dogs do, then most evaporative cooling occurs through the respiratory system. If an animal cannot pant, then the main route of evaporation is the skin. Giraffes cannot and do not pant, so most of their evaporative heat loss will be by sweating. Of the two routes, sweating from the skin is in any case likely to be more effective in giraffes simply because the surface area over which it can occur is many times larger than the surface area of the upper airways and trachea (see Table 12.1; Table 14.2). The rate of evaporation from skin is determined by exactly the same factors and by the same amount that determines the rate of convective heat loss from skin (Mitchell et al., 2018) and thus, if a giraffe were to sweat, its narrow neck and leg diameters, especially if the wind speed over them is high, will enhance evaporation just as much as they enhance convective cooling.

Evaporation of water takes place down a water vapor pressure gradient and, consequently, for sweat to evaporate, water vapor pressure on the skin has to be higher than it is in air. The water vapor pressure of air is determined by its temperature and also by its humidity, and our common experience is that the lower the humidity (and higher the wind speed) the greater is evaporation of sweat, and this would be so in giraffes. Giraffes living in their natural habitat are rarely in an environment in which the water vapor pressure of air is higher than that on their skins and, if they sweat, they always will be able to evaporate it. As likely as it is that giraffes sweat, whether they actually sweat as a way of losing heat is unknown, but they do have active sweat glands (Dimond and Montagna, 1976) concentrated in the patched areas of the skin on the legs, neck, and trunk (Table 14.5). Analysis of the factors affecting their water balance (Chapter 11) suggests that the volume of sweat an average-sized giraffe produces daily could be 10 L, which, if all were to be evaporated, would remove ~25% of the metabolic heat produced daily. Can sweating remove all metabolic heat? If sweating were the only way that all the heat generated by basal metabolism, never mind the heat generated by exercise-induced metabolism, was lost, the percentage of body water needed every hour to dissipate it would be about 0.2% of total body water every hour (Jessen, 2001). For a giraffe weighing 750 kg, that would amount to 0.8 L h^{-1} or ~20 L d^{-1}, a volume that is ~70% of daily water intake from browse and metabolism combined for that giraffe (see Table 11.1 in Chapter 11). Sweating is, however, only likely to take place during those few hours of the day around the time of solar noon when environmental conditions are such that a giraffe's ability to lose heat by convection and radiation is declining. Giraffes can also reduce the need for evaporative cooling by behavior, and by minimizing metabolic heat production by exercising little. Thus, the actual volume of sweat lost will be far less than the volume that could be lost in a worst-case scenario. Even so, about 10 L d^{-1} of water could be lost as sweat (Table 11.1) and consequently giraffes will make every effort to avoid using evaporation of sweat to dissipate body heat. In the arid environment they inhabit, giraffes will sweat only as a last resort.

The other route of evaporation is the respiratory system. Evaporation in the respiratory system occurs when inhaled air on its way to the lungs moves over the mucosa that lines the nasal cavity and trachea. As the air flows past, water from the mucosa evaporates, the air becomes saturated, and the mucosa cools. How much the mucosa cools depends on the surface area for evaporation, which is provided mainly by the anatomy of the turbinate bones in the nasal cavity and to a lesser extent the surface area of the trachea. In giraffes the nasal cavity surface area and turbinate architecture are impressively large (Figure 14.4). Langman et al. (1979) estimated the length of the giraffe nasal cavity to be 40 cm, and the perimeter of the turbinate scrolls in giraffes to be twice that of the ox. From Spinage's data (1968b), the length of the nasal chamber housing the turbinate bones in 16 male and 10 female giraffes was 70% of skull length or 48.5±3.0 cm. The surface area of the nasal mucosa (cm^2) is linearly related to the nasal cavity volume (SA = 1.4 x – 0.48; Kamau, 1991). The volume of the nasal cavity in three adult male giraffes was found by Kamau to be 5,595 cm^3 (range 3,700–7,800 cm^3) and, for comparison, he found it to be 4,550 cm^3 in an adult male camel, and in three adult male eland it was 3,550 cm^3. The body mass of the giraffes was not recorded. The surface area of the nasal mucosa in these three giraffes was on average about 0.78 m^2 (ranging from 0.5 m^2 in the smallest giraffe to 1.3 m^2 in the largest giraffe), 0.64 m^2 in the camel, and in the three elands 0.50 m^2 (0.6 m^2 to 0.8 m^2). The surface area of the trachea of adult giraffes is ~0.4 m^2 (see Table 12.1 in Chapter 12). In addition to area, the efficiency of cooling depends on the velocity of air, and air velocity depends on the narrowness of the passage through which the air is moving. The narrower it is, the higher the velocity and the greater the amount of heat lost by evaporation. Ridewood's cross-section through the skull of a giraffe (Figure 14.4) shows just how elaborate the turbinate anatomy of a giraffe can be and just how narrow the spaces between the bones can be.

Kamau (1991) measured the width of the air passages to be 15±2 mm in a giraffe (it is not clear which "air passages" were being referred to), and Langman et al. (1979) measured the distance from the center of the air flow within a turbinate coil to be 2 mm from the turbinate surface. The degree of cooling that can be achieved in the nasal cavity is

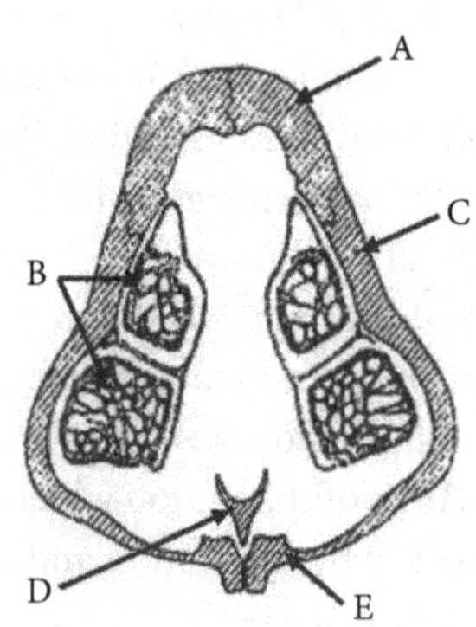

Figure 14.4. Section through the anterior part of the nose of a giraffe.
A= nasal bone; B = maxillary turbinate bones; C, E = maxillary bone; D = vomer bone.
After Ridewood, 1904, with permission of John Wiley and Sons.

illustrated by the difference between body temperature and nasal temperature measured in three giraffes (Figure 14.5). In those three giraffes, nostril temperature was 7°C colder than body temperature. The size of the temperature drop is related to ambient temperature: the colder the ambient air, the colder nasal cavity temperature becomes. A temperature gradient of 7°C occurs when ambient temperature is 24°C (Langman et al., 1979). What is also noteworthy about the data shown in Figure 14.5 is that the trachea plays only a small role in evaporative cooling. The average temperature of the air in the trachea is almost the same as rectal temperature.

How much heat is lost from the blood supplying the nasal mucosa is determined by the volume of air (L min^{-1}) that passes over the nasal mucosa. That volume is minute volume and, as Figure 14.6 shows, it varies because respiratory rate increases as body temperature increases and can double from 6 breaths per minute to about 12 breaths per minute if body temperature increases from 36°C to 39°C (Langman et al., 1982).

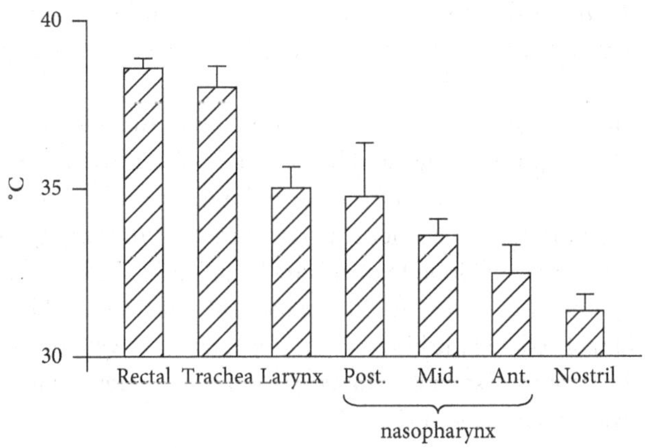

	Rectal	Trachea	Larynx	Post.	Mid.	Ant.	Nostril
N =	8	5	4	7	4	4	4
$\bar{x}$ =	38.6	38.0	35.0	34.7	33.6	32.4	31.3
SD =	0.1	0.6	0.6	1.6	0.4	0.8	0.5

Figure 14.5. Respiratory tract temperatures in three giraffes. N is the number of measurements made.

Reprinted from Mitchell and Skinner, 1993, with permission of Taylor and Francis.

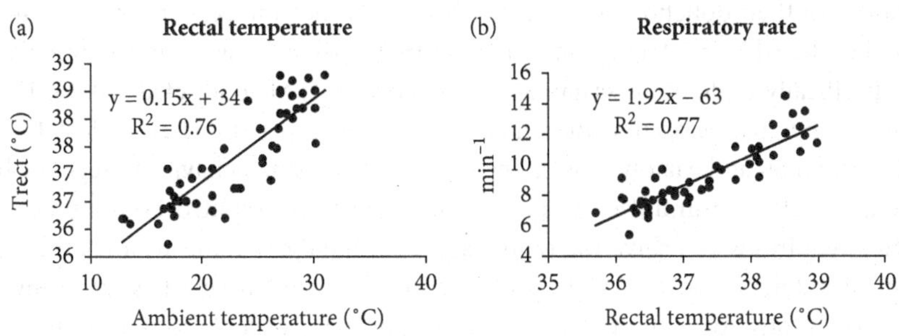

Figure 14.6. Left: The relationship between ambient temperature and rectal temperature (Trect). Right: The relationship between rectal temperature and respiratory rate in three resting giraffes.

Redrawn from data in Langman et al., 1982, with permission of Elsevier.

In an adult giraffe that is browsing quietly, minute volume can easily reach 130 L min^{-1}. The amount of heat lost by evaporation in the respiratory tract is small in a giraffe at rest (Langman et al., 1982) and amounts to 4,800 kJ d^{-1} in an average-sized giraffe, which is equivalent to the evaporation of 2 L d^{-1} of water (see Table 11.1). To that amount of heat, convective heat loss in the respiratory tract can be added. It can be calculated (Mitchell, 1974) from:

$$Heat\ loss\ (watts = joules/sec) = 0.02 \times V_E \times (Te - Ta)$$

where V_E is minute volume in L min^{-1}, Ta is inspired air temperature, Te is expired air temperature, and 0.02 is a factor that incorporates a mean density and specific heat for air (Mitchell, 1974). In a giraffe that has a minute volume of 130 L min^{-1}, and is breathing in air at 24°C and breathing out air that has a temperature of 31°C, heat lost is 18.2 watts, which translates into 1572 kJ d^{-1}. Evaporative and convective heat loss in the upper airways of a giraffe thus totals ~6,500 kJ d^{-1}. Metabolic heat produced in a 750-kg giraffe is ~115,000 kJ d^{-1}, so just 5% of metabolic heat can be lost through the respiratory system.

However, losing body heat is not the main function of the nasal heat exchanger in giraffes. More important is that the nasal blood that has been cooled by heat exchange is returned to the body via the *angularis oculi* and jugular veins. From measurements made in horses and pronghorn antelope, under some circumstances cooling of nasal blood can lower jugular vein blood temperature by up to 5°C below body temperature (Weishaupt et al., 1996; Mitchell et al., 2006; Lust et al., 2007) and, although it was not measured, presumably cooled the blood in the *angularis oculi* vein by the same degree. In effect, the nasal cooling system functions as a small refrigerator that cools the body, rather than as a way to remove body heat.

Behavioral Thermoregulation

High convective heat transfer in the neck and legs help giraffes dissipate heat only if air temperature is below skin temperature. As soon as air temperature exceeds skin temperature, heat will transfer from the environment to the animal by convection. Then the advantages of their dolichomorphic shape become disadvantages for a giraffe. Giraffes respond to the "threat" of a high air temperature by behavior, as do most animals and, more specifically, as do many savannah artiodactyls (e.g., Maloney et al., 2005). The use of behavior by giraffes occurs when their body temperature has risen by just 1.5°C to ~40°C (Langman and Maloiy, 1989). The advantage of using behavior to regulate body temperature is that it minimizes the need to sweat or shiver to control body temperature. Sweating requires water, shivering requires energy. If either can be avoided, then there is a net advantage to the animal. To get an idea of how sensitive giraffes are to environmental conditions, Anne (Innes) Dagg (1958) made many observations of giraffes in the Eastern Transvaal (now Mpumalanga) of South Africa and also on captive giraffes in the Sydney Zoo (Dagg, 1970). In the Eastern Transvaal the ambient temperature range was 0°C to 40.5°C, and in Sydney Zoo it was 12°C to 54°C. In the hottest part of the day in the Eastern Transvaal (i.e., between 12h00 and 15h00), Dagg saw that giraffes decreased their walking pace substantially and tended to lie down. However, she also saw that giraffes would lie or stand in the heat of the day, in sun or shade, facing in any direction, the direction not orientated to the position of the sun or direction of the wind. The significance of these observations is that in order to reduce radiant heat load, giraffes might be expected to seek shade, position themselves parallel to the sun's rays or lie down, or expose themselves to the wind. Dagg, in other words, could not find any behavioral pattern that could be linked to environmental conditions. In Sydney Zoo, shade seeking occurred only at very high ambient temperatures; thus, from Dagg's observations, it would be tempting to conclude that giraffes are somewhat insensitive to the temperature of their environments and do not change their behavior to minimize exposure to high ambient heat.

However, in a study of the feeding behavior of giraffes in Niger in summer (Caister et al., 2003), giraffes were not seen to feed after noon and spent most of their time resting in shade until early evening, a pattern that giraffes in Tsavo National Park in Kenya also follow (Leuthold and Leuthold, 1978b), where ambient temperature in summer reaches 49°C. In the colder temperatures of winter, the Niger giraffes foraged throughout the day. In the Northern Namib Desert, the distance traveled by giraffes during the hottest parts of the day was least, and it was highest in the cooler late evenings and early mornings (Fennessy, 2009). In Etosha National Park in Namibia, Kuntzsch and Nel (1990) observed thermoregulatory behavior of 290 giraffes and their results are summarized in Figure 14.7.

At ambient temperatures of 20°C, most (60%) of giraffes exposed the largest possible surface area of their bodies to solar radiation by standing perpendicular (lateral) to the sun's rays, but above 25°C the proportion standing perpendicular fell from 60% to 9%, and the proportion facing the sun increased from 35% to 60%, and thereby reduced the proportion of body surface exposed to solar radiation. As ambient temperature increased from 20°C to 37°C, the proportion of giraffes seeking shade increased from 5% to 35%. In the shade, radiant heat load is reduced and ambient temperature is lower and, provided that ambient temperature is below skin temperature, convective heat loss follows. Body size also affects behavior. When ambient temperatures in Etosha were

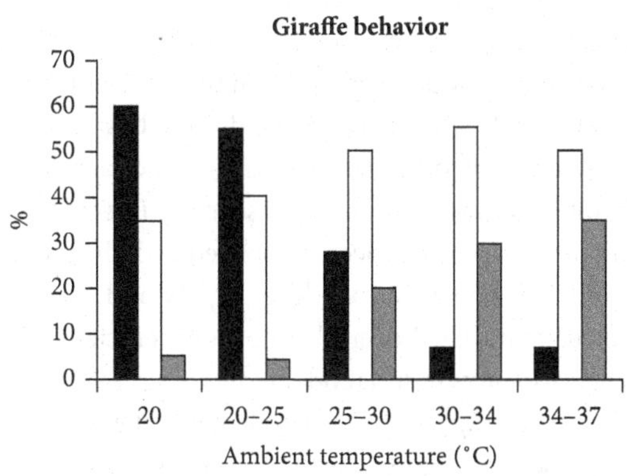

Figure 14.7. The percentage of a population of 290 giraffes in Etosha National Park, Namibia, standing perpendicular to the sun's rays (■), parallel to the sun's rays (□) or in shade (■) over a range of ambient temperatures.
Redrawn from Kuntsch and Nel, 1990.

greater than 30°C and solar radiation contemporaneously high, more than 80% of adult males (which have a body mass >1,000 kg and an average surface area of 8.2 ± 0.7 m^2) faced the sun, as did 60% of adult females (body mass >800 kg; SA = 6.5 ± 0.2 m^2) and 20% of calves (body mass <300 kg; SA = 3.7 ± 0.5 m^2), but 70% of calves selected shade, as did 32% of females and 13% of males. A reason that the calves and females sought shade more than the adult males did is because smaller giraffes have a larger relative surface area and gain radiant heat at a relatively higher rate than large ones do. Giraffe calves have a surface area:body mass ratio 1.6 times greater than that of an adult (145 cm^2 kg^{-1} vs. 91 cm^2 kg^{-1} body mass; Figure 14.2B).

The observations of Kuntsch and Nel seem conclusive, and Dagg's contrasting observations made in the bushveld of South Africa may have arisen because conditions in the Eastern Transvaal were not severe enough to elicit a clear behavior pattern, or because making consistent observations was not as easy as they were in the desert conditions of Etosha National Park. Another explanation for the differences is that behavior is related to water availability. Giraffes can live without drinking for long periods if they are eating succulent browse, but if water is available they will drink. Dagg and Foster (1976), for example, have reported that when ambient temperature was 32°C in Nairobi National Park, giraffes drank up to 45 L of water daily, and this amount decreased as ambient temperature decreased. The water available to Dagg's giraffes in South Africa was not recorded, but if they were well hydrated and water was freely available, they may have used physiological mechanisms like sweating rather than behavior to cool, with the advantage that they could continue to browse throughout the day.

The Anatomy of Giraffe Thermoregulation

Carotid Rete Mirabile

One of the great mysteries of artiodactyl evolution is why they have become such a successful, diverse group of 10 families and 90 living genera, compared with perissodactyls like horses, in which the number of genera has declined to just five that live now, from a peak of 66 in the Eocene era. Competition between the two has been ruled out as a cause of the decline of perissodactyls. One cause of the decline has been attributed to the seasonal change in the availability of food (see Chapter 7). Artiodactyls coped better with that change, but that was not the only reason for their success. The evolution of artiodactyls as a separate clade in the early Eocene has been associated with the mutations that produced a stable ankle joint and, subsequently, long legs. Agility and speed followed. Long legs themselves were followed in the late Eocene by the highly efficient selenodont molars that improved the ability to crush browse and render it digestible. In the subsequent Oligocene, those two adaptations were enough to make artiodactyls a stable, successful group, and we can imagine that they took advantage of the Oligocene to consolidate the mutations necessary for the later radiation into families. Many millions of years later, by the early part of the Miocene, when their body size had become large enough to require and accommodate it, their specialized ruminant gastrointestinal tract had evolved (Janis, 1976). The specialized tract facilitated fermentation of fibrous vegetation. Artiodactyl diversification began in earnest in the mid-Miocene about 15 Mya and gathered pace after the evolution of C4 grasses 8 Mya, a diversification that a ruminant digestive tract and agility are not enough to explain. Thus Janis (2007) wondered: "Is there something unique about bovid anatomy and physiology that enabled them to expand into the specialized grazing niches that opened up in the Pleistocene?" Evolution of an ability to grow a thick winter pelage and of behavior patterns, hibernation, migration, evaporative and convective cooling mechanisms, high metabolic rate, and mechanisms that alter the core:shell thickness ratio will all have contributed to bovid success, but none is unique to bovids. Most mammals have evolved one or more of them. What is unique to, and best developed in, bovids is a *carotid rete mirabile*—first described by Thomas Willis in 1664.

The *rete* evolved early in the evolutionary history of artiodactyls. It is present in pigs, which are the most primitive of artiodactyls, and also in dolphins, which, surprising as it may seem, are a marine artiodactyl. Pigs rely heavily on olfactory cues to detect food. In them the *rete* transfers odors and other volatile substances like pheromones from venous blood coming from the nasal mucosa to the brain (Grzegorzewski et al., 1997). The *rete*'s best-known function, first described by Willis in 1664, is "to dull the torrent of blood" going to the brain, and this is its function in dolphins (Nagel et al., 1968) and other artiodactyls (Lluch et al., 1985) and probably in giraffes (see Chapter 10). A third function and one that the data suggest evolved later is modulation of brain temperature. The anatomical position of the *rete*, lying as it does between the cooling system of the nose and the arterial blood supply to the brain, means that with little modification it could function to cool arterial blood destined for the brain. This modification most likely began its evolution in Leptomerycids during the Oligocene (Whitmore, 1952; Chapter 4) and became perfected during the Miocene. Hypertragulids probably did not have this structure (Whitmore, 1952), nor do modern Tragulids like mouse deer, but Gelocids and the artiodactyls that arose from the Gelocids do. The modifications involved rearrangement of the blood supply to the nasal mucosa and the venous drainage from it (Carlton and McKean, 1977; Johnsen et al., 1985; Mitchell et al., 2009c), so that cool venous blood in the *angularis oculi* vein draining blood from the nasal mucosa could pass the warmer arterial blood in the *rete* going to the brain.

The *rete* is well developed in giraffes (see Figure 10.8 in Chapter 10). As in all modern artiodactyls (except mouse deer), it arises from the external carotid artery as hundreds of small arterioles (up to 10 arterioles mm^{-2} of *rete* in pronghorn antelope; Mitchell et al., 2009c). Owen (1838), who was the first to describe its structure in giraffes, thought it was much less developed than it was in grazing ruminants, because giraffes browse upright and not with their heads below the level of the heart. Lawrence and Rewell (1948) revealed that the *rete* consists of two distinct halves in giraffes, each half supplying its own half of the brain. Goetz and Keen in 1957 and Kimani et al. in 1991 described its anatomy (see Figure 10.8) and Kamau (1992) measured the combined surface area of the arterioles in contact with cool venous blood. He found it to be ~260 cm^2 in giraffes, which is the same as it is in an eland, but smaller than it is in camels (325 cm^2). Thus giraffes have a *rete* of considerable size, and they also have the nasal anatomy needed to bring about cooling of blood, but do they actually use the *rete* to cool blood on its way to the brain, and is such a function necessary?

Brain metabolism produces heat continuously and the brain is sensitive to high temperature (although not more so than other organs), so removal of heat is essential.

When the brain gets too hot or too cold, the signs, unlike those for other tissues, are immediately obvious as abnormal and destructive behavior. If that consequence is to be avoided, the heat produced by brain metabolism must be lost, and usually it is, by convection, to the blood supplying the brain. The effectiveness of convective heat loss from the brain depends on the temperature gradient between the brain and blood and on the velocity of blood flow. The bigger the gradient between brain temperature and that of the blood supplying it and the faster the blood flow through the brain, the greater the amount of heat that is removed. The velocity of cerebral blood flow can vary between 25 mL 100 g tissue^{-1} min^{-1} and 80 mL 100 g tissue^{-1} min^{-1} (e.g., Mitchell, 1977). In mammals that do not have a *rete*, like us and horses, and the lesser mouse deer of Java (Fukuta et al., 2007), the temperature of brain tissue is higher than the temperature of the blood supplying it, which is at the same temperature as body temperature. This gradient between brain (at a higher temperature) and blood (at a lower temperature) is kept fairly constant at about 0.5°C±0.1°C (Figure 14.8B) by changes in blood flow rate. An increase in brain temperature produces an increase in blood flow and vice versa (within the range of 25–80 mL 100 g tissue^{-1} min^{-1}). The net outcome of the interaction between body temperature, brain blood flow rate, and brain metabolism is that in animals without a *rete* the range over which brain and body temperature varies is narrow (~37.5°C–~40°C; Figure 14.8B), and to make regulation of body temperature easier, the animals prefer to live in an environment of fairly constant temperature in the same way that we humans do. Also like humans, the animals must have access to abundant food to supply energy for heating when body temperature is low and water to allow sweating when body temperature is high, a set of circumstances that in all respects applies to those artiodactyls that do not have a *rete*, such as Java mouse deer (Whittow et al., 1977).

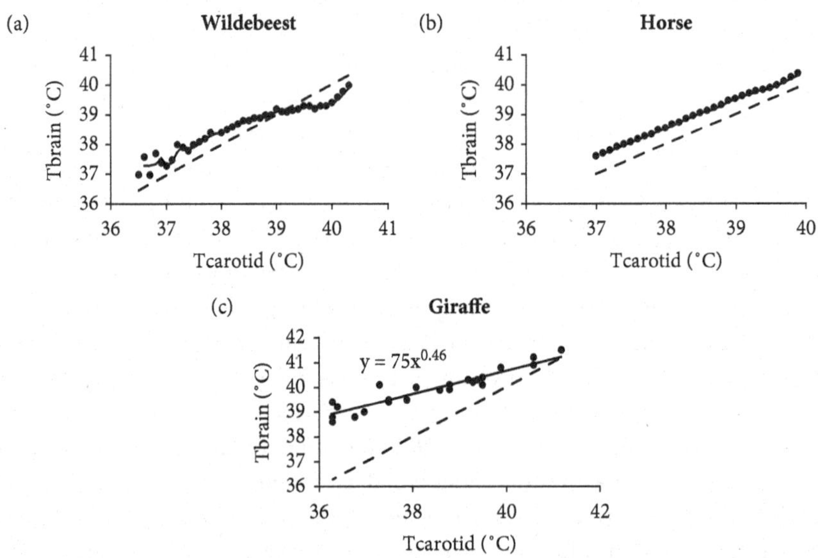

Figure 14.8. The relationship between the temperature of the hypothalamus T_{brain} (●) and the blood in the carotid artery ($T_{carotid}$) in wildebeest, horses, and a giraffe, and their temperatures in relation to the line of identical temperatures in both (dashed line).
Constructed from data in Jessen et al., 1994; Mitchell et al., 2006a; and Mitchell and Skinner, 1993.

Another way that heat can be removed from the brain by convection is by cooling the blood supplying it. Cooled blood increases the gradient between brain tissue and blood. Cooling of blood going to the brain is achieved by the *carotid rete.* The surface area of the *rete* arterioles is large, and the arterioles are bathed by blood in the venous sinuses that surround them. The sinuses are filled by cold blood returning from the nasal mucosa via the angularis oculi vein. Heat flows from the warm blood in the *rete* vessels to the cold venous blood surrounding them, and the arterial blood going to the brain can be cooled to such an extent that it can become colder than body (carotid artery) temperature (Figure 14.8A). The term "selective brain cooling" (SBC) was coined to describe this phenomenon. After the discovery of SBC by Baker and Hayward in 1968, it was assumed that the biological function of the *rete* was to protect the brain from thermal damage when body temperature unavoidably increases, for example during exercise. An ability to cool the brain in this way would prolong exercise, and would give its possessors an advantage over predators unable to cool their brains (Taylor and Lyman, 1972). This idea was compelling because it was logical and simple, but this supposed protective function of the *rete* has been disproved by studies done in Southern Hemisphere animals exposed to hot and arid conditions (Mitchell et al., 2002), and in reindeer in cold conditions (Aas-Hansen et al., 2000): brain temperature is not cooled below arterial temperature during strenuous exercise. What the *rete* can do, however, is to cool the hypothalamus, which lies at the base of the brain. The benefit of doing so is that the costs of regulating body temperature are reduced. The hypothalamus is the origin of many of the signals that increase or decrease body temperature. The signals initiate whole body warming when the hypothalamus is cold and are absent when it is hot. If the hypothalamus is cool, activation of thermoregulatory mechanisms such as sweating is subdued, even if the body itself is hot. Consequently, body temperature varies because it is less well regulated, but for example, the volume of water needed for cooling can be reduced by up to 60% of daily water requirements (Strauss et al., 2017). This type of thermoregulatory protection allows artiodactyls that have a *rete* to live within the Arctic Circle and in deserts and in equatorial forests (Mitchell and Lust, 2008). A *carotid rete mirabile* is what is unique about bovid anatomy and physiology. The *rete* enabled them to expand into specialized grazing niches, and it provides a missing, elusive mechanism sought by Janis to explain artiodactyl success.

So far only one attempt has been made to measure *rete* function in giraffes (Mitchell and Skinner, 1993). The data obtained at intervals during 50 h of recordings in a single, young giraffe were unreliable in many ways. For example, the gradient between the brain (near the hypothalamus) and the temperature of blood in the carotid artery was unusually large (1.5°C±0.8°C) when it is compared to gradients measured in other artiodactyls (Figure 14.8A). Nevertheless, when "brain" and carotid artery temperatures obtained from this giraffe are plotted in the manner shown in Figure 14.8A, what appears is the clear, unmistakable, and typical trend of brain temperature higher than carotid temperature when carotid temperature is low, and brain temperature approaching carotid temperature as carotid temperature increases (Figure 14.8C). The range of brain temperature measured near the hypothalamus was narrow (39°C–41°C); the range of the temperature of carotid artery blood (and thus body temperature) was 4.7°C (36.3°C–41°C) and thus was wider than brain temperature but similar to the daily range of body temperature found by Bligh and Harthoorn (1965) and Langman and Maloiy (1989) in giraffes. The narrowing gradient between brain and blood temperature means that the brain is being cooled as carotid temperature rises and, although the trend line suggests that brain temperature will fall below carotid temperature, no such temperatures were recorded. Although these data probably are flawed, it is likely that the relationship that exists between brain and carotid temperatures in giraffes is similar to the typical artiodactyl pattern exemplified by wildebeest (Figure 14.8A). If so, then it helps to explain why giraffes can flourish in the hot, dry savannah plains of Africa. In them, as in other artiodactyls, the *rete* probably reduces the costs of thermoregulation, especially the cost of evaporative cooling.

Ossicones and Skull

Horns, typically, are weapons and often are only found in the male of a species but, in giraffes, both sexes have them, in the form of ossicones. Consequently, Tim Ganey and colleagues (1990) suggested that ossicones may have a thermoregulatory function, a conclusion they came to after studying the anatomy of the ossicones, the blood and nerve supply to them, and the relationship between the ossicones and the sinuses overlying the brain. The sinuses enlarge as a giraffe grows, and as they enlarge, the position of the brain in the skull moves away from the roof of the skull to a more protected and insulated position under the thick overlying parietal bone (Figure 14.9; Figure 15.7). Ganey et al.'s conclusions were based on the size of the artery supplying the ossicones with blood and the path of the veins carrying the blood returning from the ossicones. The veins ran through the sinuses (Figure 14.9).

It does seem anomalous that such a large artery should be needed to supply the nutrients for the skin covering the small and unimportant ossicones and their bony core, and Ganey et al.'s explanation for the artery's size was that ossicones have a thermoregulatory function. The nerves within the substance of the ossicone could sense local ambient temperature, and if it was low enough and if cooling was needed, the nerves would direct blood through the cornual arteries to the ossicones. Heat would be lost from the warm blood going to the ossicones by convection to surrounding colder air. As a result, blood in the veins would carry cooler blood to the sinuses and a layer of cold air could form between the brain and the environment. Ganey et al. also raised the possibility that the nerves might stimulate sweating in the skin covering the ossicones and thus enhance cooling. It is also possible, of course, that because of their position, ossicones are exposed to a high radiant heat load during the day. If the heat load was high enough, the nerves would constrict the cornual arteries and thus prevent hot blood from returning through the sinuses. Contemporaneously giraffes would seek shade (see Plates 14.1 and 18.2). However, the two ossicones taken together have a very small surface area, even in a large adult male giraffe (~300 cm^2), and in an adult female their surface area is half of that, so their capacity to alter the temperature of blood flowing from them, and thus their ability to affect sinus or brain temperature and behavior, is correspondingly small and probably insignificant, especially because brain temperature can be regulated more efficiently by other means.

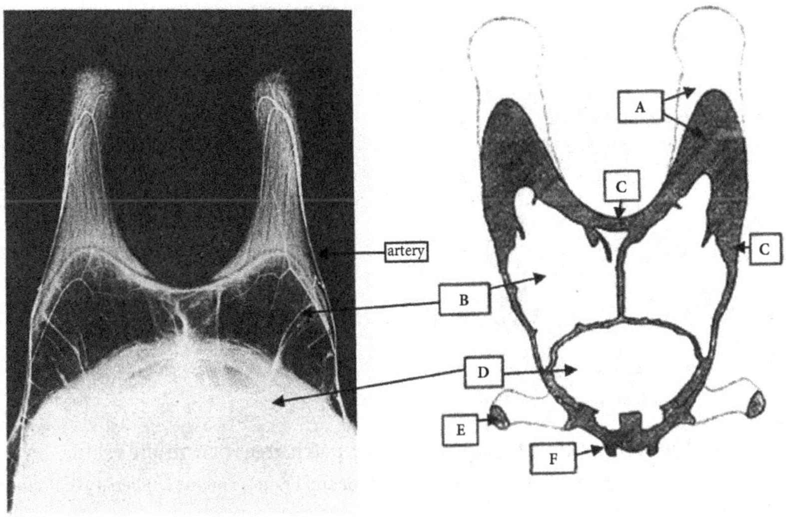

Figure 14.9. Left: The blood supply to the ossicones of a giraffe visualized by radiography. From Ganey et al., 1990 with permission of John Wiley and Sons.
Right: The positions of the ossicones and ossicone boss (A), fronto-parietal sinus (B), and the cranial cavity (D). C = parietal bone; E = squamous bone; F= sphenopalatine bone. After Ridewood, 1904, with permission of John Wiley and Sons.

The Skin and Thermal Windows

Traditionally, the skin of giraffes is regarded as having two unique functions. One is that it acts as a tight G-suit to counteract the high blood pressure present in giraffes. That function depends on the structure of the skin. The other function is camouflage. It depends on the shape and position of skin patches. Camouflage is especially important in young giraffes (see Chapter 17). But coat patches also have another function. That function, first suggested by Skinner and Smithers (1990), is that patches "act as thermal windows through which giraffes dissipate heat." This idea has its origin in an observation made by a Dutch veterinarian, L.F. De Beaufort (1927), after he carried out an autopsy on a young giraffe that died in Amsterdam Zoo and had lain on its side overnight. After death there is a strong tendency for blood to gravitate to the lower side of a recumbent animal. De Beaufort noticed that on the lower side, where blood had settled in large blood vessels in the skin, the vessels formed polygonal patches "very much like the pattern of the hide" around the patches, and that "the course of the blood vessels corresponded with the white lines" between the patches. Numerous smaller blood vessels branched off the blood vessel surrounding a patch and entered the skin underlying the patches. De Beaufort's observation was followed up 50 years later by Errol Ackermann (1976), studying under John Skinner. Ackermann devised a technique for visualizing small skin vessels, and he saw that in a giraffe fetus, although it was completely hairless, blood vessels covered the entire surface of the skin and that they "form a perfectly defined pattern that looks exactly like that of the adult's pelage": a blood vessel encircled the area of skin that would become a patch.

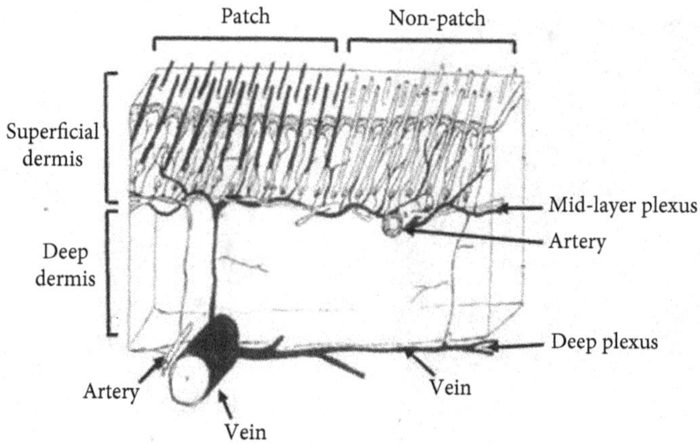

Figure 14.10. The blood supply to patch and inter-patch areas of a giraffe's skin.
After Ackermann, 1976. Reprinted with permission of Taylor and Francis and the University of Pretoria, South Africa.

Ackermann's study produced the skin blood vessel anatomy shown in Figure 14.10. Under a patch and lying about 20–30 mm below the skin surface there is an arrangement of blood vessels (a deep plexus) consisting of a small artery and a large vein. Between patches there is a mid-layer plexus that lies about 10–15 mm below the skin surface. This latter plexus consists of a large artery and a network of small veins and arteries. The large artery is muscular and confined to the junction of the deep and superficial dermis. From it, numbers of smaller arterioles branch off, more or less at right angles, and run toward the center of each patch. The large arteries in the skin bearing white hair and which surround a patch were the ones seen by De Beaufort. The whole arrangement implied to Ackermann that the blood vessels under the patches could be intermittently supplied with blood, and that blood was drained from the patch by the large vein under it. Inter-patch skin does not have this arrangement. The conclusion Ackermann drew from this anatomy is that arterial blood can be sent to the patch when body temperature is high and heat loss is needed. Patch skin temperature would rise and heat loss would be promoted. When body heat needed to be conserved, vessels underlying patches would be constricted and patch temperature would approach air temperature, thus reducing the temperature gradient between the two.

Whether these changes occur in giraffes can be judged from the temperature of patches, and that temperature has been captured in many thermographs of which one of the first taken is shown in Plate 14.2.

This image was taken on a hot day by Sabine Hilsberg, who was studying for a doctoral degree at the University of Leipzig. Patch temperatures were about 32°C–35°C and all the patches were at the same temperature (within 1°C), and the center of each patch was 1 °C–2°C hotter than the edges. Inter-patch temperatures were ~5°C colder. What happens on a cold day is unclear, but it could be expected that patch temperature would be lower because the blood vessels supplying them would constrict, inter-patch temperature would be even lower, and leg temperature colder still because the blood supply to the legs would have been reduced. This thermograph does confirm Ackermann's conclusions about how changes in blood flow to patches would contribute to heat loss, but another explanation for the clear differences between patch and inter-patch temperatures could be that the hair over patches is longer than it is over the non-patch areas. Hair can reduce heat loss by creating a layer of warm air between the skin and the environment, a trick which reduces the temperature gradient between skin and air. Wild giraffes, however, are not known to grow a thicker coat in an African winter or to change the nature of the hair over patches or non-patch areas. The length of the hair of wild giraffes is the same in both areas (Ackermann, 1976; see Chapter 16). However, Hilsberg's thermograph was taken in a zoo in Germany, and the hairs of giraffes exposed to the cold of European winters can become two and a half times longer than usual (Hagenbeck, 1910). Whether under those conditions hair length over patches differs from hair over inter-patch areas is not known, but if there is a difference, hair length may have contributed to the temperature patterns (although I think that is unlikely as the pictures were taken on a hot day, not a winter's day). It is also possible that the thermographs show nothing more than that the differences in temperatures arose because the hair on patches is dark and so absorbs more radiant heat. However, if that were the case, the patches on the back and thus exposed most to the sun would be hotter than those on the shoulder or flank, and that does not appear to be the case. The prevailing ambient temperature when Hilsberg took the picture is not known to me, but because patch temperature is high it implies that the patches were acting as thermal windows through which heat would have been lost if patch temperature was higher than ambient temperature. (Hilsberg also concluded that the hot, dependent belly was a sign of pregnancy.) In addition to this image, Hilsberg took pictures of giraffes that had been out of the sun for more than 12 hours and were walking fast in an indoor enclosure at an ambient temperature of 12°C. In those giraffes, patches over muscles were warmer than non-patch skin in the same area. If so, then the diameter of blood vessels under patches may also be controlled by local mechanisms and not only by the central signals that regulate whole-body temperature.

Patches would be even more effective as thermal windows if sweating was an additional function of them. Sweating in other large herbivores in East Africa begins at skin temperatures of 33°C–36°C (Finch, 1972), which are the temperatures of a giraffe's skin patches when they are losing heat (see Plate 14.2). Whether the skin under patches can sweat was evaluated by analysis of the skin anatomy of 10 patch and 9 non-patch areas taken from skin of the legs, trunk, and neck (Mitchell and Skinner, 2004) and the data obtained are summarized in Table 14.5. If patches do produce more sweat than non-patch areas, then they would have more sweat glands, or any sweat glands they had might produce more sweat. In either case, the glands would need a large blood supply. The data in Table 14.5 show that the area occupied by sweat glands in the skin under patches is almost twice as large as in non-patch skin and the blood vessel underlying them is larger. One way of deciding if skin blood vessels under patches are bigger is to compare their inner diameters. The bigger the inner diameter, the bigger the blood flow, and in this case the data suggest that the inner diameters of vessels under patches were on average about twice those in non-patch skin. The diameter will change depending on whether the blood vessel muscles are constricting to prevent heat loss or dilating to promote it. Sweat glands of the type found in giraffe skin release their secretions into hair follicles from where the fluid moves down the hair shaft by capillary action to the skin, so one reason for the increased sweat gland area in patches could be that patches have more hair follicles and, therefore, more glands. However, there are slightly fewer hair follicles in patch skin than there are

Table 14.5 Characteristics of Giraffe Patch and Non-Patch Skin

Site	Blood Vessel Inner Diameter (μm)	Number of Hair Follicles per Section	Sweat Gland Area (mm^2)
Trunk patch	75.6±7.4	3.3±0.6	17.0±4.8
Neck patch	31.0±1.9	3.7±1.2	13.7±4.7
Leg patch	18.7±2.5	3.0±1.6	16.8±4.6
All patch	*43.1±3.9*	*3.3±1.1*	*15.8±4.7*
Trunk non-patch	26.7±7.5	4.0±1.7	9.9±5.6
Neck non-patch	18.0±1.1	4.0±1.0	6.5±3.5
Leg non-patch	18.8±3.6	3.7±1.2	11.3±8.7
All non-patch	*21.2±5.9*	*3.9±1.3*	*9.2±5.9*

Reprinted from Mitchell and Skinner, 2004, with permission of Taylor and Francis.

in non-patch skin, and the large sweat gland area is, consequently, a function only of the size, efficiency, or number of the glands themselves. That patches on a giraffe's skin act as thermal windows therefore seems likely, and the patches are more complex structures than their conventional roles as camouflage and G-suit suggest.

A Summary

From all the studies that have been done on giraffe thermoregulation, it is possible to conclude that modern giraffes have evolved an array of heat loss mechanisms, but may have difficulty keeping warm. That is a conclusion that can be made from the life and times of ancestral *Giraffa* (Chapter 6). There are also many anecdotes of giraffes being caught and taken to Europe in the summers of the nineteenth century and dying a few months later in winter, and winter is still a trying time for giraffes living in a zoo during a Northern Hemisphere winter. Giraffes in winter in a Northern Hemisphere zoo would become hypothermic were it not for the expensive indoor accommodation built to replace the physiology that operates in their natural habitat (Clauss et al., 1999). But it is not only Northern Hemisphere winters that are a trial. Prolonged spells of cold in the Southern Hemisphere are known to be fatal to free-living giraffes (Walker et al., 1987; Chapter 18), because the effects of a cold winter are compounded by poor quantity and quality of winter browse. Severe cold is fatal because the shape of a giraffe is a disadvantage, especially so in juveniles which have a much higher relative surface area to body mass ratio than do adults: during cold weather too much body heat can be lost by convection through the neck and limbs, despite the ability to short-circuit blood destined for the lower legs and despite having well-innervated muscular arterioles in skin that regulate the amount of blood flowing to the skin and thus the amount of heat going to it. Prolonged cold, therefore, might easily have contributed to the extinction of some giraffe ancestors, with survivors being able to follow warm temperatures as much as they followed sources of nutrients.

It would have been impossible, in other words, for giraffes to have survived and flourished without sophisticated mechanisms for regulating their body temperature. A stable body temperature optimizes energy and water balance, creates an internal temperature suitable for the biochemical and physiological functions that sustain life, and allows giraffes the freedom to choose where they want to live. Giraffes like living in hot, arid habitats, and in that kind of habitat being able to lose body heat is an important advantage. When Brownlee contended that it was facilitation of heat loss that underpinned the evolution of the giraffe's dolichomorphic body shape, he offered no explanation of how it could do so. The data now available indicate that losing heat is not related to a high surface area to mass ratio. It is the slender, narrow, long necks and legs of giraffes that confer benefits for convective as well as evaporative heat loss.

In what thermal environments did giraffes evolve their dolichomorphic shape? Their morphology would have been shaped by the conditions that existed before or at the time of the appearance of *Bohlinia*, the first genus with a giraffe-like shape, and reinforced by the conditions that existed at the time of the radiation of *Giraffa*. This latter period was a time of substantial climate change (see Chapter 6) lasting many hundreds of thousands or millions of years. Ambient temperature then was much higher than it is now (Knorr et al., 2011), and occupation of open woodland then, as it does now, would have exposed ancestral giraffes to high solar radiation. Therefore, Brownlee's contention is feasible. It is, of course, quite impossible to gauge the relative contributions that the many anatomical, physiological, biochemical, and behavioral adaptations in giraffes have made to the evolution of their shape, and impossible to know for certain what evolutionary pressures existed at the time they were evolving, but it is possible to say that their elongated, dolichomorphic shape was not a disadvantage. On the contrary, the thermoregulatory advantages that flow from it are significant and may have supplemented, or been a component of, the selection pressures that resulted in the shape of giraffes. As Brownlee noted:

> *dolichomorphism is an aid in achieving heat loss—while at the same time enabling them to achieve that size and tallness which confers greater ability to evade, or defend against, predators and to reach a source of food otherwise unavailable to them.*

15
The Skeleton of Giraffes

When Robert Jacob Gordon sent a giraffe's skeleton to Holland for assembly by Arnout Vosmaer in April 1780 (see Chapter 3), the task that lay ahead was to understand how the bones joined together to form the scaffolding that supported a giraffe's shape. There was no idea then of the biology of bones beyond their functions of providing a support against gravity and a rigid framework against which muscles could contract to bring about movement. Discovery of the role of bones in blood cell production, their function as a store of calcium and phosphate, how they are built, what they are built of, their patterns of growth, and how they have evolved to prevent fractures, lay far ahead. Vosmaer was faced with the task of reassembling the skeleton of an animal that neither he nor anyone else in Europe had ever seen in life, and all he had was a skin, several boxes of bones and, helpfully, a sketch of the skeleton (see Plate 15.1) made by Johannes Schumacher, the artist who had accompanied Gordon on his journey into the interior of the Cape. The sketch was highly schematic and provided no more than a rough guide for reassembly of the bones.

Vosmaer left nothing to chance. He consulted the literature and friends and colleagues, and in the end he "followed the ideas I have gathered myself either by letters sent from the Cape or by mouth from those enlightened persons who live in those countries and that I have on occasion talked to," a group that included Gordon. Vosmaer was not an anatomist, but he had the expertise of Petrus Camper to assist him. Camper was an anatomist and had a particular interest in comparative zoology and osteology, so he was perfectly equipped for the task that lay ahead. Camper studied the bones and made the reassembly. Vosmaer focused on the skin and what the shape of the living giraffe might have looked like given the emerging shape of its skeleton.

The article that Vosmaer wrote about this giraffe in 1787 was the first scientific article ever written about giraffes. It is excellent, but it was ignored by the Comte du Buffon when he compiled his 1792 edition of the *Histoire Naturelle*, and by Owen when he wrote of his dissection of giraffes in 1838, and by Joly and Lavocat in their 1846 monograph on giraffes, and thus Vosmaer's article was forgotten. Vosmaer's article included a drawing of the giraffe as he imagined it to be in life (see Plate 3.5), as well as Camper's copious notes on how he reassembled the skeleton, and a copper engraving made by J.F. Gout in 1786 of the skeleton, labeled with letters (see Plate 15.2).

The letters were used by Camper to identify bones referred to in his description. Although the reconstruction was remarkably good given the circumstances of its assembly, Camper recognized its limitations: "The spine of the back is, in this skeleton too straight, the shoulder blades are placed too low, and the legs so much longer in the front than the rear" (Vosmaer, 1787). Gordon complained bitterly about it when he saw it for the first time, saying:

> *This is the skeleton which I sent to The Hague. I think it is badly assembled, much too low at the back and too bent in the neck which no animal has, but rather upright or at an angle forward,* (Rookmaaker, 1979)

How Giraffes Work. Graham Mitchell, Oxford University Press. © Oxford University Press 2021.
DOI: 10.1093/oso/9780197571194.003.0015

a comment that was not too dissimilar to those made by Joly and Lavocat in 1846 about Schumacher's sketch. Joly and Lavocat mistook the sketch for an anatomical drawing and wrote:

> *It is less real and complete than we can imagine. The humerus and femur are half as long as they are, and half as wide. The pelvis is enormous, the tibias ridiculous and the cervical ligament appears to be attached to the third cervical vertebra.*

In fact, Plate 15.1 is one of the very few diagrams that show the origin and insertion of the giraffe's *ligamentum nuchae* (cervical ligament), and it is accurate. The *ligamentum nuchae* attaches to the dorsal spines of all cervical vertebrae except the first (atlas), but has its main insertion on the occipital crest of the skull.

Vosmaer, as Camper's comments reveal, was preoccupied by the slope of the back. The back sloped from front to rear. His preoccupation was not new. Strabo, the squint-eyed Roman geographer, had declared in about AD 23 that it was this "lack of symmetry" that prevented giraffes from running easily (see Chapter 13) and Dio, writing of Caesar's giraffe, felt the need to comment "that it has sets of legs not of equal length. That is, its hind legs are shorter" (see Chapter 2). Mathurin Brisson had said the same. Vosmaer came to an answer. In his article he quotes several written sources and also observations made by colleagues before concluding, "All these reports old and recent agree that the posterior part of the body is lower than the front," and thus it was that in the skeleton constructed by Camper and in Schumacher's sketch the vertebral column sloped from front to back. Thirty years later, in 1809, Lamarck offered an explanation for why that might be so. He wrote that as a result of a habit long maintained of stretching to reach leaves, "the animal's forelegs have become longer than its hind legs." Buffon (1835) agreed with Lamarck, even though both Jean Allamand, who had been among the first to examine the specimens of giraffes that had been sent from South Africa, and Gordon had informed Buffon that the bones of the front and hind legs were nearly the same length. Buffon wrote:

> *We believe that in general the big difference in height which is found between the front and back of the giraffe comes from the difference in the length of the legs, so that when it is standing upright it seems that is sitting on its rump.*

But are the legs of different lengths? How long are they? And is a difference in leg length the origin of the sloping vertebral column and the slope of the back? To know whether the lengths of fore- and hindlegs are different, measurements of all of their components must be made, and so far only three have been (Table 15.1).

The bones of Vosmaer's giraffe had come from a very large giraffe that had a height of "15 feet 4 inches" (4.6 m) and thus a body mass of 1,000 kg or more. In that giraffe Camper found that the bones of the foreleg were 7 ft. 8 in. (230 cm) long, those of the hind leg 7 ft. 1 in. (212 cm) long, and that the metatarsus was a little thinner and half an inch longer (2 ft. 4 in.; 71 cm) than the metacarpus (2 ft. 3.5 in.; 70 cm). Camper wrote:

> *All observers so far say that the front legs are longer than the rear ones. This is the literal truth although the difference is not large and is not much more than 7 inches, which is not*

Table 15.1 Lengths of All Limb Bones of Giraffes

Bone	Lengths (mm)		
	Camper, 1789	Joly and Lavocat, 1846	Colbert, 1938
Humerus	483	400	430
Radius/ulna	762	690	680
Carpus	76	80	
Metacarpus	699	575	990
1st phalanx	127	120	All three phalanges combined
2nd phalanx	51	55	
3rd phalanx	102	50	
Total	*2,300*	*1,970*	*2100*
Femur	483	445	470
Tibia	533	532	520
Tarsus	127	130	
Metatarsus	711	582	1,046
1st phalanx	114	120*	combined
2nd phalanx	51	55*	
3rd phalanx	102	50*	
Total	*2,121*	*1,914*	*2,036*
Difference between forelimb and hindlimb length (mm)	*179*	*56*	*64*

* Assumed by Joly and Lavocat to be the same as the foreleg.

> *much in an object with a height of 7 feet. The difference emanates more considerably from the obliqueness of the bones of the thigh and legs, relative to the bones of the arm and forelegs.*

In other words, the legs were the same length and the back sloped, because while the front legs were straight, the hind legs were flexed (see Plate 15.2). Owen (1838) agreed but identified the scapulae as the culprit. He observed that the

> *apparent superiority in the length of the anterior extremities depends upon the nearly vertical position of the scapula upon the anterior part of the side of the deep and narrow chest and presents, perhaps, the relatively longest and narrowest scapula of all Mammalia.*

Joly and Lavocat made the second measurement of bone lengths, and in their giraffe, a three-year-old female that most likely weighed about 500 kg, the foreleg was 1,970 mm long and the hind leg 1,914 mm in length, with a difference of just 56 mm (2.25 in) between them (Table 15.1). Over 150 years later, the origin of the sloping back of giraffes was still elusive, and Edwin Colbert, preeminent delineator of giraffe phylogeny, made the third measurement of bone lengths in a giraffe whose body mass can be estimated to be 650 kg (Table 15.1). He found a difference between fore- and hindlimbs of 64 mm, thus confirming that the difference in lengths is, in fact, trivial and cannot account for the slope of the back (Colbert, 1938). Generally in ruminants, hindlegs are longer than forelegs, but after measuring the lengths of the three main limb bones in the foreleg (humerus, radius/ulna, metacarpus) and hindleg (femur, tibia/fibula, metatarsus) in members of the giraffe family, Colbert concluded that during the evolution of giraffes via Palaeotragines and Samotheres, there was gradual reduction of hindleg length toward equality until in giraffes the reverse occurred. The slope of the back was, therefore, partly because leg length differed, and leg length differed because the lower foreleg bones—the radius and metacarpus—"seem to have been elongated to a somewhat greater degree than the corresponding lower limb bones of the hind limbs." The modern view (van Sittert et al., 2015), and one that both Buffon (1835) and Colbert also had come to realize, is that while the vertebral column does slope, it does not slope as much as it seems. The illusion of slope in a living giraffe mainly results from the very long dorsal spines of the thoracic vertebrae and the quite short dorsal spines of the lumbar vertebrae (see Figure 15.4). Nevertheless, the total length of the front limb bones is longer than the total length of the hindlimb bones throughout growth (by about 40 mm in fetuses to about 200 mm in an adult, 1,500-kg giraffe (Table 15.2).

The way these differences in limb bone lengths develop is not straightforward. The humerus and femur are always the shortest of the bones, are roughly the same length, and their proportional contribution to the sum of limb bone lengths stays relatively constant at 25% for the humerus and 29% for the femur. However, as a giraffe grows, the radius in the foreleg becomes longer than the metacarpus and the proportion that the radius contributes to total bone length increases from 34% to 40%. The contribution made by metacarpus length decreases from 41% of leg bone length to 35%. In the hindlimb the tibia is always shorter than the metatarsus, whatever the size of a giraffe, but its contribution to length nevertheless increases from 31% to 34% with the

Table 15.2 The Length (mm) of the Main Limb Bones in the Fore- and Hindlegs of Giraffes (*G.c. giraffa*) at Various Growth Milestones

Bone	**50 kg (= fetus)***	**100 kg (= birth)**	**500 kg (= 3 years old)**	**1,000 kg (= adult female)**	**1,500 kg (= adult male)**
Humerus	209	254	418	518	588
Radius	287	357	605	759	940
Metacarpus	340	387	589	708	820
Total length	*836*	*998*	*1,612*	*1,988*	*2,348*
Femur	230	271	440	541	602
Tibia	246	302	507	634	723
Metatarsus	317	383	581	695	817
Total length	*793*	*957*	*1.529*	*1,870*	*2,142*

* Approximately 80% of gestation.

Adapted from van Sittert et al., 2015 with permission of John Wiley and Sons.

contribution of metatarsus length decreasing from 40% of leg bone length to 37%. Thus, in an adult giraffe weighing 1,500 kg, the difference in leg lengths of 200 mm is almost completely a consequence of the difference in lengths of the radius (940 mm) and tibia (723 mm) and not because of any great differences in metapodial length.

Number of Bones in a Giraffe's Skeleton

What the descriptions made by Camper and Joly and Lavocat of the skeletons they reassembled also reveal was the meticulous precision with which anatomists of their time measured and recorded details of their dissections. Among their measurements can be found the number of bones in a giraffe's skeleton, although the number is, in fact, uncertain and depends partly on how a bone is defined. The skull, for example, is usually understood to be a single bone, but it is made up of seven bones (occipital, sphenoid, ethmoid, interparietal, parietal, frontal, temporal) on each side of the skull (14 in total) that form the cranium, which houses the brain, and nine on each side that form the face (maxilla, premaxilla, palatine, pterygoid, nasal, lachrymal, malar, turbinates, vomer), to which the lower jaw, the mandible (which is made up of two halves that fuse together quite early in embryonic life), and the bone that supports the tongue (hyoid) must be added to the total number of skull bones. The number of bones in the vertebral column is made up of 7 cervical, 14 thoracic, and 5 lumbar vertebrae, and 4 sacral vertebrae that are fused to form a single bone (and had a length of 155 mm and a width of 145 mm, according to Joly and Lavocat in their giraffe), as well as a variable number of caudal (tail) vertebrae. Camper counted 27 caudal vertebrae, Owen 20, and Joly and Lavocat 19. The number of caudal vertebrae varies from 15 to 30 and thus so does tail length, with one source of variation being gender. The tail, which Darwin guessed functioned as a fly swatter, was measured by Cornwallis Harris (1839) to be 34 inches (86 cm) long in a giraffe he had shot, and is significantly longer in mature (>1,000 kg) males (105.1±7.5 cm; N=13) than in mature (>800 kg) females (95.2±8.1 cm; N=13; Mitchell, unpublished data). There are 14 pairs of ribs, 7 true and 7 false (although Camper reported 8 true and 6 false ribs). True ribs are attached to the first seven thoracic vertebrae and to the sternum. False ribs are attached to thoracic vertebrae 8–14 and do not connect directly to the sternum. The first five false ribs attach to the rib immediately to their front, and the last two, "floating ribs," are unattached and are buried in the muscles of the flank. In the giraffes Owen dissected, the sternum consisted of a single series of six bones and an ensiform (sword-shaped) cartilage called the xiphoid at its caudal end, while Joly and Lavocat disagreed and said it was made up of seven fused bones, presumably because they included the xiphoid in their count. Depending on how the bones are counted, the axial skeleton (skull, vertebrae and ribs) consists of a minimum of 103 bones.

The limbs are more straightforward.

Forelimbs

Each forelimb consists of a scapula (shoulder blade), a humerus (upper arm), a radius and ulna fused together (forearm), carpus (wrist), metacarpus (hand) and two digits, each digit consisting of three phalanges. In their giraffe, Joly and Lavocat found that the scapula had the shape almost of an isosceles triangle and, echoing Owen's comments, said it was distinguished from the scapula of all other mammals by its great length compared to its width and by its almost vertical position above the humerus. The ridge of bone that runs down its center (the central spine), they said, "is little elevated" (because it does not need to be taller; the infra- and supraspinatus muscles that attach to it are small and tendinous in character in giraffes; see Chapter 16). The length of the humerus, they wrote, "appears to be inverse with respect to the scapula." The scapula is long and thin, the humerus short and thick.

In the young giraffes he dissected, Owen described the ulna as becoming "less distinct" about two-thirds of the way down the radius (i.e., it became fused to the radius) but reappeared at the distal end of the radius to form, with the radius, the joint between the "arm" and the "wrist." In their young giraffe, Joly and Lavocat found the ulna to be fused to the radius throughout its length and to be visible mainly as the olecranon at its upper end (the elbow bone). The olecranon is the proximal extension of the ulna to which the extensor muscles of the arm (triceps, anconeus) are attached. In adults, as modern dissections have shown, the radius and ulna are completely fused, usually not distinct, and have the appearance of being a single bone (see Figure 15.10). The proximal end of the radius (the coronoid process) and the proximal end of the ulna (the olecranon) form a semicircular notch into which the bottom, distal end of the humerus fits tightly to form a very stable elbow joint.

Between the radius/ulna and the metacarpal bone lies the carpus, which, Camper found, consisted of six bones in two rows. The upper row contained five bones and the lower row one bone, but it was formed by the fusion of two bones, so really seven bones in all. Owen also found seven bones but in his (younger) giraffes; they consisted of an upper row of four bones and a lower row of three unfused bones. In Joly and Lavocat's giraffe, the carpus contained just six bones, also arranged in two rows, with four in the upper row and two in the lower row, all of them very irregular in shape. Clearly deciding whether bones were fused and in which row they lay, was (and is) not easy. The height of the carpus in Camper's giraffe added 75 mm to the length of the limb "taking the two rows together." The metacarpus or cannon bone is formed by fusion of what once were, in the first artiodactyls, the third and fourth metacarpal bones. Camper recorded that the two digits were noticeable for the long length of the first phalanx and the small size of the second. The first phalanx of his giraffe was 12.7 cm long, the second just 5.1 cm long, and the third, inside the hoof, was 10.2 cm long, making a total of 28.0 cm. The length of the hoof itself from front to back was 21.6 cm.

All in all, each forelimb of a giraffe consists of 17 bones, and thus the total number of forelimb bones is 34 bones for both limbs.

Hindlimbs

The hindlimbs articulate with a common and large pelvis, and each consists of a femur (thigh), a tibia (lower leg), an ankle joint (tarsus), a metatarsus (foot) and, like the front leg, two digits, each containing three phalanges. Joly and Lavocat noted that the femur was almost straight and was distinguished by its large size, but it is in fact much the same length as the humerus. Its head fits snugly into the cup-shaped acetabulum of the pelvis and in most mammals is held there by the coxo-femoral (round) ligament. During movement, that joint is exposed to great stress, and a strong bond between pelvis and femur head is required to prevent the hip joint from dislocating. The coxo-femoral ligament is absent in giraffes, and the hip is held together by external ligaments that form a fibrous capsule around the joint and which "explains the thickness and strength of this unique bond—the capsular ligament. In effect by these last conditions its mobility which, without these would be extreme, is found restrained or made solid" (Joly and Lavocat, 1846).

At its distal end, the femur articulates with the tibia to form the knee joint, a joint that is covered by the knee-cap (patella). The patella is not a true bone. It is a sesamoid bone (i.e., has the shape of a sesame seed) that lies in the single tendon formed from the four tendons that leave each part of the large quadriceps muscles of the thigh (see Chapter 16). The single tendon attaches to the tibia, and the patella functions as a pulley that increases the range over which a giraffe can extend or flex its knee. Typically, the lower hind leg of a mammal consists of a tibia and a fibula, with the fibula forming the lateral malleolus at its distal end, which supports the ankle joint. In giraffes, the fibula is absent, but the lateral malleolus is present as a single bone which encloses the external edge of the tarsus. In most ruminants the tarsus (ankle) is composed of seven bones—three in an upper row and four in a lower row, but there were just five in the box of bones sent to Camper: calcaneum (the heel bone), astragulus, a fused cuboid-scaphoid, two wedge-shaped (cuneiform) bones joined together, and a separate, third cuneiform bone. The height of the tarsus was 127 mm. Owen also counted five: an upper row of calcaneum, astragulus, a tibial bone, and the single bone that was the homologue of the malleolus, and a lower row formed by fused cuboid-scaphoid bones. Each of the two first phalanges in Camper's giraffe was 11.4 cm long, the two second phalanges 5.1 cm, and the two third phalanges contained within the hoof, 10.2 cm; a total of 26.7 cm, making a total length a little shorter than in the front limb. The hind leg hoof from back to front was 19.1 cm in length compared with 21.6 cm for the foreleg hoof.

All in all, the hind leg is made up of a pelvis and 15 other bones and thus the total number of hindlimb bones is 31 bones for both hindlimbs.

The whole skeleton therefore consists of approximately 170 bones (depending on how the counting is done), which is a formidable number to be sent from southern Africa in boxes to Europe for reassembly by people who had not seen a living giraffe.

Mass of the Skeleton

How heavy is a giraffe's skeleton, and what proportion of a giraffe's body mass is skeleton mass? Is it disproportionately heavier than it is in other mammals? It would be odd if it were not. A skeleton is the framework that supports the body. Giraffes have the highest body mass of any extant ruminant, their shape imposes unusual stresses and strains on their bones, they reach their adult height of 4–5 m in four to five years, and their skeletons must adapt to these characteristics just as their cardiovascular and other systems have to. Skeletons generally have to be strong enough to resist gravity, and as body mass increases so should skeleton mass, but does the mass of an adult giraffe's skeleton form a greater proportion of its body mass than it does in, say, a newborn giraffe? Hall-Martin and colleagues (1977a) provided a partial answer. They found that skeleton mass in 17 female giraffes with an average body mass of 792 kg was 21% of carcass mass, and in six males with an average body mass of 1,174 kg, it was 24% of carcass mass. The carcass mass of the females was 57% of body mass and in males 62% of body mass; thus the skeleton mass in a living giraffe would be ~12% of body mass in females and ~15% in males. As living bone contains ~30% water, the dry mass of the skeleton would be ~8% of body mass in females and ~11% in males. These percentages confirm the idea that evolution has produced skeletons that are heavier—proportionate to body mass—in large (male) giraffes than in smaller (female) ones, but is skeleton mass different in male and females of similar body mass? Hall-Martin et al. concluded that it was and that the difference could be attributed to differences in the two sexes of the length of their limbs and necks.

A difference in skeleton mass between giraffes of different body mass is also predicted by the few interspecific allometric equations that describe how skeleton mass increases as body mass increases in terrestrial mammals. Reynolds and Karlotski (1977), for example, analyzed the skeleton mass of 10 terrestrial mammals ranging in body mass from a 6.2-g shrew to the skeleton mass (measured by Robertson-Bullock in 1962) in African elephants weighing ~5,000 kg. They fitted the data to an allometric equation with body and skeleton mass in grams: Skeleton mass (g) = $0.03 \times Mb^{1.12\pm0.04}$. The skeleton formed ~4% of body mass in the shrew and ~20% in elephants, and the equation would predict a skeleton mass of ~12% of body mass in a 150-kg giraffe and ~15% in a 1,000-kg giraffe, proportions much the same as those found by Hall-Martin et al. (1977a). The most authoritative analysis of mammal skeleton mass (Prange et al., 1979) found that skeleton mass (kg) in 49 mammals was equal to $0.061 \times Mb^{1.09\pm0.04}$ (with body mass in kg). In 46 of these species, body mass was less than 10 kg, and in these species skeleton mass was 5.6% ±1.2% of body mass. In the other three, a dog (20 kg), a human (67.31 kg), and an elephant (6,600 kg), their skeletons were 20.6% (dog), 18.1% (human), and 27.0% (elephant) of body mass. Prange et al.'s equation would predict a skeleton mass of ~10% in a 150-kg giraffe and ~11% in a 1,000-kg giraffe. Reynolds and Karlotski and Prange et al.'s equations were developed from the relationship between skeleton mass and body mass in different species that had different body masses and thus do not necessarily represent how skeleton mass changes in a single species. So we come back to the question: does the mass of the skeleton in a single species like a giraffe form a greater proportion of body mass in an adult giraffe than it does in a juvenile giraffe that might have a body mass that is 10% of the body mass of the adult?

The only way to answer that question is to weigh each of the ~170 bones that make up a giraffe's skeleton, in several giraffes of both sexes that have body masses that range from low to high—a formidable task. Nevertheless, an attempt to do so has been made in 25 male and 26 female giraffes of different ages (van Sittert and Mitchell, unpublished). The body mass of the giraffes ranged from 147 to 1,441 kg and a total of 700 bones was collected from them. The bones included the skull and its ossicones, the mandible, 7 cervical, 14 thoracic, and 5 lumbar vertebra, and the six main limb bones (humerus, radius, femur, tibia, and two metapodials). The ribs, sternum, sacrum, caudal vertebrae, and phalanges were not weighed. Nor were the scapula and pelvis, but their mass in giraffes can be predicted from measurements made by Christiansen (2002) of limb bones in 25 species of artiodactyl, including a giraffe that had a body mass of 620 kg and three okapi weighing 228 kg, 250 kg, and 300 kg. Van Schalkwyk et al. (2004) measured the mass of the carpal and tarsal bones in mature, male giraffes. They formed 2% of skeleton mass (carpal bones) and 1.7% of skeleton mass (tarsal bones).

The mass of the individual bones analyzed and their allometric relationship with body mass in male and female giraffes are shown in Tables 15.3 and 15.4.

Analysis of all these data gives the unexpected result (given the previous conclusions about the mass of the skeleton) that the dry mass (kg) of a giraffe's skeleton increases linearly with increases in body mass in both sexes (see Figure 15.1), and not faster than body mass, as the data of Hall-Martin et al., Reynolds and Karlotski, and Prange et al. indicated it would. For giraffes of either sex, the linear equation describing dry skeleton mass is: dry skeleton mass (kg) = 0.075 ×Mb – 1.3 (R^2=0.99). The relative mass of the skeleton (g kg^{-1} body mass) of a large giraffe was the same as the relative mass of a small one. In both it was ~70 g.kg^{-1} body mass. The actual mass of the dry skeleton in male giraffes ranges from 11 kg in a neonate weighing ~150 kg to about 110 kg in a mature bull weighing ~1,500 kg. In females, skeleton mass ranged from 13 kg in a 150-kg giraffe to 67 kg in an adult weighing ~1,000 kg. The mass of the six main limb bones forms ~65% of total skeleton mass, the vertebrae ~25%, and the skull and mandible ~10%. The skeleton formed 7.4% ±0.4% of male body mass and 6.9% ±0.3% of female body mass, irrespective of actual body mass. In a living giraffe the skeleton mass can be estimated to be ~11% of body mass in males and ~10% of body mass in females, because living bones are on average about 30% water (Anderson et al., 1979). The divergence in absolute mass between males and females occurred after puberty when the giraffes had a body mass of 600–700 kg, and that divergence is a consequence mainly because of the increase in skull mass in males.

The dry mass of the skeleton in both sexes of giraffes is significantly higher than the 5.6% it forms in small mammals weighing ~10 kg or less. In other words, the skeleton of giraffes is intrinsically heavier than it is in smaller mammals, but its mass does not increase faster than its body mass does. The intrinsically heavier mass must be a consequence of natural selection and encoded in the giraffe genome, with the selective pressure for heavier skeleton mass presumably being the need to resist the forces that a higher body mass exerts on a skeleton. Thus within a species that will grow during its lifetime to have a high body mass, such as a giraffe or an elephant (in which the skeleton can form 27% of body mass), the skeleton seems to be programmed in the fetus to be heavier than it is in smaller mammals, and to be a fixed percentage of body mass, whatever that mass is. The mass of any giraffe's skeleton becomes, then, a balance between what the mass of the skeleton must be to support body mass and what the costs are of growing it and of transporting it around for a lifetime.

Table 15.3 The Allometry of the Dry Mass of the Skeleton of Male Giraffes (dry bone mass = a × Mb^b) Calculated from van Sittert and Mitchell Unpublished Data

Bone	N (bones)	N (giraffes)	Body Mass Range (kg)	Intercept a	Exponent b	R^2	Range[O] (kg)	% of skeleton[OO]	Apparent Density (g cm^{-3})
Skull*	13	13	181–1,396	0.0014	1.24±0.15	0.97	0.9–11.1	8.9±1.3	–
Mandible	13	13	181–1,396	0.004	0.89±0.12	0.96	0.3–2.2	2.2±0.3	–
All cervical vertebrae	77	11	187–1,441	0.011	1.00±0.12	0.98	1.7–14.8	12.7±0.8	1.01±0.24
All thoracic vertebrae	154	11	187–1,441	0.010	0.94±0.09	0.99	1.1–10.1	8.7±0.4	2.27±0.32
All lumbar vertebrae	55	11	187–1,441	0.005	0.84±0.06	0.99	0.4–2.3	2.3±0.2	1.46±0.34
All vertebrae	286	11	187–1,441	0.024	0.96±0.09	0.99	3.5–26.7	23.6±1.0	1.58±0.23
Humerus	8	8	77–1,343	0.003	1.03±0.11	0.99	0.6–5.0	8.9±0.5	1.51±0.16
Radius	9	9	77–1,343	0.003	1.04±0.08	0.99	0.7–6.4	11.3±0.6	1.89±0.06
Metacarpus	13	13	77–1,396	0.009	0.82±0.10	0.97	0.5–3.7	7.1±0.8	1.53±0.14
Femur	8	8	77–1,343	0.008	0.87±0.11	0.98	0.7–4.2	8.4±0.7	1.90±0.11
Tibia	8	8	77–1,343	0.004	0.96±0.09	0.99	0.6–4.6	8.7±0.4	1.62±0.12
Metatarsus	11	11	77–1,343	0.007	0.85±0.14	0.96	0.5–3.4	7.0±0.7	1.53±0.19
All limb bones [OO]	57	26	77–1,343	0.05	1.00±0.04	0.99	9.4–70.1	65.2±1.3	1.65±0.12
Whole skeleton	369		77–1,441	0.07	1.02±0.04	0.99	14.1–107.8	100	—

[O]In giraffes of the body masses shown; *includes ossicones; [OO]includes estimates of scapula and pelvis mass made from Christiansen's (2002) data, and the bones in both forelegs and the bones in both hind legs, including carpal and tarsal bones.

Table 15.4 The Allometry of the Dry Mass of the Skeleton of Female Giraffes (dry bone mass = a × Mb^b) Calculated from van Sittert and Mitchell Unpublished Data

Bone	N (bones)	N (giraffes)	Body Mass Range (kg)	Intercept a	Exponent b	R^2	Range[O] (kg)	% of Skeleton[OO]	Apparent Density ($g.cm^{-3}$)
Skull*	8	8	227–992	0.04	0.64±0.06	0.99	1.0–3.6	5.9±0.8	–
Mandible	8	8	227–992	0.004	0.87±0.18	0.96	0.3–1.8	2.5±0.1	—
All cervical vertebrae	70	10	414–1,029	0.007	1.04±0.24	0.93	1.2–9.1	11.6±0.8	0.86±0.25
All thoracic vertebrae	140	10	414–1,029	0.007	0.99±0.19	0.95	0.9–6.2	8.7±0.4	1.86±0.23
All lumbar vertebrae	50	10	414–1,029	0.003	0.90±0.20	0.93	0.3–1.7	2.5±0.1	1.39±0.28
All vertebrae	260	10	414–1,029	0.017	1.00±0.15	0.97	2.4–17.0	22.8±1.1	1.37±0.22
Humerus	7	7	147–967	0.003	1.01±0.11	0.99	0.5–3.5	9.7±0.5	1.58±0.08
Radius	6	6	147–967	0.005	0.97±0.13	0.99	0.6–4.1	11.4±0.3	1.84±0.12
Metacarpus	13	13	147–992	0.005	0.92±0.16	0.93	0.5–2.7	7.6±0.7	1.62±0.11
Femur	7	7	147–967	0.012	0.80±0.23	0.98	0.6–3.1	9.5±0.8	1.96±0.08
Tibia	7	7	147–967	0.009	0.84±0.10	0.99	0.6–3.0	9.0±0.5	1.67±0.10
Metatarsus	15	15	147–967	0.010	0.78±0.13	0.93	0.5–2.5	7.1±0.7	1.66±0.13
All limb bones[OO]	55	25	147–967	0.08	0.93±0.03	0.99	7.6–48.3	68.8±0.8	1.73±0.06
Whole skeleton	331		147–1,029	0.11	0.93±0.02	0.99	11.2–69.2	100	—

[O] In giraffes of the body masses shown; *includes ossicones; [OO]includes estimates of scapula and pelvis mass made from Christiansen's (2002) data, and the bones in both forelegs and the bones in both hind legs, including carpal and tarsal bones.

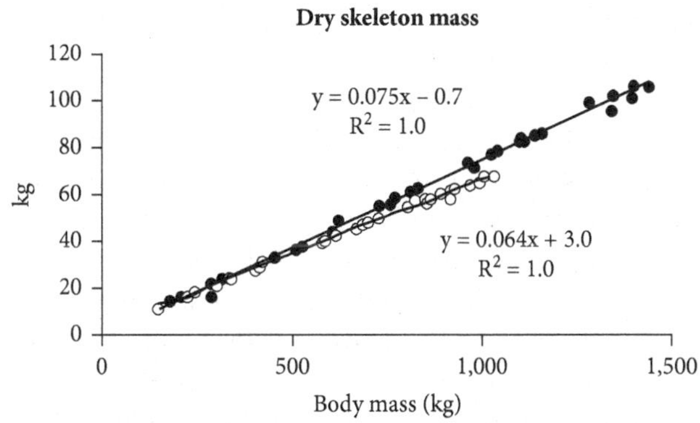

Figure 15.1. The linear relationship between dry skeleton mass and body mass in female (○) and male (●) giraffes.
Data from van Sittert and Mitchell, unpublished.

Chemical Composition of Giraffe Bones

What is the origin of the greater mass? The most parsimonious way of increasing bone mass is simply by increasing bone density. That would be a convenient way for natural selection to achieve an increase in both bone mass and strength while at the same time retaining the shape of the bones that make up a giraffe's skeleton and retaining the characteristic slender, gracile form of giraffes. Density depends on chemical composition and an average composition of a range of giraffe bones is ~33% minerals, 34% collagen, and 33% water (Bredin et al., 2008). The collagen fibers are glued together to form a flexible lattice, and they give bone the elasticity it needs to absorb the stresses imposed on it. Most of the mineral component of bones is calcium and phosphate in the form of calcium hydroxyapatite, which fills the gaps between the collagen fibers and gives bone its stiffness. It is well established that in mammalian bones density is correlated with the mineral component: reduced density is a characteristic feature of both osteoporosis (loss of calcium and collagen) and osteomalacia (loss of calcium alone, for example, in animals that have rickets). There is a strong correlation between density and mineral content of giraffe bones, with mineral content accounting for 66% of the variation in density of their bones (R^2=0.66), a correlation which, though strong, is not as strong as the relationship between density and the mineral content of the bones of Cape buffaloes (R^2=0.92; van Schalkwyk et al., 2004). This difference in correlation may simply reflect the difficulties of making accurate measurements, but it could also mean that constituents other than calcium and phosphorus contribute to bone density of giraffes. If so, they have not yet been identified. Even though a basic assumption made about bone strength is that the chemical composition of bones is similar among all terrestrial mammals (Currey and Alexander, 1985), it does seem reasonable to say that giraffes, with all the biomechanical problems created by their size, shape, and growth rate, should have bones dissimilar in chemical composition, and, therefore, different density, to the bones of other mammals.

The only attempt so far to discover whether the mineral content of giraffe bones is different was made by van Schalkwyk and colleagues in 2004. They analyzed the calcium and phosphorus composition of a range of bones collected from giraffes in the Kruger National Park in South Africa. To test if the composition in giraffes was different from other ruminants, especially a ruminant of similar body mass, they analyzed the calcium and phosphorus composition of the same bones collected from Cape buffalo cohabitant with the giraffes. Their idea was that the chemical composition of limb bones in particular would be different between the two species, and that each bone within a species would have its own specific composition. Those expectations were, however, only partially fulfilled. The composition of the bones analyzed (at least as far as calcium and phosphorus are concerned) was the same in giraffes and buffalo, but it did differ between bones in both species. The concentrations of calcium and phosphate in bones were lower in vertebrae (0.183±0.014 g/g and 0.089±0.004 g/g, respectively) compared with femur midshaft and metacarpus (0.214±0.014 and 0.103±0.004 g/g, respectively). On average, the total amount of calcium and phosphorus in the giraffe bones that were measured (20.0±1.4 g 100 g^{-1} and 9.5±1.1 g 100 g^{-1} of bone) was the same as it was in buffalo bones (20.2±0.6% and 9.5±0.2%, respectively). Surprisingly, given its function of providing the anchor for the contractions of all the muscles that propel an animal forward, the least dense bone was the pelvis (as measured in the *tuber coxae*), because of its spongy rather than solid structure, a structure that presumably has evolved as the best-suited to resist the stresses to which it is exposed. Giraffe (and other mammal) vertebrae also have a trabecular, spongy structure and they too have to cope with high stresses, and except for some thoracic vertebrae, their density, like that of the pelvis, is also relatively low (Bredin et al., 2008; Figure 15.3). The third, fourth, and fifth thoracic vertebrae are the densest bones in the skeleton with an average density of 2–3 g cm^{-3} (see Figure 15.3 later in the chapter). The densest bones in the limbs are the femur midshaft and the long, slender metacarpal bones.

Typically, the density of mammalian limb bones is given as 2.0 g cm^{-3} (Currey and Alexander, 1985), and in giraffes it was 1.9 g cm^{-3} for the compact bone of femur midshaft and metacarpus, and that density was the same in the buffalo bones that were analyzed (van Schalkwyk et al., 2004). Thus giraffes do not have uniquely dense limb bones, but bones within a skeleton have different densities and the differences depend on what function the bone has to perform. In other artiodactyls, the calcium and phosphorus content of bones increases during growth (Main and Biewener, 2004). That increase does not seem to have occurred in giraffes. If calcium and phosphorus did accumulate in giraffe bones as they grow, then so would bone density, but the density of giraffe bones does not change during growth. As John Currey, emeritus professor of Biology at York University and an expert on bone structure and mechanics, has pointed out in his excellent book on bones, if the density of a bone does not increase, then that must mean that high density brings a disadvantage. Increased density of a bone is associated with reduced resistance of the bone to the impact of a limb striking the ground. The bone is more brittle. That, perhaps, is why giraffe limb bones have never evolved to become denser than they are.

Differences in chemical composition, and thus density, give an impression that bones can range from soft to hard, but giraffe bones, like those of buffalo, are hard, and limb bones specifically are difficult to reduce to powder even by mechanical means.

Indeed giraffe bones are referred to as "bush ivory," and are used to make ornaments and implements. Even so, a hyena (*Crocuta sp*) can crack a giraffe's leg bone up to 7 cm in diameter (Tanner et al., 2008) in order to get at the marrow. To crack a giraffe leg bone is not the same as fracturing it. Bone fracture force is high. For mammals ranging in size from 0.5 to 700 kg, bones break at a force of approximately 2,000 kg cm^{-2} (Burstein et al., 1972; Biewener, 1982). To crack a giraffe bone by biting it requires much less force, but it is impressive that a hyena can do so. They have a bite force of about 70 kg cm^{-2} (compared with a human bite force of about 10 kg cm^{-2}).

The Giraffe Vertebral Column (Axial Skeleton)

Most research on the structure, function, and strength of the bones of mammals and the way bones resist fracture has been done on limb bones, but there is no doubt that the bones of the axial skeleton are exposed to high stresses (force/area, e.g., kg cm^{-2}) and also have to resist fracture. The axial skeleton is the framework that suspends the internal organs. Its basic structure is a series of vertebrae. Each vertebra has a body, articulating processes, a dorsal spine, and a vertebral (spinal) canal that provides protection for the spinal cord that runs through it (Figure 15.2). The vertebrae are separated one from the next by compressible cartilage discs. Adjacent vertebrae are held together by ligaments (interosseous ligaments). The column so formed has to be rigid enough in a quadruped to resist bending upward, a direction brought about by gravity acting downward on the weight of the head and neck, a weight that pivots over the front legs. The column also has to resist bending downward, a direction brought about by gravity acting on the weight of the internal organs, yet it has to be flexible enough to participate in movement: it is the vertebral column that converts the power generated by the muscles of the hindquarters into a whole body movement.

In his seminal analysis of the comparative biological-anatomical structure of the vertebral column and spinal musculature of mammals in 1946, a Dutch veterinarian/

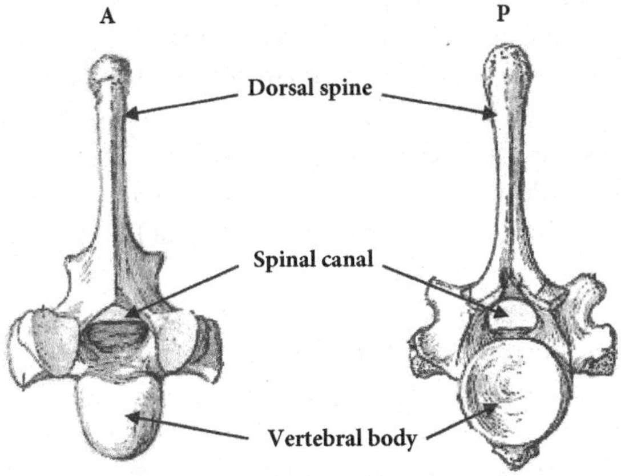

Figure 15.2. The anterior (A) and posterior (P) views of the first thoracic vertebra of a giraffe.
Redrawn from Lankester, 1908, with permission of John Wiley and Sons.

anatomist, Everhard Johannes Slijper, most renowned for his studies of cetaceans, realized that for the spine as a whole to resist the stresses it has to bear, it had to follow civil engineering principles. He was elaborating on ideas first proposed by Galen around AD 200 and much later by D'Arcy Thompson in 1917, who wrote:

> *Standing four-square upon its fore-legs and hind-legs, with the weight of the body suspended between, the quadruped at once suggests to us the analogy of a bridge, carried by its two piers with the bones as struts and the ligaments between them ties.*

D'Arcy Thompson was for many years professor of Zoology at St. Andrew's University in Scotland. His enduring fame rests on his book *On Growth and Form*, first published in 1917. It is both a work of art and an unmatchable zoological treatise that still is, over a hundred years later, essential reading for all biologists. So his opinion mattered. Nevertheless, Slijper thought that a better model was a bow and string bridge with the bow part being the vertebral column and its muscles and ligaments all tending to arch upward by the tension generated by the "string" part, consisting of the sternum, *linea alba*, and abdominal muscles, with the ends of the bow supported by the limbs. The "string" stretches from the first rib to the sacrum and pelvis. The "bow" formed by the vertebrae is connected to the "string" by the ribs and their muscles and ligaments. The mass of the abdominal organs should make the "bow" sag, but sagging is prevented by contraction of the "string" muscles, especially the abdominal muscles, assisted to a small degree by the muscles that pull the foreleg forward and those of the hindleg that pull backward, movements that, combined with contraction of the abdominal muscles, tighten the bowstring and tend to make the vertebrae arch upward.

Slijper's model works best for animals that have short necks and legs that are posted at the four corners of the trunk with the weight carried by each support fairly evenly distributed. However, in all quadrupeds the weight carried by the forequarters is greater than the weight carried by the hindquarters. In a horse that ratio is 58:42. In animals with long necks like a camel or a llama, the relative weight ratio is 67:33 (D'Arcy Thompson, 1961). Therefore, a long neck imposes a much greater weight on the forelegs and a greater stress on the spine. What effect does the long neck of the giraffe have? "In the case of the giraffe," D'Arcy Thompson wrote, "we have without doubt a very heavy load upon the forelegs, though no weighings are at hand to define the ratio," but the ratio is likely to be higher than it is in either a llama or a camel. Consequently, the stresses imposed on a giraffe's spine are asymmetrical. How do giraffes overcome the effects of that asymmetry? One way is by having forelimb bones heavier and thus stronger than hind limb bones (see Tables 15.3, 15.4), but D'Arcy Thompson had an additional answer:

> *as far as possible this disproportionate load would seem to be relieved by help of a downward as well as backwards thrust, through the sloping back to the unusually low hindquarters. The dorsal spines of the vertebrae are very high and strong, and the whole girder-system very perfectly formed. The elevated rather than protruding position of the head lessens the anterior bending moment as far as possible, but it leads to a strong compressional stress transmitted almost directly downwards through the neck: in correlation*

with which we observe that the bodies of the cervical vertebrae are exceptionally large and strong, and steadily increase in size and strength from the head downwards.

Richard Owen (1838), naturally, was the first to notice that the cervical vertebrae are the broadest of all vertebrae, but he also saw that the dimensions of the first few thoracic vertebrae (T1–T3) were more characteristic of cervical than of thoracic vertebrae, so they too contribute to the ability to resist stress.

D'Arcy Thompson, like many before him, had misunderstood the origin of the slope of the back, but that point aside, the head and neck of giraffes can be carried more vertically than they can be in any other ungulate except for llamas and alpacas. The vertical shape of a giraffe results from the relationship between the position of their forelegs and the position of the junction between the neck and trunk vertebrae (see Figure 15.5) and from their short trunk. The junction between the neck and trunk vertebrae lies directly above the forelimbs, so when the head and neck are carried vertically "the anterior bending moment" is reduced, as are its consequences for the bending stresses that act on the spine. Fielding et al. in 1976 used the analogy of a heavy weight on the top of a pole to explain the effects: little force is needed to hold the heavy weight up when the pole is vertical, but as the pole is lowered, the force needed to hold it increases. So it is with giraffes. When the head and neck are held vertically the bending stresses on the vertebrae are decreased, but compression stresses increase. Giraffes very rarely carry their heads and necks vertically. They are carried mostly at an angle midway between horizontal and vertical; thus their vertebrae are always exposed to bending stresses and, depending on the angle of carry, to more or less compressive stresses. These forces must be resisted by the vertebrae or the vertebrae would collapse. As D'Arcy Thompson supposed, those vertebrae most at risk, because of their length, position, and the forces to which they are exposed, are the cervical vertebrae.

How are stresses on vertebrae resisted? Nowhere in the skeleton is the outcome of natural selection more impressively revealed than in the shape of a giraffe's vertebrae. The volume, mass, density, cross-sectional area, width, height, and length are known for 26 vertebrae (7 cervical, 14 thoracic, 5 lumbar) of 21 giraffes (11 males, 10 females) ranging in body mass from 147 to 1,413 kg (van Sittert et al., 2010). How these six characteristics of the vertebrae change along the vertebral column and thus how they have evolved to resist stress are shown in Figure 15.3 for male and female giraffes weighing 750 kg. Males and females of equivalent body mass have vertebrae of similar length, width, cross-sectional area, and volume, but their mass and density in females were ~70% of those in males, reflecting the more slender shape of females and the lower stresses to which the vertebrae are exposed.

Slijper regarded the moment of resistance against bending as the best measure of the bending stress that the vertebral bodies must be equipped to resist in any mammal. He calculated the moment of resistance (which has the units cm^3) for the spine using an equation that defines the resistance to bending of a rectangular beam, which is $[(\text{width} \times \text{height}^2) \div 6]$, but the shape of a vertebral body of a giraffe is more circular or elliptical (Figure 15.2) and for a circular crosssection, the moment of resistance is $[(\pi \times \text{diameter}^3) \div 32]$. When the moment of resistance is calculated for each vertebra throughout the vertebral column using this latter equation, the outcome (in a 750-kg giraffe) is illustrated in Figure 15.4A. It is quite clear from the sizes of the moment of resistance

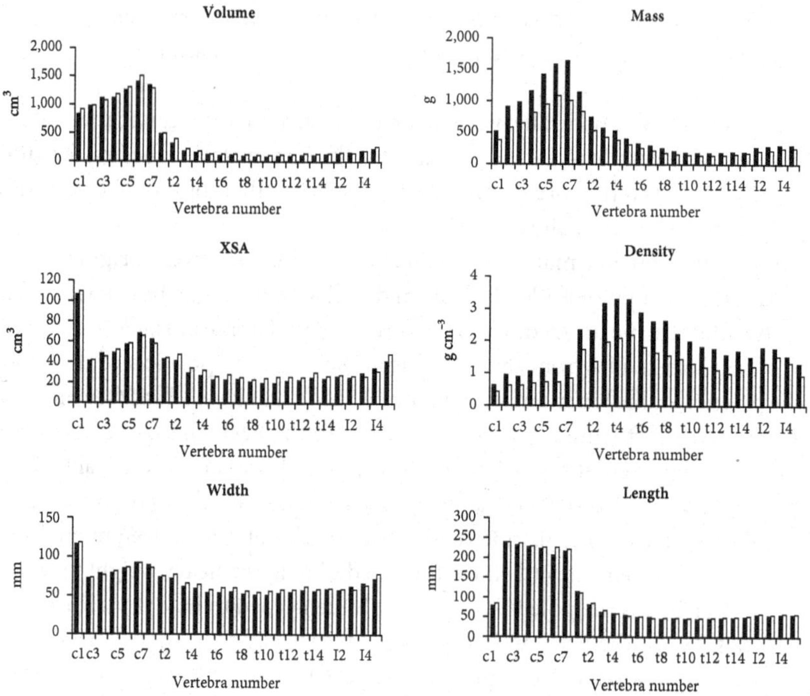

Figure 15.3. The characteristics of vertebrae for male (■) and female (□) giraffes with a body mass of 750 kg, calculated from data obtained from 11 male (body mass range 187–1,441 kg) and 10 female (body mass range 414–1,029 kg) *G.c. giraffa*.
Using data in van Sittert et al. (2010).

shown in this figure that the bending stresses on the last few cervical and first few thoracic vertebrae are the greatest. The stresses are absorbed by the width, cross-sectional area, and volume of the vertebrae, and those dimensions are perfectly adapted to resist the bending stress to which those vertebrae are exposed. Their dimensions are, on average, significantly larger than they are in the other thoracic vertebrae and in the lumbar vertebrae (Figure 15.4A), where bending stresses are lower.

D'Arcy Thompson had no data on giraffe vertebrae on which to base his conclusions about "compressional stress," but they are in fact supported by data. About 75% of the compressive strength of a vertebra depends on its spongy, trabecular structure (Mosekilde and Mosekilde, 1986), which enables it to dampen the sudden stresses that occur during the course of normal daily activities. How much compressive stress the vertebrae of quadrupeds can withstand, very surprisingly, has not been studied, but it has been in humans because of the frequency of back pain and the effects of osteoporosis on the spine. Assuming some sort of equivalence, the average maximum compressive strength of a healthy, whole, human vertebra is 6 MPa (= 61.2 kg cm^{-2}; 1 MPa = 10.2 kg cm^{-2}), and for their trabecular component it is 4.4 MPa (Mosekilde and Mosekilde, 1986). Most bones have a built-in safety factor of threefold (Biewener, 1991), so the typical or average compressive stress they should be able to resist is of the order of 2 MPa, which is equivalent to having a force of 20.4 kg cm^{-2} placed on them. What they can resist thus depends on their cross-sectional area (XSA, cm^2). The cross-sectional areas

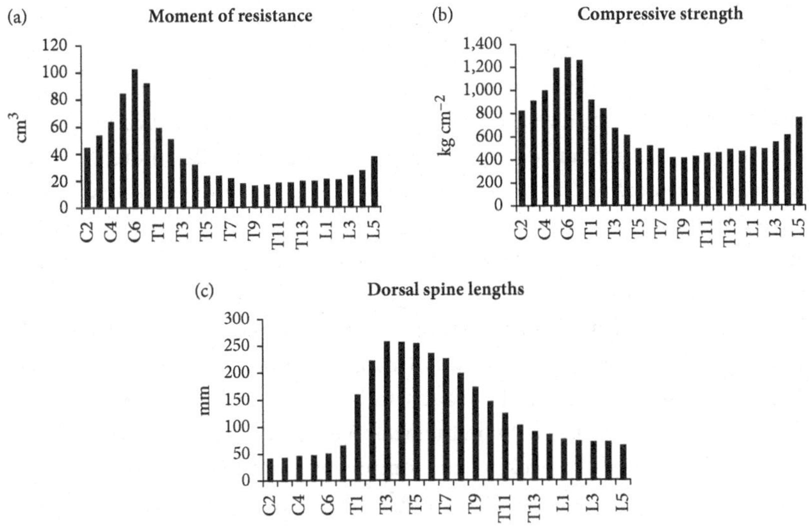

Figure 15.4. The moment of resistance (A), the compressive strength (B) and the dorsal spine lengths (C) in individual vertebrae along the spinal column calculated for a giraffe with a body mass of 750 kg.
Constructed using data in van Sittert et al., 2010.

of thoracic and lumbar vertebrae of giraffes are not significantly different at ~30 cm^2 each, but that of the cervical vertebrae is twice as large (61 cm^2). As each cm^2 can resist a mass of 20.4 kg, each of the second to seventh cervical vertebrae on average can withstand the equivalent of a 1,000-kg weight being placed on them, compared to 552 kg for each of the 14 thoracic vertebrae and 590 kg for each of the five lumbar vertebrae. These estimates may quite likely underestimate the real values in giraffes if the compressive strength of their vertebrae are ever studied in the future, but they confirm D'Arcy Thompson's principle: the vertebrae at the base of the neck are subject to the highest bending and compressive stresses (Figure 15.4B) and have evolved a shape that can resist them. The evolutionary response to both bending and compressive stresses in cervical vertebrae has been a much bigger increase in the width of the vertebrae bodies than in the less stressed thoracic and lumbar vertebrae.

Compressive strength does not depend only on cross-sectional area. It also varies with density (Carter and Hayes, 1976) and therefore it is reasonable to think that the density of the cervical vertebrae would be greater than the density of thoracic and lumbar vertebrae. But in the vertebrae measured by van Sittert et al. (2010), the density of cervical vertebrae was not greater (Figure 15.3). The average density of cervical vertebrae of giraffes (calculated as their mass divided by their displacement volume) has been reported to be 1.36±0.03 g cm^{-3} (van Schalkwyk et al., 2004). The disadvantage of the displacement method for finding density is that its accuracy depends on how porous a vertebra is. If it is porous, water is absorbed by the bone, and the density is overestimated. Another measure of density is apparent density, which is the dry mass of a bone divided by its geometric volume (Currey, 2006). The geometric volume of vertebral bodies can be estimated from their width and length, assuming that they have the shape of a cylinder, which they do. The apparent density of giraffe vertebrae calculated this way varies widely between vertebrae (Figure 15.3), with the average density in cervical vertebrae of males and females being 0.83±0.24 g cm^{-3}, a density that is much less than it is in thoracic (1.98±0.66) or lumbar vertebrae (1.44±0.26) of those giraffes. It was 1.57±0.71 g cm^{-3} for all vertebrae. The lower density of cervical vertebrae is a consequence of having a greater trabecular, spongy structure. Furthermore, the density of cervical vertebrae declined in males from 1.23 g cm^{-3} in the seventh cervical vertebra (C7) to 0.63 g cm^{-3} in C1, and in females from 0.85 g cm^{-3} in C7 to 0.42 g cm^{-3} in C1. This decline in density (and its associated decline in vertebra mass) represents an evolutionary compromise between the strength needed by the cervical vertebrae and a competing need to minimize the weight of the neck, and is compensated by an increase in width. The effect of decreasing density (mainly) in both male and female giraffes is that the mass of the cervical vertebrae decreases sharply toward the head (Figure 15.3), which reduces the effect of gravity on the neck. Vertebra density and strength are concentrated at the base of the neck where compressive and bending forces are greatest (Figure 15.4A, B). By contrast, thoracic and lumbar vertebrae have a width about two-thirds that of the cervical vertebrae, but have a density two times or more higher. Their strength comes more from their density than from their width.

The shapes of the vertebrae change as a giraffe grows, especially after birth (body mass = ~100 kg). For the most part, changes in all three dimensions correlate with body mass, which is to be expected, but it is typically width that increases most. Vertebrae bodies are, usually, broader than they are high or long (except in cervical vertebrae) in giraffes, with the exception of the fifth thoracic vertebra where width and length are, on average, exactly the same (~55 mm in a 750-kg giraffe). Elongation of the cervical vertebrae and thus of the neck is triggered at birth and perhaps by birth, presumably by activation of *Hox* gene machinery by an as yet unknown activator. (Hox genes control the shape of an animal by suppressing some genes and activating others.) As birth approaches, there is a tendency for the growth rate of all vertebrae to slow, as if in preparation for birth, with resources being reallocated to limbs or other organs (and this may well be what happens in all mammals). The reallocation produces the long legs and a relatively short neck and trunk of a newborn giraffe—an ideal combination for birth and suckling (see Figure 3.3). After birth, the growth of all vertebrae increases rapidly and at much the same rate (van Sittert et al., 2010), but elongation of cervical vertebrae is especially rapid. Despite their rapid growth, all cervical vertebrae elongate at the same rate (except for the unique first [atlas] vertebra), a growth pattern quite different from the general pattern of ungulates described by Badlangana et al. in 2009. In other ungulates, the highest elongation rate is in C2, with progressively slower rates in succeeding vertebrae so that C2 becomes longer than C3, C3 longer than C4, and so on. Not so in giraffes.

How many cervical vertebrae do giraffes have? Since Darwin's time it has been accepted that all mammals, including giraffes, have seven cervical vertebrae, except for manatees (six) and sloths of the genus *Choloepus* (six) and of the genus *Bradypus* (eight). However, in 1999 Niko Solounias proposed that giraffes had eight cervical vertebrae. His conclusion arose from the function and shape of the first thoracic vertebra, which Lankester (1908) showed had acquired the characteristics of a cervical vertebra (Figure 15.5). The defining feature of a thoracic vertebra is that it forms a joint with a rib, and the first thoracic vertebra (T1) of giraffes does articulate with a rib. Its "cervicalized"

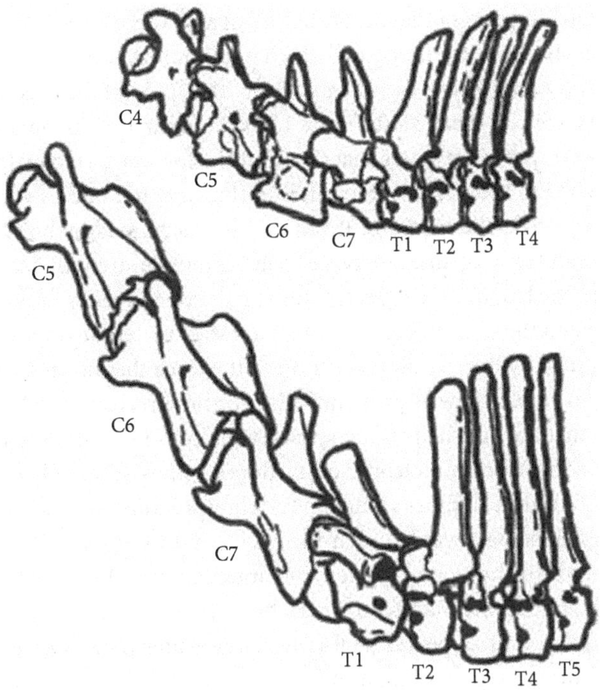

Figure 15.5. A comparison of the position of the junction between the cervical and thoracic vertebrae in giraffes and okapi. In okapi (top, and in all other mammals) the junction is between the last cervical and first thoracic vertebrae. In giraffes (below) it is between the first and second thoracic vertebrae.
From van Sittert et al., 2010, with permission of John Wiley and Sons.

shape is a consequence of the shift backward of the articulation between the cervical and thoracic vertebrae to lie between T1 and T2 rather than between the seventh cervical vertebra (C7) and T1, as is usual in all other mammals including okapi (Figure 15.5). In addition, nerves that form the brachial plexus on their way to innervating the forelegs typically emerge from the spinal cord between C7 and T1, and in giraffes they do emerge there, but they also emerge between T1 and T2. It is these unusual features that led Solounias to conclude that giraffes have eight cervical vertebrae, when they have seven like almost all other mammals. However, because the cervico-thoracic articulation has moved to between T1 and T2, the seventh cervical vertebra has been freed to elongate; thus all seven cervical vertebrae in giraffes elongate and contribute to neck length, whereas in other mammals it is mainly six that do so. The shift also moves the fulcrum of the neck backward over the forelegs and pushes the shoulder joint of giraffes forward to form a characteristic "point".

Vertebra Spines

Vertebrae not only lengthen and broaden. They also need to have the infrastructure on which muscles, tendons, and ligaments can attach. Apart from a long neck and long legs, the most eye-catching part of the infrastructure of a giraffe's skeleton is the great size of the thoracic dorsal spines, the longest of which is on the fifth thoracic vertebra in fetuses and young giraffes. However, as giraffes mature, the spine of T4 becomes the longest (Owen, 1838; van Sittert et al., 2010; Figure 15.4C) and can be 300 mm long in a mature giraffe. The dorsal spines are levers that transmit the forces exerted by the muscles and ligaments attached to them, from the spines to the vertebral bodies where the force is absorbed and resisted. The height of the spines is determined by body mass, because (counterintuitively) as body mass increases relative muscle strength decreases (see later discussion) and the length of the "levers" must increase in proportion to that decrease in order to transmit the same force. The dorsal spines of vertebrae slope. The angle of the slope is determined by the size and direction of the force that needs to be transmitted, and the most favorable slope is perpendicular to the direction in which the muscles and ligaments that are attached to the spines act because that direction produces the maximum force with minimum length of the dorsal spines. Generally in mammals, the angle of slope of the dorsal spines along the vertebral column varies. Usually at the front (T1 to T8) they slope backward, and at the back (T9 to L5) they slope forward, with the vertebra at which the slope changes (the anticlinal vertebra) pointing upright. In giraffes there is no definitive anticlinal vertebra. The first four thoracic dorsal spines point directly upward or slope a little forward. Thereafter they slope backward to the sacrum. In the lumbar region of mammals generally, the spines stand upright because the muscles and tendons of the great back muscle, the *longissimus dorsi*, which pull toward the head, are balanced by the main muscles of movement—the *gluteus*, *biceps femoris*, and *semitendinosus* muscles—which pull backward. In giraffes all the spines on lumbar vertebrae slope backward to accommodate the direction of contraction of the muscles that power movement and, simultaneously, to provide the perfect slope for the forward direction of the pull of the back muscles and the fibers that are the origin of the *ligamentum nuchae*.

The Head and Neck of Giraffes

The bending and compressive stresses on the spine are always large because the neck is long and at its end is a head nearly a meter long with a mass of as much as 45 kg in a full-grown male giraffe. In mature female giraffes, head length is about 80 cm and its mass 25 kg (Mitchell et al., 2013a; Figure 15.6A, B). A large head provides space for the tongue and teeth, for the attachments of the masseter and pterygoid muscles (see Chapter 8), and for the membranes of olfaction and temperature regulation (see Chapter 14), and the head can be used as a weapon in sparring contests (see Chapter 18).

What makes up the mass of the head? Much emphasis has been placed on the contribution that skull mass makes to head mass (Dagg, 1965; Spinage, 1968b) because a heavy skull could well be an asset if its function was to ensure supremacy during sparring. The skull can weigh 9 kg in an adult male (de Winton, 1897) or 11.3 kg (Spinage, 1968b), with female skulls weighing about half as much. But skull mass actually contributes relatively little to the mass of the head, at least in Zimbabwean *G.c. giraffa*. In 13 male and 8 female Zimbabwean *G.c. giraffa*, the mass of the components of the head have been measured, and those data indicate that although the skull of males is heavier than it is

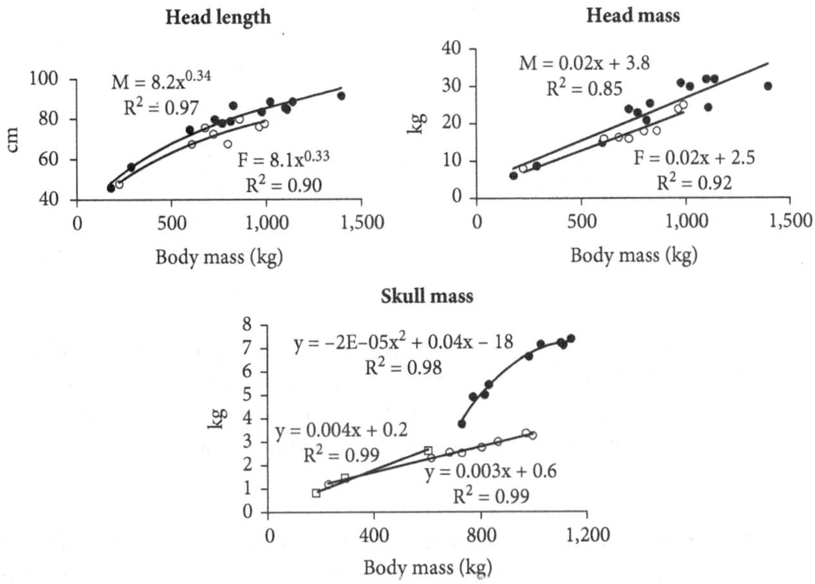

Figure 15.6. The growth patterns of head length, head mass, and skull mass including ossicones in 13 male (●, □) and 8 female giraffe *G.c. giraffa* (○).
Constructed from data in Mitchell et al., 2013a.

in females, most of the mass of the head is made up of non-skull tissue (Mitchell et al., 2013a). In males, 76% of head mass is made up of non-skull tissue and, in females, that proportion is 84%. Non-skull tissues include muscles, teeth and salivary glands (50%), the skin (16%), the mandible (6.5%), and the brain (4%). In males the heavier bones of the skull and the larger ossicones are the origin of their heavier skull. Ossicones are adornments rather than weapons (Spinage, 1970). Although they can have a circumference of 22 cm and a length of ~20 cm in males (Spinage, 1968b), they weigh only 1.5 kg in adult males and just 150 grams in an adult female. For comparison, a mature elk with a body mass of 500 kg has antlers that may weigh 18 kg and which can be 1.2 m long. Furthermore, the skull of giraffes is far less heavy than it might be. One reason that the skull is "light" is that as a giraffe grows and its skull enlarges, much of the enlargement is caused by the development of air-filled sinuses (Figure 15.7).

Skull sinuses were first evident in the Samotheres 8 Mya, and their preservation and enlargement in all subsequent giraffids implies that their benefits were great. Sinuses may act as resonance chambers for sound production (Badlangana et al., 2011) but, without the inflated sinuses, the skull (and head) would be heavier than it is and the work needed to manipulate the head and neck would be greater. The size of the sinuses of giraffes and specifically the largest—the frontal sinus (Figure 15.7)—has been compared by Badlangana et al. (2011) to its size in a range of other artiodactyls. They established that in five quite young giraffes (one of which was female), weighing between approximately 112 kg (a neonate) and 500 kg (about 3 years old), sinus volume could be predicted by the equation: Sinus volume (mL) = 2.23 × skull mass−1,100. Sinus volume in these young giraffes was ~350 mL in the smallest and 3.8 L in the largest. In mature males and females, sinus volume may be even larger if the relationship between

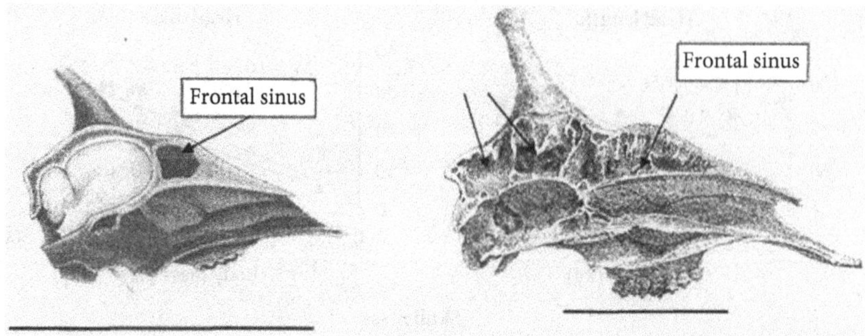

Figure 15.7. A longitudinal section through the skull of a young giraffe, five months old (left; Cobbold, 1860 with permission of John Wiley and Sons), and a mature giraffe (right; Owen, 1838, with permission of John Wiley and Sons), showing the development of the skull sinuses. The marker line is 45 cm long. The arrows indicate the sinuses.

skull mass and sinus volume holds for bigger giraffes, but whether it does is not known. For comparison, sinus volume measured by Badlangana and colleagues in the skull of an adult ox was ~225 mL. Skull mass in males diverges from the skull mass in females after puberty (ca.500 kg body mass: Figure 15.6C), when the length and diameter of the ossicones start to increase in males, and when the bones of their skulls become thicker and encrusted with bony deposits (exostoses) (Dagg, 1965; Spinage, 1968b). However, these addenda are insufficient to alter the trajectory of the increase in head mass very much (Figure 15.6B).

In the same manner that evolution has found a way to reduce the weight of a giraffe's head, evolution has found ways to make the neck as light as possible, although "light" is, again, relative. In an adult female the neck can weigh 85 kg, and in an adult male 140 kg (Mitchell et al., 2009), giving a combined head and neck mass of up to 110 kg (females) or 185 kg (males). What makes up the mass of the neck? Neck mass of *G.c. giraffa* mainly is made up of muscles (75%), skin (15%), and cervical vertebrae (10%) (Mitchell et al., 2009, 2013a) with the absolute mass of all three higher in males than in females. The absolute mass of all seven cervical vertebrae together increased in male *G.c. giraffa* from 1.0 kg at birth to 14 kg at maturity and from 0.8 to 8 kg in females. Neck muscle mass increased from about 8 kg to 115 kg in male giraffes, and from about 7 kg in a newborn female giraffe to 67 kg in a mature female. Muscle mass, vertebra mass, and head mass are interrelated. The heavier the head, the greater the muscle mass needed to maneuver it, and the greater the vertebra mass needed to support the muscles. A heavy neck has the benefit of counterbalancing body mass during a gallop, but a prerequisite for gaining access to high level browse is a light and maneuverable neck. These conflicting needs are met by neck muscle mass and vertebra mass decreasing toward the head.

The *Ligamentum Nuchae*

Despite the adaptations that reduce the mass of the head and neck of a giraffe, the effort required to manipulate the head and neck is considerable and, in addition, the mass of the head and neck generates a very large rotational (torque) force. Torque is the tendency to rotate about a pivot. The pivot in giraffes is the articulation between the cervical and thoracic vertebrae. The torque, usually acting downward, must be overcome if the head and neck are to be lifted. Overcoming the torque is partly achieved by small interspinous ligaments that join vertebrae, but mostly by the *ligamentum nuchae*, a ligament that runs from the lumbar vertebra to the head in a giraffe. A distinct *ligamentum nuchae* has been found only in the dog family (where it is a small elastic cord that links the dorsal spines of the second cervical vertebra to the spines of the first few thoracic vertebrae), in some birds (Bennett and Alexander, 1987), in horses (Gellman and Bertram, 2002), in members of the camel family, and in artiodactyls (Slijper, 1946). In horses, camels, and artiodactyls the ligament is very well developed and consists of a funicular (rope-like) part and a lamellar (leaf-like part). The funicular part is broad and flat where it attaches (very strongly) to the first four or five thoracic spines, and in giraffes it extends all the way down the back to its origin on the lumbar and sacral vertebrae (see Plate 15.1). It becomes more cord-like and narrow as it progresses up the neck toward its insertion on the occipital crest of the skull. The fibers of the lamellar part are interwoven with those of the funicular part, with the lamellar bands running downward and forward (cranio-ventrally). In giraffes they are small and inconspicuous, and their main function is to anchor the funicular element to the cervical vertebrae. The most anterior lamellar fibers insert on the dorsal spines of the second, third, and fourth cervical vertebrae. The more posterior fibers are bigger and stronger and end on the spines of the fifth, sixth, and seventh cervical vertebrae in giraffes (Owen, 1838).

The funicular part of the ligament is exceptionally well developed in giraffes and extends from the lumbar and sacral vertebrae to the head, expanding along the way to form a large sheet over the dorsal spines of the anterior thoracic vertebrae. Its size has rarely been described. Gordon was impressed by it and it featured prominently in the diagram he sent Vosmaer (see Plate 15.1). Joly and Lavocat (1846) described it as follows:

> *It is formed of a cervical portion and a dorso-lumbar portion, and is elastic along its length. The cervical part is 1.1 m long, the width of the plate at the base of the neck is 13 cm. The thickness* [i.e., diameter] *of the* (funicular) *ligament at the same point is 7 cm.*

Owen (1838) wrote:

> *The development of this elastic ligament is truly extraordinary in giraffes. The mechanical stay and support of the long neck and of the head commences from the sacral vertebrae and receives fresh accessions from each lumbar and dorsal vertebra as it advances forwards. The spines of the anterior thoracic vertebrae become greatly elongated to afford additional surface for the attachment of new portions of the ligament which appears to be inserted on a superficial dissection, in one continuous sheet into the longitudinally extended but not elevated spinous processes of the cervical vertebrae as far as the axis. The atlas, as usual, is left free for the rotatory movement of the head. The termination of the ligament passes over the atlas to end by an expanded insertion into the occipital crest.*

One hundred and fifty years later, its ability to impress remained. Thus McNeill Alexander, in 1994:

> *The ligamentum nuchae of a large mammal is an impressive structure. When I wanted to measure the extensibility of the camel's, I could not find weights heavy enough to stretch this ligament much, and performed the experiment by tying one end to a beam and asking two carefully weighed technicians to swing from the other. Even it, however, was small compared to the ligamentum nuchae of a giraffe I dissected in the field, working rather nervously because we were uncertain whether the lions who had killed it were still hiding in a nearby thicket. The ligament was about 2.5 meters (8 feet) long and its cross sectional area was no less than 50 square centimeters (8 square inches).*

Its diameter, assuming a circular crosssection, would have been about 8 cm. Alexander was referring, of course, to the funicular part, and he noted that its elasticity is the reason that the head and neck of a dead ungulate are pulled back into an unnatural but characteristic angle, a position that veterinary students are required to call "opisthotonus" (e.g., Plate 3.3).

The elasticity of the ligament can be illustrated by a comment made in 1854 by T. Spencer Cobbold, conservator of the Anatomical Museum at Edinburgh University. He recorded its length as 165 cm in situ in a young giraffe, but 107.5 cm after removal, that is, a contraction of one-third of its length. It is elastic because of its structure. In all ungulates it consists mostly (between 80% and 90%) of a highly elastic protein (elastin), with the remainder made up of collagen (Fielding et al., 1976; Bennett and Alexander, 1987; Gellman and Bertram, 2002). Elastin has an elasticity as measured by Young's modulus, (stress (MPa) ÷ strain (%)), of 8.4×10^5, compared with a collagenous tendon (1.4×10^8), so elastin is ~1,000 times more elastic than collagen is, but it ruptures more easily than collagen, which has a tensile strength of 100 MPa (equal to 1020 kg cm^{-2}; Alexander, 1989a) compared with about 1 MPa (~10 kg cm^{-2}) for elastin. The ligament cannot, of course, be made entirely of collagen, which would make it stronger than elastin, but would also make it completely non-elastic and stiff. The ligament's function is to support the head and neck, but it must be sufficiently flexible to allow drinking or feeding from the ground or, equally, from the upper branches of trees. Moreover, the elastic elements, just like those of limb tendons, can stretch and recoil, and that property smooths the movements of the head and neck. The ligament is innervated in the ox, and those nerves are thought to trigger the stretch reflex that is found in all tendons and ligaments: when a ligament is stretched too far, muscle contraction is activated that reverses the overstretch (Fielding et al., 1976).

How big is the ligament in giraffes? Despite many opportunities to do so, detailed measurements of its dimensions in male and female giraffes, old and young, have not been made. However, the prevalent mechanical force acting on the combined head and neck mass is gravity, and a basic assumption is that the ligament supports the combined mass of the head and neck against gravity, passively, when the head and neck of a giraffe are at the preferred angle of carry, an arrangement that allows free movement of the head for whatever is needed. The arrangement makes ergonomic sense. If giraffes relied on muscle power to lift the neck and head and to hold them up at the preferred angle of carry, the neck muscles would quickly fatigue. The preferred neck angle will be the angle that requires the least muscle energy to maintain, and it

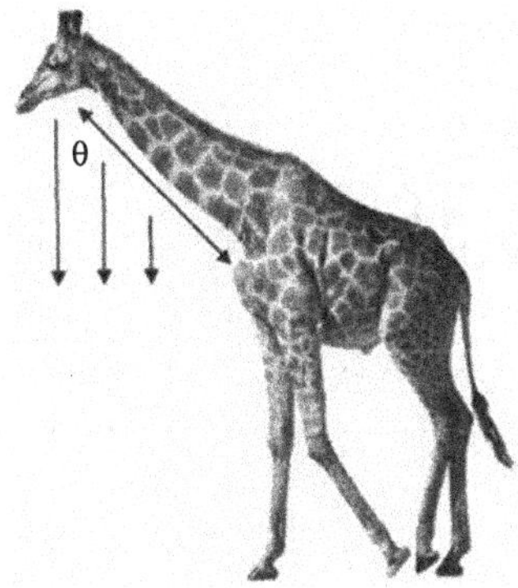

Figure 15.8. The elements of the torque generated by the head and neck. Gravitational forces exert a rotation force. The neutral position of neck carry is 55° to the horizontal and the angle (θ) at the head is then 35° to the vertical.

is presumed to be ~55° to the horizontal as that is the most common neck angle in giraffes (Figure 15.8).

The minimum size of the ligament needed to support the head and neck depends on the magnitude of the torque (newton-meters, N-m) generated at ~55° which, in turn, depends on the weight (kg) of the head and neck, the length of the neck (m), and the angle (θ) formed between the lever arm (the neck) and the vertical force applied (Figure 15.8). The torque generated by the head can be calculated by:

$$\textit{Gravity}\ (9.8\ m\ s^{-2}) \times \textit{head mass (kg)} \times \sin\theta \times \textit{neck length (m)}$$

It is sometimes mistakenly assumed that the downward force on the neck is exerted only by the head, but it is also exerted by the weight of the neck itself, and the torque the neck generates is the integral of its mass. Neck torque is calculated from:

$$\textit{Gravity} \times \sin\theta \times \textit{neck length} \times \textit{neck density} \times \pi \times [\tfrac{1}{2}R+2/3(R-r) \times (+1/4(R-r)^2]$$

where neck density is ~1,000 kg m^{-3} (Alexander, 1989a), θ, when the angle of carry is 55°, is 35°, neck length is the mean of the distances from the occipital crest to the withers and from the angle of the jaw to the *tuberculum supraglenoidale* on the scapula, *R* is the radius at the base of the neck and *r* the radius at the apex, both of which are known for a group of 60 giraffes. The size of the torque generated by the head and neck in those giraffes and the equation that describes it are shown in Figure 15.9A. On average, when θ is 35° it is 689 N-m (range 63 N-m in a 147-kg giraffe to 2,186 N-m in an adult weighing 1,441 kg). If the head and neck are lowered so that they are horizontal, then the torque exerted is at its maximum (on average 1082 N-m, range = 93 to 3,479 N-m). Consequently, if the torque generated at 90° is to be supported passively, the ligament would have to be significantly thicker than it is if torque is 689 N-m (Figure 15.9A shows the torque when the average between the torque at 35° and 90° is taken). Ligament thickness cannot change as the head is raised or lowered, so it makes much more sense from an evolutionary point of view for the thickness to increase as a giraffe grows to the size needed to match the maximum torque generated. If that had happened, then there would be little use for the extensor muscles of the neck to lift the head and head. Clearly

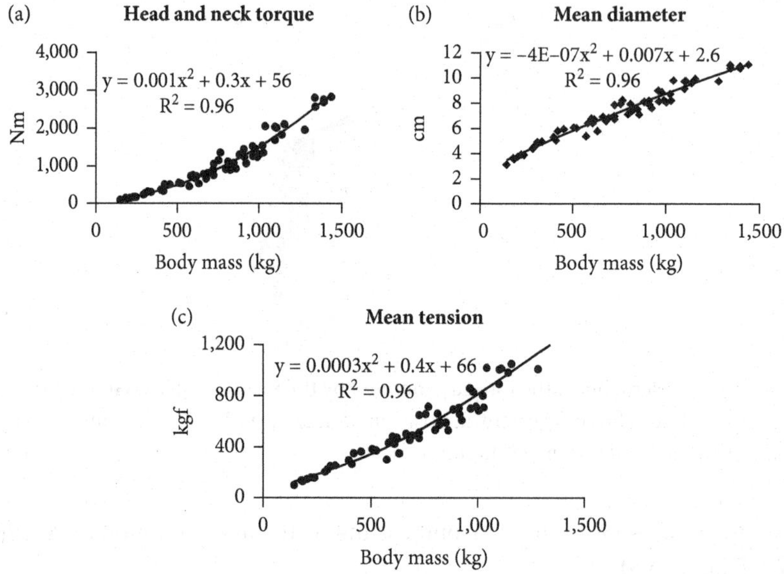

Figure 15.9. A: The torque generated by the head and neck. B: The mean diameter of the nuchal ligament. C: The average tension the ligament develops. Calculated from data for head and neck dimensions in 60 giraffes.
kgf = kilogram force where 1 kgf = 1 kg.

evolution, as it always does, has come up with a balance, and in this case the balance isbetween ligament size and the power of the extensor muscles that lift the head. If the head is raised to a vertical position and the ligament is fully relaxed, holding the head vertically depends mostly on muscle contraction.

When a giraffe lowers its head to ground level, the torque is reduced because the perpendicular distance from the fulcrum to the head is reduced, but the ligament has to stretch. Muscles have to pull the neck down against the stretch to reach that point and, consequently, the neck flexor muscles of giraffes are very well developed (see Chapter 16).

If a giraffe's ligament has the same biomechanical properties as the ligament in cattle, then stretching could elongate it by between 68% and 96% of its original length, with most elongation occurring near the head: in a camel studied by Dimery et al. (1985), ligament length at the head increased beyond initial length by 62% and at the base of the neck by 43%. Stretch and muscle strength alone, however, are insufficient to allow a giraffe's head to reach the ground and, despite the greater flexibility that the unique junction between neck and thoracic vertebrae confers (Gunji and Endo, 2016), a giraffe has to adopt its characteristic drinking posture to lower its head. What these interactions between muscles and ligaments mean for a giraffe is that when it has finished drinking, recoil of the elastic tissue of the ligament is enough to lift the head to the horizontal position without any need for a muscle contraction, after which muscles almost certainly are needed to lift the head further.

What must the size of the ligament be to support the head and neck passively against torques of the size reached in giraffes? Apart from Joly and Lavocat's and Alexander's anecdotes, the diameter of the ligament's funicular part has not been measured in giraffes. However, assuming that it is the funicular part of the ligament that resists torque, the diameter of the ligament can be calculated from its cross-sectional area, which in turn is the torque (N-m) ÷ neck length ÷ tensile strength. The tensile strength of elastin is between 1 MPa and 2 MPa (Dimery et al., 1985; Bennett and Alexander, 1987; Alexander, 1989a). Assuming that the elastin in a giraffe's ligament is that strong, the way the average diameter increases as body mass (and thus head and neck mass) increases is shown in Figure 15.9B. Its average diameter is 6.2 cm (range 3.2–11.1 cm). This average accords well with the 7-cm diameter measured by Joly and Lavocat and Alexander's diameter of 8.2 cm (which, from Figure 15.9B, means that this giraffe's body mass was ~750 kg). The calculated diameters also accord well with the ligament's diameter in an ox (5 cm at the head and 7 cm at the level of the fourth cervical vertebra) and in a camel studied by Dimery et al. (1985), the body mass of which was 402 kg. Its diameter at the head was 2.9 cm and 3.6 cm at the level of the fourth cervical vertebra. In a giraffe weighing 402 kg, the average diameter of the ligament can be calculated to be 5 cm. As a giraffe grows and its head and neck mass increase, so does the torque generated by the head and neck, and the diameter of the ligament should increase at the same rate, but it does not (Figure 15.9B). It does not because the overall load that the ligament can support is proportional to its cross-sectional area, which increases as the square of its radius. Thus, relatively small increases in diameter, say from 5 to 6 cm, increase the ligament's strength by ~50%.

Nevertheless, the calculated diameters will be sufficient only if the ligament can resist breaking when it is stretched. The ligament in cattle has a maximum tensile strength before rupture of ~3.5 MPa (= 35.7 kg.cm^2; Fielding et al., 1976), that is, it is 3.5 times stronger than its elastin component. If so, and assuming that a giraffe's ligament has the same structure as that of cattle, a ligament with the cross-sectional area of 50 cm^2, as recorded by Alexander, could carry a load of 1,750 kg before rupture, a load equivalent to the body mass of a very large giraffe and a load about nine times the combined mass of the head and neck of that very large giraffe. The average tension that develops in a giraffe's ligament can be estimated by dividing the torque generated by neck length, and how it changes as the giraffe grows is shown in Figure 15.9C. In a small giraffe weighing ~150 kg, the tension developed in the ligament is ~100 kgf, that is, a tension roughly equivalent to a weight of 100 kg being attached to it. In a mature adult male, it is ~1,300 kgf. Tensions of these sizes are far less than the predicted load that would produce rupture of the ligament. Why would evolution have produced a ligament that is far stronger than it needs to be? It is a mystery. The stresses must be greater than we think. Or, perhaps, the ligament is built with a safety factor, as are bones. Bones fracture when the load on them is 3–5 times their maximum compressive and peak bending stresses. The ligament of giraffes seems to have a similar safety ratio. The origin of the ligament's strength is equally enigmatic. One possible source of its strength is that the elastin that makes up 80–90% of its fibers has an unusual type of elastic behavior (Fielding et al, 1976). Murie (1872) reported that the outer part of the ligament in giraffes is composed of striated elastic fibers and the center of non-striated elastic fibers. He did not ascribe any functional importance to this combination of elastin fiber types, but the combination may underpin the ligament's strength in giraffes. One consequence of its elastin + collagen structure is that as a giraffe starts to bend to drink, stretch of the ligament is easy. As the head nears the ground, the ligament reaches a point where it no longer stretches easily and becomes very stiff. This anomalous behavior of the ligament is probably not related to the properties of its elastin fibers but to the 10%–20% of its fibers that are collagen. As the endpoint of elastin stretch is reached, it may well be that the collagen fibers take up the strain and resist further stretch. Collagen fibers rupture at a load 10 times that of elastin (1,020 $kg.cm^{-2}$).

The Appendicular Skeleton

The appendicular skeleton is the formal name for the limb bones; the principal limb bones are shown in Figure 15.10. To these principal bones must be added the two scapulae, the common pelvis, the carpal and tarsal bones that form the wrist and ankle, respectively, and the small bones of the phalanges. In many mammals the bones of the forearm—the radius and ulna—are separate, but in most artiodactyls they are fused to a greater or lesser degree, as are the bones of the lower leg—the tibia and fibula. In adult giraffes, as Figure 15.10 shows clearly, the radius and ulna and the tibia and fibula

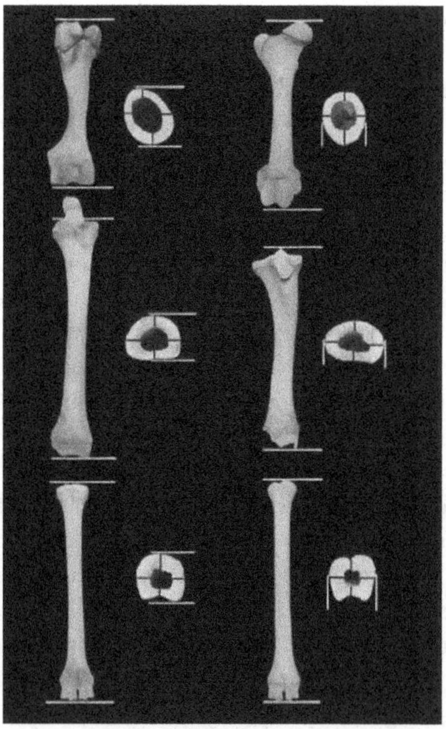

Figure 15.10. The six main limb bones of a giraffe. Left. Forelimb (from the top: humerus, radius, metacarpus). Right. Hindlimb (from the top: femur, tibia, metatarsus). The white lines show the distances (mm) over which length and, in the mid-shaft cross sections, the cranio-caudal and medio-lateral diameters, were measured. Black lines on the cross sections show the sites at which wall thickness was measured.
From van Sittert et al., 2015, with permission of John Wiley and Sons.

are completely fused. Each of the metacarpal and metatarsal bones is formed by fusion of the middle two (third and fourth) metacarpal (forelimb) and metatarsal (hindlimb) bones. Fusion of bones, over evolutionary time, has been associated with lower mass and greater strength of the product—an ideal combination.

Limb Bone Shape, Growth Patterns, and Strength

All bones are subject to forces that tend to compress them, twist them, or bend them, the limb bones most of all. Limb bones rarely fracture under compression (Selker and Carter, 1989) or torsion (twisting; Biewener, 1991). Bending stresses are the main source of fracture (Biewener, 1982, 1983) and they are produced by gravity, the impact of a foot hitting the ground, and the tension generated by muscles, especially the muscles that extend a limb, as it is they that have to do most of the work against gravity and it is they that have to develop the strongest contractions. The maximum compressive stress developed in long bones of mammals during movement on average is 550 kg cm^{-2} and peak bending stress occurs at about 400 kg cm^{-2} (Biewener, 1991), compared with the stress of approximately 2,000 kg cm^{-2} that will break them. Thus limb bones have a four- to fivefold safety factor, although safety factors of threefold have also been reported (Biewener, 1991).

The very first analysis of how limb bones avoid fracture when they are exposed to the stress generated by, say, an increase in body mass was done by Galileo in 1638. He wrote, and he might have had a giraffe in mind but, of course, giraffes would not be seen in Europe for another 200 years:

> *It would be impossible to build up the bony structures of men, horses, or other animals so as to hold together and perform their normal functions if these animals were to be increased enormously in height; for this increase in height can be accomplished only by employing a material which is harder and stronger than usual, or by enlarging the size of the bones, thus changing their shape until the form and appearance of the animals suggest a monstrosity. To illustrate briefly, I have sketched a bone whose natural length has been increased three times and whose thickness has been multiplied until, for a correspondingly large animal, it would perform the same function which the small bone performs for its small animal.* (Galileo 1638, translated by Crew H and de Salvio A, 1914)

Galileo's bone (Figure 15.11) had a length threefold longer and a diameter nine times wider in his imaginary large animal.

Of course, no modern, terrestrial animal has a bone of those monstrous proportions, and as Schmidt-Nielsen (2008) has pointed out, a threefold increase in length should be associated with 5.2-fold increase in diameter ($D \alpha L^{1.5}$), not ninefold, but the point was clear. As size and, therefore, body mass of an animal increases, the skeleton must adapt. Much, much later the problem was addressed by Thomas A. McMahon, who was professor of Applied Mechanics and Biology at Harvard University. He suggested that the most logical growth pattern to prevent fracture was a pattern of growth that preserved a constant relationship between bone dimensions and the force at which a bone will start to break, a pattern he termed "elastic similarity" (McMahon, 1975). The

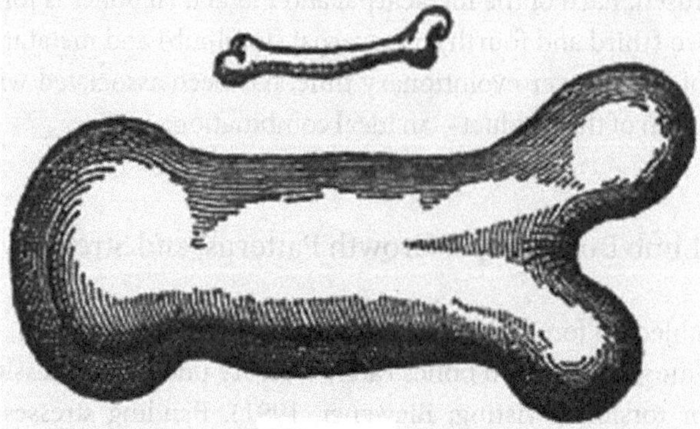

Fig. 27

Figure 15.11. Galileo's famous bones for imaginary animals differing ninefold in body mass.

From *Galileo's Dialogues*, translated by H. Crew and A. de Salvio, 1914.

advantage of an elastic growth pattern is that limb bones are resistant to fracture whatever the size of the animal. For animals that rely heavily on speed and agility to avoid predation, that advantage should outweigh any other. If a bone does grow with elastic similarity, then bone length (L) would be proportional (α) to diameter (D) to the power 0.67 ($L\ \alpha\ D^{0.67}$), $D\ \alpha\ L^{1.5}$, L α body mass $(Mb)^{0.25}$, cross-sectional area (XSA) α $Mb^{0.75}$, and $D\ \alpha\ Mb^{0.375}$ (Calder, 1996). In other words, an "elastic similarity" growth pattern is characterized by bone diameter growing faster than bone length ($D\ \alpha\ L^{1.5}$), and diameter and cross-sectional area increasing faster than increases in body mass (for isometric growth diameter would be α $Mb^{0.33}$ and XSA α $Mb^{0.67}$). The result of elastic type of growth is that bones become stockier as body mass increases, more or less as Galileo predicted. McMahon tested his idea by measuring the lengths and diameters of museum specimens of limb bones of 72 species of artiodactyls, including a giraffe, and found that $L\ \alpha\ D^{0.68}$, which was not different from the predicted $D^{0.67}$, so he concluded that resistance to "buckling and large amplitude bending" determined the relationship between length and diameter (McMahon, 1975). McMahon did not know the body mass of the animals that were the origin of his museum specimens, and so he could not test the relationship of length and diameter to body mass. However, Alexander (1977) measured the lengths and diameters of bones from artiodactyls of known body mass and found that $L\ \alpha\ Mb^{0.263\pm0.026}$ and $D\ \alpha\ Mb^{0.342\pm0.027}$, which he thought were close enough to predicted values to confirm the idea of elastic similarity.

The matter seemed settled. Limb bones grew according to elastic similarity. But in a later study, Alexander et al. (1979) measured the limb bones of 36 species of mammals ranging in body mass from a 2.9-g shrew to a 2,500-kg African elephant (but not including giraffes), and found that on average their limb bone length was α $Mb^{0.35}$ and diameter was α $Mb^{0.36}$. The relationship between length and diameter was $L\ \alpha\ D^{1.0}$ for all species, irrespective of body mass. As length and diameter increased equally as body mass increased, this type of growth pattern shows geometric similarity, and it is the pattern expected if gravity has been the selective pressure that has determined the relationship between length and diameter and body mass. (The exponents for the relationship between length and diameter and body mass are 0.33 because the units of length and diameter are cm and those of mass cm^3, and thus the exponent for equal or isometric or geometric growth is 1/3 or 0.33.) Alexander et al.'s scaling factors (exponents) were a little higher than the predicted value of 0.33; thus larger animals tend to have bones with slightly larger dimensions (Biewener, 1982) but, more importantly, it seems that the bones of terrestrial mammals generally are constructed to resist gravity. Resistance to "buckling and large amplitude bending" and elastic similarity are properties unique to the bones of terrestrial artiodactyls.

If the way bones grow in most animals follows a geometric pattern, clearly evolution usually has selected for geometric growth rather than elastic growth. The disadvantage of geometric growth is that the force required to fracture a bone decreases as body mass increases because a bone's ability to resist fracture is proportional to the bone's cross-sectional area, the size of which, if a bone grows geometrically, decreases as body mass increases (XSA α $Mb^{0.67}$; Selker and Carter, 1989): a doubling of body mass results in just a 1.5-fold increase in cross-sectional area. A corollary is that bones in older and thus larger animals should fracture more easily than they would in a younger one. However, an advantage of geometric growth is that it produces a greater relative length of bone (they do not become shorter and/or stockier as body mass increases) and thus the length of stride increases, which reduces the cost of exercise (Chapter 13). It does make some evolutionary sense to grow bones geometrically. Terrestrial mammals have to live with the cost of transporting their skeletons around for their entire lives, so that cost should be as low as possible. Large, heavy, stocky bones increase the cost of transport. Some animals cannot avoid a high cost of transport, or rather the high cost is worth it for other advantages their size might bring. This is the case in very large species like rhinoceroses and elephants (and the very large dinosaurs). Their size is/was the reason for their success, and to support their body mass their bones grow in a pattern called static similarity, in which length L α $D^{0.5}$ (D α L^2): if length doubles, diameter quadruples, and the result is greatly increased skeletal robustness and heavier bones, just as Galileo predicted.

Growth Patterns of Giraffe Limb Bones

As the strength of a bone cannot be increased by an increase in density, then strength must depend on their shape. Giraffes are artiodactyls and large, so it is reasonable to expect that their limb bones might follow the growth pattern of other artiodactyls or that of elephants or rhinoceroses. However, the limb bones of the single giraffe in McMahon's study did not fit either a geometric or elastic model: the relationship between length and diameter of the humerus and femur of the giraffe was similar to those of other artiodactyls, but the diameters of the radius, tibia, and metatarsal bones had not increased as expected as length increased, with metatarsal diameter having increased least. Similar measurements have been made by Christiansen (1999a, 1999b) of lengths and circumferences of the humerus, femur, radius, and tibia of a single giraffe with the same result. Conclusions drawn from a single giraffe are, at best, uncertain. However, an analysis of the growth patterns of the main limb bones of 10 fetuses, and 23 postnatal male and 21 postnatal female giraffes, all of known body mass, has been made by Sybrand van Sittert and colleagues (2015). The body mass of the fetuses ranged from 18 to 77 kg, in postnatal males it ranged from 187 to 1,441 kg, and in postnatal females from 147 to 1,004 kg. The bones of fetuses were included in the analysis because fetuses and their skeletons, in contrast to the bones of postnatal giraffes, are not subjected to the effects of gravity. The dimensions of fetal bones are determined mainly by genetics with the effects of intrauterine movements making a small contribution. Thus, any differences in the way fetal and postnatal giraffe bones grow would reveal the influence and extent of genetic versus physical factors on bone shape.

What the van Sittert et al. study showed was that there was no consistent pattern of growth in the limb bones of either postnatal or fetal giraffes. In fetal giraffes the average scaling factors for all six main limb bones showed a mixture of patterns: for example, $D \propto Mb^{0.45\pm0.02}$ and $XSA \propto Mb^{0.84\pm0.05}$ (~static similarity), $L \propto Mb^{0.36\pm0.03}$ (~geometric similarity), and $XSA \propto L^{2.32\pm0.21}$, $L \propto D^{0.77\pm0.04}$, $D \propto L^{1.23\pm0.09}$ (~elastic similarity). Fetal limb bones lengthen faster than the rate at which body mass increases (the exponent 0.36 is statistically greater than the expected exponent 0.33), and much faster than they lengthen in postnatal giraffes. These differences in growth patterns may partly be a consequence of growth in a gravity-free environment, but mainly exist because of a quite different biological requirement: newborn giraffes must have the athleticism to escape predation, and long, strong legs are necessary to do so. The length of a fetal giraffe's legs depends on elongation of their metapodial bones, which become longer than any of the other limb bones, and remain so until about age three (see Table 15.2).

In the limb bones of postnatal giraffes, not only did the scaling factors differ from those of fetuses, but they differed between bones within each limb and between equivalent bones of the fore- and hindlimbs (Figures 15.12, 15.13), although overall the exponents were much closer to a geometric growth pattern than to the more elastic one characteristic of fetal bones. A way of illustrating that idea is to compare the rate at which diameter and length increase in a giraffe over a lifetime. Overall, between birth and maturity the average diameter of all six bones increases 2.4 times and average lengths increase 2.3-fold. Thus both diameter and length of the main limb bones increase on average by similar proportions—geometrically. The exponents of the bones of the skeleton of postnatal giraffes are so different from those of fetuses that it must be the case that the fetal template is shaped after birth by gravity and movement, rather than only by genetics. The most conspicuous consequence of the postnatal effects is that the length of the radius becomes greater than tibia length (Figure 15.12) and the humerus becomes uniquely robust.

In the group of giraffes studied, the humerus was as long as the femur in all body mass classes but its outer diameter, wall thickness, and cross-sectional area were greater (Figure 15.10), and the rate at which its diameter increased was faster than the rate at which its length increased (Figure 15.13).

In all the other limb bones, the change in diameter was proportional to $L^{1.0}$, but for the humerus, $D \propto L^{1.3}$. The dimensions of the humerus of a postnatal giraffe arise from an elastic growth pattern, and so we have to conclude that the humerus has evolved to resist bending, whereas all the other limb bones have a more geometric growth pattern and have evolved to resist the compression stress that gravity exerts.

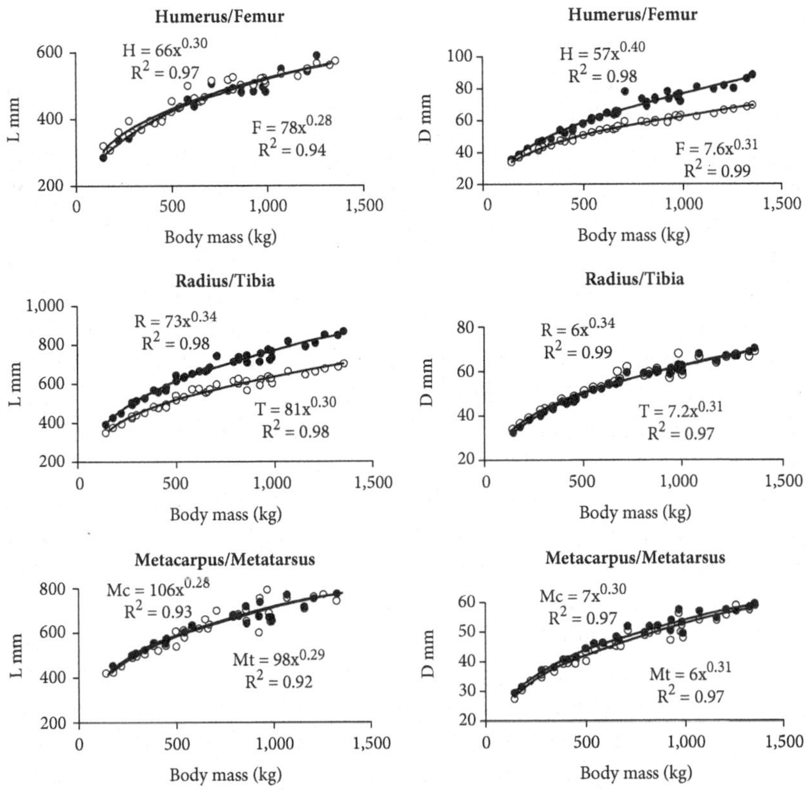

Figure 15.12. The relationship between changes in bone length or diameter and increases in body mass in 21 female and 23 male *G.c. giraffa*.

Constructed using data in van Sittert et al., 2015.

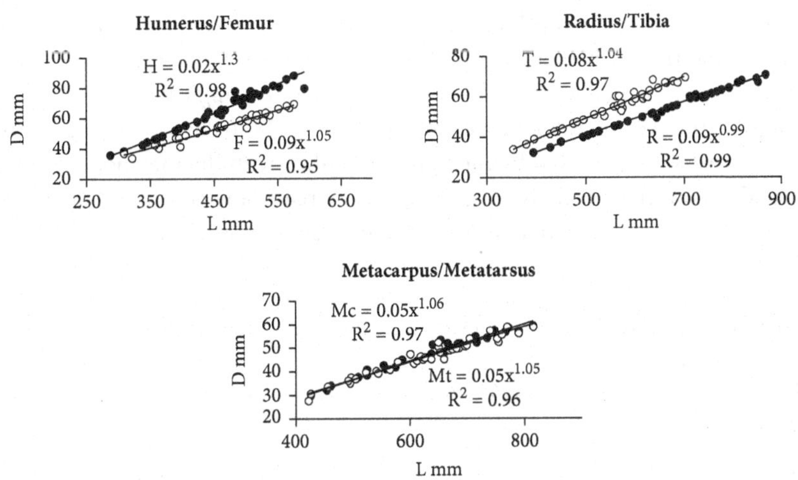

Figure 15.13. The relationship between increases in bone length and bone diameter in postnatal giraffes. Except for the humerus ($D \alpha L^{1.3}$), there is a linear relationship between diameter and length (i.e. $D \alpha L^{1}$) for all the other bones.

Constructed using data in van Sittert et al., 2015.

Limb Bone Strength

None of these patterns of bone growth is, of course, random. All are the product of natural selection with the underlying benefit of preventing fracture. All the limb bones in a skeleton must be strong enough to resist some degree of bending and compression, and the bones must also have enough extra strength to cope with accidental events like a fall (Biewener, 1989). Even so, limb bones are imperfect, as are other bones in the skeleton. John Currey has pointed out that there is an astonishingly high frequency of bone fractures in wild animals, and many of them heal (Currey, 2006) but, for the most part, easy predation follows a fracture and its possessor is removed from the gene pool. How well a limb bone can resist the stresses to which it is exposed can be estimated, in the same way that it can be in vertebrae, by applying principles of civil engineering. The critical engineering dimensions that determine the strength of a hollow, circular beam are inner diameter (ID), outer diameter (OD), wall thickness (OD–ID), length (L), and cross-sectional area (XSA). Assuming that limb bones are hollow, circular beams, the main strength indices and the equations that can be transferred from engineering and applied to giraffe bones are shown in Table 15.5.

Although there is a consensus, among those who study bones, that bone mass contributes less to limb bone strength than do the relationships between length and diameter (e.g., Selker and Carter, 1989), there is no doubt that the mass and density of giraffe limb bones are significant contributors to their strength, and that body mass and bone mass are interdependent. A good example of the interdependence is the difference in mass of male and female limb bones. Mature male giraffes are 50% heavier than are mature females, and male limb bone mass is about 50% higher than it is in females (Tables 15.3, 15.4). The relationship between the mass of a bone and its strength is defined by the ratio of its inner diameter (ID) to its outer diameter (OD), a ratio to which bone researchers have given the symbol K. Values of K can range from 0 for a completely solid bone (if it has no marrow cavity, it has no inner diameter) to nearly 1 for the very thin-walled bones found in birds (Currey and Alexander, 1985). The larger the OD and the smaller the ID the thicker the wall, the more solid the bone, the greater its mass, the greater its strength, and the smaller is K. For the *G.c. giraffa* studied by van Sittert et al. in 2015, the average K for all the limb bones was 0.48±0.05, with the humerus and femur the least solid, followed by the radius and tibia, with the metapodial bones the most solid: wall thickness relative to OD tends to increase down the limb (Figure 15.10). On average, fetal bones are even more solid than postnatal bones (mean K = 0.31±0.05)

Table 15.5 The Formulae Used to Determine the Strength of a Hollow, Circular Beam

Strength index	Formula
K	Inner diameter ÷ outer diameter (ID:OD)
I (cm^4)	$\pi/4(OD^4–ID^4)$
S (cm^2)	I÷ (length × OD)
Slenderness index (SI)	Length ÷ OD

because their marrow cavities, and hence their IDs, develop more slowly than do the ODs. Giraffe bones are significantly more solid than the average for artiodactyls generally (K = 0.49 and 0.61, respectively; $P<0.01$; see Table 15.6). In particular, wall thicknesses are greater in giraffes than in an artiodactyl of roughly equivalent body mass to the body mass of a medium-sized giraffe—the Cape buffalo (van Schalkwyk et al., 2004). Natural selection has produced thicker walls and greater strength of giraffe limb bones but also a higher mass, a compromise giraffes appear to have made during their evolution toward tallness. Throughout early postnatal life (up to 3–4 years of age, ~500–700 kg body mass), K for all six bones increases as body mass increases (the inner diameter increases faster than does the outer diameter and the wall becomes thinner) but thereafter K does not increase, becomes independent of body mass, and has an average value of 0.50±0.02 (Figure 15.14).

The capacity of a circular beam, and by extension a limb bone, to resist bending is predicted by a parameter called its second area moment of inertia (given the symbol *I*). The larger *I* is, the greater the resistance of the bone to bending. The magnitude of *I*, like that of K, is related to wall thickness (Table 15.5). *I* can be used to calculate what Selker and Carter (1989) have called the single best indicator of bone strength (S, cm^2). S is related to the bone's length and diameter (Table 15.5): the higher the resistance to bending for any specific combination of Length ×OD, the stronger the bone will be. A bone may buckle if it is exposed to too large a vertical, compression stress (as opposed to a transverse force that would cause bending). Whether a bone will buckle depends on the relationship between its length (L) and its diameter (D). The critical ratio of L:D at which buckling will occur (the slenderness index) has a threshold value of 25 (McMahon, 1973) or 26±8 (Garcia and da Silva, 2004; range 10 to 42); thus if a bone is on average 25–26 times longer than its diameter, then it is likely to buckle and fracture.

Values for these indices of strength have been calculated for the limb bones of postnatal giraffes using the measurements of length and ID and OD made by van Sittert et al. (2015). The way that S and *I* changes in all bones as body mass increases is shown in Figure 15.14. It could be expected that as body mass increases, *I* and S would increase at the same rate so that bone strength and resistance to bending and compression would stay the same. That expectation is met by the way *I* changes and, in fact, *I* increases much faster than does body mass, a pattern that implies that during postnatal life not only body mass and gravity, but also a factor like strenuous movement, affect bone dimensions. The rate at which S increases declines slowly as body mass increases (Figure 15.14), rather than increasing linearly, because the rate of increase of both bone length and diameter declines as growth proceeds. The limb bones of the fetal giraffes studied were, not unexpectedly, weaker than adult bones (*I* and S values were low), but they were more solid and had a slenderness index the same as that in postnatal bones. In both postnatal and fetal limb bones, the slenderness index fell far below the average threshold for buckling (SI=11 vs. the threshold of ~25): giraffe bones are unlikely to buckle despite their length. Strong as a postnatal giraffe's limb bones are, their relative strength (S, $cm^2\ kg^{-1}$ of body mass) declines at a rate proportional to $Mb^{-0.29}$ (Figure 15.14) and thus, in an adult giraffe, it is about half what it is in a juvenile giraffe. The relative size of *I* ($cm^4\ kg^{-1}$), on the other hand, increases at a rate proportional to $Mb^{0.36}$ (Figure 15.14), but that rate is slower than increases in body mass. Thus K, S, and *I* all decline as body mass increases, so either small giraffes have limb bones much stronger

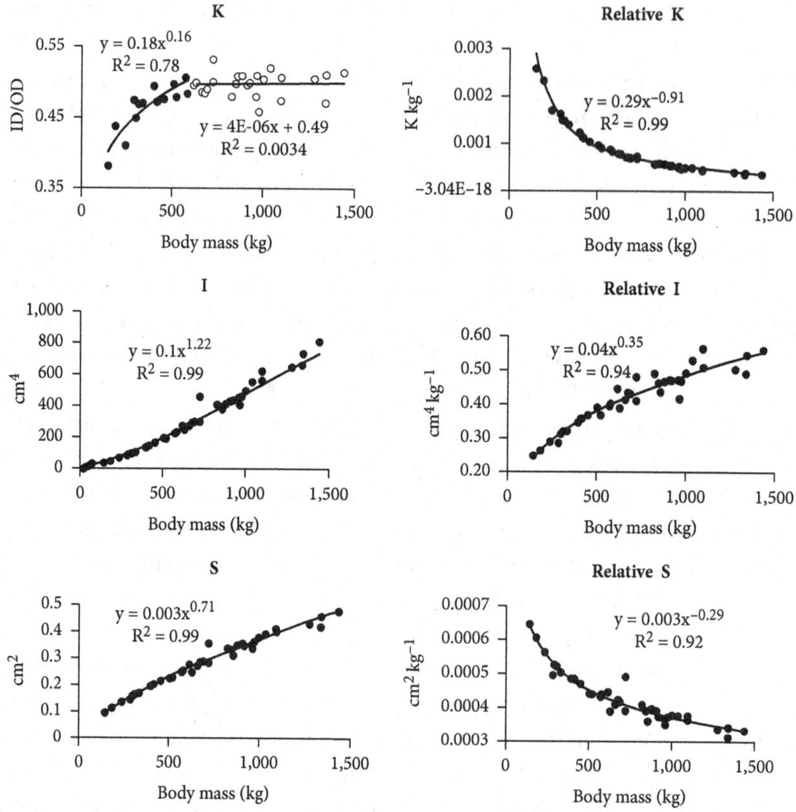

Figure 15.14. The mean strength and relative strength indices of all six limb bones combined in 39 postnatal giraffes.

Calculated from data in van Sittert et al., 2015.

than they need to be, or as body mass increases other mechanisms act to decrease the stresses on bones, or both.

Bending stress is exerted by muscle contraction, thus if the strength of contraction is reduced, the stress on limb bones can be reduced. Very large species like elephants reduce bending stresses by not being athletic (Bertram and Biewener, 1990), if that is the way to describe their ponderous movements. However, unlike elephants, giraffes are capable of extraordinary feats of exercise (see Chapter 13), although they are not agile and cannot (as I learned from a giraffe that, tripped, fell and broke its neck trying to do so) jump over a traditional, African, barbed-wire cattle fence (1–1.5 m high), but they can learn to step over it (Dagg, 1962a). The size of the force exerted by muscles when they contract is proportional to their cross-sectional area and the relative cross-sectional area (cm^2 kg^{-1} of body mass) of giraffe muscles is proportional to $Mb^{-0.23}$ (Alexander, 1977); that is, the cross-sectional area of muscles (per kg of body mass) declines as body mass increases, and in an adult giraffe it is ~60% of what it is in a juvenile giraffe, and the force the muscles can generate is also 60% of what they produce in a juvenile giraffe. Of course the actual cross-sectional area of a muscle increases throughout life, so the absolute strength of an adult giraffe's muscles is larger than that of a juvenile giraffe's muscles. However, although the relative cross-sectional area declines at $Mb^{-0.23}$, the relative strength (S) of bones declines at a rate proportional to $Mb^{-0.29}$, so relative bone strength decreases faster than muscle cross-sectional area decreases. In itself, therefore, the natural decrease in muscle cross-sectional area and strength as a giraffe ages is not enough to reduce the bending forces applied to bones through muscle contraction unless the power generated by contraction decreases simultaneously. And it may do so.

The power a muscle generates requires ATP, which requires oxygen for its synthesis. For artiodactyls, the volume of oxygen (mL kg^{-1} min^{-1}) used to move a distance (the cost of transport) decreases as body mass increases at a rate α $Mb^{-0.40}$ (see Chapter 13), and thus decreases faster than the rate that relative strength of bones declines ($Mb^{-0.29}$) and also faster than the rate at which the relative cross-sectional area of muscle declines ($Mb^{-0.23}$). The cost of transport is not necessarily a measure of the power a muscle exerts on a bone, but if the power the muscles can and do exert decreases as body mass increases, then the bending stress will reduce too. It is, however, muscle power that keeps joints stable, and if it does decrease, why do the joints of a giraffe or any other mammal not buckle as their body mass increases? Joint stability partly depends on the strength of the muscles that flex or extend them. One reason is that the strength needed to stabilize joints is small—and far smaller than the strength needed to generate a movement. A reason joints do not buckle in giraffes, and one that is peculiar to giraffes, is that they have very straight legs, and the power that muscles must exert to stabilize the joint in a straight leg is even smaller than it is in a mammal whose joints are flexed and their legs not straight. Having very straight bones also helps to reduce bending stresses. If a bone is curved along its axis, then axial compression and the power exerted by a muscle contracting along its length will tend to bend the bone, rather like a bowstring does a bow. If the bone is straight, then that bending force is reduced. As shown in Figure 15.15, giraffes have the lowest average curvature for limb bones of a variety of artiodactyls thus far measured (3.1±2.5 mm; Biewener, 1983b), and this analysis did not include the metapodials, which are completely straight in giraffes.

The forces exerted on limb bones during walking and running can also be reduced if the duty factor is increased (Alexander et al., 1977; Biewener 1983b). The duty factor refers to the time a foot spends on the ground during a stride. The longer that time is, the lower the force exerted on the limb bones by the impact of the foot hitting the ground. The length of time (seconds) a foot is on the ground during a stride increases with body

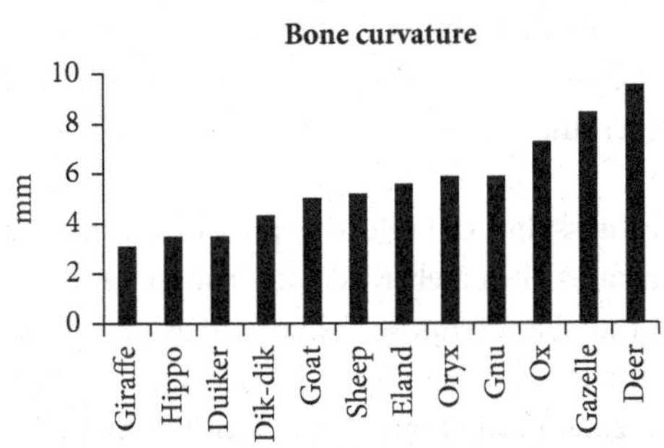

Figure 15.15. The average curvature of the humerus, radius, femur, and tibia in giraffes and other artiodactyls.
Redrawn from data in Biewener, 1983.

mass (α $Mb^{0.11}$ for forelegs and $Mb^{0.14}$ for hind legs; Alexander et al., 1977) and that time may even be longer (Garcia and da Silva, 2004) than the time predicted by Alexander at al.'s equations. Nevertheless, using Alexander et al.'s equations, the duty factor in giraffes running at their maximum speed can be estimated to range from 82 milliseconds in a giraffe weighing 150 kg to 153 milliseconds in an adult. Large giraffes do take fewer strides to cover a distance and, consequently, the length of time their hooves are on the ground during a stride is longer, the duty factor is longer than in a juvenile giraffe, and the stress generated by the impact of a hoof hitting the ground decreases, a stress further diminished by absorption of kinetic energy by their elastic tendons and perhaps by the structure of the hooves, though this last possibility has not yet been studied.

These secondary methods of reducing the bending stresses on bones (secondary, as they are unrelated to the primary structure of the bones) are clearly important and can help explain why the bones of postnatal giraffes rarely fracture. They do not explain why the bones of newborn giraffes do not fracture. The bones of a newborn giraffe are less able to resist bending stresses, their absolute strength is low, the duty factor is shorter, and relative muscle strength higher. By the second or third year of postnatal life, however, the indices that describe the strength of bones in giraffes of that age are indistinguishable from those of older postnatal animals. By then the bones of a newborn giraffe have become as strong as those of mature postnatal giraffes, but in the first few years of life, the bones of newborn and juvenile giraffes are most vulnerable to fracture, as indeed are the bones of other young, large animals like racehorses. Sometimes the bones of young racehorses do fracture, but those of giraffes seem not to. Why not? Of course, young giraffes are not forced to exercise beyond their anatomical limits, as are young racehorses, but another reason is that although absolute bone strength is low, the relative bone strength (the strength per gram of body mass) is high. It is 0.7 $cm^2\ g^{-1}$ of body mass in a newborn giraffe and 0.33 $cm^2.g^{-1}$ of body mass in a mature adult, and thus compared to the body mass they will have to carry, the bones of young giraffes are about two times stronger than they are in adults.

A Summary

Bone fractures are rare in giraffes, and so we can assume that natural selection has produced bones of just the right strength in a balance with the costs of making the bones and the costs of carrying them around. Even if they are an optimal size for giraffes, they are, if giraffe bone dimensions and strength indices are compared with those in other large artiodactyls, bigger and stronger and heavier than they are in other artiodactyls (Table 15.6).

Why? We can only guess, but the selective pressures must have been related to the high body mass that giraffes ultimately reach, their shape, and the stresses of movement that their body mass and shape impose. What makes giraffe bones strong enough to support their weight, their tallness, and the stresses of exercise? What seems to be specific for giraffes is that each of the bones in each of the limbs has a different chemical composition, different dimensions, and different characteristics and growth patterns. Each bone has evolved into the most efficient shape needed to support a giraffe and the unique ways of walking and running that giraffes have evolved. While most giraffe

Table 15.6 The Mean Dimensions of the Humerus, Radius, Femur, and Tibia (i.e., Not Including the Metacarpus and Metatarsus) Combined in a Giraffe Weighing 750 kg and in an Artiodactyl of the Same Mass, and the Strength Indices Calculated from Bone Dimensions

	Giraffe	Other Artiodactyls
Dimension	Mean	Mean
OD (mm)	54	44
ID (mm)	26	27
WT (mm)	14	9
XSA (cm^2)	27	16
Length (mm)	535	400
K	0.49	0.61
I (cm^4)	332	107
S (cm^2)	0.98	0.60

OD = outer diameter; ID = inner diameter; WT = wall thickness.

Giraffe dimensions from van Sittert et al., 2015. Artiodactyl data from Selker and Carter, 1989.

limb bones fit into either an elastic or geometric growth pattern, not all of them do, and so it seems that their individual shapes have evolved by natural selection to the stresses to which the bones have been exposed over evolutionary time and/or during a lifetime. An example of an evolutionary solution to the problem of bone strength is that with growth both the metacarpal and metatarsal bones, especially in males, develop bony columns on their caudal (ventral) surface (Figure 15.10). These columns are present in the metapodials of ancestor giraffes and so have a long evolutionary history, and they represent a specific adaptation, rather than the consequence of the bones obeying the more general patterns of growth. Other evolutionary solutions are straight legs, straight bones, limb bones that have a higher density than other parts of the skeleton, cross-sectional area and contraction characteristics of muscles that reduce the tensile stress generated by them, and a duty factor that increases as stride frequency decreases. All these factors reduce the stress on the limb bones of giraffes, and it is these other mechanisms that are as, if not more, important for preventing fractures than the geometry of bones first thought by Galileo to be the origin of limb bone strength.

16
Muscles, Nerves, and Skin

Muscles

About 700 million years ago, unicellular animals evolved preliminary contractile elements. So numerous were the benefits of an ability to move or change shape that from those modest beginnings every subsequent animal evolved its own specialized and specific muscles tailored to its lifestyle. The contractile elements are proteins and they too have become specialized, have evolved into very sophisticated sliding filaments of actin and myosin and all the complex regulatory processes that surround them without which muscles would not contract and relax. The product is muscle contractions that can range from the very fine, for example, finger muscles in a primate, to the large muscles of exercise. The whole contraction–relaxation mechanism has become packaged in the many millions of muscle cells that make up a muscle. Muscles in mammals vary in their structure and function but are of three basic types—cardiac, smooth, and skeletal (striated) muscle. In giraffes the functions and characteristics of cardiac muscle and smooth muscle meet the expectation of specialization (see Chapters 9, 10). It could be expected therefore that the skeletal muscles of giraffes will have been modified by millions of years of evolution to suit their elongated shape and will have become different from those found in mammals with a more conventional shape.

The Skeletal Muscles of Giraffes

The ways that skeletal muscles of giraffes differ from those in other mammals are mostly still a mystery because, apart from an anatomical study of forelimb muscles of a male *G.c. tippelskirchi* of undisclosed age by Kühlhorn in 1954 and very detailed studies of the hindlimb muscles made by Heinze in 1964 and another by him in 1965 of the hindlimb joints, tendons, and ligaments of a two-and-a-half-year-old female *G.c. capensis*, there have not been any detailed and complete more modern dissections that show how all giraffe skeletal muscles, not just limb muscles, might differ from those in other large herbivores. Kühlhorn and Heinze, while not correlating muscle anatomy with function, agreed that apart from "functionally unimportant slight differences," generally the limb muscles of giraffes corresponded with those found in cattle and horses but "it was not possible to determine whether some muscle parts of the giraffe show a different behavior from that of cattle and horses . . . but the exceptional proportions of the anterior extremity may correspond to peculiarities in muscle functions" (Kühlhorn, 1954). Until we have a good, modern, comparative, anatomical study of giraffe musculature, we have to rely for almost all of our knowledge of giraffe muscles on just three dissections. Richard Owen provided the first—a brief

How Giraffes Work. Graham Mitchell, Oxford University Press. © Oxford University Press 2021.
DOI: 10.1093/oso/9780197571194.003.0016

description in 1838 of, mainly, the muscles of the neck and shoulder. He was dismissive of the muscles of the body by remarking that they were for all practical purposes, the same as they were in any other animal, which is not completely accurate but, as confirmed by Kühlhorn and Heinze, is generally true at least for limb muscles. Owen did note, however, that the sheet of fibrous tissue (an aponeurosis) that is the origin of the internal and external oblique abdominal muscles had a wide area of attachment. This aponeurosis is attached to the last four ribs and all the lumbar vertebrae. The muscles attached to the aponeurosis form a sling under the belly, the muscles on each side joining in the ventral midline as the fibrous *linea alba*, to support the abdomen "which from the nature of their food requires support for the mass of the rumen" (Owen, 1838). Owen's study was followed in 1846 by that of Joly and Lavocat, who gave a very detailed and comprehensive description of all the muscles of the neck, body, and limbs, as did James Murie in 1872. Also very useful for understanding how a giraffe's muscles function, especially those of the limbs, is an ergonomic study done by P.P. Gambaryan (1974). What follows in this chapter is summary of these four studies. Joly and Lavocat's descriptions are difficult to follow, as the names they gave to the individual muscles are not in use today. Moreover, Murie complained (in 1872) that the three figures accompanying Joly and Lavocat's dissection were of such a diversified kind "that one fails to comprehend the beautiful muscular symmetry bestowed upon this towering ruminant." Murie's figures (Figures 16.1, 16.2) are, on the other hand, excellent. Despite his comment about Joly and Lavocat's dissection, Murie had the advantage of the prior dissections to guide his own, and he compared his observations with those made by Joly and Lavocat and by Owen.

Neck Muscles

Giraffes have a long and flexible but heavy neck, and supporting it against gravity and moving it around requires specific muscles. The muscles are, to put it technically, bilaterally symmetrical. Muscles on one side of the neck (or body) have a counterpart on the other side. The muscles on both sides of the neck, acting together, lift and extend the head and neck upward, but when there is a need to browse from a low shrub or drink from a pond, they lower the neck and flex it. The extensor muscles that lift the head and neck run along the dorsal or upper surface of the neck from the back of the head along the vertebrae to the shoulder. The flexor muscles that bend it down run along the ventral or lower surface of the neck from the lower jaw to the shoulder and ribs. Contracting on one side alone (i.e., on the left or right side of the neck), the neck muscles move the neck from side to side. The individual muscles have counterparts in a short-necked mammal but, in giraffes, have a uniquely complex structure and much longer tendons. The long tendons attach to the cervical vertebrae and the head, and movement is driven by contractions of the tendons' parent muscles in the lower part of the neck. Large muscles at the base of the neck near the shoulder and thin tendons near the head give the neck its tapering shape (Figure 16.1). To achieve the precise and rapid movements required, some muscles are attached at intervals to vertebrae along the neck by short tendons, and the muscles between attachments are "called upon sometimes successively or sometimes together" (Joly and Lavocat, 1846). The tendinous insertions (Figure 16.2) limit

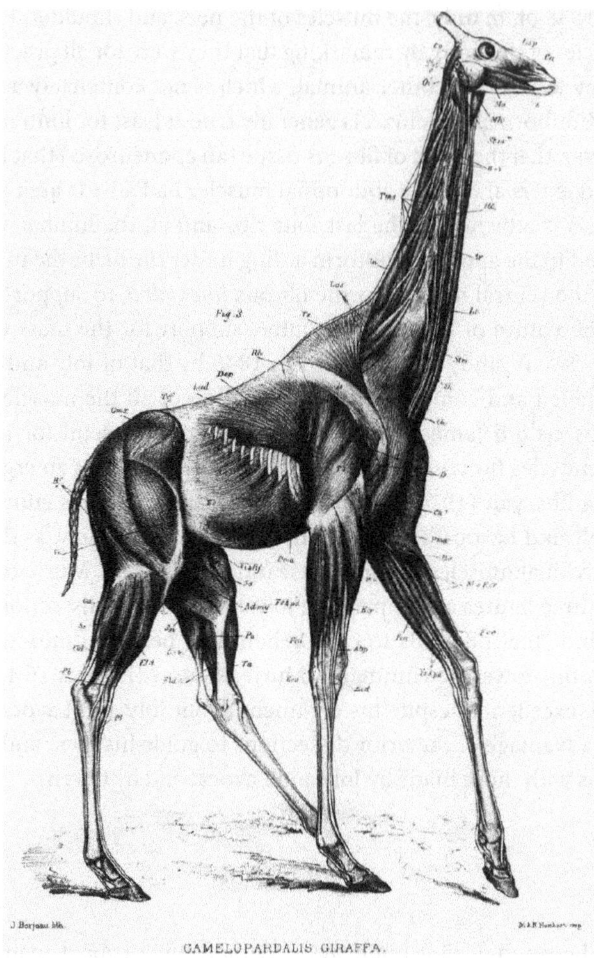

Figure 16.1. The muscles of a "giraffe in the attitude of walking to show the upper muscular layer throughout the body, neck and limbs" (52 muscles are labeled).
From Murie, 1872, with permission of Taylor and Francis

the extent or distance to which a muscle can shorten, with the effect that contractions can be more rapid and more precise than if the contractible sections were as long as the neck. Owen explained the arrangement using the muscle that controls movements of the hyoid bone of the tongue (the *sternohyoid* muscle). Its total length is about 2 m. It starts as a typical fleshy muscle, about 23 cm long, attached to the sternum, and this part ends in a tendon ~15 cm long. The tendon divides into two and each division becomes fleshy and continues as such for 23 cm, then each division again becomes tendinous for ~5 cm, then fleshy again, and so on.

> *In this alternation between contractile and non-contractile tissue is a striking example of the use of tendons in limiting the length of the fleshy contracting part of the muscle to the extent of motion required. Had the sternohyoideus been fleshy as usual from its origin*

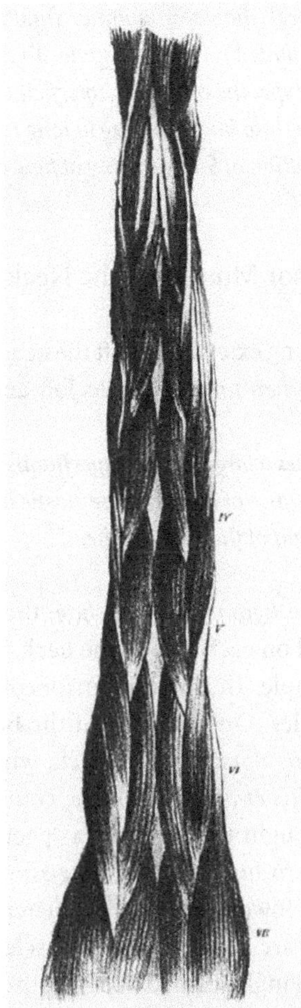

Figure 16.2. The "deep neck muscles on the lower (ventral) surface of the neck showing how they adapt themselves to the elongate character of the vertebrae and slender nuchal region. The numerals I to VI are placed opposite the cervical vertebrae."
From Murie, 1872, with permission of Taylor and Francis.

> *through the whole length of the neck to its insertion it is obvious that a great proportion of the muscle fibres would have been useless; for as these have the power of shortening themselves by their contractility only one-third of their length, if they had continued from end to end they would have been able to draw down the larynx and hyoid bone one third of the way down the neck. By the interventions of the tendons the extent of contraction is reduced, and the movements made more precise.*

Murie made the same point 40 years later.

> *Nuchal* [relating to the neck] *muscles are adapted to purpose. Just as in the legs, tendons replace flesh where power has to be transmitted with at the same time a decrease in mass*

> *of the limb or neck. In the neck, however, another function has to be subserved, viz, graceful flexion of this seven-jointed piece of the spine. To accommodate, then, all wants, and still retain the ruminant type, the ordinary superficial muscular layer appears to be absent, and the deeper ones cut into bands ending in long tendons—a method of subdivision which combines concentration of force, tenuity of figure, and pliancy.*

The Extensor Muscles of the Neck and Head

Giraffes have little need for strong extensors to lift the head because extension is brought about mainly by the elastic *ligamentum nuchae*. As Joly and Lavocat put it,

> *For the superior cervical muscles a large force is superfluous as the vertical direction of the neck and the mechanical resistance of the immense elastic ligamentum nuchae avoids an incessant fight against the weight of the neck and head.*

Despite the dominance of the *ligamentum nuchae*, there are several neck extensor muscles symmetrically placed on each side of the neck, but none is as well developed as they are in a horse, for example. The main extensors of the neck and head in giraffes are the large *complexus* muscles. On each side of the body they are made up of the continuation of the *longissimus dorsi* into the neck, where the muscle is referred to as the *longissimus (dorsi) capitis et atlantis*. Murie, commenting on the origin of the *longissimus dorsi* muscles, thought they showed a "peculiarity." The muscles start as two separate muscles, the *sacro-lumbalis* and *longissimus dorsi* muscles. These two muscles lie on one side of the lower back and are matched by two on the other side. The two muscles on one side are really a single muscle extending from the sacrum and running forward in the triangle formed between the dorsal spines and transverse processes of vertebrae on that side, all the way from the lumbar region to the last four cervical vertebrae. These two muscles are by far the longest muscles in the body, and they function to stabilize the vertebral column and prevent it from sagging downward and bending sideways. The peculiarity seen by Murie—and he was, it seems, talking about the *sacro-lumbalis* part—was that it appeared to have two halves. The upper half was tendinous (and in fact is the origin of the *ligamentum nuchae*) while the lower half was muscular.

In the upper thorax and lower neck, the *longissimus (dorsi) capitis et atlantis* is difficult to separate from two other usually distinct muscles nearby, the *semispinalis capitis* and *spinalis dorsi*, and in giraffes all are combined into one complex muscle. The *spinalis dorsi* component has unusual width because it attaches to the very long thoracic vertebra spines (Murie, 1872). The important origin of the *complexus* muscle is the first six or seven dorsal spines and transverse processes of the thoracic vertebrae. The muscle is attached to the spines underneath the sites of attachment to the spines of the *ligamentum nuchae*. The muscle's tendon is extraordinarily long, becomes thinner toward the head (Endo et al., 1997), and inserts on the occipital bone at the back of the skull below the nuchal crest of the occipital bone (to which the *ligamentum nuchae* attaches), on the mastoid process behind the ear, and on the transverse process of the atlas (1st) cervical vertebra. While Murie agreed that "as the *semispinalis* reaches the neck, it becomes much reduced in bulk corresponding therefore to the nuchal slenderness," he also thought that it inserted only on the second (axis) vertebra. He did not mention insertions on the skull, but they must be there if the muscle is to function as an extensor of the head as well as an extensor of the neck.

A second neck extensor is the *brachiocephalic* which, as its name implies, has its origin on each arm (the deltoid tuberosity and humeral crest of the humerus) and its insertion is on the skull. Its function, as Murie described it, was to "elevate and protract the fore limb or, if the shoulder and leg are fixed, the strain is applied to the lower moiety of the neck and thus drag it downwards," which implies that it flexes rather than extends the head and neck. In this he was wrong or misunderstood. When a giraffe counters the downward force on the head and neck by contracting the neck extensor muscles, contraction of the *brachiocephalic* muscle on one side draws the forelimb on that side forward and specifically pulls the humerus forward, an effect that extends and straightens the elbow joint. When the two forelimbs are kept rigid, as they are when a giraffe is standing still and taking the weight of its body on the forelegs, then the *brachiocephalic* muscle extends, and does not flex, the head and neck. Joly and Lavocat wrote: "This arrangement [of the *brachiocephalic* muscle] is unfavorable for limb action, but is all in favor of the speed and extent of the movements required by a long lever." The muscle's extensor effects are accomplished by its insertion on the mastoid process behind the ear, on the nuchal crest of the occipital bone at the back of the skull, on the transverse processes of the atlas, and also on the transverse processes of the second to sixth cervical vertebrae "whence from being narrow it broadens into a great sheet covering the shoulder sidewards and in front with an insertion into the humerus" (Murie, 1872). The extensor action of the *brachiocephalic* on each side is assisted by the *splenius* muscle (which Murie could not find), a band-like muscle (*splenius* = bandage). Its origin is the ligament that attaches the scapula to the dorsal spines of the third, fourth, and fifth thoracic vertebrae, and its tendon inserts on the same sites as the *brachiocephalic* muscle but not on the axis vertebra.

Further down and on each side of the neck are the *scalene* (= unequal) muscles. Murie, Owen, and Joly and Lavocat were impressed by their size in giraffes. Murie said they were well developed, Owen that they consisted of four parts but Joly and Lavocat found three parts, each part having a different length from the others: the *scalenus anticus* as described by Joly and Lavocat originated as two equal-sized bellies from the first rib, another from the third rib and by an aponeurosis from the sixth, and inserted on the transverse processes of the third and fourth cervical vertebrae; the *scalenus medius* arose from the first and second ribs and inserted by a tendon on the transverse processes of the seventh cervical vertebra; the *scalenus posticus* (or *s. tertius*) arose from the first rib and inserted on the sixth cervical vertebra. In other words, the scalene muscles on each side of the neck and chest arise from the first and second ribs and insert on the transverse processes of the third to seventh cervical vertebrae. The three or four parts function mainly to bend the neck to one side or the other. Extensions of the scalene muscles backward over the ribs form the supracostal musculature (Murie, 1872), which binds the scapula to the trunk. In humans the *scalene* muscles have a respiratory function. They lift the first rib when a deep inspiratory effort is made, thus increasing the volume of the thorax, but this function is very unlikely in giraffes because their first rib is immobile.

The Muscles of the Atlanto-Occipital Joint

The atlanto-occipital joint is formed between the occipital condyles at the back of a giraffe's skull and the first cervical vertebra, the atlas. The great degree to which this joint can be extended is a unique ability of giraffes and can be seen when a giraffe feeds vertically (e.g., Plate 3.6 in Chapter 3). Owen was impressed by the "extent (of the condyles) in the vertical direction; it is this structure which enables the giraffe to raise the head in line with the neck and even bend it a little way back." He should also have mentioned the other important consequence of the shape of the condyles. Their shape not only allows a giraffe to extend its head vertically when it reaches upward to browse, but also allows a giraffe to extend its head when it adopts a drinking posture, and in doing so, the degree to which the neck must be flexed so that the giraffe can reach the ground is reduced (e.g., Plate 11.1 in Chapter 11). Extension of the atlanto-occipital joint is achieved by the action of a number of small, short, but strong muscles (*obliquus*, *rectus capitis dorsalis major* and *minor*) attached to the sides of the axis and atlas. They insert on the upper part of the occipital bone of the skull. Flexion of the atlanto-occipital joint goes hand in hand with flexion of the neck.

The Flexor Muscles of the Neck and Head

Flexor muscles of the head and neck act to move the head and neck downward against the inherent elastic resistance of the *ligamentum nuchae* and are, consequently, larger than the extensor muscles. Like the extensor muscles, the main flexor muscles are duplicated on each side of the neck, an arrangement that is clearly visible in Murie's drawing (Figure 16.2). The main flexors of the head are the *rectus capitis ventralis* muscles, and in giraffes they are joined by the *rectus capitis lateralis* muscles to form a single muscle. On each side of the neck the combined muscle lies along the ventrolateral surface of the neck, where it is attached to the transverse processes of the first, third, fourth, and fifth cervical vertebrae, and inserts on the basal part of the occipital bone, well below the sites of insertion of the extensors of the head, neck, and atlanto-occipital joint. With an insertion any higher, contraction of the muscles would extend the head.

A second flexor is the *sternocephalic* muscle. Its origin is the sternum. Owen, in 1838, described the muscle as arising as a single muscle that soon divided into two bellies, with each division forming a flat muscle ~4 cm wide, and stressed the difference in structure between it and the *sternohyoideus* muscle. The divisions continued as fleshy muscles

> *the whole way to within one foot of the angle of the jaw where it terminates in a small round tendon that inserts on the inner angle of the mandible. From the freedom of*

Table 16.1 The Origins and Insertions of the Main Neck and Head Extensors and Flexors of Giraffes

	Muscle	Origin	Insertion
Neck extensors	*Complexus*	Dorsal spines of T1–T7	Occipital bone, mastoid process, atlas
	Brachiocephalic	Humerus	Occipital bone, mastoid bone, transverse processes of C1–C4
	Splenius	Dorso-scapular ligament	Occipital bone, mastoid bone, transverse processes of C1, C3, C4
	Scalene	First and second ribs	Transverse processes C3–C6
Neck flexors	*Rectus capitis ventralis*	Transverse processes C1, C3–C5	Occipital bone
	Sternocephalic	Sternum	Mandible
	Longus colli	Vertebra bodies of T1, T2, C2–C7	Transverse processes of C6, C7, bodies of C1–C5

the inflexion of the neck the head can be brought as near to the chest as the contraction of these muscles is designed to effect, the necessity for intervening tendons does not exist.

The third main neck flexor is the *longus colli*. It lies below the other flexor muscles and is composed of a succession of bundles that cover the ventral surfaces of the vertebrae from the sixth thoracic vertebra to the atlas (Figure 16.2). Each bundle originates on the body of a vertebra and inserts on the transverse processes of a more rostral one, before finally ending on the atlas by a strong tendon. Its largest bundles start on the thoracic vertebrae and they insert on the transverse processes of the sixth and seventh cervical vertebrae. Those bundles that arise on cervical vertebrae rather than on thoracic vertebrae are smaller and insert on the body of the next vertebra rather than on its transverse processes.

The main extensor and flexor muscles of the head and neck are listed in Table 16.1.

Limb Muscles

The evolution of tallness in giraffes and the very upright posture of their body were accompanied by limbs and limb bones becoming very straight. How straight is straight? That topic was studied in depth by P.P. Gambaryan. His measure of straightness was the sum of the angles formed by the elbow and wrist joints in the forelimb and by the knee and ankle joints in the hindlimb when an animal is standing. If the limbs were completely straight, then each of the two joints would form a straight line, and so the angle formed by each of them would be 180° and the sum of the two angles would be 360° (180° + 180°). When a giraffe is standing, the sum of the angles in the fore- and hindlimbs is never less than 250°, with the wrist and ankle joints always being in an almost direct line with their metapodial below or with their radius or tibia above, and the elbow and knee forming more acute angles with the humerus/femur above and the radius/tibia below. In other ungulates (like those listed in Table 16.3), the sum of the angles is usually below 200° because even when they are standing upright their joints are more flexed. Straightness in giraffes means that the elbow and wrist joints in the forelimb, and the knee and ankle joints in the hindlimb, are not flexed or bent to any great degree, and it is the straightening of a bent limb that conventionally is the source of movement. In effect the straightness of their limbs means that giraffes walk and run on stilts, and compared to smaller animals, giraffes swing their legs through smaller angles, relying on small movements at the top of the limb to bring about large movements at the bottom, a principle first recorded by Joly and Lavocat in 1846 when they described the action of the *brachiocephalic* muscle. A positive benefit of straightness is that the force that muscles must produce to maintain posture is reduced, and as a result, over evolutionary time the muscles of the shoulder of giraffes, for example, have become tendinous (meaning that muscle tissue has been replaced by collagen fibers) and less able to shorten by contraction, but much better at fixing the shoulder in positions that stabilize the upper part of the forelimbs at low energy cost. As giraffes carry most of their body weight on their front legs, having tendinous, low-energy shoulder muscles is an ideal evolutionary solution if giraffes spend most of their day standing. Do they? What does a wild giraffe do all day? Do they spend sufficient time standing each day to explain selection for tendinous muscles, rather than muscles that could produce a run?

The daily behavior patterns of modern wild giraffes have been studied quite extensively by Innes (1958), Leuthold and Leuthold (1978b), Pellew (1984b), van der Jeugd and Prins (2000), and du Toit and Yetman (2005), and presumably the same patterns existed in ancestral giraffes. Van der Jeugd and Prins's analysis of giraffe activity in Tanzania is shown in Figure 16.3; Du Toit and Yetman's data obtained in the Kruger National Park in South Africa (summarized in Table 16.2) are very similar, and both indicate that a giraffe's daily habits are the same no matter what part of Africa they inhabit.

It turns out that wild giraffes stand for ~85% of their day, walk for ~15%, and spend minutes, not hours, lying down, so throughout their day they are using muscles to stand rather than run. Giraffes in du Toit and Yetman's study ran for just a minute or two every day, and might have run about half a kilometer in that time.

Do giraffes ever rest or sleep? Generally, the amount of time an animal spends resting or sleeping is directly related to its metabolic rate because as one (sensible albeit controversial) theory holds, the purpose of sleep is for recovery from physical or mental activity, with those animals that have a high metabolic rate sleeping longer than those with low rates. Thus small animals like domestic cats sleep for 14 h a day (although mine seems to sleep for at least 20 h a day without any evidence of physical or mental overactivity), humans for ~8 h d^{-1}, and large animals like horses for just 3 h d^{-1} (e.g., Zepelin et al., 2005). Giraffes are large and have a relatively low metabolic rate, so they can be expected to sleep for short periods. Innes (1958), writing about the behavior of wild giraffes in the then Eastern Transvaal bushveld of South Africa,

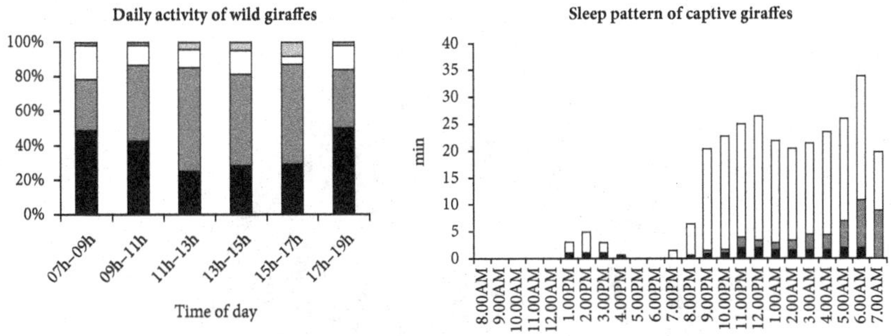

Figure 16.3. Left: Daily activity in wild giraffes. ■ = Browsing ■ = Standing □ = Moving ■ = Lying.

Redrawn from van der Jeugd and Prins, 2000, with permission of John Wiley and Sons.

Right: Periods of sleep in six captive giraffes in Emmen zoo, Germany. □ = recumbent sleep, ■ standing sleep, ■ REM sleep.

Redrawn from Tobler, 1995, with permission of Elsevier.

Table 16.2 The Daily Activities of Wild Giraffes Averaged over a Year of Observations

Activity	% of a 12-h Day	Hours per 12-h Day
Standing and feeding	58.3	7.0
Walking	18.8	2.3
Standing	9.7	1.2
Standing and ruminating	9.3	1.1
Walking and ruminating	3.5	0.42
Lying and ruminating	0.23	0.03
Running	0.13	0.02

Summarized from du Toit and Yetman, 2005, with permission of Springer Nature.

said that they lie down for between five minutes and an hour each day, a time that accounted for between 3% and 16% of their daytime activity, with lying down a more frequent behavior in the hotter parts of the year (November–December). When they were lying down, they usually did so with their heads erect (see Plate 14.1 in Chapter 14), but Innes did see one that had its neck stretched backward over its body with its head tucked beside its tail (Figure 16.4), which is the position they adopt when in deep (REM) sleep (Immelmann and Gebbing, 1962). The giraffe held that position for just five minutes. In the heat of the day giraffes also seemed to drowse standing with their heads lowered and the necks bent like a bow (Innes, 1958). Pellew found a similar pattern. Giraffes in the Serengeti spent part of the night lying down, but mostly just before dawn when feeding was at a minimum. Deep sleep with their heads resting on their flank lasted for ~ 5 minutes, just as it did in the giraffes Innes studied, so that seems to be the standard pattern in wild giraffes. Wild giraffes cannot

Figure 16.4. A giraffe in deep (REM) sleep in the "swan" posture.
From Immelmann and Gebbing, 1962, with permission of John Wiley and Sons.

afford to sleep for long periods of time. Sensory responses are depressed during sleep and the risk of predation increases. Sleep in captive giraffes is better studied (e.g., Immelmann and Gebbing, 1962; Kristal and Noonan, 1979; Tobler, 1995; Tobler and Schwierin, 1996; Siegel, 2008). In captivity, food is readily available and predator threats are nonexistent. Not surprisingly, sleep in captive giraffes is quite different from what Innes and Pellew have described for wild giraffes. For example, the sleep patterns of six captive giraffes recorded over a period of 18–25 days are summarized in Figure 16.3.

In these giraffes, total daily sleep time was ~5 h, of which deep (REM) sleep occupied ~20 minutes each night, standing sleep ~40 min each night, and lying down (recumbent sleep) occupied 3.5 to 4.3 h per night, during which the short periods of deep sleep occurred. A similar study of two captive female giraffes (Kristal and Noonan, 1979) found that they had three to eight episodes of lying down with their necks and heads held up, each episode lasting between three and 75 minutes. Periods of REM sleep during which the "swan" posture (Figure 16.4) was adopted usually occurred in these two giraffes between 02h00 and 05h00 and lasted between one and 10 minutes. Boredom can be surmised to underpin these behaviors in captive giraffes, as the length of each sleep episode increased as the night wore on and food had run out (Figure 16.3). There was nothing else for the giraffes to do except sleep.

Muscles of Movement

If a giraffe's limbs are straight and it has tendinous muscles with low contractility, what is the origin of movement, be it at walking pace or a gallop? Kühlhorn and Heinze compared giraffe limb muscles with those of cattle and horses, and the comparisons did not help to resolve this conundrum, although Heinze drew some excellent and helpful diagrams of the muscles of the hindlimb (Figures 16.5, 16.6, 16.14), but there are no diagrams yet of the forelimb muscles. Fortunately, fore- and hindlimb muscle function received the attention of Petr Petrovich Gambaryan, a distinguished Russian scholar, and Professor of Zoology at the Zoological Institute of the Russian Academy of Sciences (he died in 2017, aged 92), who also drew helpful diagrams (Figures 16.5, 16.6).

Regrettably, Gambaryan wrote (mostly) in Russian and that meant his work was and is not readily available to zoologists who, like me, cannot read Russian, or, if they can, cannot easily get access to Russian scientific journals. Fortunately, among Gambaryan's many publications was a book that was translated into English by Hilary Hardin in 1974. Both the original and the translation are monumental works. Gambaryan's research involved classification of various modes and forms of running in mammals (more than 200 species from six orders) and included the measurement of the mass of each limb muscle, as well as giving a description of its anatomy, in 25 different species of large herbivores, among them horses, Asian ungulates, the American bison, and several African ungulates including a giraffe, all aimed at working out how mammals run (Gambaryan, 1974).

Which muscles and joints are involved in giraffe movements? If giraffes are stilt walkers relying on small movements to produce larger ones by leverage, what muscles move the lever? One way of answering that question is to analyze the strength of

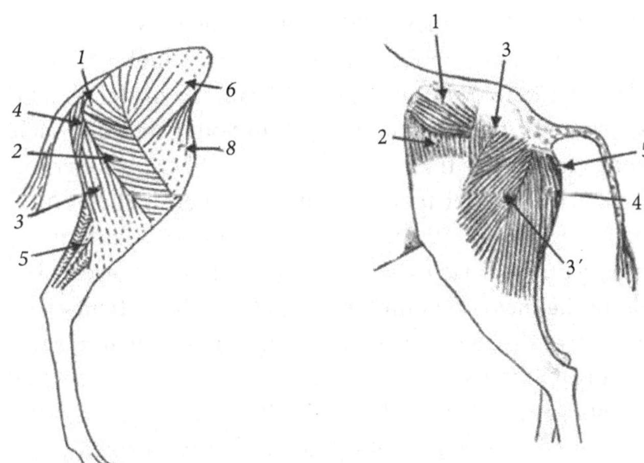

Figure 16.5. Lateral hind limb muscles of a giraffe. Gambaryan: 1, 2: Proximal and distal parts of the anterior *biceps femoris*. 3: Posterior *biceps femoris*. 4: *Semitendinosus*. 5: *Achilles* tendon. 6: *Gluteus*. 8: *Tensor fascia lata*. Heinze: 1, 3: *Gluteus medius* and *superficialis*. 2: *Tensor fascia lata*. 3′: *Biceps femoris*. 4: *Semitendinosus*. 5, *Semimembranosus*.

From Gambaryan, 1974 (left, with permission of John Wiley and Sons) and Heinze, 1964 (right, with permission of Elsevier).

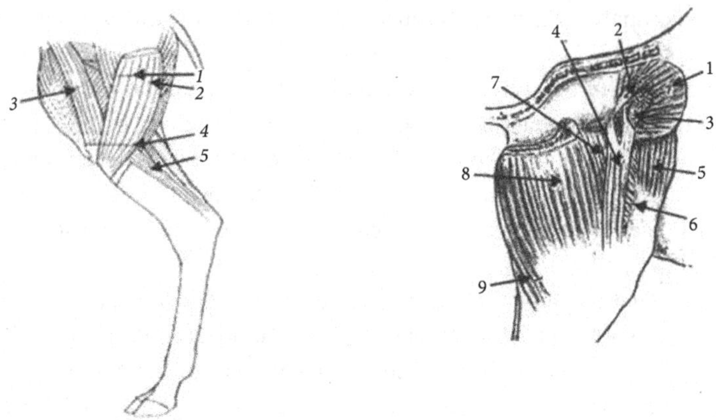

Figure 16.6. Medial hind limb muscles of a giraffe.
Gambaryan: 1: Proximal end of the *gracilis*. 2: Posterior part of the *gracilis*. 3: *Sartorius*. 4: Distal end of the *gracilis*. 5: *Achilles* tendon. Heinze: 1: *Iliacus*. 2, 3: *Psoas major* and *minor*. 4: *Sartorius*. 5: *Tensor fascia lata*. 6: *Rectus femoris*. 7: *Pectineus*. 8: *Gracilis*. 9: *Semitendinosus*.
From Gambaryan, 1974 (left, with permission of John Wiley and Sons) and Heinze, 1964 (right, with permission of Elsevier).

the muscles that move the joint. The joints that are the source of movement should have strong muscles, but although it is known in general terms that muscle strength is proportional to a muscle's cross-sectional area and muscle cross-sectional area is proportional to $Mb^{0.77}$ (Alexander, 1977), neither the cross-sectional area nor the actual muscle strength is known for any specific giraffe limb (or indeed any other) muscle. Hall-Martin (1975a) did measure the diameter of the fibers that make up the internal structure of the *longissimus dorsi* in young and old, male and female giraffes. In young giraffes less than one year old, the fiber diameter ranged from 32 to 48 nm. By maturity (greater than 10 years old) fiber diameter had increased to 53–58 nm in females and 62–72 nm in males, an increase indicative of increased strength. Gambaryan's approach was to measure relative muscle size, that is, the proportion (%) of the mass of all limb muscles that any particular muscle formed. The bigger the relative size, he argued, the stronger and more important the muscle. The actual mass of all limb muscles in mature giraffes (body mass ~1,000 kg) can be deduced from data gathered by Hall-Martin (1977). He measured the mass of the fore- and hindlimbs of male and female giraffes, and each formed ~8% of body mass in adult giraffes. Thus a limb in an adult giraffe weighing 1,000 kg will weigh ~80 kg. The bone mass in the forelimb of a 1,000-kg giraffe is ~14 kg and in the hindlimb is ~10 kg (Tables 15.3, 15.4). Apart from relatively minor contributions from skin and tendons to leg mass, muscle mass makes up the difference between total limb mass and limb bone mass. From those data the approximate mass of the main muscle groups in adult giraffes can be calculated from Gambaryan's proportions (see Tables 16.3, 16.4).

Hindlimb Muscles

The main function of the hindlimb muscles is to generate movement. A secondary function is to support body mass. Usually at the beginning of a run in most animals, including bipedal animals like humans, the first step of a run is exceptionally powerful and is brought about by extension of the hip joint. Hip joint extension should bring the femur into line with the vertebral column and the pelvis. The muscles needed to extend the hip joint have to act against the inertia of the body, and thus, typically, they are large, and the five biggest in each limb are the *gluteus maximus* (=*g. medius* + *g. minimus*), *adductor femoris*, *semimembranosus/semitendinosus* (hamstrings), *biceps femoris*, and *gracilis*, a muscle neither Murie nor Joly and Lavocat could identify, perhaps because its anatomy is confusing. In most ungulates it has two parts—anterior and posterior (Gambaryan, 1974). The anterior part is usually called the *sartorius*, which runs from the wing of the pelvis to the tibia. The posterior part runs from the pelvic symphysis to the tibia. A tendon arising from both parts joins the Achilles tendon (Figure 16.5). Together these five muscles make up over 50% of the total muscle mass of the hindlimb in a giraffe. All have their origin on the pelvis, and their insertions are on the posterior surface of the femur and/or tibia. If they are the main source of power, then the muscles should be large, but, with the exception of the *biceps femoris*, they are, taken together, somewhat smaller in giraffes than they are in other African artiodactyls, with the smallest being the hamstring muscles (*semitendinosus* and *semimembranosus*). They account for 21% of total hindlimb muscle mass in other African ungulates, compared with 12.6% in giraffes (Table 16.3).

Contraction of hip extensor muscles should change the position of the femur relative to the pelvis but, if it is assumed that Gambaryan's illustrations (see Figure 16.7) show correct angles between joints, the angle the femur forms with the pelvis hardly changes throughout the phases of a stride. What does change is the slope of the pelvis.
Its slope is determined by the action of the hip flexors—*ilio-psoas*, *superficial gluteus*, and *obturator* muscles—which are about as large as or larger in giraffes than they are in other ungulates (Table 16.3). The two psoas muscles have their origin on the ventral aspects of the last few thoracic vertebrae and on the lumbar vertebrae, and insert on the pelvis and anterior part of the femur; thus their contraction will draw the femur forward and flex the pelvis and/or hip. A giraffe cannot run with its hindlegs swinging forward parallel to its body. Its body is short and hindlegs swinging directly forward would crash into the front legs during some part of the gallop, a problem giraffes avoid by rotating the thigh outward during a stride. The muscles that achieve outward rotation are the *sartorius*, *iliacus*, *superficial gluteus*, and *obturator* muscles. The latter three muscles also flex the hip. They run from the pelvis to the lateral surface of the femur, and both form a greater proportion of hindleg muscle mass than they do in some other African ungulates (Table 16.3). The *iliacus* in giraffes is inseparably connected to the muscle fibers of the *psoas major* to form the *ilio-psoas* muscle (Heinze, 1964).

When the hip and pelvis flexors and extensors contract and relax cyclically, first flexing and then extending the hip and pelvis, the upper part of the hindleg has an easy to observe and seemingly effortless rocking motion, which, even without strong

Table 16.3 The Proportion (%) of Total Muscle Mass Formed by Each of the Hindlimb Muscles of Giraffes and Other African Ungulates

	Muscle	Wildebeest	Kudu	Eland	Gazelle	Mean % ±sd	Giraffe %	Mass (kg)[a]
Hip/pelvis flexors	*Ilio-psoas (major)*	5.2	7.7	7.2	5.5	6.4±1.2	6.9	
	Psoas minor	0.9	0.8	0.6	0.9	0.8±0.1	0.5	
	Gluteus superficialis + tensor fascia lata	2.8	3.7	3.0	2.7	3.1±0.5	4.4	
	Obturator	1.5	0.7	1.7	0.6	1.1±0.6	1.8	
		10.4	*12.9*	*12.5*	*9.7*	*11.4±1.6*	*13.6*	*10*
Hip extensors	*Pectineus*	1.0	1.5	1.6	1.7	1.5±0.3	1.6	
	Gluteus medius + minimus	13.1	9.1	10.6	8.4	10.3±2.1	7.5	
	Adductor femoris	5.3	4.4	6.5	4.5	5.2±1.0	6.3	
	Semimembranosus + semitendinosus	20.2	18.6	17.3	27.4	20.9±4.5	12.6	
	Biceps femoris	19.9	18.9	16.9	15.9	17.9±1.8	19.0	
	Gracilis/sartorius	4.3	3.8	4.4	3.4	4.0±0.5	6.0	
		63.8	*56.3*	*57.3*	*61.3*	*59.7±3.5*	*53.0*	*37*
Knee extensors	*Rectus femoris*	4.6	6.6	6.3	8.3	6.5±1.5	5.4	
	Vastus lateralis	6.6	7.9	6.7	7.7	7.2±0.7	8	
	Vastus medialis	2.9	3.7	3.9	4.1	3.7±0.5	3.1	
		14.1	*18.2*	*16.9*	*20.1*	*17.3±2.5*	*16.5*	*12*

Knee flexors	*Peroneus*	1.2	0.8	1.0	.01	1.0±0.2	1.4	
	Popliteus	0.7	0.7	1.0	0.6	0.8±0.2	1.3	
		1.9	*1.5*	*2.0*	*1.6*	*1.8±0.2*	*2.7*	*2*
Ankle extensors and flexors	*Gastrocnemius*	3.5	4.6	5.0	3.6	4.2±0.7	8.4	
	Plantaris	0.9	1.3	1.1	1.1	1.1±0.2	1.7	
	Tibialis anterior	0.3	0.5	0.5	0.1	0.4±0.2	0.5	
		4.7	*6.4*	*6.6*	*4.8*	*5.7±0.3*	*10.6*	*7*
Digits	*Extensors*	1.6	1.4	1.5	0.4	1.2±0.6	2.2	
	Flexors	3.2	2.7	2.7	1.9	2.6±0.5	2.2	

Data from Gambaryan, 1974, with permission of John Wiley and Sons.

[a] Approximate mass in a 1,000-kg giraffe.

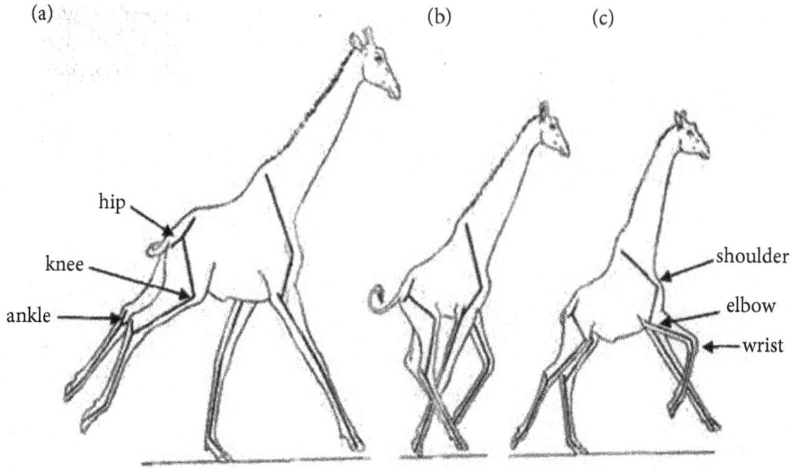

Figure 16.7. Movement of the appendicular skeleton during a gallop.
From Gambaryan, 1974, with permission of John Wiley and Sons.

propulsive thrusts, generates the power for a run. However, hip flexors and extensors are not the only source of power for movement. Another possible source of power is extension of the knee. The knee in a giraffe that is standing still is quite flexed, forming an angle between the femur above and tibia below of about 100° at the start of a stride (shown diagrammatically in Figure 16.7B). The knee is extended by the quadriceps group of muscles. Its four parts are the *rectus femoris, vastus medialis, v. intermedius* and *v. lateralis*, and the shape of the group is held by the action of the *tensor fascialata* muscle, which, as its name implies, tenses the superficial sheet of thick, strong fascia that runs over the quadriceps muscles from the sacrum to the patella. The quadriceps group of muscles have their origin on the pelvis (*rectus femoris*) and femur (the others), and their tendons insert on the patella, which in turn inserts via its patella ligament on the upper part of the tibia. In effect, therefore, the quadriceps muscles run from the hip and upper leg to the anterior surface of the tibia, via the patella. The tibia in giraffes is shorter than its counterpart bone in the forelimb: the radius. A short tibia would not have evolved unless shortness provided some sort of mechanical advantage. What that advantage might be is not clear, at least to me. The quadriceps muscles in giraffes are of similar relative mass to those in other African ungulates (Table 16.3). Contraction of them should straighten the knee and bring the femur and tibia into a straight line, and that is what they do (see Figure 13.2). However, the straightened position of the knee joint throughout the stages of a run means that it contributes very little to the power needed to run (Gambaryan, 1974).

The only other source of power for movement is extension of the ankle and digits. Extension of the ankle requires a large muscle—the *gastrocnemius*, or calf muscle—and it is twice as large in giraffes (8.4% of leg muscle mass and an absolute average mass of about 5 kg in a large giraffe, compared to 4.2% of leg muscle mass in other ungulates) (Figure 16.8; Table 16.3). The *gastrocnemius* has its origin on the distal third of the femur and inserts via the large Achilles tendon on the calcaneum (heel bone) of the ankle (Figure 16.8). Another ankle extensor is the *plantaris*, which has taken over the

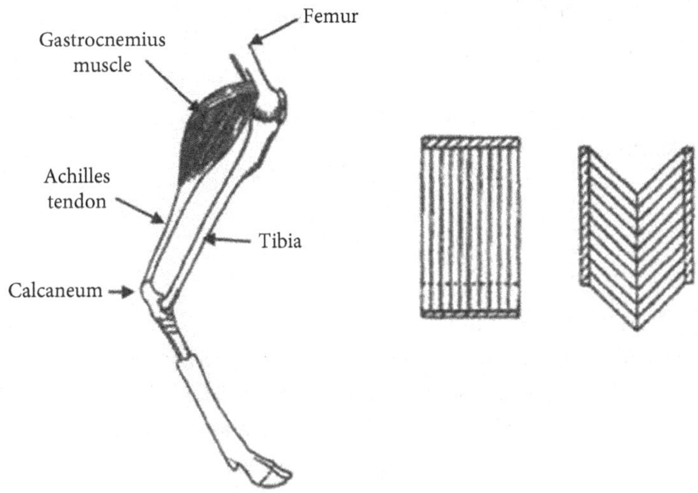

Figure 16.8. Left: Gastrocnemius muscle of a giraffe. Middle: A parallel-fibered muscle. Right: A pennate-fibered muscle.
From Gambaryan (1974), with permission of John Wiley and Sons.

postural functions of the *soleus*. In humans the *soleus* is quite large and distinct as a third calf muscle and functions together with the two heads of the *gastrocnemius* muscle to extend the ankle joint. The *soleus* has become completely incorporated into the *gastrocnemius* muscle in giraffes and so is no longer visible as a separate muscle. Joly and Lavocat thought that the *plantaris* was absent in giraffes. However, it is relatively large in giraffes (Table 16.3), and Murie described it as arising by a short tendon from the *linea aspera*, a ridge of roughened bone on the posterior surface of the femur. Its muscular part ends at about the middle of the tibia, and its tendon joins the main Achilles tendon of the gastrocnemius to insert on the calcaneum bone of the ankle joint. The fibers of the *plantaris* are mainly fatigue resistant and slow contracting. Those of the *gastrocnemius* are mainly fast contracting but fatigue quickly (Roy et al., 1988; see Chapter 13).

The relatively large size of the *gastrocnemius* and *plantaris* muscles could just reflect that they have to resist the buckling effect of a large weight bearing down on the ankle joint, rather than being a source of movement. The degree that an ankle joint can flex and extend is thought to be restricted in quadrupeds (Alexander et al., 1981), but the ankle extensors in giraffes are larger than they are in some other artiodactyls (Table 16.3) and, as Figure 13.2 (in Chapter 13) shows, the ankle joint is fully extended when a giraffe propels itself forward, so generation of movement, not just resisting buckling, must be a function of them. In addition, in giraffes the *gastrocnemius* muscles are multipennate (Gambaryan, 1974). Compared with a typical skeletal muscle in which the fibers lie in parallel with the line of contraction, in pennate muscles the fibers lie at an angle to the force-transmitting axis (Figure 16.8). Depending on the angle of the fibers, the muscle can contract quickly or slowly and generate a large or small force. Contrary to the accepted benefits of a pennate muscle (such as the masseter muscles, where a low speed of contraction is associated with a high force), Gambaryan decided that in the gastrocnemius muscle of giraffes the speed of contraction of muscle fibers is high and that high speed increases the force they exert. The extensors of the digits of giraffes are also bigger than in other artiodactyls (Table 16.3), and they act in concert with the *gastrocnemius* to produce propulsion. At the start of a stride (Figure 13.2B; Figure 16.7B), the digits are extended. At the end of a stride (Figure 13.2A, D; Figure 16.7A), the digits of the hindlimbs have flexed backward, with flexion resulting from first stretch and then recoil of the digit tendons.

Forelimb Muscles

In contrast to the functions of the hindlimb muscles, the main function of the forelimb muscles of giraffes, as it is in all quadrupedal animals, is to counter the effects of gravity and support the weight of the body. A secondary function is to generate movement. In giraffes the task of supporting the weight of the body is exaggerated by having to bear the weight of, and stresses generated by, a long and heavy neck, a combination of stresses that is complicated by the caudal displacement of the junction between the neck skeleton and the thoracic skeleton (see Figure 15.5 in Chapter 15). As a result of this shift, the proportion of the total body mass of a giraffe that is placed on the forelimbs is higher than it is, for example, in an ox or horse, or indeed in any other mammal, and over evolutionary time, the anatomy of the forelimb muscles of giraffes has evolved accordingly. It is tallness that determines how much or little the forelimbs can contribute to movement.

The shoulder and elbow joints are the forelimb joints that are most likely to buckle under a giraffe's body weight, and the task of preventing buckling falls on the shoulder and elbow joint extensor muscles. In some mammals the shoulder joint (i.e., the joint between the scapula and humerus) is stabilized by the muscles of the shoulder girdle, which attach the limb to the chest wall (*trapezius*, *rhomboid*, *serratus ventralis*, and the *pectoralis superficialis* and *profundus*), assisted by the clavicle (collarbone). Giraffes do not have a collarbone and rely completely on muscles to secure the joint. Thus it was that Owen, Joly and Lavocat, and Murie paid more attention to these muscles than they did to those of the hindlimb. It would be logical to expect that the muscles would be large, but contrary to expectation, the muscles are small relative to their size in other species (Table 16.4). A reason for their smaller size is that they are more tendinous, contain more collagen, and are therefore more inelastic, than muscle tissue itself. Their inelastic structure is perfect for supporting and suspending the trunk and for stabilizing the upper limb, but the consequence is that the muscles are less able to generate movement. Stabilizing the shoulder joint is their main purpose and depends first on fixing the scapula firmly to the chest wall so that it hardly changes position throughout the phases of walking or running. One of the muscles involved, the *rhomboid*, so-called because of its shape, according to Murie arises in giraffes from the seventh cervical to third thoracic vertebrae (a distance of ~20 cm in a large giraffe) and, according to Owen, it inserts on the broad elastic cartilage that forms the upper, vertebral edge of the scapula. It is able to contract, and when it does, it draws the scapula upward and forward, binding it closely to the ribs. It is very short, has a relative mass of one-third that of the same muscle in other ungulates, and is sufficiently small and short to be overlooked. Joly and Lavocat said it did not exist in giraffes, but Gambaryan found that it contributed 1.3% to the mass of a forelimb's muscles (Table 16.4). The smallest of the shoulder girdle muscles

Table 16.4 The Proportion (%) of Total Muscle Mass Formed by Each of the Forelimb Muscles of Giraffes and Other African Ungulates

	Muscle	Wildebeest	Kudu	Eland	Gazelle	Mean % ±sd	Giraffe %	Mass (kg)[a]
Shoulder girdle	*Trapezius*	4.9	3.7	2.1	3.3	3.5±1.2	0.3	
	Rhomboid	5.8	2.4	2.1	2.9	3.3±1.7	1.3	
	Pectorals	19.0	13.0	15.4	13.4	15.2±2.7	16.4	
	Serratus ventralis	12.3	14.9	13.2	14.5	13.7±1.2	7.5	
		42.0	*34.0*	*32.8*	*34.1*	*35.7±1.7*	*25.5*	*17*
Shoulder extensors	*Supraspinatus*	4.5	7.4	7.2	6.6	6.4±1.3	4.5	
	Infraspinatus	4.6	7.2	6.7	6.7	6.3±1.2	4.6	
	Latissimus dorsi	7.7	6.7	5.3	9.3	7.3±1.7	5.5	
	Brachiocephalic	7.1	6.6	4.5	7.1	6.3±1.2	5.1	
		8.5	*14.5*	*13.1*	*11.1*	*11.7±1.2*	*8.7*	*6*
Shoulder flexors	*Deltoid*	1.3	1.8	2.1	1.3	1.6±0.4	2.5	
	Teres minor +major	2.0	2.7	2.4	2.6	2.4±0.3	1.5	
	Coracobrachialis	0.4	0.6	0.7	0.4	0.5±0.2	1.1	
	Latissimus dorsi	7.7	6.7	5.3	9.3	7.3±1.7	5.5	
		11.4	*11.8*	*14.5*	*13.6*	*11.8±0.7*	*10.6*	*7*
Elbow extensors	*Triceps*	11.8	9.3	10.4	10.1	10.4±0.9	14.6	
	Anconeus lateralis + medialis	2.5	3.9	5.4	3.2	3.8±1.1	6.6	
		14.3	*13.2*	*15.8*	*13.3*	*14.2±1.0*	*21.2*	*14*
Elbow flexors	*Biceps*	2.1	2.7	3.0	1.6	2.4±0.6	5.0	
	Brachialis	1.4	1.9	2.5	1.2	1.8±0.6	2.7	
		3.5	*4.6*	*5.5*	*2.8*	*4.2±0.6*	*7.7*	*5*
Wrist+foot muscles	*Extensors carpus + foot*	3.6	5.2	5.9	3.7	4.6±1.1	9.7	
	Flexors carpus + foot	3.8	5.8	6.1	3.9	4.9±1.2	6.4	
		7.4	*11*	*12*	*7.6*	*9.5±2.3*	*16.1*	*11*

Data from Gambaryan, 1974, with permission of John Wiley and Sons.
[a]Approximate mass in a 1,000-kg giraffe.

is the *trapezius*. Its anatomy in giraffes is uncertain. Murie described it as "single and barely passes into the neck" and thus much smaller and "differs from the generality of heavy-necked Ruminantia and Perissodactyla, where it stretches further forward very considerably." Owen said it consisted of two distinct portions. One of them arose from the transverse process of the fifth and sixth cervical vertebrae. This part was thick and strong, expanded as it passed downward and backward to finally become lost in the strong fascia that spreads over the shoulder joint. The second part was thin and broad. It has its origin on the *ligamentum nuchae*, and its insertion is on the fascia covering the scapula. Murie thought that Owen may have mistaken the large part of the *trapezius* for the *brachiocephalic*. Gambaryan recorded the relative mass of the *trapezius* to be 0.3%, a proportion in line with Murie's assessment of its size.

The largest of the shoulder girdle muscles are the *pectoralis superficialis* and *profundus*, which have their origin along the whole length of the sternum and insert on the humerus and scapula (Owen, 1838). The muscles thus occupy the space between the lower part of the chest wall and the shoulder and arm, and function to draw a giraffe's forelegs inward toward its body and thus prevent them from splaying outward. The muscles must, however, relax when the forelegs of a giraffe are spread to facilitate drinking, but have to contract strongly to draw the legs under the chest when a giraffe lifts its head and neck to become upright again. Giraffe pectoral muscles have a relative size that is similar to the pectorals of other ungulates, but the main muscle for pulling the legs toward the trunk—the *pectoralis profundus*—in giraffes is small (its relative contribution to forelimb muscle mass is 6.5% vs. 11% in other ungulates), while the *pectoralis superficialis*, which acts to lift the trunk after a giraffe has drunk water, is more strongly developed. Its relative weight is 10% in giraffes and 4% in other artiodactyls, so while the total relative mass of the pectorals in giraffes is the same as it is in other artiodactyls, the size of its two components is different. Clearly, these two muscles have evolved in a way that accommodates both the burden of a long neck and the complications of drinking.

The *serratus ventralis* is an elastic muscle that suspends the trunk between the two scapulae. In giraffes it attaches by 13 tendons to the chest, of which 11 are attached to the ribs and the other two to cervical vertebrae. The attachments give the edge of the muscle a jagged, saw tooth appearance (that can be clearly seen in Figure 16.1), hence its name. In other ungulates the split is more even (nine thoracic and six neck tendons). Murie agreed that it arose from the 11 anterior ribs in giraffes, but did not mention any cervical serrations. In addition, and unlike all other ungulates, the tendons in giraffes attach to the proximal (upper) end of the scapula. Suspension of the trunk between the two scapulae by the *serratus ventralis* muscles on each side of the body means that the muscles, which in other animals can act to pull the trunk forward, in giraffes, together with the *rhomboid* and *trapezius*, act to keep the scapula stable during movement. Moreover, Owen wrote:

> *The fore part of the trunk is as it were slung upon the two great serrati muscles which principally support the weight of the remarkably deep chest of the giraffe. The interposition of the elastic cartilages between the upper attachments of the muscles and the capitals* [= the scapulae] *of the bony columns of the two forelegs must have the same advantage of a coach slung between springs.*

Gambaryan's description was similar: the *serrati* tendons are short and prevent forward and backward movement of the vertebral margin of the scapula, thus

> *due to the relative immobility of the vertebral margin of the scapula in the giraffe, the active work in drawing the trunk between the legs which is characteristic for this muscle in ungulates is replaced in giraffes by a passive role, that of fixing the trunk between the legs. Its relative weight in giraffes therefore constitutes 7.5% of the weight of the forelimb muscles, whereas in other ungulates it constitutes 14%* [Table 16.4].

The *serrati* are much more of a sling than an active muscle in giraffes. A continuation of the *serrati* is the *levator anguli scapulae*, which attaches to the superior (uppermost) angle (edge) of the scapula and, by contracting, also keeps the scapula close to the chest wall and rotates it inward. It has three parts: the lowest part, arising from the seventh cervical vertebra, is the largest; the other two parts, arising from the fifth and sixth cervical vertebrae, are small and "conform to the upward thinning of the neck" (Murie, 1872). The arrangement of the origins and insertions of the *serrati* muscles and the actions of the pectoral muscles help to explain how it is that a giraffe can spread its forelegs when it drinks and achieve an upright posture afterward. The scapulae are kept tightly bound to the chest, the skin between the forelegs is thin (Sathar et al., 2010), loose, flexible and folded, not thick and inflexible, and the thoracic tendons of the *serrati* arise in the upper part of the ribs near the vertebral column and not along the bottom of the ribs as they do in other ungulates. Thus, the distal end of the scapula is free to splay outward when a giraffe bends to drink, with some of the splay arising from an outward rotation of the humerus brought about by contraction of the *deltoid* muscle. After drinking, it is the pectoral muscles that contract to lift a giraffe upright.

The shoulder joint usually is straightened and extended by the *supra-* and *infra-spinatus* muscles that lie on each side of the low central spine of the scapula. Their tendons attach to the proximal, upper end of the humerus. Clearly, if these two muscles could contract in a giraffe, they would fulfill their usual function and straighten the shoulder joint and would be attached to a much higher central spine. However, they are relatively less well developed in giraffes, contributing 4.5% to forelimb muscle mass, than they are in other ungulates (6.4%; Table 16.4), because they are inelastic, fibrous, and inflexible, and their capacity to contract is limited, much as it is in elephants where they form ~4.0% of forelimb muscle mass. They function more like ligaments than muscles. The action of the two *spinatus* muscles is assisted by the *latissimus dorsi* and *brachiocephalic* muscles. The *brachiocephalic* usually is regarded as an extensor of the head and neck, but if the neck is held rigid by the action of its muscles, then the action of the *brachiocephalic* is to draw the limb forward and extend the shoulder. Consequently, and even though all the muscles of the shoulder are relatively small, their arrangement, together with the attachment of the upper border of the scapula to the ribs and vertebrae through the tendinous *rhomboid*, *trapezius*, and *levator anguli scapulae* muscles, ensures that the scapula/humerus linkage, "capitals of the bony columns," is exceptionally sturdy in giraffes, is able to counter the weight of the neck, and is able to provide the elbow and wrist joints with the platform they need to generate movement. If the action of the extensor muscles of the shoulder joint were unopposed, there would be a constant tendency for the shoulder joint to overextend. That tendency is countered by the flexors of the shoulder joint (*deltoid*, *teres*, *coracobrachialis*, and *latissimus dorsi* muscles). They form as big a proportion of forelimb muscle mass as they do in other ungulates (Table 16.4). Except for the *latissimus dorsi*, they have their origin on the scapula and they insert on the humerus. The *latissimus dorsi*, which, as its name implies, is a broad muscle, originates from the fascia that covers the dorsal spines of the lumbar and thoracic vertebrae, and inserts on the humerus. Contraction of these four muscles pulls the humerus backward and upward, and thus counters any tendency of the shoulder joint to overextend.

The muscles that provide most movement of the foreleg are the lower, distal muscles, those in other words that extend and flex the elbow, wrist, and digits. The extensors of the elbow—the *triceps/anconeus*—are large (Table 16.4). The *triceps* arises from the caudal border of the scapula, runs down the posterior surface of the humerus, and its tendon inserts on the olecranon process of the ulna. Its name suggests that it has three heads, but in giraffes it has five heads (Joly and Lavocat) arising from the scapula and the humerus but uniting to form a single tendon that inserts on the olecranon process. According to Gambaryan, it contracts by just 8% of its length in giraffes during movement, compared with 23%–47% in other ungulates, but because of the large forces to which the elbow is exposed, it is well developed. The action of the triceps is assisted by the *anconeus*, which in giraffes is twice as large as it is in other ungulates. The *anconeus* has its origin on the distal third of the humerus and also inserts on the olecranon. Throughout a movement, the *triceps/anconeus* is kept tensed and the elbow kept in extension (see Figure 13.2), much like its counterpart in the hindlimb—the *quadriceps* group—that keeps the knee in extension. Gambaryan compared the actions of the *triceps/anconeus* muscles in giraffes to those in an elephant, which has the same stiff-legged, stilt-like, walking gait with little shoulder movement, relying on the distal segments of their legs to do the walking and running. In elephants the relative weight of the elbow extensors is 21% of total forelimb muscle mass, just as it is in giraffes. The elbow flexors (*biceps brachii* and *brachialis*) are relatively less well developed in the giraffe, together weighing 7.7% of forelimb muscle mass, that is, one-third of the mass of the elbow extensors but, nevertheless, they form a greater part of limb muscle mass than they do in some other ungulates (Table 16.4). The *biceps brachii* has its origin on the scapula, lies along the anterior aspect of the humerus, and its tendon attaches to the proximal end of the radius. The *brachialis* arises on the humerus rather than the scapula, but also inserts on the radius. Contraction of these two muscles flexes the elbow in a manner very familiar to observers of posturing humans. From the size of the elbow flexors and extensors, it is clear that the effort needed to keep the elbow straight in giraffes is large: it is the forelegs which act as the fulcrum over which the body pivots during a gallop, and there can be little doubt that the muscles of the elbow joint have had to adapt accordingly.

A Summary

The distribution of relative muscle mass between fore- and hindlimbs in giraffes and between giraffes and other African ungulates differ, and the differences reveal some of the effects evolution has had on giraffe limb muscles. In the hindlimb of giraffes, the hip extensors and flexors are the biggest muscles, making up 67% of hindlimb muscle mass. Knee extensors and flexors make up 19% of muscle mass, and ankle and digit muscles the remaining 14% (Table 16.3). In the East African ungulates listed in Table 16.3, that ratio is 71:19:10. The East African ungulates rely somewhat more on hip flexors and extensors and less on ankle and digit extensors to generate movement than do giraffes. In the forelimb of giraffes, the shoulder girdle and shoulder extensors and flexors make up 45% of muscle mass, elbow extensors and flexors 29%, and wrist and digit muscles 16% (Table 16.4), compared with a ratio of 59:18:10 in East African ungulates, a ratio that shows that shoulder girdle muscles play a larger role in movement in them than they do in giraffes. These proportions nevertheless show that the power for a gallop in both giraffes and other ungulates has its origin in a powerful flexion of the pelvis and hip, followed by a powerful extension of them to complete the movement, with the lower muscles of the fore- and hindlimbs adding to, rather than originating, movement.

Nerves

As in so many other aspects of giraffe anatomy, the place to begin any analysis of giraffe nerves is with Richard Owen, but no analysis of nerves can be made without due reverence being paid to Sir Thomas Willis. He is chiefly remembered and venerated for his study of the blood vessels supplying the brain (see Chapter 10), but he also made a huge contribution to our knowledge of nerves. He was not the first to ponder their anatomy and function. Aristotle and Galen preceded him, and indeed it was expected of Willis, as Sedleian Professor of Natural Philosophy at Oxford, that he would give a series of lectures on Aristotelian and Galenical medicine (O'Connor, 2003). Willis found their explanations of signs and symptoms of diseases generally, and of the nervous system specifically, to be deficient. So it was that three decades after William Harvey's publication of *De motu cordis* had established the principle of circulation of blood, Willis began his study of the brain and

> *I determined with myself seriously to enter presently upon a new course, and to rely on this one thing, not to pin my faith on the received opinions of others, nor on the suspicions and guesses of my own mind, but for the future to believe Nature and ocular demonstrations,*

which is a dictum every scientist should obey. Willis published an account of his "ocular demonstrations" in 1660 and in it not only described the structure of the brain and the distribution of the cranial and peripheral nerves arising from it, but also considered how nerves might function. He concluded that nerve impulses "flow not within the nerves but are carried in the sides and not in the cavity," an observation that preempted by four centuries the discovery of membrane-based electrical impulses, which earned the Nobel Prize in Physiology or Medicine in 1963 for John Eccles, Alan Hodgkin, and Andrew Huxley. However, what made Willis's study truly unique was that he correlated signs and symptoms of nervous diseases with the anatomy of nerves and, in doing so, he established the discipline of neurology. Willis's studies revealed a general pattern of the structure of the nervous system that is still in use today: a central nervous system comprising brain and spinal cord and a peripheral nervous system divided into two parts—the somatic nerves that control voluntary actions like movement, and autonomic nerves that control all involuntary ones like breathing. While very little is known about the giraffe nervous system, what is known confirms that the basic anatomy of their nerves is little different from that found in any other ruminant or indeed any other mammal. However, the length of the spinal cord and peripheral nerves that supply the muscles of the limbs and the organs of the abdomen are much longer than they are in all other terrestrial mammals, except perhaps elephants. Are their central and peripheral nerves modified in some way to cope with the very elongated and large shape of giraffes?

The Brain

The anatomy of a giraffe's brain was first described by Owen in 1838. He declared that it resembles that of a deer "in the number, disposition, and depth of the convolutions." It was more depressed than in antelopes, and relatively broader and deeper than in the ox (Figure 16.9).

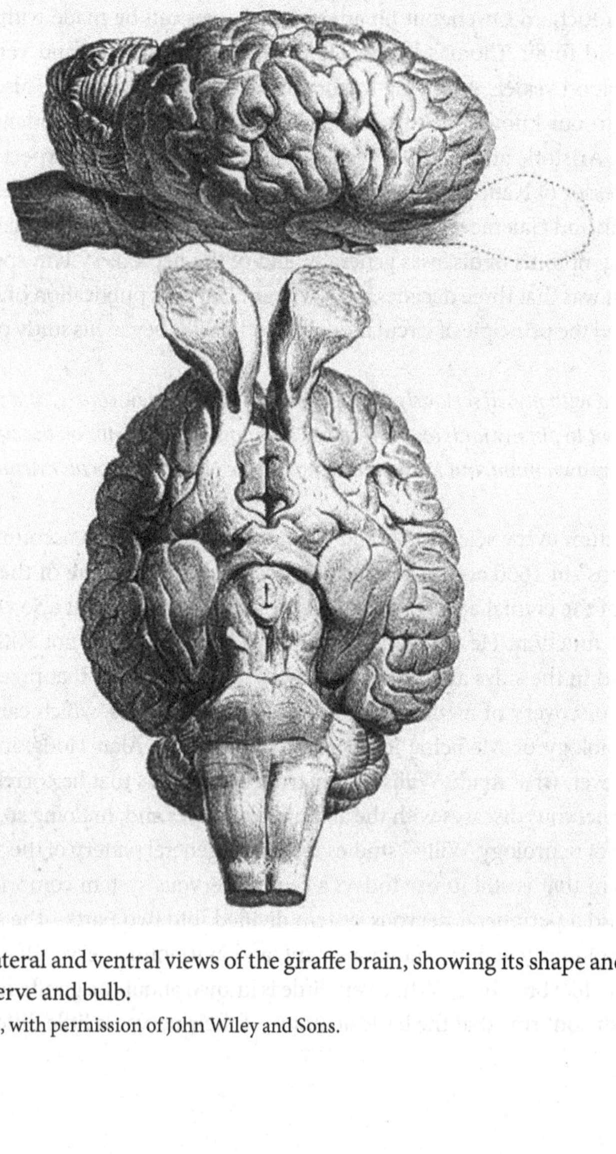

Figure 16.9 Lateral and ventral views of the giraffe brain, showing its shape and the size of the olfactory nerve and bulb.
From Owen, 1838, with permission of John Wiley and Sons.

The brain weighed 14 oz. (= 397 g) and, as the giraffe probably weighed about 300 kg, it formed just 0.12% of body mass (compared with about 2% of body mass for a human brain). The brain's cerebrum was 13.1 cm long, 6.8 cm high, and 10.6 cm wide, and its other major part, the cerebellum, was 4.6 cm long and 6.1 cm wide. Since Owen's study, the brain of giraffes has been weighed (or its weight reported) by Joly and Lavocat (1846): 710 g; Cobbold (1854): 539 g; Crisp (1864): 426 g and 390 g; Black (1915): 389 g, 420 g, 680 g; Crile and Quiring (1940): 700 g; Oboussier and Moller (1971): 655 g; Badlangana et al. (2007): 509 g; and Graïc et al. (2017): 770 g, 766 g. In most of these reports, body mass of the giraffes was not measured. It was measured (or reported) by Crisp (815 kg and 153 kg), Black (150 kg, 300 kg, 529 kg), Crile and Quiring (1220 kg), and Graïc et al. (630 kg, 800 kg). Brain mass has also been weighed in a group of 21 giraffes (*G.c. giraffa*; 13 males and eight females) for which body mass was known precisely and ranged from 179 to 1,441 kg; in other words, the giraffes were one year old to at least 10 years old (Mitchell et al., 2013a). The allometric relationship between brain mass and body mass for the 28 giraffes for which body and brain mass are known is shown in Figure 16.10. Clearly, the rate at which brain mass increases is slower than the rate at which body mass increases. Consequently, the brain's relative mass (% of body mass) declines with growth from 0.27% of body mass in a one-year-old giraffe to 0.05% in an adult (Figure 16.10), a pattern similar to that in other ruminants.

Although the relative brain mass declines with growth, its absolute mass doubles between birth and adulthood (from ~400 g at birth to ~800 g in an adult giraffe). On the other hand, a giraffe's body mass increases 14-fold (100 kg–1,400 kg). If the brain of a newborn giraffe weighs 400 g and its body mass is 100 kg, the ratio of brain mass

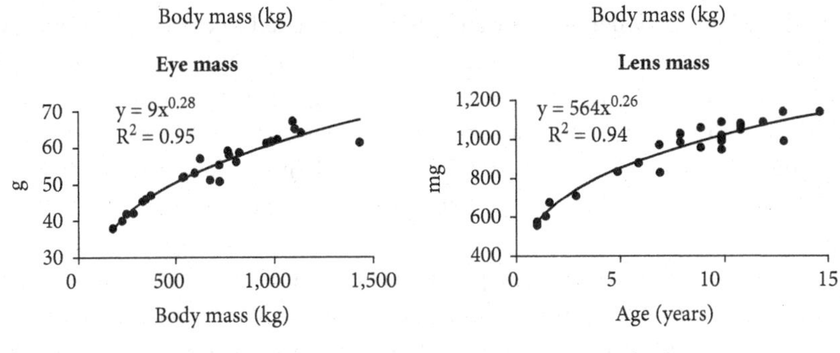

Figure 16.10. Top: Changes in absolute and relative brain mass as body mass increases.
Based on data obtained from 28 giraffes in Crisp, 1864a; Black, 1915; Crile and Quiring, 1940; Mitchell et al., 2013a; Graïc et al., 2017.
Bottom: Change in eye mass as body mass increases and in lens mass as age increases.
Redrawn using data in Hall-Martin, 1975a.

(g):body mass (g) is 1:250, and the brain is obviously able to regulate all the activity of a newborn giraffe. In an adult weighing 1,400 kg, the brain mass:body mass ratio is 1:1750, a seven-fold difference, yet the brain is able to regulate all the more complex functions that adults have. How? If brain mass does not increase linearly with body mass, then what must happen as a giraffe grows is that neural connections multiply or become modified, and it is those many more connections that allow the relatively small brain of an adult giraffe to regulate its more complicated activities. Whether this is what actually happens in giraffes is not known because, thus far, there have been very few studies of how the cells of the brain of giraffes differentiate and grow during growth. Those few studies have suggested that the types of neurons are consistent with what is known about the brains of other artiodactyls, even artiodactyls smaller than a giraffe (Bux et al., 2010; Jacobs et al., 2015). A similar conclusion was reached by Oboussier and Moller (1971), who studied the size of the neocortex in giraffes, okapi, and some unspecified African bovids. No differences were found. Another study compared the size and number of the neurons that control sleep and wakefulness in giraffes and harbor porpoises. This study was done because while porpoises have complex sleep patterns, wild giraffes sleep rarely (see earlier discussion). The relevant neurons in giraffes were larger, the porpoise had more, but no functional significance was attributed to this difference (Dell et al., 2012).

Cranial Nerves

In mammals there are 12 cranial nerves. Willis numbered them using Roman numerals, and they are still numbered that way. The first two nerves (olfactory and optic) are extensions of the cerebrum, the anterior part of the brain. The remaining 10 emerge from the brain stem, and their sites of origin in giraffes were described, but not well illustrated, in great detail by H. A. Vermeulen in 1915. The first eight cranial nerves innervate structures associated with the senses: smell (I), eyes (II, III, IV and VI), the face (V, VII), and hearing and balance (VIII). The next four innervate the tongue, salivary glands, pharynx, and carotid sinus (IX), the viscera (X), the trapezius group of muscles (XI), and the muscles of the tongue (XII). Owen was impressed by the very large olfactory nerves in giraffes, that terminated in expanded bulbs 3.8 cm long and 2.5 cm wide (Figure 16.9), a size for the bulbs similar to that recorded by Joly and Lavocat (4 cm × 2.7 cm). The olfactory acuity of giraffes is unknown, but from the size of the olfactory bulbs, odor must be an essential sense. Owen was also impressed by the large optic (II) and hypoglossal (XII) nerves, their sizes corresponding with the size of the eyes (see the following section) and the length and mobility of the tongue, respectively.

The length of the cranial nerves in giraffes is for the most part unremarkable, as the nerves innervate structures close to the brain. However, two of the nerves have exceptionally long paths. One is the vagus (X), which carries nerve fibers that innervate all the internal organs of the thorax (as well as the larynx via the superior and inferior laryngeal nerves; see Chapter 12) and the abdomen, including the intestinal tract, and thus the vagus nerve of giraffes must surely be one of the longest nerves to be found in any terrestrial mammal. The other long cranial nerve is the accessory (XI), which, generally in mammals, innervates the muscles at the base of the neck—*sternocephalic*, *brachiocephalic*, and *trapezius* muscles—muscles which in short-necked mammals are closer to the head than they are in a giraffe. The accessory nerve was once thought to be absent in giraffes, despite its origin at the base of the brain—the *nucleus acessorii*—being well developed and the nerve obvious (Vermeulen, 1915). Nevertheless, its "absence" raised the possibility that the evolution of a long neck was associated with the "saving of nervous tissue." Could the accessory nerve and its functions have been replaced during evolution by a different nerve in long-necked mammals? Solly Zuckerman (later Chief scientific officer for the UK government) and Francis Kiss were skeptical, and in 1932 investigated this idea by dissecting the pathway of the nerve in two giraffes. They found that the accessory nerve was present and did innervate the lower neck muscles, just as it did in shorter-necked artiodactyls, but the neck muscles at the base of the neck were also innervated by nerves arising nearby from the spinal cord, which is an unusual combination of innervations: the nerve had not been replaced, but it was not the only nerve supplying lower neck muscles. That opinion seemed final, but Howell and Strauss (1934) dissected a llama, could not find an accessory nerve, and concluded that there was a tendency in long-necked ungulates for the accessory nerve to be eliminated, with an intermediate step being a dual innervation. When J.J. Willemse dissected the necks of a giraffe, an okapi, a camel, and a llama in 1958, he found that the neck muscles of giraffes were supplied by both spinal and accessory nerves, just as Zuckerman and Kiss had described; in okapi the accessory nerve innervated the *trapezius* and *sternocephalic* muscles, and the *brachiocephalic* muscle was innervated by spinal nerves; in camels all the lower neck muscles were innervated by spinal nerves, and in llamas the *trapezius* and *brachiocephalic* muscles were innervated by the spinal nerves, and the *sternocephalic* by the accessory nerve. He concluded that the way lower neck muscles are innervated is family specific and not associated with evolution of a long neck, a conclusion confirmed by Angermeyer in 1966. "The muscles of the *trapezius* group" (in giraffes), Angermeyer wrote, "are supplied by the accessory nerve. The *cleidocervicalis* (*brachiocephalic*) muscle is innervated by spinal nerves," and, like others before him, he concluded that this arrangement is a familial trait and is not explained by a long neck, nor could it be, as it is also the innervation pattern found in okapis.

Eyes

Mammalian eyes are an extension of the brain in the form of a very large cranial nerve (II) innervating what once was a piece of skin sensitive to light. So numerous are the benefits of being able to detect light that, like the benefits of muscle contraction, natural selection has made sure that eyes have evolved independently in many, many different animals: in insects and snails, in octopuses, and in fish, birds, and mammals. In mammals, good vision usually depends on eye size and shape. They determine the size of the surface area of the retina and the focal length on which visual acuity depends (Howland et al., 2004). Good vision also depends on the ability of the lens to focus light onto the retina. Giraffes, as game wardens, hunters, and others have testified (see Chapter 18), have legendary, excellent vision. What is its origin?

One possibility is that the visual cortex in the brain that interprets information arriving from the eyes is better developed in giraffes than it is in other artiodactyls. However, a study of the structure of the neurons in the visual cortex of giraffes concluded that they differ little from those in other ruminants (Jacobs et al., 2015). If so, then the excellent vision that giraffes have must reside in the anatomy of their eyes, of which a surprising amount is known. The mass (but not the dimensions) of the lens of giraffes was measured in 32 giraffes by Hall-Martin in 1975. In them, lens mass increased from 389 mg at birth to 1,129 mg in an adult ~15 years old (Figure 16.10) and increases with body mass (M_L (mg) = $131*Mb^{0.28}$). The lens of giraffes has a biconvex shape like all mammalian lenses, and presumably the curvature increases when a near object is viewed and becomes flatter when a far object is viewed, as it does, for example, in humans. In 1864 Crisp recorded an eye mass of 26 g and a lens mass of 2.3 g in a young giraffe weighing 153 kg. Johnson (1901) measured the angle of the optical (orbital) axis (Figure 16.12) to be 72°; Ritland (1982) reported an axial length (the cranio-caudal diameter of the eyeball) of 38.8 mm in a giraffe that probably weighed 1,000 kg; Howland et al. (2004) reported an axial length of 48 mm in a giraffe that weighed 1,487 kg; and Burton (2006) reported an eye mass of 63.5 g in a 1,220-kg giraffe. These singular data do not give a broad idea of the size and shape of a giraffe's eye and how they change as a giraffe grows, so a systematic study of the size, shape, and growth patterns of the eye was made in a group of 27 giraffes (10 females, 17 males) with body masses that ranged from 179 to 1,441 kg (Mitchell et al., 2013b). A summary of the results of that study is shown in Table 16.5.

Table 16.5 The Size and Relations of the Eye, Skull, and Brain of Giraffes

Variable	**Value in a 750-kg Giraffe**	**Range in Giraffes with Body Mass 179–1,441 kg**	**Allometric Equation**	**R^2**
Anterior orbit width (mm)	186	124–218	$38 \times Mb^{0.24}$	0.83
Posterior orbit width (mm)	263	177–292	$47 \times Mb^{0.26}$	0.92
Orbit axis angle (deg)	59	73–46	$107 \times Mb^{-0.09}$	0.42
Eye mass (g)	58	38–67	$9.1 \times Mb^{0.28}$	0.95
Eye displacement volume (cm^3)	56	38–64	$8.8 \times Mb^{0.28}$	0.92
Eye geometric volume (cm^3)	53	37–64	$8.9 \times Mb^{0.27}$	0.93
Eye dorso-ventral diameter (mm)	47	42–50	$26 \times Mb^{0.09}$	0.93
Eye lateral diameter (mm)	49	42–51	$27 \times Mb^{0.09}$	0.90
Eye antero-posterior diameter (axial length) (mm)	44	39–48	$24 \times Mb^{0.09}$	0.88
Retina surface area (mm^2)	4,055	2,967–4,320	$1153 \times Mb^{0.19}$	0.93
Brain mass (g)	640	494–749	$237 \times Mb^{0.15}$	0.72
Eye mass:brain mass (%)	9.1	6.9–11.1	–	

Data obtained from 10 female and 17 male *G.c. giraffa* giraffes. From Mitchell et al., 2013b, with permission of Taylor and Francis.

Eye mass (g; Figure 16.10) and their displacement and geometric volumes (cm^3) in giraffes are tightly correlated (Table 16.5). Eye density is, on average, 1.01±0.03 g cm^{-3} (i.e., the same as water), which reflects the large contributions that aqueous fluid (which fills the space between cornea and lens) and vitreous fluid (which fills the space between lens and retina) make to eye mass. The maximum eye mass measured in this group of 27 giraffes was 67 g in a male that had a body mass of 1,100 kg; its displacement volume was 64 cm^3 and its geometric volume was 67 cm^3. Ideally, an eyeball should be spherical, because a spherical shape is easy to rotate within the orbit in the skull, has a small volume but a large surface area, has the smallest curvature for a given surface area and, in the case of a perfectly spherical eye, every point on the retina is equidistant from the lens. Are giraffe eyes spherical? If they were spherical, their diameter in any direction would be the same, but while the dorso-ventral and lateral (naso-temporal) diameters of the eyes in any giraffe are the same, the antero-posterior diameter is always slightly, but significantly, shorter (Table 16.5). Nevertheless, the eyeball of giraffes is indeed very close to being perfectly spherical, and is in fact more spherical than it is, for example, in sheep (Shinozaki et al., 2010). The spherical shape is maintained, as is eye shape in any mammal, by the internal, hydrostatic pressure generated by the aqueous and vitreous fluids. The eyes of giraffes are larger than they are in white rhinoceroses and elephants (Table 16.6).

Are they larger or smaller than expected for their body mass? Burton (2006) accumulated data on eye mass and body mass reported in the literature for 19 ungulates other than giraffes, and those data fit the regression equation $4 \times Mb^{0.33}$, although the correlation is weak ($R^2 = 0.40$) and 60% of eye mass is, therefore, related to factors other than body mass. Nevertheless, applying this equation to giraffes, the eye mass of a 1,100-kg giraffe can be predicted to be 40 g, compared to the 67 g that it actually was in a giraffe of that same body mass. Another way to determine the relative size of a giraffe's eye is to compare its size with the size of the brain. Eye mass was, on average, 9.4% ±4.3% of brain mass in the 19 ungulates in Burton's database. In the group of 27 giraffes it was 8.5% ±0.9% of brain mass (Table 16.5). Thus, compared to the ungulates in Burton's study, giraffe eyes are larger than expected relative to body mass, but their size is the size expected for their brain mass.

The whole purpose of an eye is to absorb light energy arriving in the form of photons in light waves arising from an object and to transform that energy into an image. Light

Table 16.6 A Comparison of Eye Mass and Focal Length in Large Land Mammals

Species	Body Mass (kg)	Eye Mass (g)	Body Mass (kg)	Axial Length (mm)
Elephant	6,654	58	4,509[a]	46[a]
Rhinoceros	764	11	764	24
Hippopotamus	947	19	1,351	29
Buffalo	666	31	759	32
Giraffe	1,220	64	1,469	48

Data on eye mass are from Burton, 2006, and on axial length from Howland et al., 2004.
[a]Mean of two measurements. Reprinted from Mitchell et al., 2013b with permission of Taylor and Francis.

energy is absorbed by eye pigments (or opsins), absorption changes the structure of the pigment, and the change in structure sets off a sequence of intracellular events that ends in the generation of a nerve impulse in the optic nerve for transmission to the visual cortex of the brain for interpretation. There is no reason to suppose that the mechanisms that turn light energy absorbed by a photoreceptor into an action potential in the optic nerve of giraffes are any different from the basic mechanisms that operate in mammals in general. In all mammals the opsins are located in photoreceptors that lie in the retina, and the photoreceptors in the retina of giraffes consist of rods and two types of cones (Schiviz et al., 2008). The best-known opsin, but not the only one, is rhodopsin, so called because it has pink-purple color. It is located in rods and absorbs the blue-green, short wavelength light that predominates at night. Its sensitivity in giraffes to blue-green light is the same as that of other savannah-adapted artiodactyls but different from that in okapis. The opsins found in okapis have evolved to become more sensitive to the gloomy light found in the closed, forest habitat that okapis occupy (Ishengoma et al., 2017). Cones endow color vision, and giraffes can distinguish red, orange, yellow, yellow-green, all long-wavelength light, and also violet (which has a short wavelength), from 34 different shades of gray (Backhaus, 1959). Cones also confer an ability to see fine detail. The average number of cones per unit area of retina is not known in giraffes, but the maximum density in parts of the retina has been measured and is 22,800 mm^{-2}. Rods are far more numerous than cones. They detect light of low intensity (but not color) and there are more in the retina of a nocturnal animal than a diurnal one. Most mammals have rod densities in the range of 200,000–400,000 mm^{-2} (Peichl, 2005). The ratio of rods to cones in sheep is 11:1 (Shinozaki et al., 2010), which, like giraffes, are diurnal and so there are probably ~250,000 rods mm^{-2} in a giraffe's retina. All impulses generated in both rods and cones converge on ganglion cells in the retina, and their axons converge to become the optic nerve. There are approximately 1.5 million ganglion cells in the retina of giraffes (Coimbra et al., 2013); therefore the ratio of photoreceptors to ganglion cells is 120:1, which is a fairly typical ratio for mammals in general.

The total number of rods and cones that can fit into a giraffe's retina depends on the retina's surface area (mm^2). The bigger the retina surface area is, the larger is the number of photoreceptors, the greater the number of photoreceptors across which an image is spread, and the greater the ability to discriminate between objects that are closely spaced. Retina surface area is proportional to eye diameter and can be calculated using Robb's formula (Robb, 1982):

$$Area = \pi \times H \times D$$

where H (mm) is the antero-posterior diameter of the eye × 0.57 (Schiviz et al., 2008), and D is the mean of the dorso-ventral (vertical) and lateral (horizontal) diameters. In giraffes with body masses of 179–1,441 kg, the retina area calculated using Robb's formula ranged from 2,900 to 4,300 mm^2 and in an average-sized giraffe weighing 750 kg, it will be ~4,000 mm^2 (Table 16.5), which is similar to the size found by Coimbra et al. (2013) in five giraffes that weighed about 500 kg (3,404±72.7 mm^2). The total number of photoreceptors in a giraffe's retina would then be around 180 million. For comparison, a human retina has about 120 million photoreceptors.

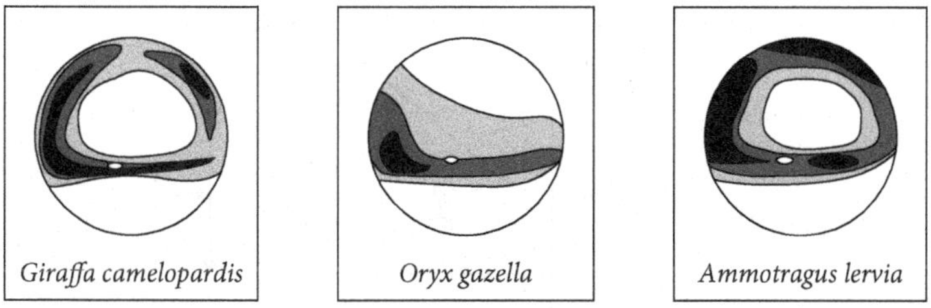

Figure 16.11. The areas of high density photoreceptors in giraffes (left) compared to that in an oryx (middle) and Barbary sheep (right). Dark areas contain more photoreceptors than lighter areas.
From Schiviz et al., 2008, with permission of John Wiley and Sons.

Although the density (mm^{-2}) of photoreceptors is much the same in giraffes as it is in other mammals, their distribution in the retina of giraffes is unique. Schiviz et al. (2008) and Coimbra et al. (2013) have found that there are three areas in the retina of giraffes where the density of the cones and their ganglion cells is very high: a horizontal streak that lies across the retina at the level of the optic disc, a temporal area that lies on the outer side of the eye, and a dorso-temporal extension on the inner side of the eye near the nose (Figure 16.11).

An artiodactyl, like an Oryx, surveys the habitat in front of it constantly to detect predators or conspecifics, for example, and it is the photoreceptors in the horizontal visual streak (Figure 16.11) that provide a broad view of the horizontal world, a world able to be surveyed with minimal movement of the head and eyes. The temporal (lateral) area of high cone density allows discrimination of fine detail in the near distance. The dorso-temporal area extends upward at both ends of the horizontal streak to form a complete arch in giraffes, reaching from the temporal to the nasal visual fields (Figure 16.11). This arch appears to have coevolved with tallness (or its proxy, the kind of life lived by Barbary sheep on the edge of precipices). Its presence means that when giraffes (or Barbary sheep) look down, images reaching the eye at a steep angle project onto areas of high cone density and so reveal detailed images of objects that otherwise would be difficult to identify (Schiviz et al., 2008). Behind the retina there is a reflecting membrane—the *tapetum lucidum.* It is found in all artiodactyls, but in giraffes its coverage is unique in that it occupies the entire dorsal hemisphere of the retina behind the areas of highest cone density (Coimbra et al., 2013). The function of the *tapetum* is to reflect any light waves that have managed to pass through the retina without being absorbed by a photoreceptor back onto the retina, thus giving the photoreceptors a second chance at detecting them. The *tapetum*, in other words, acts rather like night-vision glasses, and from its presence and shape we can speculate that giraffes can see quite well at night.

The size of the horizontal visual field that a giraffe can scan depends to a large extent on the angle that the orbit subtends with the skull midline (the orbital axis). The closer that angle is to a right angle (90°), the more the eyes are placed on the side of the head of an animal, and the more panoramic their view of the world becomes (Figure 16.12), but the ability to see and focus on objects to their front is reduced. The more acute the angle, the better is their frontal view and the less is their panoramic view.

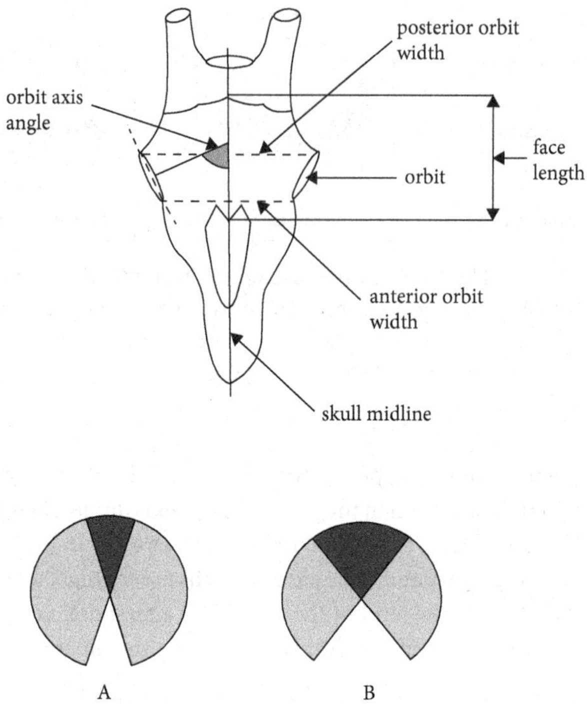

Figure 16.12. Top: A dorsal view of the skull of a giraffe showing the measurements made to determine face length and width and orbit orientation. Bottom: Diagrams that show the changes in monocular (pale gray) and binocular (dark gray) fields of vision.
A = neonate and B = mature (>1,000 kg) giraffe.
From Mitchell et al., 2013b, with permission of Taylor and Francis.

In 1901, George Lindsay Johnson, a British ophthalmologist, who, in his spare time, also made ophthalmoscopic studies of the eyes of many vertebrates other than man, reported that the divergence of the orbits in a giraffe was 72°, that is, the eyes faced almost laterally. Johnson most likely measured eye divergence in a young (<1-year-old) giraffe weighing ~240 kg, because at birth the angle of the orbital and optical axes is ~73° (Mitchell et al., 2013b). During postnatal growth, as the cranial sinuses inflate and the lachrymal bone enlarges (Owen, 1839), the face becomes broad in contrast to the long, narrow face of a newborn giraffe, which is the face-shape appropriate for birth. Broadening of the face pushes the orbit outward and forward, and its rim becomes more oval, with the horizontal diameter 1.13-fold longer than the dorso-ventral diameter. No obvious advantage arises from a change in rim shape (no doubt there is one), but broadening of the face decreases the orbit axis angle from ~73° in neonates to the more acute angle of ~ 50° in mature giraffes. The result is that the area of binocular vision increases from ~33° in neonates ((90° –73°) × 2 as both eyes are involved) to ~70° in mature adults (Figure 16.12, Figure 16.13), and the visual field changes from being mainly lateral and monocular to more frontal and binocular but the total field of vision is reduced, especially the field behind (Figure 16.12). The most acute orbit/optical axis angle so far reported is 46° in a mature female (Mb = 967 kg), which corresponds to an area of binocular overlap of 88° ((90°–46°) × 2; Mitchell et al., 2013b). For comparison, cattle have binocular overlap of 52° (Schiviz et al., 2008). The general advantage that binocular overlap confers is stereoscopic perception of distance and depth (Heesy, 2004, 2008), but the advantage specifically for giraffes is not obvious. Judgment of depth and distance is a complex process involving perspective, movement, and contrast discrimination, but it is enhanced by binocular vision (Heesy, 2009). In birds and amphibians, the effects of binocular vision are limited to a short range (<1 m; Heesy, 2008). The range over which it functions in mammals is unknown but is assumed to be limited also. The range is partly dependent on interocular distance (Heesy, 2009). The interocular distance (post-orbital width; Figure 16.12) in giraffes is comparatively very large and is up to 300 mm in a mature animal (Table 16.5), and thus the range over which depth and distance perception is possible may be longer in them. Nevertheless, if stereopsis operates only over short distances in giraffes, it, together with their ability to discriminate between closely spaced objects that the number of cones confers, will facilitate discrete selection of the browse that they require, probably supported by the temporal area of high cone density, which allows discrimination of fine detail in the near distance. We cannot ever know, of course, what a giraffe sees, but the data are consistent with such a scenario.

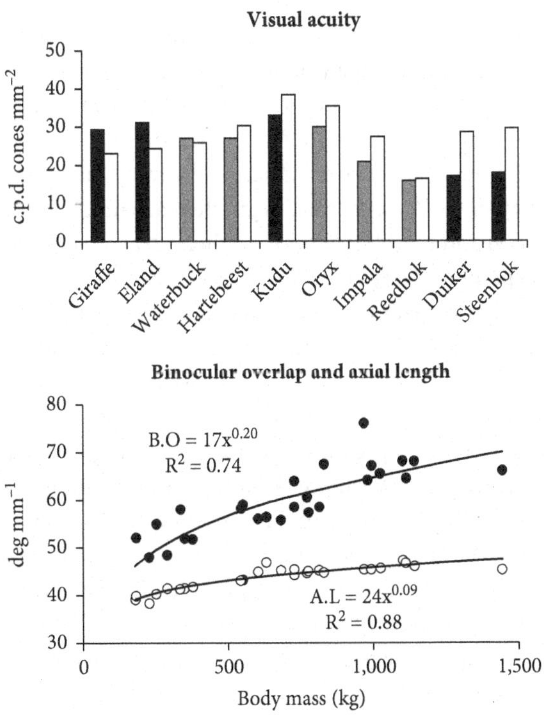

Figure 16.13. Top. The spatial resolution in cycles per degree (c.p.d.) of visual arc (■ = browsers, ■ = grazers) and the maximum number of cones (x10^3 mm^{-2}) in the retina (□), in East and South African artiodactyls, arranged in order of decreasing body mass.
Constructed using data in Schiviz et al., 2008.
Bottom: Changes in the degree of binocular overlap and the axial length (mm) as body mass increases in giraffes.
Data from Mitchell et al., 2013b.

Visual Acuity

It is one thing to have a large eye and large numbers of photoreceptors and another to have good vision. Apart from a large eye and a large retina surface area, good vision depends on the antero-posterior (cranio-caudal) diameter of the eye (the axial length; Figure 16.13), which is a measure of its focal length. The longer the focal length, the higher the magnification of an image, and higher magnification means being able to see more detail because the image covers more photoreceptors and allows more detailed analysis of the incoming information. The antero-posterior diameter of the giraffe eye is longer than it is in any other land mammal except, perhaps, elephants. Elephants have a small but elongated eye (Table 16.6). Is giraffe visual acuity good? Visual acuity is measured by the ability to distinguish between closely spaced images. The standard test of visual acuity in humans is the ability to distinguish between dark and light bars. At 6 m or 20 ft, a human with 6/6 or 20/20 vision can see lines that are ~2 mm apart, and this in visual terms is equivalent to approximately 30 cycles per degree of arc (Melin et al., 2016), a "cycle" being a pair of dark and light lines of equal width. Visual acuity calculated for giraffes from photoreceptor density and predicted focal length is between 25 cycles per degree of arc (Coimbra et al., 2013) and 29 cycles per degree of arc (Schiviz et al., 2008). A giraffe's ability to discriminate detail is thus by this measure about the same as it is in a human, and is very similar to what it is in the many other species with which giraffes share their landscape (Figure 16.13). If their visual acuity is similar to that of other animals, what is the advantage of their very large eyes? What selection pressures operated to produce a large eye? Whatever they were, they evolved at least 12 Mya, as they were present at the time *Giraffokeryx* evolved. That species also had noticeably large eyes (Chapter 6). It may be that as giraffes can run at high speed, evolution of large eyes and good vision might have allowed them to avoid collisions with objects in their path and escape predation, but Veilleux and Kirk (2014) analyzed this idea and could not find a strong correlation between maximum running speed and visual acuity in 91 species of mammals, including giraffes. There is a correlation in mammals, however, between eye size and the open habitats and daytime activity that giraffes (and presumably *Giraffokeryx*) prefer, and perhaps coping with life in an open savannah may have been among the selection pressures for a large eye in giraffe ancestors. Whatever the selection pressure may have been, we know that modern giraffes have a large, spherical eye that contains a large number of photoreceptors, a lens that focuses incoming light equally well onto the edge of the retina as its middle, a relatively long focal length, a magnified retinal image, and a retina area that provides the space to have photoreceptors distributed throughout it and not just along a narrow strip at the level of the optic disc. We have to conclude that their legendary vision, noticeable to humans as their ability to detect movement at great distances, is the result.

The Spinal Cord

The distance between the brain and, for example, the hindlimb muscles in a giraffe is as long as it gets in a terrestrial mammal, matched perhaps only by the distance in an elephant. As all nerve impulses that reach the distant parts of the body travel first in the spinal cord, its length in giraffes must be long enough to cover the length of their neck and the rest of the body. The total length of the spinal cord was, of course, first measured by Owen in 1838, who wrote:

> *The spinal cord is chiefly remarkable for the extreme length of the cervical portion. In the male dissected, which measures eight feet from muzzle to vent* [and thus the giraffe had a body mass of ~300 kg], *the length from the pyramids to the brachial plexus was three feet three inches.*

The pyramids are in the brain stem, are formed by nerve fibers that have their origin in the motor cortex of the brain, and they mark the top end of the spinal cord. The brachial plexus is the collection of nerves that have left the spinal cord to supply the forelimb. In giraffes those nerves leave the spinal cord between the seventh cervical vertebra and the first thoracic vertebra and between the first and second thoracic vertebrae (Solounias, 1999). A spinal cord has cervical, thoracic, lumbar, and sacral parts, each of which consists of segments (somites) that reflect their embryonic origin. Each somite consists of three cell layers: an ectoderm, mesoderm, and endoderm. Broadly speaking, the ectoderm is the origin of the spinal cord, its peripheral nerves and all parts of the body exposed to the environment, for example, the skin, and the lining of the intestine, and the reproductive and respiratory tracts. The mesoderm produces the vertebra, skeletal muscles and their tendons, and the endoderm is the origin of the internal organs. There are as many vertebrae as there are somites, but because of the slower growth rate of nerve tissue and/or faster growth of bones, the spinal cord segments do not correspond exactly with vertebrae segments and, consequently, the length of the spinal cord is never as long as the length of the vertebral column. The spinal cord lies in the vertebral canal (see Figure 15.2 in Chapter 15) and ends in humans, for example, at the level of the last thoracic (T12) or first lumbar vertebra (L1), which lie approximately at the level of the navel. The vertebrae extend to the coccyx. The vertebral canal below the point where the spinal cord ends contains the *cauda equina*, the collection of five or so spinal nerves arising from the lowermost lumbar and sacral segments of the spinal cord and is so called because the collection of fibers looks like the hairs of a horse's tail.

The distance the spinal cord extends down the vertebral column in ungulates is different from that in humans. In a detailed study of the length of the cord in three giraffes with body masses estimated to be 400 kg, 450 kg, and 1,500 kg (so their ages were between 2 and at least 10 years old) by Badlangana et al. (2007), the spinal cord segments were found to correspond better with each vertebra than they do in humans, but the cord extended only to the second sacral vertebra (S2) rather than the fifth (S5). Thus it ends far beyond where it ends in a human, but not as far as the full extent of the vertebrae. The *cauda equine* begins at S2 and continues into the caudal vertebrae. The cord in the giraffes studied by Badlangana et al. had a uniform diameter of 1–1.5 cm throughout its length to the level of S2, after which it tapered to end as the *conus medullaris*. The overall length of the cord was 1.7 m in the smallest giraffe and 2.6 m in the largest (Badlangana et al., 2007). Throughout life, from fetus to maturity, the proportion that each part of the spinal cord contributes to its length is consistent and, naturally, corresponds almost exactly with the proportions of the vertebral column as recorded by van Sittert et al. (2010): cervical part 55%, thoracic part 30%, lumbar part 10%, and the *conus medullaris* 5%.

The spinal cord connects the brain to all the organs of the body. Interruption of the cord results in profound and often fatal changes. The change that is most devastating is loss of movement. The nerve impulses that bring about movement arise in the motor cortex of the brain and do so in response to information being sent to the brain by the whole array of sensory systems that constantly monitor the giraffe's internal and external environments. Nerve impulses that activate contraction of the muscles of movement originate in Betz cells (named after the Ukrainian scientist Vladimir Betz, who described them in 1874) located in the motor cortex of the brain. It is logical to assume that, in giraffes, the Betz cells will be large so as to generate the energy needed to drive an impulse down a nerve that could be 3 m long. In primates, Betz cells that send impulses far down the spinal cord to the legs are larger than those that initiate impulses that go to the arms. How big the Betz cells are in giraffes is not yet clear. In one study (Badlangana et al., 2007) the Betz cells were found to be not unusually large, having a surface area of 600 μm^2 (compared to 1,000 μm^2 in primates) and resembled those found in a sheep. In another study (Jacobs et al., 2015) the cells had a surface area of 1,200 μm^2. These substantial differences may be a result of the way the cells were visualized. Whatever their size, the Betz cells must be able to generate impulses that are able to travel very long distances. Carriage of the impulse depends on electrical currents generated by diffusion of sodium ions into the nerve. Energy is required to return the sodium ions to the exterior of the nerve fiber. Usually it is the Betz cells themselves that are the source of that energy, but because the fibers (axons) of nerves are so long in giraffes and their ends far from the cell body and its nucleus and all the intracellular machinery that generates energy, it is reasonable to suppose that the long nerves of giraffes have evolved special metabolic processes to support carriage of the impulse. Badlangana et al. looked for evidence of an energy supply but could not find any. So whether and how additional energy is provided to a very long nerve fiber in a giraffe's spinal cord is, for now, an enigma. How do giraffes (and elephants and whales) maintain impulse conduction over long distances?

Peripheral Nerves

Each embryological somite, via its spinal cord segment, is the origin of two pairs of nerves, one pair on each side of the cord—the somatic nerves, which control voluntary movements. Each of the pairs has an upper, dorsal root and a lower, ventral root. The dorsal root conveys sensory information like pain and temperature to the spinal cord, and the cord carries the sensations upward to the brain, where the information is interpreted and a response, if necessary, is initiated. The sensations arise from sensory receptors in the skin or in joints, in muscles or tendons, or viscera. The lower ventral root mainly consists of nerves that convey motor impulses to the muscles. These motor nerves end on one or usually more muscle fibers to form a motor unit. The whole sequence of events from sensation to reaction is a reflex arc. Nothing happens unless all parts of the arc are activated. The system is simple and effective.

Giraffes have, of course, very long limbs and thus, theoretically, the nerves supplying the limb muscles, especially those of the hindlimb, of an adult are very long, especially if they are viewed as coming from the brain. However, cells (alpha and gamma motor neurons) in the spinal cord are the origin of the nerves going to a muscle. These spinal cord cells are in turn stimulated by impulses arriving from the motor cortex and other parts of the brain. Artiodactyls have evolved a body form such that the large muscle groups responsible for movement are close to the body core, and a consequence is that the lengths of motor nerves arising from alpha and gamma motor neurons are shorter than might be expected, and giraffes are no exception. Heinze's (1964) study of the muscles of the hindlimb and the distribution of the branches of the main motor nerve—the sciatic nerve, which supplies them, showed quite clearly that most muscles are situated above the knee and that most branches of the sciatic nerve end in them (Figure 16.14). The largest muscle of the lower leg and one of the furthest from the spinal cord is the gastrocnemius. It has its origin on the distal part of the femur, inserts on the ankle via the Achilles tendon, and lies between the knee and ankle (see Figures 16.8, 16.14).

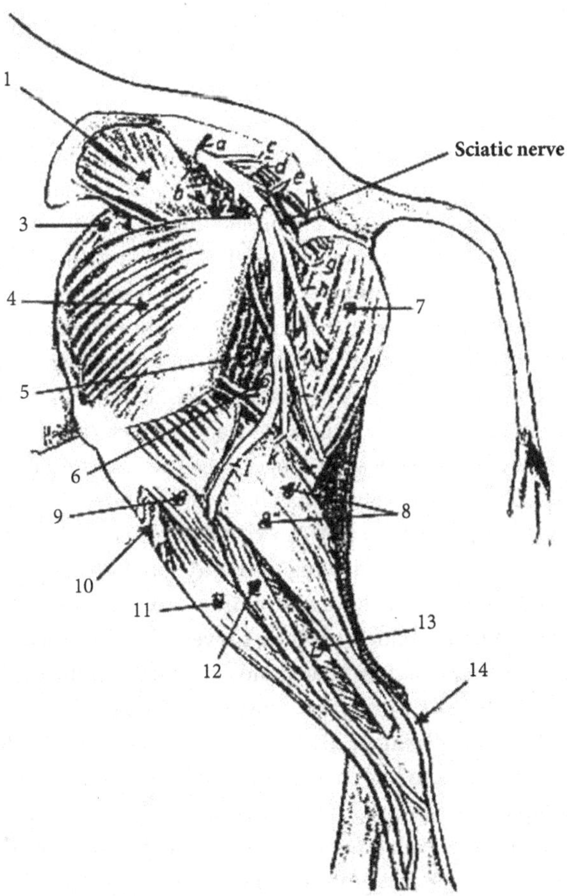

Figure 16.14. Distribution of the sciatic nerve to the hind limb of a giraffe. Numerals are muscles and tendons: 1= *gluteus*; 2 = *gluteus profundus*; 3 = *rectus femoris*; 4 = *vastus lateralis*; 5 = *adductor*; 6 = *semimembranosus*; 7 = *semitendinosus*; 8 = *gastrocnemius*; 9 = *fibularis longus*; 10 = *tibialis anterior*; 11 = *fibularis tertius*; 12 = *extensor digiti*; 13 = *flexor digiti*; 14 = *flexor tendon*. The letters are branches of the sciatic nerve.
From Heinze, 1964, with permission of Elsevier.

As the femur length in a mature giraffe is about 500–600 mm long (Table 15.2a), the distance from the spinal cord to the gastrocnemius muscle cannot be much more than that, a distance that is about the same as it is in a human. Nerves and the impulses they carry coming from sensory receptors in the feet and tendons must, however, exist. Heinze described their pathway to the feet of the hindlimb, and they are long. Thus, if a limb muscle is reacting to a stimulus arriving from a foot, such as pain or joint position or the stretch in a tendon, that stimulus in the form of a nerve impulse would have to travel up the leg to the spinal cord, from there to the brain, from the brain back to the spinal cord cell that initiates the muscle contraction, and from there to the muscle, a total distance of perhaps 5 m. A long delay could be expected between sensation and reaction when short delays are required for reflexes and escape.

Giraffes show no signs of slow reactions or of incoordination, even when running at high speeds; therefore the distances the impulses have to travel do not seem to create any problems for them. If not, why not? One reason is that a short circuit exists, in which the nerve carrying the sensory impulse first synapses on a cell in the spinal cord in the normal way. This cell sends some of the information up to the brain for interpretation and at the same time activates the cell nearby that is the origin of a motor nerve going to the appropriate muscle. We are very familiar with this short-circuit response. We react (usually by a violent muscle contraction) to a painful stimulus long before we know where to look to see what caused it. The length of this sort of reflex arc in a giraffe could be about 2.5 m long and, if so, would be half the full possible distance that a nerve impulse could travel. Another reason why giraffes may show no signs of incoordination is that the speed at which an impulse is conducted along their nerves is faster than it is in other mammals. It may be that the myelin sheath that surrounds each axon is unique in giraffes. Myelin is an electrical insulator, and myelinated nerves have higher conduction velocities than do non-myelinated nerves. Myelin sheaths are interrupted at intervals by nodes that allow the impulse to leapfrog from one node to the next. The distance between nodes can vary with longer intervals and fewer nodes, promoting faster conduction. Giraffe nerve fibers may have fewer nodes. It may also be that the number of ion channels per surface area (mm^2) of axon membrane is greater in giraffes than it is in a smaller animal. Ions flowing into, and out of, the nerve are the source of a nerve impulse. Whether giraffes have any of these modifications is not known, but if they do, then the conduction velocity of their nerves could be much faster than it is in other mammals and their well-coordinated movements would be explained. However, the velocity of conduction of an impulse mostly depends on the diameter of a nerve fiber (axon) (Swadlow and Waxman, 2012), and hence the easiest way for a giraffe to increase conduction velocity is to have nerve fibers that are larger than they are in smaller mammals. To complicate matters, giraffes have a large body and, as More et al. have pointed out (in 2013), apart from large-diameter fibers, they could or should have more sensory nerves to monitor what is going on in their bodies or the environment. Do the peripheral nerves in giraffes have large diameters, and do they carry more fibers than there are in the nerves of other equivalent-sized mammals?

As a way of answering that question, More et al. (2013) studied the number and size of axons in the sciatic nerve of giraffes and they also managed, remarkably, to obtain measurements of the conduction velocity of nerve impulses going to the gastrocnemius muscle. In the giraffes that they studied, the sciatic nerve had between 97,000 and

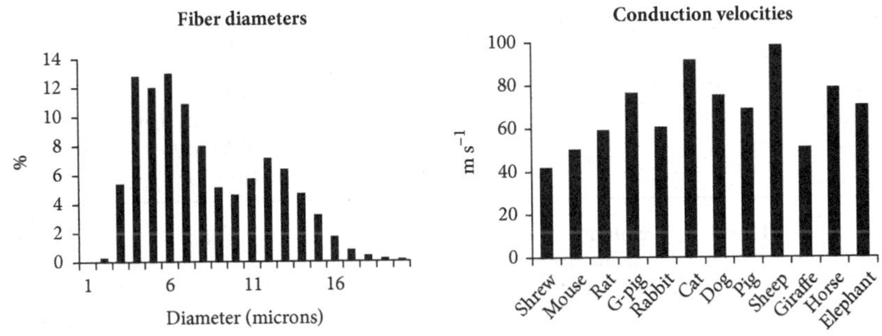

Figure 16.15. Left: Fiber diameters in the sciatic nerve of a giraffe. Right: The average combined sensory and motor nerve conduction velocities in mammals, arranged in increasing body mass.

Constructed using data in More et al., 2010, 2013.

115,000 myelinated fibers, which is slightly more than the 80,000 or so predicted from studies on other species with the same body mass of giraffes, but axon diameter was not bigger. Most of the axons had diameters that were clustered around an average of 5 μm and others that were clustered around 12 μm (Figure 16.15), which is the same size distribution that has been found in all other mammals.

Typically large peripheral nerves like the sciatic contain a mixture of axons of several different diameters and the proportion each of them contributes varies according to anatomy, lifestyle, and habitat. For example, animals that have complicated limb muscles (like humans and rats) also have a higher proportion of faster-conducting, large, motor fibers and relatively fewer sensory fibers (Gordon and Rafuse, 2001). From More et al.'s data, it is possible to identify the three main types of fiber present in the giraffe sciatic nerve, the proportion each contributes to the total, and their conduction velocity (Table 16.7). The smaller fibers, 1–5 μm and 6–12 μm in diameter, typically conduct sensory information like pain, temperature, and touch; the larger ones, 12–20 μm in diameter, carry motor impulses to muscles. From the conduction velocities measured in nerves of those sizes, and the proportion of each type of fiber in a giraffe's sciatic nerve, the expected average conduction velocity for a giraffe sciatic nerve can be calculated to be 54 m s^{-1} (Table 16.7). Average conduction velocity measured in the giraffe's sciatic nerve by More et al. (2013) was 50.4 m s^{-1} (i.e., 2 s to travel 100 m). The fiber size composition shown in Table 16.7 thus seems to be a fair estimate of the types of fiber present in the sciatic nerve of giraffes and of their conduction velocities.

Despite the very long limbs and large body mass of giraffes, none of these data indicate that giraffes have a sciatic nerve that is bigger and conducts faster, or that the number of axons in it is vastly more numerous than it is in other mammals, yet giraffes show no sensory or motor deficiencies, and their nervous system must have evolved ways of meeting the demands of an elongated shape. One way is related to the composition of their nerves. The sciatic nerve, for example, has ~100,000 axons, of which ~84% can be said to be sensory fibers and ~16% motor nerves to muscles. This distribution makes sense. The body volume and surface area of giraffes are large, and to supply all the sensory receptors that are monitoring their environment, many sensory fibers are needed. The tissue best supplied with nerves is skin, followed by muscle and

Table 16.7 Composition of the Myelinated Fibers in the Sciatic Nerves of Giraffes

FiberType	Size (μm)	Function	Expected Conduction Velocity (m s^{-1})	Giraffe Conduction Velocity (m s^{-1})*	% of Myelinated Fibers
Aα	12–20	Somatic motor/ proprioception	70–120	88	17
β	6–12	Touch, pressure, muscle spindles	15–70	37	41
γ	1–5	Pain, temperature	12–30	15	43
All	*8*		*54*	*50*	*100*

Constructed from Figure 16.14 and More et al., 2010, 2013.
*Conduction velocity of myelinated axons in cats up to 9 μ in diameter = 4.29 × diameter + 1.64 and for axons 10–20 μ in diameter = 5.24 × diameter + 6.4 (Boyd and Kalu, 1979).

then bone. The mass of the hindlimb of a mature giraffe is ~80 kg (Table 16.3), of which about 65 kg is muscle and ~15 kg is bone, skin, and all other tissues. Unlike animals that have complicated limb muscles, giraffes have few muscles requiring fine coordinated movements and many large muscles with few motor units, which a single nerve could supply without any need for more or bigger axons. If the hindlimb muscle mass is 65 kg, then each of the 16,000 motor axons in a sciatic nerve supplies ~4 g of muscle and its motor units. Assuming that all parts of a hindlimb are equally well supplied with sensory nerves, the 80 kg of tissue are supplied by 84,000 sensory axons, or one sensory axon per 1 g of tissue, which is a high density of nerves. Furthermore, sensitivity to a peripheral sensation like pain is related not only to the number of axons per unit of tissue, but also to the number and size of the receptor fields in the tissue. A receptor field contains all the sensory receptors needed to monitor the environment and is supplied by a single sensory nerve, which divides into as many branches as required to innervate all the sensory receptors in that field. The number of end branches varies according to evolutionary need, so a large number of axons is not necessarily needed to cover a large area. What is needed is just a large number of end branches. Of course, the way tissues are innervated is more complex than this description implies, but it does seem that a way evolution has kept the size of a giraffe's sciatic nerves small (and presumably other nerves too) has been by adjusting the balance between the large axons innervating muscle with much smaller axons having many branches innervating sensory receptors, in favor of more small sensory axons and fewer large motor ones.

Autonomic Nerves

If somatic nerves produce voluntary reactions to what is going on in a giraffe's external environment, the autonomic system is its involuntary counterpart for the internal environment. It is the motor system of the viscera. Most of the time we, and presumably other mammals, are not conscious that autonomic nerves are active, in contrast to the way that we are very conscious of the activity of somatic nerves. In its earliest form, the autonomic nervous system evolved to coordinate the digestive and cardiovascular systems, and this is still its dominant function, but additional functions have evolved to include reactions to information from receptors sensitive to the chemical composition of body fluids, blood pressure, pain, temperature, and so on. Like somatic nerves, autonomic nerves form a reflex arc with afferent fibers carrying information to the spinal cord, from where efferent effector fibers carry impulses to target tissues to correct imbalances. Axons of the autonomic system generally are small in diameter and non-myelinated, so their conduction velocity is slow, 2–3 m s^{-1}, compared to ~80 m s^{-1} for large motor nerves.

Two types of regulation of the internal environment are required and, accordingly, the autonomic system consists of two divisions: the parasympathetic system and the sympathetic system. Generally, activation of parasympathetic nerves promotes feeding, digestion, and absorption of food, and the regulatory parasympathetic nerves are carried in the oculomotor nerve to the eye (cranial nerve III, which controls pupil size and lens shape), in the facial nerve (VII, which supplies the salivary glands), and in the vagus nerve (X), which supplies the thoracic and abdominal viscera and controls peristalsis and the secretions of salivary and gastric glands and the pancreas. Another group of parasympathetic nerves arises from the lower part of the spinal cord as the pelvic nerve to supply all the structures associated with reproduction and excretion. Apart from the recurrent laryngeal part of the vagus nerve, none of these nerves has been studied in giraffes, and we can only assume, until shown otherwise, that they are no different from those found in all other mammals.

The other division of autonomic nerves is the sympathetic system. Typically in mammals, its axons originate from the spinal cord between the first thoracic and the sixth lumbar segments of the spinal cord. The distribution and functions of sympathetic nerves in giraffes, as they relate to the cardiovascular system, have been thoroughly studied because of their role in the control of blood pressure (see Chapters 9, 10). However, the functions of the sympathetic system extend beyond blood pressure control. It has a role in the regulation of the activity of all internal organs. The response with which we humans are most familiar is its activation in emergencies and the resulting fight-fright-flight response. Pupils dilate, heart rate, blood pressure, respiratory, and metabolic rate all increase, skin blood vessels constrict, blood glucose concentration increases, body temperature increases, and blood is diverted to the muscles that need it. In addition, the emotions of rage and aggression are activated instantly. All of these responses are caused by the release of noradrenaline and adrenaline, and anyone who has experienced a frightened giraffe knows that their sympathetic nervous system functions in the same way that it does in us and in all other vertebrates.

Skin

Evolution of the membrane that enveloped the very first self-replicating biochemical compounds, which are presumed to have been the origin of life, and separated them from their environment, was also the very first step in the long and complicated pathway that produced skin as we know it. The subsequent evolution of that primitive membrane into sophisticated epithelia and the epidermis of skin is evidence of the immense adaptive value of both of them to any organism. Epithelia line all internal body parts and cavities that are exposed to the environment, among which are the intestinal, reproductive, and respiratory tracts and, less obviously, blood vessels. Epidermis covers the external surface of the body. The functions of epithelia and especially the epidermis have multiplied from being simple but formidable barriers protecting their owner from invasive organisms and noxious chemicals. The skin has become not only a barrier but also a sense organ detecting temperature, pain, touch and pressure, an immune system detecting and responding to antigens, the source of fur, hair, and feathers, a regulator of body temperature and body fluid composition, and a surface area for gas exchange in several species of amphibians, some marine reptiles, and some bats, which have a wing membrane that provides the large surface area ideally suited for gas exchange. The functions of the skin are supported by its large blood supply. In mammals, skin receives ~10% of cardiac output and uses about 5% of all oxygen consumed. It is an active organ and is so in giraffes.

Size and Structure of the Skin

How big is a giraffe's skin? The total surface area of the skin is the sum of its head, neck, trunk, and lower leg surface areas, each of which is known with some precision in 60 giraffes (see Chapter 14). In this group of giraffes, a mature male had a skin surface area of ~12 m^2, with the head skin contributing 9% of the total, the neck 25%, the trunk 51%, and the legs 16% (Table 14.2). How much does the skin weigh? A systematic study of total skin mass in giraffes of different body masses has not been done. However, skin mass of any mammal is proportional to its surface area and its thickness (Calder, 1996). The relationship between head skin mass (kg) and skin surface area (m^2) and neck skin mass and its surface area are known for giraffes (Mitchell et al., 2013a, 2017) and their thicknesses (mm) have been measured. The surface area and thicknesses of their trunk and lower leg skin are also known (Mitchell and Skinner, 2004; Sathar et al., 2010; Mitchell et al., 2017). Assuming, not unreasonably, that the relationships between neck and head skin surface area and their mass and their thicknesses can be applied to the trunk and lower legs, the mass of the skin covering them can be estimated. Table 16.8 summarizes these data and shows the regression equations that can be used to describe them.

What emerges from the data is that total skin mass can be 130 kg in a large male and can be ~9% of its total body mass. That proportion is not correlated with body mass ($R^2 = 0.07$), and thus skin mass forms the same proportion of a newborn giraffe's body mass as it does in an adult, and it is the same proportion that a giraffe's skeleton contributes to the mass of a living giraffe (~10%; see Chapter 15), a relationship that seems to have general application to all mammals (Lindstedt and Calder, 1981). The allometric equations do indicate, however, that the mass of each square meter of skin increases as a giraffe grows, presumably because skin thickness increases with age, but other than Gordon's observation made in 1779 that the skin of a mature giraffe was 13 mm thick, and thicker than the skin of juvenile giraffes, there are no empirical data to support that contention. For dimensions as large as those shown in Table 16.8 to have

Table 16.8 Basic Dimensions of Giraffe (*G.c. giraffa*) Skin

Skin Dimension	Allometric Equation (Mb = body mass) in kg)	Value in a 750-kg Giraffe	Range for Giraffes of Mb 147–1,441 kg	R^2
Surface area (m^2)	$0.07 \times Mb^{0.70}$	7.2	2.1–11.8	0.98
Head skin mass (kg)	$0.03 \times Mb^{0.73}$	3.8	1.0–4.9	0.92
Neck skin mass (kg)	$0.02 \times Mb^{1.0}$	15	2.9–26.8	0.95
Trunk skin mass (kg)	$0.08 \times Mb^{0.95}$	43	7.9–48.5	0.92
Lower leg skin mass (kg)	$0.04 \times Mb^{0.78}$	7	3.9–7.5	0.91
Total skin mass (kg)	$0.14 \times Mb^{0.93}$	66	11–85	0.93
Skin mass:surface area (kg m^{-2})	$1.8 \times Mb^{0.24}$	8.8	6.8–7.6	0.51
Skin mass as % of body mass	$14 \times Mb^{-0.07}$	8.8	9.7–8.5	0.07

evolved, the adaptive advantages of a giraffe's skin must far outweigh the energetic and nutritional costs of growing it and carrying it around. What are those advantages? Apart from the functions that skin has in all mammals, in giraffes hair coloration, camouflage, temperature regulation (see Chapter 14), and acting as a tight G-suit to counteract high blood pressure (see Chapter 10) are among its chief features. What is its structure?

In 1976, Dimond and Montagna studied the skin of a seven-year-old female giraffe that had died in the Portland Zoological Gardens in the United States, and contemporaneously but on a different continent, Errol Ackermann studied skin structure in adult giraffes and how skin develops in giraffe fetuses. These studies were followed by those of Wood and Weldon (2002; an old male and old female), Mitchell and Skinner (2004; a young female), and Sathar et al. (2010; two young and one old male). All agree that the skin of giraffes consists of three layers—an outer epidermis, and a dermis consisting of superficial and deep layers. Thus its basic structure is no different from that of the skin of any other mammal (see Plate 16.1).

The epidermis is thin, ranging from 0.1 mm thick over most of the body to just 0.03 mm thick on the eyelids. It is covered by a layer of keratin, a fibrous protein synthesized by epidermal cells, secreted by them and deposited on the cell membrane. The keratin layer is relatively thick on the skin of the lips of giraffes, where it averages 1.6 mm, even thicker in the mouth (2.0 mm), but according to Lahkar et al. (1992), relatively thin on the tongue, ranging from 0.7 mm at the tip to 0.1 mm at the root. Elsewhere keratin thickness in giraffes varies from 0.06 mm on the back to 0.01 mm on the thinnest areas of the eyelids, but, of course, it is very thick on the hooves and on the ossicones. In all places, keratin is protective. The epidermis below the keratin layer is supplied by nerve fibers that, presumably, carry sensations of pain, touch, temperature, and pressure, as they do in other mammals. The basal layer of the epidermis contains the many melanocytes that produce the melanin that gives the skin of giraffes its uniform gray color. The superficial part of the dermis lying immediately underneath the epidermis is up to 3 mm thick and consists of a meshwork of collagen fibers in which hair follicles and sebaceous and apocrine sweat glands are embedded (see Plate 16.1).

Parasympathetic nerves run throughout the superficial dermis to hair follicles and the glands, and the glands are also surrounded by many small blood vessels arising from the well-developed vascular plexus that lies between the superficial and deep dermis (see Figure 14.10 in Chapter 14). All follicles have a sebaceous gland opening into the hair canal, except those on the nasal septum. Sebaceous glands are small and unilobular on the jaw, neck, trunk, tail, and thigh, but multilobular on the lips, nose, and eyelids. The deep dermis is made up of a network of tough collagen fibers and is thickest on the neck, trunk, and rump (see Plate 16.1, Figure 16.16). The combined superficial and deep layers of the dermis are unusually thick and, together with the epidermis, make giraffes, as Dimond and Montagna wrote a little inaccurately, one of the thickest-skinned mammals, so that its skin was once much prized for making shields and other artifacts. On average, entire skin thickness in young giraffes is 7 mm, but varies between 1.5–6.5 mm on the head and face, 5.8–18.6 mm on the neck, 3.3–12.2 mm on the trunk and over the rump, and 2.2–6.1 mm on the tail and legs. The skin of giraffes older than the one studied by Sathar and colleagues is thicker. The skin Gordon sent to Holland in 1780 was, he said, "in general about half an inch and one line" thick (~13 mm). For comparison, the average skin thickness in an elephant is 20 mm and in rhinoceroses 25 mm. In the semi-aquatic hippopotamus it is 12 mm on the belly, 22 mm in the middle of the back, and 35 mm down the sides (Luck and Wright, 1964).

Sebaceous Glands

It has long been known that old giraffe males emit a "rank and powerful smell" and, consequently, are referred to as "stink bulls." These bulls are always unmistakably darker than any others (de Winton, 1897). Selous agreed.

> *The body of an old bull giraffe gives out an excessively strong pungent odour, which can be smelt by a human being at considerable distances.* (Selous, 1908)

Why? Joly and Lavocat (1846) noted,

> *A quite remarkable peculiarity in the giraffe is the complete absence of skin muscles. They are replaced by a large and strong glossy aponeurosis, enveloping all of the body, in a manner loose enough in the skin and blending in many regions with the yellow fibrous covering indicated previously. These large aponeuroses stretch over all the muscles.*

In many mammals, the muscles of the skin (the *panniculus* muscles) allow movement of the skin independent of the movement of deeper muscle masses. Movement is triggered, for example, by painful pecks of birds and the bites of insects and has the purpose, often fruitless, of dislodging the perpetrator. Why this potentially very useful adaptation should have been lost by giraffes during their evolution is, so far, unexplained, but in its absence, giraffes have to rely on rubbing against anything nearby or on oxpeckers to remove ticks and other skin debris (Innis, 1958; see Plate 16.2), and on the secretions of the sebaceous glands for protection. The source of the powerful odor was investigated by Wood and Weldon (2002) in two captive giraffes, one a 630-kg, 12.4-year-old male and the other a 443-kg, 17.4-year-old female. They found that there were 11 major volatile compounds secreted by sebaceous glands, of which two alkaloids (indole and 3-methylindole) were the ones responsible, when secreted at high concentrations, for the intense fecal odor of male giraffes. Some of the compounds are oily and function as waterproofing, but they also have antifungal and antibacterial properties, and repel arthropods like mosquitoes and ticks. The compounds are not found in female giraffes at the same concentrations, but presumably they have the same skin-protective function in females as they do in males. Their much higher concentration in males can be attributed to testosterone, and thus their secretions and odor may be part of the wider social system of giraffes that depends on olfaction.

G-Suit

It is sometimes mistakenly assumed that the skin of giraffes is highly elastic and behaves like an elastic stocking. However, all four studies of the structure of the skin have shown that there is very little elastic tissue in it (see Plate 16.1). The superficial and deep dermises of skin consist mainly of thick bundles of collagen fibers running in different directions so that an amorphous mish-mash of fibers seems to exist (see Plate 16.1). Collagen is inelastic and provides a very tight body stocking. A skin composed entirely of collagen would, however, be very stiff and difficult to flex and, of course, giraffes have to be able to move fluently, for which elasticity of the skin is essential. Dimond and Montagna (1976) reported that the quantity of elastic fibers is directly related to the number of hair follicles and thus is limited to the superficial dermis with fine fibers arranged horizontally. There were a few elastic fibers in the deep dermis. However, the link between hair follicles and elastic tissue is not absolute. In giraffes, most elastic fibers are found over the rump, down their sides, and into the deep dermis of both fore- and hindlegs where there is greatest movement (Sathar et al., 2010), rather than where the hair follicles are most dense. There is no doubt, though, that leg skin functions as a G-suit (see Chapter 10) but if that is a main function, then the skin around the lower legs could be expected to be relatively thick. However, Sathar et al. found that total leg skin thickness was, on average, about 5 mm and the superficial dermis less than 2 mm thick (Figure 16.16), with the thinnest parts of the skin lying between the legs. Even in the absence of elastic fibers, thin skin allows movement, whereas thick skin is more rigid and difficult to fold (Jarman, 1989). To compensate for thinness, the collagen in leg skin may have unique properties that achieve its G-suit effect and simultaneously allow movement. The skin of giraffes is thickest over the shoulders and rump and on the upper and lower ridges of the neck, mainly because the deep dermis in those areas is thick (Figure 16.16). Along the top of the neck under the mane, the skin is 15–19 mm thick, along the neck's ventral surface it is 10–12 mm thick, and on its sides it varies from 8 to 10 mm thick. The trunk is covered by a large, saddle-shaped area of thick skin ranging in thickness from 10 to 19 mm. The neck and trunk are where a male giraffe is most likely to receive blows while sparring and a thick skin may act as a dermal shield by dispersing the impact of blows (Jarman, 1989). However, skin thickness in female giraffes, who never engage in bouts of sparring, does not seem to be different from its thickness in males (Mitchell and Skinner, 2004; Sathar et al., 2010) so protection from blows seems

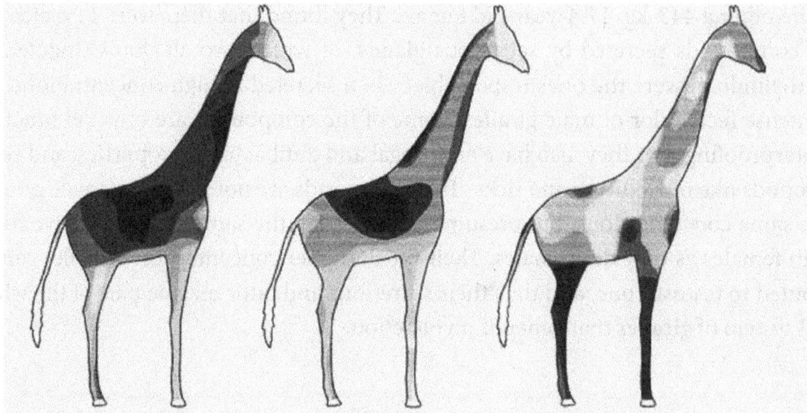

Figure 16.16. Skin thicknesses in a young giraffe. The darker the color, the thicker the skin. From Sathar et al., 2010, with permission of John Wiley and Sons.

an unlikely function. In both genders, though, thick skin could act as dermal armor to protect giraffes from thorns or attacks by lions (Sathar et al., 2010). Furthermore, thick skin over the lower neck and over the shoulders is exactly where you would expect it to be because giraffes living in woodland-savannah spend much of their day pushing past trees, and thorny ones at that. If protection against thorns is a function of their skin, then the skin of the head and face should be thick as they are exposed regularly to thorns (and ants). However, the skin over the face is the thinnest of any part, being 1.5–3.6 mm thick, and over the head where it articulates with the upper reaches of the neck, it is 3.7–5.8 mm thick. It is thin because a greater need than protection is to have free movement of the jaws and other parts of face anatomy. Moreover, the relatively thin head skin already weighs up to 5 kg, and if it were any thicker, it would add to head mass. The trade-off is between a heavier, protective head skin on the one hand and lower head mass and mobility on the other hand, and, as in so many other ways, it has been selection for whatever gives the lowest possible mass of the head that has prevailed during giraffe evolution. Giraffes rely on other mechanisms to avoid ants and the thorns in the upper levels of trees (see Chapter 7).

Skin and Patch Color

The entire skin of giraffes is hairy. Hairs are short, fine, and very dense, and in mature giraffes are 4.5 mm long and 0.035 mm in diameter (Ackermann, 1976). Follicle density is 3.6±0.3 mm^{-2} across the whole body surface (range: 3.0 mm^{-2} on the legs to 4.0 mm^{-2} on the neck and trunk). For comparison, in humans follicle density on the scalp is 1.5 mm^{-2}. Hair density is not known to vary seasonally if a giraffe is in its natural environment, where average ambient temperature varies from 10°C in winter to 21°C in summer, in the same way that hair density varies, for example, in its distant relative pronghorn antelopes that are exposed to extreme cold in winter and hot summers (−5°C to +33°C; Mitchell et al., 2009c), but if an unfortunate giraffe finds itself in a cold

Northern Hemisphere zoo, their hairs do elongate (Hagenbeck, 1910). Tail tip hair, at 1 m long (Harris, 1839), is the longest in any mammal, and once was a prized ornament. An extremely tenuous and Lamarckian attempt to establish a phylogenetic relationship between species based on hair slope patterns produced by lifestyle was made by a British physician, Walter Kidd (1900), who listed giraffids and cervids among those species with an "ordinary type" of whorl or star over the junction of the frontal and parietal bones on the forehead but, in fact, there is no correlation between habitat, phylogeny, and the arrangement of hairs (Dimond and Montagna, 1976). There is, however, a familial relationship. Walter Ridewood, a comparative anatomist who worked at the Natural History Museum in London, made a microscopic examination of hair structure in giraffes, okapi, and unspecified antelope, and established that the hairs of giraffes and okapi are similar and differed from those of antelope (Ridewood, 1901).

Skin and hair color are separate now, but in all likelihood were less so in the far-distant evolutionary past of giraffes. There is no way of knowing when the geometric hair color patterns ("patches") so characteristic of modern giraffes first evolved, but it seems unlikely that the markings of Asian giraffes, for example, would have been the same as those of African forms. An Asian coloration similar to modern okapi seems more likely if the habitat of the Asian giraffes was the tropical and paratropical forest that is assumed to have been their habitat, rather than woodland scrub/savannah as it is in East and southern Africa. While skin color of modern African giraffes is uniformly dark gray, that of their hair is highly labile. Anecdotal reports indicate that hair color can range from albino (Anon., 1971; Turner, 1969) and pure white (Butler, 1912; McDougall, 1939), to pale brown and unspotted (Arnold, 1940; Petzsch, 1950) or with a black band around the trunk (Blum, 1957), to black (Anon., 1923; Krumbiegel, 1971). Very rarely, patches and interpatch areas can lose color and become white (Muller, 2017). Reasons for the loss of color range from an absence of an enzyme needed for melanin synthesis (albinism) or a receptor for a synthesis-regulating hormone (leucism) or as Muller discovered in 2017 to a skin disease unrelated to the biochemistry of melanin production. The variations also imply that the genetic control of hair colors is supple, with, usually, adaptation to local conditions by natural selection to produce the patches with which we are so familiar today.

Patches are a defense mechanism, especially in the very young. The survival of large herbivorous species depends, partly, on the survival of the young, and there is no question that young giraffes are highly vulnerable to predation, with up to 75% dying in the first year of life despite excellent maternal care. Intuitively, selection for camouflage should contribute to the survival of young giraffes, and their survival contributes to the relative success of African giraffes. Which calves survive? The answer to that question has so far eluded solution, but the characteristic patches that have evolved would not have done so if they had no value, even if the camouflage they provide is far from foolproof for young giraffes. For camouflage to be successful, coat markings must blend into the habitat, and if survival depends on being inconspicuous, then there will be natural selection, in classic Darwinian fashion, for patch shape and color that increase the chance of survival. The high hereditability of the shape, number, and area of markings underpin this conclusion, and in giraffes is rather like that of selection for black over white peppermoths (*Biston betularia*) in the once smoggy, industrial north of England (Kettlewell, 1958, 1961). This idea is not subject to empirical analysis, but a correlation between survival and patch number and shapes in *G.c. tippelskirchi* may exist (Lee et al., 2018). Patches in *G.c. tippelskirchi* vary in size, and their edges vary from being highly fimbriated to reticulate, making analysis of correlations difficult. Lee et al. found, though, that the shapes and sizes of parches were hereditable (offspring looked like their parent), and survival in the first four months of life was improved if the calf had large spots, irrespective of their shape. If the edges of the patches were irregular, that helped too. Confusingly, best survival was among variants that had large spots and a reticulate pattern. No doubt survival of very young giraffes depends on many other factors also, and beyond four months of age, patch size, shape, and number, at least in that study, had little effect on survival. Adult giraffes are inescapably conspicuous in an open savannah, but in acacia woodland their coat markings mimic the polygonal shapes formed by the branches of the trees and break up the body outline (see Plate 16.3), especially when sunlight filters through trees and scrub, making giraffes difficult to see even when they are no more than a few meters away. Indeed, in 1908 the French zoologist Édouard Trouessart, wrote:

> *You need to have seen the giraffe in the middle of these* [Acacia] *trees which grow ordinarily isolated and also thinly, to understand the reason for the polygonal patches which give to this animal a characteristic feature. It is to have the protection of mimicry, which is explained well when we know the way the branches of the Acacia branch. These branches, with leaves only at their extremities, are whitish and curved or bifurcated at an angle more or less open, in the manner of polygonal shapes, exactly as the white lines that separate the dark patches of the coat of the giraffe. The result is that, when it is behind one of these trees, its pelage blends. From a distance, within the branches of the tree that it is in the process of browsing, it is perfectly hidden.*

It was once thought (Krumbiegel, 1971) that the appearance of dark patches was related to the tension of the skin under a patch. Krumbiegel substantiated his hypothesis by sprinkling pulverized charcoal onto stretched, thin rubber membranes and then used a brush to make artificial patches of various shapes. "A square patch," he wrote, "develops a dark centre when the membrane is tensioned on all sides, but also when it is stretched into an acute-angled rectangle. You can also get radial channels and—due to small unevenness of the tension—notches which create star-like structures" not unlike the cracks that develop "in drying clay of a creek bed, all shapes being a consequence of redistribution of melanocytes as skin tension changed." Patches can have these shapes, and so it is easy to see the logic of Krumbiegel's idea; however, no histological or other evidence has been found to confirm it. The skin is a uniform grey color, and hair color is now known to depend on the migration patterns and activity of melanocytes. Melanocytes are derived from neural crest cells in the embryo and migrate throughout the skin of the developing fetus and ultimately into its hair follicles. Their migration patterns are under strict genetic control and when their final distribution is completed, the result is the patterns seen in giraffes (and zebras; Hearing and Tsukamoto, 1991). Distribution can, of course, go completely wrong and produce an albino or a completely black giraffe, but usually patch shapes and colors are consistent among the regional variants of giraffes.

The color of the hair in a patch and between patches in giraffes depends on the types of melanin made. The final color is a mixture between brown to black eumelanin and yellow to reddish brown pheomelanin, with a wide range of colors the product. Melanin synthesis in melanocytes is controlled by enzymes, of which by far the most important is the copper-containing enzyme tyrosinase, the enzyme missing in Charicleia, the albino daughter of the Queen of Ethiopia (see Chapter 2). Tyrosinase activity, in turn, is controlled by the melanocorticin hormone system, the chief hormones of which are melanocyte stimulating hormone (MSH) and adreno-cortico-trophic hormone (ACTH; e.g., Ducrest et al., 2008). Normally these two hormones are secreted from the pituitary gland and for the first 50%–60% of fetal life, no functional pituitary gland exists, but newborn giraffes have their full complement of colors. Coloration of skin can begin only when the hormones are being secreted, and coloration of hair can occur only after hair has been formed. Ackermann (1976) determined the approximate times when these points are reached. He studied skin development in seven fetuses that were at 48, 87, 131, 148, 192, 226, and 270 days of gestation, and found that melanocytes were absent in the skin of fetuses 48 and 87 days old, but were present at 131 days. By 148 days they were evenly distributed over the fetus and the skin had begun to turn into its slate-gray color. Skin coloration was complete by day 270, when it was heavily and uniformly pigmented. Synchronously, the earliest hair follicles had appeared on the chin of those same fetuses by day 87. Those follicles ultimately produced the hairs that would become tactile hairs. By day 131, tactile follicles had also developed on lips and eyelids and ordinary (coat) hair follicles were widespread elsewhere. By 148 days, hairs had started emerging from the tactile follicles and the ordinary, coat hair follicles had begun to synthesize melanin and thus were doing so at the same time that skin melanin was being synthesized. However, those follicles did not produce hair until day 270, by which time the skin was fully pigmented and, presumably by then (it was not studied by Ackermann in fetuses), the specific arrangement of blood vessels between and under patches (see Figure 14.10) has been established. Thus it is from 270 days of gestation (60% of gestation coinciding with release of hormones from the pituitary gland) onward that most hair is produced and becomes pigmented according to the future, postnatal pattern. As gestation is over by 450 days, coloration of hair to produce its patterns occurs in the last 180 days of fetal life. The resulting pattern, while conforming closely to the pattern of the parent and subspecies, is specific for each individual giraffe, does not change after birth, and has been and is used to identify individuals (e.g., Dagg and Foster, 1976).

How is melanin production regulated so that the color of the hair in patches differs from the hair between patches, and skin color differs from hair color? Ackermann's explanation was that the epidermal melanocytes of the skin were distinct from the melanocytes in hair follicles, with the activity of each integrated but separately regulated. The two cell types, he concluded both produced melanin, a synthesis started in the embryo and was maintained throughout life. What stimulated melanin synthesis? At the time Ackermann did his study, the hormonal control of melanin synthesis was obscure but he correctly surmised that "some as yet undefined factor must act on these follicles to either stimulate or inhibit melanogenesis." It is now known that the "undefined factors" are MSH and ACTH and the regulation of melanin synthesis follows their binding to their specific receptors on each of the two cell types in hair and skin. The receptors, once activated in a giraffe, stimulate the intracellular pathways that end in synthesis of eumelanin or pheomelanin with the quantity produced determined by the sensitivity and responsiveness of the receptors to the hormones. An absence of a receptor results in leucism.

MSH and ACTH are not the only hormones that affect melanin synthesis. Testosterone does so too. Over a period of 33 years, Berry and Bercovitch (2012) studied the development of patch color in 36 male Thornycroft's giraffes resident in the Luangwa Valley of Zambia. In these giraffes, the patches darkened with age, changing from brown to black, while at the same time facial hair went gray. The starting age of change was around eight years, with the color change beginning in the center of the patch and expanding slowly outward. It took 1.8±1.1 years (i.e., between about eight months and three years) for a patch to change from brown to pitch black (see Plate 16.4). The mechanism underlying the change was not studied, but as patches in female giraffes do not darken with age, testosterone can be assumed to be the stimulus or a necessary co-factor. Completion of the change to blackness signified maturity. The youngest age that a male *G.c thornicrofti* became "mature" according to patch coloration was 8.5 years and the oldest was 10.6 years and, having reached that class of maturity, they lived another 8 to 10 years to age 20. Maximum age of females in the study area was 28 years. Their longer lives can be attributed to, among other factors, the benefits of regular infusions of stem cells that multiple pregnancies bring (e.g., O'Donoghue, 2008).

17
Reproduction and the Fetus

Mother Nature dislikes an empty womb. Persistent pregnancy is her goal, with as many progeny as possible the outcome. Only then is survival of a species in the natural world assured. What has assured the survival of giraffes? At first glance, what is known about their reproductive life cycles is not reassuring. Simply put, giraffes have a long gestation, high calf mortality, and females produce relatively few calves in a lifetime, so how they were successful in the past and in some parts of Africa still are successful is one of the intriguing mysteries of giraffe biology. The mystery is incompletely solved because part of its solution requires prolonged and difficult studies of many giraffe lifetimes, a feat achieved only once, so far, by Bercovitch and Berry in their 35-year study of *G.c. thornicrofti* in the Luangwa Valley, Zambia (Bercovitch and Berry, 2010a, 2010b, 2018; Berry and Bercovitch, 2012). On the other hand, much is known about the anatomy, physiology, biochemistry, and behavior of sex in giraffes, and it is in these that some clues to their success can be found.

The Anatomy of Giraffe Reproduction

When Owen made his dissection of the three giraffes that had died in the Regent's Park and Surrey zoos, he did not overlook the anatomy of the giraffe's reproductive system, which he rather quaintly described as the "organs of generation."

Males

Owen discovered that the giraffe penis is made up of a mixture of fibrous and elastic tissues and is bent into a sigmoid shape when retracted and is, therefore, like the penises of all ruminants. He did not make a drawing of it, but other dissections (e.g., Hall-Martin, 1975a) have confirmed that its shape and its adnexal anatomy of ducts and glands are the same as that of any ruminant. The sigmoid flexure, as it is called, of a giraffe's penis takes up half the length of the penis. The total length of the penis is 71–77 cm (Velhankar et al., 1973; Hall-Martin et al., 1978), and it is cylindrical with a circumference of 10 cm (a diameter of ~3 cm). It ends in a glans characterized by a *processus urethrae* that is 7–8 cm long (Figure 17.1).

Erection is achieved by relaxation of the *retractor penis* muscles, which are 13 cm long and are attached to the ventral surface of the sigmoid flexure. Relaxation of the retractor muscle allows the elastic elements that have been held in tension to recoil and the recoil extends the penis to its full length. Although their biochemistry has not been studied, the retractor muscles must be highly resistant to fatigue as their natural state is contraction: relaxation of them is infrequent. However, the penis has very little cavernous tissue

How Giraffes Work. Graham Mitchell, Oxford University Press. © Oxford University Press 2021.
DOI: 10.1093/oso/9780197571194.003.0017

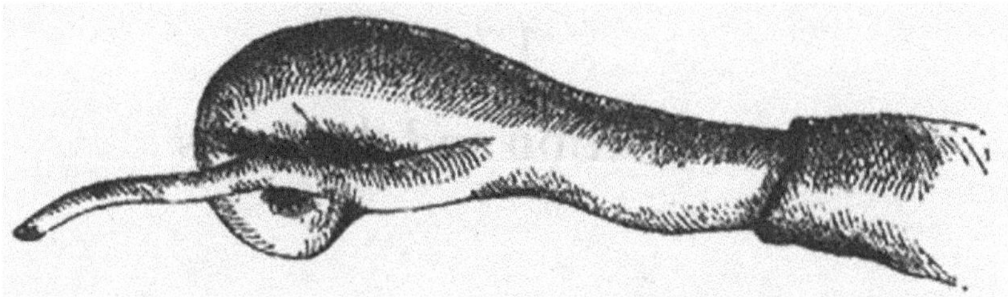

Figure 17.1. The glans of the giraffe penis showing the tube-like extension of the urethra (*processus urethrae*).
From Garrod, 1877, with permission of John Wiley and Sons.

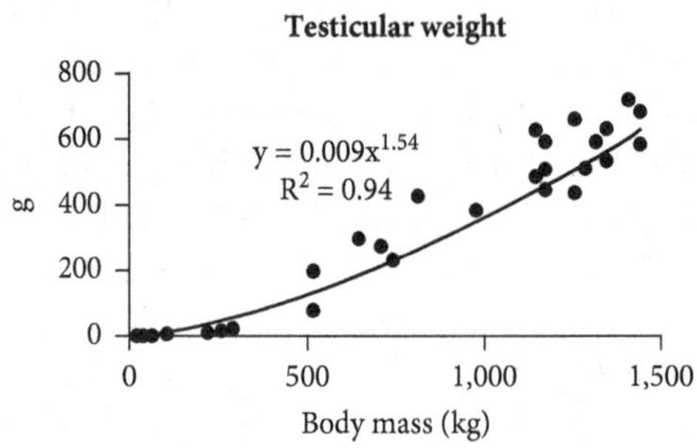

Figure 17.2. The increase in total testicle mass as a male giraffe matures and body mass increases.
Redrawn from data in Hall-Martin et al., 1978.

and thus does not have to engorge with blood to achieve erection and does not require a period of detumescence before another, thus when they are needed as many erections as necessary can be achieved in a short time. In contrast, engorgement and detumescence are required by the musculo-vascular penis of primates and horses. Their penises have large cavernous sinuses and little elastic tissue. Relatively infrequent erections are the consequence.

Owen noted that the testes of giraffes are descended at birth and carried in a short scrotum "in the situation usual in the ruminant tribe"; in other words, in the groin, and on each side of its base are rudimentary mammae. The testes, he saw, were elongated and oval and separated by a fibrous partition (mediastinum) that projected between the testes along their longitudinal axes, a position which Owen thought was most favorable for the periodical enlargement of the testis "in every direction, which takes place in giraffe and deer and antelopes." But he was wrong about giraffes. In giraffes there is no clear pattern of seasonal variation in testes mass, but their mass does increase steadily as body mass increases (Hall-Martin et al., 1975a, 1978; Figure 17.2; Table 17.1).

In an adult giraffe, each of the two testes is 10–18 cm long and about 5–8 cm wide, and they have a combined mass of up to 600 g (Velhankar et al., 1973; Hall-Martin et al., 1978; Lueders et al., 2009b). Testes mass reaches its maximum at 12 years of age. In contrast to Owen's observations of a short scrotum, Gordon (in 1779, who wrote that giraffes have "very large balls") and Hall-Martin recorded that the testes are carried in

Table 17.1. Size and Structure of Testes in Fetal, Neonatal, and Adult Giraffes

Specimen	Age	Body Mass (kg)	Testicular Mass (g)	Seminiferous Tubule Diameter (μm)	Interstitial Cells of Leydig (μm^{-2})
Fetus	270 d	16.8	2.0	-	-
Fetus	336 d	36.3	4.2	57.0	3.2
Fetus	390 d	60.3	4.4	48.2	2.8
Fetus	454 d	101.2	9.2	56.1	3.3
Neonate	2–3 d	100	6.4	57.8	2.7
Neonate	0.5 y	220	12.6	58.7	1.8
Mean of 16 adults	>8 y	1174	539	196	2.8

From Hall-Martin, 1975a, with permission of the University of Pretoria.

a pendulous scrotum, a difference most likely reflecting differences in ambient temperature at the time the observations were made. The testes internal structure is the same as it is in other mammals (Hall-Martin et al., 1978), and consists of many seminiferous tubules. The tubules are the male equivalent of the female ovary, and they are stimulated to produce spermatozoa by the same hormones that regulate ovarian activity. Each tubule is lined by primary spermatogonia, the source of spermatozoa. Between spermatogonia are Sertoli cells. They synthesize and secrete glycogen that provides the energy for the synthesis and differentiation of precursor spermatids into spermatozoa. Between the tubules lie the interstitial cells of Leydig, of which there are ~3 μm^{-2} of the interstitial space between tubules (Table 17.1). They secrete testosterone.

Male giraffes have 15 pairs of chromosomes, of which one pair is made up of the sex (XY) chromosomes (Wallace and Fairall, 1965; Buckland and Evans, 1978; Gallagher et al., 1994). When spermatogonia develop into spermatids by cell division, a process that begins at puberty, each spermatid receives one of each pair of chromosomes. At the moment of fertilization, each of these single chromosomes combines with its opposite chromosome in an ovum to form a zygote—the first step in the process that leads to an embryo. Spermatids, once formed, move into the epididymus by peristalsis. The epididymus is a 15-cm long tube in giraffes (Velhankar et al., 1973), and it is there that spermatids mature into unicellular spermatozoa. The heads of mature spermatozoa are 5.8±0.5 μm long and their tails 42±3 μm long (Velhankar et al., 1973). At ejaculation, spermatozoa are squirted into the urethra via the muscular-walled *vas deferens*, assisted by contractions of the ischio-and bulbo-cavernosus muscles that surround the urethra, along the way collecting all the other constituents of semen. The *vas deferens* is about 60 cm long (Hall-Martin et al., 1978). The urethra connects the bladder to the outside world and usually is filled with the remnants of urine. To neutralize any effect of acid or other constituents of urine that may affect the survival of spermatozoa either in the urethra or vagina, the prostate gland secretes an alkaline fluid that makes up ~20% of total semen volume. The gland consists of two separate elongated bodies in giraffes (Owen, 1838), forms part of the wall of the urethra (Hall-Martin et al., 1978), and closes off the urethra during ejaculation, an action that prevents reflux of semen into the bladder. Additional secretions are provided by a pair of egg-shaped Cowper's bulbourethral glands, each 3 cm wide and 5 cm long (Lueders et al., 2009b), and a pair of seminal vesicles. Cowper's glands have the same purpose as the prostate glands and secrete a small volume of clear alkaline fluid usually just before ejaculation. Most fluid in semen (~60%) comes from the seminal vesicles which, in giraffes, are the same size as the Cowper's glands, are not lobulated, and are smaller than they are in domestic cattle (Velhankar et al., 1973). Their important secretion is fructose, which provides the energy needed by spermatozoa to swim.

It is a common perception that mammalian spermatogenesis—the synthesis of spermatozoa—occurs best at a temperature lower than body temperature. If testes are retained within the body, as they are in cryptorchid mammals (the abnormal circumstance of one or both testes not descending into the scrotum) and in other mammals like elephants in which the testes normally are retained internally, the length of time sperm survive in the epididymus is shortened (Glover, 1973). However, spermatogenesis may not need a temperature lower than body temperature for sperm maturation and survival. What is needed is a relatively stable temperature. Whether giraffes can achieve a stable testicular temperature has not been studied in any detail, but Glover (1973) studied, in three giraffes, the course of the artery supplying blood to the testes. Arterial blood is, of course, at body temperature. In these three giraffes, Glover found that the artery was convoluted and coiled (more so than in an eland and a steenbok), especially in the region where it was surrounded by the pampiniform venous plexus, the network of veins that carries venous blood from the testes back to the body. The blood in the veins is as cool as ambient temperature. Countercurrent heat exchange occurs between venous blood and arterial blood, which cools arterial blood going to the testes in a way similar to the heat exchange between cool venous blood and warm arterial blood in the carotid *rete* (see Chapter 14). The fact that in giraffes the area of contact between artery and pampiniform plexus is larger than it is in some other artiodactyls, an enlargement that must have been produced by natural selection, implies that they rely on this method to cool arterial blood going to the testes.

The giraffe scrotum and testes are attached to the body by two muscles—the *tunica dartos* muscle of the skin of the scrotum and the *cremaster* muscle of the testes (Lueders et al., 2009b) as they are in any mammal. When these two muscles contract, they lift the testes toward the warmer body, and wrinkle the scrotal skin, reducing its surface area and thereby reducing heat loss, which in turn prevents testes temperature from falling. This mechanism can be seen to happen in giraffes on a cold morning (as it can in humans) and probably is the explanation for Owen's observation of a "short scrotum." When the testes are lifted in giraffes, the pampiniform plexus is shortened, which reduces the area of heat exchange between warm arterial blood and cool venous blood in it. Less cooling of arterial blood on its way to the testes is the outcome, and testes temperature moves closer to body temperature. The opposite happens when the testes are too hot, and explains Gordon's and Hall-Martin's observation of a "pendulous scrotum." In some artiodactyls (e.g., rams), heat can be lost from the skin of the scrotum by evaporation because scrotal skin contains sweat glands (Maloney and Mitchell, 1996). It would be a useful additional thermoregulatory mechanism if the skin of a giraffe's testicles did sweat, but whether the skin contains sweat glands (and, if it does, whether they are active) is unknown, but sweat glands may be unnecessary because the effects of skin wrinkling and the pampiniform plexus may be sufficient to keep the temperature of the testes of giraffes relatively stable.

Females

In male giraffes, the anatomy of reproductive organs is complex, but control of spermatogenesis relatively simple. In females, the anatomy is simple, but control of its functions is complex. Owen gave great details of the dimensions of the genital tract of female giraffes and emphasized differences between them and those of camels and ruminants in general, but the fact is that from a physiological point of view, anatomical differences are irrelevant. No matter what the anatomical differences are, the females of any species invariably are fertile. All female mammals, including giraffes, have two ovaries, each surrounded by the fimbriated, finger-like projections that mark the end of each of the two Fallopian tubes, a two-horned uterus, a cervix, and vagina (Figure 17.3). The purpose of the fimbriae is to guide ovulated ova into a uterine horn and, from Owen's description, the fimbriae are better developed in giraffes than in the other ruminants he studied, though they are indistinct in his illustration (Figure 17.3). The ovaries in the young giraffe Owen examined were irregularly shaped, about 4 cm long, 3 cm wide, and ~1 cm thick. Owen concluded that in an "unexcited" state the ovaries were larger than those in a camel. In a giraffe that was

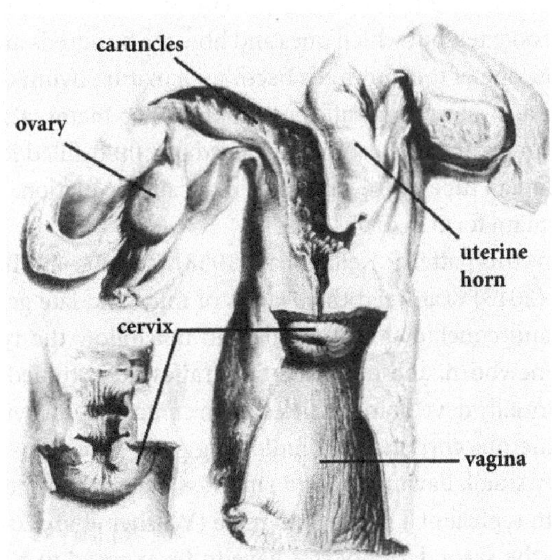

Figure 17.3. The genital tract of a female giraffe showing the vagina, cervix, the short common uterus, and the two uterine horns, their Fallopian tubes, and ovaries. The right uterine horn has been opened to show the projections (caruncles) on which cotyledons form. The cervix has been opened to show its lamellae.
From Owen, 1838, with permission of John Wiley and Sons.

11 years and 7 months old and had recently calved, an ovary described by Murie in 1872 was 7 cm long, 2.5 cm wide, and 2.5 cm thick. Each Fallopian tube is continuous with a uterine horn, and the two horns unite to form a common uterus and a vagina (Figure 17.3). Each of the uterine horns in the adult female Murie dissected was about 30 cm long and, in Owen's young giraffe, they were 20 cm long. Owen noticed four longitudinal rows of short, flattened processes (caruncles; Figure 17.3) projecting from the inner surface of the uterine horns. They are the sites of future attachment of the placenta. In Owen's giraffe, the vagina was about 15 cm long and the common uterus 5 cm long. In Murie's older giraffe, the vagina was 23 cm long and 12 cm wide and the common uterus 25 cm long. The common uterus is separated from the vagina by the cervix, which is filled by longitudinal lamellae that are lined by mucus-secreting epithelium. The lamellae converge in the midline (Figure 17.3). The mucus they secrete seals off the uterus from the external environment. All these details were confirmed in a living giraffe by sonography 170 years later by Lueders and colleagues (2009b).

The giraffe ovary, like that of other mammals, has an outer cortex and an inner medulla. The outer cortex is the germinal epithelium and contains oocytes, follicles, and corpora lutea—in short, all the components needed for reproduction. The inner medulla contains (mainly) blood vessels. In early fetal life, the germinal tissue of a giraffe's ovary is filled with many thousands of primordial follicles, each containing a primary oocyte that has the full complement of 15 pairs of chromosomes, of which one pair is the XX pair (Wallace and Fairall, 1965; Buckland and Evans, 1978; Gallagher et al., 1994). Typically in a mammal, these primary oocytes divide at puberty, when estrous cycles begin, to become secondary oocytes containing half the number of chromosomes. During a giraffe's life, several hundred of the many thousands of primary oocytes will become secondary oocytes, but which ones and how the hundreds are selected from the thousands, and how one of the hundreds becomes a mature ovum during each estrous cycle, are as much a mystery in giraffes as it is in other mammals. A secondary oocyte develops into an ovum and becomes enclosed in a fluid-filled follicle. The ovum is released into a Fallopian tube by the processes that cause ovulation. After ovulation the follicle cells that remain form a corpus luteum.

Do giraffes follow this pattern? Kellas et al. (1958), Kayanja and Blankenship (1973), and Wilsher et al. (2013) examined the ovaries of mid- and late gestation fetuses and newborn giraffes and concluded that giraffes do not follow the typical pattern. The ovaries of fetuses, newborn, and pre-pubertal giraffes they studied were not inactive. They contained normally developing follicles, each containing an ovum, and the ovaries also contained numerous corpora lutea indicating prior ovulations. The corpora lutea were, however, very small, having a combined mass of less than the mass of the single large corpus luteum typical of a pregnant female (Wilsher et al., 2013). For follicles to develop in a fetus, the secondary oocytes have to be exposed to a follicle stimulating hormone (FSH), and normally FSH is secreted by the pituitary gland only after puberty has been reached. What is the origin of FSH in a fetal giraffe? A likely origin is the placenta. In 1940 Wilkinson and de Fremery tested the urine of a pregnant giraffe and found that it contained a hormone able to stimulate follicle development in adult female rats. Urine of non-pregnant female giraffes and a male giraffe did not do so. Neither did the urine of a pregnant female for the first 240 days of gestation, but when a well-developed placenta was present on days 319 and 333 of gestation, the urine could stimulate follicle development. By that stage of gestation, the fetus has a functioning pituitary gland and thus it too could be the source of FSH. What could be the value of having ovaries functioning in female fetal giraffes as if the giraffes were post-pubertal? Gombe and Kayanja (1974) discovered that the corpora lutea of fetuses contained progesterone at the same concentration that is in the corpus luteum of a pregnant adult. Progesterone is essential for maintaining pregnancy, and secretion of it by the ovaries of a fetus may contribute to securing a successful pregnancy. If so, then stimulation of ovarian activity by a placental or pituitary hormone has a purpose. What if the fetus is male? The testes of male fetuses are small and undeveloped up to eight months of gestation and weigh just 1 g, but by late fetal life they have enlarged and in fact become larger than they are at birth (Table 17.1; Hall-Martin et al., 1978; Wilsher et al., 2013) and synthesize progesterone (Wilsher et al., 2013); thus even if the fetus is a male, pregnancy support is a possibility. However, Wilsher and colleagues did conclude that the amount of progesterone secreted by fetal testes was so small that its contribution to total progesterone synthesis could not be physiologically meaningful. The same conclusion must surely be made about the ovaries of female fetuses, and the function of follicles and corpora lutea in fetal ovaries of giraffes and the hormones they produce is an unresolved aspect of giraffe reproductive physiology.

While a placental hormone would explain the presence of follicles and corpora lutea in a fetus, a placental hormone cannot explain their presence in immature, postnatal, pre-pubertal female giraffes, which Kellas et al. and Kayanja and Blankenship had observed. Gombe and Kayanja (1974) guessed that in pre-pubertal females the origin of follicles was low-level secretion of their own stimulatory hormones. If so, then new follicles and corpora lutea would be produced regularly. However, Hall-Martin and Rowlands (1980) could find no evidence that corpora lutea were being produced afresh during the three years between birth and puberty, and they decided that those present in a pre-pubertal ovary represented nothing more than slowly regressing corpora lutea that had developed in fetal life. None of this type of corpus luteum is found after puberty. At puberty, when estrous cycles begin and normal ovulation occurs, the multiple small corpora lutea disappear and are replaced by a single large corpus luteum characteristic of an adult that has recently ovulated.

Puberty

Undoubtedly the earliest recorded observation of the age of puberty in giraffes was made by Owen (1839). In May 1836, two giraffes arrived in England. They had been caught in Kordofan, South Sudan, about six months previously. On arrival, Owen estimated they were one and a half years old, but as the height of the male was 345 cm, it was probably a little over two years old. The female's height was 330 cm, and so it too was a little over two years old (see Figure 5.3). By May 1837, each had grown 60 cm and a year later by another 20–25 cm. Owen saw seven copulations between them during March 1838, "notwithstanding the evidences of immaturity afforded by stature and dentition" when both giraffes were about 3.5 years old. Pregnancy resulted. Is that age more or less the age that all giraffes become sexually mature?

Puberty in Males

How can you tell if a male giraffe has reached puberty? If it were possible to examine the stage of development of the testes, then it would be relatively easy: the onset of puberty is marked by an increase in the diameter of the seminiferous tubules to become three- to fourfold larger than it is in a neonate (Table 17.1), and it is also marked by differentiation of spermatogonia into the spermatocytes from which spermatozoa will develop. Hall-Martin et al. (1978) detected spermatids and spermatozoa in the testes of two giraffes that weighed 501 kg and 513 kg. A body mass of 500 kg is reached at about three years of age. Sertoli cells were fully developed at that age too. The interstitial cells of Leydig, which lie between the seminiferous tubules and produce testosterone, also are able to do so from three years of age. Theoretically, therefore, an increase in testosterone concentration in blood would confirm the onset of puberty. For fairly obvious reasons, a systematic analysis of blood testosterone concentration of wild giraffes has not yet been done, but Hall-Martin et al. (1978) made an attempt to measure testosterone production by testes obtained from culled giraffes. The analysis gave very equivocal results, probably because measurement techniques available then were difficult. Whatever the reason, testosterone concentration varied and was almost undetectable in three- to six-year-old giraffes ($<0.4\ \mu g\ g^{-1}$ testis), a result which is consistent with a study done by Wolf et al. (2018a), who found that, compared with older bulls, young giraffe bulls up to four years of age secrete very little testosterone and have little interest in females. Hall-Martin et al. did conclude, however, that synthesis of testosterone had to precede spermatogenesis. The most obvious physical characteristic of sexual maturity in giraffes is testicular size and hence its mass. Testicular mass has been measured by Hall-Martin et al. (1978; Figure 17.2, Table 17.1) and from those data, an allometric equation describing testes mass (M_T, g) as a function of body mass can be derived: $M_T = 0.009 \times Mb^{1.54}$. It is clear from Figure 17.2 that combined left and right testicular mass and, therefore, their size starts increasing during fetal life, but does so more rapidly when a giraffe has reached a body mass of ~500 kg, that is, three years of age. At that age, not only do growth and expansion of the seminiferous tubules occur, but there is also initiation of spermatogenesis, and male neck mass and skull mass begin to increase (see Chapter 15). Hall-Martin et al. (1978) concluded from their data that, provided that the quality of browse available is good during their growth years, male giraffes in the wild ought to be able to breed by three years of age, but because of their small size and behavioral immaturity do not. However, captive giraffes can. Confirming Owen's observations, they recorded three cases of successful breeding in captive male giraffes at two years and eight months of age, three years and six months, and three years and nine months. Lueders et al. (2009b) used sonography to assess the development of the reproductive tract of males and reached a more general age range for sexual maturity. In their study, captive males reached sexual maturity between two and five years of age. From these relatively few studies, three years of age can be regarded as the age of puberty in male giraffes, even though, at that age, the testosterone essential for the synthesis of spermatozoa and breeding behavior is not being synthesized in the testes in significant concentrations. Nevertheless, testosterone of some origin, perhaps the cortex of the adrenal gland must be working its effects, otherwise spermatogenesis and the development of secondary sexual characters would not occur.

Puberty in Females

The time of onset of puberty and of the capacity to breed in female giraffes is a little easier to quantify. It can be estimated accurately from the date of observed first births and the assumption of a mean gestation time of 450 days. When is the milestone of a first birth reached in wild giraffes? In their study of ovary structure in 38 giraffes (*G.c. tippelskirchi*) in Kenya, Kayanja and Blankenship (1973) obtained two vital pieces of information that allows an answer to that question. First, they obtained accurate body masses of the females because data were being collected for a nutritional study. Second, 20 of the giraffes were pregnant and had a fetus of measurable mass ranging from 0.3 g to 52 kg. The gestational age of those fetuses was 16–358 days. Another seven had a small fetus, not yet implanted, and 11 were not pregnant (one of which was mature and lactating). The age range of the 27 pregnant giraffes, estimated from their body mass (see Chapter 5) was 4.1 to 8.1 years, and their average body mass was 675 kg (range = 506–792 kg). The 10 immature non-pregnant giraffes had a body mass of 393 kg (range = 212–555 kg) and on average were 2.8 years old (range = 0.8–4.7 y). The two youngest pregnant females were 4.1 and 5.2 years of age, and their age at first conception was 4.1 years (that giraffe's fetus/embryo had a mass less than 3 g and was not yet implanted) and 4.2 years (fetus mass = 52 kg and a gestational age of 358 d). In a very similar study of 31 mature female giraffes in the Timbavati Nature Reserve, Hall-Martin and Skinner (1978) concluded that the earliest conceptions occurred at 4.7 y ±3 months (compared with 3.8 y±3 months for captive giraffes). The youngest primiparous female in the Timbavati Reserve was six years old, and would have been about 4.75 years old at conception. In the Serengeti, Pellew (1983b) recorded that three adult females of known age had their first calves at 3.8, 4.3, and 4.5 years of age (46, 51, and 54 months of age, respectively) and thus conception would have occurred at 2.5, 3, and 3.2 years of age, respectively. Pellew attributed this earlier than usual onset of puberty to the high quality of browse available. Those ages are similar to the age of sexual maturity in a three-year-old captive female studied by Bercovitch et al. (2006). Metabolites of sex steroids could be detected in her fecal samples. Lueders et al. (2009b) confirmed that conclusion after using sonography to assess the degree of development of the reproductive tract: captive females in their study had also reached sexual maturity at three years of age. However, in the demanding habitat of the Luangwa Valley of Zambia, Bercovitch and Berry (2010a) observed 35 births in identifiable *G.c. thornicrofti* cows over a 35-year period and recorded that the average age at first birth was 6.4±0.7 years, with a range of six to eight years. The earliest first conception would then have been at about five years and the latest at about age six and a half years. Nevertheless, from all these studies a reasonable conclusion is that puberty can occur at around three years of age in wild female giraffes, the same age that it does in males, with the actual age of onset in individual female giraffes determined by their general health and the quality of nutrition, but the earliest conceptions can be expected to occur only about a year after puberty.

The Estrous Cycle of Giraffes

Conception cannot occur without ovulation, and ovulation is a bit like baptism: it is an inward and invisible event, with male behavior the outward and visible sign that it has occurred. Conception is orchestrated by the estrous cycles that follow the onset of puberty. In female humans the cycle has a well-described pattern that lasts roughly a month (hence called a menstrual cycle). In mice the cycle is about five days long; in several species of African ruminants (buffalo, sable, oryx, wildebeest, blesbok, eland, and suni) it is 24±3 days long; in cattle 21 days; and in elephants about 15 weeks long. There is no correlation between body mass and estrous cycle length. For the seven African ruminant species mentioned here that ranged in body mass from 5 kg (suni) to 460 kg (eland), the correlation coefficient (R^2) between estrous cycle length and body mass was 0.0006, in effect zero. Clearly estrous cycle length is governed by an array of other factors. Whatever the length of the cycle, it is characterized by pre-ovulation growth of follicles containing a growing ovum and secreting estrogens (the follicular phase), followed by ovulation and subsequent formation of a corpus luteum secreting progesterone (the luteal phase). The progesterone prepares the uterus for implantation by a fertilized ovum and maintains pregnancy. Typically, estrogen promotes follicle growth and alters female behavior to attract males. The presence of estrogen can be detected by males as changes in body odor (e.g., in humans) or in ungulates by drinking or sniffing urine. The action of urine testing by male ungulates produces the characteristic *flehmen* response of extension of the neck and curling of the upper lip ("to make other bulls think it smells bad," as the famous South African animal scientist Jan Bonsma once described it to me). The odor, taste, pheromones, or some other constituent of the urine, are detected by the vomeronasal organ in giraffes, as in any other ungulate. The vomeronasal organ is a supplementary sense organ that lies on the ventral surface of the nasal cavity. The organ is lined by an epithelium containing receptors that are activated by fat-soluble substances. Once activated, the receptors transmit the information to the very well-developed olfactory bulbs in the brain (Figure 16.9). The basic structure of the vomeronasal organ in giraffes is the same as it is in other artiodactyls, but it is richer in blood vessels. As it is the vessels that transport substances to the olfactory epithelium (Kondoh et al., 2017), and as the olfactory bulb is large in giraffes, it may be that male and female giraffes are more sensitive to odors than are those of other artiodactyl species (Kasozi and Montgomery, 2018).

The stages of an estrous cycle are controlled by the hypothalamus, which makes and secretes gonadotrophin releasing hormone (GnRH). GnRH release is inhibited by the high concentrations of estrogens and/or progesterone that are present in blood for most of the cycle, but after ovulation, and if conception fails, their concentrations fall and GnRH is secreted again to restart the cycle. GnRH in turn stimulates the release of follicle stimulating hormone (FSH) from the pituitary gland. FSH circulates to the ovary in blood and there binds to receptors on several oocytes of which one, in animals that usually have just one offspring, will develop into a mature ovum. As the ovum grows, it secretes more and more estrogen, and at high concentrations the estrogen sensitizes the pituitary gland cells that secrete luteinizing hormone (LH) to GnRH. GnRH stimulates LH release, and when the combination of hormones and the state of maturation of the ovum are right, LH acts on the mature ovum to activate the sequence of events that cause ovulation. The source of estrogen is thereby removed. A second effect of LH is to stimulate the formation of a corpus luteum from the cells remaining in the follicle, and progesterone secretion begins. If conception fails, the corpus luteum is destroyed by prostaglandins secreted from the uterus wall, progesterone concentrations fall, inhibition of GnRH secretion is lifted, and the cycle starts again.

Do giraffes fit this paradigm? At one level they do. Patton et al. (2006) injected a synthetic GnRH agonist into a female giraffe. At high doses the agonist desensitizes the pituitary gland to GnRH and inhibition of the release of FSH and LH follows. In their giraffe, estrous cycles were inhibited for 400 days, and the inhibition confirms that the main mammalian regulatory system is in place in giraffes. Whether the other parts of the regulatory mechanism are in place in giraffes can be determined from the characteristics of their estrous cycle, of which its length is, perhaps, its most revealing feature. How long is a giraffe's estrous cycle? The first observation of the pattern of estrous cycles in wild free-living giraffes was made by Hall-Martin and Skinner in the Timbavati Nature Reserve in South Africa (1978). They did not study the cycle itself but they did know, precisely to the day, the calving interval of 46 giraffes. Those 46 intervals could be divided into groups, and the groups were found to be spaced 18.1±0.9 days apart. This interval, Hall-Martin and Skinner concluded, represented the length of an estrous cycle and, if so, estrous cycle length was not all that different from the cycle length in many other artiodactyls. However, that 18-day interval was noticeably different from the length reported for the estrous cycle length of a captive female giraffe. It showed the signs of estrus six times at intervals of 16.3±0.7 days (Hall-Martin, 1975a).

As many researchers have discovered, the study of reproductive matters, and what hormones might be involved, is far easier to do in captive animals than it is in wild animals. Thus it was that Loskutoff and colleagues (1986) studied estrous cycle length by collecting daily urine samples over a period of six months from four captive giraffes and measured in them a metabolite of progesterone, pregnanediol, which accurately reflects the concentration of the parent hormone in blood. The concentration of pregnanediol increases in the urine whenever there has been an ovulation. In those giraffes, secretion of progesterone and its concentration declined quickly two to three days before birth, remained at low levels for 70 days, after which they fluctuated at regular intervals, and the intervals reflected estrous cycles (Figure 17.4). The follicular (estrogen) phase of a cycle, defined by the low progesterone concentration, averaged 4.0±0.1 days in length. The luteal (progesterone) phase lasted on average 11.5±0.4 days, and the combination of the two showed that the cycles were ~15 days long, a length much shorter than it is other artiodactyls. This pattern is identical to that in okapi (Figure 17.4) and thus a short estrous cycle has been a feature of giraffid reproductive strategy throughout their evolution and must have had and must still have selective advantages.

Bercovitch et al. (2006) confirmed this estrous cycle pattern by measuring the concentration of pregnane (another intermediary of progesterone metabolism) in fecal samples, which also accurately reflects the concentration of progesterone in blood. Fecal samples were collected over a two-year period corresponding to 46 complete

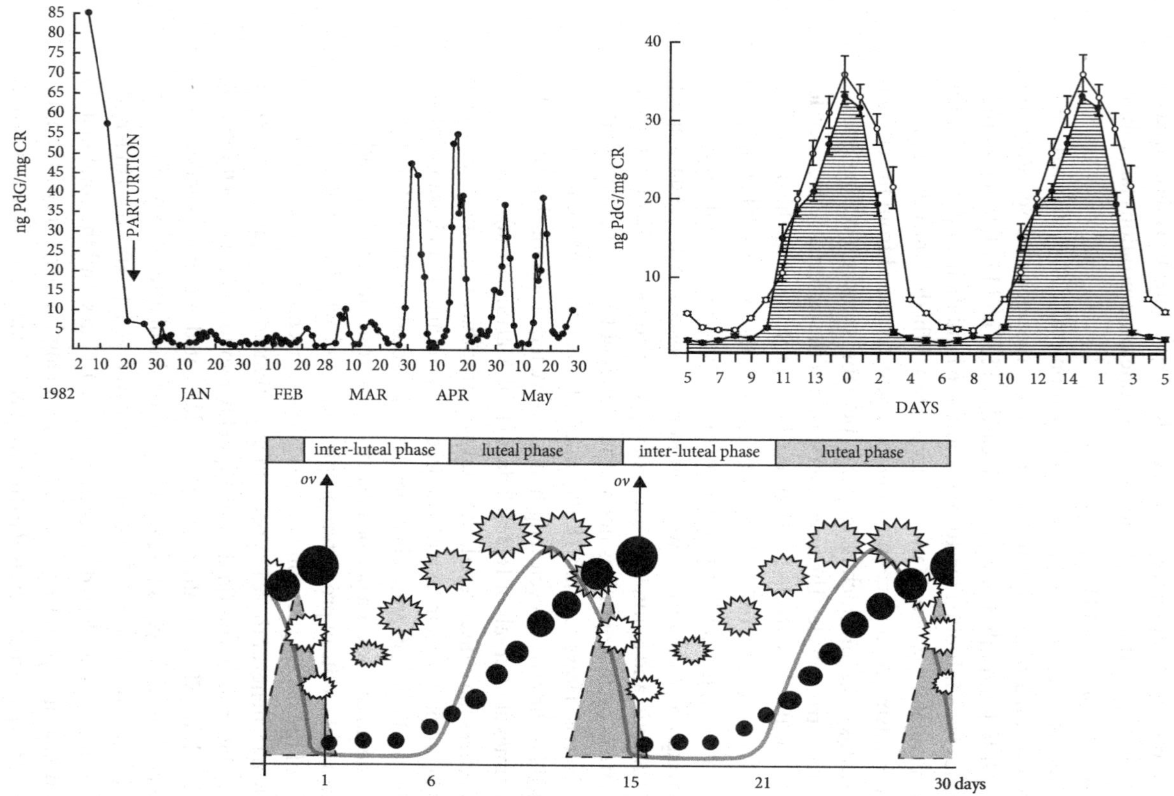

Figure 17.4. Top left: Urine concentrations of pregnanediol (ng mg^{-1} creatinine) from 10 days before birth until the reappearance of normal estrous cycles 90 days later. Top right: Urine concentrations of pregnanediol (ng mg^{-1} creatinine) during two estrous cycles in giraffes (open squares) and okapi (shaded area).

From Loskutoff et al., 1986, with permission of John Wiley and Sons.

Bottom: Schematic diagram of two estrous cycles in a giraffe. Arrows show the time of ovulation (OV), black dots show the change in size of a follicle, the gray stars show the changes in size of a corpus luteum, the gray line shows the concentration of progesterone, and the dashed line around shaded triangles shows the estrogen concentration.

From Lueders et al., 2009a, with permission of Oxford University Press.

cycles. Just as Loskutoff et al. had found, progesterone concentrations fell at or before birth, and thereafter fluctuated at regular intervals that reflected estrous cycles. The average length of those cycles was 14.7±2.8 days, with a range of 8 to 20 days. The average length of the inter-luteal (estrogen, follicular) phase was 8.8±2.3 days long, and that of the luteal (progesterone, post-ovulation) phase was 5.9±2.4 days long, a combination different from that reported by Loskutoff et al.

The most detailed study so far on the length of each phase, the mechanisms underlying the short 15-day cycles, and how they are controlled was done by Lueders et al. (2009a). Lueders and colleagues measured both progesterone and estrogen concentrations in fecal and blood samples, over a period of three years in five captive giraffes aged four to 12 years. The giraffes' reproductive status changed over the course of the study from cycling to pregnant, and one female and three male calves were born. Based on changes in progesterone concentration, the average cycle was 14.7±1.2 days long, with an inter-luteal phase of 6.4±0.9 days and a luteal phase 8.1±0.6 days long (Figure 17.4). The unusual feature of these cycles was that a corpus luteum (from the previous cycle, secreting progesterone) and a follicle (secreting estrogen) in the current cycle grew in size simultaneously, not sequentially as they do in mammals with longer cycles. How that simultaneous growth might be controlled by hormones is unknown, but it is the simultaneous (rather than sequential) development of both follicle and corpus luteum that reduces the time interval between a previous ovulation and the next, and is the reason that the cycle is compressed in giraffes to 14–15 d long. Ovulation in Lueders et al.'s giraffes occurred 1.9±0.4 days after progesterone levels reached their lowest level, and that timing implied that ovulation is precipitated by the fall in progesterone and, although it was not measured, presumably by a spike of LH secretion. It is at this time that females are most receptive to male attention, but their interest in males lasts only for about 24 hours. One day after ovulation, Lueders et al. established that a new corpus luteum started developing but only began to secrete progesterone 5.4±0.5 days after ovulation, by which time the corpus luteum of the previous cycle had disappeared. Progesterone secretion reached its peak 11–12 days after ovulation, but if no conception resulted, then secretion declined. Two days after ovulation a new follicle started forming, but estrogen secretion by the developing follicle only occurred once the follicle had reached its maximum size, thus estrogen does not seem to contribute to the growth of a follicle in giraffes in the same way that it does in other mammals. Estrogen secretion and male interest peaked as progesterone concentration started to fall. Previous studies have attributed male interest to the effects of progesterone, but Lueders and colleagues established that male giraffes are no different from the males of other mammals, including humans. Estrogen is the signal that ensures that mating behavior, ovulation, and copulation coincide exactly, as it did all those many millions of years ago when it arranged the fusion of two free-living unicellular organisms to share their DNA before dividing and thus begin the long sequence of events that has become sexual reproduction as we know it. An effect of estrogen in a mammal still is to coordinate the elaborate biology and behavior that brings about the union of the two unicellular parts of mammalian reproduction—an ovum and spermatozoon.

The studies of the length and character of a giraffe's estrous cycle have established that giraffes, by comparison with many other African artiodactyls, have a short estrous cycle. What are its advantages? Bercovitch et al. (2006) have suggested that short cycles are an advantage in non-seasonal breeders. Non-seasonal breeders have long gestations. A short estrous cycle allows them to return quickly to estrus and ovulation when environmental conditions are favorable. More important, perhaps, is that the shorter the cycles are, the greater is their frequency and the greater the number of opportunities for conception. For example, puberty in both buffaloes and giraffes occurs at about three years of age and they have equivalent life spans of 20 years (Western, 1979). Thus they have equivalent reproductive lives of ~17 years. The estrous cycle length is 23 days in buffalo and 14 days in giraffes. Over a period of 17 years, a buffalo cow could have 295 cycles and a giraffe cow 484 or 1.6-fold as many. Of course neither species has that many because for a good part of their reproductive lives they will be pregnant, but even when that is taken into account, a buffalo cow could have 44 cycles during her reproductive years and a giraffe could have 60. Short cycles are one of the mechanisms that female giraffes have evolved to maximize the number of calves they can produce in a lifetime.

Courtship and Conception

The most detailed studies of the mating behavior of male and female giraffes have been made by Innes (1958), Pratt and Anderson (1979, 1982, 1985), Hall-Martin (1975), and Bercovitch and colleagues (2006, 2010b, 2012), and it is not a simple matter, partly because male and female giraffes have very different social preferences. In the wild, and based on studies of *G.c. thornicrofti* in Zambia by Berry and Bercovitch, males seem to have relatively short reproductive lives. If a male *G.c. thornicrofti* survived the first year of life, they left their natal herd and by two years of age were independent. They dispersed between the ages of four and eight years, became solitary at seven years, and bred at 10 years (Berry and Bercovitch, 2012). They had an average life expectancy of 16 years, but could live (if they were lucky) to 21 or 22 years (Berry and Bercovitch, 2012). During their wandering years, male giraffes have large home ranges. Females, on the other hand, live in small groups, have smaller home ranges, and daughters tend to remain with their mothers. Females are attractive to males for only one or two days over the two-week period of an estrous cycle, have a short fertile window of 24 hours during those days, and have a non-seasonal breeding pattern, all of which means that the chance a bull giraffe has of finding a fertile and receptive female is low. How does a male ever find a receptive female? One possibility is that in order to make contact, males roam widely, a behavior that partly accounts for the much larger home ranges of males, and during their wanderings they come across fertile females by accident. More likely is that around the time of ovulation, a fertile female releases an attraction signal that can be detected by an errant male traveling in search of adventure. In many animals, including mammals, pheromones are one such signal. There are many types of pheromones and they have many functions, but some of those that regulate sexual behavior are those that can be dispersed by wind. They are usually volatile, lipid-soluble hydrocarbons released from body surfaces in sweat or other secretions. When they are inhaled, they attach to the membranes of the olfactory system and generate a nerve impulse. Activation of this olfactory pathway ultimately results in the secretion from the pituitary gland of the hormones that regulate sexual activity. Are pheromones or something similar secreted by female giraffes around the time of ovulation? So far no study has tested if they are, but there is circumstantial evidence to support this idea. One piece of anatomical evidence is the impressively large size of the olfactory bulbs that giraffes have (Figure 16.9). Their size, which must be a product of natural selection, tells us how important the olfactory pathway is in the social system of giraffes. Sex hormones also seem to be secreted in response to an attraction signal of some kind. In *G.c. giraffa* in Timbavati Nature Reserve, for example, testosterone synthesis and secretion are highest in June to August (Hall-Martin et al., 1975a), a time that coincides with peak births and the return to estrus in giraffe cows. The timing of the secretion of testosterone has also been studied by Seeber et al. (2013). They correlated fecal androgen concentration (as a measure of blood testosterone concentration) with male sexual behavior in 39 free-ranging male giraffes in Zimbabwe (*G.c. giraffa*), which, from the physical appearance of skull and ossicone shape and neck musculature, ranged in age from youngish to old. Assessment of maturity was subjective, but the conclusion was that sexual behavior was a characteristic of older males and was positively correlated with the appearance of androgen in fecal pellets. Sexual behavior in those males seemed to have short cycles lasting two weeks at a time, and thus as long as an estrous cycle. During the short cycles, fecal androgen concentrations were higher than at other times (39.6±16.7 μg g^{-1} of scat vs. 24.8±10.9 μg g^{-1}). Seeber et al. did not mention the frequency of the cycles or whether females were nearby, or whether the mature males were attracted to females, but if the frequency of the cycles was related to females in estrus, then a pheromonal trigger of testosterone synthesis may have been responsible for it. Wolf et al. (2018a) have also measured fecal androgen concentrations and did so in 19 male *G.c. giraffa* (five old, six young mature, and eight young and immature) in kwaZulu Natal in South Africa. The study established that the highest levels of fecal androgens and sexual activity occurred in adult males associating with females. However, that study by Wolf et al. also managed to introduce some confusion to the idea that pheromones or another attractant may play a role in courtship. In contrast to Hall-Martin's earlier study, the study showed that the lowest levels of both fecal androgen concentration and sexual activity occurred during winter months (May to August in the Southern Hemisphere) when the males tended to congregate in male-only groups. In kwaZulu Natal most (but not all) births occur in the early summer (October to January), which means that conception must have occurred in the winter of 15 months previously, when androgen concentrations were apparently lowest and the males were not actively seeking females. Moreover, while Hall-Martin (1975a) did find that testosterone synthesis and secretion are highest at a time when females are coming into estrus, he also found no seasonal variation of male fertility despite a marked drop in body condition in late winter, nor could he find a seasonal cycle in male behavior or of testes mass or seminiferous tubule diameter. Thus, if a receptive female could be identified, conception could occur at any time of the year. Hall-Martin's conclusions are more consistent with a non-seasonal pattern of births than with the social groupings and fecal androgen concentrations described by Wolf et al., but they make the role for an "attraction" signal equivocal. Nevertheless, an attraction signal of some kind must play a role, otherwise identification of a fertile female would be a random event and it does not seem to be so. How else could a female in estrous conjure the appearance of many males?

What is, however, absolutely essential is that to make a final assessment of the reproductive status of a female, a male must be in contact with the female. When in contact with a female, males assess her reproductive status by stimulating urine production and tasting or smelling it, a tactic that maximizes time spent with receptive cows and minimizes time spent on non-fertile cows (Bercovitch et al., 2006). Whatever it is in urine that males find informative is detected by the vomeronasal organ, and the information finds its way to the olfactory bulbs for interpretation and a response. Receptive females welcome the attention males give them and reciprocate by spending more time near males during the "fertile window phase" of an estrous cycle. Males do not test females that are pregnant and thus, presumably, the females are not secreting attractants. Most males are attracted to a female from the day before she has ovulated and for three days afterward (Lueders et al., 2009a), when estrogen secretion is highest. There is no jealousy between males, as several test urine at any one time, with the female urinating obligingly as necessary (Innes, 1958), a behavior that maximizes her chance of conception with least effort. Cows are not fussy, but can signal favor by choosing (or not) to urinate for a desired bull (Pratt and Anderson, 1985). A bull of sufficient size is acceptable, not necessarily the biggest (Bercovitch et al., 2006), and young cows are the preferred partner of males (Pratt and Anderson, 1985). Dominant males test and court females more frequently than do other bulls, but do not have exclusive breeding rights (Pratt and Anderson, 1985). Once a receptive female has been found, a bull will align himself behind her, tap her back legs and, as she moves forward, mount her. Ejaculation takes two seconds. At the moment of ejaculation the male emits a short guttural bleat (Owen, 1838). Coitus can occur several times (Innes, 1958). If another male is present, post-coital guarding occurs, but if not, then the bull moves off.

In almost all mammals, breeding is a seasonal affair, with births timed to maximize the use of available food either for the mother or for the offspring, or both. For herbivores, the best time to have a calf is toward the end of winter, so that by the time the calf is becoming a ruminant several months later, there is plenty of spring vegetation available for it, which will endure for the summer that follows. Giraffes do not follow this sensible pattern. They are regarded as being polyestrous (having several estrous cycles before conceiving) and non-seasonal breeders (conceiving at any time throughout a year and giving birth at any time in a year). Non-seasonal breeding means that conception, pregnancy, and births occur throughout the year and females and their calves have to manage, as best they can, the variation in the quality and quantity of browse. While a classification as non-seasonal breeders is strictly correct, Foster and Dagg (1972) were the first to notice that in Nairobi National Park there were distinct peaks of births in January and February and August and September, with most calves born in September. Gestation lasts around 450 days, thus for August and September births of a particular year, conceptions occur in May and June of the previous year, and in October and November for January and February births 15 months later. Both of those periods of conceptions coincide with the Nairobi National Park's characteristic two rainy seasons and the good nutrition that follows. Nairobi National Park, for all practical purposes, lies on the equator. In the Serengeti in Tanzania, which is two degrees of latitude (i.e., ~220 km) further south, Pellew (1983c) found an earlier, minor peak of calves in December–January and an earlier major peak in May to August. In Tsavo East National Park (3.7°S), Leuthold and Leuthold (1975) recorded 26 births over a five-year period; births were non-seasonal, but over half occurred in the months of January to May. Bercovitch and Berry (2010a) recorded 35 births in *G.c. thornicrofti* in Zambia (11°S) and they occurred in every month of the year, with most in May. In the Kaokoaveld of northern Namibia and southern Angola (19°S), the habitat of *G.c. capensis*, most conceptions occur between September and December of one year and, thus, most births in December to March 15 months later (Scheepers, 1992). Hall-Martin et al. (1975) and Skinner (1978) calculated conception dates for 123 calves and for 20 fetuses in the Timbavati Nature Reserve (24° S). Of those conceptions, 60% occurred in the four summer months of December to March, and most births occurred in March to June 15 months later. The remaining 40% of births occurred throughout the year. In the Kruger National Park, which is contiguous with the Timbavati Nature Reserve, Fairall (1968) found that 12 of 36 births (33%) occurred between February and March, and 36% between August and October. In the Willem Pretorius Game Reserve in the Orange Free State, which is situated in central South Africa (28°S), 67% of all births occur at the end of the wet season (January to April) with most conceptions therefore occurring in October to January 15 months previously, when the females were in peak condition (Kok and Opperman, 1980). An explanation for the timing of peak conceptions and, therefore, births, is that as winter becomes spring and then summer, day length increases, and longer day lengths, via their effect on the secretion of melatonin from the pineal gland, control the time of breeding, as they do in many other mammals. However, day length does not change on the equator. Therefore the effects of day length and melatonin, if any in giraffes, could not have influenced the breeding pattern reported by Foster and Dagg. On the other hand, longer day lengths and high rainfall strongly stimulate the growth of new leaves and flowers. The correlation between day length and conception is, therefore, almost certainly more to do with nutrition than with melatonin, even in the gravel-plain, sand-dune-dry-river-bed habitat of *G.c. capensis*, where annual rainfall is less than 150 mm. In that environment, conceptions correlate with an initial spring flush in vegetation (Scheepers, 1992). Thus, the common denominator of conception is good nutrition, a correlation first noted by Field and Blankenship in 1973, and as the season of abundant browse becomes more marked and earlier as distance from the equator increases, peak regional calving follows: on the equator in September (Foster and Dagg, 1972); 2°S, May to August (Pellew 1983b); 3.7°S, before June; 11°S, May (Bercovitch and Berry, 2010a); 19°S, December to April (Scheepers, 1992); 24°S, March to June (Hall-Martin et al., 1975); 28°S, January to April (Kok and Opperman, 1980). If these data are pooled, a non-seasonal pattern emerges, and the seasonal peaks first identified by Foster and Dagg become less obvious (Figure 17.5).

Pregnancy

Pregnancy follows conception, provided that a fertilized ovum becomes implanted and forms a placenta with elements of the uterus wall. The giraffe embryo's development actually proceeds without implantation for at least 30 days (Hall-Martin and Skinner, 1978; Lueders et al., 2009a, 2009b)and an embryo only becomes visible by sonography on day 37 of gestation (Lueders et al., 2009b). Implantation always occurs on the same side as ovulation (Wilsher et al., 2013) and occurs equally as often in the left as in the right ovary, although in one study 16 (60%) out of 27 ovulations occurred on the right side (Kayanja and Blankenship, 1973). Giraffes have a polycotyledonous placenta, so

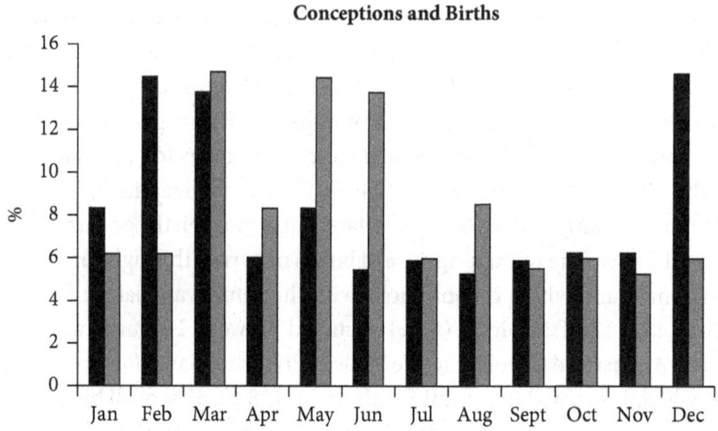

Figure 17.5. Percentage of all conceptions that occur in each month (■) and percentage of all births that occur in each month (■).
Data from Hall-Martin et al., 1975a, and Mitchell (unpublished).

called because the area of transfer of nutrients and waste products between mother and fetus is confined to discrete areas of the uterus wall—the mushroom-shaped cotyledons. The placenta is properly called syndesmochorial or epitheliochorial because the epithelium of the uterus is fused with the chorial membranes surrounding the embryo. Finger-like projections (elongated villi) grow out of the fetal membranes and insert into the equally elongated hollows (crypts) in the maternal membrane that overlies the cotyledons, and the two tissues fuse to form the epitheliochorial placenta. Each cotyledon has a rich blood supply either from the fetal (villi) or maternal (crypts) circulations. The fetal circulation is connected to the placenta by an umbilical cord, the length of which has been measured once and was ~140 cm long (Hradecky, 1983b), but Owen (1868) noted that it was "above a yard in length" and Davis (1949) also reported a length of about a meter. Those lengths are too short to survive a birth, and during birth the cord is broken.

Growth of a fetus depends on there being a large surface area for exchange of nutrients and waste products between mother and fetus. In cotyledonous placentas, it is the size of the cotyledons that matters most. Large size means many villi and crypts. The cotyledons develop on the projections (caruncles) noticed by Owen (Figure 17.3), and they are aligned in four parallel rows (Owen, 1838), or in two or three parallel rows of 60–70 caruncles each (Wilsher et al., 2013). Owen (1868) counted 180 caruncles in total; Garrod (1877) also reported 180; Hall-Martin and Skinner (1978) found 149, 183, and 191 in three placentas they examined; Deka et al. (1980) found 178 in one giraffe comprising 140 "large" cotyledons and 38 "small" cotyledons; and Hradecky (1983a) found 140 arranged in four rows in one giraffe, and in another five giraffes there were between 78 and 173, also in four rows (Hradecky, 1983b). The range is wide, but typically it seems there should be 160–180 cotyledons to support normal fetal growth. Cotyledon size varies. Their diameter, as measured by Owen (1839), ranged from 2 inches (5 cm) to 4 inches (10 cm) or 9 × 7 cm (Hradecky, 1983b) or about 8 cm in diameter (Wilsher et al., 2013) or 3.3–7.0 cm in diameter (large) and 0.5–1.9 cm

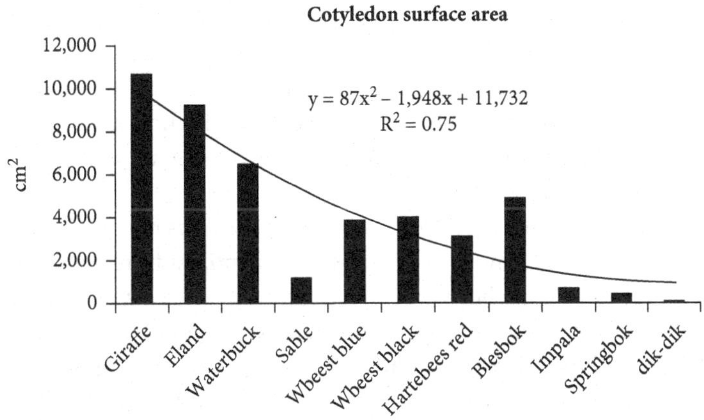

Figure 17.6. Total surface area of the cotyledons in some ungulates arranged in order of decreasing body mass, which is the independent variable.
Constructed using data in Hradecky, 1983b, and Wislocki, 1941.

(small; Deka et al., 1980). There is a reasonably strong correlation between body mass of the mother (and hence fetus size) and the total surface area of the cotyledons (number × cm^2), as illustrated by the six measurements Hradecky made in giraffes and in nine other ungulates, to which the surface area of cotyledons in the smallest artiodactyl, the dik-dik, can be added (Figure 17.6). The trend shown could be predicted. The variation implies that while cotyledon surface area is important, a more critical surface area may be the surface area provided by the villi and crypts, which may be large even if cotyledon surface area is small.

What is not as clear is how the placenta and hence the pregnancy are maintained in giraffes once conception has occurred. Growth of any placenta depends on a complex interaction between many growth factors and regulatory hormones, which is far from being fully understood in any mammal. There is no doubt, though, that the essential first step is prevention of the destruction of the corpus luteum that forms after ovulation. Without a corpus luteum to secrete progesterone, pregnancy fails and another estrous cycle will start. The details of how the corpus luteum is maintained in early pregnancy in giraffes are obscure, but from what is known, giraffes follow the general ruminant pattern as summarized by Wiltbank et al. (2016). At the time of ovulation, a mature follicle in a giraffe has to be at least 17 mm in diameter before ovulation will occur and typically is about 20 mm in diameter (Lueders et al., 2009b). The cells that remain in a follicle after ovulation form a large corpus luteum (Kayanja and Blankenship, 1973), big enough to fill the whole ovary (Lueders et al., 2009b), observations that have been confirmed in other studies of pregnancy in giraffes (e.g., Wilsher et al., 2013). If conception does not occur, then the corpus luteum is destroyed by the action of prostaglandins secreted by the uterus. Their secretion depends on the effects of estrogen synthesized by a new follicle. Estrogen stimulates synthesis of oxytocin receptors on the uterine endometrium and the natural pulses of oxytocin that occur throughout a day (roughly every four hours in humans; the frequency in giraffes is not known) by activating the oxytocin receptors, cause the uterus to secrete prostaglandins. If, however, a placenta is developing, it secretes an interferon that inhibits the effects of estrogen and, therefore, of oxytocin receptor synthesis. In this way, release of prostaglandins is suppressed. Suppression begins within the first few days after conception. The interferon effect is relatively short-lasting, however, and what prevents destruction of the corpus luteum for the rest of gestation is, thus far, unknown in any ruminant and is definitely not known in giraffes.

Hall-Martin and Rowlands (1980) thought that the corpus luteum decreased in size in the first 100 days of pregnancy and remained small until mid-pregnancy, after which, under the stimulation of the hormone luteotrophin, secreted by the placenta, its size increased gradually until birth. Kayanja and Blankenship (1973) and Lueders et al. (2009a) found a slightly different pattern: the corpus luteum was largest during the first 10 weeks of pregnancy but then decreased in size. Despite this regression, Lueders et al. found that progesterone synthesis and its blood concentration remained high. Wilsher et al. (2013) found that the ovaries increased in size for much of pregnancy. In one giraffe, the ovary containing the corpus luteum weighed 36 g after eight months (~50%) of gestation. In a second giraffe, after 11.6 months (~75%) of gestation the ovary weighed 67.6 g. Differences of this kind in data are difficult to reconcile and make the understanding of how pregnancy is maintained in giraffes just as difficult. Suffice to say that a corpus luteum is essential for the maintenance of the early part of pregnancy. Of course, a large corpus luteum does not necessarily mean that it is secreting hormones, and if it is not and if the corpus luteum does regress during middle pregnancy in giraffes, then there must be a secondary source of progesterone because the blood concentration of progesterone stays constant throughout their pregnancy (Dumonceaux et al., 2006; Isobe et al., 2007; Lueders et al., 2009a). One secondary source is the placenta itself, as it has the necessary enzymes to synthesize progesterone. The necessary enzymes also are present in the endometrium of the uterus, and so it too can be a source (Wilsher et al., 2013). In addition, ovaries of female fetuses and testes of male fetuses can produce progesterone. The amount secreted by fetuses may be as much as that produced during pregnancy (Gombe and Kayanja, 1974) or too little to be of any consequence (Wilsher et al., 2013). Even so, taken together, all these sources may produce more progesterone than the corpus luteum does, and together with the corpus luteum, provide the progesterone needed to sustain a very long gestation.

Gestation

From the moment of conception, a fetus manipulates its mother to do its bidding. The manipulation continues long after birth and, in humans, lasts into middle age and, perhaps, beyond. Of all manipulations, the most important one is when a fetus informs its mother that it wants to be born. There is nothing, of course, she can do about that, anymore than she could resist the manipulation of her behavior and body by hormones that led to conception in the first place. That was as ineluctable as the fetus's urgent demands now. Across species, but with species specificity, the moment of birth is remarkably precise, and in humans it is 40 weeks after conception or, as Plato informed us $(3^3+4^3+5^3) + (3 \times 4 \times 5)$ days. I do not suppose you can be more platonic than that, and perhaps Plato thought that human fetuses can do sums, which for all we know, they can.

How long does pregnancy last in giraffes? Is it exactly 15 months, that is, 450 days? The first record of the length of pregnancy was made, naturally, by Richard Owen in 1839. He described how he observed coitus in a male and female giraffe that had arrived in England in May 1836. On March 18, 1838, he observed a first coitus and a second on April 1, 1838, after which the female showed no further interest in the male. After a gestation which lasted 444 days after the second coitus and 457 days after the first, birth started on June 19, 1839. A male calf was produced but died. A second gestation lasted 431 days after coitus and also resulted in the birth of a male (Owen, 1839). Other observations in captive giraffes have found gestation to be 433 days long (Deka, 1980), an average of 448 days in 14 giraffes, but with a very wide range of 400–496 d (del Castillo et al., 2005), 417 days, and 438 days in two captive giraffes (Isobe, 2007), and 471±4 days also in captive giraffes (Lueders et al., 2009a). Far fewer data have been collected from wild giraffes. Skinner and Hall-Martin (1975) and Hall-Martin (1975a) calculated gestation length in three wild giraffes and in 46 captive giraffes. In the 46 captive giraffes the mean was 460±12 days and the range 440–488 days. In the wild giraffes it was 447±11 days. Pellew (1983c) found gestation to be 458±3 days in six wild giraffes in the Serengeti. A rounded and useful, easy to remember, number is 15 months, or 450 days, which is the longest of any ruminant. Two questions arise. Why is it 450 days and not some other length, and why is the range so wide? The second question is easier to answer than the first. The average is what can be expected if the pregnancy is uncomplicated. The range shows how the many factors interact in the life of a pregnant giraffe, from external factors like changes in climate or nutrition to internal factors like their general health, the health of the fetus, the size of the placenta, and the balance between hormones, any or all of which can change the length of gestation simply by altering the complicated sequence of events that time birth, not to mention the difficulties of making accurate observations. The list is endless, and what is most astonishing is that the length of gestation is so consistent pregnancy after pregnancy.

Why is it 450 days long? One answer is that it is a familial trait of the giraffids, as okapi have a gestation of 440 days. Thus its length in giraffids is an outcome of natural selection acting on ancestors over many generations many millions of years ago. What adaptive advantages could a long gestation have? It is impossible to say what they were that long ago, but from our perspective, long gestations have the effect of limiting the abundance of a particular species and so reduces competition for the browse on which they depend. Both giraffes and okapis live in habitats where competition from conspecifics for preferred browse is disadvantageous. A second advantage is that it reduces the dependence on seasonal breeding—a characteristic of short gestation species, which rely on catching the best forage for milk production and for their young when they wean. Giraffes make do, but do not necessarily do well, in any season. Answers such as these, however, ignore the biology of gestation and the requirements the mother and/or the fetus might have for giving birth to a calf that will survive. So it is that several studies have tried to predict gestation length from physiological variables. In a landmark study, David Western (1979), who has had a distinguished career in wildlife conservation in East Africa, analyzed the relationship between body mass and gestation. His hypothesis was that species that have large body masses have to have longer gestations to make sure that the larger fetus will have had time to reach a size that improves its chances of survival. Large species do tend to have long gestations—in elephants, 650 days; rhinoceroses that weigh 500 kg more than a giraffe, 490 days—but, confusingly, in hippopotamuses that weigh 1,250 kg, it is just 250 days, and in okapi that have a body mass one-quarter that of an adult female giraffe, gestation is the same length as it is in giraffes. What does seem clear is that there is a minimum gestation time for artiodactyls. For artiodactyls weighing up to 340 kg, average gestation is 219 days, with a range of 150 days in a 13-kg klipspringer to 280 days in 220-kg waterbuck and 263 days in eland weighing 340 kg. Typically, how long a mammal lives (its lifespan) or the length of gestation are partly dependent on metabolic rate, with small species (e.g., a mouse) having a high metabolic rate living short lives and having short gestation times, and the converse for large species. Biological times like these scale with body mass to the power 0.25 ($Mb^{0.25}$; Calder, 1996), and Western tested whether gestation time in, among other species, East African artiodactyls, including giraffes, scaled the same way.

In Western's data set there were 15 East African artiodactyls ranging in body mass from a 6-kg dik-dik to a 750-kg giraffe. The equation Western fitted to the relationship between gestation length and body mass was: Gestation Length (days) = 42.5 × Mb $(g)^{0.16}$ ($R^2 = 0.77$). When that equation is applied to a 750-kg giraffe, predicted gestation length is 370 days, which is far below reality. An equation that fits those data better is linear in character: Gestation Length (days) = 0.32 × Mb (kg) + 188 ($R^2 = 0.95$), and when that equation is used to predict gestation time in a 750-kg giraffe, gestation length is 427 days, which is closer to reality. A much larger set of data for gestation length in southern and East African ruminant artiodactyls can be compiled by combining data from Skinner and Smithers (1990; *N*=24 species) with Western's data (Mitchell, unpublished). The relationship between body mass and gestation length for the 36 individual species in the combined data set fitted a linear equation: Gestation Length (days) = 0.29 × Mb (kg) + 195 ($R^2 = 0.80$). That equation gives a gestation length of 412 days for a 750-kg giraffe. Both of these linear equations thus predict giraffe gestation length better than the power function used by Western, and in both linear equations, gestation length is proportional to Mb^1 and not $Mb^{0.25}$. Clauss et al. (2014) also investigated the scaling pattern of gestation and body mass. An allometric power equation they derived for 175 species of artiodactyl (with body mass in g) was 282 × $Mb^{0.02}$, which shows that gestation length could be independent of body mass but predicts gestation in a 750-giraffe to be 370 days, just as Western's equation had. Although Western's and Clauss et al.'s equations give the same gestation length for a 750-kg giraffe, the scaling exponents are very different, and why they are is not obvious. Just about the only conclusion that can be reached is that the length of gestation in large artiodactyl species is longer than it is in smaller ones. For large artiodactyls, like a giraffe, longer gestations are required, presumably because a longer time is needed for the fetus to mature, which is another way of saying that the rate at which a fetus matures is slower in large species, and more time is required by the fetus to reach a suitable birth weight. Thus it is not metabolic rate of the mother (or the fetus) that determines gestation length, but the speed of development of the fetus (Clauss et al., 2014). For giraffes this means that natural selection has come up with a combination of a slower growth rate of the fetus and a longer gestation to make sure that the fetus is mature enough to be born.

What, in turn, determines the speed of development of the fetus? Kihlström (1972) correlated gestation length with placenta type on the entirely logical grounds that the capacity for a placenta to transport materials from mother to fetus and waste products in the opposite direction is limited, and the time it takes the placenta to reach maximum transport capacity will determine gestation length. Within artiodactyls that have a syndesmochorial placenta like giraffes, Kihlström found that gestation length (days) is equal to $97.7 \times Mb^{0.17}$. If so, in a 750-kg giraffe, gestation would be just 301 days. Kihlström was wrong on the mechanism, but right in identifying the placenta as a key organ. However, because the events that lead up to birth are started late in pregnancy, Sacher and Staffeldt (1974) tested whether differences between gestation lengths among mammals were related not to the placenta, but to the rate at which body cells divide and mature. They found that all fetal tissues grow at the same (increasing) rate, except for neural tissue and in particular the brain. Neural tissue does grow slowly: a giraffe's brain at birth weighs 470 g (Mitchell et al., 2013a) and thus grows at about 1 g d^{-1} during fetal life, compared with 200 g d^{-1} for the duration of pregnancy for all the other tissues that contribute to total body mass. Sacher and Staffeldt concluded from their study that brain mass would predict the length of gestation. From a data set of 16 animals with brain masses ranging from 3 to 48 g, they derived an allometric equation: Gestation Length (days) = $28.3 \times M_{brain}\ (g)^{0.20}$. The highest of these brain masses is, of course, more than 10 times smaller than it is in a giraffe at birth, but if this equation is used to predict giraffe gestation length, it would be only 90 days. What Sacher and Staffeldt's idea did do, though, was to highlight a relationship between the rate at which the brain grows, and hence its maturation, and gestation in small mammals. If so, how is the brain involved in the timing of birth?

The first insight into how the fetal brain might be involved was that of John R. Challis (1971), who found that there is a rapid, 10-fold increase in estrogen concentration in the blood of a ewe on the day before parturition. At this same time of gestation, the concentration in blood of the progesterone that had been secreted throughout pregnancy was declining, a combination of hormone changes that is identical in every way to what happens in a female giraffe the days before birth (Loskutoff et al., 1986; Isobe et al., 2007) and is a pattern now known to be common in ruminants (Wiltbank et al., 2016). The consequences of estrogen secretion for parturition are well known—synthesis of oxytocin receptors and increased contractility of the uterus smooth muscle. The effect of estrogen on the corpus luteum of pregnancy in giraffes is clear too. Estrogen stimulates the release of prostaglandins, which destroy the corpus luteum, causing a rapid decline, three days before birth in the case of giraffes, of the progesterone that has been inhibiting oxytocin receptor synthesis and thus preventing birth (Loskutoff et al., 1986). It is one of nature's miracles that all this works almost faultlessly every pregnancy.

The origin of the estrogen that is essential for starting the birth sequence is the one surmised by Sacher and Staffeldt in 1974: maturation of the fetal brain. In sheep, and presumably something very similar in giraffes, from about midway through pregnancy, the fetal brain and its hypothalamus in particular, release corticotrophin releasing hormone (CRH). Why it does so then is a mystery, but the mechanism may be something like the slow maturation processes in the hypothalamus and pituitary gland that lead to the secretion of hormones at the time of onset of puberty many years later. CRH has the function of releasing adrenocorticotrophin hormone (ACTH) from the pituitary gland. ACTH in turn stimulates the secretion of glucocorticoid from the adrenal gland. CRH also stimulates the placenta to synthesize prostaglandin ($F_{2\alpha}$), which activates the pathways that convert progesterone into the precursors of estrogen. Consequently, the placental secretion of progesterone falls and that of estrogen increases (Challis, 1998; Challis et al., 2000), and the mix of hormone concentrations that determine the onset of birth is established. Synthesis of estrogen from its precursors requires three enzymes—3, beta-hydroxysteroid dehydrogenase and 17, 20 lyase—that stimulate and regulate testosterone synthesis from progesterone, and the third essential enzyme—aromatase—which converts testosterone into estrogen. These enzymes are not present in the placenta or adrenal tissue of giraffes up to 350 days before birth (Wilsher et al., 2014), but they are present at 400 days (of a 417-day gestation), the time when giraffes are synthesizing and secreting estrogen (Isobe et al., 2007). The timing of birth in giraffes, in other words, ultimately is determined by hormones, and their secretion in turn depends on the maturation of the fetal brain. Until the brain is ready, birth is delayed.

The Fetus

What is happening to the giraffe fetus as it spends 450 days in its mother's uterus? There has, for obvious reasons, been no systematic study of giraffe fetuses. All our knowledge of fetus anatomy and physiology has come from opportunistic measurements made in fetuses recovered from giraffe cows that have died or been culled. Nevertheless, a considerable amount of information has been collected about them. A good example is our knowledge, mentioned earlier, of fetal reproductive anatomy and the hormones that fetuses secrete. Other opportunistic studies have revealed how the shape of a fetus's body and its dimensions change throughout gestation, how the skin and its markings develop, and how the skeletons, cardiovascular, and respiratory systems do. Almost nothing is known, however, of other organs and tissues. Most of what we know has come from the study of 44 fetuses of both genders, although the gender has not been recorded in every case (Ackermann, 1976; Skinner and Hall-Martin, 1975; Mitchell et al., 2009, 2011; van Sittert et al., 2010, 2015). These 44 fetuses had body masses that ranged from 4 g (35 days old) to around 100 kg at birth. The fact that at day 35 a giraffe fetus weighs 4 g but 415 days later it weighs 100 kg is an impressive reminder of what a fetus does in a relatively very short time to become a competent newborn giraffe, able to stand and suckle within a few minutes and able to run faster than its mother can within a day or two.

Fetal Growth Patterns

Of the many possible pieces of information that could be gathered from a giraffe fetus, perhaps the most important is its age. The age of a fetus is strongly related to its body mass, and from analysis of 23 fetuses made by John Skinner and Anthony Hall-Martin in 1975, the gestational age of a giraffe fetus fits the regression equation (body mass in kg): Age (days) = $130 \times Mb^{0.26}$ (Figure 17.7). Kayanja and Blankenship (1973) also recorded the body masses of fetuses from which their gestational age (days) can be calculated to be = $130 \times Mb^{0.30}$. A combined equation derived from all fetuses is: Age (days) = $128\times Mb^{0.26}$ ($R^2 = 0.99$). In the absence of a body mass, fairly accurate estimates of gestational age can be made from linear dimensions. For example, in 1906 Frank Beddard, who was the Prosector of the Zoological Society in London, described the

Table 17.2 Body Mass and Age of Fetuses Predicted from Measured Dimensions

Dimension	*Beddard*	Predicted Body Mass (kg)	Predicted Days of Gestation	*Keen and Goetz*	Predicted Body Mass (kg)	Predicted Days of Gestation
Head length (cm)	23	15	292	18	3.5	189
Neck length (cm)	38	15	292	30	7.6	238
Back length (cm)	27	15	292	20	5.7	219
Neck:back ratio	1.41			1.5		
Crown-rump length (cm)	72	21	323			
FLW (cm)	85	18	309			
HLR (cm)	78	16	298			
Height, length	123	17	303	91	6.5	227
Neck:FLW	0.45					
Mean±sd		*17±2*	*303±12*		*6.6±1.0*	*229±13*

FLW = foreleg length from hoof to withers; HLR = hind leg length from hoof to rump.
Data from Beddard, 1906, and Keen and Goetz, 1957.

dimensions of a female giraffe fetus, which he thought was about eight months old. He did not weigh it. It had not yet developed characteristic coat markings, but growth of hairs on the head was complete and hair growth had begun elsewhere. Based on its dimensions (Table 17.2), its body mass was 17±2 kg and thus its gestational age was 303±12 days, or 10 months, and not the eight months he thought.

A similar extrapolation can be made for the fetus dissected by Keen and Goetz in 1957. It also was not weighed, and they estimated its gestational age by comparing its dimensions with those of Beddard's fetus. Keen and Goetz concluded that their fetus was about seven months old. The most reliable measurement Keen and Goetz made was from muzzle to the tip of the tail (not root of the tail, as they stated). It was 91 cm, but other dimensions can be derived from the photograph that accompanied their descriptions. From those dimensions, the body mass of their fetus was 6.6±1.0 kg and its gestational age ~230 days, so close to the seven months of age they estimated. This estimate of that fetus's gestational age is not a trivial derivation, at least not in the context of giraffe physiology. The body mass of the female from which Keen and Goetz's fetus was recovered has been recorded as ~500 kg. A body mass of 500 kg in a female giraffe correlates with an age of 3.4 years, which means that the female would have conceived the fetus at age 2.7 years, and reached puberty some months before, a combination of milestones that would be exceptional in a wild giraffe. If the calf was the female's first calf, her age was far more likely to have been ~5 years and her body mass 700 kg or even more Goetz and Budtz-Olsen, 1955). All the calculations and conclusions about how the cardiovascular and respiratory systems of giraffes work that were made by Goetz and colleagues based on a body mass of 500 kg were, therefore, incorrect. Similarly, Wilsher et al. (2013) based their analysis of placenta functions on fetus age. The body mass of their fetuses was not measured, but one of the two fetuses had a crown–rump length of 60 cm and the other 91 cm, from which their gestational ages can be estimated to have been 8.5 months (the age Wilsher et al. used) and 11.7 months (and not the 13.5 months used).

The most anomalous measurements of fetus dimensions were, surprisingly, reported by the meticulous Richard Owen in 1839. He measured the dimensions of the first giraffe calf born in Europe a few minutes after its birth at the end of a normal gestation (Figure 3.3). He did not weigh it. He recorded a back length of 112 cm and neck length of 74 cm. This back length would be found in giraffes that are four to five years old and is clearly wrong. Beddard (1906) wrote,

> *According to my own measurement the back is shorter than the neck. I feel convinced that, though I may have erred in failing to arrive at an extreme accuracy of measurement, so great an error cannot have crept in.*

Neck length as recorded by Owen was plausible, as were the lengths of the legs (140–150 cm), and from the combination of leg length and neck length, the height of that newborn giraffe can be deduced to be 224 cm. Height at birth in both wild (in South Africa) and captive giraffes (N=14) is 186±18 cm (range 170–224; Skinner and Hall-Martin, 1975) and in the Serengeti, Pellew (1983b) found mean heights at birth to be 1.90±.07 m and 1.80±0.08 m in males and females, respectively. For reasons that are completely unclear, none of the other measurements reported by Owen is consistent with the many others that have been made.

Once fetal age or body mass are known, patterns of growth can be analyzed. Of the measurements that can be made to discover the way a fetus grows, the most useful are body mass, the lengths of the neck, back, fore- and hindlegs, height, body length, and girth. Height is the distance from the sole of the front foot to the occipital crest, body length from the tip of the muzzle to the tip of the last caudal (tail) vertebra excluding tail hair, and girth is the circumference of the thorax immediately behind the forelegs. The most complete and probably unrepeatable study of how the dimensions change in early gestation was done by Errol Ackermann. His measurements were made on seven fetuses aged between 48 days (~10% of gestation) to 270 days (~60% of gestation) and are shown in Table 17.3 (Ackerman, 1976). How dimensions change during the course of a complete gestation is illustrated in Figure 17.7. The rate at which the linear dimensions of height and girth and so on increase is not directly proportional to gestation age. In all of them, as pregnancy progresses the rate at which body dimensions changes becomes faster. The relationships between the linear dimensions stay the same, however. For example, throughout pregnancy the neck is always ~1.3 times longer than the back. Another significant ratio is that between neck length and leg length. Throughout pregnancy the neck is always shorter than the legs and becomes progressively so, so that at birth a newborn giraffe has very long legs and a neck that is less than half leg length (Figure 17.7), relationships that can be seen quite clearly in the newborn giraffe drawn by Robert Hills (Figure 3.3 in Chapter 3). Body mass, on the other hand, increases rapidly as the fetus nears term (Figure 17.7). At day 35, when the fetus can be detected for the first time by sonography, it weighs a few grams, at 200 days it weighs about 3 kg, but thereafter its mass increases rapidly to its endpoint of 92.8±10.9 kg in wild giraffes

Table 17.3 Giraffe Fetal Dimensions

Variable	Days of Gestation						
	48	87	131	148	192	226	270
Body mass (kg)	0.00873	0.106	0.935	1.165	4.672	8.695	16.78
Nose to tail (cm)	10.1	24.3	49.3	58.2	94	107	127
Head length (cm)	2.6	7.9	14.8	17.3	22	25.5	27.5
Neck length (cm)	2.7	6.5	13.4	16.7	24.5	28.5	37
Tail length (cm)	0.95	3.8	8	9.4	17.5	20	22.5
Girth (cm)	4.5	11.2	23.9	28.3	42.5	49.5	66
Circumference neck base (cm)	2.25	6.2	13.8	26.1	21	27	32
Circumference neck apex (cm)	1.7	4.2	6.7	9.4	16	20	23
Withers to elbow (cm)	1.85	4.8	10.7	13	17	22	27
Elbow to carpus (cm)	0.95	3.4	7.8	9.7	17.5	18.5	24
Carpus to hoof (cm)	0.95	4.1	9.9	13.7	24	26	33
Rump to stifle (cm)	1.2	4	7.8	10.9	14.5	21	24
Stifle to tarsus (cm)	0.8	3.3	8.5	9.4	15	18	20
Tarsus to hoof (cm)	0.85	4.4	10.8	14.2	25.5	29.5	36

As measured by Ackermann (1976), with permission of the University of Pretoria.

(usually rounded to 100 kg). At birth it is 12.9% of its mother's body mass, a proportion that is greater than it is in hippos (3%), rhinoceroses (3.7%), elephants (4.8%), eland (6.6%), and Cape buffalo (9.2%)(Hall-Martin, 1975a). The mass of neonates born to captive giraffes is much lower than it is in wild giraffes (~54 kg; Skinner and Hall-Martin, 1975), for reasons that are unclear.

Fetus Skeleton

The shape of a fetal giraffe depends on the way its skeleton grows. The only measurements made of the dimensions of fetal bones have been made in 10 fetuses by van Sittert et al. (2010, 2015). From those data, allometric equations can be developed, and if those equations are applied to all 44 fetuses based on their gestational age, their growth patterns can be shown. A long neck and short back, for example, are the result of the different rates at which the cervical, thoracic, and lumbar vertebrae grow. Differences in growth rate reflect either that the intercepts (the multiplier) or slopes (the exponent), or both, of the allometric equation describing their growth are different. Which? Figure 17.8 shows that the exponents are the same for all three vertebra types (~1.2)but the intercepts are different, with the intercept of the equation for cervical vertebrae being some three

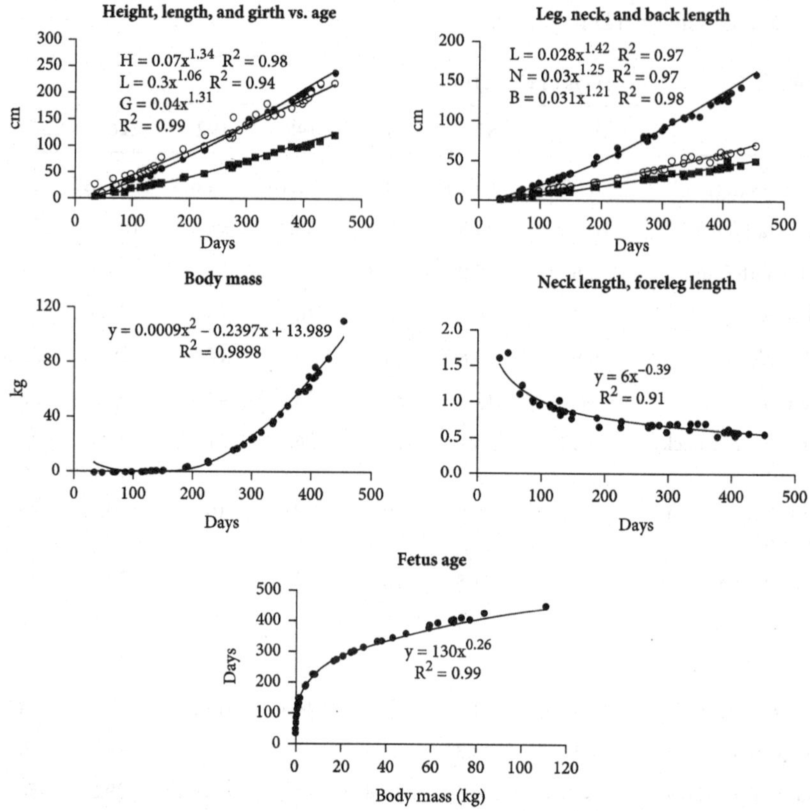

Figure 17.7. Changes in giraffe fetus dimensions during gestation. Top left. ○ = length, ■ = girth, ● = height. Top right. ○ = neck length, ■ = back length, ● = leg length.
Data from 44 fetuses: 7 in Ackermann, 1976, 23 in Skinner and Hall-Martin, 1975, and 14 from Mitchell et al., unpublished.

times larger than that of the thoracic and lumbar vertebrae (0.06 mm vs. 0.02 mm). The effect of having a larger intercept for a vertebra is not detectable in very early gestation, but its effect "multiplies" as gestation proceeds. Similarly, in a newborn giraffe an ability to run as fast as its mother, if not faster, is a useful attribute, and growing long legs is an efficient way to achieve high speed: the longer the legs, the greater the mechanical advantage. At the end of gestation, when body mass is ~100 kg, the length of the foreleg from hoof to axilla is a little more than 110 cm ($17.5 \times Mb^{0.41}$); hindleg length from hoof to groin is much the same ($18.6 \times Mb^{0.39}$). Neck length is about 70 cm ($17.1 \times Mb^{0.30}$). The origin of the long legs of a newborn giraffe lies in the elongation of the metacarpus and metatarsal bones (Figure 17.8). They are far longer than the other leg bones, constituting over 40% of leg length.

Of course fetal limb bones also have to be strong enough to resist fracture. Their strength can be estimated from measurements made by van Sittert et al. in 2015 of the outer and inner diameters of the limb bones of 10 fetuses: diameter is the main determinant of the strength of a limb bone (see Chapter 15). The outer diameter of limb bones in a fetus increases faster than the inner diameter does because the marrow cavity develops slowly in a fetus. As a result, at birth the limb bones of a fetus are more solid

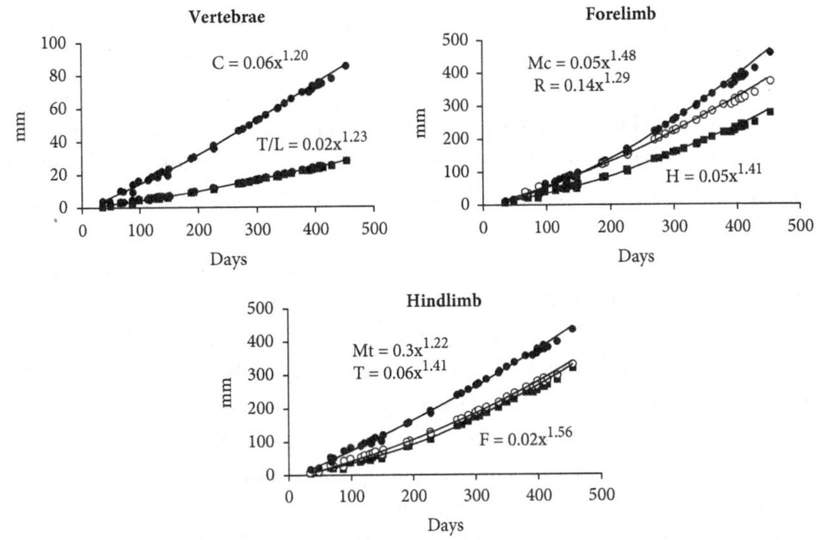

Figure 17.8. Growth patterns of limb bones in the fore- and hindlimbs and the vertebrae in giraffe fetuses ranging in age from 48 days to 454 days. Top left: C (●) = cervical, T/L (■) = thoracic and lumbar. Top right: Mc (●) = metacarpus, R (○) = radius, H (■) = humerus. Bottom: Mt (●) = metatarsus, T (○) =tibia, F (●) = femur.
Based on data in Ackermann, 1976, and van Sittert et al., 2010, 2015.

than are its mother's limb bones. Their absolute strength (S; cm^2), which depends on the relationship between diameter and length (see Table 15.5 in Chapter 15), increases linearly throughout gestation (S (cm^2) = 0.005x +0.012; R^2= 0.98; where x = age in days) but at birth, it is less than a half of the absolute strength of the bones of an adult female weighing 750 kg. Absolute strength is, however, less important than relative strength ($cm^2\ g^{-1}$ of body mass), and the relative strength of a newborn giraffe's limb bones is twice what it is in an adult giraffe (see Chapter 15) and the bones are, consequently, resistant to fracture.

Fetus Cardiovascular System

So far there has been just one study (Keen and Goetz, 1957) of the anatomy of the circulation in a giraffe fetus. It was 7–7.5 months old. Keen and Goetz found that the placenta was supplied by two thick-walled umbilical arteries and blood was carried away by two thin-walled umbilical veins. Typically in a mammalian fetus, most of the blood returning to its heart via the umbilical veins enters the fetus's inferior vena cava through the *ductus venosus*. However, in the giraffe fetus Keen and Goetz dissected, the *ductus* had closed and umbilical venous blood reached the inferior vena cava via two large hepatic veins. Blood returning to the heart from the head, neck, and upper body was carried in two jugular veins, one on each side of the neck, and two brachial veins, one draining each foreleg. These four vessels combined to form the superior vena cava. After birth, the superior and inferior venae cavae discharge their blood into the right atrium, then to the right ventricle from where it is pumped to the lungs in the pulmonary artery.

Oxygenated blood returns to the heart via pulmonary veins. It enters the left atrium first, and then the left ventricle. In a fetus the blood vessels in the nonfunctional, fluid-filled lungs offer a high resistance to blood flow, thus pulmonary blood flow is low, and if this were the only blood reaching the left ventricle from the lungs, the left ventricle would have little work to do. However, in fetuses the wall separating the left and right atria is perforated by the oval window (*foramen ovale*), allowing blood in both left and right atria to mix and to flow into both ventricles. The pulmonary artery leaving the right ventricle is connected to the aorta via the *ductus arteriosus* and in that way its blood is diverted away from the lungs. The *ductus arteriosus* in giraffe fetuses is a large and thick-walled vessel (Keen and Goetz, 1957). At birth both the *ductus arteriosus* and the *foramen ovale* close to establish the typical blood flow patterns of an adult mammal.

As the fetal organs grow, their need for blood and all the nutrients it carries increases. Fetal dimensions change rapidly in the last third of gestation, a time when the size of the placenta is hardly changing. For example, in cattle a doubling of fetus weight was associated with just a 20% increase in placenta weight (Reeves et al., 1972). Is that true for giraffes? There has been no systematic measurement of placenta mass in giraffes, but Deka et al. (1980) reported a placenta of 2.6 kg from a giraffe that was 5.3 years old after she had delivered a normal calf after 433 days of gestation. Five placentas delivered by giraffes after normal parturition weighed a very consistent 4.3±0.2 kg compared with 1.0±0.1 kg in four hartebeest and six wildebeest (Hradecky, 1983b). The body mass of the giraffes in Hradecky's study was not recorded, but presumably it would have been ~ 750 kg. The body mass of the wildebeest and hartebeest was ~180 kg (Hradecky, 1983b). Thus placentas seem to weigh about 0.6% of material body mass, which does not seem to be enough to supply nutrients to a growing giraffe fetus. Does the mass of the uterus and placenta together change? In six giraffes that had fetuses ranging from 26 to 70 kg, the mass of their uteruses and membranes was, on average, 27 kg (range 22–36 kg; Mitchell et al., unpublished), so the two combined also did not increase in size dramatically as gestation proceeded. So how does the fetus ensure that its organs and tissues get the oxygen and other nutrients they require for rapid growth if the placenta is not growing much? There are only two ways that it can. One way is by redistributing cardiac output so that an increasing proportion of the blood leaving the heart goes to the placenta. However, if there was redistribution, then blood flow to all other organs would be decreased and they would be deprived of blood, so redistribution alone is unlikely and, in fact, throughout pregnancy in other ruminants the placenta gets a fairly constant blood supply of between 40% and 50% of fetal cardiac output (e.g., Rudolph et al., 1970). The other way to increase placenta blood flow is to increase cardiac output so that flow to the placenta increases as the fetus grows, even if the proportion of cardiac output it receives does not change.

A larger cardiac output requires a larger stroke volume and/or a higher heart rate. In all mammalian fetuses thus far studied, heart rate is higher than it is after birth. A larger stroke volume requires a larger heart, and the heart does increase in size as a giraffe fetus grows. In fetuses ranging in body mass from 18 to 77 kg, heart mass increased rapidly as the end of gestation approached and thus followed the pattern of increases in fetal body mass (Figures 17.7, 17.9). In an 18-kg fetus, which would have had a gestational age of 275 days, the heart weighed just 135 g. In a 77-kg fetus it weighed about 500 g, but it is changes in the heart's relative mass rather than absolute mass that reveal how it provides the placenta with sufficient blood. As a proportion of its body mass, the fetal heart of giraffes increases in very early life from 0.55% of body mass to 0.7% in an 18-kg fetus to 0.75% by term. In a mature adult giraffe, heart mass is 0.41% of body mass, so proportional to body mass, the heart of a fetal giraffe can be nearly twice as big as the heart of an adult giraffe. The increase in mass of the fetal heart is associated with an increase in the length of the heart from base to apex more than an increase in its diameter (Figure 17.9). At the time of implantation (35 days) the heart is probably ~4 mm long and ~2 mm in diameter. After 100 days of gestation it is 2 cm long and 1.5 cm wide. By birth it is 13 cm long and has a diameter of 10 cm—the same size as the heart of an adult human.

How high is a fetal giraffe's blood pressure? Giraffe fetuses have a relatively short neck and grow in a gravity-free environment, thus both of the main factors that cause a giraffe's blood pressure to rise after birth are absent, and a reasonable conclusion is, then, that fetal blood pressures will be low—a conclusion supported by the thin walls of fetal blood vessels (see Chapter 9). The actual blood pressure of giraffe fetuses and how it changes as they grow has, understandably, never been measured, but fetal blood pressure has been measured in horses (Giussani et al., 2005), calves (Reeves et al., 1972), llamas (Llanos et al., 1998), and sheep (Faber et al., 1974). There is no reason to think that a giraffe fetus is different, and when those data are applied to giraffes, what can be predicted is that by 150 days of gestation when the fetus weighs about 100 g, its heart is generating a blood pressure of 10 mmHg, by day 300 blood pressure is 40–50 mmHg, and at birth it is ~90 mmHg. The origin of systemic arterial pressure in all postnatal mammals is the left ventricle. Their right ventricle pumps blood through the lungs, which does not require much effort, and right ventricle thickness is about one-third that of the left ventricle. However, when the thicknesses of the muscles of the ventricles were measured in 10 giraffe fetuses weighing between 18 kg and 77 kg (Mitchell et al., 2009), the thicknesses showed that it was the right ventricle that carried most of the burden of pumping blood around the body at the start of fetal life. Snelling et al. (2015) used ventricle mass rather than wall thickness of the hearts of fetal kangaroos to reach the same conclusion. A right ventricle wall bigger than the left ventricle wall in early fetal life is consistent with the way the heart develops embryologically (e.g., Moorman et al., 2003) and with the way blood enters the fetal heart on its return from the placenta and organs, and with the way it enters the systemic circulation. However, the left ventricle starts to take over the task of generating pressure and circulating blood to the body quite

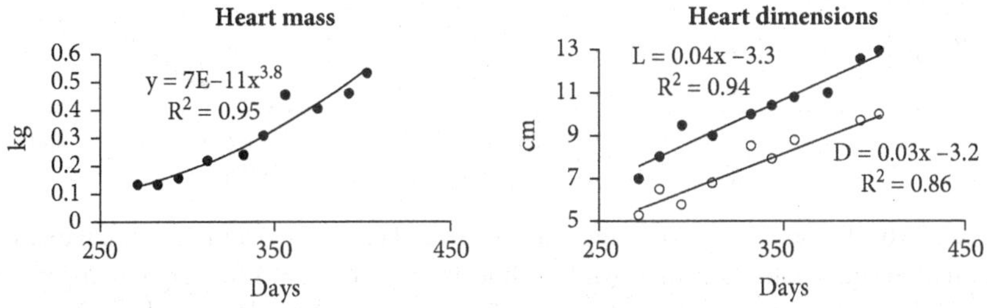

Figure 17.9. Changes in the mass and shape of the heart in 10 giraffe fetuses as gestation progresses.

quickly. At birth the left ventricle wall is about twice as thick as the right ventricle wall, and it becomes three times thicker than the right ventricle wall in an adult.

Fetus Respiration and Metabolism

Metabolic rate has not been measured in fetal giraffes. Many measurements of metabolic rate in the fetuses of sheep, cows, and horses (reviewed by Battaglia and Meschia, 1978) have been made, but mostly in fetal sheep (Clapp et al., 1971). In those species metabolic rate measured as oxygen consumption was constant at around 8 mL O_2 min^{-1} kg^{-1} body mass throughout pregnancy. There is no reason to think that the giraffe fetuses are different, and if this is true, then fetal metabolic rate will increase as body mass increases (Figure 17.10), will be about four times the basal metabolic rate of an adult female giraffe (2 mL O_2 min^{-1} kg^{-1}; see Chapter 8), and daily oxygen consumption will increase from 200 L d^{-1} in a fetus weighing 18 kg to (if the extrapolation is accurate) 1,200 L d^{-1} at term (an adult human at rest consumes about 400 L of oxygen a day). Most oxygen used by a fetus is used to oxidize the glucose provided by its mother and much, but not all, of the energy produced is used for growth, as fetuses do not have to spend energy to move against or resist gravity or to spend very much trying to keep their body temperature warm. Combustion of a liter of oxygen produces 20 kJ of energy and thus, over the course of gestation, daily energy requirements increase from perhaps 1 kJ d^{-1} (equivalent to 0.5 g of glucose daily) to 25,000 kj d^{-1}, needing 1.25 kg of glucose daily. This energy, and oxygen, glucose, and other nutrients, all have to be

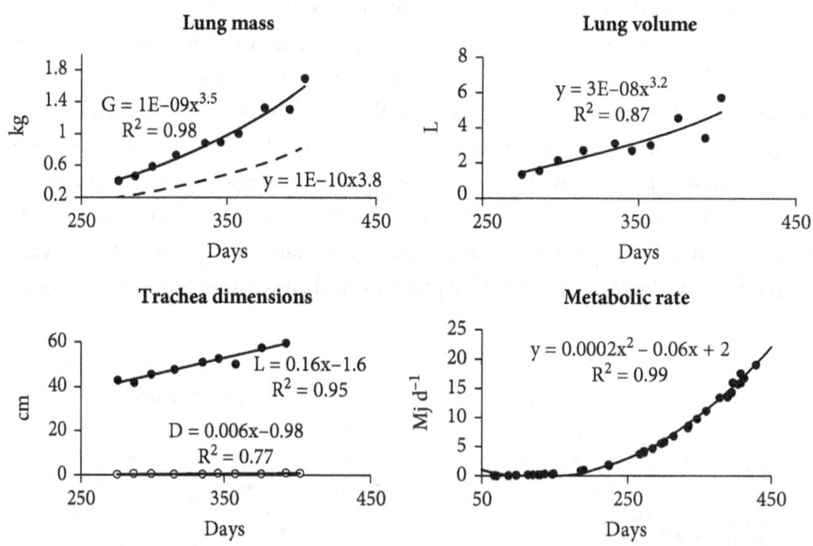

Figure 17.10. Top left: Measured lung mass in 10 fetal giraffes (•) and functional mass in mammals of equivalent body mass (dashed line; from Stahl, 1967). Top right: Lung volume calculated from lung mass after subtraction of fluid mass. Bottom left: Trachea length and diameter in 10 giraffe fetuses. Bottom right: Estimated metabolic rate of 44 giraffe fetuses assuming that the mass-specific metabolic rate was 8 mL O_2 min^{-1} kg^{-1}.

provided by the mother. Although fetuses do not have to allocate much energy to the task of keeping warm, fetal metabolism does generate heat. To lose that heat there must be a temperature gradient between the fetus and its mother. In those mammals in which fetal temperature has been measured (e.g., Laburn et al., 2002) it is about 0.5°C higher than its mother's body temperature, and that gradient is sufficient to establish the balance between heat generation and heat loss that keeps fetal body temperature constant. Most (85%) of the heat lost is removed by convection into the maternal circulation via the placenta. The remainder is lost by conduction through fetal fluids to maternal body fluids, and both that heat and the heat transferred via the placenta are lost via maternal heat loss mechanisms (Schröder and Power, 1997).

The source of oxygen for a giraffe fetus is, of course, the placenta, but the source immediately after birth has to be the lungs. Thus the sizes of the lungs and airways have to increase during gestation to reach the sizes needed after birth, and they do so in giraffes (Figure 17.10). Trachea length and diameter have been measured in 10 giraffe fetuses with gestational ages of 275–405 days. Trachea length increased linearly with days of gestation to match elongation of the neck. Trachea diameter in those same fetuses ranged from 5 mm at 275 days to 1.5 cm at birth. At birth, the trachea immediately inflates and its diameter doubles to 2–3 cm (Mitchell and Skinner, 2011). Figure 17.10 also shows the measured mass of the lungs of a giraffe fetus and, from equations developed by Stahl (1967), the mass of functional, that is, fully inflated, lungs in mammals of equivalent body mass to the body mass of giraffe fetuses. The difference in mass between giraffe fetal lungs (higher) and the lungs of a normally breathing mammal (lower) is mostly due to the fluid that is present in fetal lungs. It amounts to 0.1 mL in giraffes at the start of gestation, but it is as much as 1.1 L at birth, when fluid makes up more than half of fetal lung mass (57% ±7%). All this fluid is absorbed from the lungs into blood within a few minutes after birth, when spontaneous breathing starts. The volume of a giraffe's lung at birth, calculated from lung mass after removal of fluid, is 6 L, which is the same size as the lungs of an adult human.

Birth

Many, many observations have been made of birth in giraffes, almost all of them captive (Dagg and Foster, 1976). All tell the same story. The first observation of a giraffe birth was made, of course, by Owen in 1839. No one at that time had any idea of how long gestation was in a giraffe, but

> *it may be confidently expected that the period of gestation and the condition of the new-born animal—facts which are of essential importance in the natural history of every species—may soon be recorded in the Proceedings of the Zoological Society.* (Owen, 1839)

With all the anxiety of an expectant father, he wrote,

> *Many months elapsed before any change in the form of the abdomen could be perceived, but at length the expansion of the lumbar region became manifest. Notwithstanding*

> *these unequivocal symptoms of pregnancy, when more than a year had elapsed without a symptom of parturition, many began to doubt the fact.* (Owen, 1839)

Eventually, after 444 days, parturition commenced at noon on June 19, 1839. The mother stood and the forelegs of the calf appeared first as far as the carpal joints, then the muzzle appeared, and birth followed quickly with the mother standing. It took a minute before the calf took its first breath, and then once it had started, respiratory rate was 43 breaths min^{-1}. In half an hour the calf tried to rise and managed to do so within an hour. It was a male. Its markings were clear, hooves soft, and testes descended. By 10 hours it was walking and by June 21, two days after its birth, it was running. Sadly in this case, the mother rejected the calf and it did not survive. Davis (1949) described another birth in a captive giraffe in the National Zoo in Washington, DC. The first stages of birth began at 09:00. By 10:00 the fetus's forelegs were exposed to the extent of the carpus, followed by the head and part of the humeri. At this point labor appeared to stop, and keepers intervened by pulling on the fetus. At 10:40 the entire fetus was expelled, dropping a distance of 1.7 m. The umbilical cord was wrapped around the body and was about 1 m long. By 12:35 the neonate had found the teats and was drinking. The placenta was expelled at 14:00, 5 hours after birth began.

Far fewer observations have been made and recorded of births in wild giraffes. In one observed birth in a wild giraffe, nine female giraffes formed a circle around a female that had just given birth (Anon, 1965). After 25 minutes the calf was mobile. All the midwives then nuzzled the calf. Three male giraffes did not approach nearer than 50 yards (45 m). The sequence of events described by Davis in 1949 was almost identical to those made by Rudi van Aarde, a long-time observer of free-living mammals and now Director of the Conservation Ecology and Research Unit at the University of Pretoria in South Africa. He has provided the most complete record of a birth in a wild giraffe so far. In 1976 in the Jack Scott Nature Reserve in South Africa at 10h20 one morning, van Aarde observed a giraffe cow alone. At 10h25 intermittent contractions of the abdomen could be seen and the forelegs of the calf protruded. By 10h38 the front legs, head, and neck were emerging, and at 10h50, in a renewed effort, the cow pushed the calf's shoulders clear. Within a few seconds the calf landed "with a thud" on the ground. Birth had taken just 35 minutes. Van Aarde did not say so, but during the fall the umbilical cord will have broken as the distance the calf falls (~1.8 m) is greater than the length of the cord (~1 m). The remnant of the cord remains attached to the calf for 6–8 weeks after birth (Leuthold and Leuthold, 1975). Immediately after the birth, van Aarde wrote, the cow turned, bent down, and licked the calf, but at intervals stood upright and looked around. 11h30: the calf raised its head and struggled to a sitting position. 11h52: the calf stood leaning against its mother's legs for support, and 20 minutes later attempted to suckle. 12h25: both mother and calf walked slowly toward some nearby Acacia trees and the cow started browsing. The calf gazed around. Although van Aarde watched the pair for another three hours, the placenta was not expelled. The placenta is attractive to scavengers, so delay of its expulsion while the calf is still immobile has evolutionary value. When is it expelled? Davis (1949) reported that the placenta was expelled after five hours, Lightfoot (1957) reporting a birth in a wild giraffe in Zimbabwe also recorded that the placenta was expelled after five hours, as did Owen (1939), while Dagg and Foster (1976) have said that expulsion may not occur for up to nine hours after birth. The placenta is not eaten by the mother, as it often is in ungulates. The birth recorded by van Aarde occurred in the morning, which, as van Aarde noted, was the best time, and indeed most births recorded even in captive giraffes occur in the morning, with the evolutionary benefit that the calf is mobile by nightfall when predators become more active.

Lactation

Providing enough milk for her newborn calf is the most formidable physiological task facing the cow after the birth. The calf will grow at its fastest rate in its first six months of life (see Figure 5.4 in Chapter 5). The time of peak conceptions is when conditions are optimum for the cow, but calves often are born in the dry season, a consequence that can only be mitigated by a long lactation and provision by the mother of good quality and quantity of milk. In the wild, calves suckle up to 10 months of age, and milk has been found in the stomach of a one-year-old (Hall-Martin et al., 1977). Typically, pregnant females lactate for the first 150 days of a subsequent gestation, which, when added to the time between birth and re-conception (~100 days), means that lactation lasts at least 8 months but can last as long as 12–13 months (Hall-Martin, 1975a). Thus, because the time to re-conception (~3 months) is shorter than the period of lactation, most female giraffes will conceive while lactating, which is unusual in mammals, but despite some reports to the contrary (e.g., Deacon et al., 2015), is normal for giraffes and is yet another way that giraffes optimize the frequency of pregnancy. If conception occurred only after lactation has ended, the interval between calves would be prolonged and the number of calves a female could have would be reduced. A calf is not, of course, solely dependent on milk throughout the period of lactation. Calves start nibbling browse in the first few days of life (Pratt and Anderson, 1979). They can get no nutritional benefit from doing so but learn the habit early. Browsing, not grazing, is innate in giraffes. The calves nibble with greater purpose by three weeks of age (Owen, 1839; Innes, 1958; Reason, 1995), are drinking water by two and a half months, and are able to digest vegetation by four months (Owen, 1839; Pratt and Anderson, 1979). Newborn giraffes quickly develop the necessary teeth to do so. Owen studied the early development and succession of teeth in the calf born in 1839. The four middle deciduous incisors began to cut the gum one week after birth and had emerged completely by four weeks of age. By then the crowns of the first and second deciduous premolars were visible. At two months the third incisor had cut the gum, and at three months the third premolar was in place, as was the first permanent molar (Owen, 1840).

Usually, no calves older than six months suckle successfully, and calves are fully established ruminants at six to eight months of age (Langman, 1977). Mothers do not permit calves other than their own to nurse (Pratt and Anderson, 1979). In the many, many hours of observations made of cow–calf interactions in the wild, only six cases of "cross-sucking," by which is meant that a calf drinks from a giraffe that is not its mother, have been recorded (Pratt and Anderson, 1979; Saito and Idani, 2018). In all of them, once discovered, they were immediately terminated, so "cross-sucking" is not a part of normal giraffe behavior. On the other hand, cross-sucking is frequently seen in captive giraffes (Gloneková et al., 2016), and compared with wild giraffe behavior, it is abnormal but presumably it is an instinctive, adaptive, behavioral response that enables the survival of a calf in the unnatural environment of captivity. The decision to suckle is determined by the mother more than the calf, although

> *occasionally a calf would dispense with the preliminaries: catching its browsing mother unawares, it would approach quickly and dart in for a brief unauthorized suck—an invariably unproductive tactic.* (Pratt and Anderson, 1979)

This "sneaky-sucker" tactic increases with age, but peaks when the calf is three to four months old, is becoming a ruminant, and is being weaned by its mother. Thirsty calves can be demanding by pushing on their mother's rump with their chest or using their head and neck to hit her rump, behaviors often accompanied by a downward stamping of a foreleg (Langman, 1977), tactics not unfamiliar to human mothers. A calf's behavior is of course a consequence of their dependence on milk for much of their water and energy for at least four months of postnatal life. Their mother has to provide it. They obviously do. How?

Giraffes have two pairs of inguinal teats and thus their udder is divided into four quarters, as it is in cattle. Udder mass has been reported to range from 1.8 kg (in a pregnant, non-lactating, 10-year-old female) to 7.3 kg (in a 14-year-old lactating female; Hall-Martin et al., 1977). Its mass increases rapidly in the last 50 days of gestation (Hall-Martin, 1975a). How much milk is produced? The largest volume of milk in a single sample obtained by Hall-Martin et al. (1977) was 2.25 L, but most samples ranged between 100 mL and 800 mL. Aschaffenburg et al. (1962) collected 600 mL from a 16-year-old anesthetized female that had a calf of five months old, so that volume represented milk production on the 150th day of lactation. Hall-Martin et al. (1977) concluded that, conservatively, giraffes provide at least 500–2,000 mL of milk per suckling bout and because calves could suckle every five hours, 24-hour milk yield could range between 2.5 L and 10 L, if not more. Duration of suckling varies between a few seconds and four minutes (Langman, 1977; Pratt and Anderson, 1979). Frequency of suckling partly depends on age (Pratt and Anderson, 1979). In the first week of life, frequency is once per hour (12 d^{-1}), falling to 8 d^{-1} by the second week, then dropping to 4 d^{-1} and then 2–3 d^{-1} until weaning. The composition of milk in captive giraffes has been reported by Greed (1961), using an analysis done by Aschaffenburg et al. by Aschaffenburg et al. themselves in 1962 and also by Ben-Shaul (1962). Milk composition of wild giraffes has been analyzed by Hall-Martin et al. (1977) and, more recently, by Osthoff et al. (2017). The results of these analyses, and for comparison the composition of cow's milk and human milk, are shown in Table 17.4. Differences between wild and captive giraffe data can be attributed to differences in diet. Milk is, of course, mostly water, but compared to cow and human milk, giraffe milk generally has higher concentrations of all the main constituents—protein, fat, lactose, and minerals. Its composition, at least in wild giraffes, does not change with season or with age, but as lactation proceeds, the milk becomes more dilute.

How much milk must a cow provide to meet the nutritional needs of her calf? The milk must provide, as a minimum, enough energy and is the only source of water for a calf for at least a month or more. The daily turnover rate of water (L d^{-1}) of newborn giraffes weighing 100–130 kg is about 5–6 L (Chapter 11). Milk is 70%–90% water, so

Table 17.4 Composition of Giraffe, Cow, and Human Milks

Constituent	Prelactation[a]	Early[b] <3 mth	Early[c]	Mid[e] 3 mth	Late[b] >3 mth	Late[c]	Late[d]	Cow	Human
Water ($mL\ L^{-1}$)	680	700	800	835	840	870	700	880	880
Fat ($g\ L^{-1}$)	167	151	125	79	72	47	125	37	38
Protein ($g\ L^{-1}$)	129	133	59	49	63	29	104	33	68
Lactose ($g\ L^{-1}$)	12	24	34	41	30	54	58	49	12
Minerals ($mg\ L^{-1}$)	800	1300	1000	—	900	700	602	700	117
Calcium ($mg\ L^{-1}$)	—	80	—	—	70	—	154	125	33
Phosphorus ($mg\ L^{-1}$)	—	99	—	—	102	—	104	100	15
Magnesium ($mg\ L^{-1}$)	—	6.0	—	—	5	—	8.0	11	3
Potassium ($mg\ L^{-1}$)	—	134	—	—	96	—	100	150	55
Sodium ($mg\ L^{-1}$)	—	31	—	—	40	—	100	44	15
Iron ($\mu g\ L^{-1}$)	—	8	—	—	137	—	2	31	76
Manganese ($\mu g\ L^{-1}$)	—	20	—	—	20	—	—	30	37
Copper ($\mu g\ L^{-1}$)	—	28	—	—	35	—	—	11	36
Zinc ($\mu g\ L^{-1}$)	—	570	—	—	580	—	—	390	295
B vitamins ($\mu g\ mL^{-1}$ of milk)	—	—	—	—	—	—	7	2.8	2.3
Vits A and E ($\mu g\ g^{-1}$ of fat)	—	—	—	—	—	—	9	3.4	5.3

Data from: [a]Hall-Martin 1975a; [b]Hall-Martin et al., 1977b; [c]Ben Shaul, 1962; [d]Aschaffenburg et al., 1962; [e]Osthoff et al., 2017.

if milk is the only source of water for the first four months of life, milk volume must be about 7 L daily to supply the water needed. The amount of energy needed can be estimated from the Field Metabolic Rate (FMR; kJ d^{-1}) of calves, and in a newborn giraffe it can be estimated to be ~25,500 kJ d^{-1}. Using the composition of early milk of wild giraffes (Table 17.4), 25,545 kJ can be provided by just 3 L of milk containing 453 g of fat, 399 g of protein, and 72 g of lactose. The quantity of milk that can be produced thus exceeds the minimum by a large margin, and that difference ensures robust growth. Approaching weaning at four months of age, a young giraffe has a body mass of about 130 kg and its FMR is ~32,000 kJ d^{-1}. If milk were its only source of food and if the milk had the composition found in late lactation, milk yield would need to be 4 L a day, containing 500 g of fat, 416 g of protein, and 232 g of lactose. A daily volume of 4 L is well within the maximum production of milk by a giraffe cow and, of course, much less would actually be needed because, by that age, browse would be supplying an increasing proportion of the calf's daily nutrients and the calf would be drinking surface water.

The energy (and water and other nutrients) required by the calf comes from the cow and has to be added to the cow's energy and water balance budgets. If the cow has a typical body mass of 750 kg, its FMR is ~115,000 kJ d^{-1} and it needs to eat a minimum of 14 kg of dry matter (DM) daily to supply that FMR, and it has a water intake of 25–30 L d^{-1} (Chapter 11). To these requirements the calf's early needs of about 25,000 kJ d^{-1} and 6 L of water must be added, as well as the energy cost of actually making milk. This latter cost is not known for giraffes, but for dairy cows it is about 3,000 kJ L^{-1} of milk produced. For the first few weeks of a calf's life, the giraffe cow would, therefore, need to increase her daily DM intake by about 4 kg d^{-1} to ~18 kg d^{-1}, change her sources of browse to improve its quality (Pellew, 1984a, 1984b), and avoid any tannin-rich or other browse that would make the milk unpalatable to the calf (Caister et al., 2003). Unless a high quantity of succulent browse is available, almost certainly the cow would have to supplement the water taken in through browse and generated by metabolism by drinking surface water. By the time the calf is four months old, its mother might have to be eating 19 kg DM d^{-1} if the calf were fully dependent on her, and by that time the cow is very likely to be pregnant again.

The additional energy burden of a fetus is negligible at first, but once the fetus is 200 days old, its metabolism requires 2,000 kJ d^{-1} and rises rapidly from then until birth when it is 25 500 kJ d^{-1}. Throughout the last 150 days of pregnancy, daily energy requirements of a fetus average 13 000 kJ d^{-1}, necessitating an increase in dry matter consumption by the mother of another 1.5 kilograms a day, unless her previous calf has been weaned. This amount is relatively easy to achieve if the calf is born in the growing season, but less easy if the calf is born outside the growing season, when many are. A reduction in DM intake can, theoretically, be achieved if a cow has been able to accumulate body fat. A kilogram of body fat provides about 40,000 kJ of energy. The amount of energy needed to feed a calf throughout its first four months of dependency would require the energy stored in about 95 kg of "dry" fat, or, because about 20% of stored fat is water, about 119 kg of body fat. However, Hall-Martin (1975a) has estimated that even when a giraffe is in peak condition, fat forms only 1.5% of body mass, or 11 kg in a 750-kg giraffe. If this estimate of body fat is correct, while she is lactating the cow's only options are to eat more (stomach-fill of lactating cows is higher than it is in non-lactating cows;

Hall-Martin, 1975a), and select for high-quality, high-energy browse, which is what they do (Pellew, 1984c).

Giraffe calves need micronutrients like iron and zinc just as much as protein, fat, lactose, and water, and to obtain the micronutrients newborn calves lick at soil, eat dead twigs, bark, and tree stumps, and bones, all of which provide experiential learning and a nutritional benefit. These activities cannot, though, supply the calcium and phosphate needed for the growth of the calf's skeleton, and of all the minerals present in milk it is calcium and phosphorus that are the most important for calf growth. They too have to come from the mother. The dry mass of a giraffe's skeleton at birth is about 8 kg, of which 2.5 kg is minerals, and of the minerals 1.6 kg is calcium and 900 g is phosphorus. Thus over 450 days of gestation the fetus requires on average about 3.5 g of calcium and 1.8 g of phosphorus daily, amounts that have to be supplied by the mother from what she eats or from her skeleton. The daily drain on her skeleton, if that is a main source of the minerals, could be greater as some calcium and phosphorus is lost in urine each day (Table 11.5 in Chapter 11). In the months immediately after birth, skeletal growth in a neonatal giraffe is rapid, and they need on average 6 g of calcium and 3 g of phosphorus daily, amounts that continue to be required until giraffes have stopped growing at 10 years of age (Chapter 7). To provide sufficient calcium and phosphorus for the calf after birth, at least 5 L of milk are needed daily, more than is required to supply energy but less than that required to provide water. Langman (1977) has reported that during the two years that he studied calf behavior in the Timbavati Nature Reserve in South Africa, weaning typically occurred at 6–8 months of age, and between birth and weaning, providing for its calf drained the cow's resources of calcium and phosphorus. Thus it is that osteophagia is more common in female than in male giraffes (see Chapter 7). Weaning is, perhaps, a matter of survival for the cow, not a lifestyle choice.

Maternal Care

Survival of calves of any ruminant species depends on more than just an adequate supply of milk. It also depends on being protected from predators. How do giraffe cows do that? One answer is not very well, as calf mortality is very high, with 50%–75% of calves dying or being killed in the first year of life, of which 90% are killed by lions (Hall-Martin, 1975a), proportions that suggest that maternal care is poor, that the bond between cow and calf weak, or that causes of death are beyond the cow's influence. That the bond between cow and calf is weak was the conclusion reached by Innes (1958) after studying giraffe behavior in the Eastern Transvaal of South Africa, a view subsequently supported by studies she made with J. Bristol Foster in Nairobi National Park (Foster, 1966; Dagg, 1971; Foster and Dagg, 1972). Foster and Dagg thought the bond was not only weak but even unnecessary because weaning seemed to occur at one month, cow and calf stayed together for just the first six weeks postpartum after which the calf wandered off on its own, and there was no correlation between cow–calf bond and calf mortality. Animals do use different strategies for rearing their young, partly dependent on how mature the young are when they are born. The greater the maturity, the less intensive the care needed. The young of ruminants are mature and so it is quite understandable that a female giraffe would concentrate on the next birth rather than the current one. However, as Langman (1977) has pointed out, there is no selective advantage of a weak cow–calf bond. Calves at six weeks of age and even up to two years of age do not have the physical attributes needed, other than speed, to avoid predation: calves six weeks old have a body mass not much more than 100 kg and a height of about 1.8 m, and juveniles one to two years of age have a height of 2.7–3.3 m and weigh 180–350 kg. It is giraffes that survive to three years of age and weighing in excess of 500 kg that are likely to reach reproductive age (Leuthold and Leuthold, 1978a; Pellew, 1983c).

For a giraffe calf to reach three years of age does, therefore, require maternal care, and exhaustive studies (very entertainingly described) by Vaughan Langman of giraffes in South Africa (1977) and Pratt and Anderson of giraffes in Tanzania (1979) (and thus, presumably, can be applied to wild giraffes elsewhere) have established that giraffes have a strong bond with their calves, especially female calves (Pratt and Anderson, 1979), that may last several years, but they use progressively less protective behavior to weaken the bond and promote independence. In the first three weeks of life the calf is hidden and kept isolated from other calves and from other adults by the cow. During this period of its life, sleep occupies at least 25% of the calf's day, and 90% of that sleep time is spent in deep sleep with its head folded back onto its rump (Langman, 1977; Kristal and Noonan, 1979; Figure 16.4 in Chapter 16). This low profile, together with its coat markings, makes the calf almost invisible. How the cow selects the hiding place for her calf is uncertain, but it does in some way become etched into her memory, or she can communicate with her calf perhaps by sounds because, having hidden the calf, she moves off to browse or drink but returns up to 12 times a day to the precise spot where the calf was left. It is during this time that the calf learns who its mother is. The distance between cow and calf depends as much on a need to protect it as on the quality and succulence of browse and surface water available for the cow. In the first few weeks after birth, if the cow can obtain all her nutritional needs from nearby, then she will move no more than 20–25 m away from the calf, but distances of 1 km (Kok, 1982), 3 km (Langman, 1977), and 4.2 km (Pratt and Anderson, 1979) have been observed, and cows can leave their calves for 4–5 hours. This early hiding phase progresses to a nursery phase comprising more calves than juveniles (in a ratio of 3:1) where two or more calves and juvenile giraffes and their mothers move and browse together, and then to a calf pool stage that contains more juveniles than calves (in a ratio of 2:1) when cows leave their calves with one or two females while they move to drink or to browse, a trait common among ruminants, and in other mammals. Calving pools last a single day, with young collected by their mothers by evening, which is a pattern very familiar to modern urban human mothers. By their second year, most female calves still remain within 50 m of their mother but males move off, foreshadowing the largely solitary life of an adult male (Pratt and Anderson, 1979; Le Pendu and Ciofolo, 1999; Berry and Bercovitch, 2012).

Populations and Reproduction

The whole purpose of reproduction is to ensure the survival of a species. Survival of a species requires, at the very least, replacement of reproductively capable members that have died. An increase in the number of members of a species requires a potential to expand beyond replacement in favorable and even in unfavorable times. That combination implies that a safety factor has evolved too—a basic replication rate and a concealed excess available on demand. Giraffes have the first, but there is very little evidence of a capacity for the second. An ideal combination that ensures that an adequate number of offspring will reach maturity is early puberty, early pregnancy, and longevity, combined with a high birth rate and low calf mortality. The earlier puberty and the greater longevity are, the more calves are produced over the lifetime of a female and the more calves there are to secure the success of the species. For individual females, the more calves she has in her lifetime, the more likely her genes will be represented in the gene pool. Thus it is that evolution of early puberty and longevity are subject to natural selection. However, there is no evidence that the onset of puberty is becoming earlier in giraffes, or indeed that it has shortened over evolutionary time, so we have to assume that the onset of puberty is as early as it can possibly be. Puberty occurs at about three years of age in both males and females with some small variation. Its variability is not genetic in origin but a consequence of local environmental circumstances and, in addition, some of the variability can be attributed to the difficulties of measurement rather than to giraffe biology. Does a longer life expectancy contribute to population stability and growth? David Western (1979) derived an allometric relationship between body mass and life expectancy for African ruminant artiodactyls: Life expectancy (years) = $5.94 \times Mb^{0.22\pm0.02}$ ($R^2 = 0.85$) with body mass in kg. If applied to a giraffe with the average body mass (750 kg) typical of adult females, its predicted lifespan is 22–28 years, and this range of ages is not uncommon in females in the wild, despite the hazards of nature (Hall-Martin and Skinner, 1978; Berry and Bercovitch, 2012).

What advantages does longevity confer on a species? Why has it evolved? Evolution of longevity in large mammals or, as in humans, mammals with large brains, is often assumed to have occurred because of the benefits conferred on younger members of a population by elders. If elderly animals are a selective benefit, then there will be more in a population. This relationship is well established for many human populations and also for African elephant populations (Lindeque, 1991), even though in both examples the ability of the elderly to contribute to population growth by having offspring is declining or has ended. The contribution that older giraffes make to a population is less obvious because of their loose herd structure, but the same trend between longevity and large size exists in them. This trend implies that there are benefits of longevity for giraffes too, possibly of wisdom, but more likely because longevity gives a longer time to have calves, particularly as female giraffes do not seem to undergo menopause (Hall-Martin et al., 1978). Giraffe herd structure is replete with the elderly. Those explanations for the structure of giraffe populations may well be valid, but there is a far more simple (and obvious) explanation. A population dominated by the elderly would also be the consequence of differential survival. Young giraffes have a high mortality. Giraffes older than three years of age have a high survival. This sort of differential survival produces more older giraffes in a population than there are younger ones (and will do so in any mammalian population that shows age-dependent survival, including human populations). The result for giraffe populations is the typical structure of a healthy, natural population unaltered by human predation (a rare circumstance these days, and is shown in Figure 18.5 in Chapter 18). In the Serengeti and Nairobi National Parks, immature adults 1–5 years old make up ~30% of the populations, mature adults make up ~55%, and the fewest giraffes are those less than a year old (~15%). In the Timbavati Nature Reserve in South Africa the same trend has been recorded. The Timbavati is an undisturbed area of 210 square miles and had, during the period 1964–1967, a resident giraffe population of about 1,500 members. Poaching was minimal, the population reasonably stable, and 38% of the population consisted of adult male giraffes, 46% of adult females (adults thus making 84% of the total), 8% of subadults of both sexes, and 7% of juveniles less than a year old (Hirst, 1969).

Across species, large size is balanced by a decline in birth rate because large size is associated with longer gestation times. This combination is common in biology—the smaller the body mass, the shorter the lifespan, the more the number of eggs or the larger the size of the litter produced—but it is not as readily recognizable in animals as large as giraffes that usually produce just one calf. For any mammal, birth rate (BR) is defined as the percentage of births per annum in a population, and its relationship with body mass in African ungulates is described by BR (%) = $153.3 \times Mb^{-0.36}$ (Western, 1979). If that equation is applied to giraffes with an average body mass of 750 kg, then the number of calves born each year would be 14% of the total population. The actual birth rate in a population of giraffes in Tsavo National Park (Leuthold and Leuthold, 1978) was found to be 17–27 per 100 females y^{-1} and, as reproductive females formed ~30% of the population, that rate amounted to a birth rate of 5%–8% of the members of the total population. Birth rate was 35–40 per 100 females y^{-1} (10%–12% of the population) in Nairobi National Park (Foster and Dagg, 1976), and 66 per 100 females y^{-1} (21%) in the Serengeti (Pellew, 1983b). The higher birth rate in the Serengeti at the time of Pellew's study was attributed by him to the excellent quantity and quality of browse available. It is of course exceptionally difficult to discover how many calves a giraffe female in the wild might have in her lifetime. Bercovitch and Berry (2010a) who have made a long-term study of *G.c. thornicrofti* in Zambia found that one female definitely had 10 calves and may have had an 11th, with the oldest female having a calf when she was 24 years old. The number of calves per cow is easier to establish in captive giraffes, and a 24-year-old in Taronga Zoo in Australia is also reported to have had 10 calves. In South Africa, a 20-year-old giraffe in the wild had had eight calves, or one every two years after puberty had been reached (Hall-Martin and Skinner, 1978). In the Tarangire Ecosystem of Northern Tanzania, giraffe females have five or six calves in a lifetime (Lee et al., 2017). As an expected life span of a female in the wild is not much more than 20 years, the eight calves reported by Hall-Martin and Skinner (1978) is probably close to the maximum for wild giraffes. Of those calves, just two or perhaps three survive to adulthood—not very many, in other words, but clearly enough to ensure the survival of giraffes in the wild.

Calf Mortality

A low birth rate is successful only if the calves survive and in giraffes, like many other ungulates, few do. In wild populations of giraffes, calf mortality is extraordinarily high, despite the strong cow-calf bond, with most deaths caused by predation (Hall-Martin, 1975a), but disease (Leuthold and Leuthold, 1978) and drought contribute (Mitchell et al., 2010). The first assessment of the high calf mortality phenomenon was made in the Nairobi National Park, where mortality rates were found to be 73% in the calf's first year and 7% in its second year, with just 20% of calves surviving to puberty (Foster and Dagg, 1972). In the Serengeti, Pellew (1983b) found that 50% of calves died in the first year of life; in Timbavati, 48% did (Hall-Martin, 1975a), and 45% did in the Luangwa Valley (Bercovitch and Berry, 2010a). In the semi-arid conditions of Tsavo National Park, if adulthood is reached, survival is good, with>80% surviving for at least three years after first being identified as an adult (Leuthold and Leuthold, 1978). A similar proportion survives after reaching three years in the Serengeti (Pellew, 1983c). However, for younger giraffes in Tsavo it is very different. Leuthold and Leuthold (1978) followed the lives of 15 calves: 12 of those calves survived for 1 month, 10 survived for 3 months, 7 for 6 months, 5 for 12 months (so 10 [67%] died in the first year), and just 4 survived to 24 months (27%). In the south of Arusha National Park, Tanzania, 20% of calves survive to adulthood (Pratt and Anderson, 1982). These rates of calf mortality, while high, are actually similar to those in buffalo (30%–60%), blue wildebeest (80%), and zebra (40%–80%), so survival of all these species has been accomplished despite high calf mortality. The giraffe calves that survive may be better camouflaged (Chapter 15) but, more generally, survivors have a close association with their mothers for 12–18 months (Langman, 1977; Leuthold and Leuthold, 1978) or live in areas of few large predators (Pratt and Anderson, 1982) or where there is reduced predation because predators, and specifically lions, are distracted by easier prey such as that offered by migrating herds of wildebeest and zebras (Lee et al., 2016). There may also be a small effect of season of birth, with calves born just before or early in a rainy season having a slightly higher probability of surviving beyond three months, although no numerical data were reported to support that inference (Lee et al., 2017). A possible reason for the effect of season is that early in a rainy season growth of vegetation has yet to begin and predators are more conspicuous and more easily avoided. With such high mortality, it would be useful if twins were the norm in giraffes, as they are, for example, in their ancient sisters, pronghorn antelopes, but giraffe twins are exceptionally rare. Hall-Martin and Skinner (1978) recorded the only properly documented case of twins in the wild (with a photograph; Hall-Martin, 1975a), and they were stillborn twins in the Timbavati Nature Reserve. Some 30 or so cases of twins in captive giraffes have been recorded and, in most if not all of those cases, both calves did not survive.

Calving Interval

A corollary of high calf mortality is that the number of offspring a female giraffe has in her lifetime is critical for the survival of the species. That number is determined by the age at puberty and the length of reproductive life, but also by the length of time it takes for a female to become pregnant again after having a calf. The length of time between one calf and the next is the calving interval, and one analysis of its length in 74 giraffes (23 wild giraffes in Africa and 51 captive giraffes in Europe, North America, and Australia) is summarized in Figure 17.11.

In that group of giraffes the most common interval was 16–17 months (Hall-Martin, 1975a). The weighted average for all 74 intervals is 19 months. Hall-Martin and Skinner (1978) analyzed the calving intervals of 32 captive giraffes in Australia and

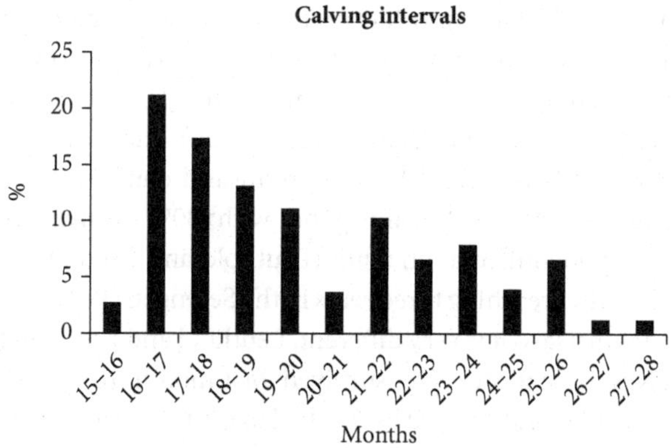

Figure 17.11. Frequency distribution of calving intervals of 23 wild and 51 captive giraffes. Constructed using data in Hall-Martin (1975a).

35 from Europe and America. The average calving interval for those 67 intervals was between 21.2 and 21.7 months. The shortest times to re-conception were 19, 23, and 27 days. A calving interval has two parts. The first part is the length of gestation which is ~450 days long and, as we have seen, it varies remarkably little. The second part is the time between the birth of a calf and conception of the next one. This second period can vary enormously. Theoretically, it should last about 90 days because on average there are six exploratory estrous cycles between birth and the next conception, and each cycle is 15 days long (Figure 17.4). That theoretical interval is roughly correct. For the calving intervals of 16–17 months shown in Figure 17.11, the time to re-conception was 109 ± 11 days (Hall-Martin 1975a), and for all the calving intervals shown in Figure 17.11 the average interval between birth and re-conception was ~120 days. Age, social hierarchy, the availability of males but not mate preference, calf mortality, and also the health of the cow all affect the calving interval length in captive giraffes (del Castillo et al., 2005; Bercovitch et al., 2006; Lueders et al., 2009b). The effect of male availability can be seen in the birth history of a captive female giraffe reported by Pournelle (1955): she had a calf and then six more at intervals of 27 months, 22 months, 30 months, 24 months, 23 months, and 28 months. She conceived whenever a captive male was present. But the length also depends much on the individual. Owen, for example, in 1839 recorded that it took Zaida nearly 9 months to re-conceive after the birth of her first calf.

How long it takes for a wild female giraffe to re-conceive after giving birth is far more difficult to establish than it is in a captive one, but the average time to re-conception can be estimated by subtracting average gestation time from calving interval. It is much longer than it is in captive giraffes. For 13 wild giraffes in the Timbavati, the calving interval was 19.9–22.1 months (597 to 674 days; Hall-Martin 1975a; Hall-Martin & Skinner, 1978), and thus the time between birth and re-conception was 5 to 7 months (150 to 217 days). In Tsavo National Park, Leuthold and Leuthold (1978) found the delay to conception to be usually 8–8.5 months, with a minimum of 1.5–2.5 months. The average calving interval was 22.6 months (678 days). In the Serengeti Pellew recorded an average calving interval of 18.4±1.0 months, with the longest postpartum anestrus being 9.2 months, or 267 days. In *G.c. thornicrofti* in the Luangwa Valley of Zambia, the calving intervals recorded over a 35 year period were on average 607±77 days

(19.6±2.5 months) long, and thus the time to re-conception was ~160 days (Bercovitch and Berry, 2010a).

The variation in the length of calving intervals in wild giraffes can be attributed to seasonal variation in the quality and quantity of browse. If nutrition is good, then the nutritional stress of lactation is reduced and the duration of postpartum infertility is shorter, because an important factor affecting return to estrus is the condition of the cow. Figure 17.5 shows, for example, that the re-conception rate of wild giraffes that have given birth between March and June is low for the 150 days between June and December. By December (in the Southern Hemisphere) summer rains have stimulated production of good quality and quantities of browse and cows have had time to recover from the deficits of the dry season, and a spike of conceptions occurs between December and March: calves born in March to June were conceived between December and March 15 months previously, when the cows were in peak condition. The length of the calving interval also depends on whether the calf survives. If a newborn calf dies early after birth, return to estrus is quick, although in Zaida's case her first calf died nine days after birth and she did not re-conceive for nine months (Owen, 1839). The shortest interval between birth and conception in wild giraffes so far recorded is 24 days (Pellew 1983b). Kok (1982) has described how a female gave birth in the marginal habitat (for giraffes) of the Willem Pretorius Reserve in South Africa. The calf died four days later, and 49 days postpartum the cow conceived again. Early return to conception after a calf death is also the pattern reported by Foster and Dagg (1972): the disappearance of a calf within a month led to a second calf 17 months later, and as 15 of the 17 months were gestation, conception occurred after 2 months. Survival of a calf for three months led to a calf 21–22 months later, so conception occurred at about 6 months after the previous birth. A relationship between calf death and time to estrus for wild giraffes has been disputed (Bercovitch and Berry, 2010a), but Pellew's data (Pellew 1983c) obtained in the Serengeti are convincing (at least for that population at that time). In the Serengeti if a calf died within a week of birth, the next calf was born ~500 days afterward. For calves surviving for one month after birth, the interval increased to 519 days, survival for up to 3 months increased the interval to 555 days, and survival for up to 6 months increased it to 573 days. Survival of a calf for more than a year increased the interval to 591 days: the longer a calf survived, the longer the calving interval became because re-conception was delayed.

A common explanation for the absence of estrous cycles, the failure to reconceive during the early part of lactation, and the lengthening of the calving interval is that prolactin, a hormone released from the pituitary gland by suckling and which controls milk synthesis, inhibits the release of GnRH. Prolactin's effects (if any) have not been studied in giraffes, but if it inhibits GnRH secretion in giraffes in the same way that it does in other mammals, the benefit would be suppression of postpartum estrous cycles, delay of the next pregnancy, and a long calving interval. The advantage of a delay like that is that it allows the mother to divert all her resources to her newborn calf. However, a strong effect of prolactin in giraffes would be a disadvantage, as delay of re-conception would reduce the number of calves a female giraffe could have. If prolactin does have an effect, then the theoretical 90-day postpartum period between birth and the next conception would be explained. However, estrous cycles do occur in that period (Figure 17.4) but seemingly without the delicate hormonal balance needed to achieve ovulation and conception. A complete absence of estrous cycles would be a stronger indicator of an effect of prolactin. In captive giraffes studied by Loskutoff et al. (1986), estrous cycles were absent for three months postpartum. A more mixed picture emerges from Lueders et al.'s (2009b) sonographic study of ovarian function. In one (of three) captive giraffes, a follicle had developed by 8 days after birth. Her calf was removed (for veterinary reasons), but it was not until 5 months later that she re-conceived, never having lactated. In a second giraffe, that nursed a calf, follicles developed but none ovulated, and it took 10 months before normal estrous cycles returned. In a third giraffe that also nursed a calf, no follicular activity occurred, and partly because of her poor condition it was a year before estrous cycles returned. On balance, the stress of lactation does seem to affect the frequency of estrous cycles, but a role for prolactin is uncertain. What is more certain is that if the stress of lactation is for any reason absent or reduced, then ovarian activity returns quickly. It is the earliest possible re-conception that is the biological response of giraffes to a birth and it contributes to their survival as a species.

Population Dynamics

Giraffes have flourished for many millions of years despite changing climates and changing habitats and the consequences that those two disruptors have for food supply and all that follows. How, if reproductive success is nicely balanced, have they flourished? On what parts of reproduction have their populations relied for stability? What we can say is that population stability, or even growth, is not limited by male availability. There always is an overabundance of adult males. By contrast, much depends on the number of adult females. Their proportion in a typical giraffe population is higher, although not much higher, than the proportion of males, but their long gestation is a disadvantage. How is the number of calves maximized, and is that number sufficient to balance births and deaths? One way that it could be maximized would be an early onset of puberty, but the onset of puberty in giraffes occurs at much the same time in their lives as it does, for example, in humans, that is, after approximately the first one-eighth of their expected lifespan has been completed. Another way, and the way that female giraffes have evolved (and uniquely so for such a large mammal), is short estrous cycles, combined with a capacity to return to estrus as soon as possible after calving. Both result in more opportunities for pregnancy than is the case in other large herbivores that have a longer estrous cycle or, because of the contraceptive effects of lactation stress, a delayed return to estrus after giving birth. Underlying reproductive success is, of course, good nutrition. Robin Pellew (1983b), who studied changes in the population of giraffes in the Serengeti in the 1970s, emphasized that aspect of giraffe populations: as food supply increased, so did the population; the numbers then exceeded food supply, and the population declined to a point of relative stability where the number of its members was higher than it was at first but less than maximum number. That pattern is replicated by all species and has been replicated in many human populations, but whether the pattern's origin in giraffes is an oversupply of food, as Pellew emphasized, or a lack of it, is less certain. In temperate regions of our planet (rather than the tropical or subtropical climes enjoyed by wild giraffes), for example, healthy, natural herbivore populations—natural in the sense that they are regulated by natural cycles and not human interventions—are characterized by high and stable adult survival and variable juvenile survival, and these characteristics have been shown to be independent of the density of the population, independent of food supply, and independent of causes of mortality (Gaillard et al., 1998, 2000). Which of these two patterns is true of giraffes? Are their adult populations relatively stable, birth rates constant, and the survival of juvenile giraffes variable?

Whatever the type of pattern for the size of a population to remain stable, birth rate must match death rate. Pellew (1983b) calculated birth rates and mortality rates in a sub-population of 711 giraffes in the Serengeti and found that birth rate was 20.7 per 100 head per year, and mortality 14.3 per 100 head per year. The difference was a 6.4% per year increase in population size. However, in most giraffe populations the average birth rate for giraffes does not vary much from the 14% predicted by Western's analysis in 1979, nor have any studies shown that the average proportion of adult females in a natural giraffe population varies. Thus in any population of 100 giraffes in which there will be about 30 females older than five years of age, each having a calf every two years, 15 calves will be produced each year, of which three will reach adulthood. Any circumstances that increase adult mortality to more than three per 100 adults per year, such as reduced food resources, diseases, or an increase in the number of predators, both human and natural, or circumstances that reduce the number of calves surviving, will cause the population to decline. Strauss et al., for example, in 2015 studied the same population that Pellew had studied 40 years previously in the Serengeti (Strauss et al., 2015). That population had declined from 16,483 in the 1970s (Pellew 1983c) to 3,520 in 2008–2010. Causes of the decline included proliferation of unpalatable browse species that changed the food supply but, and more significantly, adult survival had been reduced by poaching. The number of calves reaching adulthood each year had also declined, not because calf survival had declined, but because poor quality of browse had prolonged the calving intervals and fewer calves per female was the result. Consequently, the population had become particularly sensitive to adult survival, and poaching of adults was (and will be) its nemesis. On the other hand, in the population of giraffes that inhabits the Kruger National Park in South Africa where poaching of giraffes is minimal, Owen-Smith and Mason (2005) established that it was the highly variable survival of juvenile giraffes that contributed more to the rate of population growth than did adult survival, a pattern consistent with the stable adult–variable juvenile survival pattern characteristic of large herbivore populations in temperate regions (Gaillard et al., 1998; 2000). Interesting as these patterns are, it may be fair to conclude that the factors that produce stable or growing giraffe populations are not mysterious. Stable giraffe populations depend on the survival of both adults and juveniles. Growth of a population of giraffes will occur if either exceeds expectations.

Growth of populations must have been the norm during the radiation of giraffes many millions of years ago. What can the expected growth rate be now? In a typical, natural, giraffe population in Africa consisting of 100 individuals, we can assume there will be about 30 female giraffes of reproductive age, 25 adult male giraffes, 30 juvenile giraffes one to five years of age, and 15 neonates less than one year old (e.g., Dagg and Foster, 1976; Pellew 1983c). Under normal circumstances, each of the 30 females will produce a calf every two years. Three of the 15 calves born each year will reach adulthood because eight of them can be expected to die in the first year of life and four will die before reaching three years of age. Two adults will also die (Pellew, 1983c). Extrapolating those proportions to the current total African wild giraffe population of ~100,000 means that each year ~3,000 juvenile giraffes will reach adulthood and ~2,000 adults will die, and the total population would grow by 1,000 individuals (1%) each year. It is a delicate and sobering balance, but presumably it is at least that balance which existed 3 Mya in East Africa at the time of giraffe radiation. Small increases in juvenile survival to adulthood and small decreases in adult mortality would tip the balance toward more rapid growth. The converse, however, would tip populations into decline.

There is no doubt that high adult mortality (and therefore fewer births each year) has happened and is happening in Sudan and Tanzania and Kenya. According to the latest IUCN report (2016-3), over the last 40 years the number of giraffes in those three countries has declined from around 125,000 to around 42,000, with the most catastrophic decline in Sudan (20,000 to 650). That overall decline is equivalent to an annual average decrease of 3%: just three deaths of adults per 100 giraffes y^{-1} more than the number of calves surviving to adulthood each year. In southern Africa the balance is the opposite. In southern Africa the population of giraffes has increased from approximately 30,000 to 55,000 over 40 years. An increase of 25,000 giraffes in 40 years corresponds to a compound interest rate of 2%: each year for the 40-year period the average annual number of juveniles surviving to adulthood per 100 giraffes y^{-1} has been two greater than the average number of adult deaths. A reason given for this increase in population size is an increase in the number of privately owned nature conservancies and the much better protection of giraffes that they provide (Deacon and Tutchings, 2018). It does not take much, in other words, to ensure the survival and growth of a giraffe population, provided that humans show care and the habitat can provide an adequate diet. Habitats now and in the future will provide the diet. Ensuring the survival of giraffes is not ranked high on the list of humankind's priorities, so care by humans is far less certain.

18
A Shape to Die For?

When all is said and done about how giraffes work, what remains is their shape. Enigmatic, beautiful, unmatched in nature. No matter the variant or subspecies, natural selection has balanced the advantages and disadvantages of their shape to produce an identical shape in all of them. Their shape's most striking visible features are elongated legs and neck and distinctive coat markings. Its cryptic attributes are as numerous as they are remarkable. Many are adaptations to a high blood pressure, but giraffes also have a skeleton that has become reorganized to support their elevated shape, a gastrointestinal tract suited to the needs of a concentrate selector, a large intestinal surface area, a range of shape-dependent thermoregulatory tools that is impressive, a respiratory system adjusted to fit an elongated neck, kidneys that are well-adapted to the East and southern African habitats in which modern giraffes evolved and in which they now live, and a reproductive life that, in the normal order of things, sustains their populations. Their capacity for endurance exercise and high-speed gallops is substantial, and giraffes have sufficient skin surface area to dissipate the metabolic heat produced by that exercise. They have a skull enlarged by sinuses that increase head volume without increasing its weight, and an elongated head that adds, in a large male giraffe, an extra meter to the length of the neck and increases the height to which it can reach browse. Their pointed muzzle is perfect for reaching browse in narrow spaces and also provides the surface area for a role in temperature regulation and for olfaction, and provides the room for a long prehensile tongue. Absence of upper incisors facilitates extension of the tongue. Robust, rugose, brachydont, and selenodont molars provide a broad grinding surface that cuts and crushes leaves and fruits of giraffes' favored browse. Their eyes are large, their field of vision panoramic, their visual acuity excellent, and vision periscopic. All these idiosyncratic cryptic attributes make the complex shape of giraffes a harmonious whole and are the origin of its complete and perfect integration into the habitats it occupies. It is a shape to die for.

A shape endures in the absence of any reason not to. It is new factors, new habitats, new environments that bring about change. The attributes giraffes have evolved have done so because of the vegetation, habitat, and climate changes to which modern giraffes, their ancestors, and their antecedents have been exposed. Giraffes have had their share of those, and the giraffes we see today are the product of at least 15 million years of accumulated mutations, polished and shaped by intervals of punctuated and gradualistic evolution, and natural selection in the best of Darwinian traditions. Their lineage can be traced from *Diacodexis* to the present via the Hypertragulids and Leptomerycids of the Eocene, the Gelocids and Palaeomerycines of the Oligocene and early Miocene, the Canthumerycids, Palaeotragines, and Samotheres of the middle Miocene, and then *Bohlinia* of the late Miocene and early Pliocene. From *Bohlinia* arose first the Asian *Giraffa* and then the African *Giraffa*. After *Bohlinia*, much of *Giraffa* evolution was influenced by changes in climate and the habitat and vegetation

How Giraffes Work. Graham Mitchell, Oxford University Press. © Oxford University Press 2021.
DOI: 10.1093/oso/9780197571194.003.0018

changes that followed. The most influential climate changes seem to have begun about 10 Mya, with a second phase in the uplift of the Himalaya–Tibetan plateau and retreat of the Paratethys Sea, the great inland sea that once filled a vast area behind what is now the Italian Peninsula and the remnants of which are the Caspian and Black Seas. North of the Himalayas, precipitation became seasonal and the Tibetan plateau drier. Summer and winter monsoons intensified. South of the Himalayas, summer temperatures and precipitation increased. Over eastern and northern Africa and over western India, tropical vegetation disappeared, was replaced by arid-adapted plants, and a savannah belt developed. Forests were replaced by smaller, small-leafed trees and C3 shrubs. Sea floor spreading, silicate weathering, and sediment burial lowered atmospheric carbon dioxide concentrations. By 9 Mya those lower carbon dioxide concentrations had created the conditions for the evolution and growth of C4 grasses. The transition from habitats dominated by C3 shrubs to those dominated by C4 grasses began in earnest 8 Mya and took 1–1.5 million years to reach the grassland-dominated, woodland-savannah vegetation of today. Dramatic changes in fauna resulted. At the same time that vegetation changed, so did migratory routes from Europe to Asia and from Europe to Africa. Migration out of Europe followed. The first of the true giraffes, *Giraffa punjabiensis*, appeared in India about 7 Mya and, slightly later, *Giraffa jumae* did in Africa. Between 6 Mya and 2.6 Mya there was further uplift of the Himalayas, continental icesheets expanded, and the world, at intervals, became colder. The change to an almost pure C4 flora in Asia and China, and the aridity and cold, brought extinction to the Asian *Giraffa* by 1 Mya. In East Africa variation between sparse and denser tree- and shrub-filled habitats over the period from 5 to 1 Mya provided the habitats for African *Giraffa* to evolve and flourish, reaching an apogee of species or variants in the East African Pliocene, when all four classic, ancestral subspecies coexisted.

Allometry and the Shape of Giraffes

The shape of adult, modern giraffes, which has resulted from their long evolutionary history and the effects of climate and vegetation, is defined by its dimensions, for example, the width of their body or the length of their neck. Achieving an adult shape requires time. How does the shape of a giraffe change over the time of a lifespan, rather than over evolutionary time, to become an adult giraffe? Like many of the analyses of how giraffes work, their growth patterns can be described best by allometry, because, as D'Arcy Thompson explained in 1917, although "we begin by describing the shape of an object in the simple words of common speech: we end by defining it in the precise language of mathematics." Allometry is the best mathematical language to describe the shape of giraffes, especially when the data available cover a wide range of sizes and ages from the very young to mature adults. Most of what we know about the way a giraffe's shape changes as it grows comes from studies of just two subspecies—*G.c. giraffa* of southern Africa and *G.c. capensis* of Namibia—between which physical differences are almost indistinguishable. Are the dimensions of these two subspecies the same as they are in other subspecies/variants? Any descriptive equation will, of course, be most accurate when applied to individuals from which the measurements

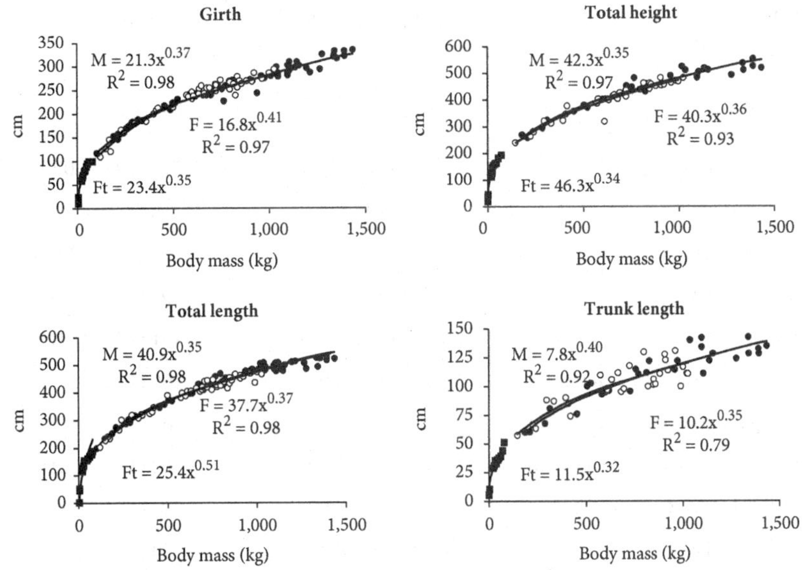

Figure 18.1. Body growth patterns in male (M, ●) and female (F, ○) *G.c. giraffa*, and *G.c. giraffa* fetuses (Ft, ■) from South Africa and Zimbabwe. Girth is the circumference of the thorax measured immediately behind the front legs; trunk length is the distance from the withers to the base of the tail; total length is the sum of head, neck, and trunk lengths and tail length to the tip of the last tail vertebra; total height is the distance from the sole of front hoof to the occipital crest.

Data obtained from studies by Hall-Martin, 1975a and Mitchell et al., 2009b.

were obtained, but there are no statistically significant differences between the cranial measurements of *G.c. giraffa* and *G.c. capensis* and those of *G.c. thornicrofti*, for example (Hall-Martin, 1975a), or between the dimensions of the limb bones of all other subspecies/variants (Singer and Boné, 1960), or between those of modern giraffes and ancestral giraffes (Chapter 6); therefore it is not unreasonable to assume that the growth patterns of all subspecies are similar. The basic dimensions of giraffes—their length, height and girth—and how they change as body mass increases from fetal life to adulthood are shown in Figure 18.1, and actual sizes in postnatal giraffes are summarized in Table 18.1.

When linear dimensions are correlated with body mass, as shown in Figure 18.1, what is strikingly obvious is that the growth patterns of males and females are identical, except that the final body mass of females is approximately 800–1,000 kg compared with 1,300–1,500 kg in males. Despite the difference in final body mass, the differences between the sexes in girth, lengths, and height are small. Trunk length of a mature female is just 3 cm shorter than that of a mature male. The girth, total length, and total height of mature females are 20–70 cm less than they are in males, and average thorax diameter is 83 cm in males and 76 cm in females. Females are, simply, more slender, graceful, and feminine versions of males. The growth of limbs is also similar in both genders (Figure 18.2).

The legs of males and females elongate at the same rate, but the legs of females end up 15–20 cm shorter than they do in males, on average. At birth (100-kg body mass),

Table 18.1 The Body Dimensions of an Average-Sized Giraffe (*G.c. giraffa*) Weighing 750 kg and the Range in 30 Male and 30 Female Giraffes from Zimbabwe

Dimension	Males	Females
Body mass (kg)	750 (181–1,441)	750 (147–1,029)
Girth (cm)	251 (148–332)	240 (128–292)
Trunk length (cm)	114 (61–133)	101 (58–130)
Total length (cm)	431 (204–525)	399 (231–472)
Total height (cm)	439 (261–547)	401 (244–479)
Foreleg length to the axilla (cm)	173 (108–209)	160 (112–189)
Foreleg length to the withers (cm)	289 (175–347)	262 (167–307)
Hindleg length to the groin (cm)	176 (109–210)	162 (118–188)
Hindleg length to the rump (cm)	237 (141–287)	224 (144–260)
Head length (cm)	79 (46–97)	70 (43–84)
Head mass (kg)	25 (6–45)	17 (6–25)
Neck length (cm)	168 (92–217)	155 (74–202)
Neck mass (kg)	78 (15–138)	56 (14–87)

Data extracted from Figures 18.1, 18.2, 18.3.

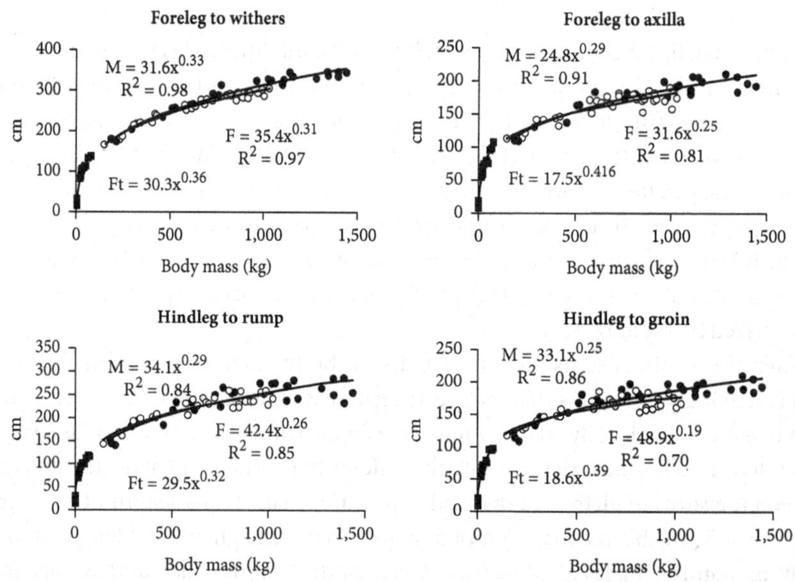

Figure 18.2. Growth patterns of the fore- and hindlegs of 30 male (M, ●) and 30 female (F, ○) *G.c. giraffa*, and 10 *G.c. giraffa* fetuses from Zimbabwe (Ft, ■). Foreleg length is the perpendicular distance from the sole of the hoof to the axilla or to the withers, hindleg length the perpendicular distance from the sole of the hind hoof to the groin or to the rump. Data from studies by Mitchell et al., 2009b, and van Sittert et al., 2010, 2015).

foreleg length to the axilla in males is 95 cm on average and in females 100 cm. At maturity they are ~210 cm and ~190 cm, respectively, and have doubled in length since birth. Hindleg length does the same. At birth, neck length is on average 75 cm in males and 65 cm in females. At maturity, neck length in males and females is ~215 cm and ~200 cm on average, respectively—a threefold increase. The necks of female giraffes start shorter, but grow faster than their legs do, and faster than the necks of males do, so the ratio of neck length to leg length is higher in mature females than in mature males (Figure 18.3), giving females almost equal access to high-level browse, despite their smaller, more delicate frame. Equality of leg and neck length (neck length:foreleg length ratio= 1) occurs at a body mass of ~700 kg in females and ~1,000 kg in males (Figure 18.3) so that, at full maturity, necks are longer than forelegs in both sexes.

The shape of giraffes and the adaptations that support it are, and are not, fortuitous. They are fortuitous because natural selection operates by chance. They are not

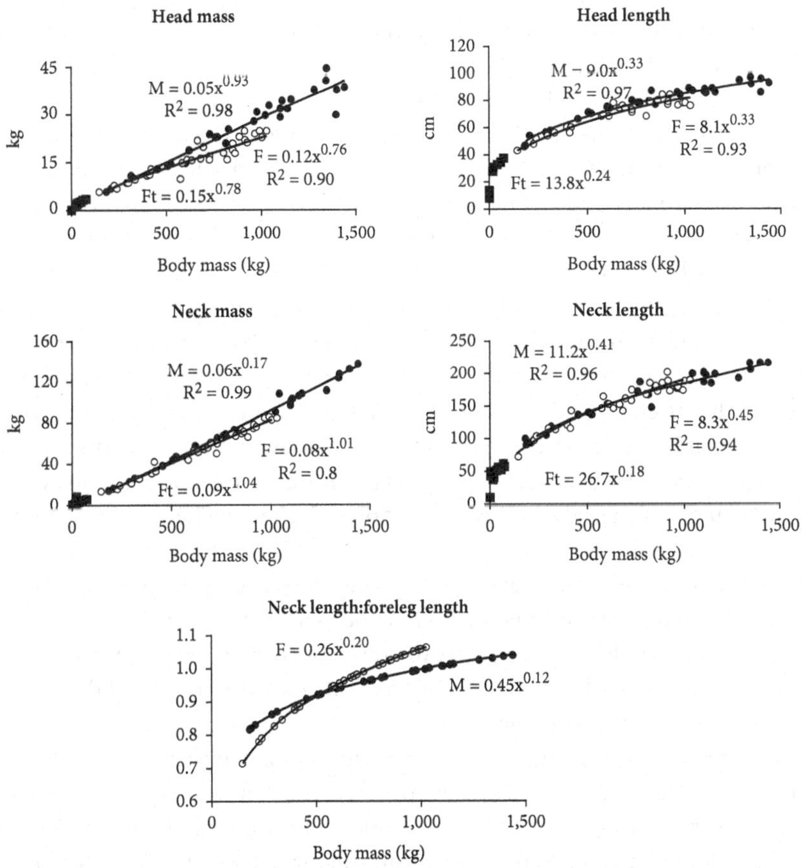

Figure 18.3. Growth patterns of neck and head mass of 30 male (M, ●) and 30 female (F, ○) *G.c. giraffa*, and 10 *G.c. giraffa* fetuses (Ft, ■) from Zimbabwe, head length and neck length, and the neck length to leg length ratio. Neck length is the mean of the distances from the occipital crest to the withers and from the angle of the jaw to the point of the shoulder.
Data from studies by Mitchell et al., 2009b, 2013a.

fortuitous because the shape and adaptations that persist are those that are the best suited to the vegetation and climate, in short the environment, which giraffes inhabit. When the word "environment" is mentioned, the thoughts that first come to mind are of ambient temperature or of rainfall or of vegetation, and changes in them are what giraffes adapt to by natural selection. However, perhaps the most overlooked factor in the environment generally, and for giraffes in particular, that has affected their shape is gravity. Gravity does not change, of course, but it is to the constraints of its effects that all other adaptations are subservient. For example, one analysis has proposed that the maximum size a land animal can reach is 20 tons, because land mammals of that mass cannot develop a skeleton that is both a reasonable proportion of body mass and, at the same time, is strong enough to resist the force gravity imposes (Economos, 1981). Another analysis (Hokkanen, 1986) arrived at the higher upper limit of 100 tons. Skeletons of animals as heavy as 20 or 100 tons compete for space with organs that support metabolism or produce movement. The balance, in other words, between skeleton mass and all other needs is narrow. With some notable exceptions like *Brachiosaurus*, which is estimated to have had a body mass of ~50 tons, and the recently discovered *Argentinosaurus* that might have weighed up to 100 tons, very few terrestrial animals have ever had a body mass of 20 tons. The largest modern land mammal is an African elephant (*Loxodonta africana*). It can weigh up to ~6 tons (6,000 kg) and has a skeleton that weighs perhaps 1,500 kg (25% of body mass). Giraffes reach no more than a quarter of that mass and have a skeleton that weighs ~10% of its body mass. The upper limit of a giraffe's body size is linked to its capacity to harvest and digest sufficient nutrients, but its size is also constrained by the biomechanics of the skeleton and by the capacity of the heart to pump blood to the brain, both of which are affected by gravity. Thus, the degree to which giraffes have been able to adapt to other environmental pressures has been offset by the effects of gravity, so it is not unreasonable to say that gravity has been the ultimate determinant of the shape of a giraffe. If gravity had been less than it is, then giraffes might have evolved a larger body mass, thinner bones, longer legs, and smaller hearts. If gravity had been larger, limb bones would be thicker and shorter and a stronger/larger heart or a shorter neck might have evolved. Body size of ancestral giraffes increased when food energy was plentiful and energy expenditure was allocated to growth rather than, say, keeping warm, and it might have continued to increase beyond what modern giraffes have become, but, even if food shortages did not occur, gravity would always be there to prevent growth to a size too large. In short, the shape and maximum body size of modern giraffes are the product of the combined influences of its habitat and gravity.

What advantages accrue to giraffes as a consequence of their shape? A granddaughter tells me that they have a long neck because their feet smell. It is an explanation for a long neck that has the merit of causality, not creation, but the question remains. How have all the physiological functions and anatomical structures related to their shape of which we are aware come to pass, what are their advantages, and how do they fit in with the life and evolution of giraffes? What selective pressures, if any, operated such that any ancestor that had a neck longer than their legs left more progeny than did those with a neck shorter than their legs? What factors and pressures played on natural selection to produce those progeny best suited to the habitat they chose to inhabit? What made *Giraffa* a relatively successful genus? There is little consensus. Why not?

An Accidental Shape?

One reason is that all of the several hypotheses assume that the current use and advantages of long legs and a long neck mirror the selective pressures that produced the shape. However, as Stephen J. Gould pointed out (in 1996), current use is not evidence for a selective pressure no matter how obvious the relationship. He had a point. The current pervasive use of the human brain for frivolous twitterings on Facebook cannot have been the selective pressure that led to the enlargement of our brains 2 Mya at the time of *Homo habilis*. It is worth considering, therefore, if a way other than natural selection could have produced the shape of giraffes. Could it have come about accidentally, as an add-on to other changes? It could. And the possibility that it did is attractive, as we can never know what the selection pressures were at the time giraffes were evolving. They can only be guessed. An explanation that removes guesswork would be compelling. Moreover, as Bertrand Russell (1931) has remarked, there is nothing in the process of evolution that demands the hypothesis of a purpose. Evolution is governed by nothing other than the laws of physics and chemistry. Thus, one origin of a long neck, like any other overspecialized structure, as Gould explained in 1974 when he analyzed the origin of the massive antlers of the now-extinct Irish elk, is that it could simply be an allometric product of increases in body size. In postnatal *G.c. giraffa* we know that the allometric relationship between body mass and neck length is described by the equation: neck length = $11.2 \times Mb^{0.41}$ in males and $8.3 \times Mb^{0.45}$ in females and thus, in growing giraffes, neck length increases faster than does body mass (for an isometric, equivalent increase in both, the exponent would be 0.33). Neck mass also increases faster than body mass in males ($\alpha\ Mb^{1.07}$; for isometry, the exponent would be 1.0) and does so in some females but, on average, in females neck mass increases in parallel with increases in body mass ($\alpha\ Mb^{1.01}$; Figure 18.3). Positive allometry of this kind in large animals may be a consequence of the trait having an advantage—an easily detectable signal of masculinity, for example—that is greater in large animals than it is in smaller ones (Bonduriansky and Day, 2003; Kodric-Brown et al., 2006). Thus, it is entirely possible that, somewhere in the long prehistory of giraffes, it was not neck dimensions that responded to selection pressures, but it was body mass that did. If body mass increased, then the disproportionate increases in neck length and neck mass would have followed passively, simply because of the link between them and body mass. All that would have been required was a mutation that allowed an increase in neck length, rather like the one that produced a long neck in the camel family—llamas, alpacas, and camels themselves—from the very beginning of their evolution, or the one in *Palaeotragus* that initiated neck elongation in giraffids, and which ended up as a long neck in *Bohlinia*. A mutation like that automatically would produce a neck much longer than expected if body mass increased, and the long neck of giraffes would have evolved by accident. Of course, if a long neck had evolved and turned out to be a useful acquisition with no selective disadvantages such as higher mortality that might have acted against its elongation, then natural selection would have preserved and even enhanced it.

The body mass of Palaeotragines and Samotheres was ~600 kg, that of *Bohlinia* ~900 kg, and that of modern male giraffes is up to 1,500 kg. What selection pressures might have led to that increase in body mass? What advantages might have accrued from a large size? An appealing advantage is resistance to predation. Healthy modern adult giraffes are hardly ever killed by predators other than humans. Lions account for most deaths by predation, although a pack of hyaenas can kill a weakened giraffe by working as a team. Giraffes are formidable prey animals and are killed by lions mainly during the winter when because of loss of condition they are weak and are easier prey (Hirst, 1969). Even so, in two studies done in the Timbavati Nature Reserve in South Africa (Hirst, 1969; Hall-Martin, 1975b), in a habitat in which lions were plentiful, just 4%–6% of a large giraffe population was killed by lions, and of those fewer than half were adults. Adults are too big a foe and are able to defend themselves. Using their front hooves—legend has it but evidence to support the legend is scarce, so the legend verges on myth—adult giraffes are able to deliver a fatal blow to a lion. If larger giraffes survived to produce more offspring, and size was a heritable trait, natural selection would have led to offspring with a larger size and potentially greater resistance to predation. And it may well be that large body size is attractive to female giraffes and thus a desirable character subject to sexual selection—thus the selective pressure for body size. Over evolutionary time, body size would have reached a balance between the effects of gravity, the size needed to resist predation, and the quality and quantity of browse required to sustain the size. Theoretically, body size and its associated shape could increase *ad infinitum*, but size is constrained by size itself: the greater the size, the greater the effects of gravity, the greater the mass of the skeleton needed, the greater the requirement for food, and the greater the vulnerability to food shortages. Thus, the very largest giraffes may have been selected against, leaving the more moderately sized modern giraffes to survive and produce progeny, but those that did survive had acquired a long neck. This explanation, attractive as it is, is an uncomfortable one, at least for me—not least because if any trait, like neck length, increases faster than any other trait, like body mass, unconstrained by the forces of natural selection that would resist the change, then an imbalance results and the shape of a giraffe would depart from the adaptive optimum needed for its habitat. Extinction might follow, as it did in giant elks. Furthermore, too many other adaptations are needed to support the shape of a giraffe, and a simple link between neck dimensions and body mass is, perhaps, too unlikely, as it does not take into account, in the way that natural selection can, all the subtle attributes needed for success. If so, then what other selective pressures might have been involved to produce the familiar shape of giraffes?

The Shape of Giraffes and the Browsing Hypothesis

If the selection pressures themselves will always be mysterious, what can be done is to see what measurable benefits, if any, a modern giraffe's shape confers on its daily life. An obvious modern benefit is that their shape gives access to browse above a height that other browsers can reach, or browse that is at the center of low trees and shrubs (see Plate 18.1).

Long necks in many taxa, for example in the herbivorous *Apatosaurus*, have evolved to increase the volume of food available (Wilkinson and Ruxton, 2011), and it has been assumed that throughout the evolution of *Giraffa* sp., access to high browse has been a prominent selection pressure for tallness. But was it? The first recorded connection between shape and browsing was made by Jean Baptiste Pierre Antoine de Monet (the

Chevalier de Lamarck), in 1809, almost two decades before the first living giraffes arrived in Europe, but when numerous more-or-less accurate illustrations of what they might look like were available, starting with Arnout Vosmaer's drawing (Plate 3.5 in Chapter 3). Lamarck thought the origin of their shape was obvious. It is worth quoting his thoughts. He wrote,

> *It is interesting to observe the result of habit in the peculiar shape and size of the giraffe: this animal, the largest of the mammals, is known to live in the interior of Africa in places where the soil is nearly always arid and barren, so that it is obliged to browse on the leaves of trees and to make constant efforts to reach them. From this habit long maintained in all its race, it has resulted that the animal's fore-legs have become longer than its hind legs and that its neck is lengthened to such a degree that the giraffe, without standing up on its hind legs, attains a height of six metres (nearly 20 feet). The numerous repetitions of these organized activities strengthen, stretch, develop and even create the organs necessary to them. We have only to watch attentively what is happening all around us, to be convinced that this is the true course of organic development and changes. It is not the shape either of its body or its parts which give rise to the habits of animals and their mode of life; but that it is, on the contrary, the habits, mode of life, and all the other influences of the environment which have in the course of time built up the shape of the body and of parts of animals. With new shapes, new faculties have been acquired, and little by little nature has succeeded in fashioning animals such as we actually see them.*

Though he had not seen a giraffe, he must have read the literature then current on giraffes, but, if he had, he seemingly ignored it, especially Vosmaer's monograph published 20 years earlier, which undid the fallacy of markedly different leg lengths. Nevertheless, it is easy to be seduced by the logic of Lamarck's ideas, mainly because we are familiar with the many examples we see daily about how human bodies can be shaped by experience: the abnormal development of muscles in laborers and athletes, the change in shape of bones in response to stresses and strains, thickened skin on the feet, a suntan, all are caused by variable expression of genes in response to environmental stimuli. Frequent use of a body part does influence gene expression of that part but, regrettably, changes in gene expression do not alter the genome and so are not hereditable: acquired characteristics cannot be passed to descendants in the way imagined by Lamarck. Darwin was not dismissive of Lamarck's ideas, but together with Wallace came up with an alternative explanation for the "perpetuation of varieties." In his famous presentation to the Linnaean Society in London in 1858, Alfred Russel Wallace wrote on the "hypothesis of Lamarck":

> *The powerful retractile talons of the falcon—and cat—tribes have not been produced or increased by the volition of those animals; but among the different varieties which occurred in the earlier forms, those always survived longest which had the greatest facilities for seizing their prey. Neither did the giraffe acquire its long neck by desiring to reach the foliage of more lofty shrubs, and constantly stretching its neck for the purpose, but because any varieties which occurred among its antitypes with a longer neck than usual at once secured a fresh range of pasture over the same ground as their shorter-necked companions, and on the first scarcity of food were thereby enabled to outlive them.*

Darwin did not mention giraffes in his presentation to the Linnaean Society in 1858, nor in the first edition of *The Origin of Species* in 1859, but concurred and paraphrased Wallace's words later.

The comments of Lamarck and Wallace were the origin of the idea that the shape of giraffes confers a feeding advantage. Were there benefits other than access to browse? Darwin did not think so. He was reluctant to be drawn into a debate about the evolution of tallness in giraffes, but eventually did so in 1872 to answer a critique by St. George Mivart, who questioned the ability of natural selection to produce new species. Mivart, who tried (but failed) to reconcile Darwin's theories with the doctrine of the Catholic Church, was a "distinguished zoologist," according to Darwin, who had "recently collected all the objections which have ever been advanced by myself and others against the theory of natural selection" (Darwin, 1872). Darwin felt obliged to respond to Mivart's objections which Mivart had published in 1871 in *On the Genesis of Species* and did so at great length in the sixth edition of *The Origin of Species* (published in 1872), which included the famous passage that has become the standard explanation for the evolution of giraffe tallness:

> *The giraffe, by its lofty stature, much elongated neck, forelegs, head and tongue, has its whole frame beautifully adapted for browsing on the higher branches of trees. It can thus obtain food beyond the reach of other Ungulata or hoofed animals inhabiting the same country; and this must be a great advantage to it during dearths. . . . So under nature with nascent giraffe, the individuals which were the highest browsers and were able during dearths to reach even an inch or two above the others, will often have been preserved.*

Darwin too was deceived by the optical illusion of foreleg length, but his summary of the benefits of tallness is persuasive and its details are supported by modern research. Edible biomass is concentrated at high levels where light intensity is greatest: a cubic meter of high-level canopy can contain three times more browse than does a cubic meter at levels of less than a meter, and high browse can be succulent and rich in protein and minerals. Added to the advantage of reduced competition, the better quality and quantity stimulates faster growth rates, shorter calving intervals, a younger age of first conception and, therefore, more calves per female lifetime and a shorter intergenerational gap, all of which favor survival of the species. All attempts so far made to confirm or refute Darwin's hypothesis have studied giraffe feeding patterns and behavior, and the results have been mixed, although they have provided some confirmation. For example, giraffes do not browse at high levels all the time (see Plate 18.1). They often feed at heights at which other browsers also feed (du Toit, 1990a), prefer to feed at shoulder height most of the time (Leuthold and Leuthold, 1972; Pellew, 1984b; Young and Isbell, 1991), even in the dry season, feeding from the tops of trees only when new shoots are available (Pellew, 1984b) and the quality of browse good (Van Aarde and Skinner, 1975). In addition, there is a vigilance cost to feeding at high levels. Nevertheless, a long neck is an overwhelming advantage when the lower leaves of trees have been eaten by shorter browsers or have been depleted by drought (see Plate 18.2).

Mivart, however, spotted the problem and challenged Darwin on the cost of browsing. He asked "whether the disadvantages thence arising would not in times of scarcity, more than counterbalance the advantages." Darwin responded that "a supply of food left untouched by other hoofed quadrupeds of the country, would have been of some advantage." Moreover, "nor must we overlook the fact, that increased bulk would act as protection against beasts of prey." Mivart was right and Darwin's response weak and inconclusive. Rapid growth and large body mass do mean that nutritional demands increase, and acquiring sufficient nutrients, the fossil record implies, was not easy. We cannot know for sure, but it seems very likely that good-quality browse was, once, too scarce to allow a larger body size to evolve in ancestral giraffes. Any attempt then to evolve a larger body size would have been fruitless, as not enough browse would have been available to support a large size and provide enough for all the other browsers eating it. The first descendants of *G. jumae* were medium-body mass giraffes weighing ~600 kg. Evolution of the East African ecosystem, and the change from tropical forest into the savannah–acacia-woodland biome—made abundant browse of good quality available, but at a height of two or more meters. Relatively suddenly, a whole world of food became available that had not been there before. The tallest giraffes in a herd would have been able to reach it. Tallness would have survived and flourished and removed a bottleneck restriction on size. Giraffes already had the genes for long legs. Selection for long necks was decisive. Size could increase until a new balance between size, tallness, and availability of high-quality browse was reached. Tallness provided food, the food sustained the tallness, and the cycle led to a progressive increase in size until the size and shape of small ancestral giraffes became indistinguishable from that of modern giraffes (Chapter 6).

Browsing and Survival

While much of the modern evidence is consistent with access to browse as the main selection pressure that led to the shape of giraffes, a better test of Darwin's browsing hypothesis is whether, in a drought, taller giraffes survive better than do shorter ones. Intermittent droughts have been a part of *Giraffa* evolution from its very beginning through to modern times, and it could be predicted from Darwin's explanation that during a "dearth" the survivors would be the tallest members of any population, and there is support for this expectation. A severe drought in Nairobi National Park during 1973–1974 killed 2,623 ruminants, just four of which were giraffes (Hillman and Hillman 1977). Most deaths were of grazer ruminants. Giraffes were almost unaffected by a prolonged drought in the Kruger National Park and surrounding conservancies in 1987 (Walker et al., 1987). In the already extreme environment of the Kaokoveld in the far north of Namibia, no giraffe deaths were recorded during a drought that also killed hundreds of grazers (Scheepers, 1992). However, drought was a factor in reducing the size of the Niger population of *G.c. peralta* (Le Pendu and Ciofolo, 1999), and in reducing the size of the population in Timbavati Nature Reserve (Hall-Martin, 1975a), and there is some evidence that during a "dearth" the large size of giraffes can be a fatal disadvantage (Mitchell et al., 2010). It is, of course, impossible to arrange a drought to test Darwin's explanation, but in 2008 a severe drought did come to the Bubye Valley Conservancy in southeastern Zimbabwe (21°42′S, 29°54′E), a habitat of *G.c. giraffa*. The principal source of browse for *G.c. giraffa* there is *Acacia* spp., and they defoliate in the latter parts of winter. When rainfall in late summer (January to May in the Southern

Hemisphere) is good, the resulting high water table stimulates the *Acacia* trees to produce nutritious buds and flowers in August. They sustain giraffes until the next summer rains arrive. In drought years, a lower water table delays both flowering and leaf budding at a time when ambient temperature is still relatively low and demand for food energy is high: in the Bubye Conservancy, average minimum winter (August) temperatures (10.1° ±3.0°C) are less than half those in summer (20.7° ±1.2°C). In the summer months of 2006 and 2007, 461 mm (2006) and 440 mm (2007) of rain were recorded at Bubye, with 248 mm (2006) and 158 mm (2007) falling between January and May. In 2008, summer rainfall was 236 mm and just 55 mm fell in late summer, precipitating a collapse in browse production. By August large numbers of giraffes had succumbed to cold and malnutrition.

From the skeletons of giraffes that had died and from the robust allometric relationships between the length of bones and age, size, and gender, the demography of the giraffes that had died could be established, and compared to the typical demography of giraffe populations in which the ratio of adults:sub-adults is ~2.5:1 and sub-adults to juveniles ~3:1 (Fennessy et al., 2003; Figure 18.4).

The demography of the giraffes that had died was not representative of the demography of typical giraffe populations, and differences between them were statistically significant. No skeletons of giraffes less than one year old were collected. Most skeletons collected came from animals that fell into two main groups. The first group consisted of young animals, typically two years old, weighing 200–400 kg and having a height of

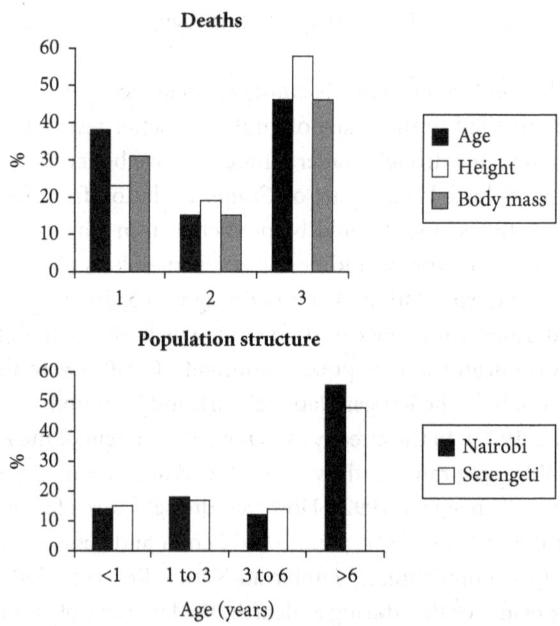

Figure 18.4. Top: Frequency distribution of deaths during a drought by age, height, and body mass. Age categories are 1 = 1–3 years; 2 = 3–6 years; 3 = >6 years. Height categories are 1 = <3 m; 2= 3–4 m; 3 = >4 m. Body mass categories are 1 = 200–400 kg, 2= 400–800 kg, 3 = >800 kg. Bottom: Typical (historical) frequency distribution of giraffe populations in Nairobi National Park (Dagg and Foster, 1976) and Serengeti (Pellew, 1983b).

less than 3 m. This group accounted for 32% of deaths, but it constitutes only ~15% of a typical, healthy population. The second group comprised animals that were mature, tall, and heavy. Typically, these animals were greater than 10 years old, weighed ~1,200 kg or more, and were at least 4.5 meters tall. This group accounted for 54% of all deaths, a proportion that reflects the frequency of mature giraffes in a population, rather than the lower frequency expected if the advantages conferred by their height were protective. In addition, in a giraffe population there are 1.3 to 1.5 females for each male (Hall-Martin, 1975a, 1975b; Fennessy et al., 2003), and the reason sometimes given for the bias is that more males than females are killed by lions or humans (Hall-Martin, 1975a, 1975b; Marealle et al., 2010). At birth the ratio of males to females is 1:1 (Hall-Martin, 1975b), although in one study of East African giraffe populations, males outnumbered females by an extraordinary ratio of 1.6 males:1.0 females at birth (Bourliere, 1961). Nevertheless, more-or-less equal numbers of adult males and females should be found in a population and more-or-less equal numbers of males and females should have been represented in the skeletons, but they were not. In mature giraffes, the sex of the skeleton can be established from the size and shape of the skull, and of the 15 skeletons of mature giraffes collected, 13 were male, for a male to female ratio of 6.5:1, quite different from that in the general population. Hence, the giraffes most at risk in this "dearth" were large males.

The demography of deaths in this drought is consistent with the demography of deaths in the Timbavati Nature Reserve in South Africa over the years of 1964 to 1967 (Hirst, 1969). During that period there were 1,618 deaths of ungulates from starvation, predation, and accidents, of which 317 were giraffes. Most (88% or 279/317) of all giraffe deaths occurred in winter (June to November in the Southern Hemisphere) and most of those in the winter of 1965 following a long period of low summer rainfall. There were no deaths of disease. Of the 317 giraffe deaths, 103 were the result of predation (6% of all deaths and 32% of giraffe deaths) and 182 by starvation (11% of all deaths and 57% of giraffe deaths). Starvation and predation were linked. Giraffes in a weakened condition at the end of winter are easier prey than when they are in full condition. The age and sex of the giraffes that had died was determined from skull characteristics, and virtually all giraffes that had died fell into the prime and old age classes, with males in those classes forming 65% of deaths and females 35%.

An explanation for these departures from Darwin's predictions hinges on the availability of browse and the food energy and water it provides. The food energy required by giraffes every day (kJ d^{-1}) is related to its body mass and can be predicted from the regression equation $860 \times Mb^{0.74}$. In the mature, 1,200-kg or more giraffes that died in this drought, it would have been ~160,000 kJ.d^{-1}. The daily water needed by a 1,200-kg giraffe is also related to its body mass, and it can be predicted from the regression equation $0.21 \times Mb^{0.74}$, and would have been ~40 L. All of a giraffe's food energy and a large part of its daily water requirements under normal circumstances come from eating fresh, succulent browse, and the amount of fresh browse required can be predicted by the regression equation $0.34 \times Mb^{0.74}$. For a 1200-kg giraffe it is 65 kg d^{-1}. In a drought, of course, neither fresh browse nor surface water is available. A 200-kg, one-year-old giraffe requires 17 kg of fresh browse daily, containing 5 kg d^{-1} of dry matter, ~45 000 kJ d^{-1} of energy, and 12 L of water. Thus, small giraffes have much lower needs and, if they could get access to good-quality browse free of competition from other browsers, should survive best, but their skeletons were over-represented. The reason that more juveniles of one to three years old died than expected was because they could not avoid competing for browse with other browsers such as kudu and eland, which can browse up to a height of two meters (du Toit, 1990). Adults of those species also die disproportionately during droughts (Hirst, 1969; Knight, 1995). The reason that the biggest giraffes died was simply because the amount of browse available to them during the drought was well below the amount needed to maintain their metabolism. Fewer adult females died than males because they are smaller but not much shorter and their metabolic demands are less, but the consequences of drought for them is that their fetuses die and are resorbed or aborted, or lactation ceases and their unweaned calves then have no source of nutrients and die. Aborted remains and dead neonates are highly desired by scavengers, easily dismembered, and quickly devoured or dispersed. The resorption of fetuses and scavenging explains the absence of skeletons of giraffes less than one year old. In a census done after the drought, in a sample of 314 giraffes, 85% were adults (of which 75% were female), 12% were young adults, and just 2% were yearlings. Thus the giraffes that survived this drought were not the tallest and most able to reach "even an inch or two above the others," which is what Darwin's hypothesis would have predicted. A greater determinant of survival appears to be the ratio of metabolic demand to available browse. The survivors were young adults or adult females in their reproductive years. As young cows are the preferred partner of bulls, they are those most likely to breed successfully when favorable conditions return, to ensure survival of the species.

The conclusions that can be drawn from this drought are not quite what Darwin had in mind when he was at last forced to comment on the utility of the shape of the giraffe, and the causes of its preservation. When he wrote his famous "standard" story, he might have changed its emphasis had he taken note of what his archenemy, Richard Owen, had written some 20 years before. In 1851 Owen wrote:

> *In proportion to the bulk of a species is the difficulty of the contest which, as a living organized whole, the individual of each species has to maintain against the surrounding agencies that are ever tending to dissolve the vital bond, and subjugate the living matter to the ordinary chemical and physical forces. Any changes therefore, in such external conditions as a species may have been originally adapted to exist in, will militate against that existence in a degree proportionate, perhaps in a geometric ratio, to the bulk of the species. If a dry season be greatly prolonged, the large mammal will suffer from the drought sooner than the small one; if any alteration of climate affects the quantity of vegetable food, the bulky herbivore will be the first to feel the effects of stinted nourishment.*

It turns out that for *G.c. giraffa* living in Zimbabwe, Owen was right. Darwin was wrong. In that drought, the biggest and tallest giraffes did not survive. They died from a lack of sufficient browse.

A Shape for Drinking?

The browsing hypothesis is a classic example of the idea that current use mirrors selective pressures, and even if its accuracy cannot be known, it is an intuitively sensible conclusion to make about the long neck and long head of giraffes because large modern giraffes can and do acquire more browse than they could if their necks and heads were shorter. However, Chapman Pincher (1949), much better known as an investigative journalist and spy-hunter than as a biologist, although he had graduated with a degree in biology (he died in 2014, aged 100), made one of the first objections to the traditional story. He concluded that a Darwinian dearth, severe and long-lasting enough and/or frequent enough for natural selection to operate to produce a long neck, would cause the recurrent wastage of young giraffes, and thus lead to extinction of the species rather than its evolution. A.R. Brownlee, in his 1963 analysis of giraffe evolution, also came to the conclusion that preferential access to nutrients could not be the only or even main evolutionary stimulus for a long neck. He argued, as had Pincher, that a long neck served adult males best and a benefit for them alone was unlikely to be a benefit for the species generally. Brownlee's proposal was that their shape conferred a thermoregulatory benefit, which it does for giraffes of all ages, although it is unlikely that better thermoregulation was the main evolutionary stimulus for a long neck (see Chapter 14). Pincher also wondered why it was that "dearths" did not encourage selection for long necks in other ungulates, and why it was that only giraffes achieved this distinction, or as St. George Mivart had put it over 80 years earlier, "if high browsing be so great an advantage, why has not any other hoofed quadruped acquired a long neck and lofty stature?" If a long neck had some general utility or advantage, then its evolution should have initiated an impressive radiation of forms, but did not and has not. Pincher went further. Adult males, he noted, are on average taller than adult females, which in turn are taller than their young, and a life punctuated by dearths, he argued, would place all less tall members of the species at a permanent disadvantage, intermittently exacerbated by drought, and extinction would be inevitable.

Darwin's answer to Mivart was that

> *in every district some of one kind of animal will almost certainly be able to browse higher than the others and it is almost equally certain that this one kind alone could have its neck elongated for this purpose through natural selection and* [using Lamarck's words] *the effects of increased use. Why in other quarters of the world various animals belonging to this same order have not acquired an elongated neck cannot be distinctly answered; but it is as unreasonable to expect a distinct answer to such a question, as why some event in the history of mankind did not occur in one country but did in another. We are ignorant with respect to the conditions which determine the numbers and range of each species—for instance some areas support singularly few large quadrupeds, for instance S. America, though it is so luxuriant; whilst S. Africa abounds with them to an unparalleled degree. Whatever the causes may have been, we can see that certain districts and times would have been much more favourable than others for the development of so large a quadruped as the giraffes.*

In the end he conceded that . . .

> *we cannot explain why, in many quarters of the world, hoofed quadrupeds have not acquired much elongated necks or other means for browsing on the higher branches of trees.*

Pincher's explanation for tallness had nothing to do with trees. He proposed that the most extraordinary feature of a giraffe is the length of its forelegs, which, in his view, conferred a long stride and a speed of 32 miles per hour (~50 km h^{-1}). This view was immediately contested on the (erroneous) grounds that if speed was the purpose, then the hind legs should also have lengthened proportionately, but they had not (Robinson, 1949)—erroneous because as we now know, there is very little difference between foreleg and hind leg lengths. Robinson also assumed that giraffes were altruistic, and thus "it should be remembered that many animals will deliberately leave the best and most easily obtained food for their young, the same law holding good between male and female," so the idea that males would survive a drought but females would not had to be wrong. Pincher's idea also was supported by Frederic Wood-Jones in a briefer but more pertinent letter to *Nature*, also in 1949, in which he commented that Pincher's explanation had been "advanced with perhaps more cogency by myself" in his book *Hallmarks of Mankind* (Wood-Jones, 1948). Wood-Jones was a peripatetic comparative anatomist who filled (for short intervals) the Chairs of distinguished anatomy departments all over the world. His last academic appointment was at the Royal College of Surgeons in London, as Professor of Human and Comparative Anatomy and Conservator of the Hunterian Museum, the latter position once held by Richard Owen. One of Wood-Jones's tasks was to rebuild the collection that had been started by John Hunter. His support added considerable weight to Pincher's arguments, not least because Wood-Jones wrote beautiful and persuasive prose, but also because in his letter he mentioned that another, perhaps even more famous anatomist, Robert Broom, most renowned for his discovery of *Australopithecus africanus* in South Africa, agreed with his analysis. Wood-Jones, like Owen, argued ceaselessly against Darwin's ideas on evolution, especially the evolution of mankind, and ascribed all adaptations to an ill-defined "directive force" and the Lamarckian principles that "function is the creator of structure and not its offspring" and of "satisfaction of a need." Accordingly he wrote that the long legs of giraffes were "created for rapid progression in open spaces" but they also created a need—a need to be able to obtain water and "this need is met by the harmonious elongation of the neck." Speed is one of the selective benefits of long legs, but with respect to the benefit of a long neck, Pincher (as well as Wood-Jones and Broom) was following an idea first analyzed by Edwin Colbert 10 years earlier, in 1938. Colbert had also speculated that as a response to increasing limb length there had to be concomitant elongation of the neck if a giraffe was to be able reach the ground.

This explanation for the development of a long neck is, as Wood-Jones put it, a cogent one, but it can be dismissed on empirical grounds. If a long neck did evolve to offset long legs, then during giraffid evolution, as leg length increased there should have been a concomitant increase in neck length. In the first giraffid, *Canthumeryx*, leg length was ~800 mm and thus similar to the length of the legs of modern-day buffalo and wildebeest (Figure 18.5). For the succeeding many millions of years, in concert with increases in body mass, leg lengths increased from the 800 mm in *Canthumeryx* to 1,200 mm in that offshoot of *Palaeotragus*, the okapi, to ~1,600 mm in *Palaeotragus sp.* proper, to ~1,700 mm in *Samotherium sp.* and to ~2,100 mm in modern giraffes. What was neck length in each? In *Canthumeryx*, neck length was about 550 mm (Badlangana et al., 2009), and thus similar to the length of buffalo and wildebeest necks, for a ratio of leg length:neck length of 1.5:1. Leg lengths of the giraffids that followed *Canthumeryx* were,

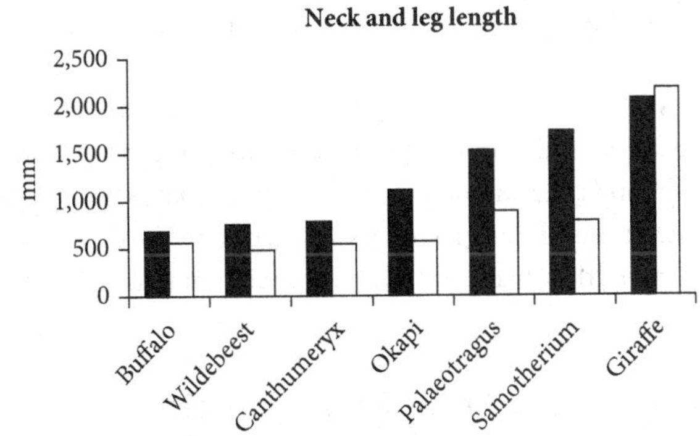

Figure 18.5. Increases in neck (□) and leg (■) length during the evolution of giraffids, compared to their lengths in extant adult grazing buffalo (*Syncerus caffer*; body mass = 434±69 kg) and wildebeest (*Connochaetes sp*; body mass = 225±15 kg).
Data from Colbert, 1938; Badlangana et al., 2009; Mitchell and Skinner, unpublished; Roberts and Mitchell, unpublished.

on average, almost twice as long as neck length (leg length:neck length = 1.8±0.1:1; Figure 18.5). In other words, neck length did not increase until the evolution of *Bohlinia* and *Giraffa* as species, where, for the first time, neck length exceeded leg length to become 1.21-fold longer in female *G.c. giraffa* and 1.14-fold longer in male *G.c. giraffa* (Figure 18.3). Thus Pincher's hypothesis that the neck of giraffes had elongated during evolution because their legs did, with the selective advantage that a long neck enabled giraffes to reach the ground, is not supported by analysis of the evolutionary evidence. Besides, the contortions modern giraffes go through to drink is ample evidence that while a long neck (and long head) does make drinking from a pond possible, it does not make drinking all that much easier. Leg length initially increased in artiodactyls to provide speed and agility, but in giraffids, all of them, except for *Canthumeryx*, principally browsers, long legs were the primary development that allowed access to browse at higher levels. Contrary to Darwin's view that they had not acquired "other means for browsing on the higher branches of trees," they had.

There is a good biomechanical explanation for why, in *Bohlinia* and *Giraffa*, an elongated neck evolved, rather than a continuation of the trend for elongated legs. It is that there is a limit to the length that legs can become. That limit is determined by the strength of the limb bones that are needed to support body mass. The strength of limb bones mostly is related to their length and diameter, with, typically, increases in length compensated for by proportional increases in diameter. Among artiodactyls generally, diameter does not increase as rapidly as length, and fracture will occur, for example, if length is 25–26 times greater than diameter. The ratio between length and diameter for the limb bones of giraffes is remarkably constant in all ages and is on average 11.0±0.6 (length ÷ diameter). Bones, however, have a safety factor of 3–5, and so *in situ* bone length should, at least theoretically, not exceed bone diameter by more than a factor of 8 (25/3) and 5 (25/5) to be within safety limits. Giraffes have evolved limb bones that have features that prevent fracture, but their limb bones cannot become increasingly slender during growth without a concurrent increase in the risk of fracture. Limb bones of modern giraffes are as slender as they can be (van Sittert et al.,

2015). It is the biomechanics of bone strength that limit limb bone and hence leg elongation. If natural selection prevents further leg elongation for biomechanical reasons as a "means for browsing on the higher branches of trees," then selection for those giraffes having a longer neck in addition to their already long legs would have achieved that goal. Evolution of a long neck in *Bohlinia* and its descendants was not because they had long legs and needed to reach the ground to drink, but because a longer neck was one of the selective advantages they had and have over their competitors.

A Shape for Vigilance?

A.R. Brownlee in 1963 had concluded that a long neck was "an aid in achieving heat loss," but he was careful to put that conclusion in perspective. Achieving heat loss was not its only benefit. Long necks, he wrote, also enabled giraffes "to achieve that size and tallness which confers greater ability to evade, or defend against, predators and to reach a source of food otherwise unavailable to them." Brownlee's thought that a long neck would confer an ability to evade predators had been pointed out in 1872 by Darwin: "its tall neck—and the taller the better—would serve as a watchtower." There is no doubt that tallness, combined with good vision, confers a unique advantage on giraffes. The ancient Egyptians used giraffes as the hieroglyph meaning "to foretell" because they had seen that giraffes had (and have) an uncanny ability to detect changes occurring in their environment long before other species sharing their habitat had done so. The vision of giraffes is and has been legendary among hunters and game wardens. Hendrik Wikar, writing in 1779, said, "If you manage to, by stalking, get within shot of this animal it is by great good fortune, and it seldom happens, for, being so tall and always eating the tree tops, it sees you from afar." Selous, writing of his experiences, declared that "owing to their great height, their keenness of sight and great range of vision they are able to see anything approaching them, when still a long way off" (Selous, 1908). Sir Harry Johnston, the "discoverer" of okapis, wrote:

> *Towering above the trees, the giraffe with its large eyes can scan the surrounding country from an altitude of twenty feet above the ground, and in this way during the daytime, and possibly on nights that are not too dark, can detect the approach of a troop of lions, the only creature besides man which can do it any harm.* (Johnston, 1902)

Good vision in giraffes could be expected. The preferred habitat of giraffes is woodland-savannah, group sizes are small, home ranges are much larger than predicted from metabolic needs, and distances between group members are large. Within that habitat, giraffes must be able to detect movement, shapes, and coat patch patterns at relatively long distances if they are to avoid predators, recognize calves, detect competitors, find a mate, or keep in touch with members of their group. Most herds of giraffes these days (75%) comprise fewer than six individuals, and distances between individuals that make up the group can be a kilometer or more. Small herd size means that the "many eyes" vigilance that a large herd provides is not available to giraffes (if it ever existed in them; see Chapter 7), and when intraspecific distances within a group are large, good vision is a useful asset (Kiltie, 2000). Indeed, distances between conspecifics increase as visual acuity increases (Kiltie, 2000) because they can. Hearing and smell also help. Giraffes' auditory acuity is unknown, but observation suggests it is acute. They are very likely able to make sounds that are understood by other giraffes, but what those understandable sounds are, is unresolved. Their olfactory acuity also is unknown, but as both sexes have very large olfactory bulbs, odor must be an important signal. But it is on their eyes that they rely most. The origin of their excellent vision is the large size of their eyes and the anatomy and dimensions of the retina, but their visual acuity also depends on the long axial length that is the consequence of large eye size, and the degree of binocular overlap.

The size of an adult giraffe's binocular overlap and the way it changes with age is shown in Figure 18.6, where it is compared with the overlap in some other species. Plains-adapted animals like wildebeest and rabbits have very little overlap (and thus very large lateral visual fields), and humans and other predators have poor lateral vision but excellent frontal vision. Giraffes have good overlap and thus good frontal vision, but they also have something more than just a larger overlap. As Figure 18.6 shows, as their neck elongates, both binocular overlap and axial length increase. In other words, the benefits of the good vision that giraffes have evolved have been magnified by a periscope-like ability to see above tree level, and together the long neck and good vision provide an advantageous adaptation for the open woodland-savannah environment that giraffes inhabit, and which *Giraffokeryx* inhabited: it was in *Giraffokeryx* that a large eye first evolved in the giraffe lineage. The visual anatomy of ancestral giraffes started

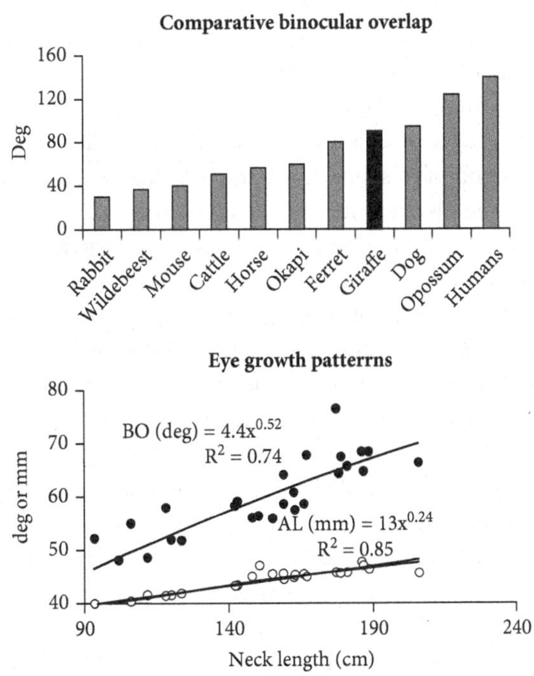

Figure 18.6. Top: Comparison of the degrees of binocular overlap in giraffes and some other animals. Bottom: The change in binocular overlap (BO, ●, deg) and axial length (AL, ○, mm) in giraffes as neck length increases.
Constructed from data in Heesy, 2004; Mitchell et al., 2013b; and Roberts and Mitchell, unpublished.

by *Giraffokeryx* could have been emphasized by the conditions that existed at the time when giraffes were evolving rapidly. Coincident with the evolution of *Bohlinia* and the subsequent radiation of *Giraffa* there was a shift from tropical vegetation to the now familiar arid, open, woodland-savannah, a habitat that was congenial not only for giraffes but also for other species. Immigration of large predators followed (Janis, 1993). The habitat allowed the size of the home range of giraffes to increase, and vigilance, communication, and detection and avoidance of predators became correspondingly difficult. In this scenario, the evolution of good, periscope-like vision could easily have been one of the factors underpinning natural selection for a neck as long as that of *Bohlinia* and even longer, to become as long it is in modern giraffes. Whether browsing, temperature regulation, or vigilance were primary, secondary, or just coincidental, contemporaneous adjustments to body shape is difficult to gauge, but each would have reinforced the others' selective advantages.

The Shape of Giraffes and the Sexual Selection Hypothesis

In his book *The Descent of Man and Selection in Relation to Sex*, published in 1871, Darwin developed the theory of sexual selection to account for sexual dimorphism as shown by exaggerated secondary sexual characters like the antlers of stags or the tails of peacocks. Ronald Fisher, best known for his monumental contributions to statistics, but equally influential for his twentieth-century revisions of the theory of evolution, proposed in 1930 "a runaway hypothesis" to account for the "rapid" evolution of sexually selected characters and decided that they arose because of persistent female selection of males that had them. The species would benefit through propagation of the "best" genes, assuming (without much evidence to do so) that beneficial genes go along with those for exaggerated secondary characters. No doubt human females (and males) practice selection of this kind too, although what it is that they find attractive, from what an interested observer sees, is not as obvious as a peacock's tail. Characters that have evolved by sexual selection always are more exaggerated in one sex than the other and, consequently, sexual dimorphism for a character has become a criterion for deciding if sexual selection is its origin. Any character that ends up as being sexually selected starts as one shared equally by both sexes, then, over evolutionary time it becomes more highly developed in one of the sexes, often but not only in males, either because of their different role in reproduction and/or because the advantages the trait confers on that sex means it is able to use resources better (Lande, 1980) or, indeed, for other yet to be discovered reasons. Whatever the cause of excess development, for example in a male, female preference for it follows and the character finally becomes limited to one sex, able only to be expressed by that sex, as are the primary sex characters that define gender. It is a process of intra-species speciation such that only one of the sexes ends up with the genetic makeup to express the character.

While the character may increase breeding success, it usually has a cost. Darwin, in 1874, identified a survival cost (a higher mortality rate) as a result, simply because sexually selected characters do not evolve as an adaptation to the environment and thus they reduce their bearer's "fitness optimum" (Lande, 1980). Consequently, there is a built-in constraint against the development of sexually selected characters: their expression increases until balanced by natural selection operating under environmental stimuli. Thus characters as obvious as a peacock's tail or the grossly exaggerated antlers of the now-extinct Irish elk occur rarely—indeed, so rarely that since Darwin's exposition of the theory, the defining criteria for what constitutes a sexually selected character have had to be extended beyond dimorphism and a "cost." Senter (2007), for example, writing about the origin of long necks in dinosaurs, which, he argued (controversially; Taylor et al., 2011), were a result of sexual selection, added that sexually selected characters should be used in dominance contests and, during their phylogenetic history, should have increased in size independently of any increases in size of other parts of the body. Fisher's runaway hypothesis also implies that a sexually selected character that is distinguished by its size will increase in size faster than, say, the rate at which body mass increases. In other words, the allometric equations that describe its growth rate relative to increases in body mass will usually, although not always (Bonduriansky, 2007), have an exponent greater than 1 ($Mb^{>1}$; Kodric-Brown et al., 2006).

Is the short body and the tall elongated shape of giraffes a result of sexual selection? It is not obvious that they are. Like all the other possible selection pressures that have produced the shape of giraffes, sexual selection for a heavy, long neck would have had to operate first in *Bohlinia*, as it was in *Bohlinia* that a shape the same as it is in modern giraffes first arose. Very few fossil remains of *Bohlinia* have been recovered so far (many of which are listed in Table 6.3 in Chapter 6), and they do not indicate that there was significant sexual dimorphism. There may well have been sexual dimorphism in the Asian *G. sivalensis* (Chapter 6), but that dimorphism could just have been a consequence of the way that secondary sexual characters develop to produce the typical differences in size and shape of male and female mammals. Secondary sexual characters begin to develop at puberty and depend for their development on the secretion of sex steroid hormones, which are among the most primitive of hormones (Roth et al., 1982; Thornton, 2001). Thus, presumably, the external, visible features of male and female *Bohlinia* would have been affected by puberty as well. Growth in modern giraffes is definitely affected by puberty. Figure 18.3, for example, shows that the masses of the head and neck in modern male giraffes diverge from those in females around a body mass of 500 kg, which is the body mass at the onset of puberty. Figure 15.6 shows how skull mass changes in males after puberty. Furthermore, as much of the data recorded in Figures 18.1, 18.2, 18.3, 18.7, and 18.8 and in previous chapters of this book show, growth of all body parts continues in male giraffes for longer, and ends at a much higher body mass than it does in females, and that kind of growth pattern is typical of large, terrestrial, herbivorous mammals (Reeve and Fairbairn, 1996). That pattern is regulated by sex steroid hormones and specifically by testosterone in males (Badyaev, 2002). The endproducts of hormone action are the differences in the shape and size of male and female mammals generally and giraffes in particular, differences that are not necessarily the endproduct of sexual selection.

Even though we cannot know for certain (for now) whether male and female *Bohlinia* were sexually dimorphic, or whether any differences between them were initiated by puberty, we can analyze the shape and growth patterns, particularly of neck length and mass, in male and female modern giraffes to see if they meet the criteria for sexually selected characters, especially in males. If they do, then it is likely that sexual selection started in *Bohlinia*. In modern giraffes, both sexes rather than just one sex have elongated necks, and in both a long neck has advantages that increase, not decrease, their "fitness." Furthermore, the increase in neck length and size in *Bohlinia* were associated with an increase in body mass, so Senter's criterion of unassociated growth of a sexually selected character can be set aside. Nevertheless, after studying the way that neck mass (rather than length) and head mass increased in a population of Namibian giraffes (*G.c. capensis*), Robert Simmons and Lue Scheepers, writing in 1996, concluded that the necks of giraffes were an example of sexual selection. They found that in Namibian giraffes the neck mass of male giraffes, but not female giraffes, increased faster than did the rate of increase in body mass. Moreover, while head and neck mass increased throughout life in male *G.c. capensis*, their masses reached a plateau in females at a quite low body mass. Are there costs of a heavy head and neck for male *G.c. capensis*? Simmons and Scheepers identified large hearts and high blood pressure as maladaptive costs for giraffes. However, those "costs," it is now recognised, are not "costs" for giraffes. Giraffe hearts are not larger than the typical size predicted for mammalian hearts from body mass, their high blood pressure is as much a "cost" for females as for males, has survival value, and has no known deleterious consequences for giraffes. Simmons and Scheepers also identified a higher predation rate as a cost for large males. Predation rates of male giraffes are higher than they are for females, but rather than to a long neck, that cost can be attributed to the very familiar solitary lifestyle that mature male giraffes follow, to their preference for thickly wooded habitats in which vigilance is compromised, and to their habit of reducing vigilance behavior when alone (Foster, 1966; Hall-Martin, 1975b; Pratt and Anderson, 1985; Cameron and du Toit, 2005). In addition, absence of a continuous supply of high-quality browse leads to malnutrition and, sometimes, death in male more so than in female giraffes, but those excess deaths also are a cost of large size, not of a long neck. Nevertheless, the hyper-allometric increases in neck and head mass in male *G.c. capensis*, their plateau in females, together with the discomforting uncertainties of the browsing hypothesis, led Simmons and Scheepers to the conclusion that the necks and heads of male giraffes were a product of sexual selection. That conclusion has been supported by the doyenne of giraffe biology, Anne Dagg, who wrote in 2014 that males with larger necks more often win combats and females prefer such males and, thus, male shape fulfills the criterion of female preference for a sexually selected character.

Are giraffe bulls that have heavy necks and heads preferred by females, and are those heavy necks and heads used to win contests that decide mating rights? If they were, the benefit for the best-endowed males would be more frequent access to females. Size would matter. The corollary would be that females would choose a well-endowed male, and Fisher's runaway hypothesis would have another example. In our competitive, ruthless world this is a recognizable, but not very attractive, idea. In nature, red in tooth and claw, a heavy neck and head seem to be ideal attributes for dealing with predators or rivals. Giraffes, of course, use hooves to defend against predators, not necks, but what about rivals? Extensive analysis of this plausible assumption has been made by Coe (1967) and Pratt and Anderson (1982, 1985) and they have shown that "sparring," or "necking" occurs in strictly bachelor groups comprising adults and young males (Coe, 1967), but it may also occur between neonates and yearlings (Innes, 1958). Large mature bulls were not seen to participate in any of the many contests observed by Pratt and Anderson, an observation corroborated by

Wolf et al. (2018a). When sparring or a rare fight occurs between mature bulls that could be regarded as being rivals, defense of territory is its likely origin, especially when home ranges are large and the intruder is an unknown bull (van der Jeugd and Prins, 2000). Sparring between young giraffes is more typical of play, sometimes has sexual overtones (Innis, 1958), but does allow careful assessment of fighting ability. Its purpose is to establish a dominance hierarchy, which, once established, lasts a lifetime. Sparring is a stressful activity, at least as determined by the concentration of glucocorticoid stress-hormones in fecal pellets (Wolf et al., 2018b). The fecal pellets of mature bulls with a well-established hierarchical position have significantly lower concentrations of glucocorticoids than do the pellets of younger males. Clearly, sparring is not a matter of dealing with rivals and does not have the purpose of gaining access to females, a conclusion reached in several studies (Pratt and Anderson, 1982, 1985; Bercovitch and Deacon, 2015; Wolf et al., 2018a). High social rank determined by sparring does give dominant bulls breeding preference, but not exclusive breeding rights. Moreover, while females can show a preference for a specific male, they do not select males with the heaviest necks. Any male of adequate size will do (Bercovitch et al., 2006), but not necessarily any male.

How frequently do fights (rather than sparring) occur? In their study of the social interaction and behavior of 304 male giraffes over a period of 10 months and encompassing 3,264 hours of observation, Pratt and Anderson (1985) recorded 166 contests, of which 164 were bouts of sparring and only two of which they classified as fighting. Fighting thus was rare in this group of giraffes, and there is no reason to think that this group was unique: necking contests never, for example, escalated into fights among giraffes in the Lake Manyara National Park (van der Jeugd and Prins, 2000). Serious fights leading to injury are rarer; injury is to be avoided, as severe injury has the potential for, and undesirable cost of, removing an adult from a population. In any case, the young immature males that engage in sparring are not equipped to inflict serious damage. They have heads and necks the dimensions of which are not significantly different from those of females of equivalent body mass (Figure 18.3). The ossicones on a male skull are, though, much bigger than those on a female's skull, but if ossicones had evolved specifically to be a weapon, they would have to be much larger than they are to withstand the force that could be generated by the body mass and neck mass of a large giraffe (Spinage, 1970). If the ossicones had become bigger, then the neck would have had to be more muscular, and giraffes would have looked more like a Sivathere (Figure 4.10 in Chapter 4), but of course they do not. Winners of the contests watched by Pratt and Anderson also did not necessarily have the longest and strongest necks, but if there was a difference in size between two contestants, the winners often were the larger. Nevertheless, Pratt and Anderson described how "a small but evidently old and experienced bull with determination and high self-esteem" drove off three younger but much larger bulls. Standing as upright as possible and looking down his nose with as much hauteur as possible was enough for this bull. It projected its rank simply by display, without needing to spar. All that was required was an appropriate demeanor, not a clash of necks, although as Jarman (1989) has noted, "bluff must be backed up by genuine fighting ability." Thus, winning dominance contests, obtaining sexual favors, and head and neck morphology are not strongly linked and are unlikely to have driven sexual selection for characters such as a heavy neck or head. Females do not die for the

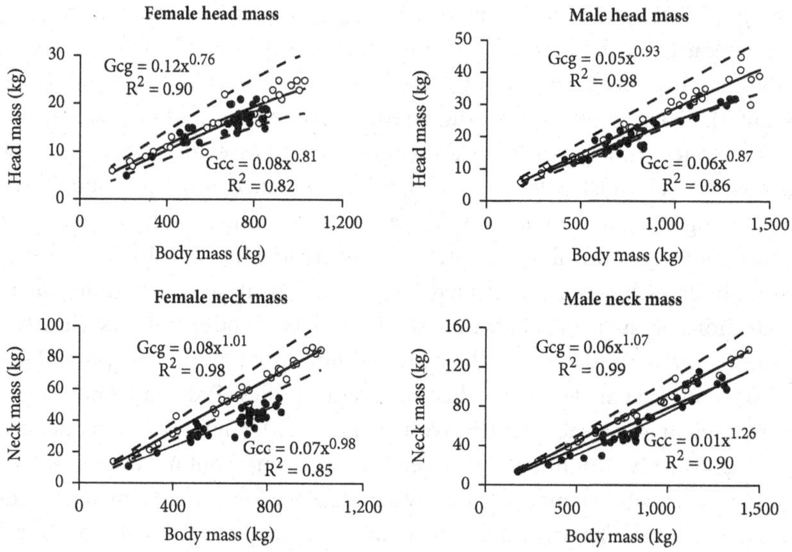

Figure 18.7. A comparison of the growth rate relative to body mass of heads and necks in 30 male and 30 female *G.c. giraffa* (○) from Zimbabwe and 35 male and 41 female *G.c. capensis* (●) from Namibia. The dark solid lines show the best fit line for *G.c. giraffa*, and the dashed lines show the 95% confidence intervals of those lines. The lighter solid lines show the best fit lines for *G.c. capensis*.

From Mitchell et al., 2013a, with permission of John Wiley and Sons.

shape of large males. Dominant bulls do not fight with each other over females and show no jealousy.

Darwin's primary criterion for a sexually selected trait like a peacock's tail was sexual dimorphism for that trait. Simmons and Scheepers drew their conclusions from a study of giraffes resident in Namibia (*G.c. capensis*), in which the neck mass of male giraffes, but not female giraffes, increased faster than did the rate of increase in body mass, and while head and neck mass increased throughout life in male *G.c. capensis*, their masses reached a plateau in females at quite low body mass. Those growth patterns are consistent with the sexual selection hypothesis, and if all subspecies showed the same differences, sexual selection would be a possibility. Are they the same in other giraffe subspecies/variants? Measurements of the girth and total body length in 24 female and 27 male *G.c. giraffa* by Hall-Martin (1975a) in the Timbavati Nature Reserve, and those same measurements, together with others, in 30 female and 30 male *G.c. giraffa* in Zimbabwe (Mitchell et al., 2009b) have been made, and the way they change with growth in males and females are shown in Figures 18.1, 18.2, and 18.3. At least in those giraffes there was little evidence of sexual dimorphism, nor is there when the growth patterns of head and neck mass in *G.c. capensis* and *G.c. giraffa* are compared (Figure 18.7).

Head mass in both sexes and subspecies increases throughout life, with the rate of increase declining as maturity approaches, as did neck mass in females. What the comparisons in Figure 18.7 and in Figure 18.8 also show, though, is that head and neck mass in both male and female *G.c. capensis* are smaller than they are in *G.c.*

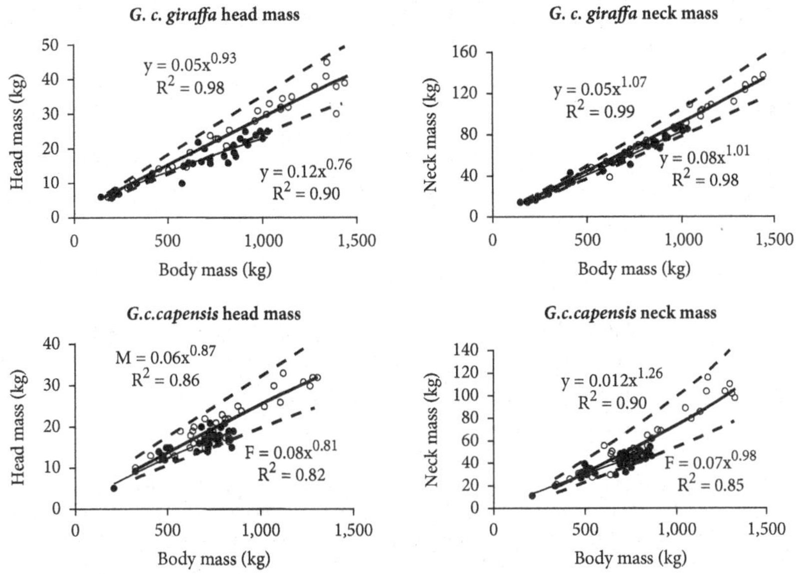

Figure 18.8. A comparison of the growth rate relative to body mass of head and neck mass in male (○) and female (●) *G.c. giraffa* and *G.c. capensis*. The dark solid lines show the best fit line for males, and the dashed lines show the 95% confidence intervals for males. The lighter solid lines show the best fit lines for female *G.c. giraffa* and *G.c. capensis*.
From Mitchell et al., 2013a, with permission of John Wiley and Sons.

giraffa, and are especially smaller and stop growing earlier in female *G.c. capensis* than they do in *G.c. giraffa*.

The difference in size between *G.c. giraffa* and *G.c. capensis* can be attributed to the harsh habitat of Namibia, which can be arid and forbidding and the vegetation sparse. A decrease in body mass is a selective adaptation to an arid environment. The feral horses of the Namib Desert, for example, are also significantly smaller than are the horses from which they are descended (Sneddon et al., 1991) and *G. stillei*, the first and much smaller offshoot of *G. jumae*, evolved in East Africa at a time of aridity. Nevertheless, the much smaller size of female *G.c. capensis* does give an impression that the two sexes of *G.c. capensis* are different. Furthermore, Figures 18.7 and 18.8 do show that the neck mass of males of both subspecies has a hyper-allometric growth pattern (the exponents are greater than 1), a pattern which would fulfill Kodric-Brown et al.'s criterion for a sexually selected trait if that growth pattern produced significantly different (in the statistical sense) consequences for males and females. Figure 18.8 illustrates, though, that the trajectories of head and neck mass growth in male and female *G.c. giraffa* and *G.c. capensis* are not different, and the growth patterns are not consistent, therefore, with sexual dimorphism. Indeed, so similar are the allometric equations that describe the growth patterns of the necks and heads in the two sexes of each subspecies that it is possible to develop generic equations to describe how necks and heads grow irrespective of sex in them. Those combined equations, shown in Table 18.2, have high coefficients of determination (R^2), small standard errors of prediction and, consequently, head or neck mass can be accurately predicted in each subspecies from the body mass of either sex. And, extrapolating

Table 18.2 The Allometry of Head and Neck Mass in Male and Female *G.c. giraffa* and in Male and Female *G.c. capensis*

Subspecies	Variable	*N*	Allometric Equation	R^2	STEYX*
G.c. giraffa	Male and female neck mass (kg)	60	$0.06 \times Mb^{1.06}$	0.93	5.1 kg
G.c. giraffa	Male and female head mass (kg)	60	$0.05 \times Mb^{0.90}$	0.95	3.1 kg
G.c. capensis	Male and female neck mass (kg)	75	$0.08 \times Mb^{0.97}$	0.80	8.1 kg
G.c. capensis	Male and female head mass (kg)	68	$0.09 \times Mb^{0.79}$	0.87	2.4 kg

Calculated from data in Figures 18.7, 18.8. The body mass of male *G.c. giraffa* giraffes ranged from 181 to 1,441 kg, and of male *G.c. capensis* from 333 to 1,324 kg. For female *G.c. giraffe*, body mass ranged from 147 to 1,029 kg, and for female *G.c. capensis* from 211 to 863 kg.
*Standard error of the prediction of Y from X. *N*= number of giraffes.

backward through time, I think we can assume the same about the growth patterns of male and female *Bohlinia*.

While these analyses do not support the sexual selection hypothesis as an origin of the shape of giraffes, in wondering how giraffes evolved to reach their current size and shape, how giraffes work, how they grow, how they get through their day, year after year, generation after generation, it is impossible and indeed would be wrong to eliminate an element of sexual selection in the evolution of the size and shape of giraffes. Present in all liaisons is an element of sexual selection for the visible, like body mass, and the invisible, like odor, and that is as true for humans as it is for all other mammals and perhaps all other animals. Female giraffes are selective of their mates, although what they select for is not well-defined. Male giraffes, we can be sure, are just as selective, and what they find attractive is equally obscure. It is also impossible to judge how much each of the many known and unknown anatomical, physiological, and biochemical adaptations, behavior patterns, and mechanisms have contributed not just to the survival of giraffes, but to their unique shape. No glib summary is possible, but Darwin, as usual, was able to summarize the dilemma neatly: "The preservation of each species can rarely be determined by any one advantage, but by the union of all, great and small." For giraffes, this summary seems particularly apt.

A Shape to Die For?

The dominant theme of giraffe evolution and the emergence of modern giraffes from their plains-dwelling, okapi-like ancestors is that it has depended on, and has been stimulated by, the evolution of a woodland-savannah. That biome is rich in leguminous browse that provides the nutrients that giraffes need, and as that browse grows to a height of around two meters, that food source is unavailable to most other browsers.

It is a biome to which giraffes are adapted and to which they seem irretrievably bound. When a particular habitat has contributed so much to the success of a species, there is a suspicion that the species can occupy no other biome successfully. By this criterion, giraffes have evolved into a dead end from which there is no escape, no evolutionary way that an animal with its shape and adaptations can progress. What if selection for a long neck has produced a structure that could become a disadvantage? In other words, could the anatomy, physiology, and behavior patterns that giraffes have evolved set them on a path to extinction?

Specializations like those found in giraffes always have short-term advantages, otherwise they would not have evolved. The most obvious advantages of the specializations giraffes have evolved are access to browse free of competition, some thermoregulatory and vigilance benefits, and an adult size that is virtually predator-proof. Giraffes can pick and choose what they eat in a large home range and have few costs in doing so. However, they have become accustomed to selecting browse from a few sources and their diets have become narrow. Their size and shape now are highly dependent on a plentiful supply of the succulent browse that they prefer. If the underlying preferred resources diminish, increased vulnerability to extinction follows, a sequence well established for small and large mammals (Safi and Kerth, 2004; Clauset and Erwin, 2008), whereas generalist, less-fussy bulk feeders like buffaloes can tolerate change with greater equanimity. It is natural selection that explains the correlation between specialized diets and the risk of extinction. Natural selection favors species best adapted to their environment, specialization is the consequence, and specialization ineluctably progresses to overspecialization. That such a sequence has occurred in giraffes is obvious to us: giraffes alive today are descended from less-specialized ancestors with, for example, shorter necks. The perilous cost of overspecialization for giraffes would be an inability, or at least a slowness, to adapt to any change that reduces the quality and/or quantity of the browse on which they rely. Overspecialization is also associated with reduced genetic variability and, thus, a reduced capacity to diversify and persist as a species: specialists generate fewer species than do generalists (Vrba, 1987) because genetic variation is low. Only if the specialist species is still capable of speciation can that species avoid extinction (Day et al., 2016). Genetic variation between the current recognized subspecies is very low (Chapter 1), and their ability to generate new species is, therefore, also likely to be low.

One of the more common causes of the reduction or change in a supply of food is habitat fragmentation. In the prehistory of modern giraffes, habitat fragmentation was a consequence of tectonic events, and new habitats in home ranges of almost infinite size resulted. Speciation followed to produce the ancestral giraffes. Nowadays fragmentation of habitats mostly flows from expansion of human activities. Home ranges have shrunk and opportunities for migration and speciation have reduced rather than expanded: the distribution range of giraffes in Africa is now 45% of what it was in the 1970s (Martínez-Freiría et al., 2016), and within that distribution they occupy just 11% of their historical habitat (Ripple et al., 2015). In a home range reduced to that extent, competition between giraffes for browse, and between giraffes and other browsers able to reach it, increases, and even modest declines in browse abundance can cause populations to crash, with the most vulnerable being the largest, a biological relationship noticed as long ago as 1851 by Richard Owen.

Does the large size of giraffes itself increase the risk of extinction? In the past this idea was accepted without question. For example, in a throwaway remark made by Guy Pilgrim, that eminent describer of the giraffids of the Siwaliks, he wrote (Pilgrim, 1934), referring to a sheep-like antelope that lived in Mongolia, that it was of "a larger size than any form hitherto (and) is, therefore, probably a terminal species." Progression of any species to a very large size, in other words, is a cul-de-sac. The underlying reasons for that conclusion have since become clearer. Using a very large database of ~4,700 vertebrates that, according to the International Union for Conservation of Nature (IUCN), are threatened with extinction, Ripple et al. (2017) found a direct correlation between body mass/size and extinction risk, with degree of habitat loss and/or diet specialization secondary factors. Large animals are more susceptible to human persecution (Lyons et al., 2004), with direct killing the dominant threat and, currently, persecution is the cause of the decline and slow extirpation of giraffes in the Serengeti and the Sudan. Large size also is associated with prolonged development time, longer gestation times, and longer generation intervals and, therefore, fewer lifetime offspring. A low reproductive rate reduces the ability to recover from population declines, as well as reducing the rate of gene recombinations and the potential evolution of adaptations to environmental change that recombinations promise. The risk of extinction is correspondingly higher. Elisabeth Vrba (1987) summarized the dilemma of specialization:

> *While specialist species may proliferate during the vast time periods of background evolution, they pay for it when rare and massive environmental disturbances strike. It is the specialists that are shaved off the Tree of Life by mass extinction.*

She could have had the shape and habits of giraffes in mind. Giraffes are highly adapted, overspecialized, and have little room for further evolution. In the longer term they have come to an evolutionary dead end. Their shape (see Plate 18.2) is potentially fatal.

The shape of giraffes is also fatal in quite different ways. A case has been made that their charismatic shape and presence in private conservancies and public National Parks and their ubiquitous use as a marketing tool gives an illusion of abundance, whereas scarcity, in parts of Africa, is the reality (Courchamp et al., 2018). The illusion leads to indifference toward the need for conservation, and it can be predicted that indifference, if it persists, will lead to a further decline in their number. Their charismatic shape and demeanor also make giraffes animals that zoos find essential to keep. People pay to see them. Keeping a giraffe in captivity in a zoo is difficult because it is impossible to replace their complex natural diet with the simple artificial diets that zoos are obliged to use (Hofmann and Matern, 1988; Clauss et al., 2002), or to recreate their natural environments and ambient temperatures (Clauss et al., 1999). Giraffes tend, therefore, to have short lives in zoos, and the reasons fill many pages of learned pathological journals (e.g., Clauss et al., 2002). Zoos too grossly restrict a giraffe's natural tendency to meander. In their natural world they have a home range of about 70 km^2. In a small cage in a zoo, their space for meandering may be no more than a few hundred square meters. That some giraffes seem able to survive in a small cage is, I suppose, a testament to their tenacity and will to live, albeit in a much diminished way. It is hard to imagine anything more soul-destroying for a giraffe. Isak Dinesen in *Out of Africa* perfectly captured the essence of a life in captivity for giraffes when she poignantly described the horror of giraffes being sent to a zoo:

> *In the harbour of Mombasa lay a rusty German cargo-steamer, homeward bound. I passed her in Ali bin Salim's rowing boat with his Swaheli rowers, on my way to the Island and back. Upon the deck there stood a tall wooden case, and above the edge of the case rose the heads of two giraffes. They were, Farah, who had been on board the boat, told me, coming from Portuguese East Africa, and were going to Hamburg, to a travelling menagerie. The giraffes turned their delicate heads from the one side to the other, as if they were surprised, which they might well be. They had not seen the sea before. They could only just have room to stand in the narrow case. The world had suddenly shrunk, changed and closed around them. They could not know or imagine the degradation to which they were sailing. For they were proud and innocent creatures, gentle amblers of the great plains; they had not the least knowledge of captivity, cold, stench, smoke, and mange, nor of the terrible boredom in a world in which nothing is ever happening. Crowds in dark smelly clothes, will be coming in from the wind and sleet of the streets to gaze on the giraffes, and to realize man's superiority over the dumb world. They will point and laugh at the long slim necks when the graceful, patient, smoky-eyed heads are raised over the railings of the menagerie. The children will be frightened at the sight and cry, or they will fall in love with the giraffes, and hand them bread, then the fathers and mothers will think the giraffes nice beasts, and believe they are giving them a good time. In the long years before them will the giraffes sometimes dream of their lost country? Where are they now, where have they gone to, the grass and the thorn-trees, the rivers and waterholes and the blue mountains? The high sweet air over the plains has lifted and withdrawn. Where have the other giraffes gone to, that were side by side with them when they set going, and cantered over the undulating land? They have left them, they have all gone, and it seems they are never coming back.*

Alphabetical References

Aas-Hansen Ø, Folkow LP, Blix AS. 2000. Panting in reindeer (*Rangifer tarandus*). *American Journal of Physiology* 279: R1190–R1195.

Abercrombie HR. 1952. The secret history of South Africa, or sixty-five years in the Transvaal. Central News Agency, Johannesburg.

Abdalla MA, Abdalla O. 1979. Morphometric observations on the kidney of the camel, *Camelus dromedarius*. *Journal of Anatomy* 129: 45–50.

Abdulrazak SA, Fujihara T, Ondiek JK, Ørskov ER. 2000. Nutritive evaluation of some Acacia tree leaves from Kenya. *Animal Feed Science and Technology* 85:
89–98.

Ackerman E. 1976. The histogenesis of hair follicles in the zebra and giraffe with special reference to pigmentation and cutaneous vasculature. MSc Dissertation, University of Pretoria, South Africa.

Agaba M, Ishengoma E, Miller WC, McGrath BC, Hudson CN, Reina OCB, Rattan A, Burhans R, Chikhi R, Medvedev P, Praul CA, Wu-Cavener L, Wood B, Robertson H, Penfold L, Cavener DR. 2016. Giraffe genome sequence reveals clues to its unique morphology and physiology. *Nature Communications* 7: 11519.

Agatharchides of Cnidus. *On the Erythraean Sea.* Translated by Stanley M. Burstein, 1989, book 5, fragment 73a.

Agusti J, Anton M. 2002. *Mammoths, sabretooths and hominids.* Columbia University Press, New York.

Agusti J, Moya Sola S. 1991. Spanish Neogene mammal succession and its bearing on continental biochronology. *Newsletters on Stratigraphy* (Leiden) 25: 91–114.

Agusti J, Rook L, Andrews P. 1999. *The evolution of Neogene terrestrial ecosystems in Europe.* Cambridge University Press, Cambridge.

Alexander RM. 1977. Allometry of the limbs of antelopes (Bovidae). *Journal of Zoology* 183: 125–146.

Alexander RM. 1981. Allometry of leg muscles of mammals. *Journal of Zoology* 194: 539–552.

Alexander RM. 1984. Elastic energy stores in running vertebrates. *American Zoologist* 24: 85–94.

Alexander RM. 1989a. Mechanics of fossil vertebrates. *Journal of the Geological Society, London* 146: 41–52.

Alexander RM. 1989b. On the synchronization of breathing with running in wallabies (*Macropus* spp.) and horses (*Equus caballus*). *Journal of Zoology* 218: 69–85.

Alexander RM. 1994. *Bones.* Macmillan, New York.

Alexander RM, Bennet-Clark HC. 1977. Storage of elastic strain energy in muscle and other tissues. *Nature* 265: 114–117.

Alexander RM, Jayes AS, Maloiy GMO, Wathuta EM. 1979. Allometry of the limb bones of mammals from shrews (*Sorex*) to elephant (*Loxodonta*). *Journal of Zoology* 189: 305–314.

Alexander RM, Langman VA, Jayes AS. 1977. Fast locomotion of some African ungulates. *Journal of Zoology (London)* 183: 291–300.

Allard MW, Miyamoto MM, Jarecki L, Kraus F, Tennant MR. 1992. DNA systematics and evolution of the artiodactyl family Bovidae. *Proceedings of the National Academy of Sciences, USA* 89: 3972–3976.

Allen JA. 1877. The influence of physical conditions in the genesis of species. *The Radical Review* 1: 108–140.

Allin M. 1998. *Zarafa.* Headline, London.

Altman PL, Dittmer DS. 1971. *Respiration and circulation.* FASEB, Bethesda, MD.

Altringham JD, Young IS. 1991. Power output and the frequency of oscillatory work in mammalian diaphragm muscle: The effects of animal size. *Journal of Experimental Biology* 157: 381–389.

Alvarez W. 1997. *T. rex and the crater of doom*. Princeton University Press, Princeton, NJ.

Amoroso EC, Edholm OG, Rewell RE. 1947. Venous valves in the giraffe, okapi, camel, and ostrich. *Proceedings of the Zoological Society, London* 117: 435–440.

Anderson JF, Rahn H, Prange HD. Scaling of supportive tissue mass. *Quarterly Review of Biology* 54: 139–148.

Angermeyer M. 1966. Zur Kenntnis der Halsmuskeln und Halsnerven von *Giraffa camelopardalis* (L). *Zoologischer Anzeiger* 177: 188–200.

Anonymous. 1827. The Camelopardalis, or, Giraffe (being an account of the arrival of George IV's giraffe in England on 11 August 1827. *The London Literary Gazette*. August 25: 553–555.

Anonymous. 1923. A black giraffe, seen by "a competent European observer," was reported to the Zoological Society, London by Mr. Caldwell of the Game Warden's office, Nairobi. *Proceedings of the Zoological Society* 1923: 179.

Anonymous. 1965. Giraffe midwives at a birth. *African Wildlife* 19: 323.

Anonymous. 1971. Giraffe albino. *Africana* 4 (8): 39.

Anonymous. 1974. The giraffe's parameter. *Lancet* 2: 1491–1492.

Arambourg C. 1947. Contribution a l'etude geologique et palaeontologie du bassin du lac Rudolph et de la basse vallee de l'Omo. *Musee d'Histoire Naturel, Paris* 1: 232–562.

Arambourg C. 1959. Vertebres continentaux du Miocene superieur de l'Afrique du Nord. Publications du service de la Carte Geologique de l'Algerie (Nouvelle Serie), Palaeontologie, *Memoires Algiers* 4: 1–159.

Arambourg C. 1963. Continental vertebrate faunas of the tertiary of North Africa. In *African ecology and human evolution*. Clark Howell F, Bourliere F (Eds.). Aldine, Chicago, pp. 55–63.

Archer D, Sanson G. 2002. Form and function of the selenodont molar in southern African ruminants in relation to their feeding habits. *Journal of Zoology (London)* 257: 13–26.

Aristotle. 384–322 BC [1987]. *History of animals*. Translated by Richard Cresswell. George Bell and Sons, London.

Arnold R. 1940. Eine ungefleckte Giraffe in Nord-Ost-Afrika? *Zeitschrift fur Saugetierkunde* 15: 306–311.

Artemidorus of Ephesus. The Eighth book of Artemidorus's Geography. A commentary by Irene Pajón Leyra in *Historia: Zeitschrift für Alte Geschichte*. Bd. 61, H. 3 (2012), pp. 336–361, 366–367.

Aschaffenburg R, Gregory ME, Rowland SJ, Thompson SY. 1962. The composition of the milk of the giraffe (*Giraffa camelopardalis reticulata*). *Proceedings of the Zoological Society (London)* 139: 359–363.

Axelson E. 1954. *South African explorers*. Oxford University Press, Oxford.

Axmacher H, Hofmann RR. 1988. Morphological characteristics of the masseter muscle of 22 ruminant species. *Journal of Zoology (London)* 215: 463–473.

Backhaus D. 1959. Experimentelle Prüfung des Farbsehvermögens einer Massai-giraffe (*Giraffa camelopardalis tippelskirchi* Matschie 1898). *Zeitschrift fur Tierpsychologie* 16: 468–477.

Badeer HS. 1986. Does gravitational pressure of blood hinder flow to the brain of the giraffe. *Comparative Biochemistry and Physiology A* 83: 207–211.

Badeer HS. 1988. Haemodynamics of the jugular vein in the giraffe. *Nature* 332: 788–789.

Badeer HS. 1997. Is the flow in the giraffe's jugular vein a "free" fall? *Comparative Biochemistry and Physiology A* 118: 573–576.

Badlangana NL, Adams JW, Manger PR. 2009. The giraffe (*Giraffa camelopardalis*) cervical vertebral column: A heuristic example in understanding evolutionary processes? *Zoological Journal of the Linnean Society* 155: 736–757.

Badlangana NL, Adams JW, Manger PR. 2011. A comparative assessment of the size of the frontal air sinuses in the giraffe (*Giraffa camelopardalis*). *Anatomical Record* 294: 931–940.

Badlangana NL, Bhagwandin A, Fuxe K, Manger PR. 2007a. Distribution and morphology of putative catecholaminergic and serotonergic neurons in the medulla oblongata of a sub-adult giraffe, *Giraffa camelopardalis*. *Journal of Chemical Neuroanatomy* 34: 69–79.

Badlangana NL, Bhagwandin A, Fuxe K, Manger PR. 2007b. Observations on the giraffe central nervous system related to the corticospinal tract, motor cortex and spinal cord: What difference does a long neck make? *Neuroscience* 148: 522–534.

Badyaev AV 2002. Growing apart: An ontogenetic perspective on the evolution of sexual size dimorphism. *Trends in Ecology and Evolution* 17: 369–378.

Baker MA. 1979. A brain cooling system in mammals. *Scientific American* 240: 130–139.

Baker MA, Hayward JN. 1968. The influence of the nasal mucosa and carotid rete upon hypothalamic temperature in sheep. *Journal of Physiology (London)* 198: 561–579.

Baker RJ, Bradley RD. 2006. Speciation in mammals and the genetic species concept. *Journal of Mammalogy* 87: 643–662.

Baldwin BA, Bell FR. 1963. The anatomy of the cerebral circulation in the sheep and the ox: The dynamic distribution of the blood supplied by the carotid and vertebral arteries to cranial regions. *Journal of Anatomy* 97: 203–215.

Balsdon JPVD. 1969. *Life and leisure in ancient Rome.* McGraw-Hill, New York.

Baotic A, Sicks F, Stoeger AS. 2015. Nocturnal "humming" vocalizations: Adding a piece to the puzzle of giraffe vocal communication. *BMC Research Notes* 8: 425–434.

Barrow J. 1801. *An account of travels into the interior of Southern Africa in the years 1797 and 1798.* T. Cadell and W. Davis, London.

Barton N, Partridge L. 2000. Limits to natural selection. *BioEssays* 22: 1075–1084.

Bassett DR, Howley ET. 2000. Limiting factors for maximum oxygen uptake and determinants of endurance performance. *Medicine & Science in Sports & Exercise* 32: 70–84.

Basu C, Falkingham PL, Hutchinson JR. 2016. The extinct, giant giraffid *Sivatherium giganteum:* Skeletal reconstruction and body mass estimation. *Biology Letters* 12: 20150940.

Bate DMA. 1951. The mammals from Singa and Abu Hugar. *Fossil Mammals of Africa* 2: 1–28.

Battaglia FC, Meschia G. 1978. Principal substrates of fetal metabolism. *Physiological Reviews* 58: 499–527.

Beddard JE. 1902a. Note on crooked neck in giraffe. *Proceedings of the Zoological Society (London)* 1: 52–54.

Beddard JE. 1902b. *The Cambridge natural history.* Volume 10. *Mammalia.* Macmillan, London.

Beddard JE. 1906. Description of the external characters of an unborn foetus of a giraffe (*Giraffa camelopardalis wardii*). *Proceedings of the Zoological Society (London)* 1906: 463–468.

Beekman JH, Prins HHT. 1989. Feeding strategies of sedentary large herbivores in East Africa, with emphasis on the African buffalo, *Syncerus caffer. African Journal of Ecology* 27: 129–147.

Beintema JJ, Gaastra W, Munniksma J. 1979. Primary structure of Pronghorn pancreatic ribonuclease: Close relationship between giraffe and pronghorn. *Journal of Molecular Evolution* 13: 305–316.

Belon P. 1553. *Les observations de plusieurs singularitez et choses memorables trouvées en Grèce, Asie, Judée, Egypte, Arabie et autres pays étrangèrs.* Gilles Corrozet, Paris.

Belozerskaya M. 2006. *The Medici giraffe.* Little, Brown, New York.

Ben Shaul DM. 1962. The composition of the milk of wild animals. *International Zoo Yearbook* 4: 333–342.

Bennett MB, Alexander RM. 1987. Properties and function of extensible ligaments in the necks of turkeys (*Meleagris gallopavo*) and other birds. *Journal of Zoology, London* 212: 275–281.

Bercovitch FB, Bashaw MJ, del Castillo SM. 2006. Sociosexual behavior, male mating tactics, and the reproductive cycle of giraffe *Giraffa camelopardalis. Hormones and Behavior* 50: 314–321.

Bercovitch FB, Berry PSM. 2010a. Reproductive life history of Thornicroft's giraffe in Zambia. *African Journal of Ecology* 48: 535–538.

Bercovitch FB, Berry PSM. 2010b. Ecological determinants of herd size in the Thornicroft's giraffe of Zambia. *African Journal of Ecology* 48: 962–971.

Bercovitch FB, Berry PSM. 2018. Social and demographic influences on the feeding ecology of giraffe in the Luangwa Valley, Zambia: 1973–2014. *African Journal of Ecology* 56: 254–261.

Bercovitch FB, Berry PSM, Dagg A, Deacon F, Doherty JB, Lee DE, Mineur F, Muller Z, Ogden R, Seymour R, Shorrocks B, Tutchings A. 2017. How many species of giraffe are there? *Current Biology* 27: R136–R137.

Bercovitch FB, Deacon F. 2015. Gazing at a giraffe gyroscope: Where are we going? *African Journal of Ecology* 53: 135–146.

Berman A. 2003. Effects of body surface area estimates on predicted energy requirements and heat stress. *Journal of Dairy Science* 86: 3605–3610.

Berner RA. 1997. The rise of plants and their effect on weathering and atmospheric CO2. *Science* 276: 544–546.

Berry PSM. 1978. Range movements of giraffe in the Luangwa Valley, Zambia. *East African Wildlife Journal* 16: 77–83.

Berry PSM, Bercovitch FB. 2012. Darkening coat colour reveals life history and life expectancy of male Thornicroft's giraffes. *Journal of Zoology (London)* 287: 157–160.

Bertram JEA, Biewener AA. 1990. Differential scaling of the long bones in the terrestrial Carnivora and other mammals. *Journal of Morphology* 204: 157–169.

Beuchat CA. 1990. Body size, medullary thickness, and urine concentrating ability in mammals. *American Journal of Physiology* 258: R298–R308.

Bhatti ZH, Khan MA, Akhtar M, Khan AM, Ghaffar A, Iqbal M, Siddiq MK. 2012. *Giraffa punjabiensis* (Giraffidae: Mammalia) from the Middle Siwaliks of Pakistan. *Pakistan Journal of Zoology* 44: 1689–1695.

Bibi F, Kiessling W. 2015. Continuous evolutionary change in Plio-Pleistocene mammals of eastern Africa. *Proceedings of the National Academy of Sciences* 112: 10623–10628.

Biewener AA. 1982. Bone strength in small mammals and bipedal birds: Do safety factors change with body size. *Journal of Experimental Biology* 98: 289–301.

Biewener, AA. 1983. Allometry of quadrupedal locomotion: The scaling of duty factor, bone curvature and limb orientation to body size. *Journal of Experimental Biology* 105: 147–171.

Biewener AA. 1989. Scaling body support in mammals: Limb posture and muscle mechanics. *Science* 245: 45–48.

Biewener AA. 1991. Musculoskeletal design in relation to body size. *Journal of Biomechanics* 24 (Suppl. 1): 19–29.

Bigalke R. 1951. The discovery of the giraffe in South Africa. *Fauna and Flora (Transvaal)* 2: 24–28.

Bininda-Emonds ORP, Cardillo M, Jones KE, MacPhee RDE, Beck RMD, Grenyer R, Price SA, Vos RA, Gittleman JL, Purvis A. 2007. The delayed rise of present-day mammals. *Nature* 446: 507–512.

Black D. 1915. A study of the endocranial casts of Ocapia, Giraffa and Samotherium, with special reference to the convoluted pattern in the family Giraffidae. *Journal of Comparative Neurology* 28: 329–360.

Bligh J, Harthoorn AM. 1965. Continuous radiotelemetric records of the deep body temperature of some restrained African mammals under near-natural conditions. *Journal of Physiology* 176: 145–162.

Blixen K. 1938. *Out of Africa*. 1st edition. Random House, New York.

Blum L. 1957. A photograph of a young giraffe with a black band around its body. *African Wildlife* 11: 328.

Bochardt S. 1663. *Hierozoicon sive bipartitum opus de animalibus sacrae scripturae*. Vol. 1. *De animalibus in genere et de quadrupedibus viviparous et oviparis*. Thomas Roycroft, London.

Bock F, Fennessy J, Bidon T, Tutchings A, Marais A, Deacon F, Janke A. 2014. Mitochondrial sequences reveal a clear separation between Angolan and South African giraffe along a cryptic rift valley. *BMC Evolutionary Biology* 14: 219 (1–12).

Bohlin B. 1926. Die Familie Giraffidae mit besondere Berucksichtiging der fossilen Formen aus China. *Palaeontologica Sinica, Series C* 4: 1–179.

Bohlin B. 1935. Some remarks on fossil Giraffidae. *Bulletin of the Geological Society of China* 14: 83–89.

Bond WJ, Lofell D. 2001. Introduction of giraffe changes acacia distribution in a South African savanna. *African Journal of Ecology* 39: 286–294.

Bonduriansky R. 2007. Sexual selection and allometry: A critical reappraisal of the evidence and ideas. *Evolution* 61: 838–849.

Bonduriansky R, Day T. 2003. The evolution of static allometry in sexually selected traits. *Evolution* 57: 2450–2458.
Bourgaud F, Gravot A, Milesi S, Gontier E. 2001. Production of plant secondary metabolites: A historical perspective. *Plant Science* 161: 839–851.
Bourlière F. 1961. Le sex-ratio de la giraffe. *Mammalia* 25: 467–471.
Bourlière F. 1963. Observations on the ecology of some large African mammals. In *African ecology and human evolution.* Clark Howell F, Bourliere F (Eds.). Aldine, Chicago.
Boyd IA, Kalu KU. 1979. Scaling factor relating conduction velocity and diameter for myelinated afferent nerve fibres in the cat hind limb. *Journal of Physiology* 289: 277–297.
Bradley RD, Baker RJ. 2001. A test of the genetic species concept: Cytochrome-*b* sequences and mammals. *Journal of Mammalogy* 82: 960–973.
Bramble DM, Carrier DR. 1983. Running and breathing in mammals. *Science* 219: 251–256.
Bredin I, Skinner JD, Mitchell G. 2008. Can osteophagia provide giraffes with phosphorus and calcium? *Onderstepoort Journal of Veterinary Research* 75: 1–9.
Brenton LCL. 1851. *The Septuagint with Apocrypha.* Samuel Bagster & Sons, London.
Brisson MJ. 1762. *Regnum animale in classes IX distributum.* Lugduni Batavorum, Apud Theodorum Haak, pp. 37–38.
Britten J. 1914. Robert Jacob Gordon. *Journal of Botany (London)* 52: 75–77.
Brondum E, Hasenkam JM, Secher NH, Bertelsen MF, Grondhal C, Peterson KK, Buhl R, Aalkaer C, Baandrup U, Nygaard H, Smerup M, Stegmann F, Sloth E, Østergaard KH, Nissen P, Runge M, Pitsillides K, Wang T. 2009. Jugular venous pooling during lowering of the head affects blood pressure of the anesthetized giraffe. *American Journal of Physiology* 297: R1058–1065.
Brook BS, Pedley TJ. 2002. A model for time-dependent flow in (giraffe jugular veins): Uniform tube properties. *Journal of Biomechanics* 35: 95–107.
Brown DM, Brenneman RA, Koepfli K-P, Pollinger JP, Mila B, Georgiadis, NJ, Louis EE, Grether GF, Jacobs DK, Wayne RK. 2007. Extensive population genetic structure in the giraffe. *BMC Biology* 5: 57 (1–13).
Brown MB, Wells E. 2020. Skeletal dysplasia-like syndromes in wild giraffe. *BMC Research Notes* 13: 569.
Brown WL. 1960. Ants, acacias and browsing mammals. *Ecology* 41: 587–592.
Brownlee A. 1963. Evolution of the giraffe. *Nature* 200: 1022.
Buckland RA, Evans HJ. 1978. Cytogenetic aspects of phylogeny in the Bovidae 1. G -banding. *Cytogenetics and Cell Genetics* 21: 42–63.
Buffon G-LL. 1781. *Historia naturale: General y particular*, Vol XIV. Published by Viuda de Joaquim Ibarra, Madrid. Translated by William Smellie.
Buffon G-LL. 1792 and 1807. *Buffon's Natural History, containing a theory of the earth, a general history of man, of the brute creation, and of vegetables, minerals, &c. &c.* Vol. VIII : 260–274. Translated from French by JS Barr. JS Barr, London.
Buffon G-LL. 1796. *Historia natural, general y particular.* Tomo 14, published by Viuda de Don Joaquin Ibarra, Madrid, Figure 223. *Astas de Girafa.*
Buffon G-LL. 1828. *A natural history.* Thomas Kelly, London. Vol. 1: 649–653. Translated by William Smellie.
Buffon G-LL. 1835. *Oeuvres completes de Buffon avec les supplements augmentées de la classification de G Cuvier, et accompagnèes de 700 vignettes gravées sur acier représentant au monis 900 animaux.* Vol. VI. Ed. P Dumènil. Rue des Beaux-arts, Paris, pp. 186–196.
Burchell W. 1822. *Travels in the interior of Southern Africa.* Vol 1. Longman, Hurst, Rees, Orme and Brown, London.
Burckhardt JL. 1819. *Travels in Nubia.* John Murray, London.
Burne RH. 1939. Description of the stomach, intestine, liver, and pancreas of the okapi, *Okapia johnstoni. Proceedings of the Zoological Society B* 109: 451–479.
Burstein AH, Currey JD, Frankel VH, Reilly DT. 1972. The ultimate properties of bone tissue: The effects of yielding. *Journal of Biomechanics* 5: 35–44.
Burton RF. 2006. A new look at the scaling of size in mammalian eyes. *Journal of Zoology* 269: 225–232.

Bush M, Custer RS, Whitla JC. 1980. Hematology and serum chemistry profiles for giraffes (Giraffa camelopardalis): Variations with sex, age, and restraint. *Journal of Zoo Animal Medicine* 11: 122–129.

Bush M, Grobler DG, Raath JP, Phillips LG, Stamper MA, Lance WR. 2001. Use of medetomidine and ketamine for immobilization of free-ranging giraffes. *Journal of the American Veterinary Medical Association* 218: 245–249.

Butler AL. 1912. A white giraffe. *Field* 119: 596.

Bux F, Bhagwandin A, Fuxe K, Manger PR. 2010. Organization of cholinergic, putative catecholaminergic and serotonergic nuclei in the diencephalon, midbrain, and pons of sub-adult male giraffes. *Journal of Chemical Neuroanatomy* 39: 189–203.

Caister LE, Shields WM, Gosser A. 2003. Female tannin avoidance: A possible explanation for habitat and dietary segregation of giraffes (*Giraffa camelopardalis peralta*) in Niger. *African Journal of Ecology* 41: 201–210.

Calder WA. 1981. Scaling of physiological processes in homeothermic animals. *Annual Reviews of Physiology* 43: 301–322.

Calder WA. 1996. *Size, function, and life history*. Dover Publications, Mineola, NY.

Calder WA, Braun EJ. 1983. Scaling of osmotic regulation in mammals and birds. *American Journal of Physiology* 244: R601–RR606.

Cameron EZ, du Toit JT. 2005. Social influences on vigilance behaviour in giraffes, *Giraffa camelopardalis*. *Animal Behaviour* 69: 1337–1344.

Cameron EZ, du Toit JT. 2007. Winning by a neck: Tall giraffes avoid competing with shorter browsers. *The American Naturalist* 169: 130–135.

Cano I, Perez W. 2009. Quantitative anatomy of the trachea of the giraffe (*Giraffa camelopardalis rothschildi*). *International Journal of Morphology* 27: 905–908.

Carlton C, McKean T. 1977. The carotid and orbital retia of the pronghorn deer and elk. *Anatomical Record* 189: 91–108.

Carroll RL. 1988. *Vertebrate palaeontology and evolution*. WH Freeman, New York, pp. 506–520.

Carter DR, Hayes WC. 1976. Bone compressive strength: The influence of density and strain rate. *Science* 194: 1174–1176.

Carteret P. 1770. A letter on a Camelopardalis found about the Cape of Good Hope, from Capt Carteret to Mathew Maty, M.D. Sec. R.S. *Philosophical Transactions of the Royal Society* 60: 27–29.

Castigli G, Sacchi R. 1941. L'apparato di conduzione nel cuore della giraffe. *Rivista di biologia* 32: 71–80.

Cautley PT. 1838. Note on a fossil ruminant genus allied to Giraffidae, in the Siwalik hills. *Journal of the Asiatic Society of Bengal* 7: 658–660.

Cavagna GA, Heglund NC, Taylor CR. 1977. Mechanical work in terrestrial locomotion: Two basic mechanisms for minimizing energy expenditure. *American Journal of Physiology* 233: R243–R261.

Cavagna GA, Saibene FP, Margaria R. 1964. Mechanical work in running. *Journal of Applied Physiology* 19: 249–256.

Cave AJE. 1950. On the liver and gall-bladder of the Giraffe. *Proceedings of the Zoological Society, London* 120: 381–393.

Cerling TE, Ehleringer JR, Harris JM. 1998. Carbon dioxide starvation, the development of C4 ecosystems, and mammalian evolution. *Philosophical Transactions of the Royal Society. London B* 353: 159–171.

Cerling TE, Harris JM, Leakey MG. 2005. Environmentally driven dietary adaptations in African mammals. In *A history of atmospheric carbon dioxide and effects on plants animals and ecosystems*, Vol. 177, ECOLSTUD, Chapter 12, pp. 258–272.

Cerling TE, Harris JM, Ambrose SH, Leakey MG, Solounias N. 1997a. Dietary and environmental reconstruction with stable isotope analyses of herbivore tooth enamel from the Miocene locality of Fort Ternan, Kenya. *Journal of Human Evolution* 33: 635–650.

Cerling TE, Harris JM, MacFadden BJ, Leakey MG, Quade J, Eisenmann V, Ehleringer JR. 1997b. Global vegetation change through the Miocene/Pliocene boundary. *Nature* 389: 153–158.

Cerling TE, Quade J, Ambrose SH, Sikes NE. 1991. Fossil soils, grasses, and carbon isotopes from Fort Ternan, Kenya: Grassland or woodland? *Journal of Human Evolution* 21: 295–306.

Cerling TE, Wang Y, Quade J. 1993. Expansion of C4 ecosystems as an indicator of global ecological change in the late Miocene. *Nature* 361: 344–345.

Cerling TE, Wynn JG, Andanje SA, Bird MI, Kofir DK, Levin NE, Mace W, Macharia AN, Quade J, Remien CH. 2011. Woody cover and hominin environments in the past 6 million years. *Nature* 476: 51–56.

Challis JRG. 1971. Sharp increase in free circulating oestrogens immediately before parturition in sheep. *Nature* 229: 208.

Challis JRG. 1998. The placental clock. *Placenta* 19 (Suppl 2): 361–374.

Challis JRG, Matthews SG, Gibb W, Lye SJ. 2000. Endocrine and paracrine regulation of birth at term and preterm. *Endocrine Reviews* 21: 514–550.

Chapman HC. 1875. Notes on the great blood vessels of the giraffe. *Proceedings of the Academy of Natural Science (Philadelphia)* 1875: 401–402.

Chapman HC. 1877. Note on the great blood vessels of the giraffe. *Proceedings of the Academy of Natural Science (Philadelphia)* 1877: 37–38.

Christiansen P. 1999a. Scaling of the limb long bones to body mass in terrestrial mammals. *Journal of Morphology* 239: 167–190.

Christiansen P. 1999b. Scaling of mammalian long bones: Small and large mammals compared. *Journal of Zoology (London)* 247: 333–348.

Christiansen P. 2002. Mass allometry of the appendicular skeleton in terrestrial mammals. *Journal of Morphology* 251: 195–209.

Churcher CS. 1970. Two new upper Miocene Giraffids from Fort Ternan, Kenya, East Africa. *Fossil Vertebrates of Africa* 2: 1–105.

Churcher CS. 1978. Giraffidae. In *Evolution of African mammals*. Maglio VJ, Cooke HBS (Eds.). Harvard University Press, Cambridge, MA, pp. 509–535.

Cifelli RL. 1981. Patterns of evolution among the Artiodactyla and Perissodactyla (Mammalia). *Evolution* 35: 433–440.

Ciofolo I. 1995. West Africa's last giraffes: The conflict between development and conservation. *Journal of Tropical Ecology* 11: 577–588.

Clapp JF, Abrams RM, Caton D, Cotter JR, Barron DH. 1971. Fetal oxygen consumption in late gestation. *Quarterly Journal of Experimental Physiology* 56: 137–146.

Clauset A, Erwin DH. 2008. The evolution and distribution of species body size. *Science* 321: 399–401.

Clauss M, Deutsch A, Lechner-Doll M, Flach E, Tack C. 1998. Passage rate of fluid and particle phase in captive giraffe (*Giraffa camelopardalis*). *Advances in Ethology* 33: 98.

Clauss M, Dierenfeld ES. 2008. The nutrition of "browsers." *Zoo and Wild Animal Medicine Current Therapy* 6: 444–454.

Clauss M, Dittman MT, Müller DWH, Zerbe P, Codron D. 2014. Low scaling of a life history variable: Analyzing Eutherian gestation periods with and without phylogeny-informed statistics. *Mammalian Biology* 79: 9–16.

Clauss M, Frey R, Kiefer B, Lechner-Doll M, Loehlein W, Polster C, Rössner GE, Streich WJ. 2003a. The maximum attainable size of herbivorous mammals: Morphophysiological constraints on foregut, and adaptations of hindgut fermenters. *Oecologia* 136: 14–27.

Clauss M, Hofmann RR, Hummel J, Adamczewski J, Nygren K, Pitra C, Streich WJ, Reese S. 2006a. Macroscopic anatomy of the omasum of free-ranging moose (*Alces alces*) and muskoxen (*Ovibos moschatus*) and a comparison of the omasal laminal surface in 34 ruminant species. *Journal of Zoology* 270: 346–358.

Clauss M, Hofmann RR, Streich WJ, Fickel J, Hummel J. 2006b. Convergence in the macroscopic anatomy of the reticulum in wild ruminant species of different feeding types and a new resulting hypothesis on reticular function. *Journal of Zoology* 281: 26–38.

Clauss M, Hofmann RR, Streich WJ, Fickel J, Hummel J. 2008. Higher masseter muscle mass in grazing than in browsing ruminants. *Oecologia* 157: 377–385.

Clauss M, Hume DI, Hummel J. 2010. Evolutionary adaptations of ruminants and their potential relevance for modern production systems. *Animal* 7: 979–992.

Clauss MA, Hummel JBC, Streich WJD. 2006c. The dissociation of the fluid and particle phase in the forestomach as a physiological characteristic of large grazing ruminants: An evaluation of available, comparable ruminant passage data. *European Journal of Wildlife Research* 52: 88–98.

Clauss MA, Lechner-Doll MB, Flach EJC, Tack CD, Hatt J-ME. 2001. Comparative use of four different marker systems for the estimation of digestibility and low food intake in a group of captive giraffes (*Giraffa camelopardalis*). *Zoo Biology* 20: 315–329.

Clauss M, Lechner-Doll M, Flach EJ, Wiser J, Hatt J-M. 2002. Digestive tract pathology of captive giraffe (*Giraffa camelopardalis*): A unifying hypothesis. *European Association of Zoo and Wildlife Veterinarians. Proceedings of the 14th Scientific Meeting*, pp. 99–107.

Clauss M, Lechner-Doll M, Streich WJ. 2003b. Ruminant diversification as an adaptation to the physicomechanical characteristics of forage: A reevaluation of an old debate and a new hypothesis. *Oikos* 102: 253–262.

Clauss M, Lechner-Doll M, Streich WJ. 2004. Differences in the range of faecal dry matter content between feeding types of captive wild ruminants. *Acta Theriologica* 49: 259–267.

Clauss M, Nunn C, Fritz J, Hummel J. 2009. Evidence for a tradeoff between retention time and chewing efficiency in large mammalian herbivores. *Comparative Biochemistry and Physiology A* 154: 376–382.

Clauss M, Schwarm A, Ortmann S, Streich WJ, Hummel J. 2007. A case of non-scaling in mammalian physiology? Body size, digestive capacity, food intake, and ingesta passage in mammalian herbivores. *Comparative Biochemistry and Physiology A* 148: 249–265.

Clauss M, Steuer P, Muller DWH, Codron D, Hummel J. 2013. Herbivory and body size: Allometries of diet quality and gastrointestinal physiology, and implications for herbivore ecology and dinosaur gigantism. *PLOS One* 8: 1–14.

Clauss M, Suedmeyer K, Flach EJ. 1999. Susceptibility to cold in captive giraffe (*Giraffa camelopardalis*). *Proceedings of the American Association of Zoo Veterinarians* 1999: 183–186.

Clemens ET, Maloiy GMO. 1983. Digestive physiology of East African wild ruminants. *Comparative Biochemistry and Physiology A* 76: 319–333.

Cobbold TS. 1854. On the anatomy of the giraffe (*Camelopardalis giraffa, Linn.*) *Annals and Magazine of Natural History*, Edinburgh. 1854: 1–5, and republished in the *Annals* June 1854: 484–488.

Cobbold TS. 1860. Contributions to the anatomy of the giraffe. *Proceedings of the Zoological Society (London)* 1860: 99–105.

Coe MJ. 1967. "Necking" behaviour in the giraffe. *Journal of Zoology (London)* 151: 313–321.

Coimbra JP, Hart N, Collin SP, Manger PR. 2013. Scene from above: Retinal ganglion cell topography and spatial resolving power in the giraffe (*G camelopardalis*). *Journal of Comparative Neurology* 521: 2042–2057.

Colbert EH. 1933. A skull and mandible of *Giraffokeryx punjabiensis* (Pilgrim). *American Museum Novitates* 632: 1–14.

Colbert EH. 1935a. Distributional and phylogenetic studies in Indian fossil mammals. V. The classification and the phylogeny of the Giraffidae. *American Museum Novitates* 800: 1–15.

Colbert EH. 1935b. Siwalik mammals in the American Museum of Natural History. *Transactions of the American Philosophical Society* 26: 323–375.

Colbert EH. 1936a. Was the extinct giraffe (*Sivatherium*) known to early Sumerians. *American Anthropologist* 38: 605–608.

Colbert EH. 1936b. *Palaeotragus* in the Tungur formation of Mongolia. *American Museum Novitates* 874: 1–17.

Colbert EH. 1938. The relationships of the okapi. *Journal of Mammalogy* 19: 47–50.

Colbert EH. 1941. The osteology and relationships of *Archaeomeryx*, an ancestral ruminant. *American Museum Novitates* 1135: 1–24.

Cooke HBS. 1978. Africa: The physical setting. In *Evolution of African mammals*. Maglio VJ, Cooke HBS (Eds.). Harvard University Press, Cambridge, MA.

Cooper SM, Owen-Smith N, Bryant JP. 1988. Foliage acceptability to browsing ruminants in relation to seasonal changes in leaf chemistry of woody plants in a South African savanna. *Oecologia* 75: 336–342.

Courchamp F, Jaric I, Albert C, Meinard Y, Ripple WJ, Chapron G. 2018. The paradoxical extinction of the most charismatic animals. *PLOS Biol* 16: e2003997.

Cramer MD, Mazel, AD. 2007. The past distribution of giraffe in Kwazulu-Natal. *South African Journal of Wildlife Research* 37: 197–201.

Cresswell R. 1878. *Aristotle's history of animals.* George Bell and Sons, London.

Crile G, Quiring DP. 1940. A record of the body weights and certain organ and gland weights of 3690 animals. *The Ohio Journal of Science* 40: 219–259.

Crisp E. 1853. Note on the splenic vein of the giraffe. *Proceedings of the Zoological Society (London)* 1853: 99.

Crisp E. 1864a. Contributions to the anatomy of the giraffe, with an account of the length of the alimentary canal of many other ruminants. *Proceedings of the Zoological Society (London)* 1864: 63–68.

Crisp E. 1864b. Further contributions to the anatomy of the giraffe and the nylghau. *Proceedings of the Zoological Society (London)* 1864: 269–271.

Crisp E. 1866. Further observations relating to the anatomy of the giraffe. *Proceedings of the Zoological Society (London)* 1866: 563–566.

Crompton AW, Owerkowicz T, Bhullar B-AS, Musinsky C. 2017. Structure of the nasal region of non-mammalian cynodonts and mammaliforms: Speculations on the evolution of mammalian endothermy. *Journal of Vertebrate Paleontology.* e1269116–2.

Cronin MA, Stuart R, Pierson BJ, Patton JC. 1996. K-casein gene phylogeny of higher ruminants (Pecora, Artiodactyla). *Molecular Phylogenetics and Evolution* 6: 295–311.

Crutzen PJ, Aselmann I, Seiler W. 1986. Methane production by domestic animals, wild ruminants, other herbivorous fauna, and humans. *Tellus* 38B: 271–284.

Cumming DHM, Cumming GS. 2003. Ungulate community structure and ecological processes: Body size, hoof area and trampling in African savannas. *Oecologia* 134: 560–568.

Currey, JD. 2006. *Bones.* Princeton University Press, Princeton, NJ, and Oxford.

Currey JD, Alexander RM. 1985. The thickness of the walls of tubular bones. *Journal of Zoology (London) A* 206: 453–468.

Curry JW, Hohl R, Noakes TD, Kohn TA. 2012. High oxidative capacity and type IIX fibre content in springbok and fallow deer skeletal muscle suggest fast sprinters with a resistance to fatigue. *Journal of Experimental Biology* 215: 3997–4005.

Cushing H. 1901. Concerning a definitive regulatory mechanism of the vaso-motor centre which controls blood pressure during cerebral compression. *Bulletin of Johns Hopkins Hospital* 12: 290–292.

Czerkawski JW. 1980. A novel estimate of heat produced by the rumen. *British Journal of Nutrition* 43: 239–243.

Dagg AI. 1960. Food preferences of the giraffe. *Proceedings of the Zoological Society (London)* 135: 640–642.

Dagg AI. 1962a. The role of the neck in the movements of the giraffe. *Journal of Mammalogy* 43: 88–97.

Dagg AI. 1962b. Giraffe movement and the neck. *Natural History* 2: 44–51.

Dagg AI. 1962c. The subspeciation of the giraffe. *Journal of Mammalogy* 43: 550–552.

Dagg AI. 1965. Sexual differences in giraffe skulls. *Mammalia* 29: 610–612.

Dagg AI. 1968. External features of giraffe. *Mammalia* 32: 657–669.

Dagg AI. 1970. Preferred environmental temperatures of some captive mammals. *International Zoo Yearbook* 10: 127–130.

Dagg AI. 1971. Giraffa camelopardalis. *Mammalian Species* 5: 1–8.

Dagg AI. 2014. *Giraffe: Biology, behaviour, and conservation.* Cambridge University Press, Cambridge, UK.

Dagg AI, Foster JB. 1976. *The giraffe: Its biology, behaviour, and ecology.* Van Nostrand Reinhold, New York.

Damkjaer M, Wang, T, Brondum E, Østergaard KH, Baandrup U, Horlyck A, Hasenkam JM, Smerup M, Funder J, Marcussen N, Danielsen CC, Bertelsen MF, Grondal C, Pedersen M, Agger P, Candy G, Aalkjaer C, Bie P. 2015. The giraffe kidney tolerates high arterial blood pressure by high renal interstitial pressure and low glomerular filtration rate. *Acta Physiologica* 214: 497–510.

Dangerfield JM, Modukanele B. 1996. Overcompensation by *Acacia erubescens* in response to simulated browsing. *Journal of Tropical Ecology* 12: 905–908.

Danowitz M, Domalski R, Solounias N. 2015. The cervical anatomy of *Samotherium*, an intermediate-necked giraffid. *Royal Society Open Science* 2: 150521.

Danowitz M, Vasilyev A, Kortlandt V, Solounias N. 2015. Fossil evidence and stages of elongation of the *Giraffa camelopardalis* neck. *Royal Society Open Science* 2: 150393.Dapper O. 1668. *Naukeurige Beschrijvinge der Afrikaensche Gewesten*. Jacob van Meurs, Amsterdam.

Darwin CR. 1869. *On the origin of species by means of natural selection, or the preservation of favoured races in the struggle for life*. John Murray, London.

Darwin CR. 1871. *The descent of man and selection in relation to sex*. John Murray, London.

Darwin CR. 1872. *The origin of species by means of natural selection, or the preservation of favoured races in the struggle for life*. John Murray, London.

Darwin CR. 1888. *The origin of species: By means of natural selection, the preservation of favoured races in the struggle for life*. John Murray, London.

Darwin CR, Wallace AR. 1858. On the tendency of species to form varieties, and on the perpetuation of varieties and species by natural means of selection. *Proceedings of the Linnaean Society (London)* 3: 45–63.

Davis M. 1949. Parturition of the Nubian giraffe (Giraffa camelopardalis). *Journal of Mammalogy* 30: 306–309.

Davis RB. 1827. Of the Giraffe (being an account of the habits of George IV's giraffe). *The London Literary Gazette*, December 1: 778–779.

Dawkins R. 2009. *The greatest show on earth: The evidence for evolution*. Transworld, London, p. 361.

Dawson EA, Secher NH, Dalsgaard MK, Ogoh S, Yoshiga CC, González-Alonso J, Steensburg A, Raven PB. 2004. Standing up to the challenge of standing: A siphon does not support cerebral blood flow in humans. *American Journal of Physiology* 287: R911–R914.

Dawson WR. 1927. The earliest records of giraffe. *Annals and Magazine of Natural History (Edinburgh)* Series 9, 19: 478–485.

Dawson WR. 1934. Deer in Ancient Egypt. *Journal of the Linnean Society* 39: 137–145.

Day EH, Hua X, Bromham L. 2016. Is specialization an evolutionary dead end? Testing for differences in speciation, extinction and trait transition rates across diverse phylogenies of specialist and generalists. *Journal of Evolutionary Biology* 29: 1257–1267.

Deacon F, Bercovitch FB. 2018. Movement patterns and herd dynamics among South African giraffes (*Giraffa camelopardalis giraffa*). *African Journal of Ecology* 56: 620–628. doi: 10.1111/aje.12514.

Deacon F, Nel PJ, Bercovitch FB. 2015. Concurrent pregnancy and lactation in wild giraffes (*Giraffa camelopardalis*). *African Zoology* 50: 331–334.

Deacon F, Smit N. 2017. Spatial ecology and habitat use of giraffe (*Giraffa camelopardalis*) in South Africa. *Basic and Applied Ecology* 21: 55–65. doi: 10.1016.j.baae.2017.04.003.

Deacon F, Tutchings A. 2018. The South African giraffe *Giraffa camelopardalis giraffa*: A conservation success story. *Fauna and Flora International* 53: 45–48. doi: 10.1017/50030605317001612.

De Beaufort LF. 1927. On a case of correlation between blood vessels and colour pattern in a giraffe. *Tijdschrift der Nederlandsche Dierkundige Vereeniging* 3: 31–32.

Dehority BA, Odenyo AA. 2003. Influence of diet on the rumen protozoal fauna of indigenous African wild ruminants. *Journal of Eukaryotic Microbiology* 50: 220–223.

Deka BC, Nath KC, Borghain BN. 1980. Clinical note on the morphology of the placenta in a giraffe. *Journal of Zoo Animal Medicine* 11: 117–118.

del Castillo SM, Bashaw MJ, Patton ML, Rieches RR, Bercovitch FB. 2005. Fecal steroid analysis of female giraffe (*Giraffa camelopardalis*) reproductive condition and the impact of endocrine status on daily time budgets. *General and Comparative Endocrinology* 141: 271–281.

Dell L-A, Patzke N, Bhagwandin A, Bux F, Fuxe K, Barber G, Siegel JM, Manger PR. 2012. Organization and number of orexinergic neurons in the hypothalamus of two species of Cetartiodactyla: A comparison of giraffe (*Giraffa camelopardalis*) and harbour porpoise (*Phocoena phocoena*). *Journal of Chemical Neuroanatomy* 44: 98–109.

deMenocal PB. 1995. Plio-pleistocene African climate. *Science* 270: 53–59.

deMenocal PB. 2004. African climate change and faunal evolution during the Pliocene-Pleistocene. *Earth and Planetary Science Letters* 200: 3–24.

deMenocal PB. 2011. Climate and human evolution. *Science* 331: 540–542.

deMenocal, PB. 2015. End of the African humid period. *Nature Geoscience* 8: 86–87.

deMenocal P, Ortiz J, Guilderson T, Adkins J, Sarnthein M, Baker L, Yarusinsky M. 2000. Abrupt onset and termination of the African Humid Period: Rapid climate responses to gradual insolation forcing. *Quaternary Science Reviews* 19: 347–361.

Demment MW. 1982. The scaling of ruminoreticulum size with body weight in East African ungulates. *African Journal of Ecology* 20: 43–47.

Demment MW, Van Soest PJ. 1985. A nutritional explanation for body-size patterns of ruminants and non-ruminant herbivores. *American Naturalist* 125: 641–672.

de Monet JBPA. 1809. (The Chevalier de Lamarck). *Philosophie zoologique.* Paris. Translated by HS Elliot. Macmillan, London.

Denton DA, Blair-West JR, McKinley MJ, Nelson JF. 1986. Physiological analysis of bone appetite (osteophagia). *Bioessays* 4: 40–42.

Deuteronomy 14: verses 4–5.

de Winton WE. 1897. Remarks on the existing forms of giraffe. *Proceedings of the Zoological Society* 18: 273–283.

Dickinson CJ. 1965. *Neurogenic hypertension. A monograph.* Blackwell Scientific Publications, Oxford.

Dietrich WO. 1942. Ältesquartäre Säugetiere aus den südlichen Serengeti, Deutsch-Ost-afrika. *Paleontographica* 94: 43–133.

Dimery NJ, Alexander RM, Deyst KA. 1985. Mechanics of the ligamentum nuchae of some artiodactyls. *Journal of Zoology, London* 206: 341–351.

Dimond RL, Montagna W. 1976. The skin of the giraffe. *Anatomical Record* 185: 63–76.

Dinesen I (Karen Blixen). 1938. *Out of Africa.* Random House, New York.

Dio Cassius Cocceianus (Dio). AD 200–229 [1905]. *Annals of Rome.* Translated from the Greek by HB Foster. Pafraet's, New York.

Dietrich WO. 1942. Ältesquartäre Säugetiere aus den südlichen Serengeti, Deutsch-Ost-afrika. *Paleontographica* 94: 43–133.

Dobson J. 1959. John Hunter's giraffe. *Annals of the Royal College of Surgeons of England* 24: 124–128.

Doepp F, Schreiber SJ, von Munster T, Rademacher J, Klingebiel R, Valdueza JM. 2004. How does the blood leave the brain? A systematic ultrasound analysis of cerebral venous drainage patterns. *Neuroradiology* 46: 565–570.

Dougall HW, Drysdale VM, Glover PE. 1964. The chemical composition of Kenya browse and pasture herbage. *East African Wildlife Journal* 2: 86–121.

Doyle JT, Patterson JL, Warren JV, Detweiler DK. 1960. Observations on the circulation of domestic cattle. *Circulation Research* 8: 4–15.

Drevemo S, Grootenhuis JG, Karstad L. 1974. Blood parameters in wild ruminants in Kenya. *Journal of Wildlife Diseases* 10: 327–333.

Driver SR. 1902. *A critical and exegetical commentary on Deuteronomy.* T and T Clark, Edinburgh.

Ducrest A-L, Keller L, Roulin A. 2008. Pleiotropy in the melanocortin system, coloration and behavioural syndromes. *Trends in Ecology and Evolution* 23: 502–510.

Du Bois D, du Bois EF. 1916. A formula to estimate the approximate surface area if height and weight be known. *Archives of Internal Medicine* 17: 863–871.

Dumonceaux GA, Bauman JE, Camilo GR. 2006. Evaluation of progesterone levels in feces of captive reticulated giraffe (*Giraffa camelopardalis reticulata*). *Journal of Zoo and Wildlife Medicine* 37: 255–261.

du Toit JT. 1990a. Feeding-height stratification among African browsing ruminants. *African Journal of Ecology* 28: 55–61.

du Toit JT. 1990b. Giraffe feeding on *Acacia* flowers: Predation or pollination? *African Journal of Ecology* 28: 63–68.

du Toit JT. 1990c. Home range: Body mass relations: A field study on African browsing ruminants. *Oecologia* 85: 301–303.

du Toit JT, Bryant JP, Frisby K. 1990. Regrowth and palatability of Acacia shoots following pruning by African savanna browsers. *Ecology* 71: 149–154.

du Toit JT, Owen-Smith N. 1989. Body size, population metabolism, and habitat specialization among large African herbivores. *The American Naturalist* 133: 736–740.

du Toit JT, Yetman CA. 2005. Effects of body size on the diurnal activity budgets of African browsing ruminants. *Oecologia* 143: 317–325.

Eastman CR. 1915. Early representations of the giraffe. *Nature* 95: 672–673.

Eastman CR. 1917. Chinese and Persian giraffe paintings. *Nature* 99: 334.

Eckenhoff JE. 1971. The vertebral venous plexus. *Canadian Anaesthetists' Society Journal* 18: 487–495.

Economos AC. 1981. The largest land mammal. *Journal of Theoretical Biology* 89: 211–215.

Edelman NH, Epstein P, Cherniak NS, Fishman AP. 1972. Control of cerebral blood flow in the goat: Role of the carotid rete. *American Journal of Physiology* 223: 615–619.

Edvinsson L. 1975. Neurogenic mechanisms in the cerebrovascular bed. *Acta Physiologica Scandinavica Suppl* 427: 1–35.

Edwards EJ, Osborne CP, Stromberg AE, Smith SA. 2010. The origin of C4 grasslands: Integrating evolutionary and ecosystem science. *Science* 328: 587–591.

Edwards NA. 1975. Scaling of renal functions in mammals. *Comparative Biochemistry and Physiology A* 52: 63–66.

Eldredge N. 1971. The allopatric model and phylogeny in Paleozoic invertebrates. *Evolution* 25: 156–167.

Eldredge N, Gould SJ. 1972. Punctuated equilibria: An alternative to phyletic gradualism. In *Models in Paleobiology*. Schopf TJM (Ed.). Freeman Cooper, San Francisco, pp. 82–115.

Elsner J. 2016. Backgrounds and shadows in the Artemidorus Papyrus. In *Intorno al Papiro di Artemidoro III. I designi*. Milano, LED, pp. 37–43.

Elting EC. 1926. A formula for estimating surface area of dairy cattle. *Journal of Agricultural Research* 33: 269–278.

Endo H, Yamagiwa D, Fujisawa M, Kimura J, Kurohmaru M, Hayashi Y. 1997. Modified muscular system of the giraffe (*Giraffa camelopardalis*). *Annals of Anatomy* 179: 481–485.

Erdogan S, Perez W. 2013. Anatomical characteristics of the larynx in giraffe (*Giraffa camelopardalis*). *Journal of Morphological Science* 30: 266–271.

Faber JJ, Green TJ, Thornburg KL. Arterial blood pressure in the unanaesthetized fetal lamb after changes in fetal blood volume and haematocrit. *Quarterly Journal of Experimental Physiology and Cognate Medical Sciences* 59: 241–255.

Fairall N. 1968. The reproductive seasons of some large mammals in the Kruger National Park. *Zoologica Africana* 3: 189–210.

Faith JT. 2014. Late Pleistocene and Holocene mammal extinctions on continental Africa. *Earth-Science Reviews* 128: 105–121.

Falconer H. 1868. *Palaeontological memoirs and notes of the late Hugh Falconer*. Edited by C Murchison. Hardwicke, London.

Falconer H, Cautley PT. 1843–1844. On some fossil remains of *Anoplotherium* and giraffe, from the Sewalik Hills, in the north of India. *Proceedings of the Geological Society, London* 4: 235–249.

Falster DS, Westoby M. 2003. Plant height and evolutionary games. *Trends in Ecology and Evolution* 18: 337–343.

Fancy SG, Blanchard JM, Holleman DF, Kokjer KJ, White, RG. 1986. Validation of doubly labeled water using a ruminant. *The American Journal of Physiology* 251: R143–R149.

Fedak MA, Heglund NC, Taylor CR. 1982. Energetics and mechanics of terrestrial locomotion. II. Kinetic energy changes of the limbs and body as a function of speed and body size in birds and mammals. *Journal of Experimental Biology* 97: 23–40.

Fenn WO. 1930. Work against gravity and work due to velocity changes in running. *American Journal of Physiology* 93: 433–462.

Fennessy JT. 2004. Ecology of desert-dwelling giraffe *Giraffa camelopardalis* in northwestern Namibia. PhD thesis. University of Sydney, Australia.

Fennessy J. 2009. Home range and seasonal movements of *Giraffa camelopardalis angolensis* in the northern Namib desert. *African Journal of Ecology* 47: 318–327.

Fennessy J, Bidon T, Reuss F, Kumar V, Elkan P, Nilsson MA, Vamberger M, Fritz U, Janke A. 2016. Multi-locus analyses reveal four giraffe species instead of one. *Current Biology* 26: 1–7.

Fennessy J, Bock FG, Tutchings A, Brenneman R, Janke A. 2013. Mitochondrial DNA analyses show that Zambia's South Luangwa Valley giraffe (*Giraffa camelopardalis thornicrofti*) are genetically isolated. *African Journal of Ecology* 51: 635–640.

Fennessy JT, Leggett KEA, Schneider S. 2003. Distribution and status of the desert-dwelling giraffe (*Giraffa camelpardalis angolensis*) in northwestern Namibia. *African Zoology* 38: 184–188.

Fernandez, MH, Vrba ES. 2005. A complete estimate of the phylogenetic relationships in Ruminantia: A dated species-level supertree of the extant ruminants. *Biological Reviews* 80: 269–302.

Ferreira P. 2003. Report of a giraffe with a bone stuck in its mouth. *Stalking Magazine*. Winter, p. 4.

Field CR, Blankenship LH. 1973. Nutrition and reproduction of Grant's and Thomson's gazelles, Coke's hartebeest and giraffe in Kenya. *Journal of Reproduction and Fertility* Suppl. 19: 287–301.

Fielding JW, Burstein AH, Frankel VH. 1976. The nuchal ligament. *Spine* 1: 3–14.

Finlay R. 2008. The voyages of Zheng He: Ideology, state power, and maritime trade in Ming China. *The Journal of the Historical Society* 8: 327–347.

Finch VA. 1972. Thermoregulation and heat balance of the East African eland and hartebeest. *American Journal of Physiology* 222: 1374–1379.

Fisher RA. 1930. *The genetical theory of natural selection*. Clarendon Press, Oxford.

Flatt JP. 1995. Use and storage of carbohydrate and fat. *American Journal of Clinical Nutrition* 61: 952S–959S.

Fleming PA, Hofmeyr SD, Nodson SW, du Toit JT. 2006. Are giraffes pollinators or flower predators of *Acacia nigrescens* in Kruger National Park, South Africa? *Journal of Tropical Ecology* 22: 247–253.

Foley WJ, Moore BD. 2005. Plant secondary metabolites and vertebrate herbivores—from physiological regulation to ecosystem function. *Current Opinion in Plant Biology* 8: 430–435.

Folkow B. 1991. Giraffe, rats, and man: What is the importance of the "structural factor" in normo- and hypertensive states? *Clinical and Experimental Pharmacology and Physiology* 18: 3–11.

Forbes VS. 1970. Some scientific matters in early writings on the Cape. In *A History of scientific endeavour in South Africa*. Royal Society of South Africa, Cape Town, South Africa, pp. 33–59.

Forsyth Major CJ. 1891. On the fossil remains of species of the family *Giraffidae*. *Proceedings of the Zoological Society, London* 1891: 315–326.

Forsyth Major CJ. 1902. On a specimen of the okapi lately received at Brussels. *Proceedings of the Zoological Society, London* 1902: 339–345.

Foster JB. 1966. The giraffe of Nairobi national park: Home range, sex ratios, the herd, and food. *East African Wildlife Journal* 4: 139–148.

Foster JB. 1977. Africa's gentle giants. *National Geographic* 152: 402–416.

Foster JB, Dagg AI. 1972. Notes on the biology of the giraffe. *East African Wildlife Journal* 10: 1–16.

Frackowiak H, Jakubowski H. 2008. Arterial vascularization in the giraffe brain. *Annales Zoologici Fennici* 45: 353–359.

Frank A. 1960. Die glandula thyreoidea und die glandulae thyreoideae accessoriae bei Giraffa camelopardalis L. *Acta Anatomica* 42: 267.

Franklin KJ, Haynes F. 1927. The histology of the giraffe's carotid functionally considered. *Journal of Anatomy* 62: 115–117.

Frutos P, Hervás G, Giráldez FJ, Mantecon AR. 2004. Review. Tannins and ruminant nutrition. *Spanish Journal of Agricultural Research* 2: 191–202.

Fukuta K, Kudo H, Sasaki M, Kimura J, Ismail DB, Endo H. 2007. Absence of carotid rete mirabile in small tropical ruminants: Implications for the evolution of the arterial system in artiodactyls. *Journal of Anatomy* 21: 112–116.

Furstenburg D, van Hoven W. 1994. Condensed tannin as anti-defoliate agent against browsing by giraffe (*Giraffa camelopardalis*) in the Kruger National Park. *Comparative Biochemistry and Physiology A* 107: 425–431.

Furuuchi K, Koyabe D, Mori K, Endo H. 2013. Physiological cross-sectional area of the masticatory muscles in the giraffe (*Giraffa camelopardalis*). *Mammal Study* 38: 67–71.

Gaillard J-M, Festa-Bianchet M, Yoccoz NG. 1998. Population dynamics of large herbivores: Variable recruitment with constant adult survival. *Trends in Ecology and Evolution* 13: 58–63.

Gaillard J-M, Festa-Bianchet M, Yoccoz NG, Loison A, Toïgo C. 2000. Temporal variation in fitness components and population dynamics of large herbivores. *Annual Review of Ecology and Systematics* 31: 367–393.

Galileo. 1638 [1914]. *Dialogues concerning two new sciences.* Translated by Crew H & de Salvio A. McMillan, New York.

Gallagher DS, Derr JN, Womack JE. 1994. Chromosome conservation among the advanced pecorans and determination of the primitive bovid karyotype. *Journal of Heredity* 85: 204–210.

Gambaryan PP. 1974. *How mammals run.* John Wiley & Sons, New York. Translated from Russian by Hilary Hardin.

Ganey T, Ogden J, Olsen J. 1990. Development of the giraffe horn and its blood supply. *Anatomical Record* 227: 497–507.

Garcia GJM, da Silva JKL. 2004. On the scaling of mammalian long bones. *Journal of Experimental Biology* 207: 1577–1584.

Gardiner A. 1927. *Egyptian grammar.* Clarendon Press, London.

Garland T. 1983. The relation between maximal running speed and body mass in terrestrial mammals. *Journal of Zoology* 199: 157–170.

Garrod AH. 1877. Notes on the visceral anatomy and osteology of the ruminants. *Proceedings of the Zoological Society (London)* 1877: 2–18.

Gehr P, Mwangi DK, Ammann A, Maloiy GMO, Taylor CR, Weibel ER. 1981. Design of the mammalian respiratory system. V. Scaling morphometric pulmonary diffusing capacity to body mass: Wild and domestic mammals. *Respiration Physiology* 44: 61–86.

Geiser DR, Morris PJ, Adair HS. 1992. Multiple anesthetic events in a reticulated giraffe (*Giraffa camelopardalis*). *Journal of Zoo and Wildlife Medicine* 23: 189–196.

Gellman KS, Bertram JEA. 2002. The equine nuchal ligament 1: Structural and material properties. *Veterinary and Comparative Orthopaedics and Traumatology* 15: 1–6.

Gentry AW. 1970. The Bovidae (Mammalia) of the Fort Ternan fossil fauna. *Fossil Vertebrates of Africa* 2: 243–323.

Gentry AW. 1994. The Miocene differentiation of Old World Pecora (Mammalia). *Historical Biology* 7: 115–158.

Gentry AW, Hooker JJ. 1988. The phylogeny of the Artiodactyla. In *The phylogeny and classification of the tetrapods*, Volume 2: *Mammals.* Benton MJ (Ed.). Clarendon Press, Oxford, pp. 235–272.

Geraads D. 1979. Les Giraffinae (Giraffidae, Mammalia) du Miocène supérieur de la région de Thessalonique (Grèce). *Bulletin du Museum National d'Histoire Naturelle* C 1: 377–389.

Geraads D. 1986. Remarques sur la systematique et la phylogenie des Giraffidae (Artiodactyla, Mammalia). *Geobios* 19: 465–477.

Geraads D. 1998. Le gisement de vertébrés pliocènes de Calta, Ankara, Turquie. 9. Cervidae et Giraffidae. *Geodiversitas* 20: 455–465.

Geraads D. 2006. The late Pliocene locality of Ahl al Oughlam, Morocco: Vertebrate fauna and interpretation. *Transactions of the Royal Society of South Africa* 61: 97–101.

Geraads D. 2009. Giraffidae (Mammalia) de la fin du Néogène de la République de Macédoine (ARYM). *Geodiversitas* 31: 893–908.

Geraads D, Reed K, Bobe R. 2013. Pliocene Giraffidae (Mammalia) from the Hadar formation of Hadar and Ledi-Geraru, Lower Awash, Ethiopia. *Journal of Vertebrate Palaeontology* 33: 470–481.

Geraads D, Spassov N, Kovachev D. 2005. Giraffidae (Artiodactyla, Mammalia) from the late Miocene of Kalimantsi and Hadjidimovo, Southwestern Bulgaria. *Geologica Balcanica* 35: 11–18.

Gibbon E. 1897. *The history of the decline and fall of the Roman Empire*. Bury JB (Ed.). Methuen, London.

Giesecke D, van Gylswyk NO. 1975. A study of feeding types and certain rumen functions in six species of South African wild ruminants. *Journal of Agricultural Science, Cambridge* 85: 75–83.

Ginnett TF, Demment MW. 1997. Sex differences in giraffe foraging behaviour at two spatial scales. *Oecologia* 110: 291–300.

Ginnett TF, Demment MW. 1999. Sexual segregation by Masai giraffes at two spatial scales. *African Journal of Ecology* 37: 93–106.

Giussani DA, Forhead AJ, Fowden AL. 2005. Development of cardiovascular function in the horse fetus. *Journal of Physiology* 565.3: 1019–1030.

Gloneková M, Brandlová K, Jan Pluháček J. 2016. Stealing milk by young and reciprocal mothers: High incidence of allonursing in giraffes, *Giraffa camelopardalis*. *Animal Behaviour* 113: 113–123.

Glover TD. 1973. Aspects of sperm production in some East African mammals. *Journal of Reproduction and Fertility* 35: 45–53.

Godee Molsbergen EC. 1916. *Reizen in Zuid-Afrika in de Hollandse Tijd. Eerste Deel, Tochten naar het Noorden, 1652–1686*. 's Gravenhage. In Bigalke R. 1951. The discovery of the giraffe in South Africa. *Fauna and Flora (Transvaal)* 2: 24–28.

Goetz RH. 1955. Preliminary observations on the circulation of the giraffe. *Transactions of the American College of Cardiologists* 5: 39–48.

Goetz RH. 1988. Early surgical research at the University of Cape Town. *South African Journal of Surgery* 26: 140–147.

Goetz RH, Budtz-Olsen O. 1955. Scientific safari: The circulation of the giraffe. *South African Medical Journal* 29: 773–776.

Goetz RH, Keen EN. 1957. Some aspects of the cardiovascular system in the giraffe. *Angiology* 8: 542–564.

Goetz RH, Warren JV, Gauer OH, Patterson JL, Doyle JT, Keen EN, McGregor M. 1960. Circulation of the giraffe. *Circulation Research* 8: 1049–1058.

Goldberg JL. 2003. How do axons grow? *Genes and Development* 17: 941–958.

Gombe S, Kayanja FIB. 1974. Ovarian progestins in Masai giraffe (*Giraffa camelopardalis*). *Journal of Reproduction and Fertility* 40: 45–50.

Goodman PS, Tomkinson AJ. 1987. The past distribution of giraffe in Zululand and its implications for reserve management. *South African Journal of Wildlife Research* 17: 28–32.

Gordon IJ, Illius AW. 1994. The functional significance of the browser-grazer dichotomy in African ruminants. *Oecologia* 98: 167–175.

Gordon IJ, Illius AW. 1996. The nutritional ecology of African ruminants: A reinterpretation. *Journal of Animal Ecology* 65: 18–28.

Gordon RJ. 1777–1779 [1993]. *The journals of Robert Jacob Gordon*. Translated from Dutch by PR Cullinan. Brenthurst Library, Johannesburg.

Gould SJ. 1974. The origin and function of "bizarre" structures: Antler size and skull size in the "Irish Elk" *Megaloceros giganteus*. *Evolution* 28: 191–220.

Gould SJ. 1996. The tallest tale. *Natural History* 5: 18–23, 54–57.

Graïc J-M, Peruffo A, Ballarin C, Cozzi B. 2017. The brain of the giraffe (*Giraffa camelopardalis*): Surface configuration encephalization quotient, and analysis of the existing literature. *The Anatomical Record* 300: 1502–1511.

Greed RE. 1961. Composition of the milk of giraffe. *International Zoo Yearbook* 2: 106.

Grimsdell JJR, Field CR. 1976. Grazing patterns of buffaloes in the Rwenzori National Park, Uganda. *East African Wildlife Journal* 14: 339–344.

Grzegorzewski WJ, Skipor J, Wasows B, Krzymowski T. 1997. Countercurrent transfer of I-125-LHRH in the perihypophyseal cavernous sinus–carotid rete vascular complex, demonstrated on isolated pig heads perfused with autologous blood. *Domestic Animal Endocrinology* 14: 149–160.

Gunji M, Endo H. 2016. Functional cervicothoracic boundary modified by anatomical shifts in the neck of giraffes. *Royal Society Open Science* 3: 150604.

Gutsche T. 1979. *There was a man: The life and times of Sir Arnold Theiler K.C.M.G of Onderstepoort.* Howard Timmins, Cape Town.

Hagenbeck C. 1910. *Beasts and men.* Translated from German by Elliot HSR, Thacker AG. Longman's Green, London.

Hahn CN, Mayhew IG. 1999. Do giraffe roar? *Veterinary Record* 145: 28.

Haller JD, Olearchyk AS. 2002. Cardiology's 10 greatest discoveries. *Texas Heart Institute Journal* 29: 342–344.

Hall-Martin AJ. 1974a. Food selection by Transvaal Lowveld giraffe as determined by analysis of stomach contents. *Journal of the South African Wildlife Management Association* 4: 191–202.

Hall-Martin AJ. 1974b. A note on the seasonal utilization of different vegetation types by giraffe. *South African Journal of Science* 70: 122–123.

Hall-Martin AJ. 1975a. Studies on the biology and productivity of the giraffe *Giraffa camelopardalis*. DSc thesis, University of Pretoria, South Africa.

Hall-Martin AJ. 1975b. Aspects of the ecology and sociality of the giraffe (*Giraffa camelopardalis giraffa*). *Publications of the University of Pretoria (New Series)* 97: 48–56.

Hall-Martin AJ. 1976. Dentition and age determination of the giraffe, *Giraffa camelopardalis. Journal of Zoology, London* 180: 263–289.

Hall-Martin AJ. 1977. Giraffe weight estimation using dissected leg weight and body measurements. *Journal of Wildlife Management* 41: 740–745.

Hall-Martin AJ, Basson WD. 1975. Seasonal chemical composition of the diet of Transvaal Lowveld giraffe. *Journal of the South African Wildlife Management Association* 5: 19–21.

Hall-Martin AJ, Rowlands IW. 1980. Observations on ovarian structure and development of the southern giraffe, *Giraffa camelopardalis giraffa. South African Journal of Zoology* 15: 217–221.

Hall-Martin AJ, Skinner JD. 1978. Observations on puberty and pregnancy in female giraffe (*Giraffa camelopardalis*). *South African Journal of Wildlife Research* 8: 91–94.

Hall-Martin AJ, Skinner JD, Hopkins BJ. 1978. The development of the reproductive organs of the male giraffe, *Giraffa camelopardalis. Journal of Reproduction and Fertility* 52: 1–7.

Hall-Martin AJ, Skinner JD, Smith A. 1977b. Observations on lactation and milk composition of the giraffe, *Giraffa camelopardalis. South African Journal of Wildlife Research* 7: 67–71.

Hall-Martin AJ, Skinner JD, van Dyk JM. 1975. Reproduction in the giraffe in relation to some environmental factors. *East African Wildlife Journal* 13: 237–248.

Hall-Martin AJ, von la Chevallerie M, Skinner JD. 1977a. Carcass composition of the giraffe, *Giraffa camelopardalis giraffa. South African Journal of Animal Science* 7: 55–64.

Hamilton HC, Falconer W. 1892. *Geographica of Strabo.* G. Bell & Sons, London.

Hamilton WR. 1973. The lower Miocene ruminants of Gebel Zelten, Libya. *Bulletin of the British Museum of Natural History* 21: 73–150.

Hamilton WR. 1978a. Fossil giraffes from the Miocene of Africa and a revision of the phylogeny of the Giraffoidea. *Philosophical Transactions of the Royal Society B* 283: 165–229.

Hamilton WR. 1978b. Cervidae and Palaeomerycidae. In *Evolution of African Mammals.* Maglio VJ, Cooke HBS (Eds.). Harvard University Press, Cambridge, MA, pp. 496–508.

Hampton C. 2002. Carnivorous giraffe, or natural phenomenon? *Endangered Wildlife* 40: 22–25.

Hanks J. 1972. Reproduction of the elephant (*Loxodonta africana*) in the Luangwa Valley, Zambia. *Journal of Reproduction and Fertility* 30: 13–26.

Hargens AR, Gershuni DH, Danzig LA, Millard RW, Pettersson K. 1988. Tissue adaptations to gravitational stress: Newborn versus adult giraffes. *Physiologist* 31: S110–S113.

Hargens AR, Millard RW, Pettersson K, van Hoven W, Gershuni DH, Johansen K. 1987a. Transcapillary fluid balance in giraffe. *Advances in Microcirculation* 13: 195–202.

Hargens AR, Millard RW, Pettersson K, Johansen K. 1987b. Gravitational haemodynamics and oedema prevention in the giraffe. *Nature* 329: 59–60.

Harper F. 1940. The nomenclature and type localities of certain old world mammals. *Journal of Mammalogy* 21: 322–332.

Harris JM. 1976a. Pliocene Giraffoidea (Mammalia, Artiodactyla) from the Cape Province. *Annals of the South African Museum* 69: 325–353.

Harris JM. 1976b. Giraffidae from the East Rudolf succession. In *Earliest man and environments in the Lake Rudolph Basin*. Coppens Y, Howell FC, Isaac GL, Leakey RE (Eds.). Chicago University Press, Chicago, pp. 264–267.

Harris JM. 1976c. Pleistocene Giraffidae (Mammalia, Artiodactyla) from East Rudolf, Kenya. *Fossil Vertebrates of Africa* 4: 283–332.

Harris JM. 1987. Fossil Giraffidae and Camelidae from Laetoli. In *Laetoli, a Pliocene site in Northern Tanzania*. Leakey MD, Harris JM (Eds.). Oxford University Press, Oxford, pp. 358–377.

Harris JM. 1991. Family Giraffidae. *Koobi Fora Research Project* 3: 93–138.

Harris JM, Solounias N, Geraads D. 2010. Giraffoidea. In *Cenozoic mammals of Africa*. Werdelin L, Saunders WJ (Eds.). University of California Press, Oakland, CA, pp. 797–811.

Harris Capt. Sir WC. 1839. *The Wild Sports of Southern Africa: Being the narrative of an expedition from the Cape of Good Hope, through the territories of the Chief Moselekatse, to the Tropic of Capricorn*. John Murray, London.

Harris Capt. Sir WC. 1852 [1963]. *Wild Sports of Southern Africa*. 5th edition. Henry G. Bohn, London. Facsimile reprint, C. Strunk, Cape Town, South Africa.

Harrison DFN. 1980. Biomechanics of the Giraffe larynx and trachea. *Acta Otolaryngology* 89: 258–264.

Harrison DFN. 1981. Fibre size frequency in the recurrent laryngeal nerves of man and giraffe. *Acta Otolaryngology* 91: 383–389.

Harrison DFN. 1986. Victor Negus: 57 years later. *Annals of Otology, Rhinology and Laryngology* 95: 561–566.

Harrison DFN, Denny S. 1985. The possible influence of laryngeal and tracheal size on the running speed of mammals. *Acta Otolaryngology (Stockholm)* 99: 229–235.

Haskell JP, Ritchie ME, Olff H. 2002. Fractal geometry predicts varying body size relationships for mammal and bird home ranges. *Nature* 418: 527–530.

Haskins CH. 1922. Science at the court of the Emperor Frederick II. *The American Historical Review* 27: 669–694.

Hassanin A, Douzery EJ. 1999. The tribal radiation of the family Bovidae (Artiodactyla) and the evolution of the mitochondrial cytochrome b gene. *Molecular Phylogenetics and Evolution* 13: 227–243.

Hassanin A, Douzery EJP. 2003. Molecular and morphological phylogenies of ruminantia and the alternative position of the Moschidae. *Systematic Biology* 52: 206–228.

Hassanin A, Ropiquet A, Gourmand A-L, Chardonnet B, Rigoulet J. 2007. Mitochondrial DNA variability in *Giraffa camelopardalis*: Consequences for taxonomy, phylogeography and conservation of giraffes in west and central Africa. *Comptes Rendus Biologies* 330: 265–274.

Hasselquist F. 1766. *Voyages and travels in the Levant, 1749–1752*. Published by Charles Linnaeus. Translated by A. Foster from the Swedish for Davis & Reymers, London.

Haughton SH. 1922. A note on some fossils from the Vaal River gravels. *Transactions of the Geological Society of South Africa* 24: 11–16.

Hawkey CM. 1975. *Comparative mammalian haematology*. William Heineman Medical Books, London.

Hawkey CM, Bennett PM, Gascoyne SC, Hart MG, Kirkwood JK. 1991. Erythrocyte size, number and haemoglobin content in vertebrates. *British Journal of Haematology* 77: 392–397.

Hays JD, Imbrie J, Shackleton NJ. 1976. Variations in the Earth's orbit: Pacemaker of the ice ages. *Science* 194: 1121–1132.

Hayssen V, Lacy RC. 1985. Basal metabolic rates in mammals: Taxonomic differences in the allometry of BMR and body mass. *Comparative Biochemistry and Physiology, Part A* 81: 741–754.

Hearing VJ, Tsukamoto K. 1991. Enzymatic control of pigmentation in mammals. *FASEB Journal* 5: 2902–2909.

Heesy CP. 2004. On the relationship between orbit orientation and binocular visual field overlap in mammals. *Anatomical Record* 281A: 1104–1110.

Heesy CP. 2008. Ecomorphology of orbit orientation and the adaptive significance of binocular vision in primates and other mammals. *Brain Behavior and Evolution* 71: 54–67.

Heesy CP. 2009. Seeing in stereo: The ecology and evolution of primate binocular vision and stereopsis. *Evolutionary Anthropology* 18: 21–35.

Heglund NC, Fedak MA, Taylor CR, Cavagna GA. 1982. Energetics and mechanics of terrestrial locomotion. IV. Total mechanical energy changes as a function of speed and body size in birds and mammals. *Journal of Experimental Biology* 97: 57–66.

Heglund NC, Taylor CR. 1988. Speed, stride frequency and energy cost per stride: How do they change with body size and gait. *Journal of Experimental Biology* 138: 301–318.

Heinze W. 1964. Die Muskulatur der Hintergliedmaße von *Giraffa camelopardalis angoliensis* mit einigen Hinweisen zur Blutgefäß-und Nervenversorgung. *Anatomischer Anzeiger Bd* 115: 476–496.

Heinze W.1965. Gelenke und Bänder der Hintergliedmasse der Giraffe (*Giraffa camelopardalis angolensis*). *Anatomischer Anzeiger Bd* 116: 39–58.

Heintz E. 1975. Origine, migration et paleobiogeographie des Paleotraginae (Giraffidae, Artiodactyla) Antevallesiens de l'Ancien Monde. *Colloque International C.N.R.S.* 218: 723–730.

Heliodorus. AD 250 [1967]. *An Aethiopian history*. Translated from Joly and Lavocat, and from T. Underdowne's translation of 1587. Reprinted by AMS Press, New York.

Helm C, Cawthra H, Cowling R, De Vynck J, Marean C, McCrea R, Rust R. 2018. Palaeoecology of giraffe tracks in Late Pleistocene aeolianites on the Cape south coast. *South African Journal of Science* 114: 1–8.

Henderson DM, Naish D. 2010. Predicting the buoyancy, equilibrium and potential swimming ability of giraffes by computational analysis. *Journal of Theoretical Biology* 265: 151–159.

Hendey QB. 1970. The age of the fossiliferous deposits at Langebaanweg, Cape Province. *Annals of the South African Museum* 56: 119–131.

Hendey QB. 1978. Preliminary report on the Miocene vertebrates from Arrisdrift, South West Africa. *Annals of the South African Museum* 76: 1–41.

Hendey QB. 1981. Geological succession at Langebaanweg, Cape Province, and global events of the late Tertiary. *South African Journal of Science* 77: 33–38.

Hendey QB. 1982. *Langebaanweg: A record of past life*. South African Museum, Cape Town, South Africa.

Hetem RS, Maloney SK, Fuller A, Mitchell D. 2016. Heterothermy in large mammals: Inevitable or implemented? *Biological Reviews* 91: 187–205.

Hevelius J. 1690. *Firmamentum sobiescianum, sive uranographia.* Johannes Zacharia Stoll, Gdansk.

Hicks JW, Badeer HS. 1989. Siphon mechanisms in collapsible tubes: Application to circulation of the giraffe head. *American Journal of Physiology* 256: R567–R571.

Hicks JW, Munis JR. 2005. The siphon controversy counterpoint: The brain need not be "baffling." *American Journal of Physiology* 289: R629–R632.

Hillman JC, Hillman KK. 1977. Mortality of wildlife in Nairobi National Park, during the drought of 1973–1974. *East African Wildlife Journal* 15: 1–8.

Hilsberg, S. 2000. Aspekte klinischer Anwendung der Infrarot: Thermographie in der Zoo-und Wildtiermedizin. PhD dissertation. University of Leipzig, Germany.

Hirst SM. 1969. Populations in a Transvaal Lowveld Nature Reserve. *Zoologica Africana* 4: 199–230.

Hitching F. 1982. *The neck of the giraffe, or, where Darwin went wrong*. Pan Books, London.

Historia Augusta, Volume II. 1924. Loeb Classical Library. Translated by D Magie. Harvard University Press, Cambridge, MA.

Hofmann RR. 1989. Evolutionary steps of ecophysiological adaptation and diversification of ruminants: A comparative view of their digestive system. *Oecologia* 78: 443–457.

Hofmann RR, Matern B. 1988. Changes in gastrointestinal morphology related to nutrition in giraffes *Giraffa camelopardalis*: A comparison of wild and zoo specimens. *International Zoo Yearbook* 27: 168–176.

Hofmann RR, Streich WJ, Fickel J, Hummel J, Clauss M. 2008. Convergent evolution in feeding types: Salivary gland mass differences in wild ruminant species. *Journal of Morphology* 269: 240–257.

Hogan AG, Skouby CI. 1923. Determination of the surface areas of cattle and swine. *Journal of Agricultural Research* 25: 419–430.

Hokkanen, JEI. 1986. The size of the largest land animal. *Journal of Theoretical Biology* 118: 491–499.

Holbourn A, Kuhnt W, Kochhann KGD, Andersen N, Meier KJS. 2015. Global perturbation of the carbon cycle at the onset of the Miocene Climatic Optimum. *Geology* 43: 123–126.

Holbourn A, Kuhnt W, Schulz M, Erienkeuser H. 2005. Impacts of orbital forcing and atmospheric carbon dioxide on Miocene ice-sheet expansion. *Nature* 438: 483–487.

Holt JP. 1959. Flow of liquids through collapsible tubes. *Circulation Research* 7: 342–353.

Holt JP, Rhode EA. 1976. Similarity of renal glomerular hemodynamics in mammals. *American Heart Journal* 92: 465–472.

Holt JP, Rhode EA, Kines H. 1968. Ventricular volumes and body weight in mammals. *American Journal of Physiology* 215: 704–715.

Home E. 1828. On the peculiarities of the tongue of the Xariffa and on the muscular structure of tongues in general. *Lectures on Comparative Anatomy* 5: 244–249.

Home E. 1830. A report on the stomach of the Zariffa. *Philosophical Transactions of the Royal Society* 120: 85–86.

Hoppeler H, Mathieu O, Weibel ER, Krauer R, Lindstedt SL, Taylor CR. 1981. Design of the mammalian respiratory system. VIII. Capillaries in skeletal muscle. *Respiration Physiology* 44: 129–150.

Hoppeler H, Weibel ER. 1998. Limits for oxygen and substrate transport in mammals. *Journal of Experimental Biology* 201: 1051–1064.

Hopwood AT. 1934. New fossil mammals from Olduvai (Tanganyika Territory). *Annals and Magazine of Natural History* 14: 546–550.

Horace. (Quintus Horatius Flaccus). 65 BC–8 AD [1983]. *Epistle 2.1* in *The complete works of Horace*. Translated based on CE Passage, Frederick Unger, New York.

Hoskins GA. 1835. *Travels in Ethiopia, above the second cataract of the Nile.* Longman, Rees, Orme, Brown, Green, & Longman, London.

Howell AB, Strauss WL. 1934. Note on the spinal accessory nerve of long-necked ungulates. *Proceedings of the Zoological Society (London)* 1: 29–32.

Howland HC, Merola S, Basarab JR. 2004. The allometry and scaling of the size of vertebrate eyes. *Vision Research* 44: 2043–2065.

Hradecký P. 1983a. Some uterine parameters in antelopes and a giraffe. *Theriogenology* 20: 491–498.

Hradecký P. 1983b. Placental morphology in African antelopes and giraffes. *Theriogenology* 20: 725–734.

Hsiu KJ, Montadert L, Bernoulli D, Cita MB, Erickson A, Garrison RE, Kidd RB, Melieres F, Muller C, Wright R. 1977. History of the Mediterranean salinity crisis. *Nature* 267: 399–403.

Hugh-Jones P, Barter CE, Hime JM, Rusbridge MM. 1978. Dead space and tidal volume of the giraffe compared with some other mammals. *Respiration Physiology* 35: 53–58.

Hummel J, Südekum K-H, Streich WJ, Clauss M. 2006. Forage fermentation patterns and their implications for herbivore ingesta retention times. *Functional Ecology* 20: 989–1002.

Huxley JS. 1932. *Problems of relative growth.* Lincoln MacVeagh, Dial Press, New York.

Iason G. 2005. The role of plant secondary metabolites in mammalian herbivory: Ecological perspectives. *Proceedings of the Nutrition Society* 64: 123–131.

Illius AW, Gordon IJ. 1992. Modelling the nutritional ecology of ungulate vertebrates: Evolution of body size and competitive interactions. *Oecologia* 89: 428–434.

Imbrie J, Berger A, Boyle EA, Clemens SC, Duffy A, Howard WR, Kukla G, Kutzbach J, Martinson DG, McIntyre A, Mix AC, Molfino B, Morley JJ, Peterson LC, Pisias NG, Prell WL, Raymo ME, Shackleton NJ, Toggweiler JR. 1993. On the structure and origin of major glaciation cycles 2. The 100,000-year cycle. *Paleoceanography* 8: 699–735.

Imbrie J, Imbrie KP. 1979. *Ice ages: Solving the mystery.* Harvard University Press, Cambridge, MA.

Immelmann K, Gebbing H. 1962. Schlaf bei Giraffiden. *Zeitschrift fur Tierpsychologie* 19: 84–92.

Innis AC. 1958. The behaviour of the giraffe, *Giraffa camelopardalis*, in the Eastern Transvaal. *Proceedings of the Zoological Society (London)* 131: 245–278.

Irwin DM, Kocher TD, Wilson AC. 1991. Evolution of the cytochrome b gene of mammals. *Journal of Molecular Evolution* 32: 128–144.

Ishengoma E, Agaba M, Cavener DR. 2017. Evolutionary analysis of vision genes identifies potential drivers of visual differences between giraffes and okapi. *PeerJ* 5: 1–26.

Isobe N, Nakao T, Shimada M, Fukumoto Y, Watanabe H, Minami S, Noda A, Yoshimura Y. 2007. Fecal progestagen and estrone during pregnancy in a giraffe: A case report. *Journal of Reproduction and Development* 53: 159–164.

Jacobs B, Harland T, Kennedy D, Schall M, Wicinski B, Butti C, Hof PR, Sherwood CC, Manger PR. 2015. The neocortex of cetartiodactyls. II. Neuronal morphology of the visual and motor cortices in the giraffe (*Giraffa camelopardalis*). *Brain Structure and Function* 220: 2851–2872.

Janis CM. 1976. The evolutionary strategy of the Equidae and the origins of rumen and cecal digestion. *Evolution* 30: 757–774.

Janis CM. 1982. Evolution of horns in ungulates: Ecology and palaeo-ecology. *Biological Reviews* 57: 261–318.

Janis CM. 1993. Tertiary mammal evolution in the context of changing climates, vegetation, and tectonic events. *Annual Review of Ecology and Systematics* 24: 467–500.

Janis CM. 2007. Artiodactyl paleoecology and evolutionary trends. In *The evolution of artiodactyls.* Prothero DR, Foss SE (Eds.). Johns Hopkins University Press, Baltimore, MD, pp. 292–315.

Janis CM, Damuth J, Theodor JM. 2000. Miocene ungulates and terrestrial primary productivity: Where have all the browsers gone? *Proceedings of the National Academy of Sciences* 97: 7899–7904.

Janis, CM, Damuth J, Theordor JM. 2004. The species richness of Miocene browsers and implications for habitat type and primary productivity in the North American grassland biome. *Palaeogeography, Palaeoclimatology, Palaeoecology* 207: 371–398.

Janis CM, Scott KM. 1988. The phylogeny of the Ruminantia (Artiodactyla, Mammalia). In *The phylogeny and classification of the tetrapods,* Volume 2: *Mammals.* Benton MJ (Ed.). Clarendon Press, Oxford, pp. 273–282.

Jardine W. 1838. Giraffidae. *The Naturalist's Library* XXI: 186–189.

Jarman PJ. 1974. The social organisation of antelope in relation to their ecology. *Behaviour* 48: 215–267.

Jarman PJ. 1989. On being thick-skinned: Dermal shields in large mammalian herbivores. *Biological Journal of the Linnean Society* 36: 169–191.

Jefferson JW. 1971. Electrocardiographic and phonocardiographic findings in a reticulated giraffe. *Journal of the American Veterinary Medical Association* 159: 602–604.

Jessen C. 2001. *Temperature regulation in humans and other mammals.* Springer, Berlin.

Jetz W, Carbone C, Fulford J, Brown JH. 2004. The scaling of animal space use. *Science* 306: 266–268.

Johnsen HK, Blix AS, Jorgensen L, Mercer JB. 1985. Vascular basis for regulation of nasal heat exchange in reindeer. *American Journal of Physiology* 249: R617–R623.

Johnson GL. 1901. Contributions to the comparative anatomy of the mammalian eye, chiefly based on ophthalmoscopic examination. *Philosophical Transactions of the Royal Society of London, Series B* 194: 1–82.

Johnston H. 1902. *The Uganda Protectorate,* Volume 1. Hutchinson, London.

Joleaud L. 1937. Remarques sur les giraffids fossiles d'afrique. *Mammalia* 1: 85–96.

Joly N, Lavocat A. 1846. Recherches historiques, zoologiques, anatomiques et palaeontologiques sur la giraffe. *Memoirs de la Societe des Sciences Naturale de Strasbourg* III: 1–112.
Jones HL. 1930. *The geography of Strabo.* Vol. VII, Book 16, p. 337. William Heinemann, London.
Jones AM, Vanhatalo A, Burnley M, Morton RH, Poole DC. 2010. Critical power: Implications for determination of VO2max and exercise tolerance. *Medicine & Science in Sports & Exercise* 42: 1876–1890.
Jones ML. 1993. Longevity of ungulates in captivity. *International Zoo Yearbook* 32: 159–169.
Jones RJ, Meyer JHF, Bechaz FM, Stoltz MA, Palmer B, van der Merwe G. 2001. Comparison of rumen fluid from South African game species to digest tanniniferous browse. *Australian Journal of Agricultural Research* 52: 453–460.
Joost-Gaugier, CL. 1987. Lorenzo the Magnificent and the Giraffe as a symbol of power. *Artibus et Historiae* 8: 91–99.
Kamau JMZ. 1991. Morphogenetic analysis of nasal passages in ungulates. *Ongulès/Ungulates* 91: 91–94.
Kamau JMZ. 1992. A comparative approach to the structural basis for brain cooling in ungulates with observation on *in vitro* heat exchange. *Discovery and Innovation* 2: 95–102.
Kasozi H, Montgomery RA. 2018. How do giraffes locate one another? A review of visual, auditory, and olfactory communication among giraffes. *Journal of Zoology* 306: 139–146. doi: 10.1111/jzo.12604.
Kayanja FIB, Blakenship HL. 1973. The ovary of the giraffe, *Giraffa camelopardalis. Journal of Reproduction and Fertility* 34: 305–313.
Kearney CC. 2005. Effects of dietary physical form and carbohydrate profile on captive giraffe. MSc thesis, University of Florida.
Keen EN, Goetz RH. 1957. Cardiovascular anatomy of a foetal giraffe. *Acta Anatomica* 31: 562–571.
Kellas LM, van Lennep EW, Amoroso EC. 1958. Ovaries of some foetal and prepubertal giraffes. *Nature* 181: 487–488.
Kendrick KM, Hinton MR, Atkins K, Haupt MA, Skinner JD. 1998. Mothers determine sexual preferences. *Nature* 395: 229–230.
Kent DV, Olsen PE, Rasmussen C, Lepre C, Mundil R, Irmis RB, Gehreis GE, Giesler D, Geissman JW, Parker WG. 2018. Empirical evidence for stability of the 405-kiloyear Jupiter-Venus eccentricity cycle over hundreds of millions of years. *Proceedings of the National Academy of Sciences* 115: 6153–6158.
Kerr RA. 1980. Changing global sea levels as a geologic index. *Science* 209: 483–486.
Kettlewell HBD. 1958. A survey of the frequency of *Biston betularia* (L)(Lep) and its melanic forms in Great Britain. *Heredity* 12: 51–72.
Kettlewell HBD. 1961. The phenomenon of industrial melanism in Lepidoptera. *Annual Review of Entomology* 6: 245–262.
Kidd W. 1900. The significance of hair-slope in certain mammals. *Proceedings of the Zoological Society, London* 1900: 676–686.
Kihlström JE. 1972. Period of gestation and body weight in some placental mammals. *Comparative Biochemistry and Physiology A* 43: 673–679.
Kiltie RA. 2000. Scaling of visual acuity with body size in mammals and birds. *Functional Ecology* 14: 226–234.
Kimani JK. 1981a. Structural evidence for insertion of collagen fibres to smooth muscle cells in the carotid arterial system of the giraffe (*Giraffa camelopardalis*). *Cell and Tissue Research* 214: 225–238.
Kimani JK. 1981b. Subendothelial fibrillar laminae in the carotid arteries of the giraffe (*Giraffa camelopardalis*). *Cell and Tissue Research* 219: 441–443.
Kimani JK. 1983a. The structural organization of the tunica intima in the carotid arteries of the giraffe (*Giraffa camelopardalis*). *African Journal of Ecology* 21: 300–315.
Kimani JK. 1983b. The structural organization of the carotid arterial system of the Giraffe (*Giraffa camelopardalis*). *African Journal of Ecology* 21: 317–324.

Kimani JK. 1987. Structural organization of the vertebral artery in the giraffe (*Giraffa camelopardalis*). *Anatomical Record* 217: 256–262.
Kimani JK, Mbuva RN, Kinyama RM. 1991a. Sympathetic innervation of the hindlimb arterial system in the giraffe (*Giraffa camelopardalis*). *Anatomical Record* 229: 103–108.
Kimani K, Mungai JM. 1983. Observations on the structure and innervation of the presumptive carotid sinus area in the giraffe. *Acta Anatomica* 115: 117–133.
Kimani JK, Opole IO. 1991. The structural organization and adrenergic innervation of the carotid arterial system of the giraffe (*Giraffa camelopardalis*). *Anatomical Record* 320: 369–377.
Kimani JK, Opole IO, Ogengo JA. 1991b. Structure and sympathetic innervation of the intracranial arteries in the giraffe (*Giraffa camelopardalis*). *Journal of Morphology* 208: 193–203.
Kinzelbach RK. 2012. A cassowary *Casuarius casuarius* (Linnaeus, 1758): Record from Alexandria, Egypt, in 20 BC. (Aves, Ratitae, Casuaridae). *The Open Ornithology Journal* 5: 26–31.
Kleiber M. 1947. Body size and metabolic rate. *Physiological Review* 27: 511–541.
Kleiber M. 1961. *The fire of life*. Wiley, New York.
Kleynhans CJ, van Hoven W. 1976. Rumen protozoa of the giraffe with a description of two new species. *East African Wildlife Journal* 14: 203–214.
Knight HH. 1995. Drought-related mortality of wildlife in the southern Kalahari and the role of man. *African Journal of Ecology* 33: 377–394.
Knorr G, Butzin M, Micheels A, Lohmann G. 2011. A warm Miocene climate at low atmospheric CO2 levels. *Geophysical Research Letters* 38: L20701.
Knottnerus-Meyer T. 1910. Eine neue Giraffe aus dem südlichien Abessinien. *Zoologischer Anzeiger* (May): 797–800.
Kodric-Brown A, Sibly RM, Brown JH. 2006. The allometry of ornaments and weapons. *Proceedings of the National Academy of Sciences* 103: 8733–8738.
Koeslag JH. 1990. Koinophylia groups sexual creatures into species, promotes stasis, and stabilizes social behaviour. *Journal of Theoretical Biology* 144: 15–35.
Kohn TA. 2014. Insights into the skeletal muscle characteristics of three southern African antelope species. *Biology Open* 3: 1037–1044. doi:10.1242/bio.20149241.
Kok OB. 1982. Cannibalism and post-partum return to oestrus of a female Cape giraffe. *African Journal of Ecology* 20: 141–143.
Kok OB, Opperman DPJ. 1980. Voedingsgedrag van kameelperde, *Giraffa camelopardalis*, in die Willem Pretorius-Wildtuin, Oranje-Vrystaat. *South African Journal Wildlife Research* 10: 45–55.
Kondoh D, Nakamura KG, Ono YS, Yuhara K, Bando G, Watanabe K, Horiuchi N, Kobayashi Y, Saski M, Kitamura N. 2017. Histological features of the vomeronasal organ in the giraffe, *Giraffa camelopardalis*. *Microscopy Research and Technique* 80: 652–656.
Konstantinov IE. 2001. Robert H Goetz: The surgeon who performed the first successful clinical coronary artery bypass operation. *Einstein Quarterly Journal of Biology and Medicine* 18: 73–78.
Kostopoulos DS. 2009. The late Miocene mammals faunas of the Mytilinii Basin, Samos Island, Greece: New collection. 13. *Giraffidae. Beiträge zur Paläontologie, Wien* 31: 299–343.
Kostopoulos DS, Koliadimou KK, Koufos GD. 1996. The Giraffids (Mammalia, Artiodactyla) from the late Miocene mammalian localities of Nikiti (Macedonia, Greece). *Palaeontographica Abteilung A* 239: 61–88.
Kraus F, Miyamoto MM. 1991. Rapid cladogenesis among the pecoran ruminants: Evidence from mitochondrial DNA sequences. *Systematic Zoology* 40: 117–130.
Krieg H. 1944. Der Schädel einer Giraffe. *Naturwissenschaften* 32: 148–156.
Kristal MB, Noonan M. 1979. Note on sleep in captive giraffes (Giraffa camelopardalis reticulata). *South African Journal of Zoology* 14: 108.
Krogh A. 1920. A contribution to the physiology of capillaries. Nobel Lecture, December 11, 1920.
Krogh A. 1924. *The anatomy and physiology of capillaries*. Yale University Press, New Haven, CT, pp. 225–226.
Krumbiegel I. 1951. Giraffenmischlinge. *Der Zoologische Garten* 18: 109–114.

Krumbiegel I. 1971. *Die Giraffe*. A Ziemsen, Wittenberg, Germany.

Kühlhorn F. 1954. Ursprung und Insertion der wichtigsten Muskeln der Vorderextremitat der Massaigiraffe (Giraffa camelopardalis tippelskirchi Mtsch. 1898). *Anatomischer Anzeiger* 101: 237–242.

Kuntsch V, Nel JAJ. 1990. Possible thermoregulatory behaviour in *Giraffa camelopardalis*. *Zeitschrift fur Saugetierkunde* 55: 60–62.

Kurten B. 1957. Mammal migrations, cenozoic stratigraphy, and the age of Peking man and the Australopithecines. *Journal of Palaeontology* 31: 215–227.

Lagueux O. 2003. Geoffroy's giraffe: The hagiography of a charismatic mammal. *Journal of the History of Biology* 36: 225–247.

Lande R. 1980. Sexual dimorphism, sexual selection, and adaptation in polygenic characters. *Evolution* 34: 292–305.

Langman VA. 1973a. The immobilization and capture of giraffe. *South African Journal of Science* 69: 200–203.

Langman VA. 1973b. Radio-tracking giraffe for ecological studies. *Journal of the South African Wildlife Management Association* 3: 75–78.

Langman VA. 1977. Cow-calf relationships in giraffe (*Giraffa camelopardalis giraffa*). *Zeitschrift fur Tierpsychologie* 43: 264–286.

Langman VA. 1978. Giraffe pica behaviour and pathology as indicators of nutritional stress. *Journal of Wildlife Management* 42: 141–147.

Langman VA, Bamford OS, Maloiy GMO. 1982. Respiration and metabolism in the giraffe. *Respiration Physiology* 50: 141–152.

Langman VA, Maloiy GMO. 1989. Passive obligatory heterothermy of the giraffe. *Journal of Physiology* 415: 89P.

Langman VA, Maloiy MO, Schmidt-Nielsen K, Schroter RC. 1979. Nasal heat exchange in the giraffe and other large mammals. *Respiration Physiology* 37: 325–333.

Lankester ER. 1901. On *Okapia johnstoni* (Giraffidae). *Proceedings of the Zoological Society (London)* 1901: 279–281, 472–474.

Lankester ER. 1907. The origin of the lateral horns of the giraffe in foetal life on the area of the parietal bones. *Proceedings of the Zoological Society (London)* 1: 100–115.

Lankester R. 1908. On certain points in the structure of the cervical vertebrae of the okapi and the giraffe. *Proceedings of the Zoological Society (London)* 1908: 320–334.

Latham RE. 1958. *The travels of Marco Polo*. Penguin Books, London.

Laufer B. 1928. *The giraffe in history and art*. Field Museum of Natural History, Leaflet 27, Chicago.

LaRiviere JP, Ravelo AC, Crimmins A, Dekens PS, Ford HL, Lyle M, Wara MW. 2012. Late Miocene decoupling of oceanic warmth and atmospheric carbon dioxide forcing. *Nature* 486: 97–100.

Lawrence WE, Rewell RE. 1948. The cerebral blood supply in the Giraffidae. Proceedings of the Zoological Society (London) 118: 202–212.

Leakey LSB. 1965. *Olduvai, 1951–1961*. Cambridge University Press, Cambridge, UK.

Leakey LSB. 1970. Additional information on the status of *G. jumae* from East Africa. *Fossil Vertebrates of Africa* 2: 325–330.

Leblanc MJ, Leduc C, Stagnitti F, van Oevelen PJ, Jones C, Mofor LA, Razack M, Favreau G. 2006. Evidence for Megalake Chad, north-central Africa, during the late Quaternary from satellite data. *Palaeogeography, Palaeoclimatology, Palaeoecology* 230: 230–242.

Lechner I, Barboza P, Collins W, Günther D, Hattendorf B, Hummel J, Clauss M. 2009. No "bypass" in adult ruminants: Passage of fluid ingested vs fluid inserted into the rumen in fistulated muskoxen (*Ovibos moschatus*), reindeer (*Rangifer tarandus*) and moose (*Alces alces*). *Comparative Biochemistry and Physiology A* 154: 151–156.

Lee DE, Bond ML, Bolger DT. 2017. Season of birth affects juvenile survival of giraffe. *Population Ecology* 59: 45–54.

Lee DE, Cavener DR, Bond ML. 2018. Seeing spots: Quantifying mother-offspring similarity and assessing fitness consequences of coat pattern traits in a wild population of giraffes (*Giraffa camelopardalis*). *Peer J* 6: e5690.

Lee DE, Kissui BM, Kiwango YA, Bond ML.2016. Migratory herds of wildebeests and zebras indirectly affect calf survival of giraffes. *Ecology and Evolution* 6: 8402–8411.

Le Pendu Y, Ciofolo I. 1999. Seasonal movements of giraffes in Niger. *Journal of Tropical Ecology* 15: 341–353.

Le Pendu Y, Ciofolo I, Gosser A. 2000. The social organization of giraffes in Niger. *African Journal of Ecology* 38: 78–85.

Lesson R-P. 1842. *Nouveau tableau du règne animal*. Arthus Bertrand, Paris, pp. 168–169.

Leuthold BM, Leuthold W. 1972. Food habits of giraffe in Tsavo National Park, Kenya. *East African Wildlife Journal* 10: 129–141.

Leuthold W, Leuthold BM. 1975. Temporal patterns of reproduction in ungulates of Tsavo East National Park, Kenya. *East African Wildlife Journal* 13: 159–169.

Leuthold BM, Leuthold W. 1978a. Ecology of the giraffe in Tsavo East national Park, Kenya. *East African Wildlife Journal* 16: 1–20.

Leuthold BM, Leuthold W. 1978b. Daytime activity patterns of gerenuk and giraffe in Tsavo National Park, Kenya. *East African Wildlife Journal* 16: 231–243.

LeVaillant F. 1795. *Second voyage dans l'interieur de l'Afrique par le Cap de Bonne-Esperance, dans les annes 1783, 84, 85*. HJ Jansen, Paris.

Leyra IP. 2012. Artemidorus behind Artemidorus: Geographic aspects in the zoological designs of the Artemidorus papyrus. *Historia: Zeitschrift für Alte Geschichte* 61(3): 336–367.

Lightfoot CJ. 1957. Birth of a giraffe. *Wild Rhodesia* 7: 28–30.

Lindeque M. 1991. Age structure of the elephant population in the Etosha National Park, Namibia. *Madoqua* 18: 27–32.

Lindstedt SL, Calder WA. 1981. Body size, physiological time, and longevity of homeothermic animals. *Quarterly Review of Biology* 56: 1–16.

Lindstedt SL, Miller BJ, Buskirk SW. 1986. Home range, time, and body size in mammals. *Ecology* 67: 413–418.

Linnaeus C. 1758. *Systema naturae per regna tria naturae, secundum classes, ordines, genera, species, cum characteribus, differentiis synonymis, locis*. Tomus 1, Edition 10, p. 66.

Linnaeus C. 1766. *Voyages and Travels in the Levant in the years 1749, 50, 51, 52, by Frederick Hasselquist*. Translated by A. Foster from the Swedish for Davis & Reymers, London, 1766.

Linton RAF, Taylor PM, Linton NWF, Flach EJ, O'Brien TK, Band DM. 1999. Cardiac output measurement in an anaesthetized giraffe. *Veterinary Record* 145: 498–499.

Lisiecki LE. 2010. Links between eccentricity forcing and the 100,000-year glacial cycle. *Nature Geoscience* 3: 349–352.

Lisiecki LE, Raymo ME. 2005. A Pliocene-Pleistocene stack of 57 globally distributed benthic D18O records. *Paleoceanography* 20: PA1003, doi: 10.1029/2004PA001071.

Llanos A, Riquelme R, Sanhueza E, Gaete C, Cabello G, Parer J. 1996. Cardiorespiratory responses to acute hypoxemia in the chronically catheterized fetal llama at 0.7–0.9 of gestation. *Comparative Biochemistry and Physiology A* 119: 705–709.

Lluch S, Dieguez G, Garcia A L, Gomez B. 1985. Rete mirabile of goat: Its flow-damping effect on cerebral circulation. *American Journal of Physiology* 249: R482–R489.

Loskutoff NM, Walker L, Ott-Joslin JE, Raphael BL, Lasley BL. 1986. Urinary steroid evaluations to monitor ovarian function in exotic ungulates: II. Comparison between the giraffe (*Giraffa camelopardalis*) and the Okapi (*Okapia johnstoni*). *Zoo Biology* 5: 331–338.

Luck CP, Wright PG. 1964. Aspects of the anatomy and physiology of the skin of the hippopotamus (*H. amphibius*). *Quarterly Journal of Experimental Physiology* 49: 1–14.

Lueders I, Hildebrandt TB, Pootoolal J, Rich P, Gray CS, Niemuller CA. 2009a. Ovarian ultrasonography correlated with fecal progestins and estradiol during the estrous cycle and early pregnancy in giraffes (*Giraffa camelopardalis rothschildi*). *Biology of Reproduction* 81: 989–995.

Lueders I, Niemuller C, Pootoolal J, Rich P, Gray C, Streich WJ, Hildebrandt TB. 2009b. Sonomorphology of the reproductive tract in male and pregnant and non-pregnant female Rothschild's giraffes (*Giraffa camelopardalis rothschildi*). *Theriogenology* 72: 22–31.

Lukhele MS, van Ryssen JBJ. 2003. The chemical composition and potential nutritive value of the foliage of four subtropical tree species in southern Africa for ruminants. *South African Journal of Animal Science* 33: 132–141.

Lust A, Fuller A, Maloney SK, Mitchell D, Mitchell G. 2007. Thermoregulation in Pronghorn antelope (*Antilocapra americana, Ord*) in the summer. *Journal of Experimental Biology* 210: 2444–2452.

Lydekker R. 1876. Notes on the fossil mammalian faunae of India and Burma. In *Records of the Geological Survey of India*. Trubner, London, pp. 86–105.

Lydekker R. 1883. Siwalik camelopardalidae. *Indian Tertiary and Post-tertiary Vertebrata* 2: 99–142.

Lydekker R. 1891. The giraffe and its allies. *Nature* 44: 524–526.

Lydekker R. 1901. Giraffidae. In *Library of natural history*. Saalfield, London. Vol. II, pp. 918–925.

Lydekker R. 1904a. On old pictures of Giraffes and Zebras. *Proceedings of the Zoological Society (London)* l: 339–345.

Lydekker R. 1904b. On the subspecies of Giraffa camelopardalis. *Proceedings of the Zoological Society (London)* 1: 202–227.

Lydekker R. 1911. Two undescribed giraffes. *Nature* 87: 484.

Lyons SK, Smith FA, Brown JH. 2004. Of mice, mastodons and men: Human-mediated extinctions on four continents. *Evolutionary Ecology Research* 6: 339–358.

MacInnes DG. 1936. A new genus of fossil deer from the Miocene of Africa. *Journal of the Linnaean Society* 39: 521–530.

Madden D, Young TP. 1992. Symbiotic ants as an alternative defense against giraffe herbivory. *Oecologia* 91: 235–238.

Main RP, Biewener AA. 2004. Ontogenetic patterns of limb loading, *in vivo* bone strains and growth in the goat radius. *Journal of Experimental Biology* 207: 2577–2588.

Maglio VJ, Cooke HBS (Eds.). 1978. *Evolution of African mammals*. Harvard University Press, Cambridge, MA.

Makhabu SW. 2005. Resource partitioning within a browser guild in a key habitat, the Chobe Riverfront, Botswana. *Journal of Tropical Ecology* 21: 641–649.

Makkar HPS. 2003. Effects of tannins in ruminant animals, adaptation to tannins, and strategies to overcome detrimental effects of feeding tannin-rich feeds. *Small Ruminant Research* 49: 241–256.

Maloiy GMO, Clemens ET, Kamau JMZ. 1982. Aspects of digestion and *in vitro* rumen fermentation rate in six species of East African wild ruminants. *Journal of Zoology (London)* 197: 345–353.

Maloney SK, Mitchell D. 1996. Regulation of ram scrotal temperature during heat exposure, cold exposure, fever, and exercise. *Journal of Physiology* 496: 421–430.

Maloney SK, Moss G, Mitchell D. 2005. Orientation to solar radiation in black wildebeest (*Connochaetes gnou*). *Journal of Comparative Physiology A* 191: 1065–1077.

Maluf NSR. 2002. Kidney of giraffes. *Anatomical Record* 267: 94–111.

Marealle WN, Fossøy F, Holmern T, Stokke BG, Røskaft E. 2010. Does illegal hunting skew Serengeti wildlife sex ratios? *Wildlife Biology* 16 (4): 419–429.

Marina N, Christie IN, Korsak A, Doronin M, Brazhe A, Hosford PS, Wells JA, Sheikhbahaei S, Humoud I, Paton JFR, Lythgoe MF, Semyanov A, Kasparov S, Gourine AV. 2019. Astrocytes monitor cerebral perfusion and control systemic circulation to maintain brain blood flow. *Nature Communications* 11: 1–9. doi.org/10.1038/s41467-019-13956-y.

Marra AC, Solounias N, Carone G, Rook L. 2011. Palaeogeographic significance of the giraffid remains (Mammalia, Artiodactyla) from Cessaniti (Late Miocene, Southern Italy). *Geobios* 44: 189–197.

Martínez-Freiría F, Tarroso P, Rebelo H, Brito JC. 2016. Contemporary niche contraction affects climate change predictions for elephants and giraffes. *Diversity and Distributions* 22: 432–444.

Maslin MA. 2017. Forty years of linking orbits to ice ages. *Nature* 540: 208–210.

Maslin MA, Brierley CM, Milner AM, Shultz S, Trauth MH, Wilson KE. 2014. East African climate pulses and early human evolution. *Quaternary Science Reviews* 101: 1–17.

Mathieu O, Krauer R, Hoppeler H, Gehr P, Lindstedt SL, Alexander RM, Taylor CR, Weibel ER. 1981. Design of the mammalian respiratory system. VII. Scaling of mitochondrial volume in skeletal muscle to body mass. *Respiration Physiology* 44: 113–128.
Matthew WD. 1929. Critical observations upon Siwalik mammals. *Bulletin of the American Museum of Natural History* 56: 437–560.
Matthew WD. 1934. Phylogenetic chart of the artiodactyla. *Journal of Mammalogy* 15: 207–209.Mayr E. 1996. What is a species, and what is not? *Philosophy of Science* 63: 262–277.
McBryde FD, Malpas SC, Paton J. 2017. Intracranial mechanisms for preserving brain blood flow in health and disease. *Acta Psychologica* 219: 274–287.
McCalden TA, Borsook D, Mendelow AD, Shimell CJ, de Vos V, Pieterse PC, de Klerk B. 1977. Autoregulation and haemodynamics of giraffe carotid blood flow. *South African Journal of Science* 73: 278–279.
McCalden TA, Mendelow AD. 1980. Carotid haemodynamics in the giraffe. *South African Journal of Science* 76: 246–247.
McCrindle JW. 1897. *The Christian topography of Cosmas, an Egyptian monk*. Hakluyt Society, London.
McDougall DSA. 1939. A white giraffe. *Field* 174: 1003.
McEvedy C, Jones R. 1978. *Atlas of world population history*. Penguin Books, London.
McFadden PL. 1979. Palaeomagnetism and the age of the Makapansgat hominid site. *Earth and Planetary Science Letters* 44: 373–382.
McMahon TA. 1973. Size and shape in biology. *Science* 179, 1201–1204.
McMahon TA. 1975. Allometry and biomechanics: Limb bones in adult ungulates. *The American Naturalist* 109: 547–563.
McNab BK. 1963. Bioenergetics and the determination of home range size. *The American Naturalist* 97: 133–140.
McNab BK. 2002. *The physiological ecology of vertebrates*. Cornell University Press, Ithaca, NY.
Melin AD, Kline DW, Hiramatsu C, Caro T. 2016. Zebra stripes through the eyes of their predators, zebras and humans. *PLOS One* 11: e0145679.
Merceron G, Colyn M, Geraads D. 2018. Browsing and non-browsing extant and extinct giraffids: Evidence from dental microwear textural analysis. *Palaeogeography, Palaeoclimatology, Palaeoecology* 505: 128–139.
Midgley JJ, Mclean P, Botha M, Balfour D. 2001. Why do some African thorn trees (*Acacia spp*) have a flat top: A grazer-plant mutualism hypothesis? *African Journal of Ecology* 39: 226–228.
Milewski AV, Young TP, Madden D. 1991. Thorns as induced defenses: Experimental evidence. *Oecologia* 86: 70–75.
Mitchell D. 1974. Convective heat transfer from man and other animals. In *Heat loss from animals and man*. Monteith JL, Mount LE (Eds.). Butterworths, London, pp. 59–76.
Mitchell D, Maloney SK, Jessen C, Laburn HP, Kamerman P, Mitchell G, Fuller A. 2002. Adaptive heterothermy and selective brain cooling in arid-zone mammals. *Comparative Biochemistry and Physiology B* 131: 571–585.
Mitchell D, Snelling E, Hetem R, Maloney S, Strauss W, Fuller A. 2018. Revisiting concepts of thermal physiology: Predicting responses of mammals to climate change. *Journal of Animal Ecology* 87: 956–973.
Mitchell G. 1977. *Studies on the control of cerebral blood flow*. PhD thesis, University of the Witwatersrand, Johannesburg, South Africa.
Mitchell G. 2008. Robert Goetz: A pioneer of the study of the physiology of animals in the wild. *Transactions of the Royal Society of South Africa* 63: 182–185.
Mitchell, G. 2009. The origin of the scientific study and classification of giraffes. *Transactions of the Royal Society of South Africa* 64: 1–13.
Mitchell G, Bobbitt JP, Devries S. 2008. Cerebral perfusion pressure in giraffe: Modelling the effects of head raising and lowering. *Journal of Theoretical Biology* 252: 98–108.
Mitchell G, Fuller A, Maloney SK, Rump N, Mitchell D. 2006a. Guttural pouches, brain temperature and exercise in horses. *Biology Letters* 2: 475–477.

Mitchell G, Hattingh J. 1993. Giraffe colloid osmotic pressure and oedema. *South African Journal of Science* 89: 569–570.

Mitchell G, Hepburn HR, Mitchell D, Shimell C, Borsook D, Ford DM, Klugman K, Rogers G. 1980. Carotid haemodynamics in the giraffe. *South African Journal of Science* 76: 246.

Mitchell G, Lust, A. 2008. The carotid rete and artiodactyl success. *Biology Letters* 4: 415–418.

Mitchell G, Maloney SK, Mitchell D, Keegan DJ. 2006b. The origin of mean arterial and jugular venous blood pressures in giraffes. *Journal of Experimental Biology* 209: 2515–2524.

Mitchell G, Roberts D, van Sittert S, Skinner JD. 2013a. Growth patterns and masses of the heads and necks of male and female giraffes. *Journal of Zoology (London)* 290: 49–57.

Mitchell G, Roberts DG, van Sittert SJ. 2015. The digestive morphophysiology of wild, free-living, giraffes. *Comparative Biochemistry and Physiology A* 187: 119–129.

Mitchell G, Roberts DG, van Sittert SJ. 2016. Corrigendum to "The digestive morphophysiology of wild, free-living, giraffes." *Comparative Biochemistry and Physiology A* 191: 226.

Mitchell G, Roberts DG, van Sittert SJ, Skinner JD. 2013b. Orbit orientation and eye morphometrics in giraffes (*Giraffa camelopardalis*). *African Zoology* 48: 333–339.Mitchell G, Skinner JD. 1993. How giraffe adapt to their extraordinary shape. *Transactions of the Royal Society of South Africa* 48: 207–218.

Mitchell G, Skinner JD. 2003. On the origin, evolution and phylogeny of giraffes *Giraffa camelopardalis*. *Transactions of the Royal Society of South Africa* 58: 51–73.

Mitchell G, Skinner JD. 2004. Giraffe thermoregulation: A review. *Transactions of the Royal Society of South Africa* 59: 49–58.

Mitchell G, Skinner JD. 2009. An allometric analysis of the giraffe cardiovascular system. *Comparative Biochemistry and Physiology A* 154: 523–529.

Mitchell G, Skinner JD. 2011. Lung volumes in giraffes, *Giraffa camelopardalis*. *Comparative Biochemistry and Physiology, Part A* 158: 72–78.

Mitchell G, van Schalkwyk OL, Skinner JD. 2005. The calcium and phosphorus content of giraffe (*Giraffa camelopardalis*) and buffalo (*Syncerus caffer*) skeletons. *Journal of Zoology (London)* 264: 307–315.

Mitchell G, van Sittert S, Roberts D, Mitchell D. 2017. Body surface area and thermoregulation in giraffes. *Journal of Arid Environments* 145: 35–42.

Mitchell G, van Sittert S, Skinner JD. 2009a. The structure and function of giraffe jugular vein valves. *South African Journal of Wildlife Research* 39: 175–180.

Mitchell G, van Sittert S, Skinner JD. 2009b. Sexual selection is not the origin of long necks in giraffes. *Journal of Zoology (London)* 278: 281–286.

Mitchell G, van Sittert S, Skinner JD. 2010. The demography of giraffe deaths in a drought. *Transactions of the Royal Society of South Africa* 65: 165–168.

Mitchell JA, Mitchell G, Lust A. 2009c. Thermoregulatory anatomy of pronghorn (*Antilocapra americana*). *European Journal of Wildlife Research* 55: 23–31.

Mitchell PC. 1905. On the intestinal tract of mammals. *Transactions of the Zoological Society* 17: 474–476.

Mivart St. G. 1871. *On the genesis of species*. Macmillan, London and New York.

Monod J, Changeux JP, Jacob F. 1963. Allosteric proteins and cellular control systems. *Journal of Molecular Biology* 6: 306–329.

Moorman A, Webb S, Brown NA, Lamers W, Anderson RH. 2003. Development of the heart: (1) formation of the cardiac chambers and arterial trunks. *Heart* 89: 806–814.

Morales J, Soria D, Pickford M. 1999. New stem giraffoid ruminants from the early and middle Miocene of Namibia. *Geodiversitas* 21: 229–253.

More HL, Hutchinson JR, Collins DF, Weber DJ, Aung SKH, Donelan JM. 2010. Scaling of sensorimotor control in terrestrial mammals. *Proceedings of the Royal Society B* 277: 3563–3568.

More HL, O'Connor SM, Brøndum E, Wang T, Bertelsen MF, Grøndahl C, Kastberg K, Hølyck A, Funder J, Donelan JM. 2013. Sensorimotor responsiveness and resolution in the giraffe. *Journal of Experimental Biology* 216: 1003–1011.

Morgan ME, Kingston JD, Marino BD. 1994. Carbon isotopic evidence for the emergence of C4 plants in the Neogene from Pakistan and Kenya. *Nature* 367: 162–165.

Morris P. 1972. A review of mammalian age determination methods. *Mammal Review* 2: 69–104.
Mortola JP, Wilfong D. 2017. Hematocrit of mammals (Artiodactyla, Carnivora, Primates) at 1500m and 2100m altitudes. *Zoology* 125: 10–23.
Mosekilde L, Mosekilde L. 1986. Normal vertebral body size and compressive strength: Relations to age and to vertebral and iliac trabecular bone compressive strength. *Bone* 7: 207–212.
Mossop EE (Ed.). 1935. *The Journals of Wikar, Coetsé and van Reenen.* Translated from the Dutch by van der Horst AW and Mossop EE. The Van Riebeeck Society, Cape Town.
Mossop EE. (Ed.). 1947. *The Journals of Brink and Rhenius: Being the journal of Carel Frederick Brink of the journey into Great Namaqualand (1761-2) and the journal of Ensign Johannes Tobias Rhenius (1724).* Van Riebeeck Society, Cape Town.
Muller CFJ. 1971. *Five hundred years: A history of South Africa.* 2nd edition. Academica, Pretoria and Cape Town.
Müller DWH, Codron D, Meloro C, Munn A, Schwarm A, Hummel J, Clauss M. 2013. Assessing the Jarman-Bell principle: Scaling of intake, digestibility, retention time and gut fill with body mass in mammalian herbivores. *Comparative Biochemistry and Physiology A* 164: 129–140.
Muller Z. 2017. White giraffes: The first record of vitiligo in a wild adult giraffe. *African Journal of Ecology* 55: 118–123.
Muller Z, Cuthill IC, Harris S. 2018. Group sizes of giraffes in Kenya: The influence of habitat, predation and the age and sex of individuals. *Journal of Zoology* 306: 77–87.
Munis JR, Lozada LJ. 2000. Giraffes, siphons, and Starling resistors. Cerebral perfusion pressure revisited. *Journal of Neurological Anesthesiology* 12: 290–296.
Murie J. 1872. On the horns, viscera and muscles of the giraffe, with a record of the post-mortem examination of two specimens killed by a fire. *The Annals and Magazine of Natural History, Edinburgh*, Ser. 4, IX: 177–195.Nagel E L, Morgane PJ, McFarlane WL, Galliano RE. 1968. Rete mirabile of dolphin: Its pressure damping effect on cerebral circulation. *Science* 161: 898–900.
Nagy KA. 1987. Field metabolic rate and food requirement scaling in mammals and birds. *Ecological Monographs* 57: 111–128.
Nagy KA. 1994. Field bioenergetics of mammals: What determines field metabolic rates. *Australian Journal of Zoology* 42: 43–53.
Nagy KA. 2005. Field metabolic rate and body size. *Journal of Experimental Biology* 208: 1621–1625.
Nakakuki S. 1983. The bronchial ramification and lobular division of the giraffe lung. *Anatomischer Anzeiger* 154: 313–317.
Nesbit-Evans EM. 1970. The reaction of a group of Rothschild's giraffe to a new environment. *East African Wildlife Journal* 8: 53–62.
Nilsson O, Booj S, Dahlstrom A, Hargens AR, Millard RW, Pettersson K. 1988. Sympathetic innervation of the cardiovascular system in the giraffe. *Blood Vessels* 25: 299–307.
Noirot-Timothée C. 1963. Sur les cilies du rumen de *Giraffa camelopardalis. Comptes rendus de l'Académie des Sciences* 256: 5400–5401.
Nyman G, Röken B, Hedin E, Hedenstierna G. 2016. Case studies in physiology: Ventilation and perfusion in a giraffe—does size matter? *Journal of Applied Physiology* 121: 1374–1378.
Oates LG. 1972. Food preferences of giraffe in Transvaal lowveld Mopane woodland. *Journal of the South African Wildlife Management Association* 2: 21–23.
Oboussier H, Moller G. 1971. Zur kenntnis des Gehirns der Giraffidae (Pecora, Artiodactyla, Mammalia): Ein Verleich der Neocortex-oberflachengrosse. *Zeitschrift fur Saugetierk* 36: 291–296.
O'Brien HD, Bourke J. 2015. Physical and computational fluid dynamics models for the hemodynamics of the artiodactyl rete. *Journal of Theoretical Biology* 386: 122–131.
O'Brien HD, Gignac PM, Hieronymus TL, Witmer, LM. 2016. A comparison of postnatal arterial pressure patterns in a growth series of giraffe (Artiodactyla: *Giraffa camelopardalis*). *PeerJ* 4: e1696.
Ochs M, Nyengaard JR, Jung A, Knudsen L, Voigt M, Wahlers T, Richter J, Gundersen HJG. 2004. The number of alveoli in the human lung. *American Journal of Respiratory and Critical Care Medicine* 169: 120–124.

O'Connor, JPB. 2003. Thomas Willis and the background to *Cerebri Anatome*. *Journal of the Royal Society of Medicine* 96: 139–143.

Odenyo AA, McSweeney CS, Palmer B, Negassa D, Osuji PO. 1999. *In vitro* screening of rumen fluid samples from indigenous African ruminants provides some evidence for rumen fluid with superior capacities to digest tannin-rich fodders. *Australian Journal of Agricultural Research* 50: 1147–1157.

O'Donoghue K. 2008. Fetal microchimerism and maternal health during and after pregnancy. *Obstetric Medicine* 1: 56–64.

Ogilby J. 1670. *Africa: Being an accurate description of the Regions of Aegypt, Barbary, Lybia, and Billedulgerid, the Land of the Negroes . . . Collected and Translated from most Authentick Authors, and augmented from later observations: Illustrated with Notes, and Adorned with peculiar maps, and proper Sculptures*. A Translation of Dapper, 1668 [see earlier entry]. Thomas Johnson, London.

O'Kane CAJ, Duffy KJ, Page BR, Macdonald DW. 2011. Overlap and seasonal shifts in use of woody plant species amongst a guild of savanna browsers. *Journal of Tropical Ecology* 27: 249–258.

Oken L. 1802. *Grundriss der Naturphilosophie, der Theorie der Sinne, mit der darauf gegründeter Classification der Thiere*. Oken, Frankfurt.

Østergaard KH, Baandrup UT, Wang T, Bertelsen MF, Andersen JB, Smerup M, Nyengaard JR. 2013. Left ventricular morphology of the giraffe heart examined by stereological methods. *The Anatomical Record* 296: 611–621.

Østergaard KH, Bertelsen MF, Brondum ET, Aalkjaer C, Hasenkam JM, Smerup M, Wang T, Nyengaard JR, Baandrup U. 2011. Pressure profile and morphology of the arteries along the giraffe limb. *Journal of Comparative Physiology B* 181: 691–698.

Osthoff C, Hugo A, Madende M, Deacon F, Nel PJ. 2017. Milk composition of free-ranging red hartebeest, giraffe, Southern reedbuck and warthog and a phylogenetic comparison of the milk of African Artiodactyla. *Comparative Biochemistry and Physiology A* 204: 93–103.

Owen R. 1838. Notes on the anatomy of the Nubian giraffe. *Transactions of the Zoological Society (London)* 2: 217–248.

Owen R. 1839. Notes on the birth of the Giraffe at the Zoological Gardens, and description of the foetal membranes and some of the natural and morbid appearances observed in the dissection of the young animal. *Transactions of the Zoological Society of London* 3: 21–28.

Owen R. 1840–1845. *Odontography, or a treatise on the comparative anatomy of the teeth*. Bailliere, London, pp. 527–543.

Owen R. 1848. Description of teeth and proportion of jaws of two extinct Anthracotheroid quadrupeds (*Hyopotamus vectianus and Hyopotamus bovinus*) discovered by the Marchioness of Hastings in the Eocene deposits of the N.W. coast of the Isle of Wight: With an attempt to develop Cuvier's idea of the classification of pachyderms by the number of their toes. *Quarterly Journal of the Geological Society of London* 4: 103–141.

Owen R. 1851. On Dinornis (Part IV). *Transactions of the Zoological Society* 4: 1–20.

Owen R. 1868. *On the anatomy of vertebrates*. Vol. 3, *Mammals*. Longmans, Green, London.

Owen RB, Muiruri VM, Lowenstein TK, Renaut RW, Rabideaux N, Luo S, Deino AL, Sier MJ, Dupont-Nivet G, McNulty EP, Leet K, Cohen A, Campisano C, Deocampo D, Shen C-C, Billingsley A, Mbuthia A. 2018. Progressive aridification in East Africa over the last half million years and implications for human evolution. *Proceedings of the National Academy of Sciences* 115: 11174–11179.

Owen-Smith N. 1997. Distinctive features of the nutritional ecology of browsing versus grazing ruminants. *Zeitschrift fur Saugetierkunde* 62: 176–191.

Owen-Smith N, Mason DR. 2005. Comparative changes in adult vs juvenile survival affecting population trends of African ungulates. *Journal of Animal Ecology* 74: 762–773.

Pagani M, Zachos JC, Freeman KH, Tipple B, Bohaty S. 2005. Marked decline in atmospheric carbon dioxide concentrations during the Paleogene. *Science* 309: 600–603.

Pallas PS. 1778. *The Natural history of strange animals, in which mainly new and unknown animal species are explained by copper engravings, descriptions and explanations*. GA Lange, Berlin. In German.

Pander C, D'Alton E. 1823. *Die Skelete der Wiederkäuer*. Eduard Weber, Bonn.

Parker DM, Bernard RTF, Colvin SA. 2003. The diet of a small group of extralimital giraffe. *African Journal of Ecology* 41: 245–253.

Paterson W. 1789. *A narrative of four journeys into the country of the Hottentots and Caffraria in the years 1777–1789.* J Johnson, London.

Paton JFR, Dickinson C J, Mitchell G. 2009. Harvey Cushing and the regulation of blood pressure in giraffe, rats and man: Introducing the Cushing's mechanism. *Experimental Physiology* 94: 11–17.

Patterson JL, Goetz RH, Doyle JT, Warren JV, Gauer OH, Detweiler DK, Said SI, Hoernicke H, McGregor M, Keen EN, Smith MH, Hardie EL, Reynolds M, Flatt WP, Waldo DR. 1965. Cardiorespiratory dynamics in the ox and giraffe, with comparative observations on man and other mammals. *Annals of the New York Academy of Sciences* 127: 393–413.

Patterson JL, Warren JV. 1952. Mechanisms of adjustments in the cerebral circulation upon assumption of the upright position. *Journal of Clinical Investigation* 31: 653.

Patterson JL, Warren JV, Doyle JT, Gauer O, Keen EN, Goetz RH. 1957. Circulation and respiration in the giraffe. *Journal of Clinical Investigation* 36: 919.

Patton ML, Bashaw MJ, del Castillo SM, Jochle W, Lamberski N, Rieches R, Bercowitz FB. 2006. Long-term suppression of fertility in female giraffe using GnRH agonist deslorelin as a long-acting implant. *Theriogenology* 66: 431–438.

Pearson PN, Palmer MR. 2000. Atmospheric carbon dioxide for the past 60 Myr. *Nature* 406: 695–699.

Pedley TJ. 1987. How giraffes prevent oedema. *Nature* 329: 13–14.

Pedley TJ, Brook BS, Seymour RS. 1996. Blood pressure and flow rate in the giraffe jugular vein. *Philosophical Transactions of the Royal Society (London) B* 351: 855–866.

Peichl L. 2005. Diversity of mammalian photoreceptor properties: Adaptations to habitat and lifestyle. *Anatomical Record* 287A: 1001–1012.

Pellew RA. 1983a. The impact of elephant, giraffe, and fire upon the *Acacia tortillas* woodlands of the Serengeti. *African Journal of Ecology* 21: 41–74.

Pellew RA. 1983b. The giraffe and its food resource in the Serengeti. I. Composition, biomass, and production of available browse. *African Journal of Ecology* 21: 241–267.

Pellew RA. 1983c. The giraffe and its food resource in the Serengeti. II. Response of the giraffe population to changes in the food supply. *African Journal of Ecology* 21: 269–283.

Pellew RA. 1984a. Food consumption and energy budgets of the giraffe. *Journal of Applied Ecology* 21: 141–159.

Pellew RA. 1984b. The feeding ecology of a selective browser, the giraffe (*Giraffa camelopardalis tippelskirchi*). *Journal of Zoology (London)* 202: 57–81.

Pennycuick CJ. 1975. On the running of the Gnu (*Connochaetes taurinus*) and other animals. *Journal of Experimental Biology* 63: 775–799.

Perez W, Lima M, Clauss M. 2009. Gross anatomy of the intestine in the giraffe (*Giraffa camelopardalis*). *Journal of Veterinary Medicine: Anatomia, Histologia, Embryologia* 38: 432–435.

Perez W, Michel V, Hassen J, Vazquez N. 2012. Anatomy of the mouth of the Giraffe (*Giraffa camelopardalis rothschildi*). *International Journal of Morphology* 30: 322–329.

Perry TW. 2012. *Beef cattle feeding and nutrition*. Academic Press, Cambridge, MA.

Petersen KK, Horlyck A, Østergaard KH, Andresen J, Broegger T, Skovgaard N, Telinius N, Laher I, Bertelsen MF, Grondahl C, Smerup M, Secher NH, Brondum E, Hasenkam JM, Wang T, Baandrup U, Aalkjaer C. 2013. Protection against high intravascular pressure in giraffe legs. *American Journal of Physiology* 305: R1021–R1030.

Pettersson K, Hargens AR, Millard RW, Johansen K, Gershuni DH, Burroughs R, Meltzer DGA, van Hoven W. 1986. Dependent hypertension and arterial wall hypertrophy without interstitial edema in the giraffe. *Proceedings of the International Union of Physiological Sciences* 16: 411.

Petzold A, Hassanin A. 2020. A comparative approach for species delimitation based on multiple methods of multi-locus DNA sequence analysis: A case study of the genus *Giraffa* (Mammalia, Cetartiodactyla). *PLOS One* 15 (2): e0217956.

Petzsch H. 1950. Zur Frage des Vorkommens ungefleckter albinotischer Giraffen. *Der Zoologische Garten* 17: 44–47.

Pickering J. 1997. William J Burchell's South African mammal collection, 1810–1815. *Archives of Natural History* 24: 311–326.

Pickford M. 1975. Late Miocene sediments and fossils from the northern Kenya Rift Valley. *Nature* 256: 279–284.

Pienaar U de V. 1969. Predator prey relationships amongst the larger mammals of the Kruger National Park. *Koedoe* 12: 108–176.

Pilgrim GE. 1911. The fossil giraffidae of India. *Palaeontologia Indica* 4: 1–29.

Pilgrim GE. 1934. Two new species of sheep-like antelope from the Miocene of Mongolia. *American Museum Novitates* 716: 1–29.

Pincher C. 1949. Evolution of the giraffe. *Nature* 164: 29–30.

Pliny the Elder (Gaius Plinius Secundus). AD 64. *Natural history*: Book VIII. Translated from Latin by Bostock J, Riley HT, George Bell & Sons, London, 1900: 276–277, and by Rackham H, 1943. *Pliny: Natural history*, Volume III, Books 8–11. Loeb Classical Library No. 353. Harvard University Press, Cambridge, MA.

Pomel A. 1892. Sur le Libytherium maurusium, grand ruminant du terrain Pliocene plaisancier de l'Algerie. *Comptes rendus de l'Acadedmie des Sciences (Paris)* 115: 100–102.

Pontzer H. 2007. Effective limb length and the scaling of locomotor cost in terrestrial animals. *Journal of Experimental Biology* 210: 1752–1761.

Potts R. 1998. Environmental hypotheses of hominin evolution. *Yearbook of Physical Anthropology* 41: 93–136.

Pournelle GH. 1955. Notes on the reproduction of a Baringo giraffe. *Journal of Mammalogy* 36: 574.

Prange HD, Anderson JF, Rahn, H. 1979. Scaling of skeletal mass to body mass in birds and mammals. *American Naturalist* 113: 103–122.

Pratt DM, Anderson VH. 1979. Giraffe cow-calf relationships and social development of the calf in the Serengeti. *Zeitschrift fur Tierpsychologie* 51: 233–251.

Pratt DM, Anderson VH. 1982. Population, distribution, and behaviour of giraffe in the Arusha National Park, Tanzania. *Journal of Natural History* 16: 481–489.

Pratt DM, Anderson VH. 1985. Giraffe social behavior. *Journal of Natural History* 19: 771–781.

Prothero J. 1979. Heart weight as a function of body weight in mammals. *Growth* 43: 139–150.

Prothero DR, Foss SE. 2007. *The evolution of Artiodactyls*. Johns Hopkins University Press, Baltimore, MD.

Rackham, H. 1943. *Pliny: Natural history*, Volume III, Books 8–11. Loeb Classical Library No. 353. Harvard University Press, Cambridge, MA.

Ramstein G, Fluteau F, Besse J, Joussaume S. 1997. Effect of orogeny, plate motion and land-sea distribution on Eurasian climate change over the past 30 million years. *Nature* 386: 788–795.

Raper PE, Boucher M (Eds.). 1988. *Robert Jacob Gordon: Cape travels, 1777 to 1786*. Volume 1. Brenthurst Press, Johannesburg, South Africa.

Ray J. 1693. *Synopsis methodica animalium quadrupedum et serpentini generis*. Robert Southwell, London.

Raymo ME, Ruddiman WF. 1992. Tectonic forcing of late Cenozoic climate. *Nature* 359: 117–122.

Reason RC. 1995. Development of feeding behaviours in giraffe (*Giraffa camelopardalis*) calves at Brookfield Zoo, Chicago. *International Zoo Yearbook* 34: 222–224.

Reeve JP, Fairbairn DJ. 1996. Sexual size dimorphism as a correlated response to selection on body size: An empirical test of the quantitative genetic model. *Evolution* 50: 1927–1938.

Reeves JT, Daoud FS, Gentry M. 1972. Growth of the fetal calf and its arterial pressure, blood gases, and hematological data. *Journal of Applied Physiology* 32: 240–244.

Retallack GJ, Dugas DP, Bestland EA. 1990. Fossil soils and grasses of a middle Miocene East African grassland. *Science* 247: 1325–1328.

Reynolds WW, Karlotski, WJ. 1977. The allometric relationship of skeleton weight to body weight in teleost fishes: A preliminary comparison with birds and mammals. *Copeia* 1977: 160–163.

Rhodes JA. 1975. Biochemical profiles in exotic captive animals. *Annual Proceedings of the American Association of Zoo Veterinarians* 1975: 81–111.

Richiardi S. 1880. Sull'anatomia della giraffa (Anatomy of the giraffe). *Zoologische Anzeiger* 3: 92–93.

Ridewood WG. 1901. A note on the hairs of antelope, giraffe and okapi. *Proceedings of the Zoological Society, London* 1901: 3.

Ridewood WG. 1904. Some observations on the skull of the giraffe. *Proceedings of the Zoological Society, London* 1: 150–157.

Rigoulet J. 2012. Histoire de Zarafa, la giraffe de Charles X. *Bulletin de l'Académie Vétérinaire de France* 165: 169–176.

Ringmar E. 2006. Audience for a giraffe: European expansionism and the quest for the exotic. *Journal of World History* 17: 375–397.

Rios M, Danowitz M, Solounias N. 2016. First comprehensive morphological analysis on the metapodials of Giraffidae. *Palaeontologia Electronica* 19.3.52A: 1–39.

Ripple WJ, Newsome TM, Wolf C, Dirzo R, Everatt KT, Galetti M, Hayward MW, Kerley GIH, Levi T, Lindsey PA, Macdonald DW, Malhi Y, Painter LE, Sandom CJ, Terborgh J, Van Valkenburgh B. 2015. Collapse of the world's largest herbivores. *Science Advances* 1: e1400103.

Ripple WJ, Wolf C, Newsome TM, Hoffman M, Wirsing AJ, McCauley DJ. 2017. Extinction risk is most acute for the world's largest and smallest vertebrates. *Proceedings of the National Academy of Sciences* 114 (40): 10678–10683.

Ritland S. 1982. The allometry of the vertebrate eye. PhD dissertation, University of Chicago.

Roaf R. 1961. Giraffe-necked women: A myth exploded. *Journal of Bone and Joint Surgery* 43: 114–115.

Robb RM. 1982. Increase in retinal surface area during infancy and childhood. *Journal of Pediatric Ophthalmology and Strabismus* 19: 16–20.

Robbins CT, Spalinger DE, van Hoven W. 1995. Adaptation of ruminants to browse and grass diets: Are anatomical-based browser-grazer interpretations valid? *Oecologia* 103: 208–213.

Robertson-Bullock, W. 1962. The weight of the African Elephant *Loxodonta africana*. *Proceedings of the Zoological Society* 138: 133–135.

Robin ED, Corson JM, Dammin GJ. 1960. The respiratory dead space of the giraffe. *Nature* 186: 24–26.

Robinson CA. 2011. Giraffidae. In *Paleontology and Geology of Laetoli: Human evolution in context.* Harrison T (Ed.). Series Vertebrate paleobiology and paleoanthropology. Springer, New York, pp. 339–362.

Robinson E. 1949. Evolution of the giraffe. *Nature* 164 (4164): 323.

Roggenbuck M, Sauer C, Poulsen M, Bertelsen MF, Sorensen SJ. 2014. The giraffe (*Giraffa camelopardalis*) rumen microbiome. *FEMS Microbiology Ecology* 90: 237–246.

Rookmaaker LC. 1979. Robert Jacob Gordon (1743–1795) *en zijn bijdrae tot de zoölogie van de Kaap de Goede Hoop.* PhD thesis, University of Utrecht.

Rookmaaker LC. 1983. The observations of Robert Jacob Gordon (1743–1795) on giraffes (*Giraffa camelopardalis*) found in Namaqualand. *Journal of the South West Africa Scientific Society* 36–37: 71–90.

Rose KD. 1982. Skeleton of *Diacodexis*, oldest known artiodactyl. *Science* 216: 621–623.

Rossof AH. 1972. An electrocardiographic study of the giraffe. *American Heart Journal* 83: 142–143.

Roth J, LeRoith D, Shiloach J, Rosenzweig JL, Lesniak MA, Havrankova J. 1982. The evolutionary origins of hormones, neurotransmitters, and other extracellular chemical messengers. *New England Journal of Medicine* 306: 523–527.

Rowan J, Locke EM, Robinson JR, Campisano CJ, Wynn JG, Reed KE. 2017. Fossil Giraffidae (Mammalia, Artiodactyla) from Lee Adoyta, Ledi-Geraru, and Late Pliocene dietary evolution in Giraffids from the Lower Awash Valley, Ethiopia. *Journal of Mammal Evolution* 24: 359–371. doi: 10.1007/s10914-016-9343-z.

Roy RR, Graham S, Peteson JA. 1988. Fiber type composition of the plantar flexors of giraffes (*Giraffa camelopardalis*) at different post-natal stages of development. *Comparative Biochemistry and Physiology A* 91: 347–352.

Ruddiman WF 2006. Orbital changes and climate. *Quaternary Science Reviews* 25: 3092–3112.

Rudolph AM, Heymann MA. 1970. Circulatory changes in the fetal lamb. *Circulation Research* 26: 289–299.

Ruggieri G, Rocca AR. 2010. Analysis of past and present methods of measuring and estimating body surface area and the resulting evaluation of its doubtful suitability to universal application. *Blood Purification* 30: 296–305.

Rüppell E. 1829. *Reisen in Nubien, Kordofan und dem petraeischen Arabien vorzueglich in geographisch-statisticher Hinsicht.* Friedrich Wilmans, Frankfurt.

Russell B. 1931. *The scientific outlook.* George Allen & Unwin, London.

Ryan SJ, Jordaan W. 2005. Activity patterns of African buffalo *Syncerus caffer* in the lower Sabie region, Kruger National Park, South Africa. *Koedoe* 48: 117–124.

Sacher GA, Staffeldt EF. 1974. Relation of gestational time to brain weight for placental mammals: Implications for the theory of vertebrate growth. *American Naturalist* 108: 593–615.

Safi K. Kerth G. 2004. A comparative analysis of specialization and extinction risk in temperate-zone bats. *Conservation Biology* 18: 1293–1303.

Saito M, Idani G. 2018. Suckling and allosuckling behaviour in wild giraffe (*Giraffa camelopardalis tippelskirchi*). *Mammalian Biology* 93: 1–4.

Sarrus P, Rameaux J-F. 1839. Application des sciences accessories et principalment des mathématiques á la physiologie général. *Bulletin de l'Académie Royale de Médicine* 3: 1904–1100.

Sasaki M, Endo H, Kitamura N, Yamada J, Yamamoto M, Arishima K, Hayashi Y. 2001. The structure of the masseter muscle in the giraffe (*Giraffa camelopardalis*). *Anatomia, Histologia, Embryologia* 30: 313–319.

Sathar F, Badlangana L, Manger PR. 2010. Variations in the thickness and composition of the skin of the giraffe. *Anatomical Record* 293: 1615–1627.

Sauer C,.Bertelsen MF, Lund P, Weisbjerg MR, Clauss M. 2015. Quantitative macroscopic anatomy of the giraffe (*Giraffa camelopardalis*) digestive tract. *Anatomia, Histologia, Embryologia* 45: 338–349.

Sauer JJC. 1983a. Food selection by giraffes in relation to changes in chemical composition of the leaves. *South African Journal of Animal Science* 13: 40–43.

Sauer JJC. 1983b. A comparison between *Acacia* and *Combretum* leaves utilized by giraffe. *South African Journal of Animal Science* 13: 43–44.

Sauer JJC, Skinner JD, Neitz AWH. 1982. Seasonal utilization of leaves by giraffes, *Giraffa camelopardalis*, and the relationship of the seasonal utilization to the chemical composition of the leaves. *South African Journal of Zoology* 17: 210–219.

Sauer JJC, Theron GK, Skinner JD. 1977. Food preferences of giraffe *Giraffa amelopardalis* in the arid bushveld of the western Transvaal. *South African Journal of Wildlife Research* 7: 53–59.

Savage RJG, Long MR. 1986. *Mammal evolution: An illustrated guide.* British Museum (Natural History), London.

Scafetta N, Mirandola A, Bianchini A. 2017. Natural climate variability, part 1: Observations versus model predictions. *International Journal of Heat and Technology* 35: S9–SS17.

Scheepers JL. 1992. Habitat selection and demography of a giraffe population in northern Namib desert, Namibia. *Ongulès/Ungulates* 91: 223–228.

Scheffer M, Vergnon R, Cornelissen JHC, Hantson S, Holmgren M, van Nes EH, Xu C. 2014. Why trees and shrubs but rarely trubs? *Trends in Ecology and Evolution* 29: 433–434.

Scherren H. 1908. On certain errors with reference to George the Fourth's Giraffe. *Proceedings of the Zoological Society, London* 1908: 403–405.

Schiviz AN, Ruf T, Kueberr-Heiss A, Schubert C, Ahnelt PK. 2008. Retinal cone topography of artiodactyl mammals: Influence of body height and habitat. *Journal of Comparative Neurology* 507: 1336–1350.

Schmidt DA, Ball RL, Grobler D, Ellersleck MR, Griffin ME, Citino SB, Bush M. 2007. Serum concentrations of amino-acids, fatty acids, lipoproteins, Vitamins A and E, and minerals in apparently healthy, free-ranging southern giraffe (*Giraffa camelopardalis giraffe*). *Zoo Biology* 26: 13–25.

Schmidt-Nielsen K. 1983. *Animal physiology: Adaptation and environment*. Cambridge University Press, Cambridge.

Schmidt-Nielsen K. 2008. *Why is animal size so important?* Cambridge University Press, Cambridge.

Schoene B, Samperton KM, Eddy MP, Keller G, Adatte T, Bowring SA, Khadri SFR. 2015. U-Pb geochronology of the Deccan Traps and relation to the end-Cretaceous mass extinction. *Science* 347: 182–184.

Schreider E. 1950. Geographical distribution of the body-weight/body-surface ratio. *Nature* 165: 286.

Schreider E. 1951. Anatomical factors of body heat regulation. *Nature* 167: 823–824.

Schröder HJ, Power GG. 1997. Engine and radiator: Fetal and placental interactions for heat dissipation. *Experimental Physiology* 82: 403–414.

Schroter RC. 1978. Convective flows in mammalian lungs. In *Comparative physiology*. Schmidt-Nielsen K, Bolis L, Maddrell S (Eds.). Cambridge University Press, Cambridge, pp. 303–325.

Sclater PL. 1901. On a new African mammal. *Proceedings of the Zoological Society, London* 1901: 3–6.

Scott KM, Janis CM. 1993. Relationships of the Ruminantia (Artiodactyla) and an analysis of the characters used in ruminant taxonomy. In *Mammal phylogeny: Placental*. Szalay FS, Novacek MJ, McKenna MC (Eds.). Springer-Verlag, London, pp. 282–302.

Scott WB. 1913. *A history of land mammals of the Western Hemisphere*. MacMillan, New York.

Seeber PA, Duncan P, Fritz H, Ganswindt A. 2013. Androgen changes and flexible rutting behaviour in male giraffes. *Biology Letters* 9: 20130396.

Seeber PA, Ndlovu HT, Duncan P, Ganswindt A. 2012. Grazing behavior of the giraffe in Hwange National Park, Zimbabwe. *African Journal of Ecology* 50: 247–250.Selker F, Carter DR. 1989. Scaling of long bone fracture strength with animal mass. *Journal of Biomechanics* 22: 1175–1183.

Selous FC. 1881. *A hunter's wanderings in Africa: Being a narrative of nine years spent amongst the game of the far interior of South Africa*. Richard Bentley, London.

Selous FC. 1908. *African nature notes and reminiscences*. Macmillan, London.

Senter P. 2007. Necks for sex: Sexual selection an explanation for sauropods dinosaur neck elongation. *Journal of Zoology (London)* 271: 45–53.

Seymour RS. 2000. Model analogues in the study of cephalic circulation. *Comparative Biochemistry and Physiology A* 125: 517–524.

Seymour RS, Blaylock AJ. 2000. The principle of LaPlace and scaling of ventricular wall stress and blood pressure in mammals and birds. *Physiological and Biochemical Zoology* 73: 389–405.

Seymour RS, Hargens AR, Pedley TJ. 1993. The heart works against gravity. *American Journal of Physiology* 265: R715–R720.

Seymour RS, Johansen K. 1987. Blood flow up hill and downhill: Does a siphon facilitate circulation above the heart. *Comparative Biochemistry and Physiology A* 88: 167–170.

Shackleton NJ. 2000. The 100,000 year ice-age cycle identified and found to lag temperature, carbon dioxide and orbital eccentricity. Science 289: 1897–1902.

Sharkey HJ. 2015. *La Belle Africaine*: The Sudanese giraffe who went to France. *Canadian Journal of African Studies* 49: 39–65.

Shinozaki A, Hosaka Y, Imagawa T, Uehara M. 2010. Topography of ganglion cells and photoreceptors in the sheep retina. *Journal of Comparative Neurology* 518: 2305–2315.

Siegel JM. 2008. Do all animals sleep? *Trends in Neurosciences* 31: 208–213.

Sikes SK. Habitat and cardiovascular disease: Observations made on elephants (*Loxodonta africana*) and other free-living animals in East Africa. 1969. *Transactions of the Zoological Society (London)* 32: 1–104.

Silanikove N. 1994. The struggle to maintain hydration and osmoregulation in animals experiencing severe dehydration and rapid rehydration: The story of ruminants. *Experimental Physiology* 79: 281–300.
Simmons RE, Altwegg R. 2010. Necks-for-sex or competing browsers? A critique of ideas on the evolution of giraffe. *Journal of Zoology* 282: 6–12.
Simmons RE, Scheepers L. 1996. Winning by a neck: Sexual selection in the evolution of giraffe. *The American Naturalist* 148: 771–786.
Simons EL, Pilbeam D, Boyer SJ. 1971. Appearance of *Hipparion* in the tertiary of the Siwalik Hills of north India, Kashmir and west Pakistan. *Nature* 229: 408–409.
Singer MA, Morton AR. 2000. Mouse to elephant: Biological scaling and Kt/V. *American Journal of Kidney Diseases* 35: 306–309.
Singer R, Bonè EL. 1960. Modern giraffes and the fossil Giraffids of Africa. *Annals of the South African Museum* 45: 375–548.
Sisson S, Grossman JD. 1967. *The anatomy of the domestic animals*. 4th edition, WB Saunders, Philadelphia and London.
Shackleton NJ. 2000. The 100,000-year ice-age cycle identified and found to lag temperature, carbon dioxide, and orbital eccentricity. *Science* 289: 1897–1902.
Skead CJ. 1958. Mammals of the Uitenhage and Cradock Districts Cape Province in recent times. *Koedoe* 1: 19–38.
Skead CJ. 1980. *Historical mammal incidence in the Cape Province*. Volume 1. Department of Nature and Environmental Conservation, Cape of Good Hope, Cape Town, South Africa.
Skinner JD. 1978. Breeding cycles in three species of African ungulates. In *Proceedings in Life Sciences: Environmental Endocrinology*. Springer-Verlag, Berlin, pp. 64–72.
Skinner JD, Hall-Martin AJ. 1975. A note on foetal growth and development of the giraffe, *Giraffa camelopardalis giraffa*. *Journal of Zoology, London* 177: 73–79.
Skinner JD, Mitchell G. 2011. Family Giraffidae (Giraffe and Okapi). In *Handbook of mammals of the world*. Vol. 2, *Hoofed mammals*. Wilson DE, Mittermeier RA (Eds.). Lynx Edicions, Barcelona, pp. 788–802.
Skinner JD, Smithers RHN. 1990. Giraffidae. In *The mammals of the Southern African Sub-region*. 2nd edition. University of Pretoria, South Africa, pp. 604–607.
Slijper EJ. 1946. Comparative biological-anatomical investigation on the vertebral column and spinal musculature of mammals. *Proceedings of the Koninklike Nederlandse Akademie van Wetenschappen, Amsterdam* 42: 1–128.
Smerup M, Damkjaer M, Brondum E, Baandrup UT, Kristiansen SB, Nygaard H, Aalkjaer C, Sauer C, Buchanan R, Bertelsen MF, Østergaard K, Grondahl C, Candy G, Hasenkam JM, Secher NH, Bie P, Wang T.2016. The thick left ventricular wall of the giraffe heart normalizes wall tension, but limits stroke volume and cardiac output. *Journal of Experimental Biology* 219: 457–463.
Smith JR. 1883. Observations on Texas cattle: Age, weight, temperature, liver, and spleen. *Public Health Paper Report* 9: 99–110.
Smithers RHN, Skinner JD. 1983. *The mammals of the Southern African Sub-region*. 1st edition. University of Pretoria, South Africa.
Sneddon JC, Argenzio RA. 1998. Feeding strategy and water homeostasis in equids: The role of the hind gut. *Journal of Arid Environments* 38: 493–509.
Sneddon J, van der Walt J, Mitchell G. 1991. Water homeostasis in desert dwelling horses. *Journal of Applied Physiology* 71: 112–117.
Snelling EP, Taggart DA, Maloney SK, Farrell AP, Seymour RS. 2015. Biphasic allometry of cardiac growth in the developing Kangaroo *Macropus fuliginosus*. *Physiological and Biochemical Zoology* 88: 216–225.
Snyder GK, Sheafor BA. 1999. Red blood cells: Centerpiece in the evolution of the vertebrate circulatory system. *American Zoologist* 39: 189–198.
Solounias N. 1988. Prevalence of ossicones in Giraffidae (Artiodactyla, Mammalia). *Journal of Mammalogy* 69 (4): 845–848.

Solounias N. 1999. The remarkable anatomy of the giraffe's neck. *Journal of Zoology (London)* 247: 257–268.

Solounias N. 2007. Family Giraffidae. In *The evolution of artiodactyls*. Prothero DR, Foss SE (Eds.). Johns Hopkins University Press, Baltimore, MD, pp. 257–277.

Solounias N, Moelleken SMC. 1993. Dietary adaptation of some extinct ruminants determined by premaxillary shape. *Journal of Mammalogy* 74: 1059–1071.

Solounias N, Tang N. 1990. The two types of cranial appendages in *Giraffa camelopardalis* (Mammalia, Artiodactyla). *Journal of Zoology (London)* 222: 293–302.

Sonnenblick EH, Parmley WW, Buccino RA, Spann JF. 1968, Maximum force development in cardiac muscle. *Nature* 219: 1056–1058.

Sonntag CF. 1922. The comparative anatomy of the tongues of the Mammalia. VII. Cetacea, Sirenia, and Ungulata. *Proceedings of the Zoological Society* 92: 639–657.

Spinage CA. 1968a. *The book of the giraffe*. Collins, London.

Spinage CA. 1968b. Horns and other bony structures of the skull of the giraffe, and their functional significance. *East African Wildlife Journal* 6: 53–61.

Spinage CA. 1970. Giraffid horns. *Nature* 227: 735–736.

Spinage CA. 1993. The median ossicone of *Giraffa camelopardalis*. *Journal of Zoology (London)* 230: 1–5.

St.-Hilaire E. Geoffroy. 1827. Quelques considerations sur la girafe. *Annales des sciences naturelles (Paris)* 11: 210–223.

Stahl, WR. 1965. Organ weights in primates and other mammals. *Science* 150: 1039–1042.

Stahl WR. 1967. Scaling of respiratory variables in mammals. *Journal of Applied Physiology* 22: 453–460.

Stavorinus JS. 1798. *Voyages to the East Indies*. Translated from Dutch by SH Wilcocke. GG and J Robinson. Pater-Noster Row, London.

Stirton RA. 1944. Comments on the relationships of the cervoid family Palaeomerycidae. *American Journal of Science* 242: 633–655.

Stott K. 1959. Giraffe intergradation in Kenya. *Journal of Mammalogy* 40: 251.

Strauss MKL, Kilewo M, Rentsch D, Packer C. 2015. Food supply and poaching limit giraffe abundance in the Serengeti. *Population Ecology* 57: 505–516.

Strauss WM, Hetem RS, Mitchell D, Maloney SK, O'Brien HD, Meyer LCR, Fuller A. 2017. Body water conservation through selective brain cooling by the carotid rete: A physiological feature for surviving climate change? *Conservation Physiology* 5: 1–15.

Stromer E. 1907. Fossile Wirbeltier-Reste aus dem Uadi Faregh und Uadi Natrum in Aegypten. *Abhandlungen der Senckenbergischen Naturforschenden Gesellschaft* 29: 99–132.

Sullivan K, van Heugten E, Ange-van Heugten K, Poore MH, Dierenfeld ES, Wolfe B. 2010. Analysis of nutrient concentrations in the diet, serum, and urine of giraffe from surveyed North American Zoological Institutions. *Zoo Biology* 29: 457–469.

Swadlow HA, Waxman SG. 2012. Axonal conduction delays. *Scholarpedia* 7: 1451–1463.

Swainson W. 1835. *A treatise on the geography and classification of animals*. Longman, Rees, London.

Talbot WJ. 1970. Pathfinders and pioneers, explorers and scientists 1487–1976. In *A history of scientific endeavour in South Africa*. Royal Society of South Africa, Cape Town, South Africa, pp. 1–32.

Tanner JB, Dumont ER, Sakai ST, Lundrigan BL, Holekamp KE. 2008. Of arcs and vaults: The biomechanics of bone-cracking in spotted hyenas (*Crocuta crocuta*). *Biological Journal of the Linnean Society* 95: 246–255.

Taylor CR, Heglund NC, Maloiy GMO. 1982. Energetics and mechanics of terrestrial locomotion. I. Metabolic energy consumption as a function of speed and body size in birds and mammals. *Journal of Experimental Biology* 97: 1–21.

Taylor CR, Heglund NC, McMahon TA, Looney TR. 1980. Energetic cost of generating muscular force during running: A comparison of large and small animals. *Journal of Experimental Biology* 86: 9–18.

Taylor CR, Karas RH, Weibel ER, Hoppeler H. 1987. Adaptive variation in the mammalian respiratory system in relation to demand: II. Reaching the limits to oxygen flow. *Respiration Physiology* 69: 7–26.
Taylor CR, Lyman, CP. 1967. A comparative study of the environmental physiology of an East African antelope, the Eland, and the Hereford steer. *Physiological Zoology* 40: 280–295.
Taylor CR, Lyman CP. 1972. Heat storage in running antelopes: Independence of brain and body temperatures. *American Journal of Physiology* 222: R114–R117.
Taylor CR, Maloiy GMO, Weibel ER, Langman VA, Kamau JMZ, Seeherman HJ, Heglund NC. 1981. Design of the mammalian respiratory system. III. Scaling maximum aerobic capacity to body mass: Wild and domestic animals. *Respiration Physiology* 44: 25–37.
Taylor CR, Weibel ER. 1981. Design of the mammalian respiratory system. 1. Problem and strategy. *Respiration Physiology* 44: 1–10.
Taylor MP, Hone DWE, Wedel MJ, Naish D. 2011. The long necks of sauropods did not evolve primarily through sexual selection. *Journal of Zoology* 285: 150–161.
Tenney SM, Bartlett D. 1967. Comparative quantitative morphology of the mammalian lung: Trachea. *Respiration Physiology* 3: 130–135.
Tenney SM, Remmers JE. 1963. Comparative quantitative morphology of the mammalian lung: Diffusing area. *Nature* 197: 54–56.
Theiler A, du Toit PJ, Viljoen PR, Green HH, Meier H, Robinson EM. 1927. Lamsiekte (Parabotulism) in cattle in South Africa: Osteophagia and phosphorus deficiency in relation to lamsiekte. *11th and 12th Reports of the Director of Veterinary Education and Research*, Union of South Africa, pp. 1211–1248.
Theiler AK, Green HH, du Toit PJ. 1924. Phosphorus in the livestock industry. *South African Department of Agriculture Journal* 8: 460–504.
Thenius E. 1992. The okapi (*Mammalia, Artiodactyla*) from Zaire: A living fossil or a secondary rainforest inhabitant. *Zeitschrift fur Zoologische Systematik und Evolutionsforschung* 30: 163–179.
Theodorou GE, Roussiakis SJ, Athanassiou A, Filippidi A. 2010. Mammalian remains from a site near the classical locality of Pikermi (Attica, Greece). *Proceedings of the XIX CBGA Congress, Thessalonika, Greece* 99: 109–119.
Thomas EM. 1959. *The harmless people*. Random House, New York.
Thomas H. 1984. Les Giraffoidea et les Bovidae Miocenes de la formation Nyakach (rift Nyanza, Kenya). *Palaeontographica, Stuttgart A* 183: 64–89.
Thompson D'Arcy W. 1961. *On growth and form*. Cambridge University Press, Cambridge, UK.
Thornton JW. 2001. Evolution of vertebrate steroid receptors from and ancestral estrogen receptor by ligand exploitation and serial genome expansions. *Proceedings of the National Academy of Sciences* 98: 5671–5676.
Tobias PV. 1955. Teeth, jaws, and genes. *Journal of the Dental Association of South Africa* 10: 88–104.
Tobler I. 1995. Is sleep fundamentally different between mammalian species? *Behavioural Brain Research* 69: 35–41.
Tobler I, Schwierin B. 1996. Behavioural sleep in the giraffe (*Giraffa camelopardalis*) in a zoological garden. *Journal of Sleep Research* 5: 21–32.
Topsell E. 1607. *Historie of Foure-footed Beasts*. William Jaggard, London.
Trouessart E. 1908. La nouvelle giraffe du Museum et les differentes variétés de l'espèce. *La Nature (Paris)* 36: 339–342.
Tucker MA, Ord TJ, Rogers TL. 2014. Evolutionary predictors of mammalian home range size: Body mass, diet and the environment. *Global Ecology and Biogeography* 23: 1105–1114.
Turner M. 1969. Partial albino giraffe. *Africana* 3: 45.
T-W-Fiennes RN. 1960. Report of the society's pathologist for the year 1958. *Proceedings of the Zoological Society (London)* 134: 297–308.
Tweedale M. (Ed.). Deuteronomy *ca* 1200 BC Ch 14 verses 4–5. *Biblia Sacra juxta Vulgatum Clementinam*. London, 2005. Editio Electronica.
Udroiu I. 2014. Estimation of erythrocyte surface area in mammals. doi: 10.1101/012815.

van Aarde RJ. 1976. A note on the birth of a giraffe. *South African Journal of Science* 72: 307.

van Aarde RJ, Skinner JD. 1975. The food and feeding behaviour of the giraffe (*Giraffa camelopardalis*) in the Jack Scott Nature Reserve. *Publications of the University of Pretoria (New Series)* 97: 59–69.

van Citters RL, Kemper WS, Franklin DL. 1966. Blood pressure responses of wild giraffes studied by radio telemetry. *Science* 152: 384–386.

van Citters RL, Kemper WS, Franklin DL. 1968. Blood flow and pressure in the giraffe carotid artery. *Comparative Biochemistry and Physiology* 24: 1035–1042.

van Citters RL, Franklin DL, Vatner SF, Patrick T, Warren JV. 1969. Cerebral haemodynamics in the giraffe. *Transactions of the Association of American Physicians* 82: 293–304.

van Dam JA, Aziz HA, Sierra MAA, Hilgen FJ, van den Hoek Ostende LW, Lourens LJ, Mein P, van der Meulen AJ, Pelaez-Campomanes P. 2006. Long-period astronomical forcing of mammal turnover. *Nature* 443: 687–691.

van den Berghe L, Boné G. 1944. De la parenté sérologique de l'okapi et de la girafe. *Bulletin de la Classe des Sciences Academie Royale de Belgique* 30: 239–244.

Van der Jeugd HP, Prins HHT. 2000. Movements and group structure of giraffe (*Giraffa camelopardalis*) in Lake Manyara National park, Tanzania. *Journal of Zoology* 251: 15–21.

VanderWaal KL, Wang H, McCowan B, Fushing H, Isbell LA. 2014. Multilevel social organization and space use in reticulated giraffe (*Giraffa camelopardalis*). *Behavioral Ecology* 25: 17–26.

van Schalkwyk OL. 2004. *Bone density and Ca and P content of the giraffe skeleton*. MSc dissertation, University of Pretoria, South Africa.

van Schalkwyk OL, Skinner JD, Mitchell G. 2004. A comparison of the bone density and morphology of giraffe (*Giraffa camelopardalis*) and buffalo (*Syncerus caffer*) skeletons. *Journal of Zoology (London)* 264: 307–315.

van Sittert SJ. 2015. Ontogenetic allometry of the postcranial skeleton of the giraffe (Giraffa camelopardalis), with application to giraffe life history, evolution and *palaeontology*. PhD thesis, University of Pretoria, South Africa.

van Sittert SJ, Mitchell G. 2015. On reconstructing *Giraffa sivalensis* an extinct giraffid from the Siwalik Hills, India. *PeerJ* 3: e1135.

van Sittert SJ, Skinner JD, Mitchell, G. 2010. From fetus to adult: An allometric analysis of the giraffe vertebral column. *Journal of Experimental Zoology B* 314: 469–479.

van Sittert SJ, Skinner JD, Mitchell G. 2015. Scaling of the appendicular skeleton of the giraffe (*Giraffa camelopardalis*). *Journal of Morphology* 276: 503–516.

van Soest PJ. 1996. Allometry and ecology of feeding behavior and digestive capacity in herbivores: A review. *Zoo Biology* 15: 455–479.

van Wyk P. 1984. *Field guide to the trees of the Kruger National Park*. Struik, Cape Town, South Africa.

van Zinderen-Bakker EM, Mercer JH. 1986. Major late Cenozoic climatic events and palaeoenvironmental changes in Africa viewed in a worldwide context. *Palaeogeography, Palaeoclimatology, Palaeoecology* 56: 217–235.

Veilleux CC, Kirk EC. 2014. Visual acuity in mammals: Effects of eye size and ecology. *Brain Behavior and Evolution* 83: 43–53.

Velhankar DP, Hukeri VB, Deshpande BR, Sane CR. 1973. Biometry of the genitalia and the spermatazoa of male giraffe. *The Indian Veterinary Journal* 50: 789–792.

Vermeulen HA. 1915. The vagus-area in Camelopardalus Giraffe. *Proceedings of the Royal Academy (Amsterdam)* 18: 647–690.Volodina EV, Volodin IA, Chelysova EV, Frey R. 2018. Hiss and snort call types of wild-living giraffes *Giraffa camelopardalis*: Acoustic structure and context. *BMC Research Notes* 11: 1–7.

Vosmaer A. 1787. *Description du Chameau-Leopard (Camelopardalis)*. P Meyers & P Warner, Amsterdam.

Vrba ES. 1985a. Environment and evolution: Alternative causes of the temporal distribution of evolutionary events. *South African Journal of Science* 81: 229–236.

Vrba ES. 1985b. African Bovidae: Evolutionary events since the Miocene. *South African Journal of Science* 81: 263–266.

Vrba ES. 1987. Ecology in relation to speciation rates: Some case histories of Miocene-Recent mammal clades. *Evolutionary Ecology* 1: 283–300.

Walker BH, Emslie RH, Owen-Smith RN, Scholes RJ. 1987. To cull or not to cull: Lessons from a southern African drought. *Journal of Applied Ecology* 24: 381–401.

Wallace C, Fairall N. 1965. Chromosome analysis in the Kruger National Park with special reference to the chromosomes of the giraffe (*Giraffa camelopardalis giraffa*, Boddaert). *Koedoe* 8: 97–103.

Ward D, Young TP. 2002. Effects of large mammalian herbivores and ant symbionts on condensed tannins. *Journal of Chemical Ecology* 28: 913–929.

Warren JV. 1974. The physiology of the giraffe. *Scientific American* 231: 96–105.

Warren JV, Patterson JL, Doyle JT, Gauer OH, Keen EN, McGregor M, Goetz RH. 1957. Circulation and respiration in the giraffe. *Circulation* 16: 947–948.

Webb SD, Taylor BE. 1980. The phylogeny of hornless ruminants and a description of the cranium of *Archaeomeryx*. *Bulletin of the American Museum of Natural History* 167: 117–158.

Weibel ER, Bacigalupe LD, Schmitt B, Hoppeler H. 2004. Allometric scaling of maximal metabolic rate in mammals: Muscle aerobic capacity as determinant factor. *Respiratory Physiology and Neurobiology* 140: 115–132.

Weibel ER, Taylor CR, Gehr P, Hoppeler H, Mathieu O, Maloiy GMO. 1981. Design of the mammalian respiratory system. IX. Functional and structural limits for oxygen flow. *Respiration Physiology* 44: 151–164.

Weishaupt MA, Stämpfli H, Billeter R, Straub R. 1996. Temperature changes during strenuous exercise indifferent body compartments of the horse. *Pferdeheilkunde* 12: 450–454.

Western D. 1971. Giraffe chewing a Grant's gazelle carcass. *East African Wildlife Journal* 9: 156–157.

Western D. 1979. Size, life history and ecology in mammals. *African Journal of Ecology* 17: 185–204.

Wheeler PE. 1991. The thermoregulatory advantages of hominid bipedalism in open equatorial environments: The contribution of increased convective heat loss and cutaneous evaporative cooling. *Journal of Human Evolution* 21: 107–115.

White CR, Seymour RS. 2002. Mammalian basal metabolic rate is proportional to body mass$^{2/3}$. *Proceedings of the National Academy of Sciences* 100: 4046–4049.

White CR, Seymour RS. 2013. The role of gravity in the evolution of mammalian blood pressure. *Evolution* 68: 901–908.

Whitmore FC. 1952. Cranial morphology of some Oligocene artiodactyla. *Geological Survey Professional Paper* 243-H, pp. 117–160.

Whittow GC, Scamell CA, Leong M, Rand D. 1977. Temperature regulation in the smallest ungulate, the Lesser Mouse deer (*Tragulus javanicus*). *Comparative Biochemistry and Physiology A* 56: 23–26.

Wikar HJ. 1779 [1935]. In *The Journals of Wikar, Coetsé and van Reenen*. Mossop EE (Ed.). Translated from the Dutch by van der Horst AW and Mossop EE. The van Riebeeck Society, Cape Town.

Wilkinson DM, Ruxton GD. 2011. Understanding selection for long necks in different taxa. *Biological Reviews* 87: 616–630.

Wilkinson JF, de Fremery P. 1940. Gonadotropic hormones in the urine of the giraffe. *Nature* 146: 491.

Willemse JJ. 1958. The innervation of the muscles of the trapezius-complex in giraffe, okapi, camel and llama. *Archives Nederlandaises de Zoologie* 12: 532–536.

Williams DC, Murison PJ, Hill CL. 2007. Dystocia in a Rothschild giraffe leading to a Caesarean section. *Journal of the Veterinary Medical Association* 55: 199–202.

Williams E. 2010. *Giraffe*. Reaction Books, London.

Williamson JR, Vogler NJ, Kilo C. 1971. Regional variations in the width of the basement membrane of muscle capillaries in man and giraffe. *American Journal of Pathology* 63: 359–370.

Willis T. 1664 [1978]. *Cerebri anatome: Qui accessit nervorum descriptio et uses*. Roycroft, London. Reprinted by McGill-Queens University Press, Birmingham, AL.

Willis T. 1681. *The remaining medical works of the famous and renowned physician Dr Thomas Willis*. Dring, Harper, Leigh, and Martyn, London.

Wilsher S, Stansfield F, Greenwood RES, Trethowan PD, Anderson RA, Wooding FBW, Allen WR. 2013. Ovarian and placental morphology and endocrine functions in the pregnant giraffe (*Giraffa camelopardalis*). *Reproduction* 145: 541–554.

Wiltbank, MC, Median R, Ochoa J, Baez GM, Giordano JO, Ferreira JCP, Sartori R. 2016. Maintenance or regression of the corpus luteum during multiple decisive periods of bovine pregnancy. *Animal Reproduction* 13: 217–233.

Winter S, Fennessy J, Janke A. 2018. Limited introgression supports division of giraffe into four species. *Ecology and Evolution* 8: 10156–10165.

Wislocki GB. 1941. The placentation of an antelope (*Rhyncotragus kirkii nyikae,* Heller). *Anatomical Record* 81: 221–241.

Wolf TE, Bennett NC, Burroughs R, Ganswindt A. 2018b. The impact of age-class and social context on fecal glucocorticoid metabolite levels in free-ranging male giraffes. *General and Comparative Endocrinology* 255: 16–31.

Wolf TE, Schaebs FS, Bennett NC, Burroughs R, Ganswindt A. 2018a. Age and socially related changes in fecal androgen metabolite concentrations in free-ranging male giraffes. *General and Comparative Endocrinology* 255: 19–25.

Wolfe BA, Sladsky KK, Loomis MR. 2000. Obstructive urolithiasis in a reticulated giraffe (*Giraffa camelopardalis reticulata*). *Veterinary Record* 146: 260–261.

Wood WF, Weldon PJ. 2002. The scent of the reticulated giraffe (*Giraffa camelopardalis reticulata*). *Biochemical Systematics and Ecology* 30: 913–917.

Woodall PF, Skinner JD. 1993. Dimensions of the intestines, diet and faecal water loss in some African antelope. *Journal of Zoology (London)* 229: 457–471.

Wood-Jones F. 1948. *Hallmarks of mankind.* Baillière, Tindall and Cox, London.

Wood-Jones F. 1949. Evolution of the giraffe. *Nature* 164 (4164): 323.

Woolnough AP, du Toit JT. 2001. Vertical zonation of browse quality in tree canopies exposed to a size-structures guild of African browsing ungulates. *Oecologia* 129: 585–590.

Worthington J, Young IS, Altringham JD. 1991. The relationship between body mass and ventilation rate in mammals. *Journal of Experimental Biology* 161: 533–536.

Wrobel KH. 1965. Das Nierenbecken der Giraffe. *Zeitschrift fur Saugetierk* 30: 233–241.

Wyatt JR. 1971. Osteophagia in Masai giraffe. *East African Wildlife Journal* 9: 157.

Yeakel JD, Pires MM, Rudolf L, Dominy NJ, Koch PL, Guimarães PR, Gross T. 2014. Collapse of an ecological network in Ancient Egypt. *Proceedings of the National Academy of Sciences* 111: 14472–14477.

Young TP. 1987. Increased thorn length in *Acacia depranolobium*: An induced response to browsing. *Oecologia (Berlin)* 71: 436–438.

Young TP, Isbell LA. 1991. Sex differences in giraffe feeding ecology: Energetic and social constraints. *Ethology* 87: 79–89.

Zachos J, Pagani M, Sloan L, Thomas E, Billups K. 2001. Trends and rhythms and aberrations in global climate 65 Ma to present. *Science* 292: 686–693.

Zanazzi A, Kohn MJ, MacFadden BJ, Terry DO. 2007. Large temperature drop across the Eocene-Oligocene transition in central North America. *Nature* 445: 639–642.

Zepelin HJ, Siegel M, Tobler I. 2005. Mammalian sleep. *Principles and Practice of Sleep Medicine* 4: 91–100.

Zhisheng A, Kutzbach JE, Prell WL, Porter SC. 2001. Evolution of Asian monsoons and phased uplift of the Himalaya-Tibetan plateau since late Miocene times. *Nature* 411: 62–66.

Zhu Z (+31 others). 2016. Greening of the earth and its drivers. *Nature Climate Change* 6: 791–795.

Zinn AD, Ward D, Kirkman K. 2007. Inducible defences in *Acacia sieberiana* in response to giraffe browsing. *African Journal of Range and Forage Science* 24: 123–129.

Zivotofsky D, Zivotofsky AZ, Amar Z. 2002/3. Giraffe: A halakhically oriented dissection. *The Torah u-Madda Journal* 11: 203–219.

Zuckerman S, Kiss F. 1932. The spinal accessory nerve of the giraffe. *Proceedings of the Zoological Society (London)* 2: 767–769.

Giraffe-Specific References in Chronological Order

~1200 BC:
Deuteronomy 14, verses 4–5.

384–322 BC:
Aristotle. *History of animals*. Translated by Richard Cresswell. George Bell and Sons, London, 1891.

~250 BC:
Septuagint. *The Septuagint with Apocrypha*. Translated by Brenton LCL. Samuel Bagster & Sons, London, 1851.

~169 BC:
Agatharchides of Cnidus. *On the Erythraean Sea*. Translated by Stanley M. Burstein, 1989, book 5, fragment 73a.

~100 BC:
Artemidorus of Ephesus. *The Eighth book of Artemidorus's Geography. A commentary by Irene Pajón Leyra in Historia: Zeitschrift für Alte Geschichte*. Bd. 61, H. 3 (2012), pp. 336–361, 366–367.

65 BC–AD 8:
Quintus Horatius Flaccus (Horace). *Epistle 2.1* in *The complete works of Horace*. Translated based on CE Passage. Frederick Unger, New York, 1983.

~ AD 1:
Strabo. *The geography of Strabo*. Vol .VII, Book 16, p. 337. Translated by Hamilton HC, Falconer W. 1892. G. Bell & Sons, London, and by Jones HL. 1930. William Heinemann, London.

64:
Gaius Plinius Secundus (Pliny the Elder). *Natural History*: Book VIII. Translated from Latin by Bostock J, Riley HT, George Bell &Sons, London, 1900: 276–277, and by Rackham H. 1943. *Natural History*, Volume III, Books 8–11. Loeb Classical Library No. 353. Harvard University Press, Cambridge, MA.

200–229:
Dio Cassius Cocceianus (Dio). *Annals of Rome*. Translated from the Greek by Foster HB. Pafraet's, New York, 1905.

250:
Heliodorus. *An Aethiopian history*. Translated from Joly and Lavocat, and from Underdowne T's translation of 1587. Reprinted by AMS Press, New York, 1967.

~400:
St. Jerome's Vulgate. Tweedale M. (Ed.) Deuteronomy *ca* 1200 BC Ch 14 verses 4–5. *Biblia Sacra juxta Vulgatum Clementinam*. London, 2005. Editio Electronica.

1553:
Belon P. 1553. Les observations *de plusieurs singularitez et choses memorables trouvées en Grèce, Asie, Judée, Egypte, Arabie et autres pays étrangèrs*. Gilles Corrozet, Paris.

1607:

Topsell E. 1607. *Historie of Foure-footed Beasts*. William Jaggard, London, pp. i–xli, 1–759.

1663.

Bochardt S. 1663. *Hierozoicon sive bipartitum opus de animalibus sacrae scripturae*. Vol. 1. *De animalibus in genere et de quadrupedibus viviparous et oviparis*. Thomas Roycroft, London.

1668:

Dapper O. 1668. *Naukeurige Beschrijvinge der Afrikaensche Gewesten*. Jacob van Meurs, Amsterdam.

1670:

Ogilby J. 1670. *Africa: Being an accurate description of the Regions of Aegypt, Barbary, Lybia, and Billedulgerid, the Land of the Negroes . . . Collected and Translated from most Authentick Authors, and augmented from later observations: Illustrated with Notes, and Adorned with peculiar maps, and proper Sculptures*. Thomas Johnson, London.

1690:

Hevelius J. 1690. *Firmamentum sobiescianum, sive uranographia*. Johannes Zacharia Stoll, Gdansk.

1758:

Linnaeus, C. *Systema naturae per regna tria naturae, secundum classes, ordines, genera, species, cum characteribus, differentiis synonymis, locis*. Tomus 1, Editio 10, p. 66.

1762:

Brisson MJ. Regnum animale in classes IX distributum. Lugduni Batavorum, Apud Theodorum Haak, pp. 37–38.

1766:

Hasselquist F. *Voyages and travels in the Levant in the years 1749–1752*. Published by Charles Linnaeus. Translated by A. Foster from the Swedish for Davis & Reymers, London, 1766.

1770:

Carteret P. 1770. A letter on a Camelopardalis found about the Cape of Good Hope, from Capt Carteret to Mathew Maty, M.D. Sec. R.S. *Philosophical Transactions of the Royal Society* 60: 27–29.

1778:

Pallas PS. 1778. *The Natural history of strange animals, in which mainly new and unknown animal species are explained by copper engravings, descriptions and explanations*. GA Lange, Berlin.

1777–1779.

Gordon RJ. 1777–1779. *The Journals of Robert Jacob Gordon*. Translated from Dutch by PR Cullinan. Brenthurst Library, Johannesburg, 1993.

1779:

Wikar HJ. 1779. In *The Journals of Wikar, Coetsé and van Reenen*. Mossop EE (Ed.). Translated from the Dutch by van der Horst AW and Mossop EE. The van Riebeeck Society, Cape Town, 1935.

1781.

Buffon G-LL. 1781. *Historia Naturale: general y particular*. Vol XIV. Published by Viuda de Joaquim Ibarra, Madrid. Translated by William Smellie.

1787:

Vosmaer, A. 1787. *Description d'un recueil exquis des animaux rares*. Amsterdam.

1789:

Paterson W. 1789. *A narrative of four journeys into the country of the Hottentots and Caffraria in the years 1777–1789*. J Johnson, London.

1792:

Buffon G-LL. 1792 and 1807. *Buffon's Natural History, containing a theory of the earth, a general history of man, of the brute creation, and of vegetables, minerals, &c. &c.* Vol. VIII: 260–274. Translated from French by JS Barr. JS Barr, London.

1795:

LeVaillant F. 1795. *Second voyage dans l'interieur de l'Afrique par le Cap de Bonne-Esperance, dans les annes 1783, 84, 85.* HJ Jansen, Paris.

1796:

Buffon G-LL. 1796. *Historia Naturale, general y particular Tomo XIV*, pp 193–226. Figure 223 *Astas de Girafa*. Published by viuda de Joaquim Ibarra, Madrid.

1801:

Barrow J. 1801. *An account of travels into the interior of Southern Africa in the years 1797 and 1798.* T Cadell and W Davis, London.

1802:

Oken L. 1802. *Grundriss der Naturphilosophie, der Theorie der Sinne, mit der darauf gegründeter Classification der Thiere.* Oken, Frankfurt.

1809:

de Monet JBPA. (The Chevalier de Lamarck). 1809. *Philosophie zoologique.* Paris. Translated by HS Elliot. Macmillan, London.

1819:

Burckhardt JL. 1819. *Travels in Nubia.* John Murray, London.

1822:

Burchell W. 1822. *Travels in the interior of Southern* Africa. Vol 1. Longman, Hurst, Rees, Orme and Brown, London.

1823:

Pander C, D'Alton E. 1823. *Die Skelete der Wiederkäuer.* Eduard Weber, Bonn.

1827:

Davis RB. 1827. Of the Giraffe (being an account of the habits of George IV's giraffe). The *London Literary Gazette.* December 1: 778–779.

St. Hilaire E Geoffroy. 1827. Quelques considerations sur la girafe. *Annales des sciences naturelles (Paris)* 11: 210–223.

Anonymous. 1827. The Camelopardalis, or, Giraffe (being an account of the arrival of George IV's giraffe in England on 11 August 1827. *The London Literary Gazette.* August 25, 553–555.

1828:

Home E. 1828. On the peculiarities of the tongue of the Xariffa, and on the muscular structure of tongues in general. *Lectures on Comparative Anatomy* 5: 244–249.

Buffon G-LL. 1828. *A natural history.* Thomas Kelly, London. Vol. 1: 649–653. Translated by William Smellie.

1829:

Rüppell E. 1829. *Reisen in Nubien, Kordofan und dem petraeischen Arabien vorzueglich in geographisch-statisticher Hinsicht.* Frankfurt, Friedrich Wilmans.

1830:

Home E. 1830. A report on the stomach of the Zariffa. *Philosophical Transactions of the Royal Society* 120: 85–86.

1835:

Buffon G-LL. 1835. *Oeuvres completes de Buffon avec les supplements augmentées de la classification de G Cuvier, et accompagnèes de 700 vignettes gravées sur acier représentant au monis 900 animaux*. Vol. VI. Ed. P Dumènil, Rue des Beaux-arts, 10. 1835.

Hoskins GA. 1835. *Travels in Ethiopia, above the second cataract of the Nile*. Longman, Rees, Orme, Brown, Green, & Longman, London.

Swainson W. 1835. *A treatise on the geography and classification of animals*. Longman, Rees, London.

1838:

Cautley, PT. 1838. Note on a fossil ruminant genus allied to Giraffidae, in the Siwalik hills. *Journal of the Asiatic Society of Bengal* 7: 658–660.

Jardine W. 1838. Giraffidae. *The Naturalist's Library* XXI: 186–189.

Owen R. 1838. Notes on the anatomy of the Nubian giraffe. *Transactions of the Zoological Society, London* 2: 217–248.

1839:

Harris Capt. Sir WC. 1839. *The Wild Sports of Southern Africa: Being the narrative of an expedition from the Cape of Good Hope, through the territories of the Chief Moselekatse, to the Tropic of Capricorn*. John Murray, London.

Owen R. 1839. Notes on the birth of the Giraffe at the Zoological Gardens, and description of the foetal membranes and some of the natural and morbid appearances observed in the dissection of the young animal. *Transactions of the Zoological Society, London* 3: 21–28.

1840–1845:

Lesson R-P. 1842. *Nouveau tableau du règne animal*. Arthus Bertrand, Paris. pp. 168–169.

Owen R. 1840–1845. *Odontography, or a treatise on the comparative anatomy of the teeth*. London, Bailliere.

1843–1844:

Falconer H, Cautley PT. 1843–1844. On some fossil remains of *Anoplotherium* and giraffe, from the Sewalik Hills, in the north of India. *Proceedings of the Geological Society, London* 4 (ii): 235–249.

1846:

Joly N, Lavocat A. 1846. Recherches historiques, zoologiques, anatomiques et palaeontologiques sur la giraffe. *Memoirs de la Societe des Sciences Naturale de Strasbourg* Vol. III: 1–112.

1848:

Owen R. 1848. Description of teeth and proportion of jaws of two extinct Anthracotheroid quadrupeds (*Hyopotamus vectianus and Hyopotamus bovinus*) discovered by the Marchioness of Hastings in the Eocene deposits of the N.W. coast of the Isle of Wight: with an attempt to develop Cuvier's idea of the classification of pachyderms by the number of their toes. *Quarterly Journal of the Geological Society of London* 4: 103–141.

1851:

Owen R. 1851. On Dinornis (Part IV). *Transactions of the Zoological Society* 4: 1–20.

1852:

Harris, Capt. Sir WC. *Wild Sports of Southern Africa*. 5th edition. Henry G. Bohn, London, Facsimile reprint, C. Strunk, Cape Town, 1963.

1853:

Crisp E. 1853. Note on the splenic vein of the giraffe. *Proceedings of the Zoological Society, London* 1853: 99.

1854:

Cobbold TS. 1854. On the anatomy of the giraffe. *Annals and Magazine of Natural History, Edinburgh* June 1854: 484–488.

1858:

Darwin C, Wallace AR. 1858. On the tendency of species to form varieties and on the perpetuation of varieties and species by natural means of selection. *Proceedings of the Linnaean Society (London)* 3: 45–63.

1860:

Cobbold TS. 1860. Contributions to the anatomy of the giraffe. *Proceedings of the Zoological Society, London* 1860: 99–105.

1864:

Crisp E. 1864a. Contributions to the anatomy of the giraffe, with an account of the length of the alimentary canal of many other ruminants. *Proceedings of the Zoological Society, London* February 9: 63–68.

Crisp E. 1864b. Further contributions to the anatomy of the giraffe and the nylghau. *Proceedings of the Zoological Society, London* June 14: 269–271.

1866:

Crisp E. 1866. Further observations relating to the anatomy of the giraffe. *Proceedings of the Zoological Society, London* December 13: 563–566.

1868:

Falconer H. 1868. *Palaeontological memoirs and notes of the late Hugh Falconer*. Ed. C Murchison. Hardwicke, London.

Owen R. 1868. *On the anatomy of vertebrates*. Vol. 3. *Mammals*. Longmans, Green, and Co., London.

1869:

Darwin CR. 1869. *On the origin of species by means of natural selection, or the preservation of favoured races in the struggle for life*. John Murray, London.

1871:

Darwin CR. 1871. *The descent of man and selection in relation to sex*. John Murray, London.

Mivart St. G. 1871. *On the genesis of species*. Macmillan, London and New York.

1872:

Darwin CR. 1872. *The origin of species by means of natural selection, or the preservation of favoured races in the struggle for life*. John Murray, London.

Murie J. 1872. On the horns, viscera and muscles of the giraffe, with a record of the post-mortem examination of two specimens killed by a fire. *The Annals and Magazine of Natural History, Edinburgh*, Ser. 4, IX: 177–195.

1875:

Chapman HC. 1875. Notes on the great blood vessels of the giraffe. *Proceedings of the Academy of Natural Science (Philadelphia)* 1875: 401–402.

1876:

Lydekker R. 1876. Notes on the fossil mammalian faunae of India and Burma. In *Records of the Geological Survey of India*, Trubner, London, pp. 86–105.

1877:

Chapman HC. 1877. Note on the great blood vessels of the giraffe. *Proceedings of the Academy of Natural Science (Philadelphia)* 1877: 37–38.

Garrod AH. 1877. Notes on the visceral anatomy and osteology of the ruminants. *Proceedings of the Zoological Society (London)* 1877: 2–18.

1880:

Richiardi S. 1880. Sull'anatomia della giraffa (Anatomy of the giraffe). *Zoologischer Anzeiger* 3: 92–93.

1881:

Selous FC. 1881. *A hunter's wanderings in Africa: Being a narrative of nine years spent amongst the game of the far interior of South Africa.* Richard Bentley, London.

1883:

Lydekker R. 1883. Siwalik camelopardalidae. *Indian Tertiary and Post-tertiary Vertebrata* 2: 99–142.

1888:

Darwin C. 1888. *The origin of species: By means of natural selection, the preservation of favoured races in the struggle for life.* 6th edition. London, John Murray, pp. 157, 177–181.

1891:

Forsyth Major, CJ. 1891. On the fossil remains of species of the family *Giraffidae. Proceedings of the Zoological Society, London* 1891: 315–326.

Lydekker, R. 1891. The giraffe and its allies. *Nature* 44 (1144): 524–526.

1892:

Pomel, A. 1892. Sur le Libytherium maurusium, grand ruminant du terrain Pliocene plaisancier de l'Algerie. *Comptes rendus de l'Academie des Sciences (Paris)* 115: 100–102.

1897:

de Winton, WE. 1897. Remarks on the existing forms of giraffe. *Proceedings of the Zoological Society* 18: 273–283.

Gibbon E. 1897. *The history of the decline and fall of the Roman Empire.* Bury JB (Ed.). Methuen, London.

McCrindle JW. 1897. *The Christian topography of Cosmas, an Egyptian monk.* Hakluyt Society, London.

1900:

Kidd, W. 1900. The significance of hair-slope in certain mammals. *Proceedings of the Zoological Society, London* 1900: 676–686.

1901:

Johnson, GL. 1901. Contributions to the comparative anatomy of the mammalian eye, chiefly based on ophthalmoscopic examination. *Philosophical Transactions of the Royal Society of London, Series B* 194: 1–82.

Lankester, ER. 1901. On *Okapia johnstoni* (Giraffidae). *Proceedings of the Zoological Society, London* 1901: 279–281, 472–474.

Lydekker R. 1901. Giraffidae. In *Library of natural history.* Saalfield, London. Vol. II, pp. 918–925.

Ridewood, WG. 1901. A note on the hairs of antelope, giraffe and okapi. *Proceedings of the Zoological Society, London* 1901: 3.

Sclater, PL. 1901. On a new African mammal. *Proceedings of the Zoological Society, London* 1901: 3–6.

1902:

Beddard JE. 1902a. Note on crooked neck in giraffe. *Proceedings of the Zoological Society (London)* 1: 52–54.

Driver SR. 1902. *A critical and exegetical commentary on Deuteronomy.* T and T Clark, Edinburgh.

Forsyth Major, CI. 1902. On a specimen of the okapi lately received at Brussels. *Proceedings of the Zoological Society, London* 1902: 339–345.

Johnston H. 1902. *The Uganda Protectorate,* Volume 1. Hutchinson, London.

1904:

Lydekker R. 1904a. On old pictures of Giraffes and Zebras. *Proceedings of the Zoological Society, London* 1: 339–345.

Lydekker R. 1904b. On the subspecies of Giraffa camelopardalis. *Proceedings of the Zoological Society, London* 1: 202–227.

Ridewood WG. 1904. Some observations on the skull of the giraffe. *Proceedings of the Zoological Society, London* 1: 150–157.

1905:

Mitchell PC. 1905. On the intestinal tract of mammals. *Transactions of the Zoological Society* 17: 474–476.

1906:

Beddard JE. 1906. Description of the external characters of an unborn foetus of a giraffe (Giraffa camelopardalis wardii). *Proceedings of the Zoological Society, London* 1906: 463–468.

1907:

Lankester ER. 1907. The origin of the lateral horns of the giraffe in foetal life on the area of the parietal bones. *Proceedings of the Zoological Society, London* 1: 100–115.

Stromer E. 1907. Fossile Wirbeltier-Reste aus dem Uadi Faregh und Uadi Natrum in Aegypten. *Abhandlungen der Senckenbergischen Naturforschenden Gesellschaft* 29: 99–132.

1908:

Lankester R. 1908. On certain points in the structure of the cervical vertebrae of the okapi and the giraffe. *Proceedings of the Zoological Society, London*. 1908: 320–334, 1908.

Scherren H. 1908. On certain errors with reference to George the Fourth's Giraffe. *Proceedings of the Zoological Society, London* 1908: 403–405.

Selous FC. 1908. *African nature notes and reminiscences*. Macmillan, London.

Trouessart E. 1908. La nouvelle giraffe du Museum et les differentes variétés de l'espèce. *La Nature (Paris)* 36: 339–342.

1910:

Hagenbeck C. 1910. *Beasts and men*. Translated from German by Elliot HSR and Thacker AG. Longman's Green, London.

Knottnerus-Meyer T. 1910. Eine neue Giraffe aus dem südlichien Abessinien. *Zoologischer Anzeiger* (May): 797–800.

1911:

Lydekker R. 1911. Two undescribed giraffes. *Nature* 87: 484.

Pilgrim GE. 1911. The fossil giraffidae of India. *Palaeontologia Indica* 4: 1–29.

1912:

Butler AL. 1912. A white giraffe. *Field* 119: 596.

1914:

Britten J. 1914. Robert Jacob Gordon. *Journal of Botany (London)* 52: 75–77.

1915:

Black D. 1915. A study of the endocranial casts of Ocapia, Giraffa and Samotherium, with special reference to the convoluted pattern in the family Giraffidae. *Journal of Comparative Neurology* 28: 329–360.

Eastman CR. 1915. Early representations of the giraffe. *Nature* 95: 672–673.

Vermeulen HA. 1915. The vagus-area in Camelopardalus Giraffe. *Proceedings of the Royal Academy (Amsterdam)* 18: 647–690.

1917:

Eastman CR. Chinese and Persian giraffe paintings. *Nature* 99: 334.

1922:

Haskins CH. 1922. Science at the court of the Emperor Frederick II. *The American Historical Review* 27: 669–694.

Haughton SH. 1922. A note on some fossils from the Vaal River gravels. *Transactions of the Geological Society of South Africa* 24: 11–16.

Sonntag CF. 1922. The comparative anatomy of the tongues of the Mammalia. VII. Cetacea, Sirenia, and Ungulata. *Proceedings of the Zoological Society* 92: 639–657.

1923:

Anonymous. 1923. A black giraffe, seen by "a competent European observer," was reported to the Zoological Society, London by Mr. Caldwell of the Game Warden's office, Nairobi. *Proceedings of the Zoological Society* 1923: 179.

1924:

Krogh A. 1924. *The anatomy and physiology of capillaries*. Yale University Press, New Haven, CT, pp. 225–226.

Bohlin B. 1926. Die Familie Giraffidae mit besondere Berucksichtiging der fossilen Formen aus China. *Palaeontologica Sinica*, Series C 4: 1–179.

Historia Augusta, Volume II, 1924. Loeb Classical Library. Translated by D Magie. Harvard University Press, Cambridge, MA.

1926:

Elting EC. 1926. A formula for estimating surface area of dairy cattle. *Journal of Agricultural Research* 33: 269–278.

1927:

De Beaufort LF. 1927. On a case of correlation between blood vessels and colour pattern in a giraffe. *Tijdschrift der Nederlandsche Dierkundige Vereeniging* 3: 31–32.

Franklin KJ, Haynes F. 1927. The histology of the giraffe's carotid functionally considered. *Journal of Anatomy* 62: 115–117.

1928:

Laufer B. 1928. *The giraffe in history and art*. Field Museum of Natural History, Leaflet 27, Chicago. 100 pp.

1929:

Matthew WD. 1929. Critical observations upon Siwalik mammals. *Bulletin of the American Museum of Natural History* 56: 437–560.

1932:

Zuckerman S, Kiss F. 1932. The spinal accessory nerve of the giraffe. *Proceedings of the Zoological Society, London* 2: 767–769.

1933:

Colbert EH. 1933. A skull and mandible of *Giraffokeryx punjabiensis* (Pilgrim). *American Museum Novitates* 632: 1–14.

1934:

Dawson WR. 1934. Deer in ancient Egypt. *Journal of the Linnean Society* 39: 137–145.

Hopwood AT. 1934. New fossil mammals from Olduvai (Tanganyika Territory). *Annals and Magazine of Natural History* 14: 546–550.

Howell AB, Straus WL. 1934. Note on the spinal accessory nerve of long-necked ungulates. *Proceedings of the Zoological Society (London)* 1: 29–32.

Matthew WD. 1934. Phylogenetic chart of the artiodactyla. *Journal of Mammalogy* 15: 207–209.

1935:

Bohlin B. 1935. Some remarks on fossil Giraffidae. *Bulletin of the Geological Society of China* 14: 83–89.

Colbert EH. 1935a. Distributional and phylogenetic studies in Indian fossil mammals. V. The classification and the phylogeny of the Giraffidae. *American Museum Novitates* 800: 1–15.

Colbert EH. 1935b. Siwalik mammals in the American Museum of Natural History. *Transactions of the American Philosophical Society* 26: 323–375.

Mossop EE (Ed.). 1935. *The Journals of Wikar, Coetsé and van Reenen*. Translated from the Dutch by van der Horst AW and Mossop EE. The Van Riebeeck Society, Cape Town.

1936:

Colbert EH. 1936a. Was the extinct giraffe (*Sivatherium*) known to early Sumerians. *American Anthropologist* 38: 605–608.

Colbert EH. 1936b. *Palaeotragus* in the Tungur formation of Mongolia. *American Museum Novitates* 874: 1–17.

MacInnes DG. 1936. A new genus of fossil deer from the Miocene of Africa. *Journal of the Linnaean Society* 39: 521–530.

1937:

Joleaud L. 1937. Remarques sur les giraffids fossiles d'afrique. *Mammalia* 1: 85–96.

1938:

Colbert EH. 1938. The relationships of the okapi. *Journal of Mammalogy* 19: 47–50.

Dinesen I. (Karen Blixen). 1938. *Out of Africa*. 1st edition. Random House, New York.

1939:

Burne RH. 1939. Description of the stomach, intestine, liver, and pancreas of the okapi, *Okapia johnstoni*. *Proceedings of the Zoological Society B* 109: 451–479.

McDougall DSA. 1939. A white giraffe. *Field* 174: 1003.

1940:

Arnold R. 1940. Eine ungefleckte Giraffe in Nord-Ost-Afrika? *Zeitschrift fur Saugetierkunde* 15: 306–311.

Crile G, Quiring DP. 1940. A record of the body weights and certain organ and gland weights of 3690 animals. *The Ohio Journal of Science* 40: 219–259.

Harper F. 1940. The nomenclature and type localities of certain old world mammals. *Journal of Mammalogy* 21: 322–332.

Wilkinson JF, de Fremery P. 1940. Gonadotropic hormones in the urine of the giraffe. *Nature* 146: 491.

1941:

Castigli G, Sacchi, R. L'apparato di conduzione nel cuore della giraffe. *Rivista di biologia* 32: 71–80, 1941.

Colbert EH. 1941. The osteology and relationships of *Archaeomeryx*, an ancestral ruminant. *American Museum Novitates* 1135: 1–24.

1942:

Dietrich WO. 1942. Ältesquartäre Säugetiere aus den südlichen Serengeti, Deutsch-Ost-afrika. *Paleontographica* 94: 43–133.

1943:

Rackham, H. 1943. *Pliny: Natural history*, Volume III, Books 8–11. Loeb Classical Library No. 353. Harvard University Press, Cambridge, MA.

1944:

Krieg H. 1944. Der Schädel einer Giraffe. *Naturwissenschaften* 32: 148–156.

Stirton RA. 1944. Comments on the relationships of the cervoid family Palaeomerycidae. *American Journal of Science* 242: 633–655.

Van den Berghe L, Boné G. 1944. De la parenté sérologique de l'okapi et de la girafe. *Bulletin de la Classe des Sciences Academie Royale de Belgique* 30: 239–244.

1946:

Slijper, EJ. 1946. Comparative biological-anatomical investigation on the vertebral column and spinal musculature of mammals. *Proceedings of the Koninklike Nederlandse Akademie van Wetenschappen, Amsterdam* 42: 1–128.

1947:

Amoroso EC, Edholm OG, Rewell RE. Venous valves in the giraffe, okapi, camel, and ostrich. *Proceedings of the Zoological Society, London* 117: 435–440.

Arambourg C. 1947. Contribution a l'etude geologique et palaeontologie du bassin du lac Rudolph et de la basse vallee de l'Omo. *Musee d'Histoire Naturel, Paris* 1: 232–562.

Mossop EE. (Ed.). 1947. *The Journals of Brink and Rhenius. Being the journal of Carel Frederick Brink of the journey into Great Namaqualand (1761–2) and the journal of Ensign Johannes Tobias Rhenius (1724).* Van Riebeeck Society, Cape Town.

1948:

Lawrence WE, Rewell RE. 1948. The cerebral blood supply in the Giraffidae. *Proceedings of the Zoological Society, London* 118: 202–212.

Wood-Jones F. 1948. *Hallmarks of mankind.* Baillière, Tindall and Cox, London.

1949:

Davis M. 1949. Parturition of the Nubian giraffe (*Giraffa camelopardalis*). *Journal of Mammalogy* 30: 306–309.

Pincher C. 1949. Evolution of the giraffe. *Nature* 164 (4157): 29–30.

Robinson E. 1949. Evolution of the giraffe. *Nature* 164 (4164): 323, 1949.

Wood-Jones F. 1949. Evolution of the giraffe. *Nature* 164 (4164): 323, 1949.

1950:

Cave AJE. 1950. On the liver and gall-bladder of the Giraffe. *Proceedings of the Zoological Society, London* 120: 381–393.

Petzsch H. 1950. Zur Frage des Vorkommens ungefleckter albinotischer Giraffen. *Der Zoologische Garten* 17: 44–47.

1951:

Bate DMA. 1951. The mammals from Singa and Abu Hugar. *Fossil Mammals of Africa* 2: 1–28.

Bigalke R. 1951. The discovery of the giraffe in South Africa. *Fauna and Flora (Transvaal)* 2: 24–28.

Godee Molsbergen EC. *Reizen in Zuid-Afrika in de Hollandse Tijd. Eerste Deel, Tochten naar het Noorden, 1652–1686. 's Gravenhage, 1916.* In Bigalke, R. 1951. The discovery of the giraffe in South Africa. *Fauna and Flora (Transvaal)* 2: 24–28.

Krumbiegel I. 1951. Giraffenmischlinge. *Der Zoologische Garten* 18: 109–114.

1952:

Patterson JL, Warren JV. 1952. Mechanisms of adjustments in the cerebral circulation upon assumption of the upright position. *Journal of Clinical Investigation* 31: 653.

1954:

Kühlhorn F. 1954. Ursprung und Insertion der wichtigsten Muskeln der Vorderextremitat der Massaigiraffe (Giraffa camelopardalis tippelskirchi Mtsch. 1898). *Anatomischer Anzeiger* 101: 237–242.

1955:

Goetz RH. 1955. Preliminary observations on the circulation of the giraffe. *Transactions of the American College of Cardiologists* 5: 39–48.

Goetz RH, Budtz-Olsen O. 1955. Scientific safari: The circulation of the giraffe. *South African Medical Journal* 29: 773–776.

Pournelle GH. 1955. Notes on the reproduction of a Baringo giraffe. *Journal of Mammalogy* 36: 574.

1957:

Blum L. 1957. A photograph of a young giraffe with a black band around its body. *African Wildlife* 11: 328.

Keen EN, Goetz RH. 1957. Cardiovascular anatomy of a foetal giraffe. *Acta Anatomica* 31: 562–571.

Goetz RH, Keen EN. 1957. Some aspects of the cardiovascular system in the giraffe. *Angiology* 8: 542–564.

Lightfoot CJ. 1957. Birth of a giraffe. *Wild Rhodesia* 7: 28–30.

Patterson JL, Warren JV, Doyle JT, Gauer O, Keen EN, Goetz RH. 1957. Circulation and respiration in the giraffe. *Journal of Clinical Investigation* 36: 919.

Warren JV, Patterson JL, Doyle JT, Gauer OH, Keen EN, McGregor M, Goetz RH. 1957. Circulation and respiration in the giraffe. *Circulation* 16: 947–948.

1958:

Innis AC. 1958. The behaviour of the giraffe, *Giraffa camelopardalis*, in the Eastern Transvaal. *Proceedings of the Zoological Society (London)* 131: 245–278.

Kellas LM, van Lennep EW, Amoroso EC. 1958. Ovaries of some foetal and prepubertal giraffes. *Nature* 181: 487–488.

Latham RE. 1958. *The travels of Marco Polo.* Penguin Books, London.

Skead CJ. 1958. Mammals of the Uitenhage and Cradock Districts Cape Province in recent times. *Koedoe* 1: 19–38.

Willemse JJ. 1958. The innervation of the muscles of the trapezius-complex in giraffe, okapi, camel and llama. *Archives Nederlandaises de Zoologie* 12: 532–536.

1959:

Arambourg C. 1959. Vertebres continentaux du Miocene superieur de l'Afrique du Nord. *Publications du service de la Carte Geologique de l'Algerie (Nouvelle Serie), Palaeontologie, Memoires Algiers* 4: 1–159.

Backhaus D. 1959. Experimentelle Prufung des Farbsehvermogens einer Massai-giraffe (*Giraffa camelopardalis tippelskirchi* Matschie 1898). *Zeitschrift fur Tierpsychologie* 16: 468–477.

Dobson J. 1959. John Hunter's giraffe. *Annals of the Royal College of Surgeons of England.* 24: 124–128.

Stott K. 1959. Giraffe intergradation in Kenya. *Journal of Mammalogy* 40: 251.

1960:

Brown WL. 1960. Ants, acacias and browsing mammals. *Ecology* 41 (3): 587–592.

Dagg AI. 1960a. Food preferences of the giraffe. *Proceedings of the Zoological Society (London)* 135: 640–642.

Dagg AI. 1960b. Gaits of the giraffe and okapi. *Journal of Mammalogy* 41: 282.

Doyle JT, Patterson JL, Warren JV, Detweiler DK. 1960. Observations on the circulation of domestic cattle. *Circulation Research* 8: 4–15.

Frank A. 1960. Die glandula thyreoidea und die glandulae thyreoideae accessoriae bei Giraffa camelopardalis L. *Acta Anatomica* 42: 267.

Goetz RH, Warren JV, Gauer OH, Patterson JL, Doyle JT, Keen EN, McGregor M. 1960. Circulation of the giraffe. *Circulation Research* 8: 1049–1058.

Robin ED, Corson JM, Dammin GJ. 1960. The respiratory dead space of the giraffe. *Nature* 186: 24–26.

Singer R, Bonè, EL. 1960. Modern giraffes and the fossil Giraffids of Africa. *Annals of the South African Museum* 45: 375–548.

T-W-Fiennes RN. 1960. Report of the society's pathologist for the year 1958. *Proceedings of the Zoological Society (London)* 134: 297–308.

1961:

Bourliere F. 1961. Le sex-ratio de la giraffe. *Mammalia* 25: 467–471.

Greed RE. 1961. Composition of the milk of giraffe. *International Zoo Yearbook* 2: 106.

Thompson, D'Arcy W. 1961. *On growth and form.* Cambridge University Press, Cambridge, UK.

1962:

Aschaffenburg R, Gregory ME, Rowland SJ, Thompson SY. 1962. The composition of the milk of the giraffe (Giraffa camelopardalis reticulata). *Proceedings of the Zoological Society (London)* 139: 359–363.

Ben Shaul DM. 1962. The composition of the milk of wild animals. *International Zoo Yearbook* 4: 333–342.

Dagg AI. 1962a. The role of the neck in the movements of the giraffe. *Journal of Mammalogy* 43 (1): 88–97.

Dagg AI. 1962b. Giraffe movement and the neck. *Natural History* 2: 44–51, 1962.

Dagg AI. 1962c. The subspeciation of the giraffe. *Journal of Mammalogy* 43: 550–552.

Immelmann K, Gebbing H. 1962. Schlaf bei Giraffiden. *Zeitschrift fur Tierpsychologie* 19: 84–92.

1963:

Arambourg C. 1963. Continental vertebrate faunas of the tertiary of North Africa. In *African ecology and human evolution*. Clark Howell F, Bourliere F (Eds.) Aldine, Chicago, pp. 55–63.

Bourlière, F. 1963. Observations on the ecology of some large African mammals. In *African ecology and human evolution*. Clark Howell F, Bourliere F (Eds.). Aldine, Chicago.

Brownlee, A. 1963. Evolution of the giraffe. *Nature* 200: 1022.

Noirot-Timothée C. 1963. Sur les cilies du rumen de *Giraffa camelopardalis*. *Comptes rendus de l'Académie des Sciences* 256: 5400–5401.

1964:

Dougall HW, Drysdale VM, Glover PE. 1964. The chemical composition of Kenya browse and pasture herbage. *East African Wildlife Journal* 2: 86–121.

Heinze W. 1964. Die Muskulatur der Hintergliedmaße von *Giraffa camelopardalis angoliensis* mit einigen Hinweisen zur Blutgefäß-und Nervenversorgung. *Anatomischer Anzeiger Bd* 115: 476–496.

1965:

Anonymous. 1965. Giraffe midwives at a birth. *African Wildlife* 19 (4): 323.

Bligh J, Harthoorn AM. 1965. Continuous radiotelemetric records of the deep body temperature of some restrained African mammals under near-natural conditions. *Journal of Physiology* 176: 145–162.

Dagg AI. 1965. Sexual differences in giraffe skulls. *Mammalia* 29: 610–612.

Heinze W. 1965. Gelenke und Bänder der Hintergliedmasse der Giraffe (*Giraffa camelopardalis angolensis*). *Anatomischer Anzeiger Bd* 116: 39–58.

Leakey LSB. 1965. *Olduvai, 1951–1961*. Cambridge University Press, Cambridge, UK.

Patterson JL, Goetz RH, Doyle JT, Warren JV, Gauer OH, Detweiler DK, Said SI, Hoernicke H, McGregor M, Keen EN, Smith MH, Hardie EL, Reynolds M, Flatt WP, Waldo DR. 1965. Cardiorespiratory dynamics in the ox and giraffe, with comparative observations on man and other mammals. *Annals of the New York Academy of Sciences* 127: 393–413.

Wallace C, Fairall N. 1965. Chromosome analysis in the Kruger National Park with special reference to the chromosomes of the giraffe (*Giraffa camelopardalisgiraffa*, Boddaert). *Koedoe* 8: 97–103.

Wrobel KH. 1965. Das Nierenbecken der Giraffe. Zeitschrift fur Saugetierk. 30: 233–241.

1966:

Angermeyer M. 1966. Zur Kenntnis der Halsmuskeln und Halsnerven von *Giraffa camelopardalis* (L). *Zoologischer Anzeiger* 177: 188–200.

van Citters RL, Kemper WS, Franklin DL. 1966. Blood pressure responses of wild giraffes studied by radio telemetry. *Science* 152: 384–386.

1967:

Coe MJ. 1967. "Necking" behaviour in the giraffe. *Journal of Zoology (London)* 151: 313–321.

1968:

Dagg AI. 1968. External features of giraffe. *Mammalia* 32: 657–669, 1968.

Fairall N. 1968. The reproductive seasons of some large mammals in the Kruger National Park. *Zoologica Africana* 3: 189–210.

Spinage CA. 1968. Horns and other bony structures of the skull of the giraffe, and their functional significance. *East African Wildlife Journal* 6: 53–61.

Spinage CA. 1968. *The book of the giraffe*. Collins, London.

van Citters RL, Kemper WS, Franklin DL. 1968. Blood flow and pressure in the giraffe carotid artery. *Comparative Biochemistry and Physiology* 24: 1035–1042.

1969:

Balsdon JPVD. 1969. *Life and leisure in ancient Rome*. McGraw-Hill, New York.

Hirst SM. 1969. Populations in a Transvaal Lowveld Nature Reserve. *Zoologica Africana* 4: 199–230.

Pienaar U de V. 1969. Predator prey relationships amongst the larger mammals of the Kruger National Park. *Koedoe* 12: 108–176.

Sikes SK. 1969. Habitat and cardiovascular disease: Observations made on elephants (Loxodonta africana) and other free-living animals in East Africa. *Transactions of the Zoological Society (London)* 32: 1–104.

Turner M. 1969. Partial albino giraffe. *Africana* 3: 45.

van Citters RL, Franklin DL, Vatner SF, Patrick T, Warren JV. 1969. Cerebral haemodynamics in the giraffe. *Transactions of the Association of American Physicians* 82: 293–304.

1970:

Churcher CS. 1970. Two new upper Miocene Giraffids from Fort Ternan, Kenya, East Africa. *Fossil Vertebrates of Africa* 2: 1–105.

Dagg AI. 1970. Preferred environmental temperatures of some captive mammals. *International Zoo Yearbook* 10: 127–130.

Forbes VS. 1970. Some scientific matters in early writings on the Cape. In *A history of scientific endeavour in South Africa*. Royal Society of South Africa, Cape Town, South Africa, pp. 33–59.

Gentry AW. 1970. The Bovidae (Mammalia) of the Fort Ternan fossil fauna. *Fossil Vertebrates of Africa* 2: 243–323.

Hendey QB. 1970. The age of the fossiliferous deposits at Langebaanweg, Cape Province. *Annals of the South African Museum* 56: 119–131.

Leakey LSB. 1970. Additional information on the status of *G. jumae* from East Africa. *Fossil Vertebrates of Africa* 2: 325–330.

Nesbit-Evans EM. 1970. The reaction of a group of Rothschild's giraffe to a new environment. *East African Wildlife Journal* 8: 53–62.

Spinage CA. 1970. Giraffid horns. *Nature* 227: 735–736.

Talbot WJ. 1970. Pathfinders and pioneers, Explorers and Scientists 1487–1976. In *A history of scientific endeavour in South Africa*. Royal Society of South Africa, Cape Town, South Africa, pp. 1–32.

1971:

Altman PL, Dittmer DS. 1971. *Respiration and circulation*. FASEB, Bethesda, MD.

Anonymous. 1971. Giraffe albino. *Africana* 4: 39.

Dagg AI. 1971. Giraffa camelopardalis. *Mammalian Species* 5: 1–8.

Jefferson JW. 1971. Electrocardiographic and phonocardiographic findings in a reticulated giraffe. *Journal of the American Veterinary Medical Association* 159: 602–604.

Krumbiegel I. 1971. *Die Giraffe*. A Ziemsen, Wittenberg, Germany.

Oboussier H, Moller G. 1971. Zur kenntnis des Gehirns der Giraffidae (Pecora, Artiodactyla, Mammalia): Ein Verleich der Neocortex-oberflachengrosse. *Zeitschrift fur Saugetierk* 36: 291–296.

Western D. 1971. Giraffe chewing a Grant's gazelle carcass. *East African Wildlife Journal* 9: 156–157.

Williamson JR, Vogler NJ, Kilo C. 1971. Regional variations in the width of the basement membrane of muscle capillaries in man and giraffe. *American Journal of Pathology* 63: 359–370.

1972:

Foster JB, Dagg AI. 1972. Notes on the biology of the giraffe. *East African Wildlife Journal* 10: 1–16.

Leuthold BM, Leuthold W. 1972. Food habits of giraffe in Tsavo National Park, Kenya. *East African Wildlife Journal* 10: 129–141.

Oates LG. 1972. Food preferences of giraffe in Transvaal lowveld Mopane woodland. *Journal of the South African Wildlife Management Association* 2: 21–23.

Rossof AH. 1972. An electrocardiographic study of the giraffe. *American Heart Journal* 83: 142–143.

1973:

Glover TD. 1973. Aspects of sperm production in some East African mammals. *Journal of Reproduction and Fertility* 35: 45–53.

Hamilton WR. 1973. The lower Miocene ruminants of Gebel Zelten, Libya. *Bulletin of the British Museum of Natural History* 21: 73–150.

Kayanya FIB, Blakenship HL. 1973. The ovary of the giraffe, *Giraffa camelopardalis. Journal of Reproduction and Fertility* 34: 305–313.

Langman VA. 1973a. The immobilization and capture of giraffe. *South African Journal of Science* 69: 200–203.

Langman VA. 1973b. Radio-tracking giraffe for ecological studies. *Journal of the South African Wildlife Management Association* 3: 75–78.

Velhankar DP, Hukeri VB, Deshpande BR, Sane CR. 1973. Biometry of the genitalia and the spermatazoa of male giraffe. *The Indian Veterinary Journal* 50: 789–792.

1974:

Anonymous. 1974. The giraffe's parameter. *Lancet* 2: 1491–1492.

Drevemo S, Grootenhuis JG, Karstad L. 1974. Blood parameters in wild ruminants in Kenya. *Journal of Wildlife Diseases* 10: 327–333.

Gambaryan PP. 1974. *How mammals run.* John Wiley & Sons, New York. Translated from Russian by Hilary Hardin.

Gombe S, Kayanja FIB. 1974. Ovarian progestins in Masai giraffe (*Giraffa camelopardalis*). *Journal of Reproduction and Fertility* 40: 45–50.

Hall-Martin AJ. 1974a. Food selection by Transvaal Lowveld giraffe as determined by analysis of stomach contents. *Journal of the South African Wildlife Management Association* 4: 191–202.

Hall-Martin AJ. 1974b. A note on the seasonal utilization of different vegetation types by giraffe. *South African Journal of Science* 70: 122–123.

Warren JV. 1974. The physiology of the giraffe. *Scientific American* 231: 96–105.

1975:

Hall-Martin AJ. 1975a. Studies on the biology and productivity of the giraffe *Giraffa camelopardalis*. DSc thesis, University of Pretoria, South Africa.

Hall-Martin AJ. 1975b. Aspects of the ecology and sociality of the giraffe (*Giraffa camelopardalis giraffa*). *Publications of the University of Pretoria (New Series)* 97: 48–56.

Hall-Martin AJ, Basson WD. 1975. Seasonal chemical composition of the diet of Transvaal Lowveld giraffe. *Journal of the South African Wildlife Management Association* 5: 19–21.

Hall-Martin AJ, Skinner JD, van Dyk JM. 1975. Reproduction in the giraffe in relation to some environmental factors. *East African Wildlife Journal* 13: 237–248.

Hawkey CM. 1975. *Comparative mammalian haematology*. William Heineman Medical Books, London.

Heintz E. 1975. Origine, migration et paleobiogeographie des Paleotraginae (Giraffidae, Artiodactyla) Antevallesiens de l'Ancien Monde. *Colloque International C.N.R.S.* 218: 723–730.

Leuthold W, Leuthold BM. 1975. Temporal patterns of reproduction in ungulates of Tsavo East National Park, Kenya. *East African Wildlife Journal* 13: 159–169.

McMahon TA. 1975. Allometry and biomechanics: Limb bones in adult ungulates. *The American Naturalist* 109: 547–563.

Pennycuick CJ. 1975. On the running of the Gnu (*Connochaetes taurinus*) and other animals. *Journal of Experimental Biology* 63: 775–799.

Pickford M. 1975. Late Miocene sediments and fossils from the northern Kenya Rift Valley. *Nature* 256: 279–284.

Rhodes JA. 1975. Biochemical profiles in exotic captive animals. *Annual Proceedings of the American Association of Zoo Veterinarians* 1975: 81–111.

Skinner JD, Hall-Martin AJ. 1975. A note on foetal growth and development of the giraffe, *Giraffa camelopardalis giraffa*. *Journal of Zoology, London* 177: 73–79.

van Aarde RJ, Skinner JD. 1975. The food and feeding behaviour of the giraffe (*Giraffa camelopardalis*) in the Jack Scott Nature Reserve. *Publications of the University of Pretoria (New Series)* 97: 59–69.

1976:

Ackerman E. 1976. The histogenesis of hair follicles in the zebra and giraffe with special reference to pigmentation and cutaneous vasculature. MSc dissertation, University of Pretoria, South Africa.

Dagg AI, Foster JB. 1976. *The giraffe: Its biology, behaviour, and ecology*. Van Nostrand Reinhold, New York.

Dimond RL, Montagna W. 1976. The skin of the giraffe. *Anatomical Record* 185: 63–76.

Hall-Martin AJ. 1976. Dentition and age determination of the giraffe, *Giraffa camelopardalis*. *Journal of Zoology, London* 180: 263–289.

Harris JM. 1976a. Pliocene Giraffoidea (Mammalia, Artiodactyla) from the Cape Province. *Annals of the South African Museum* 69: 325–353.

Harris JM. 1976b. Giraffidae from the East Rudolf succession. In *Earliest man and environments in the Lake Rudolph Basin*. Coppens Y, Howell FC, Isaac GL, Leakey RE (Eds.). Chicago University Press, Chicago, pp. 264–267.

Harris JM. 1976c. Pleistocene Giraffidae (Mammalia, Artiodactyla) from East Rudolf, Kenya. *Fossil Vertebrates of Africa* 4: 283–332.

Kleynhans CJ, van Hoven W. 1976. Rumen protozoa of the giraffe with a description of two new species. *East African Wildlife Journal* 14: 203–214.

van Aarde RJ. 1976. A note on the birth of a giraffe. *South African Journal of Science* 72: 307.

1977:

Alexander RM, Bennet-Clark HC. 1977. Storage of elastic strain energy in muscle and other tissues. *Nature* 265: 114–117.

Alexander RM, Langman VA, Jayes AS. 1977. Fast locomotion of some African ungulates. *Journal of Zoology (London)* 183: 291–300.

Foster JB. 1977. Africa's gentle giants. *National Geographic* 152: 402–416.

Hall-Martin AJ. 1977. Giraffe weight estimation using dissected leg weight and body measurements. *Journal of Wildlife Management* 41: 740 745.

Hall-Martin AJ, Skinner JD, Smith A. 1977b. Observations on lactation and milk composition of the giraffe Giraffa camelopardalis. *South African Journal of Wildlife Research* 7: 67–71.

Hall-Martin AJ, von la Chevallerie M, Skinner JD. 1977. Carcass composition of the giraffe, *Giraffa camelopardalis giraffa*. *South African Journal of Animal Science* 7: 55–64.

Hillman JC, Hillman KK. 1977. Mortality of wildlife in Nairobi National Park, during the drought of 1973–1974. *East African Wildlife Journal* 15: 1–8.

Langman VA. 1977. Cow-calf relationships in giraffe (*Giraffa camelopardalis giraffa*). *Zeitschrift fur Tierpsychologie* 43: 264–286.

McCalden TA, Borsook D, Mendelow AD, Shimell CJ, de Vos V, Pieterse PC, de Klerk B. Autoregulation and haemodynamics of giraffe carotid blood flow. *South African Journal of Science* 73: 278–279.

Sauer JJC, Theron GK, Skinner JD. 1977. Food preferences of giraffe *Giraffa camelopardalis* in the arid bushveld of the western Transvaal. *South African Journal of Wildlife Research* 7: 53–59.

1978:

Berry PSM. 1978. Range movements of giraffe in the Luangwa Valley, Zambia. *East African Wildlife Journal* 16: 77–83.

Buckland RA, Evans HJ. 1978. Cytogenetic aspects of phylogeny in the Bovidae 1. G—banding. *Cytogenetics and Cell Genetics* 21: 42–63.

Churcher CS. 1978. Giraffidae. In *Evolution of African Mammals*. Maglio VJ, Cooke HBS (Eds.). Harvard University Press, Cambridge, MA, pp. 509–535.

Cooke HBS. 1978. Africa: The physical setting. In *Evolution of African Mammals*. Maglio VJ, Cooke HBS (Eds.). Harvard University Press, Cambridge, MA, pp. 17–45.

Hall-Martin AJ, Skinner JD. 1978. Observations on puberty and pregnancy in female giraffe (*Giraffa camelopardalis*). *South African Journal of Wildlife Research* 8: 91–94.

Hall-Martin AJ, Skinner JD, Hopkins BJ. 1978. The development of the reproductive organs of the male giraffe, *Giraffa camelopardalis*. *Journal of Reproduction and Fertility* 52: 1–7.

Hamilton WR. 1978a. Fossil giraffes from the Miocene of Africa and a revision of the phylogeny of the Giraffoidea. *Philosophical Transactions of the Royal Society B* 283: 165–229.

Hamilton WR. 1978b. Cervidae and Palaeomerycidae. In *Evolution of African Mammals*. Maglio VJ, Cooke HBS (Eds.). Harvard University Press, Cambridge, MA, pp. 496–508.

Hendey QB. 1978. Preliminary report on the Miocene vertebrates from Arrisdrift, South West Africa. *Annals of the South African Museum* 76: 1–41.

Hugh-Jones P, Barter CE, Hime JM, Rusbridge MM. 1978. Dead space and tidal volume of the giraffe compared with some other mammals. *Respiration Physiology* 35: 53–58.

Langman VA. 1978. Giraffe pica behaviour and pathology as indicators of nutritional stress. *Journal of Wildlife Management* 42: 141–147.

Leuthold BM, Leuthold W. 1978a. Ecology of the giraffe in Tsavo East national Park, Kenya. *East African Wildlife Journal* 16: 1–20.

Leuthold BM, Leuthold W. 1978b. Daytime activity patterns of gerenuk and giraffe in Tsavo National Park, Kenya. *East African Wildlife Journal* 16: 231–243.

Maglio VJ, Cooke HBS (Eds.). 1978. *Evolution of African Mammals*. Harvard University Press, Cambridge, MA.

Schroter RC. 1978. Convective flows in mammalian lungs. In *Comparative physiology*. Schmidt-Nielsen K, Bolis L, Maddrell S (Eds.). Cambridge University Press, Cambridge, UK, pp. 303–325.

Skinner JD. 1978. Breeding cycles in three species of African ungulates. In *Proceedings in Life Sciences: Environmental Endocrinology*. Springer-Verlag, Berlin, pp. 64–72.

1979:

Beintema JJ, Gaastra W, Munniksma J. 1979. Primary structure of Pronghorn pancreatic ribonuclease: close relationship between giraffe and pronghorn. *Journal of Molecular Evolution* 13: 305–316.

Geraads D. 1979. Les Giraffinae (Giraffidae, Mammalia) du Miocène supérieur de la région de Thessalonique (Grèce). *Bulletin du Museum National d'Histoire Naturelle C* 1: 377–389.

Kristal MB, Noonan M. 1979. Note on sleep in captive giraffes (Giraffa camelopardalis reticulata). *South African Journal of Zoology* 14: 108.

Langman VA, Maloiy MO, Schmidt-Nielsen K, Schroter RC. 1979. Nasal heat exchange in the giraffe and other large mammals. *Respiration Physiology* 37: 325–333.

Pratt DM, Anderson VH. 1979. Giraffe cow-calf relationships and social development of the calf in the Serengeti. *Zeitschrift fur Tierpsychologie* 51: 233–251.

Rookmaaker LC. 1979. Robert Jacob Gordon (1743–1795) *en zijn bijdrae tot de zoölogie van de Kaap de Goede Hoop*. PhD thesis, University of Utrecht.

Western D. 1979. Size, life history and ecology in mammals. *African Journal of Ecology* 17: 185–204.

1980:

Bush M, Custer RS, Whitla JC. 1980. Hematology and serum chemistry profiles for giraffes (*Giraffa camelopardalis*): Variations with sex, age, and restraint. *Journal of Zoo Animal Medicine* 11: 122–129.

Deka BC, Nath KC, Borghain BN. 1980. Clinical note on the morphology of the placenta in a giraffe. *Journal of Zoo Animal Medicine* 11: 117–118.

Hall-Martin AJ, Rowlands IW. 1980. Observations on ovarian structure and development of the southern giraffe, *Giraffa camelopardalis giraffa*. *South African Journal of Zoology* 15: 217–221.

Harrison DFN. 1980. Biomechanics of the Giraffe larynx and trachea. *Acta Otolaryngology* 89: 258–264.

Kok OB, Opperman DPJ. 1980. Voedingsgedrag van kameelperde, *Giraffa camelopardalis*, in die Willem Pretorius-Wildtuin, Oranje-Vrystaat. *South African Journal Wildlife Research* 10: 45–55.

McCalden TA, Mendelow AD. 1980. Carotid haemodynamics in the giraffe. *South African Journal of Science* 76: 246–247.

Mitchell G, Hepburn HR, Mitchell D, Shimell C, Borsook D, Ford DM, Klugman K, Rogers G. 1980. Carotid haemodynamics in the giraffe. *South African Journal of Science* 76: 246.

Skead CJ. 1980. Giraffe distribution. In *Historical mammal incidence in Cape Province* 1: 433–443. Department of Nature and Environmental Conservation, Cape of Good Hope, South Africa.

Webb SD, Taylor BE. 1980. The phylogeny of hornless ruminants and a description of the cranium of *Archaeomeryx*. *Bulletin of the American Museum of Natural History* 167: 117–158.

1981:

Gehr P, Mwangi DK, Ammann A, Maloiy GMO, Taylor CR, Weibel ER. 1981. Design of the mammalian respiratory system. V. Scaling morphometric pulmonary diffusing capacity to body mass: wild and domestic mammals. *Respiration Physiology* 44: 61–86.

Harrison DFN. 1981. Fibre size frequency in the recurrent laryngeal nerves of man and giraffe. *Acta Otolaryngology* 91: 383–389.

Hendey QB. 1981. Geological succession at Langebaanweg, Cape Province, and global events of the late Tertiary. *South African Journal of Science* 77: 33–38.

Kimani JK. 1981a. Structural evidence for insertion of collagen fibres to smooth muscle cells in the carotid arterial system of the giraffe (*Giraffa camelopardalis*). *Cell and Tissue Research* 214: 225–238.

Kimani JK. 1981b. Subendothelial fibrillar laminae in the carotid arteries of the giraffe (Giraffa camelopardalis). *Cell and Tissue Research* 219: 441–443.

1982:

Hendey QB. 1982. *Langebaanweg: A record of past life*. South African Museum, Cape Town, South Africa.

Hitching F. 1982. *The neck of the giraffe, or, where Darwin went wrong*. Pan Books, London.

Kok OB. 1982. Cannibalism and post-partum return to oestrus of a female Cape giraffe. *African Journal of Ecology* 20: 141–143.

Langman VA, Bamford OS, Maloiy GMO. 1982. Respiration and metabolism in the giraffe. *Respiration Physiology* 50: 141–152.

Maloiy GMO, Clemens ET, Kamau JMZ. 1982. Aspects of digestion and in vitro rumen fermentation rate in six species of East African wild ruminants. *Journal of Zoology (London)* 197: 345–353.Pratt DM, Anderson VH. 1982. Population, distribution, and behaviour of giraffe in the Arusha National Park, Tanzania. *Journal of Natural History* 16: 481–489.

Ritland S. 1982. *The allometry of the vertebrate eye*. PhD dissertation, University of Chicago, Ann Arbor.

Sauer JJC, Skinner JD, Neitz AWH. 1982. Seasonal utilization of leaves by giraffes, *Giraffa camelopardalis*, and the relationship of the seasonal utilization to the chemical composition of the leaves. *South African Journal of Zoology* 17: 210–219.

Taylor CR, Heglund NC. 1982. Energetics and mechanics of terrestrial locomotion. *Annual Reviews of Physiology* 44: 97–107.

1983:

Biewener AA. 1983. Allometry of quadrupedal locomotion: The scaling of duty factor, bone curvature and limb orientation to body size. *Journal of Experimental Biology* 105: 147–171.

Clemens ET, Maloiy GMO. 1983. Digestive physiology of East African wild ruminants. *Comparative Biochemistry and Physiology A* 76: 319–333.

Garland T. 1983. The relation between maximal running speed and body mass in terrestrial mammals. *Journal of Zoology* 199: 157–170.

Hradecký P. 1983a. Some uterine parameters in antelopes and a giraffe. *Theriogenology* 20: 491–498.

Hradecký P. 1983b. Placental morphology in African antelopes and giraffes. *Theriogenology* 20: 725–734.

Kimani JK. 1983a. The structural organization of the tunica intima in the carotid arteries of the giraffe (*Giraffa camelopardalis*). *African Journal of Ecology* 21: 300–315.

Kimani JK. 1983b. The structural organization of the carotid arterial system of the giraffe (Giraffa camelopardalis). *African Journal of Ecology* 21: 317–324.

Kimani K, Mungai JM. 1983. Observations on the structure and innervation of the presumptive carotid sinus area in the giraffe. *Acta Anatomica* 115: 117–133.

Nakakuki S. 1983. The bronchial ramification and lobular division of the giraffe lung. *Anatomischer Anzeiger* 154: 313–317.

Pellew RA. 1983a. The impact of elephant, giraffe, and fire upon the *Acacia tortillas* woodlands of the Serengeti. *African Journal of Ecology* 21: 41–74.

Pellew RA. 1983b. The giraffe and its food resource in the Serengeti. I. Composition, biomass, and production of available browse. *African Journal of Ecology* 21: 241–267.

Pellew RA. 1983c. The giraffe and its food resource in the Serengeti. II. Response of the giraffe population to changes in the food supply. *African Journal of Ecology* 21: 269–283, 1983.

Rookmaaker LC. 1983. The observations of Robert Jacob Gordon (1743–1795) on giraffes (*Giraffa camelopardalis*) found in Namaqualand. *Journal of the South West Africa Scientific Society* 36–37: 71–90.

Sauer JJC. 1983a. Food selection by giraffes in relation to changes in chemical composition of the leaves. *South African Journal of Animal Science* 13: 40–43.

Sauer JJC. 1983b. A comparison between *Acacia* and *Combretum* leaves utilized by giraffe. *South African Journal of Animal Science* 13: 43–44.

Smithers RHN, Skinner JD. 1983. *The mammals of the Southern African Sub-region*. 1st edition. University of Pretoria, South Africa.

1984:

Pellew RA. 1984a. Food consumption and energy budgets of the giraffe. *Journal of Applied Ecology* 21: 141–159.

Pellew RA. 1984b. The feeding ecology of a selective browser, the giraffe (*Giraffa camelopardalis tippelskirchi*). *Journal of Zoology (London)* 202: 57–81.

Thomas H. 1984. Les Giraffoidea et les Bovidae Miocenes de la formation Nyakach (rift Nyanza, Kenya). *Palaeontographica, Stuttgart A* 183: 64–89.

van Wyk P. 1984. *Field guide to the trees of the Kruger National Park*. Struik, Cape Town, South Africa.

1985:

Harrison DFN, Denny S. 1985. The possible influence of laryngeal and tracheal size on the running speed of mammals. *Acta Otolaryngology (Stockholm)* 99: 229–235.

Pratt DM, Anderson VH. 1985. Giraffe social behavior. *Journal of Natural History* 19: 771–781.

1986:

Badeer HS. 1986. Does gravitational pressure of blood hinder flow to the brain of the giraffe. *Comparative Biochemistry and Physiology A* 83: 207–211.

Geraads D. 1986. Remarques sur la systematique et la phylogenie des Giraffidae (Artiodactyla, Mammalia) *Geobios* 19: 465–477.

Harrison DFN. 1986. Victor Negus: 57 years later. *Annals of Otology, Rhinology and Laryngology* 95: 561–566.

Loskutoff NM, Walker L, Ott-Joslin JE, Raphael BL, Lasley BL. 1986. Urinary steroid evaluations to monitor ovarian function in exotic ungulates: II. Comparison between the giraffe (*Giraffa camelopardalis*) and the Okapi (*Okapia johnstoni*). *Zoo Biology* 5: 331–338.

Pettersson K, Hargens AR, Millard RW, Johansen K, Gershuni DH, Burroughs R, Meltzer DGA, van Hoven W. 1986. Dependent hypertension and arterial wall hypertrophy without interstitial edema in the giraffe. *Proceedings of the International Union of Physiological Sciences* 16: 411.

Savage RJG, Long MR. 1986. *Mammal evolution: An illustrated guide*. British Museum of Natural History, London.

1987:

Goodman PS, Tomkinson AJ. 1987. The past distribution of giraffe in Zululand and its implications for reserve management. *South African Journal of Wildlife Research* 17: 28–32.

Hargens AR, Millard RW, Pettersson K, van Hoven W, Gershuni DH, Johansen K. 1987a. Transcapillary fluid balance in giraffe. *Advances in Microcirculation* 13: 195–202.

Hargens AR, Millard RW, Pettersson K, Johansen K. 1987b. Gravitational haemodynamics and oedema prevention in the giraffe. *Nature* 329: 59–60.

Harris JM. 1987a. Fossil Giraffidae and Camelidae from Laetoli. In *Laetoli, a Pliocene site in Northern Tanzania*. Leakey MD, Harris JM (Eds.). Oxford University Press, Oxford, pp. 358–377.

Joost-Gaugier, CL. 1987. Lorenzo the Magnificent and the Giraffe as a symbol of power. *Artibus et Historiae* 8: 91–99.

Kimani JK. 1987. Structural organization of the vertebral artery in the giraffe (*Giraffa camelopardalis*). *Anatomical Record* 217: 256–262.

Pedley TJ. 1987. How giraffes prevent oedema. *Nature* 329: 13–14.

Seymour RS, Johansen K. 1987. Blood flow up hill and downhill: Does a siphon facilitate circulation above the heart. *Comparative Biochemistry and Physiology A* 88: 167–170.

Walker BH, Emslie RH, Owen-Smith RN, Scholes RJ. 1987. To cull or not to cull: lessons from a southern African drought. *Journal of Applied Ecology* 24: 381–401.

Young TP. 1987. Increased thorn length in *Acacia depranolobium*: An induced response to browsing. *Oecologoa (Berlin)* 71: 436–438.

1988:

Badeer HS. 1988. Haemodynamics of the jugular vein in the giraffe. *Nature* 332: 788–789.

Gentry AW, Hooker JJ. 1988. The phylogeny of the Artiodactyla. In *The phylogeny and classification of the tetrapods*, Volume 2: *Mammals*. Benton MJ (Ed.). Clarendon Press, Oxford, pp. 235–272.

Goetz RH. 1988. Early surgical research at the University of Cape Town. *South African Journal of Surgery* 26: 141–147.

Hargens AR, Gershuni DH, Danzig LA, Millard RW, Pettersson K. 1988. Tissue adaptations to gravitational stress: newborn versus adult giraffes. *Physiologist* 31: S110–S113.

Hofmann RR, Matern B. 1988. Changes in gastrointestinal morphology related to nutrition in giraffes *Giraffa camelopardalis*: A comparison of wild and zoo specimens. *International Zoo Yearbook* 27: 168–176.

Janis CM, Scott KM. 1988. The phylogeny of the Ruminantia (Artiodactyla, Mammalia). In *The phylogeny and classification of the tetrapods*, Volume 2: *Mammals*. Benton MJ (Ed.). Clarendon Press, Oxford, pp. 273–282.

Nilsson O, Booj S, Dahlstrom A, Hargens AR, Millard RW, Pettersson K. 1988. Sympathetic innervation of the cardiovascular system in the giraffe. *Blood Vessels* 25: 299–307.

Raper PE, Boucher M (Eds.). 1988. *Robert Jacob Gordon: Cape travels, 1777 to 1786*. Volume 1. Brenthurst Press, Johannesburg, South Africa.

Roy RR, Graham S, Peteson JA. 1988. Fiber type composition of the plantar flexors of giraffes (*Giraffa camelopardalis*) at different post-natal stages of development. *Comparative Biochemistry and Physiology A* 91: 347–352.

Solounias N. 1988. Prevalence of ossicones in Giraffidae (Artiodactyla, Mammalia). *Journal of Mammalogy* 69: 845–848.

1989:

du Toit JT, Owen-Smith N. 1989. Body size, population metabolism, and habitat specialization among large African herbivores. *The American Naturalist* 133 (5): 736–740.

Hicks JW, Badeer HS. 1989. Siphon mechanisms in collapsible tubes: application to circulation of the giraffe head. *American Journal of Physiology* 256: R567–R571.

Hofmann RR. Evolutionary steps of ecophysiological adaptation and diversification of ruminants: a comparative view of their digestive system. *Oecologia* 78: 443–457.

Langman VA, Maloiy GMO. 1989. Passive obligatory heterothermy of the giraffe. *Journal of Physiology* 415: 89P.

1990:

du Toit JT. 1990a. Feeding-height stratification among African browsing ruminants. *African Journal of Ecology* 28: 55–61.

du Toit JT. 1990b. Giraffe feeding on *Acacia* flowers: predation or pollination? *African Journal of Ecology* 28: 63–68.

du Toit JT. 1990c. Home range: Body mass relations: A field study on African browsing ruminants. *Oecologia* 85: 301–303.

du Toit JT, Bryant JP, Frisby K. 1990. Regrowth and palatability of Acacia shoots following pruning by African savanna browsers. *Ecology* 71 (1): 149–154.

Ganey T, Ogden J, Olsen J. 1990. Development of the giraffe horn and its blood supply. *Anatomical Record* 227: 497–507.

Kuntsch V, Nel JAJ. 1990. Possible thermoregulatory behaviour in *Giraffa camelopardalis*. *Zeitschrift fur Saugetierkunde* 55: 60–62.

Skinner JD, Smithers RHN. 1990. Giraffidae. In *The mammals of the Southern African Sub-region*. 2nd edition. University of Pretoria, South Africa, pp. 604–607.

Solounias N, Tang N. 1990. The two types of cranial appendages in *Giraffa camelopardalis* (Mammalia, Artiodactyla). *Journal of Zoology (London)* 222: 293–302.

1991:

Agusti J, Moya Sola S. 1991. Spanish Neogene mammal succession and its bearing on continental biochronology. *Newsletters on Stratigraphy (Leiden)* 25: 91–114.

Folkow B. 1991. Giraffe, rats, and man: What is the importance of the "structural factor" in normo- and hypertensive states? *Clinical and Experimental Pharmacology and Physiology* 18: 3–11.

Harris JM. 1991. Family Giraffidae. *Koobi Fora Research Project* 3: 93–138.

Hawkey CM, Bennett PM, Gascoyne SC, Hart MG, Kirkwood JK. 1991. Erythrocyte size, number and haemoglobin content in vertebrates. *British Journal of Haematology* 77: 392–397.

Hearing VJ, Tsukamoto K. 1991. Enzymatic control of pigmentation in mammals. *FASEB Journal* 5: 2902–2909.

Irwin DM, Kocher TD, Wilson AC. 1991. Evolution of the cytochrome b gene of mammals. *Journal of Molecular Evolution* 32: 128–144.

Kamau JMZ. 1991. Morphogenetic analysis of nasal passages in ungulates. *Ongulès/Ungulates* 91: 91–94.

Kimani JK, Mbuva RN, Kinyama RM. 1991a. Sympathetic innervation of the hindlimb arterial system in the giraffe (*Giraffa camelopardalis*). *Anatomical Record* 229: 103–108.

Kimani JK, Opole IO. 1991. The structural organization and adrenergic innervation of the carotid arterial system of the giraffe (*Giraffa camelpardalis*). *Anatomical Record* 320: 369–377.

Kimani JK, Opole IO, Ogengo JA. 1991b. Structure and sympathetic innervation of the intracranial arteries in the giraffe (*Giraffa camelopardalis*). *Journal of Morphology* 208: 193–203.

Milewski AV, Young TP, Madden D. 1991. Thorns as induced defenses: experimental evidence. *Oecologia* 86: 70–75.

Worthington J, Young IS, Altringham JD. 1991. The relationship between body mass and ventilation rate in mammals. *Journal of Experimental Biology* 161: 533–536.

Young TP, Isbell LA. 1991. Sex differences in giraffe feeding ecology: Energetic and social constraints. *Ethology* 87: 79–89.

1992:

Geiser DR, Morris PJ, Adair HS. 1992. Multiple anesthetic events in a reticulated giraffe (*Giraffa camelopardalis*). *Journal of Zoo and Wildlife Medicine* 23: 189–196.

Kamau JMZ. 1992. A comparative approach to the structural basis for brain cooling in ungulates with observation on in vitro heat exchange. *Discovery and Innovation* 2: 95–102.

Madden D, Young TP. 1992. Symbiotic ants as an alternative defense against giraffe herbivory. *Oecologia* 91: 235–238.

Scheepers JL. 1992. Habitat selection and demography of a giraffe population in northern Namib desert, Namibia. *Ongulès/Ungulates* 91: 223–228.

Thenius E. 1992. The okapi (Mammalia, Artiodactyla) from Zaire: A living fossil or a secondary rain-forest inhabitant. *Zeitschrift fur Zoologische Systematik und Evolutionsforschung* 30: 163–179.

1993:

Jones ML. Longevity of ungulates in captivity. *International Zoo Yearbook* 32: 159–169.

Mitchell G, Hattingh J. 1993. Giraffe colloid osmotic pressure and oedema. *South African Journal of Science* 89: 569–570.

Mitchell G, Skinner JD. 1993. How giraffe adapt to their extraordinary shape. *Transactions of the Royal Society of South Africa* 48: 207–218.

Scott KM, Janis CM. 1993. Relationships of the Ruminantia (Artiodactyla) and an analysis of the characters used in ruminant taxonomy. In *Mammal phylogeny: Placental.* Szalay FS, Novacek MJ, McKenna MC (Eds.). Springer-Verlag, London, pp. 282–302.

Seymour RS, Hargens AR, Pedley TJ. 1993. The heart works against gravity. *American Journal of Physiology* 265: R715–R720.

Solounias N, Moelleken SMC. 1993. Dietary adaptation of some extinct ruminants determined by premaxillary shape. *Journal of Mammalogy* 74: 1059–1071.

Spinage CA. 1993. The median ossicone of *Giraffa camelopardalis*. *Journal of Zoology (London)* 230: 1–5.

1994:

Alexander RM. 1994. *Bones*. Macmillan, New York, p. 103.

Furstenburg D, van Hoven W. 1994. Condensed tannin as anti-defoliate agent against browsing by giraffe (*Giraffa camelopardalis*) in the Kruger National Park. *Comparative Biochemistry and Physiology A* 107: 425–431.

Gallagher DS, Derr JN, Womack JE. 1994. Chromosome conservation among the advanced pecorans and determination of the primitive bovid karyotype. *Journal of Heredity* 85: 204–210.

Gentry AW. 1994. The Miocene differentiation of Old World Pecora (Mammalia). *Historical Biology* 7: 115–158.

Gordon IJ, Illius AW. 1994. The functional significance of the browser-grazer dichotomy in African ruminants. *Oecologia* 98: 167–175.

1995:

Ciofolo I. 1995. West Africa's last giraffes: the conflict between development and conservation. *Journal of Tropical Ecology* 11: 577–588.

Reason RC. 1995. Development of feeding behaviours in giraffe (*Giraffa camelopardalis*) calves at Brookfield Zoo, Chicago. *International Zoo Yearbook* 34: 222–224.

Robbins CT, Spalinger DE, van Hoven W. 1995. Adaptation of ruminants to browse and grass diets: are anatomical-based browser-grazer interpretations valid? *Oecologia* 103: 208–213.

Tobler I. 1995. Is sleep fundamentally different between mammalian species? *Behavioural Brain Research* 69: 35–41.

1996:

Cronin MA, Stuart R, Pierson BJ, Patton JC. 1996. K-casein gene phylogeny of higher ruminants (Pecora, Artiodactyla). *Molecular Phylogenetics and Evolution* 6: 295–311.

Gordon IJ, Illius AW. 1996. The nutritional ecology of African ruminants: A reinterpretation. *Journal of Animal Ecology* 65: 18–28.

Gould SJ. 1996. The tallest tale. *Natural History* 5: 18–23, 54–57.

Kostopoulos DS, Koliadimou KK, Koufos GD. 1996. The Giraffids (Mammalia, Artiodactyla) from the late Miocene mammalian localities of Nikiti (Macedonia, Greece). *Palaeontographica Abteilung A* 239: 61–88.

Pedley TJ, Brook BS, Seymour RS. 1996. Blood pressure and flow rate in the giraffe jugular vein. *Philosophical Transactions of the Royal Society (London) B* 351: 855–866.

Simmons RE, Scheepers L. 1996. Winning by a neck: Sexual selection in the evolution of giraffe. *The American Naturalist* 148: 771–786.

Tobler I, Schwierin B. 1996. Behavioural sleep in the giraffe (*Giraffa camelopardalis*) in a zoological garden. *Journal of Sleep Research* 5: 21–32.

Van Soest PJ. 1996. Allometry and ecology of feeding behaviour and digestive capacity in herbivores: a review. *Zoo Biology* 15: 455–479.

1997:

Badeer HS. 1997. Is the flow in the giraffe's jugular vein a "free" fall? *Comparative Biochemistry and Physiology A* 118: 573–576.

Endo H, Yamagiwa D, Fujisawa M, Kimura J, Kurohmaru M, Hayashi Y. 1997. Modified muscular system of the giraffe (*Giraffa camelopardalis*). *Annals of Anatomy* 179: 481–485.

Ginnett TF, Demment MW. 1997. Sex differences in giraffe foraging behaviour at two spatial scales. *Oecologia* 110: 291–300.

Owen-Smith N. 1997. Distinctive features of the nutritional ecology of browsing versus grazing ruminants. *Zeitschrift fur Saugetierkunde* 62: 176–191.

Pickering J. 1997. William J Burchell's South African mammal collection, 1810–1815. *Archives of Natural History* 24: 311–326.

1998:

Allin M. 1998. *Zarafa*. Headline, London.

Cerling TE, Ehleringer JR, Harris JM. 1998. Carbon dioxide starvation, the development of C4 ecosystems, and mammalian evolution. *Philosophical Transactions of the Royal Society, London B* 353: 159–171.

Clauss M, Deutsch A, Lechner-Doll M, Flach E, Tack C. 1998. Passage rate of fluid and particle phase in captive giraffe (*Giraffa camelopardalis*). *Advances in Ethology* 33: 98.

Geraads D. 1998. Le gisement de vertébrés pliocènes de Calta, Ankara, Turquie. 9. Cervidae et Giraffidae. *Geodiversitas* 20: 455–465.

1999:

Christiansen P. 1999a. Scaling of the limb long bones to body mass in terrestrial mammals. *Journal of Morphology* 239: 167–190.

Christiansen P. 1999b. Scaling of mammalian long bones: Small and large mammals compared. *Journal of Zoology (London)* 247: 333–348.

Clauss M, Suedmeyer K, Flach EJ. 1999. Susceptibility to cold in captive giraffe (*Giraffa camelopardalis*). *Proceedings of the American Association of Zoo Veterinarians* 1999: 183–186.

Ginnett TF, Demment MW. 1999. Sexual segregation by Masai giraffes at two spatial Scales. *African Journal of Ecology* 37: 93–106.

Hahn CN, Mayhew IG. 1999. Do giraffe roar? *Veterinary Record* 145: 28.

Linton RAF, Taylor PM, Linton NWF, Flach EJ, O'Brien TK, Band DM. 1999. Cardiac output measurement in an anaesthetized giraffe. *Veterinary Record* 145: 498–499.

Morales J, Soria D, Pickford M. 1999. New stem giraffoid ruminants from the early and middle Miocene of Namibia. *Geodiversitas* 21: 229–253.

Odenyo AA, McSweeney CS, Palmer B, Negassa D, Osuji PO. 1999. In vitro screening of rumen fluid samples from indigenous African ruminants provides some evidence for rumen fluid with superior capacities to digest tannin-rich fodders. *Australian Journal of Agricultural Research* 50: 1147–1157.

Solounias N. 1999. The remarkable anatomy of the giraffe's neck. *Journal of Zoology (London)* 247: 257–268.

Von Muggenthaler E, Baes C, Hill D, Fulk R, Lee A. 1999. Infrasound and low frequency vocalizations from the giraffe. Presented at the American Acoustical Society Conference. Published in the Proceedings of Riverbanks Consortium.

2000:

Hilsberg S. 2000. Aspekte klinischer Anwendung der Infrarot: Thermographie in der Zoo-und Wildtiermedizin. PhD dissertation, University of Leipzig.

Le Pendu Y, Ciofolo I, Gosser A. 2000. The social organization of giraffes in Niger. *African Journal of Ecology* 38: 78–85.

Munis JR, Lozada LJ. 2000. Giraffes, siphons, and Starling resistors: Cerebral perfusion pressure revisited. *Journal of Neurological Anesthesiology* 12: 290–296.

Seymour RS. 2000. Model analogues in the study of cephalic circulation. *Comparative Biochemistry and Physiology A* 125: 517–524.

Seymour RS, Blaylock AJ. 2000. The principle of LaPlace and scaling of ventricular wall stress and blood pressure in mammals and birds. *Physiological and Biochemical Zoology* 73: 389–405.

Singer MA, Morton AR. 2000. Mouse to elephant: Biological scaling and Kt/V. *American Journal of Kidney Diseases* 35: 306–309.

Van der Jeugd HP, Prins HHT. 2000. Movements and group structure of giraffe (*Giraffa camelopardalis*) in Lake Manyara National park, Tanzania. *Journal of Zoology* 251: 15–21.

Wolfe BA, Sladsky KK, Loomis MR. 2000. Obstructive urolithiasis in a reticulated giraffe (*Giraffa camelopardalis reticulata*). *Veterinary Record* 146: 260–261.

2001:

Bond WJ, Lofell D. 2001. Introduction of giraffe changes acacia distribution in a South African savanna. *African Journal of Ecology* 39: 286–294.

Bush M, Grobler DG, Raath JP, Phillips LG, Stamper MA, Lance WR. 2001. Use of medetomidine and ketamine for immobilization of free-ranging giraffes. *Journal of the American Veterinary Medical Association* 218: 245–249.

Clauss MA, Lechner-Doll MB, Flach EJC, Tack CD, Hatt J-ME. 2001. Comparative use of four different marker systems for the estimation of digestibility and low food intake in a group of captive giraffes (*Giraffa camelopardalis*). *Zoo Biology* 20: 315–329.

Jones RJ, Meyer JHF, Bechaz FM, Stoltz MA, Palmer B, van der Merwe G. 2001. Comparison of rumen fluid from South African game species to digest tanniniferous browse. *Australian Journal of Agricultural Research* 52: 453–460.

Midgley JJ, Mclean P, Botha M, Balfour D. 2001. Why do some African thorn trees (*Acacia spp*) have a flat top: A grazer-plant mutualism hypothesis? *African Journal of Ecology* 39: 226–228.

Sasaki M, Endo H, Kitamura N, Yamada J, Yamamoto M, Arishima K, Hayashi Y. 2001. The structure of the masseter muscle in the giraffe (*Giraffa camelopardalis*). *Anatomia, Histologia, Embryologia* 30: 313–319.

Woolnough AP, du Toit JT. 2001. Vertical zonation of browse quality in tree canopies exposed to a size-structures guild of African browsing ungulates. *Oecologia* 129: 585–590.

2002:

Agusti J, Anton M. 2002. *Mammoths, sabretooths and hominids.* Columbia University Press, New York.

Brook BS, Pedley TJ. 2002. A model for time-dependent flow in (giraffe jugular veins): uniform tube properties. *Journal of Biomechanics* 35: 95–107.

Christiansen P. 2002. Mass allometry of the appendicular skeleton in terrestrial mammals. *Journal of Morphology* 251: 195–209.

Clauss M, Lechner-Doll M, Flach EJ, Wiser J, Hatt J-M. 2002. Digestive tract pathology of captive giraffe (*Giraffa camelopardalis*): A unifying hypothesis. *European Association of Zoo and Wildlife Veterinarians. Proceedings of the 14th Scientific Meeting*, pp. 99–107.

Hampton C. 2002. Carnivorous giraffe, or natural phenomenon? *Endangered Wildlife* 40: 22–25.

Maluf NSR. 2002. Kidney of giraffes. *Anatomical Record* 267: 94–111.

Ward D, Young TP. 2002. Effects of large mammalian herbivores and ant symbionts on condensed tannins. *Journal of Chemical Ecology* 28: 913–929.

Wood WF, Weldon PJ. 2002. The scent of the reticulated giraffe (*Giraffa camelopardalis reticulata*). *Biochemical Systematics and Ecology* 30: 913–917.

Zivotofsky D, Zivotofsky AZ, Amar Z. 2002/3. Giraffe: A halakhically oriented dissection. *The Torah u-Madda Journal* 11: 203–219.

2003:

Caister LE, Shields WM, Gosser A. 2003. Female tannin avoidance: A possible explanation for habitat and dietary segregation of giraffes (Giraffa camelopardalis peralta) in Niger. *African Journal of Ecology* 41: 201–210.

Clauss M, Frey R, Kiefer B, Lechner-Doll M, Loehlein W, Polster C, Rössner GE, Streich WJ. 2003. The maximum attainable size of herbivorous mammals: Morphophysiological constraints on foregut, and adaptations of hindgut fermenters. *Oecologia* 136: 14–27.

Clauss M, Lechner-Doll M, Streich WJ. 2003. Ruminant diversification as an adaptation to the physicomechanical characteristics of forage: A reevaluation of an old debate and a new hypothesis. *Oikos* 102: 253–262.

Cumming DHM, Cumming GS. 2003. Ungulate community structure and ecological processes: body size, hoof area and trampling in African savannas. *Oecologia* 134: 560–568.

Dehority BA, Odenyo AA. 2003. Influence of diet on the rumen protozoal fauna of indigenous African wild ruminants. *Journal of Eukaryotic Microbiology* 50: 220–223.

Fennessy JT, Leggett KEA, Schneider S. 2003. Distribution and status of the desert-dwelling giraffe (*Giraffa camelpardalis angolensis*) in northwestern Namibia. *African Zoology* 38: 184–188.

Ferreira P. 2003. Report of a giraffe with a bone stuck in its mouth. *Stalking Magazine*. Winter, p. 4.

Hassanin A, Douzery EJP. 2003. Molecular and morphological phylogenies of ruminantia and the alternative position of the Moschidae. *Systematic Biology* 52: 206–228.

Lagueux O. 2003. Geoffroy's Giraffe: The hagiography of a charismatic mammal. *Journal of the History of Biology* 36: 225–247.

Mitchell G, Skinner JD. 2003. On the origin, evolution and phylogeny of giraffes *Giraffa camelopardalis*. *Transactions of the Royal Society of South Africa* 58: 51–73.

Parker DM, Bernard RTF, Colvin SA. The diet of a small group of extralimital giraffe. *African Journal of Ecology* 41: 245–253.

2004:

Clauss M, Lechner-Doll M, Streich WJ. 2004. Differences in the range of faecal dry matter content between feeding types of captive wild ruminants. *Acta Theriologica* 49: 259–267.

Fennessy JT. 2004. Ecology of desert-dwelling giraffe *Giraffa camelopardalis* in northwestern Namibia. PhD thesis. University of Sydney, Australia.

Howland HC, Merola S, Basarab JR. 2004. The allometry and scaling of the size of vertebrate eyes. *Vision Research* 44: 2043–2065.

Mitchell G, Skinner JD. 2004. Giraffe thermoregulation: a review. *Transactions of the Royal Society of South Africa* 59: 49–57.

van Schalkwyk OL. 2004. Bone density and Ca and P content of the giraffe skeleton. MSc dissertation, University of Pretoria, South Africa.

van Schalkwyk OL, Skinner JD, Mitchell G. 2004. A comparison of the bone density and morphology of giraffe (*Giraffa camelopardalis*) and buffalo (*Syncerus caffer*) skeletons. *Journal of Zoology (London)* 264: 307–315.

2005:

Cameron EZ, du Toit JT. 2005. Social influences on vigilance behaviour in giraffes, *Giraffa camelopardalis*. *Animal Behaviour* 69: 1337–1344.

del Castillo SM, Bashaw MJ, Patton ML, Rieches RR, Bercovitch FB. 2005. Fecal steroid analysis of female giraffe (*Giraffa camelopardalis*) reproductive condition and the impact of endocrine status on daily time budgets. *General and Comparative Endocrinology* 141: 271–281.

du Toit JT, Yetman CA. 2005. Effects of body size on the diurnal activity budgets of African browsing ruminants. *Oecologea* 143: 317–325.

Fernandez MH, Vrba ES. 2005. A complete estimate of the phylogenetic relationships in Ruminantia: A dated species-level supertree of the extant ruminants. *Biological Reviews* 80: 269–302.

Geraads D, Spassov N, Kovachev D. 2005. Giraffidae (Artiodactyla, Mammalia) from the late Miocene of Kalimantsi and Hadjidimovo, Southwestern Bulgaria. *Geologica Balcanica* 35 (1–2): 11–18.

Kearney CC. 2005. Effects of dietary physical form and carbohydrate profile on captive giraffe. MSc thesis, University of Florida.

Makhabu SW. 2005. Resource partitioning within a browser guild in a key habitat, the Chobe Riverfront, Botswana. *Journal of Tropical Ecology* 21: 641–649.

Mitchell G, van Schalkwyk OL, Skinner JD. 2005. The calcium and phosphorus content of giraffe (*Giraffa camelopardalis*) and buffalo (*Syncerus caffer*) skeletons. *Journal of Zoology (London)* 264: 307–315.

Owen-Smith N, Mason DR. 2005. Comparative changes in adult vs juvenile survival affecting population trends of African ungulates. *Journal of Animal Ecology* 74: 762–773.

Tweedale M. (Ed.) 2005. Deuteronomy *ca* 1200 BC Ch 14 verses 4–5. Biblia Sacra juxta Vulgatum Clementinam, London. Editio Electronica.

Zepelin HJ, Siegel M, Tobler I. 2005. Mammalian sleep. *Principles and Practice of Sleep Medicine* 4: 91–100.

2006:

Belozerskaya M. 2006. *The Medici giraffe*. Little, Brown, New York, pp. 86–129.

Bercovitch FB, Bashaw MJ, del Castillo SM. 2006. Sociosexual behavior, male mating tactics, and the reproductive cycle of giraffe *Giraffa camelopardalis*. *Hormones and Behavior* 50: 314–321.

Burton RF. 2006. A new look at the scaling of size in mammalian eyes. *Journal of Zoology* 269: 225–232.

Clauss M, Hofmann RR, Hummel J, Adamczewski J, Nygren K, Pitra C, Streich WJ, Reese S. 2006. Macroscopic anatomy of the omasum of free-ranging moose (*Alces alces*) and muskoxen (*Ovibos moschatus*) and a comparison of the omasal laminal surface in 34 ruminant species. *Journal of Zoology* 270: 346–358.

Clauss M, Hofmann RR, Streich WJ, Fickel J, Hummel J. 2006. Convergence in the macroscopic anatomy of the reticulum in wild ruminant species of different feeding types and a new resulting hypothesis on reticular function. *Journal of Zoology* 281: 26–38.

Clauss MA, Hummel JBC, Streich WJD. 2006. The dissociation of the fluid and particle phase in the forestomach as a physiological characteristic of large grazing ruminants: An evaluation of available, comparable ruminant passage data. *European Journal of Wildlife Research* 52: 88–98.

Dumonceaux GA, Bauman JE, Camilo GR. 2006. Evaluation of progesterone levels in feces of captive reticulated giraffe (*Giraffa camelopardalis reticulata*). *Journal of Zoo and Wildlife Medicine* 37: 255–261.

Fleming PA, Hofmeyr SD, Nodson SW, du Toit JT. 2006. Are giraffes pollinators or flower predators of *Acacia nigrescens* in Kruger National Park, South Africa? *Journal of Tropical Ecology* 22: 247–253.

Geraads D. 2006. The late Pliocene locality of Ahl al Oughlam, Morocco: Vertebrate fauna and interpretation. *Transactions of the Royal Society of South Africa* 61: 97–101.

Mitchell G, Maloney SK, Mitchell D, Keegan DJ. 2006. The origin of mean arterial and jugular venous blood pressures in giraffes. *Journal of Experimental Biology* 209: 2515–2524.

Patton ML, Bashaw MJ, del Castillo SM, Jochle W, Lamberski N, Rieches R, Bercowitz FB. 2006. Long-term suppression of fertility in female giraffe using GnRH agonist deslorelin as a long-acting implant. *Theriogenology* 66: 431–438.

Ringmar E. 2006. Audience for a giraffe: European expansionism and the quest for the exotic. *Journal of World History* 17: 375–397.

2007:

Badlangana NL, Bhagwandin A, Fuxe K, Manger PR. 2007a. Distribution and morphology of putative catecholaminergic and serotonergic neurons in the medulla oblongata of a sub-adult giraffe, *Giraffa camelopardalis*. *Journal of Chemical Neuroanatomy* 34: 69–79.

Badlangana NL, Bhagwandin A, Fuxe K, Manger PR. 2007b. Observations on the giraffe central nervous system related to the corticospinal tract, motor cortex and spinal cord: What difference does a long neck make? *Neuroscience* 148: 522–534.

Brown DM, Brenneman RA, Koepfli K-P, Pollinger JP, Mila B, Georgiadis, NJ, Louis EE, Grether GF, Jacobs DK, Wayne RK. 2007. Extensive population genetic structure in the giraffe. *BMC Biology* 5: 57 (1–13).

Cameron EZ, du Toit JT. 2007. Winning by a neck: Tall giraffes avoid competing with shorter browsers. *The American Naturalist* 169: 130–135.

Clauss M, Schwarm A, Ortmann S, Streich WJ, Hummel J. 2007. A case of non-scaling in mammalian physiology? Body size, digestive capacity, food intake, and ingesta passage in mammalian herbivores. *Comparative Biochemistry and Physiology A* 148: 249–265.

Cramer MD, Mazel AD. 2007. The past distribution of giraffe in Kwazulu-Natal. *South African Journal of Wildlife Research* 37: 197–201.

Hassanin A, Ropiquet A, Gourmand A-L, Chardonnet, B, Rigoulet J. 2007. Mitochondrial DNA variability in Giraffa camelopardalis: consequences for taxonomy, phylogeography and conservation of giraffes in West and central Africa. *Comptes Rendus Biologies* 330: 265–274.

Isobe N, Nakao T, Shimada M, Fukumoto Y, Watanabe H, Minami S, Noda A, Yoshimura Y. 2007. Fecal progestagen and estrone during pregnancy in a giraffe: A case report. *Journal of Reproduction and Development* 53: 159–164.

Pontzer H. 2007. Effective limb length and the scaling of locomotor cost in terrestrial animals. *Journal of Experimental Biology* 210: 1752–1761.

Schmidt DA, Ball RL, Grobler D, Ellersleck MR, Griffin ME, Citino SB, Bush M. 2007. Serum concentrations of amino-acids, fatty acids, lipoproteins, Vitamins A and E, and minerals in apparently healthy, free-ranging southern giraffe (*Giraffa camelopardalis giraffe*). *Zoo Biology* 26: 13–25.

Solounias N. 2007. Family Giraffidae. In *The evolution of artiodactyls*. Prothero DR, Foss SE (Eds.). Johns Hopkins University Press, Baltimore, MD, pp. 257–277.

Williams DC, Murison PJ, Hill CL. 2007. Dystocia in a Rothschild giraffe leading to a Caesarean section. *Journal of the Veterinary Medical Association* 55: 199–202.

Zinn AD, Ward D, Kirkman K. 2007. Inducible defences in *Acacia sieberiana* in response to giraffe browsing. *African Journal of Range and Forage Science* 24: 123–129.

2008:

Bredin I, Skinner JD, Mitchell G. 2008. Can osteophagia provide giraffes with phosphorus and calcium? *Onderstepoort Journal of Veterinary Research* 75: 1–9.

Clauss M, Dierenfeld ES. 2008. The nutrition of "browsers." *Zoo and Wild Animal Medicine Current Therapy* 6: 444–454.

Clauss M, Hofmann RR, Streich WJ, Fickel J, Hummel J. 2008. Higher masseter muscle mass in grazing than in browsing ruminants. *Oecologia* 157: 377–385.

Finlay R. 2008. The voyages of Zheng He: Ideology, State power, and maritime trade in Ming China. *The Journal of the Historical Society* 8: 327–347.

Frackowiak H, Jakubowski H. 2008. Arterial vascularization in the giraffe brain. *Annales Zoologici Fennici* 45: 353–359.

Hofmann RR, Streich WJ, Fickel J, Hummel J, Clauss M. 2008. Convergent evolution in feeding types: Salivary gland mass differences in wild ruminant species. *Journal of Morphology* 269: 240–257.

Mitchell G. 2008. Robert Goetz: A pioneer of the study of the physiology of animals in the wild. *Transactions of the Royal Society of South Africa* 63 (2): 182–185.

Mitchell G, Bobbitt JP, Devries S. 2008. Cerebral perfusion pressure in giraffe: Modelling the effects of head raising and lowering. *Journal of Theoretical Biology* 252: 98–108.

Siegel JM. 2008. Do all animals sleep? *Trends in Neurosciences* 31: 208–213.

Schiviz AN, Ruf T, Kueberr-Heiss A, Schubert C, Ahnelt PK. 2008. Retinal cone topography of artiodactyl mammals: Influence of body height and habitat. *Journal of Comparative Neurology* 507: 1336–1350.

Tanner JB, Dumont ER, Sakai ST, Lundrigan BL, Holekamp KE. 2008. Of arcs and vaults: The biomechanics of bone-cracking in spotted hyenas (*Crocuta crocuta*). *Biological Journal of the Linnean Society* 95: 246–255.

2009:

Badlangana NL, Adams JW, Manger PR. 2009. The giraffe (*Giraffa camelopardalis*) cervical vertebral column: A heuristic example in understanding evolutionary processes? *Zoological Journal of the Linnean Society* 155: 736–757.

Brondum E, Hasenkam JM, Secher NH, Bertelsen MF, Grondhal C, Peterson KK, Buhl R, Aalkaer C, Baandrup U, Nygaard H, Smerup M, Stegmann F, Sloth E, Østergaard KH, Nissen P, Runge M, Pitsillides K, Wang T. 2009. Jugular venous pooling during lowering of the head affects blood pressure of the anesthetized giraffe. *American Journal of Physiology* 297: R1058–1065.

Cano I, Perez W. 2009. Quantitative anatomy of the trachea of the giraffe (*Giraffa camelopardalis rothschildi*). *International Journal of Morphology* 27: 905–908.

Clauss M, Nunn C, Fritz J, Hummel J. 2009. Evidence for a tradeoff between retention time and chewing efficiency in large mammalian herbivores. *Comparative Biochemistry and Physiology A* 154: 376–382.

Dawkins R. 2009. *The greatest show on earth: The evidence for evolution*. Transworld, London, p. 361.

Fennessy J. 2009. Home range and seasonal movements of *Giraffa camelopardalis angolensis* in the northern Namib desert. *African Journal of Ecology* 47: 318–327.

Geraads D. 2009. Giraffidae (Mammalia) de la fin du Néogène de la République de Macédoine (ARYM). *Geodiversitas* 31: 893–908.

Kostopoulos DS. 2009. The late Miocene mammals faunas of the Mytilinii Basin, Samos Island, Greece: new collection. 13. Giraffidae. *Beiträge zur Paläontologie, Wien* 31: 299–343.

Lueders I, Hildebrandt TB, Pootoolal J, Rich P, Gray CS, Niemuller CA. 2009a. Ovarian ultrasonography correlated with fecal progestins and estradiol during the estrous cycle and early pregnancy in giraffes (*Giraffa camelopardalis rothschildi*). *Biology of Reproduction* 81: 989–995.

Lueders I, Niemuller C, Pootoolal J, Rich P, Gray C, Streich WJ, Hildebrandt TB. 2009b. Sonomorphology of the reproductive tract in male and pregnant and non-pregnant female Rothschild's giraffes (*Giraffa camelopardalis rothschildi*). *Theriogenology* 72: 22–31.

Mitchell, G. 2009. The origin of the scientific study and classification of giraffes. *Transactions of the Royal Society of South Africa* 64: 1–13.

Mitchell G, Skinner JD. 2009. An allometric analysis of the giraffe cardiovascular system. *Comparative Biochemistry and Physiology A* 154: 523–529.

Mitchell G, van Sittert S, Skinner JD. 2009a. The structure and function of giraffe jugular vein valves. *South African Journal of Wildlife Research* 39: 175–180.

Mitchell G, van Sittert S, Skinner JD. 2009b. Sexual selection is not the origin of long necks in giraffes. *Journal of Zoology (London)* 278: 281–286.

Paton JFR, Dickinson CJ, Mitchell G. 2009. Harvey Cushing and the regulation of blood pressure in giraffe, rats and man: introducing the Cushing's mechanism. *Experimental Physiology* 94: 11–17.

Perez W, Lima M, Clauss M. 2009. Gross anatomy of the intestine in the giraffe (*Giraffa camelopardalis*). *Journal of Veterinary Medicine: Anatomia, Histologia, Embryologia* 38: 432–435.

2010:

Bercovitch FB, Berry, PSM. 2010a. Reproductive life history of Thornicroft's giraffe in Zambia. *African Journal of Ecology* 48: 535–538.

Bercovitch FB, Berry, PSM. 2010b. Ecological determinants of herd size in the Thornicroft's giraffe of Zambia. *African Journal of Ecology* 48: 962–971.

Bux F, Bhagwandin A, Fuxe K, Manger PR. 2010. Organization of cholinergic, putative catecholaminergic and serotonergic nuclei in the diencephalon, midbrain, and pons of subadult male giraffes. *Journal of Chemical Neuroanatomy* 39: 189–203.

Clauss M, Hume DI, Hummel J. 2010. Evolutionary adaptations of ruminants and their potential relevance for modern production systems. *Animal* 7: 979–992.

Harris JM, Solounias N, Geraads D. 2010. Giraffoidea. In *Cenozoic mammals of Africa*. Werdelin L, Saunders WJ (Eds.). University of California Press, Oakland, pp. 797–811.

Henderson DM, Naish D. 2010. Predicting the buoyancy, equilibrium and potential swimming ability of giraffes by computational analysis. *Journal of Theoretical Biology* 265: 151–159.

Marealle WN, Fossøy F, Holmern T, Stokke BG, Røskaft E. 2010. Does illegal hunting skew Serengeti wildlife sex ratios? *Wildlife Biology* 16 (4): 419–429.

Mitchell G, van Sittert S, Skinner JD. 2010. The demography of giraffe deaths in a drought. *Transactions of the Royal Society of South Africa* 65: 165–168.

Sathar F, Badlangana L, Manger PR. 2010. Variations in the thickness and composition of the skin of the giraffe. *Anatomical Record* 293: 1615–1627.

Simmons RE, Altwegg R. 2010. Necks-for-sex or competing browsers? A critique of ideas on the evolution of giraffe. *Journal of Zoology* 282: 6–12.

Sullivan K, van Heugten E, Ange-van Heugten K, Poore MH, Dierenfeld ES, Wolfe B. 2010. Analysis of nutrient concentrations in the diet, serum, and urine of giraffe from surveyed North American Zoological Institutions. *Zoo Biology* 29: 457–469.

Theodorou GE, Roussiakis SJ, Athanassiou A, Filippidi A. 2010. Mammalian remains from a site near the classical locality of Pikermi (Attica, Greece). *Proceedings of the XIX CBGA Congress, Thessalonika, Greece* 99: 109–119.

van Sittert SJ, Skinner JD, Mitchell, G, 2010. From fetus to adult: An allometric analysis of the giraffe vertebral column. *Journal of Experimental Zoology B* 314: 469–479.

2011:

Badlangana NL, Adams JW, Manger PR. 2011. A comparative assessment of the size of the frontal air sinuses in the giraffe (*Giraffa camelopardalis*). *Anatomical Record* 294: 931–940.

Marra AC, Solounias N, Carone G, Rook L. 2011. Palaeogeographic significance of the giraffid remains (Mammalia, Artiodactyla) from Cessaniti (Late Miocene, Southern Italy). *Geobios* 44: 189–197.

Mitchell G, Skinner JD. 2011. Lung volumes in giraffes, *Giraffa camelopardalis*. *Comparative Biochemistry and Physiology A* 158: 72–78.

Østergaard KH, Bertelsen MF, Brondum ET, Aalkjaer C, Hasenkam JM, Smerup M, Wang T, Nyengaard JR, Baandrup U. 2011. Pressure profile and morphology of the arteries along the giraffe limb. *Journal of Comparative Physiology B* 181: 691–698.

Robinson CA. 2011. Giraffidae. In *Paleontology and geology of Laetoli: Human evolution in context.* Harrison T (Ed.). Series Vertebrate paleobiology and paleoanthropology. Springer, New York, pp. 339–362.

Skinner JD, Mitchell G. 2011. Giraffidae. In *Handbook of world mammals*, Vol. 2.Wilson, DE, Mittermeier RA (Eds.). Lynx Edicions, Barcelona, pp. 788–802.

Taylor MP, Hone DWE, Wedel MJ, Naish D. 2011. The long necks of sauropods did not evolve primarily through sexual selection. *Journal of Zoology* 285: 150–161.

Wilkinson DM, Ruxton GD. 2011. Understanding selection for long necks in different taxa. *Biological Reviews* 87: 616–630.

2012:

Berry PSM, Bercovitch FB. 2012. Darkening coat colour reveals life history and life expectancy of male Thornicroft's giraffes. *Journal of Zoology (London)* 287: 157–160.

Bhatti ZH, Khan MA, Akhtar M, Khan AM, Ghaffar A, Iqbal M, Siddiq MK. 2012. *Giraffa punjabiensis* (Giraffidae: Mammalia) from the Middle Siwaliks of Pakistan. *Pakistan Journal of Zoology* 44: 1689–1695.

Dell L-A, Patzke N, Bhagwandin A, Bux F, Fuxe K, Barber G, Siegel JM, Manger PR. 2012. Organization and number of orexinergic neurons in the hypothalamus of two species of

Cetartiodactyla: A comparison of giraffe (*Giraffa camelopardalis*) and harbour porpoise (*Phocoena phocoena*). *Journal of Chemical Neuroanatomy* 44: 98–109.

Kinzelbach RK. 2012. A cassowary *Casuarius casuarius* (Linnaeus, 1758): Record from Alexandria, Egypt, in 20 BC. (Aves, Ratitae, Casuaridae). *The Open Ornithology Journal* 5: 26–31.

Leyra IP. 2012. Artemidorus behind Artemidorus: Geographic aspects in the zoological designs of the Artemidorus papyrus. *Historia: Zeitschrift für Alte Geschichte* 61(3): 336–367.

Perez W, Michel V, Hassen J, Vazquez N. 2012. Anatomy of the mouth of the Giraffe (*Giraffa camelopardalis rothschildi*). *International Journal of Morphology* 30: 322–329.

Perry TW. 2012. *Beef cattle feeding and nutrition*. Academic Press, Cambridge, MA.

Rigoulet J. 2012. Histoire de Zarafa, la giraffe de Charles X. *Bulletin de l'Académie Vétérinaire de France* 165: 169–176.

Seeber PA, Ndlovu HT, Duncan P, Ganswindt A. 2012. Grazing behavior of the giraffe in Hwange National Park, Zimbabwe. *African Journal of Ecology* 50: 247–250.

2013:

Clauss M, Steuer P, Muller DWH, Codron D, Hummel J. 2013. Herbivory and body size: allometries of diet quality and gastrointestinal physiology, and implications for herbivore ecology and dinosaur gigantism. *PLOS One* 8: 1–14.

Coimbra JP, Hart N, Collin SP, Manger PR. 2013. Scene from above: Retinal ganglion cell topography and spatial resolving power in the giraffe (*G camelopardalis*). *Journal of Comparative Neurology* 521: 2042–2057.

Erdogan S, Perez W. 2013. Anatomical characteristics of the larynx in giraffe (*Giraffa camelopardalis*). *Journal of Morphological Science* 30: 266–271.

Fennessy J, Bock FG, Tutchings A, Brenneman R, Janke A. 2013. Mitochondrial DNA analyses show that Zambia's South Luangwa Valley giraffe (*Giraffa camelopardalis thornicrofti*) are genetically isolated. *African Journal of Ecology* 51: 635–640.

Furuuchi K, Koyabe D, Mori K, Endo H. 2013. Physiological cross-sectional area of the masticatory muscles in the giraffe (*Giraffa camelopardalis*). *Mammal Study* 38: 67–71.

Geraads D, Reed K, Bobe R. 2013. Pliocene Giraffidae (Mammalia) from the Hadar formation of Hadar and Ledi-Geraru, Lower Awash, Ethiopia. *Journal of Vertebrate Palaeontology* 33: 470–481.

Mitchell G, Roberts D, van Sittert S, Skinner JD. 2013a. Growth patterns and masses of the heads and necks of male and female giraffes. *Journal of Zoology (London)* 290: 49–57.

Mitchell G, Roberts DG, van Sittert SJ, Skinner JD. 2013. Orbit orientation and eye morphometrics in giraffes (*Giraffa camelopardalis*). *African Zoology* 48: 333–339.

More HL, O'Connor SM, Brøndum E, Wang T, Bertelsen MF, Grøndahl C, Kastberg K, Hølyck A, Funder J, Donelan JM. 2013. Sensorimotor responsiveness and resolution in the giraffe. *Journal of Experimental Biology* 216: 1003–1011.

Müller DWH, Codron D, Meloro C, Munn A, Schwarm A, Hummel J, Clauss M. 2013. Assessing the Jarman-Bell principle: Scaling of intake, digestibility, retention time and gut fill with body mass in mammalian herbivores. *Comparative Biochemistry and Physiology A* 164: 129–140.

Østergaard KH, Baandrup UT, Wang T, Bertelsen MF, Andersen JB, Smerup M, Nyengaard JR. 2013. Left ventricular morphology of the giraffe heart examined by stereological methods. *The Anatomical Record* 296: 611–621.

Petersen KK, Horlyck A, Østergaard KH, Andresen J, Broegger T, Skovgaard N, Telinius N, Laher I, Bertelsen MF, Grondahl C, Smerup M, Secher NH, Brondum E, Hasenkam JM, Wang T, Baandrup U, Aalkjaer C. 2013. Protection against high intravascular pressure in giraffe legs. *American Journal of Physiology* 305: R1021–R1030.

Seeber PA, Duncan P, Fritz H, Ganswindt A. 2013. Androgen changes and flexible rutting behaviour in male giraffes. *Biology Letters* 9: 20130396.

Wilsher S, Stansfield F, Greenwood RES, Trethowan PD, Anderson RA, Wooding FBW, Allen WR. 2013. Ovarian and placental morphology and endocrine functions in the pregnant giraffe (*Giraffa camelopardalis*). *Reproduction* 145: 541–554.

2014:

Bock F, Fennessy J, Bidon T, Tutchings A, Marais A, Deacon F, Janke A. 2014. Mitochondrial sequences reveal a clear separation between Angolan and South African giraffe along a cryptic rift valley. *BMC Evolutionary Biology* 14: 219 (1–12).

Clauss M, Dittman MT, Müller DWH, Zerbe P, Codron D. 2014. Low scaling of a life history variable: analyzing Eutherian gestation periods with and without phylogeny-informed statistics. *Mammalian Biology* 79: 9–16.

Dagg AI. 2014. *Giraffe: Biology, behaviour, and conservation.* Cambridge University Press, Cambridge, UK.

Roggenbuck M, Sauer C, Poulsen M, Bertelsen MF, Sorensen SJ. 2014. The giraffe (*Giraffa* camelopardalis) rumen microbiome. *FEMS Microbiology Ecology* 90: 237–246.

VanderWaal KL, Wang H, McCowan B, Fushing H, Isbell LA. 2014. Multilevel social organization and space use in reticulated giraffe (*Giraffa camelopardalis*). *Behavioral Ecology* 25: 17–26.

Veilleux CC, Kirk EC. 2014. Visual acuity in mammals: Effects of eye size and ecology. *Brain Behavior and Evolution* 83: 43–53.

Yeakel JD, Pires MM, Rudolf L, Dominy NJ, Koch PL, Guimarães PR, Gross T. 2014. Collapse of an ecological network in Ancient Egypt. *Proceedings of the National Academy of Sciences* 111: 14472–14477.

2015:

Baotic A, Sicks F, Stoeger AS. 2015. Nocturnal "humming" vocalizations: Adding a piece to the puzzle of giraffe vocal communication. *BMC Research Notes* 8: 425–434.

Bercovitch FB, Deacon F. 2015. Gazing at a giraffe gyroscope: Where are we going? *African Journal of Ecology* 53: 135–146.

Damkjaer M, Wang, T, Brondum E, Østergaard KH, Baandrup U, Horlyck A, Hasenkam JM, Smerup M, Funder J, Marcussen N, Danielsen CC, Bertelsen MF, Grondal C, Pedersen M, Agger P, Candy G, Aalkjaer C, Bie P. 2015. The giraffe kidney tolerates high arterial blood pressure by high renal interstitial pressure and low glomerular filtration rate. *Acta Physiologica* 214: 497–510.

Danowitz M, Domalski R, Solounias N. 2015. The cervical anatomy of *Samotherium*, an intermediate-necked giraffid. *Royal Society Open Science* 2: 150521.

Danowitz M, Vasilyev A, Kortlandt V, Solounias N. 2015. Fossil evidence and stages of elongation of the *Giraffa camelopardalis* neck. *Royal Society Open Science* 2: 150393.Deacon F, Nel PJ, Bercovitch FB. 2015. Concurrent pregnancy and lactation in wild giraffes (*Giraffa camelopardalis*). *African Zoology* 50: 331–334.

Jacobs B, Harland T, Kennedy D, Schall M, Wicinski B, Butti C, Hof PR, Sherwood CC, Manger PR. 2015. The neocortex of cetartiodactyls. II. Neuronal morphology of the visual and motor cortices in the giraffe (*Giraffa camelopardalis*). *Brain Structure and Function* 220: 2851–2872.

Mitchell G, Roberts DG, van Sittert SJ. 2015. The digestive morphophysiology of wild, free-living, giraffes. *Comparative Biochemistry and Physiology A* 187: 119–129.

Ripple WJ, Newsome TM, Wolf C, Dirzo R, Everatt KT, Galetti M, Hayward MW, Kerley GIH, Levi T, Lindsey PA, Macdonald DW, Malhi Y, Painter LE, Sandom CJ, Terborgh J, Van Valkenburgh B. 2015. Collapse of the world's largest herbivores. *Science Advances* 1: e1400103.

Sauer C, Bertelsen MF, Lund P, Weisberg MR, Clauss M. 2015. Quantitative macroscopic anatomy of the giraffe (*Giraffa camelopardalis*) digestive tract. *Anatomia, Histologia, Embryologia* 45: 338–349.

Sharkey HJ. 2015. *La Belle Africaine*: The Sudanese giraffe who went to France. *Canadian Journal of African Studies* 49: 39–65.

Strauss MKL, Kilewo M, Rentsch D, Packer C. 2015. Food supply and poaching limit giraffe abundance in the Serengeti. *Population Ecology* 57: 505–516.

van Sittert SJ. 2015. Ontogenetic allometry of the postcranial skeleton of the giraffe (Giraffa camelopardalis), with application to giraffe life history, evolution and *palaeontology*. PhD thesis, University of Pretoria, South Africa.

van Sittert SJ, Mitchell G. 2015. On reconstructing *Giraffa sivalensis* an extinct giraffid from the Siwalik Hills, India. *PeerJ* 3: e1135.

van Sittert SJ, Skinner JD, Mitchell G. 2015. Scaling of the appendicular skeleton of the giraffe (*Giraffa camelopardalis*). *Journal of Morphology* 276: 503–516.2016:

Agaba M, Ishengoma E, Miller WC, McGrath BC, Hudson CN, Reina OCB, Rattan A, Burhans R, Chikhi R, Medvedev P, Praul CA, Wu-Cavener L, Wood B, Robertson H, Penfold L, Cavener DR. 2016. Giraffe genome sequence reveals clues to its unique morphology and physiology. *Nature Communications* 7: 11519.

Basu C, Falkingham P, Hutchinson J. 2016. The extinct giant giraffid *Sivatherium giganteum*: Skeletal reconstruction and body mass estimation. *Biology Letters* 12: 20150940.

Elsner J. 2016. Backgrounds and shadows in the Artemidorus Papyrus. In *Interno al Papiro di Artemidoro III. I designi.* Milano, LED.

Fennessy J, Bidon T, Reuss F, Kumar V, Elkan P, Nilsson MA, Vamberger M, Fritz U, Janke A. 2016. Multi-locus analyses reveal four giraffe species instead of one. *Current Biology* 26 : 1–7.

Gloneková M, Brandlová K, Jan Pluháček J. 2016. Stealing milk by young and reciprocal mothers: High incidence of allonursing in giraffes, *Giraffa camelopardalis*. *Animal Behaviour* 113: 113–123.

Gunji M, Endo H. 2016. Functional cervicothoracic boundary modified by anatomical shifts in the neck of giraffes. *Royal Society Open Science* 3: 150604.

Lee DE, Kissui BM, Kiwango YA, Bond ML. 2016. Migratory herds of wildebeests and zebras indirectly affect calf survival of giraffes. *Ecology and Evolution* 6: 8402–8411.

Martínez-Freiría F, Tarroso P, Rebelo H, Brito JC. 2016. Contemporary niche contraction affects climate change predictions for elephants and giraffes. *Diversity and Distributions* 22: 432–444.

Mitchell G, Roberts DG, van Sittert SJ. 2016. Corrigendum to "The digestive morphophysiology of wild, free-living, giraffes." *Comparative Biochemistry and Physiology A* 191: 226.

Nyman G, Röken B, Hedin E, Hedenstierna G. 2016. Case studies in physiology: Ventilation and perfusion in a giraffe—does size matter? *Journal of Applied Physiology* 121: 1374–1378.

O'Brien HD, Gignac PM, Hieronymus TL, Witmer, LM. 2016. A comparison of postnatal arterial pressure patterns in a growth series of giraffe (Artiodactyla: *Giraffa camelopardalis*). *PeerJ* 4: e1696.

Rios M, Danowitz M, Solounias N. 2016. First comprehensive morphological analysis on the metapodials of Giraffidae. *Palaeontologia Electronica* 19.3.52A: 1–39.

Rowan J, Locke EM, Robinson JR, Campisano CJ, Wynn JG, Reed KE. 2017. Fossil Giraffidae (Mammalia, Artiodactyla) from Lee Adoyta, Ledi-Geraru, and Late Pliocene dietary evolution in Giraffids from the Lower Awash Valley, Ethiopia. *Journal of Mammal Evolution* 24: 359–371. doi: 10.1007/s10914-016-9343-z.

Smerup M, Damkjaer M, Brondum E, Baandrup UT, Kristiansen SB, Nygaard H, Aalkjaer C, Sauer C, Buchanan R, Bertelsen MF, Østergaard K, Grondahl C, Candy G, Hasenkam JM, Secher NH, Bie P, Wang T. 2016. The thick left ventricular wall of the giraffe heart normalizes wall tension, but limits stroke volume and cardiac output. *Journal of Experimental Biology* 219: 457–463.

2017:

Bercovitch FB, Berry, PSM, Dagg A, Deacon F, Doherty JB, Lee DE, Mineur F, Muller Z, Ogden R, Seymour R, Shorrocks B, Tutchings A. 2017. How many species of giraffe are there? *Current Biology* 27: R136–R137.

Deacon F, Smit N. 2017. Spatial ecology and habitat use of giraffe (*Giraffa camelopardalis*) in South Africa. *Basic and Applied Ecology* 21: 55–65. doi: 10.1016.j.baae.2017.04.003.

Graïc J-M, Peruffo A, Ballarin C, Cozzi B. 2017. The brain of the giraffe (*Giraffa camelopardalis*): Surface configuration encephalization quotient, and analysis of the existing literature. *The Anatomical Record* 300: 1502–1511.

Ishengoma E, Agaba M, Cavener DR. 2017. Evolutionary analysis of vision genes identifies potential drivers of visual differences between giraffes and okapi. *PeerJ* 5: 1–26.

Kondoh D, Nakamura KG, Ono YS, Yuhara K, Bando G, Watanabe K, Horiuchi N, Kobayashi Y, Saski M, Kitamura N. 2017. Histological features of the vomeronasal organ in the giraffe, *Giraffa camelopardalis. Microscopy Research and Technique* 80: 652–656.

Lee DE, Bond ML, Bolger DT. 2017. Season of birth affects juvenile survival of giraffe. *Population Ecology* 59: 45–54.

Mitchell G, van Sittert S, Roberts D, Mitchell D. 2017. Body surface area and thermoregulation in giraffes. *Journal of Arid Environments* 145: 35–42.

Muller Z. 2017. White giraffes: The first record of vitiligo in a wild adult giraffe. *African Journal of Ecology* 55: 118–123.

Osthoff C, Hugo A, Madende M, Deacon F, Nel PJ. 2017. Milk composition of free-ranging red hartebeest, giraffe, Southern reedbuck and warthog and a phylogenetic comparison of the milk of African Artiodactyla. *Comparative Biochemistry and Physiology A* 204: 93–103.

2018:

Bercovitch FB, Berry PSM. 2018. Social and demographic influences on the feeding ecology of giraffe in the Luangwa Valley, Zambia: 1973–2014. *African Journal of Ecology* 56: 254–261.

Courchamp F, Jaric I, Albert C, Meinard Y, Ripple WJ, Chapron G. 2018. The paradoxical extinction of the most charismatic animals. *PLOS Biol* 16: e2003997.

Deacon F, Bercovitch FB. 2018. Movement patterns and herd dynamics among South African giraffes (*Giraffa camelopardalis giraffa*). *African Journal of Ecology* 56: 620–628. doi: 10.1111/aje.12514.

Deacon F, Tutchings A. 2018. The South African giraffe *Giraffa camelopardalis giraffa*: A conservation success story. *Fauna and Flora International* 53: 45–48. doi: 10.1017/S0030605317001612.

Helm C, Cawthra H, Cowling R, De Vynck J, Marean C, McCrea R, Rust R. 2018. Palaeoecology of giraffe tracks in Late Pleistocene aeolianites on the Cape south coast. *South African Journal of Science* 114: 1–8.

Kasozi H, Montgomery RA. 2018. How do giraffes locate one another? A review of visual, auditory, and olfactory communication among giraffes. *Journal of Zoology* 306: 139–146. doi: 10.1111/jzo.12604.

Lee DE, Cavener DR, Bond ML. 2018. Seeing spots: Quantifying mother-offspring similarity and assessing fitness consequences of coat pattern traits in a wild population of giraffes (*Giraffa camelopardalis*). *Peer J* 6: e5690.

Merceron G, Colyn M, Geraads D. 2018. Browsing and non-browsing extant and extinct giraffids: evidence from dental microwear textural analysis. *Palaeogeography, Palaeoclimatology, Palaeoecology* 505: 128–139.

Muller Z, Cuthill IC, Harris S. 2018. Group sizes of giraffes in Kenya: The influence of habitat, predation and the age and sex of individuals. *Journal of Zoology* 306: 77–87.

Saito M, Idani G. 2018. Suckling and allosuckling behaviour in wild giraffe (*Giraffa camelopardalis tippelskirchi*). *Mammalian Biology* 93: 1–4.

Volodina EV, Volodin IA, Chelysova EV, Frey R. 2018. Hiss and snort call types of wild-living giraffes *Giraffa camelopardalis*: acoustic structure and context. *BMC Research Notes* 11: 1–7.

Winter S, Fennessy J, Janke A. 2018. Limited introgression supports division of giraffe into four species. *Ecology and Evolution* 8: 10156–10165.

Wolf TE, Bennett NC, Burroughs R, Ganswindt A. 2018b. The impact of age-class and social context on fecal glucocorticoid metabolite levels in free-ranging male giraffes. *General and Comparative Endocrinology* 255: 16–31.

Wolf TE, Schaebs FS, Bennett NC, Burroughs R, Ganswindt A. 2018a. Age and socially related changes in fecal androgen metabolite concentrations in free-ranging male giraffes. *General and Comparative Endocrinology* 255: 19–25.

2020:

Brown MB, Wells E. 2020. Skeletal dysplasia-like syndromes in wild giraffe. *BMC Research Notes* 13: 569.

Petzold A, Hassanin A. 2020. A comparative approach for species delimitation based on multiple methods of multi-locus DNA sequence analysis: A case study of the genus *Giraffa* (Mammalia, Cetartiodactyla). *PLOS One* 15 (2): e0217956.

Index